Modern Industrial Statistics

STATISTICS IN PRACTICE

Series Advisors

Human and Biological Sciences
Stephen Senn
CRP-Santé, Luxembourg

Earth and Environmental Sciences
Marian Scott
University of Glasgow, UK

Industry, Commerce and Finance
Wolfgang Jank
University of Maryland, USA

Founding Editor
Vic Barnett
Nottingham Trent University, UK

Statistics in Practice is an important international series of texts which provide detailed coverage of statistical concepts, methods and worked case studies in specific fields of investigation and study.

With sound motivation and many worked practical examples, the books show in down-to-earth terms how to select and use an appropriate range of statistical techniques in a particular practical field within each title's special topic area.

The books provide statistical support for professionals and research workers across a range of employment fields and research environments. Subject areas covered include medicine and pharmaceutics; industry, finance and commerce; public services; the earth and environmental sciences, and so on.

The books also provide support to students studying statistical courses applied to the above areas. The demand for graduates to be equipped for the work environment has led to such courses becoming increasingly prevalent at universities and colleges.

It is our aim to present judiciously chosen and well-written workbooks to meet everyday practical needs. Feedback of views from readers will be most valuable to monitor the success of this aim.

A complete list of titles in this series appears at the end of the volume.

Modern Industrial Statistics

with Applications in R, MINITAB and JMP

Third Edition

RON S. KENETT

Chairman and CEO, the KPA Group, Raanana, Israel
Research Professor, University of Turin, Turin, Italy, and
Senior Research Fellow, Samuel Neaman Institute for National Policy Research, Technion, Israel

SHELEMYAHU ZACKS

Distinguished Professor,
Binghamton University, Binghamton, USA

With contributions from
DANIELE AMBERTI
Turin, Italy

This edition first published 2021

Edition History
John Wiley and Sons, Inc. (2e, 2014), Cengage (1e, 1980)

Registered Offices
John Wiley & Sons, Inc., 111 River Street, Hoboken, NJ 07030, USA
John Wiley & Sons Ltd, The Atrium, Southern Gate, Chichester, West Sussex, PO19 8SQ, UK

Editorial Office
9600 Garsington Road, Oxford, OX4 2DQ, UK

For details of our global editorial offices, customer services, and more information about Wiley products visit us at www.wiley.com.

Wiley also publishes its books in a variety of electronic formats and by print-on-demand. Some content that appears in standard print versions of this book may not be available in other formats.

Library of Congress Cataloging-in-Publication Data

Names: Kenett, Ron, author. | Zacks, Shelemyahu, 1932- author. | Amberti,
    Daniele, author.
Title: Modern industrial statistics : with applications in R, MINITAB and
    JMP / Ron S. Kenett, Chairman and CEO, the KPA Group, Raanana, Israel
    Research Professor, University of Turin, Turin, Italy, and Senior
    Research Fellow, Samuel Neaman Institute for National Policy Research,
    Technion, Israel, Shelemyahu Zacks, Distinguished Professor, Binghamton
    University, Binghamton, USA ; with contributions from Daniele Amberti,
    Turin, Italy.
Description: Third edition. | Hoboken, NJ : Wiley, 2021. | Series:
    Statistics in practice | Includes bibliographical references and index.
Identifiers: LCCN 2020051133 (print) | LCCN 2020051134 (ebook) | ISBN
    9781119714903 (hardback) | ISBN 9781119714927 (adobe pdf) | ISBN
    9781119714965 (epub)
Subjects: LCSH: Quality control – Statistical methods. | Reliability
    (Engineering) – Statistical methods. | R (Computer program language) |
    Minitab. | JMP (Computer file)
Classification: LCC TS156 .K42 2021 (print) | LCC TS156 (ebook) | DDC
    620/.00452–dc23
LC record available at https://lccn.loc.gov/2020051133
LC ebook record available at https://lccn.loc.gov/2020051134

Cover Design: Wiley
Cover Image: © merrymoonmary/E+/Getty Images

Set in 10/12pt, TimesLTStd by SPi Global, Chennai, India.
Printed and bound by CPI Group (UK) Ltd, Croydon, CR0 4YY

C9781119714903_220421

To my wife Sima, our children Dolav, Ariel, Dror, Yoed and their spouses, and our grandchildren Yonatan, Alma, Tomer, Yadin, Aviv, Gili, Matan, Eden, Ethan.
RSK

To my wife Hanna, our sons Yuval and David, and their families with love.
SZ

To Anna and my family.
DA

Contents

Also available on book's website: www.wiley.com/go/modern_industrial_statistics

Appendix I: Introduction to R
Appendix II: Introduction to MINITAB and Matrix Algebra for Statistics
Appendix III: List of R scripts, by chapter
Appendix IV: Data sets as csv Files
Appendix V: MINITAB macros
Appendix VI: JMP scripts

Preface to Third Edition

This third edition of modern industrial statistics with applications in R, MINITAB, and JMP consists of a reorganization of the material in the second edition and four new chapters. It includes two parts. Part I, titled *Modern Statistics: A Computer-Based Approach* consists of eight chapters including three new chapters: (1) an introduction to computer age statistics and analytics (Chapter 1), (2) a chapter on time series analysis and predictions (Chapter 7), and (3) a chapter on modern analytic methods with an introduction to decision trees, clustering methods, naive Bayes classifiers, functional data analysis, and text analytics (Chapter 8). Part I provides modern foundations to Part II that is dedicated to industrial statistics methods and applications. Part I can be used as textbook in introductory courses in statistics and analytics.

Part II, titled *Modern Industrial Statistics: Design and Control of Quality and Reliability* builds on parts II–V in the second edition and is particularly well suited to address challenges posed by the 4th industrial revolution. Part II covers statistical process control, the design of experiments, computer experiments, reliability analysis, including Bayesian reliability, and acceptance sampling, also involving sequential methods. Such methods are used, for example, in A/B testing of web applications. Bayesian inference is treated extensively in the 3rd edition. Part II can be used in courses in industrial statistics at both the introductory and the advanced levels.

Like the 2nd edition, the book refers to three software platforms, R, MINITAB, and JMP. It includes specially designed applications that provide several simulation options. We believe that teaching modern industrial statistics, with simulations, provides the right context for gaining sound hands on experience. This is especially accentuated in the context of blended education combining on-line and offline interactions with students. To take full advantages of this book, both students and instructors must be willing to try out different models and methods and spend time "learning by doing." To achieve this, we provide over 50 data sets representing real-life case studies which are typical of what one finds while performing statistical work in business and industry.

The book comes with a set of appendices available for download from the book's website. Appendix I (online material) is an introduction to R contributed by Stefano Iacus, a member of the R core team, and we thank him for that. Appendix II (online material) is about basic MINITAB Commands and a review of matrix algebra for statistics. With this third edition, the R *mistat* application you can download from the CRAN website has been updated and so has Appendix III (online material)R_scripts which contains files with R scripts. With *mistat* installed, these files reproduce all the R analysis and outputs you see in the chapters, the script file number corresponding to the book's chapter. In preparing this third edition, we got feedback and comments from many experts in industry and academia. In particular, we would like to gratefully acknowledge the inputs of Alessandro Di Bucchianico, Shirley Coleman, Ian Cox, Fred Faltin, Peter Gedeck, Blan Godfrey, Anan Halabi, Ran Jin, Anne Milley, Malcolm Moore, Mojgan Naeeni, Matt Rowson, Jean Michel Poggi, Naomi Rom, Fabrizio Ruggeri, Galit

Shmueli, David Steinberg, Amit Teller, Lina Tepper, Bill Woodall, Emmanuel Yashchin, and Avigdor Zonnenshain.

We obviously look forward to feedback, comments and suggestions from students, teachers, researchers, and practitioners and hope the book will help these different target groups achieve concrete and significant impact with tools and methods of industrial statistics.

Ron S. Kenett
Raanana and Technion, Haifa, Israel

Shelemyahu Zacks
McLean, VA, USA
2020

Preface to Second Edition

This book is about modern industrial statistics and its application using R, MINITAB, and JMP. It is an expanded second edition of a book entitled *Modern Industrial Statistics: Design and Control of Quality and Reliability*, Duxbury/Wadsworth Publishing, 1998. In this second edition, we provide examples and procedures in the now popular R language and also refer to MINITAB and JMP. Each of these three computer platforms carry unique advantages. Exercises are provided at the end of each chapter, in order to create opportunities to learn and test your knowledge. R is an open source programming language and software environment for statistical computing and graphics based on the S programming language created by John Chambers while at Bell Labs in 1976. It is now developed by the R Development Core Team, of which Chambers is a member. MINITAB is a statistics package developed at the Pennsylvania State University by researchers Barbara Ryan, Thomas Ryan, Jr., and Brian Joiner in 1972. MINITAB began as a light version of OMNITAB, a statistical analysis program developed at the National Bureau of Standards now called the National Institute of Standards and Technology (NIST). JMP was originally written in 1989 by John Sall and others to perform simple and complex statistical analyses by dynamically linking Statistics with graphics to interactively explore, understand, and visualize data. JMP stands for John Macintosh Project and it is a division of SAS Institute Inc. (SAS Institute Inc. 1983).

A clear advantage of R is that it is free open source software. It requires, however, knowledge of command language programming. MINITAB is a popular statistical software application providing extensive collaboration and reporting capabilities. JMP, a product of the SAS company, is also very popular and carries advanced scripting features and high-level visualization components. Both R and JMP have fully compatible versions for Mac OS. We do not aim to teach programming in R or using MINITAB or JMP, our goal is to provide students, researchers, and practitioners of modern industrial statistics with examples of what can be done with these software platforms. An Appendix (online material) provides an introduction to R. Installations of JMP and MINITAB include effective tutorials with introductory material. Such tutorials have not been replicated in this text.

The three software platforms we use also provide simulation options. We believe that teaching modern industrial statistics, with simulations, provides the right context for gaining sound hands-on experience. We aim at the middle road target, between theoretical treatment and how to approach. To take full advantages of this book, both students and instructors must be willing to try out different models and methods and spend time "learning by doing." To achieve this, we provide over 30 data sets representing real-life case studies which are typical of what one finds while performing statistical work in business and industry. Figures in the book have been produced with R, MINITAB, and JMP.

The material on the book website should be considered part of the book. We obviously look forward to feedback, comments, and suggestions from students, teachers, researchers, and practitioners and hope the book will help these different target groups achieve concrete and significant impact with tools and methods of industrial statistics.

Ron S. Kenett
Raanana, Israel and Turin, Italy

Shelemyahu Zacks
Binghamton, NY, USA
2014

Preface to First Edition

Modern Industrial Statistics provides the tools for those who drive to achieve perfection in industrial processes. Learn the concepts and methods contained in this book and you will understand what it takes to measure and improve world-class products and services.

The need for constant improvement of industrial processes, in order to achieve high quality, reliability, productivity, and profitability, is well recognized. Further management techniques, such as total quality management and business process reengineering, are insufficient in themselves without the strong backing by management of specially tailored statistical procedures.

Statistical procedures designed for solving industrial problems are called Industrial Statistics. Our objective in writing this book was to provide statistics and engineering students, as well as practitioners, the concepts, applications, and practice of basic and advanced industrial statistical methods, which are designed for the control and improvement of quality and reliability.

The idea of writing a text on industrial statistics developed after several years of collaboration in industrial consulting, teaching workshops, and seminars, and courses at our universities. We felt that no existing text served our needs in both content and approach, so we decided to develop our notes into a text. Our aim was to make the text modern and comprehensive in terms of the techniques covered, lucid in its presentation, and practical with regard to implementation.

Ron S. Kenett
Binghamton, NY, USA and Raanana, Israel

Shelemyahu Zacks
Binghamton, NY, USA
1998

List of Abbreviations

ANOVA   analysis of variance
ANSI   American National Standard Institute
AOQ   average outgoing quality
AOQL   average outgoing quality limit
AQL   acceptable quality level
ARIMA   autoregressive integrated moving average
ARL   average run length
ASN   average sample number
ASQ   American Society for Quality
ATI   average total inspection
BECM   Bayes estimation of the current mean
BI   business intelligence
BIBD   balanced incomplete block design
BN   Bayesian network
BP   bootstrap population
c.d.f. cumulative distribution function
CAD   computer aided design
CADD   computer aided drawing and drafting
CAM   computer aided manufacturing
CART   classification and regression trees
CBD   complete block design
CED   conditional expected delay
cGMP   current good manufacturing practices
CHAID   chi-square automatic interaction detector
CIM   computer integrated manufacturing
CLT   central limit theorem
CMM   coordinate measurement machines
CMMI   capability maturity model integrated
CNC   computerized numerically controlled
CPA   circuit pack assemblies
CQA   critical quality attribute
CUSUM   cumulative sum
DACE   design and analysis of computer experiments
DAG   directed acyclic graph
DFIT   difference in fits distance
DLM   dynamic linear model
DoE   Design of Experiments
DTM   document term matrix
EBD   empirical bootstrap distribution

ETL   Extract−Transform−Load
EWMA   exponentially weighted moving average
FDA   Food and Drug Administration
FDA   functional data analysis
FPCA   functional principal component analysis
FPM   failures per million
GFS   Google File System
GRR   gage repeatability and reproducibility
HPD   highest posterior density
HPLC   high-performance liquid chromatography
i.i.d.   independent and identically distributed
IDF   inverse document frequency
InfoQ   Information Quality
IPO   initial public offering
IPS   inline process control
IQR   inter quartile range
ISC   short circuit current of solar cells (in ampere)
KS   Kolmogorov−Smirnov test
LCL   lower control limit
LLN   law of large numbers
LQL   limiting quality level
LSL   lower specification limit
LTPD   lot tolerance percent defective
LWL   lower warning limit
m.g.f.   moment generating function
MLE   maximum likelihood estimator
MSD   mean squared deviation
MTBF   mean time between failures
MTTF   mean time to failure
NID   normal independently distributed
OAB   one-armed bandit
OC   operating characteristic
p.d.f.   probability density function
PCA   principal component analysis
PERT   project evaluation and review technique
PFA   probability of false alarm
PL   product limit estimator
PPM   defects in parts per million
PSE   practical statistical efficiency
QbD   Quality by Design
QMP   quality measurement plan
QQ-plot   quantile versus quantile plot
RCBD   randomized complete block design
Regex   regularized expression
RMSE   root mean squared error
RSWOR   random sample without replacement
RSWR   random sample with replacement

SE	standard error
SL	skip lot
SLOC	source lines of code
SLSP	skip lot sampling plans
SPC	statistical process control
SPRT	sequential probability ratio test
SR	Shiryaev–Roberts
SSE	sum of squares of errors
SSR	sum of squares around the regression model
SST	total sum of squares
STD	standard deviation
SVD	singular value decomposition
TAB	two-armed bandit
TTC	time till censoring
TTF	time till failure
TTR	time till repair
TTT	total time on test
UCL	upper control limit
USL	upper specification limit
UWL	upper warning limit
WLLN	weak law of large numbers
WSP	wave soldering process

Part I

Modern Statistics: A Computer-Based Approach

1

Statistics and Analytics in Modern Industry

1.1 Analytics, big data, and the fourth industrial revolution

During the past decade, industries in advanced economies have experienced significant changes in processes, technologies, and their engineering and manufacturing practices. This phenomenon is often referred to as the fourth industrial revolution or Industry 4.0. It is based on advanced manufacturing and engineering technologies, massive digitization, big data analytics, robotics, adaptive automation, additive and precision manufacturing (e.g., 3D printing), modeling and simulation, artificial intelligence, and nano-engineering of materials (Kenett et al. 2020). This revolution presents challenges and opportunities to analytics, systems and manufacturing engineering, and material science disciplines. The term "Industry 4.0," or Industrie 4.0, as originally coined by a 2011 initiative of the Federal German Ministry of Education and Research, epitomizes the convergence of traditional manufacturing technologies with the exploding information and communication technologies. As noted in *Bundesministerium fur Bildung und Forschung* (BMBF), 2016 (translated from German): "The economy is on the threshold of the fourth industrial revolution. Driven by the Internet, real and virtual worlds grow together into one Internet of Things (IoT). With the project Industry 4.0 we want to support this process."

The fourth industrial revolution is fueled by data from sensors and IoT devices and driven by increasing computer power and flexible manufacturing capabilities. Information technology, telecommunications, and manufacturing are merging, and production is increasingly autonomous. Futurists talk of machines that organize themselves, delivery chains that automatically assemble themselves, and applications that feed customer orders directly into production. In this context, industrial statistics and data analytics play an expanded role in industrial applications. Early signs of this expanded role of data analysis were available half a century ago. To quote John Tukey (1962): "For a long time I have thought I was a statistician, interested in inferences from the particular to the general. But, as I have watched mathematical statistics evolve, I have had cause to wonder and to doubt. ... All in all, I have come to feel that

Modern Industrial Statistics: With Applications in R, MINITAB and JMP, Third Edition.
Ron S. Kenett and Shelemyahu Zacks.
© 2021 John Wiley & Sons, Ltd. Published 2021 by John Wiley & Sons, Ltd.

my central interest is in data analysis, which I take to include, among other things: procedures for analyzing data, techniques for interpreting the results of such procedures, ways of planning the gathering of data to make its analysis easier, more precise or more accurate, and all the machinery and results of (mathematical) statistics which apply to analyzing data.", a insight providing a vision for current industrial statistics, statistics, and data science in general.

These are only a few examples of problem areas where the tools of industrial statistics are used within modern industrial and service organizations. In order to provide more specific examples, we first take a wide angle look at a variety of industries. Later, we discuss examples from such industries.

There are basically three types of production systems: (1) continuous flow production, (2) job shops, and (3) discrete mass production. Examples of continuous flow production include steel, glass, and paper making, thermal power generation, and chemical transformations. Such processes typically involve expensive equipment that is very large in size, operates around the clock, and requires very rigid manufacturing steps. Continuous flow industries are both capital-intensive and highly dependent on the quality of the purchased raw materials. Rapid customizing of products in a continuous flow process is extremely challenging, and new products are introduced using complex scale-up procedures. We expand on these types of production processes in Chapter 9. With this background on modern industry, in the next section, we provide a broad overview of computer age analytics. Part I of the book is about statistical analysis and analytics. Part II is about industrial statistics and analytics in modern industry.

1.2   Computer age analytics

There are three main goals in analyzing data (Efron and Hastie 2016):

1. *Prediction*: To predict the response to future input variables.
2. *Estimation*: To infer how response variables are associated with input variables.
3. *Explanation*: To understand the relative contribution of input variables to response values.

Predictive modeling is the process of applying statistical models or data mining algorithms to data, for the purpose of predicting new or future observations. In other words, a predictive model is a model generating forecasts, whatever the method or algorithm used. In contrast, explanatory models aim to explain the relationship between the independent variables and the dependent variable. Classical statistics focuses on modeling the stochastic system generating the data. Statistical learning, or computer age statistics, builds on big data and the modeling of the data itself. If the former aimed at properties of the model, the latter is looking at the properties of computational algorithms. These are two different cultures (Breiman 2001).

Frequentism (or "objectivism") is based on the probabilistic properties of a procedure applied to the output from using it on observed data. This provides us with an assessment of bias and variance. The frequentists interpretation is based on a scenario where the same situation is repeated, endlessly. The Neyman–Pearson lemma provides an optimum hypothesis testing algorithm, where a black-and-white decision is made. You either reject the null hypothesis in testing for an alternative hypothesis, or not. This offers an apparently simple and effective way to conduct statistical inference and can be scaled up to an industrialized type application of statistics. Confidence intervals (CI), in contrast to hypothesis testing, are considered by many as more informative. However, they are also the object of criticism, like p-value hacking where self-selection biases the analysis. Barnett and Wren (2019) demonstrate the wide prevalence of

CI hacking. We discuss hypothesis testing and CI methods in Chapter 4. Alternatively, statistical analysis can be conducted within a Bayesian framework by transforming a prior distribution on the parameters of interest, to a posterior, using the observed data. In this probabilistic framework, one often calculates the Bayes factor, a ratio of the marginal likelihood of two competing hypotheses, usually a null and an alternative. The Bayes factor is a sort of Bayesian alternative to classical hypothesis testing. Bayesian inference is introduced in Section 4.8 and discussed at length in the context of Bayesian reliability in Chapter 17. Bayesian methods used in testing websites are introduced in Section 18.5.

In computer age analytics, like in frequentist or Bayesian statistics, one distinguishes between algorithms aiming at estimation, prediction, and explanations of structure in the data. Estimation is assessed by accuracy of estimators, prediction-by-prediction error, and explanations are based on variable selection using variance bias trade offs, penalized regression, and regularization criteria. These three types of goals and utility functions are considered in the information quality framework described in Section 1.4.

Overall, Part I of this book (Chapters 1–8) is an introduction to statistics in modern industry and to statistical analytics in frequentist, Bayesian, or computer age variants. Typical industrial problems are described, and basic statistical concepts and tools are presented through case studies and computer simulations. To help focus on data analysis and interpretation of results, we refer to three leading software platforms for statistical analysis, R, MINITAB, and JMP. R is an open source programming language available at the Comprehensive R Archive Network (http://cran.r-project.org/). MINITAB and JMP are commercial statistical packages widely used in business and industry (www.minitab.com, www.jmp.com) with 30 days fully functional free downloads. Chapter 2 presents basic concepts and tools for describing variability. It emphasizes data visualization techniques used to explore and summarize variability in observations. The chapter introduces the reader to R and provides examples in MINITAB and JMP. Chapter 3 is an introduction to probability models, including a comprehensive treatment of statistical distributions that have applicability to industrial statistics. The chapter provides a reference to fundamental results and basic principles used in later chapters. Chapter 4 is dedicated to statistical inference. It covers hypothesis testing, as used prevalently in industry, and provides an introduction to Bayesian decision procedures and to nonparametric computer intensive bootstrapping methods. Bayesian methods transform, with data, prior distributions to posterior distributions generating credibility intervals. The chapter presents basic methods that rely on computer-intensive methods with examples in R, MINITAB, and JMP. The chapter shows how statistical procedures are used in making inference from a sample to a population. Chapter 5 deals with variability in several dimensions and regression models. It begins with graphical techniques that handle observations taken on several variables, simultaneously. Linear and multiple regression models are covered including diagnostics and prediction intervals. Categorical data and contingency tables are also analyzed. Chapter 6 presents foundations for estimation of finite population quantities using samples. It shows how to account for stratified data and known covariates. This is especially important in big data applications, where stationarity cannot be validated and variability in the data structure over time is to be expected. Chapter 7 is about time series analysis with emphasis on forecasting and prediction. Chapter 8 provides a brief introduction to computer age analytics such as supervised and unsupervised learning, text analytics, and functional data analysis. Supervised learning applies to data with one or more target responses (Y variables) and several covariate inputs (X variables). As an example, decision trees are used to create a model that predicts the value of a target variable based on several input variables. Trees are "learned" by splitting the data into subsets based on prioritized input variables. This is repeated in a recursive manner called recursive partitioning. The recursion is completed when the subset at a node has a unique value of the target variable, or when splitting no longer adds value to the predictions. Other approaches, based on decision trees, are random forests and boosting. In random forests, one grows many decision trees to randomized versions of the training data and averages them. In boosting one repeatedly grows trees and builds up an additive model consisting of a sum of trees. Unsupervised learning is performed when you do not have target variables, only covariates you want to cluster. Some of the clustering methods include hierarchical clusters and K-means clusters. We also introduce functional data analysis where functions are analyzed, for example, when your data consists of profiles over time. Text analytics deals with unstructured data. A document is represented by the presence or count of words in a dictionary. This leads to wide document-term matrices (DTM). We present such techniques with examples and for more details refer the interested reader to Hastie et al. (2009).

1.3   The analytics maturity ladder

In observing organizations implementing statistics and analytics, we identify five types of behavior:

- *Level 1*: Random reports to be delivered yesterday.
- *Level 2*: A focus on descriptive statistics.
- *Level 3*: Capturing variability with models and statistical distributions.
- *Level 4*: Planning interventions and experiments for data gathering.
- *Level 5*: A holistic view of data analytics aimed at learning and discovery.

Level 1 reflects a basic level of maturity of organizations, so chaotic that people cannot think beyond the short-term. Level 1 organizations work on quick-fix solutions. The work is frantic. In such environments, the data analyst or statistician, is asked for reports to be urgently produced. These reports are typically shallow and lack insights. They are shallow because they focus only on immediately available data, and they lack insights because management does not give analysts an opportunity to reflect on the findings.

Level 2 companies routinely produce reports with descriptive statistics and run charts in tables and figures. These are often based on business intelligence (BI) applications and focused on productivity and quality indicators. An example is *in line process control* (IPC) applications that screen defective products and track the yield of a production process. When collected over time, such data provides a rear-view mirror perspective. The typical report in such organizations is based on dashboards with basic visuals such as run charts, bar charts, or pie charts. Just as it is impossible to drive a car by looking in the mirror, it is also difficult to run an organization with only a perspective on historical data. Drivers, and managers, need to look ahead, through the windshield.

Organizations at Level 3 deploy forward-looking capabilities. The science of monitoring took a leap forward with the control chart, invented by Walter Shewhart at Bell Laboratories in the 1920s. Shewhart explicitly embraced variation in his formulation of control limits. The chart triggers an alarm, signaling that a process is out-of-control, when observations go beyond the control limits typically set up at three standard deviations above and below the mean. This approach is based on a statistical distribution representing the performance of a stable process. The alarm indicates that the underlying distribution has changed and the process has gone "out-of-control." Instead of relying on product inspection, the control chart lets you control

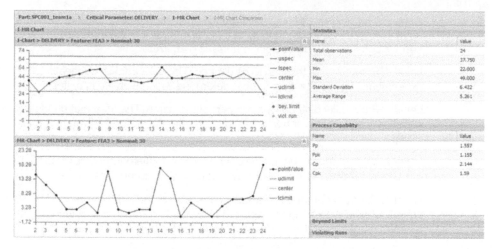

Figure 1.1   *Control chart on individual measurements (upper chart tracks the measurements, lower chart their moving range representing variability). The right panel provides summary statistics (figure obtained by permission from the SPCLive system by KPA Ltd.).*

the production process itself (Shewhart 1926). Control charts are presented in Chapters 10–12 of Part II. Figure 1.1 is an example of a control chart from a modern cloud hosted web-based system using data from probes measuring dimensions using coordinate measurement machines (CMM). The application integrates measurements with work orders and information on critical parameters available in engineering design systems. Operator interventions are annotated, and sophisticated diagnostic and prognostic models provide to the plant manager opportunities for enhanced quality and efficiencies. A brief video on the system can be seen at https://vimeo.com/285180512. It is an example of Industry 4.0 applications.

Ambitious companies move up to the next level of the data analytics maturity curve, Level 4. Here managers extend forward-looking thinking into the design of products and services. This stage requires statistically designed experiments, robust design, and other methods to ensure that the product/service meets customer needs and performs well, even when the raw materials are of uneven quality and environmental conditions vary. Genichi Taguchi and Joseph M. Juran played key roles in developing this approach. In the 1980s, Taguchi introduced the West to methods he developed in Japan in the 1950s (Taguchi 1987). Juran described a structured approach for quality planning that starts with examining customer needs and ensuring those needs were met in the final product (Juran 1988). In a Level 4 organization, the data analyst understands the role of experimental design and is proactive in planning improvement experiments. This is a precursor of A/B testing (Kohavi and Thomke 2017) where web application designers direct customers to alternative designs to learn which works best, using behavioral data such as click-through-rates. A complementary aspect is the design and evaluation of product and system reliability. Chapters 13–19 in Part II cover these topics.

Neither Taguchi nor Juran anticipated the Big Data era, with data coming from all directions, including social media, web clicks, "connected devices" (e.g., the Internet of Things), personal trackers, and so forth. This new age poses new challenges and opportunities and suggests yet another maturity level, level 5.

Level 5: This is where attention is paid to information quality. Data from different sources is integrated. Chronology of Data and Goal and Generalization is a serious consideration in designing analytic platforms.

Level 4: Experimental thinking is introduced. The data analyst suggests experiments, like A/B testing, to help determine which website is better.

Level 3: Probability distributions are part of the game. The idea that changes are statistically significant, or not, is introduced. Some attention is given to model fitting.

Level 2: Management asks to see dashboards with histograms, bar charts, and run charts. Models are not used, data is analyzed in rather basic ways.

Level 1: Random demand for reports that address questions such as: How many components of type X did we replaced last month or how many customers in region Y required service?

Figure 1.2   *The analytics maturity ladder.*

Level 5 organizations are focused on extracting knowledge from data and generating information quality. Figure 1.2 presents the five maturity levels with a brief description of how data is used at each level. We expand on an information quality framework in Section 1.4.

In proposing this maturity ladder, we emphasize the need for organizations to move up to the learning and discovery phase characterizing Level 5. Senge 1990 also emphasized the importance of doing so.

1.4   Information quality

Kenett and Shmueli (2014, 2016) define information quality as an approach to assess the level of knowledge generated by analyzing data with specific methods, given specific goals.

Formally, let:

- g A specific analysis goal
- X The available dataset
- f An empirical analysis method
- U A utility measure

A goal could be to keep a process under control. We mentioned earlier three types of utility functions matching the predictive, estimation, and explanation goals of statistical analysis. In process control applications, the available data could be a rational sample of size 5, collected every 15 minutes. In such a case, the analysis method can be a an Xbar-R or Xbar-S control chart (see Chapter 10).

Information quality (InfoQ) is defined in Kenett and Shmueli (2014, 2016) as: $InfoQ(U, f, X, g) = U(f(X|g))$, i.e., the utility U derived by conducting a certain analysis, f, on a given dataset X, conditioned on a goal g. In terms of our example, it is the economic value derived from applying an Xbar-R or Xbar-S chart on the sample data in order to keep the process under control.

To achieve high InfoQ we consider eight dimensions:

1. *Data resolution*: This is determined by measurement granularity, measurement uncertainty, and level of data aggregation, relative to the task at hand. The concept of rational sample is such an example.
2. *Data structure*: This relates to the data sources available for the specific analysis. Comprehensive data sources combine structured quantitative data with unstructured, semantic-based data derived from video images or machine logs.
3. *Data integration*: Properly combining data sources is sometimes not a trivial task. Information system experts apply Extract–Transform-Load (ETL) technologies to integrate data sources with aliased nomenclature and varying time stamps.
4. *Temporal relevance*: A data set contains information collected during a certain time window. The degree of relevance of the data in that time window, to the current goal at hand, must be assessed. Data collected a year ago might not be relevant anymore in characterizing process capability.
5. *Generalizability*: Statistical generalizability refers to inferring from a sample to a target population. Sampling from a batch implies that decisions based on the sample apply to the whole batch. Generalizability is also conducted using first principles and expert opinion.
6. *Chronology of data and goal*: This is obvious. If a control chart is updated once a month, ongoing responsive process control cannot be conducted.
7. *Operationalization*: Construct operationalization is about determining how to measure abstract elements. Action operationalization is about deriving concrete actions from analyzing data. Quoting Deming (1986): "An operational definition is a procedure agreed upon for translation of a concept into measurement of some kind."
8. *Communication*: If the information does not reach the right person at the right time, then the quality of information is necessarily poor. Data visualization is directly related to the quality of information. Poor visualization of findings can lead to degradation of the information quality contained in the analysis performed on the data.

A JMP add in for summarizing the information quality assessment is available from https://community.jmp.com/kvoqx44227/attachments/kvoqx44227/add-ins/338/1/InfoQ.jmpaddin. Examples showing how JMP is used to generate information quality are available from https://www.jmp.com/en_us/whitepapers/book-chapters/infoq-support-with-jmp.html. Kenett and Shmueli (2016) provide many examples on how to derive an overall InfoQ score. In considering the various tools and methods of industrial statistics presented in this book, one should keep in mind that the ultimate objective is high-information quality or InfoQ. Achieving high InfoQ is however necessary but not sufficient to make the application of industrial statistics both effective and efficient. InfoQ is about using statistical and analytic methods effectively so that they generate the required knowledge. Efficiency is related to the organizational impact of the methods used. Level 5 organizations achieve high efficiency by generating information quality.

The chapters in Part I cover general tools and methods of statistics and data analytics that will get you to move up the analytic maturity ladder in Figure 1.2. Part II is about the tools and methods of industrial statistics including control charts, the design of experiments, models of reliability, and sequential tests. The goal in Part II is to help organization go up a quality ladder introduced in Chapter 9.

1.5   Chapter highlights

This chapter introduces Part I of the book which is about methods and tools of statistics and computer age analytics. It starts with a description of the advanced manufacturing ecosystem, sometimes labeled Industry 4.0 or the fourth industrial revolution. This creates opportunities and challenges for analytics. Part I is a general introduction to analytics, starting from basic descriptive methods, followed by a review of probability distributions, classical inference and bootstrapping, Bayesian analysis, regression models and multivariate techniques, time series forecasting and a review of statistical learning methods such as decision trees, clustering, functional data analysis, and text analytics. It refers to an example of a state-of-the-art web-based statistical process control system used to monitor processes in metalwork companies. To deploy such systems, requires adequate analytic maturity. An analytic maturity ladder is described in Section 1.3. Level 5 is reached by organizations effectively generating information quality. A framework for assessing information quality is also presented.

The main terms and concepts introduced in this chapter include

- Advanced Manufacturing
- Industry 4.0
- Analytics in Manufacturing Applications
- Computer Age Analytics
- The Analytics Maturity Ladder
- Information Quality

1.6   Exercises

1.1 Identify a friend or relative who works in a manufacturing plant. Present to him (or her) the five levels of the analytic maturity ladder and determine jointly the position of that plant on the ladder.

1.2 Evaluate the InfoQ dimensions of the case study on predicting days with unhealthy air quality in Washington DC. Several tour companies' revenues depend heavily on favorable weather conditions. This study looks at air quality advisories, during which people are advised to stay indoors, within the context of a tour company in Washington DC. https://www.galitshmueli.com/data-mining-project/tourism-insurance-predicting-days-unhealthy-air-quality-washington-dc.

1.3 Evaluate the InfoQ dimensions of the case study on quality-of-care factors in US nursing homes. Thousands of Americans reside in nursing homes across the United States, with facilities spanning a wide range. This study looks at the quality of care in nursing homes in the United States. https://www.galitshmueli.com/data-mining-project/quality-care-factors-us-nursing-homes.

1.4 Evaluate the InfoQ dimensions of a case study of corporate earnings in relation to an existing theory of business forecasting developed by Joseph H. Ellis (former research analyst at Goldman Sachs). http://www.galitshmueli.com/data-mining-project/predicting-changes-quarterly-corporate-earnings-using-economic-indicators.

1.5 Zillow.com is a free real estate service that calculates an estimated home valuation ("Zestimate") as a starting point for anyone to see for most homes in the United States. Evaluate the InfoQ dimensions of a case study that looks at the accuracy of Zestimates. http://www.galitshmueli.com/data-mining-project/predicting-zillowcom-s-zestimate-accuracy.

1.6 An Initial Public Offering (IPO) is the first sale of stock by a company to the public. Evaluate the InfoQ dimensions of a case study that looks at the first-day returns on IPOs of Japanese companies. http://www.galitshmueli.com/data-mining-project/predicting-first-day-returns-japanese-ipos.

2

Analyzing Variability: Descriptive Statistics

2.1 Random phenomena and the structure of observations

Many phenomena which we encounter are only partially predictable. It is difficult to predict the weather or the behavior of the stock market. In this book, we focus on industrial phenomena, like performance measurements from a product which is being manufactured or the sales volume in a specified period of a given product model. Such phenomena are characterized by the fact that measurements performed on them are often not constant but reveal a certain degree of variability. The objective of this chapter is to present methods for analyzing this variability, in order to understand the variability structure and enhance our ability to control, improve, and predict future behavior of such phenomena. We start with a few simple examples.

Example 2.1. A piston is a mechanical device that is present in most types of engines. One measure of the performance of a piston is the time it takes to complete one cycle. We call this measure **cycle time**. In Table 2.1, we present 50 cycle times of a piston operating under fixed operating conditions (a sample data set is stored in file **CYCLT.csv**). We provide with this book code in R and JMP for running a piston software simulation. If you installed JMP, download from the book website the file **com.jmp.cox.ian.piston.jmpaddin** and double click on it. You will get access to a piston simulator with seven factors that you can change using interactive sliders (see Figure 15.1). Set the number of samples to 50, leave the sample size as 1 and click *Run* at the bottom of the screen. This will generate 50 cycle times that are determined by how you set up the factors on the sliders and your computer random number generator. JMP produces an output file with several graphical displays and summary statistics that we will discuss later. We will use this simulator when we discuss Statistical Process Control (Chapters 10–12) and the Design of Experiments (Chapters 13–15). We continue at a pedestrian pace by recreating Table 2.1 using R. All the R applications referred to in this

Modern Industrial Statistics: With Applications in R, MINITAB and JMP, Third Edition.
Ron S. Kenett and Shelemyahu Zacks.
© 2021 John Wiley & Sons, Ltd. Published 2021 by John Wiley & Sons, Ltd.

book are contained in a package called `mistat` available in the CRAN R repository. The following R commands will install the `mistat` package, read the cycle time data, and print them on your monitor:

```
> # This is a comment
> install.packages("mistat",          # Install mistat package
                   dependencies=TRUE) # and its dependencies
>                      #
> library(mistat)      # A command to make our datasets
>                      # and functions available
>                      #
> data(CYCLT)          # Load specified data set
>                      # CYCLT is a vector of values
>                      #
> help(CYCLT)          # Read the help page about CYCLT
>                      #
> CYCLT                # Print CYCLT to Console
```

Notice that functions in R have parenthesis. The `library()` function loads an additional package to extend R functionalities and CYCLT is an object containing a simple vector of values.

The differences in cycle times values is quite apparent, and we can make the statement "cycle times are varying." Such a statement, in spite of being true, is not very useful. We have only established the existence of variability – we have not yet characterized it and are unable to predict and control future behavior of the piston.

Table 2.1 *Cycle times of piston (in seconds) with control factors set at minimum levels*

1.008	1.117	1.141	0.449	0.215
1.098	1.080	0.662	1.057	1.107
1.120	0.206	0.531	0.437	0.348
0.423	0.330	0.280	0.175	0.213
1.021	0.314	0.489	0.482	0.200
1.069	1.132	1.080	0.275	0.187
0.271	0.586	0.628	1.084	0.339
0.431	1.118	0.302	0.287	0.224
1.095	0.319	0.179	1.068	1.009
1.088	0.664	1.056	1.069	0.560

Example 2.2.   Consider an experiment in which a coin is flipped once. Suppose the coin is fair in the sense that it is equally likely to fall on either one of its faces. Furthermore, assume that the two faces of the coin are labeled with the numbers "0" and "1." In general, we cannot predict with certainty on which face the coin will fall. If the coin falls on the face labeled "0," we assign to a variable X the value 0; if the coin falls on the face labeled "1," we assign to X the value 1. Since the values which X will obtain in a sequence of such trials cannot be

predicted with certainty, we call X a **random variable**. A typical random sequence of 0, 1 values that can be generated in this manner might look like the following:

$$0, 1, 1, 0, 1, 0, 1, 0, 1, 1, 1, 1, 1, 0, 1, 0, 1, 1, 1, 1,$$
$$0, 0, 1, 1, 1, 1, 1, 0, 0, 1, 0, 1, 1, 1, 0, 0, 0, 1, 0, 1.$$

In this sequence of 40 random numbers, there are 15 0's and 25 1's. We expect in a large number of trials, since the coin is unbiased, that 50% of the random numbers will be 0's and 50% of them will be 1's. In any particular short sequence, the actual percentage of 0's will fluctuate around the expected number of 50%.

At this point, we can use the computer to **"simulate"** a coin tossing experiment. There are special routines for generating random numbers on the computer. We will illustrate this by using the R environment. The following commands generate a sequence of 50 random binary numbers (0 and 1). A similar sequence can be created with *Random Functions* in JMP or *Random Data* in MINITAB.

```
> X <-                       # Assign to object X
    rbinom(n = 50,           # 50 pairs of binomial variates
           size = 1,         # set the number of trials
           prob = 0.5)       # set the probability of success
>                            #
> X                          # Equivalent to command print(X)
>                            #
> ls()                       # List the available objects
>                            #
> rm(X)                      # Remove object X
```

The command uses a binomial function (`rbinom`) with the number of trials (argument size) set to 1, the first argument n specifies the number of observations and `prob` is the probability of 1s. Execute this command to see another random sequence of 50 0s and 1s. Compare this sequence to the one given earlier.

Example 2.3. Another example of a random phenomenon is illustrated in Figure 2.1 where 50 measurements of the length of steel rods are presented. These data are stored in file **STEELROD.csv**. To generate Figure 2.1, in R type at the prompt the following commands:

```
> data(STEELROD)                        # STEELROD is a vector
>                                        #
> plot(STEELROD,                         # Plot vector STEELROD
       ylab = "Steel rod Length",        # set y axis title
       xlab = "Index")                   # set x axis title
```

Steel rods are used in the car and truck industry to strengthen vehicle structures. Automation of assembly lines has created stringent requirements on the physical dimensions of parts. Steel rods supplied by Urdon Industries for Peugeot car plants are produced by a process adjusted to obtain rods of length 20 cm. However, due to natural fluctuations in the production process, the actual length of the rods varies around the nominal value of 20 cm. Examination

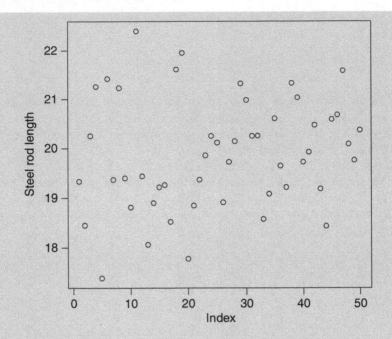

Figure 2.1   *Length of 50 Steel Rods (in cm).*

of this sequence of 50 values does not reveal any systematic fluctuations. We conclude that the deviations from the nominal values are random. It is impossible to predict with certainty what the values of additional measurements of rod length will be. However, we shall learn later that with further analysis of this data, we can determine that there is a high likelihood that new observations will fall close to 20 cm.

It is possible for a situation to arise in which, at some time, the process will start to malfunction, causing a **shift** to occur in the average value of the process. The pattern of variability might then look like the one in Figure 2.2. An examination of Figure 2.2 shows

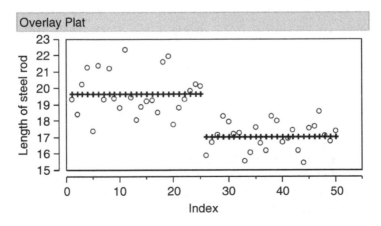

Figure 2.2   *Level shift after the first 25 observations (JMP).*

that a significant shift has occurred in the level of the process after the 25th observation and that the systematic deviation from the average value of the process has persisted constantly. The deviations from the nominal level of 20 cm. are first just random and later systematic and random. The steel rods obviously became shorter. A quick investigation revealed that the process got accidentally misadjusted by a manager who played with machine knobs while showing the plant to important guests.

In formal notation, if X_i is the value of the i-th observation, then

$$X_i = \begin{cases} O + E_i & i = 1, \ldots, 25 \\ N + E_i & i = 26, \ldots, 50, \end{cases}$$

where $O = 20$ is the original level of the process, $N = 17$ is its new level after the shift, and E_i is a random component. Note that O and N are fixed and, in this case, constant nonrandom levels. Thus, a random sequence can consist of values which have two components: a **fixed** component and a **random** component. A fixed-nonrandom pattern is called a **deterministic** pattern. As another example, in Figure 2.3, we present a sequence of 50 values of

$$X_i = D_i + E_i, \quad i = 1, \ldots, 50,$$

where the D_i's follow a sinusoidal pattern shown on Figure 2.3 by dots, and E_i's are random deviations having the same characteristics as those of Figure 2.1. The sinusoidal pattern is $D_i = \sin(2\pi i/50)$, $i = 1, \ldots, 50$. This component can be determined exactly for each i, and is

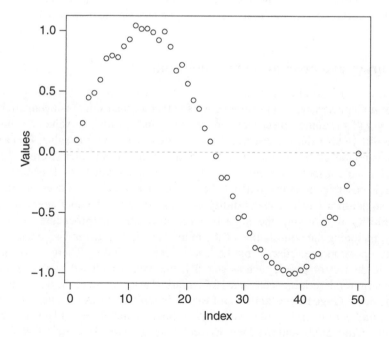

Figure 2.3   *Random variation around a systematic trend.*

therefore called deterministic while E_i is a random component. In R, we can construct such a sequence and plot it with the following commands:

```
> X <-  seq(from=1,          # Assign to X a sequence from 1
            to=50,           # to 50
            by=1)            # increment of sequence
>                            #
> # Equivalent to:
> # X <- 1:50                # Integer sequence from 1 to 50
>                            #
> X <- sin(                  # Reassign X with sine of
    X*(2*pi)/50)             # X*2*pi/50
>                            #
> X <- X + rnorm(n=length(X),# Add to X a random normal
              mean=0,        # component with mean 0
              sd=0.05)       # and standard deviation 0.05
>                            #
> plot(X,
      ylab="Values")
>                            #
> abline(h=0,                # Add a horizontal line at y=0
        lty="dashed",        # set line type
        col="lightgray")     # set line color
```

If the random component could be eliminated, we would be able to predict exactly the future values of X_i. For example, by following the pattern of the D_i's, we can determine that X_{100} would be equal to 0. However, due to the existence of the random component, an exact prediction is impossible. Nevertheless, we **expect**, that the actual values will fall around the deterministic pattern. In fact, certain prediction limits can be assigned, using methods which will be discussed later.

2.2   Accuracy and precision of measurements

Different measuring instruments and gages or gauges (such as weighing scales, voltmeters) may have different characteristics. For example, we say that an instrument is **accurate** if repetitive measurements of the same object yield an average equal to its true value. An instrument is **inaccurate** if it yields values whose average is different from the true value. **Precision**, on the other hand, is related to the dispersion of the measurements around their average. In particular, small dispersion of the measurements reflects high precision, while large dispersion reflects low precision. It is possible for an instrument to be inaccurate but precise, or accurate but imprecise. Precision, sometimes called **Repeatability** is a property of the measurement technology. **Reproducability** is assessing the impact of the measurement procedure on measurement uncertainty, including the contribution of the individuals taking the measurement. Differences between lab operators are reflected by the level of reproducibility. There are other properties of measuring devices or gages, such as stability, linearity, which will not be discussed here. A common term for describing techniques for empirical assessment of the uncertainty of a measurement device is **Gage Repeatability and Reproducability** (GR&R). These involve repeated testing of a number of items by different operators. Such methods are available in the MINITAB *Quality Tools Gage Study* and the JMP *Measurement Systems Analysis* pull-down windows. In addition to a (GR&R) assessment, to ensure proper accuracy, measuring instruments need to be calibrated periodically relative to an external standard. In the United States, the National Institute of Standards and Technologies (NIST) is responsible for such activities.

Example 2.4.   In Figure 2.4, we present weighing measurements of an object whose true weight is 5 kg. The measurements were performed on three instruments, with 10 measurements on each one. We see that instrument *A* is accurate (the average is 5.0 kg.), but its dispersion is considerable. Instrument *B* is not accurate (the average is 2.0 kg.), but is more precise than *A*. Instrument *C* is as accurate as *A* but is more precise than *A*.

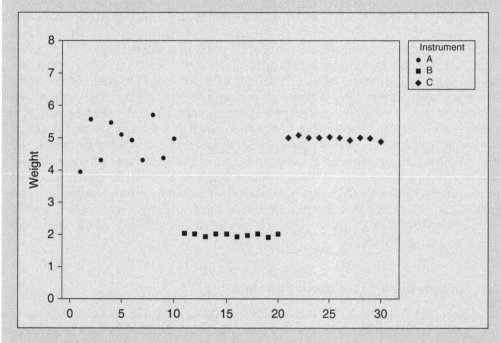

Figure 2.4   *Samples of 10 measurements on three different instruments (MINITAB).*

As a note, it should be mentioned that Repeatability and Reproducibility are also relevant in the wider context of research. For a dramatic failure in reproducibility, see the article by Nobel winner Paul Krugman on the research of Harvard economists, Carmen Reinhart, and Kenneth Rogoff, that purported to identify a critical threshold or tipping point, for government indebtedness. Their findings were flawed because of self-selected data points and coding errors in Excel (http://www.nytimes.com/2013/04/19/opinion/krugman-the-excel-depression.html?_r=0). Another dramatic example of irreproducible research is a Duke university genomic study which proposed genomic tests that looked at the molecular traits of a cancerous tumor and recommended which chemotherapy would work best. This research proved flawed because of errors such as moving a row or a column over by one in a giant spreadsheet and other more complex reasons (http://www.nytimes.com/2011/07/08/health/research/08genes.html). Repeatability in microarray studies is related to identifying the same set of active genes in large and smaller studies. These topics are however beyond the scope of this book.

2.3   The population and the sample

A **statistical population** is a collection of units having a certain common attribute. For example, the set of all the citizens of the USA on January 1, 2010, is a statistical population. Such a population is comprised of many subpopulations, e.g., all males in the age group of 19–25 living in Illinois, etc. Another statistical population is the collection of all concrete cubes of specified dimensions that can be produced under well-defined conditions. The first example of all the citizens of the USA on January 1, 2010, is a **finite** and **real** population, while the population of all units that can be produced by a specified manufacturing process is **infinite** and **hypothetical**.

A **sample** is a subset of the elements of a given population. A sample is usually drawn from a population for the purpose of observing its characteristics and making some statistical decisions concerning the corresponding characteristics of the whole population. For example, consider a lot of 25,000 special screws which were shipped by a vendor to factory *A*. Factory *A* must decide whether to accept and use this shipment or reject it (according to the provisions of the contract). Suppose it is agreed that, if the shipment contains no more than 4% defective items, it should be accepted and, if there are more than 6% defectives, the shipment should be rejected and returned to the supplier. Since it is impractical to test each item of this population (although it is finite and real), the decision of whether or not to accept the lot is based on the number of defective items found in a **random sample** drawn from the population. Such procedures for making statistical decisions are called **acceptance sampling** methods. Chapter 18 is dedicated to these methods. Chapter 6 provides the foundations for estimation using samples from finite populations including random sample with replacement (RSWR) and random sample without replacement (RSWOR). Chapter 4 includes a description of a technique called bootstrapping, which is based on RSWR.

2.4   Descriptive analysis of sample values

In this section, we discuss the first step for analyzing data collected in a sampling process. One way of describing a distribution of sample values, which is particularly useful in large samples, is to construct a **frequency distribution** of the sample values. We distinguish between two types of frequency distributions, namely, frequency distributions of (i) **discrete** variables; and (ii) **continuous** variables.

A random variable, X, is called discrete if it can assume only a finite (or at most a countable) number of different values. For example, the number of defective computer cards in a production lot is a discrete random variable. A random variable is called continuous if, theoretically, it can assume all possible values in a given interval. For example, the output voltage of a power supply is a continuous random variable.

2.4.1   Frequency distributions of discrete random variables

Consider a random variable, X, that can assume only the values $x_1, x_2, \ldots, x_k$, where $x_1 < x_2 < \ldots < x_k$. Suppose that we have made n different observations on X. The frequency of x_i ($i = 1, \ldots, k$) is defined as the number of observations having the value x_i. We denote the frequency of x_i by f_i. Notice that

$$\sum_{i=1}^{k} f_i = f_1 + f_2 + \cdots + f_k = n.$$

The set of ordered pairs

$$\{(x_1, f_1), (x_2, f_2), \ldots, (x_k, f_k)\}$$

constitutes the frequency distribution of X. We can present a frequency distribution in a tabular form as follows:

Value	Frequency
x_1	f_1
x_2	f_2
$\vdots$	$\vdots$
x_k	f_k
Total	n

It is sometimes useful to present a frequency distribution in terms of the proportional or **relative frequencies** p_i, which are defined by

$$p_i = f_i/n \quad (i = 1, \ldots, k).$$

A frequency distribution can be presented graphically in a form which is called a **bar-diagram**, as shown in Figure 2.5. The height of the bar at x_j is proportional to the frequency of this value.

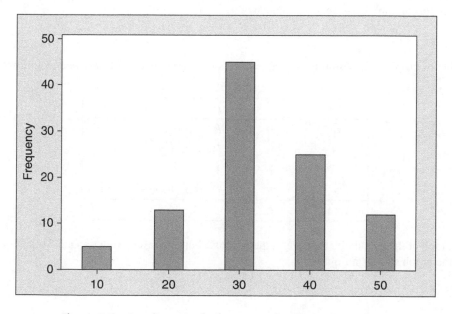

Figure 2.5   *Bar-diagram of a frequency distribution (MINITAB).*

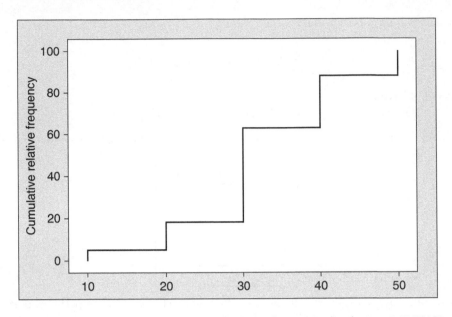

Figure 2.6   *Step function of a cumulative relative frequency distribution (MINITAB).*

In addition to the frequency distribution, it is often useful to present the cumulative frequency distribution of a given variable. The **cumulative frequency** of x_i is defined as the sum of frequencies of values less than or equal to x_i. We denote it by F_i, and the proportional cumulative frequencies or cumulative relative frequency by

$$P_i = F_i/n.$$

A table of proportional cumulative frequency distribution could be represented as follows:

Value	p	P
x_1	p_1	$P_1 = p_1$
x_2	p_2	$P_2 = p_1 + p_2$
$\vdots$	$\vdots$	$\vdots$
x_k	p_k	$P_k = p_1 + \cdots + p_k = 1$
Total	1	

The graph of the cumulative relative frequency distribution is a step function and looks typically like the graph shown in Figure 2.6.

Example 2.5. A US manufacturer of hybrid microelectronic components purchases ceramic plates from a large Japanese supplier. The plates are visually inspected before screen printing. Blemishes will affect the final product's electrical performance and its overall yield. In order to prepare a report for the Japanese supplier, the US manufacturer decided to characterize the variability in the number of blemishes found on the ceramic plates. The following measurements representing the number of blemishes found on each of 30 ceramic plates:

$$0, 2, 0, 0, 1, 3, 0, 3, 1, 1, 0, 0, 1, 2, 0$$
$$0, 0, 1, 1, 3, 0, 1, 0, 0, 0, 5, 1, 0, 2, 0.$$

Here the variable X assumes the values $0, 1, 2, 3$, and 5. The frequency distribution of X is displayed in Table 2.2.

Table 2.2 *Frequency distribution of blemishes on ceramic plates*

x	f	p	P
0	15	0.50	0.50
1	8	0.27	0.77
2	3	0.10	0.87
3	3	0.10	0.97
4	0	0.00	0.97
5	1	0.03	1.00
Total	30	1.00	

We did not observe the value $x = 4$, but since it seems likely to occur in future samples, we include it in the frequency distribution, with frequency $f = 0$. For pedagogical purposes, we show next how to calculate a frequency distribution and how to generate a bar diagram in R:

```
> data(BLEMISHES)   # BLEMISHES is a matrix-like structure
```

The object BLEMISHES is not a simple vector like CYCLT – it is called a data frame, i.e., a matrix-like structure whose columns (variables) may be of differing types. Because of this, we access this object with the subsetting operator square brackets [i, j] specifying elements to extract. Square brackets can be used also on vectors. Type help("[") at the command prompt for additional information. Below are the first few rows of the object BLEMISHES.

```
> BLEMISHES[1:3, ] # Return rows 1 to 3, all columns

        plateID count
Plate 1       1     0
Plate 2       2     2
Plate 3       3     0
```

```
> head(BLEMISHES,   # Equivalently head returns part of an object
      n=3)          # set number of elements
```

Like the previously introduced subsetting operator [i,j] the $ operator extracts a whole column by name.

```
> X <- factor(BLEMISHES$count, # Encode count vector as a factor
    levels=0:5)                 # specify labels for the levels
>                               #
> X <- table(X                  # Reassign to X a frequency table
           )                    # of encoded variable count
>                               #
> X <- prop.table(X)            # Reassign with proportional table
>                               #
> barplot(X,                    # Plot a Bar diagram
        width=1,                # set width of bars
        space=4,                # set space between bars
        col="grey50",
        ylab="Proportional Frequency")
```

The bar-diagram and cumulative frequency step-function are shown in Figures 2.7 and 2.8, respectively. Figure 2.8 is obtained in MINITAB by going to the Graph > Empirical CDF window.

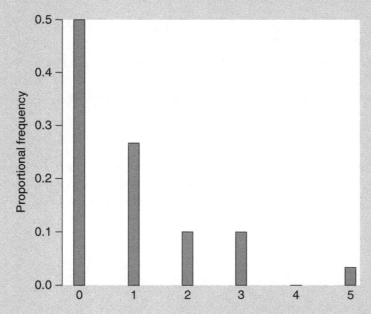

Figure 2.7   *Bar-diagram for number of blemishes on ceramic plates.*

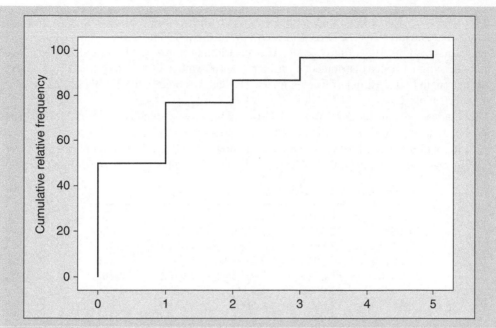

Figure 2.8 *Cumulative relative frequency distribution for number of blemishes on ceramic plates (MINITAB).*

2.4.2 Frequency distributions of continuous random variables

For the case of a continuous random variable, we partition the possible range of variation of the observed variable into k subintervals. Generally speaking, if the possible range of X is between L and H, we specify numbers $b_0, b_1, b_r, \ldots, b_k$ such that $L = b_0 < b_1 < b_2 < \cdots < b_{k-1} < b_k = H$. The values $b_0, b_1, \ldots, b_k$ are called the limits of the k subintervals. We then classify the X values into the interval (b_{i-1}, b_i) if $b_{i-1} < X \leq b_i$ $(i = 1, \ldots, k)$. (If $X = b_0$, we assign it to the first subinterval.) Subintervals are also called **bins**, **classes** or **class-intervals**.

In order to construct a frequency distribution, we must consider the following two questions:

(i) How many subintervals should we choose?
   and
(ii) How large should the width of the subintervals be?

In general, it is difficult to give to these important questions exact answers which apply in all cases. However, the general, recommendation is to use between 10 and 15 subintervals in large samples, and apply equal width subintervals. The frequency distribution is given then for the subintervals, where the midpoint of each subinterval provides a numerical representation for that interval. A typical frequency distribution table might look like the following:

Subintervals	Mid-Point	Freq.	Cum. Freq.
$b_0 - b_1$	$\overline{b}_1$	f_1	$F_1 = f_1$
$b_1 - b_2$	$\overline{b}_2$	f_2	$F_2 = f_1 + f_2$
$\vdots$			
$b_{k-1} - b_k$	$\overline{b}_k$	f_k	$F_k = n$

Example 2.6. Nili, a large fiber supplier to United States, South America, and European textile manufacturers, has tight control over its yarn strength. This critical dimension is typically analyzed on a logarithmic scale. This logarithmic transformation produces data that is more symmetrically distributed. Consider $n = 100$ values of $Y = \ln(X)$, where X is the yarn-strength [lb./22 yarns] of woolen fibers. The data is stored in file **YARNSTRG.csv** and shown in Table 2.3

The smallest value in Table 2.3 is $Y = 1.1514$ and the largest value is $Y = 5.7978$. This represents a range of $5.7978 - 1.1514 = 4.6464$. To obtain approximately 15 subintervals, we need the width of each interval to be about $4.6464/15 = 0.31$. A more convenient choice for this class width might be 0.50. The first subinterval would start at $b_0 = 0.75$ and the last

Table 2.3   *A sample of 100 log (yarn strength)*

2.4016	1.1514	4.0017	2.1381	2.5364
2.5813	3.6152	2.5800	2.7243	2.4064
2.1232	2.5654	1.3436	4.3215	2.5264
3.0164	3.7043	2.2671	1.1535	2.3483
4.4382	1.4328	3.4603	3.6162	2.4822
3.3077	2.0968	2.5724	3.4217	4.4563
3.0693	2.6537	2.5000	3.1860	3.5017
1.5219	2.6745	2.3459	4.3389	4.5234
5.0904	2.5326	2.4240	4.8444	1.7837
3.0027	3.7071	3.1412	1.7902	1.5305
2.9908	2.3018	3.4002	1.6787	2.1771
3.1166	1.4570	4.0022	1.5059	3.9821
3.7782	3.3770	2.6266	3.6398	2.2762
1.8952	2.9394	2.8243	2.9382	5.7978
2.5238	1.7261	1.6438	2.2872	4.6426
3.4866	3.4743	3.5272	2.7317	3.6561
4.6315	2.5453	2.2364	3.6394	3.5886
1.8926	3.1860	3.2217	2.8418	4.1251
3.8849	2.1306	2.2163	3.2108	3.2177
2.0813	3.0722	4.0126	2.8732	2.4190

Table 2.4   *Frequency distribution for log yarn-strength data*

$b_{i-1} - b_i$	$\bar{b}_i$	f_i	p_i	F_i	P_i
0.75–1.25	1.0	2	0.02	2	0.02
1.25–1.75	1.5	9	0.09	11	0.11
1.75–2.25	2.0	12	0.12	23	0.23
2.25–2.75	2.5	26	0.26	49	0.49
2.75–3.25	3.0	17	0.17	66	0.66
3.25–3.75	3.5	17	0.17	83	0.83
3.75–4.25	4.0	7	0.07	90	0.90
4.25–4.75	4.5	7	0.07	97	0.97
4.75–5.25	5.0	2	0.02	99	0.99
5.25–5.75	5.5	0	0.00	99	0.99
5.75–6.25	6.0	1	0.01	100	1.00

subinterval would end with $b_k = 6.25$. The frequency distribution for this data is presented in Table 2.3.

A graphical representation of the distribution is given by a **histogram** as shown in Figure 2.9. Each rectangle has a height equal to the frequency (f) or relative frequency (p) of the corresponding subinterval. In either case, the area of the rectangle is proportional to the frequency of the interval along the base. The cumulative frequency distribution is presented in Figure 2.10.

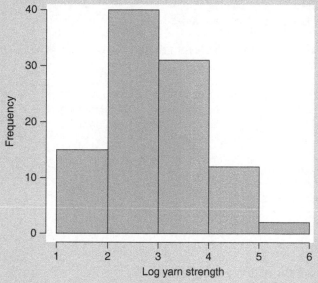

Figure 2.9 *Histogram of log yarn strength (Table 2.4).*

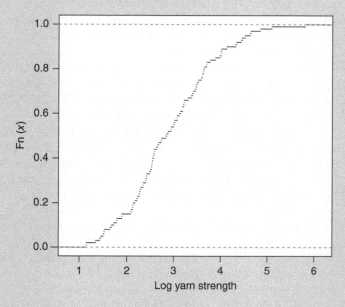

Figure 2.10 *Cumulative relative distribution of log yarn strength.*

Computer programs select a default midpoint and width of class intervals but provide the option to change these choices. The shape of the histogram depends on the number of class intervals chosen. You can experiment with the data set **YARNSTRG.csv**, by choosing a different number of class intervals, starting with the default value.

Apply the following R commands:

```
> data(YARNSTRG)   # YARNSTRG is a vector of values
>                  #
> hist(YARNSTRG)   # Plot an histogram of the given data values
```

This produces an histogram with 5 class intervals. The commands below produced the histogram presented in Figures 2.9 and 2.10.

```
> hist(YARNSTRG,   # Plot an histogram of the given data values
       breaks=6,   # set the number of cells for the histogram
       main="",    # set the main title to void
       xlab = "Log yarn strength")
>                  #
> plot.ecdf(YARNSTRG,  # Plot empirical cumulative distribution
            pch=NA,    # set no symbol in steps
            main="",
            xlab="Log Yarn Strength")
```

2.4.3   Statistics of the ordered sample

In this section, we identify some characteristic values of a sample of observations that have been sorted from smallest to largest. Such sample characteristics are called **order statistics**. In General, **statistics** are computed from observations and are used to make an inference on characteristics of the population from where the sample was drawn. Statistics that do not require to sort observation are discussed in Section 2.4.4.

Let $X_1, X_2, \ldots, X_n$ be the observed values of some random variable, as obtained by a random sampling process. For example, consider the following 10 values of the shear strength of welds of stainless steel (lb./weld): 2385, 2400, 2285, 2765, 2410, 2360, 2750, 2200, 2500, 2550. What can we do to characterize the variability and location of these values?

The first step is to sort the sample values in increasing order. That is, we rewrite the list of sample values as follows: 2200, 2285, 2360, 2385, 2400, 2410, 2500, 2550, 2750, 2765. These ordered values are denoted by $X_{(1)}, X_{(2)}, \ldots, X_{(n)}$, where $X_{(1)} = 2200$ is the smallest value in the sample, $X_{(2)} = 2285$ is the second smallest, and so on. We call $X_{(i)}$ the i-**the order statistic** of the sample. For convenience, we can also denote the average of consecutive order statistics by

$$X_{(i.5)} = (X_{(i)} + X_{(i+1)})/2 = X_{(i)} + 0.5(X_{(i+1)} - X_{(i)}). \tag{2.4.1}$$

For example, $X_{(2.5)} = (X_{(2)} + X_{(3)})/2$. We now identify some characteristic values that depend on these order statistics, namely the sample minimum, the sample maximum, the sample range, the sample median, and the sample quartiles. The **sample minimum is** $X_{(1)}$ and the **sample maximum** is $X_{(n)}$. In our example, $X_{(1)} = 2200$ and $X_{(n)} = X_{(10)} = 2765$. The **sample range** is the difference $R = X_{(n)} - X_{(1)} = 2765 - 2200 = 565$. The "middle" value in the ordered sample is called the **sample median**, denoted by M_e. The sample median is defined as $M_e = X_{(m)}$, where $m = (n+1)/2$. In our example, $n = 10$ so $m = (10+1)/2 = 5.5$. Thus,

$$M_e = X_{(5.5)} = (X_{(5)} + X_{(6)})/2 = X_{(5)} + 0.5(X_{(6)} - X_{(5)})$$

$$= (2400 + 2410)/2$$

$$= 2405.$$

The median characterizes the center of dispersion of the sample values, and is therefore called a **statistic of central tendency**, or **location statistic**. Approximately, 50% of the sample values are smaller than the median. Finally, we define the **sample quartiles** as $Q_1 = X_{(q_1)}$ and $Q_3 = X_{(q_3)}$, where

$$q_1 = \frac{(n+1)}{4}$$

and   (2.4.2)

$$q_3 = \frac{3(n+1)}{4}.$$

Q_1 is called the **lower quartile** and Q_3 is called the **upper quartile**. These quartiles divide the sample so that approximately one-fourth of the values are smaller than Q_1, one-half are between Q_1 and Q_3, and one-fourth are greater than Q_3. In our example, $n = 10$ so

$$q_1 = \frac{11}{4} = 2.75$$

and

$$q_3 = \frac{33}{4} = 8.25.$$

Thus, $Q_1 = X_{(2.75)} = X_{(2)} + 0.75 \times (X_{(3)} - X_{(2)}) = 2341.25$ and $Q_3 = X_{(8.25)} = X_{(8)} + 0.25 \times (X_{(9)} - X_{(8)}) = 2600$.

These sample statistics can be obtained from a frequency distribution using the cumulative relative frequency as shown in Figure 2.11 which is based on the log yarn-strength data of Table 2.3.

Using linear interpolation within the subintervals, we obtain $Q_1 = 2.3$, $Q_3 = 3.6$, and $M_e = 2.9$. These estimates are only slightly different from the exact values $Q_1 = X_{(0.25)} = 2.2789$, $Q_3 = X_{(0.75)} = 3.5425$, and $M_e = X_{(0.5)} = 2.8331$.

The sample median and quartiles are specific forms of a class of statistics known as sample quantiles. The **p-th sample quantile** is a number that exceeds exactly $100p\%$ of the sample values. Hence, the median is the 0.5 sample quantile, Q_1 is the 0.25th quantile and Q_3 is the 0.75th sample quantile. We may be interested, for example, in the 0.9 sample quantile. Using linear interpolation in Figure 2.11, we obtain the value 4.5, while the value of $X_{(0.9)} = 4.2233$. The p-th sample quantile is also called the $100p$-th sample percentile.

The R commands

```
> quantile(CYCLT,            # Sample quantiles
          probs = seq(from=0,   # set given probabilities
                     to=1,      #
                     by=0.25),  #
          type=6               # set algorithm to be used:
  )                            # 6 for MINITAB like one
```

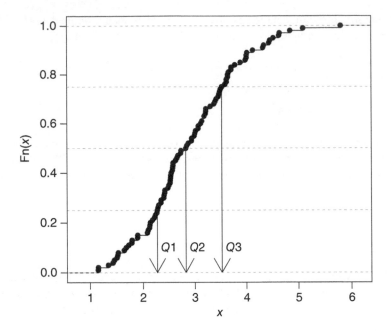

Figure 2.11 *Cumulative relative distribution function with linear interpolation lines at quartiles.*

```
       0%       25%       50%       75%      100%
0.17500  0.29825  0.54550  1.07175  1.14100

>                                #
> mean(CYCLT,                    # Arithmetic mean
       trim=0.0,                 # set fraction of observations
                                 # to be trimmed from each end
       na.rm=TRUE                # set whether NA values should
  )                              # be stripped out

[1] 0.65246

>                                #
> # summary(CYCLT)               # As above but uses R default
>                                # type algorithm
```

yields the following statistics of the data: Median, Min, Max, Q_1, and Q_3. Applying this command on the piston cycle time of file **CYCLT.csv**, we find $X_{(1)} = 0.1750$, $Q_1 = 0.2982$, $M_e = 0.5455$, $\overline{X} = 0.6525$, $Q_3 = 1.0718$, and $X_{(50)} = 1.1410$.

2.4.4   Statistics of location and dispersion

Given a sample of n measurements, $X_1, \ldots, X_n$, we can compute various statistics to describe the distribution. The **sample mean** is determined by the formula

$$\overline{X} = \frac{1}{n}\sum_{i=1}^{n} X_i. \tag{2.4.3}$$

Like the sample median, $\overline{X}$ is a measure of central tendency. In Physics, the sample mean represents the "center of gravity" for a system consisting of n equal-mass particles located on the points X_i on the line.

As an example consider the following measurements, representing component failure times in hours since initial operation

$$45, \; 60, \; 21, \; 19, \; 4, \; 31.$$

The sample mean is

$$\overline{X} = (45 + 60 + 21 + 19 + 4 + 31)/6 = 30.$$

To measure the spread of data about the mean, we typically use the **sample variance** defined by

$$S^2 = \frac{1}{n-1} \sum_{i=1}^{n} (X_i - \overline{X})^2, \tag{2.4.4}$$

or the **sample standard deviation**, given by

$$S = \sqrt{S^2}.$$

The sample standard deviation is used more often since its units (cm., lb.) are the same as those of the original measurements. In the next section, we will discuss some ways of interpreting the sample standard deviation. Presently, we remark only those data sets with greater dispersion about the mean that will have larger standard deviations. The computation of S^2 is illustrated in Table 2.5 using the failure time data.

The sample standard deviation and sample mean provide information on the variability and central tendency of observation. For the data set (number of blemishes on ceramic plates) in Table 2.2, one finds that $\overline{X} = 0.933$ and $S = 1.258$. Looking at the histogram in Figure 2.7, one notes a marked asymmetry in the data. In 50% of the ceramic plates, there were no blemishes and in 3% there were five blemishes. In contrast, consider the histogram of Log Yarn-Strength which shows remarkable symmetry with $\overline{X} = 2.9238$ and $S = 0.93776$. The difference in shape is obviously not reflected by $\overline{X}$ and S. Additional information pertaining to the shape of a distribution of observations is derived from the **sample skewness** and **sample kurtosis**. The sample

Table 2.5 *Computing the sample variance*

X	$(X - \overline{X})$	$(X - \overline{X})^2$
45	15	225
60	30	900
21	−9	81
19	−11	121
4	−26	676
31	1	1
Sum 180	0	2004

$\overline{X} = 180/6 = 30$
$S^2 = 2004/5 = 400.8$

skewness is defined as the index

$$\beta_3 = \frac{1}{n}\sum_{i=1}^{n}(X_i - \overline{X})^3/S^3. \tag{2.4.5}$$

The sample kurtosis (steepness) is defined as

$$\beta_4 = \frac{1}{n}\sum_{i=1}^{n}(X_i - \overline{X})^4/S^4. \tag{2.4.6}$$

These indices can be computed in R using package e1071 and in MINITAB with Stat > Basic Statistics > Display Descriptive Statistics.

In JMP the output from the *Analyze Distribution* window looks like this:

Distributions

Log Yarn Strength

Quantiles			Summary Statistics	
100.0%	maximum	5.7978	Mean	2.923843
99.5%		5.7978	Std Dev	0.9377595
97.5%		4.96125	Std Err Mean	0.0937759
90.0%		4.30186	Upper 95% Mean	3.1099148
75.0%	quartile	3.57325	Lower 95% Mean	2.7377712
50.0%	median	2.83305	N	100
25.0%	quartile	2.27895	Sum Wgt	100
10.0%		1.68344	Sum	292.3843
2.5%		1.2533	Variance	0.8793929
0.5%		1.1514	Skewness	0.4164256
0.0%	minimum	1.1514	Kurtosis	-0.007958
			CV	32.07284
			N Missing	0
			N Zero	0
			N Unique	99
			Uncorrected SS	941.94568
			Corrected SS	87.059893
			Autocorrelation	-0.014168
			Median	2.83305
			Mode	3.186
			5% Trimmed Mean	2.8982144
			Geometric Mean	2.7709747
			Range	4.6464
			Interquartile Range	1.2943
			Median Absolute Deviation	0.63425

Skewness and kurtosis are provided by most statistical computer packages. If a distribution is symmetric (around its mean), then skewness $= 0$. If skewness > 0, we say that the distribution is positively skewed or skewed to the right. If skewness < 0, then the distribution is negatively skewed or skewed to the left. We should also comment that in distributions which are positively skewed $\overline{X} > Me$, while in those which are negatively skewed $\overline{X} < Me$. In symmetric distributions, $\overline{X} = Me$.

The steepness of a distribution is determined relative to that of the normal (Gaussian) distribution, which is described in the next section and specified in Section 3.4.2. In a normal

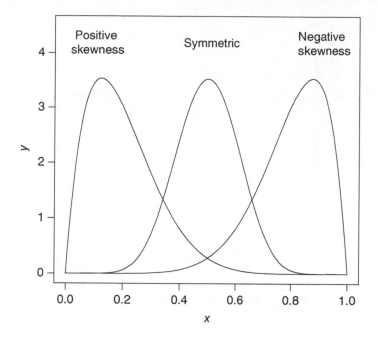

Figure 2.12   *Symmetric and asymmetric distributions.*

distribution, kurtosis = 3. Thus, if kurtosis > 3, the distribution is called steep. If kurtosis < 3, the distribution is called flat. A schematic representation of shapes is given in Figures 2.12 and 2.13.

To illustrate these statistics, we computed $\overline{X}$, S^2, S, skewness, and kurtosis for the log yarn-strength data of Table 2.3, We obtained

$$\overline{X} = 2.9238$$

$$S^2 = 0.8794 \quad S = 0.93776$$

$$\text{Skewness} = 0.4040 \quad \text{Kurtosis} = -0.007958.$$

The sample mean is $\overline{X} = 2.9238$, for values on a logarithmic scale. To return to the original scale [lb/22 yarns], we can use the measure

$$G = \exp\{\overline{X}\}$$

$$= \left(\prod_{i=1}^{n} Y_i\right)^{1/n} = 18.6119, \tag{2.4.7}$$

where $Y_i = \exp(X_i)$, $i = 1, \ldots, n$. The measure G is called the **geometric mean** of Y. The geometric mean, G, is defined only for positive valued variables. It is used as a measure of central tendency for rates or change and index numbers such as the desirability function implemented in MINITAB and JMP. One can prove the following general result:

$$G \leq \overline{X}.$$

Equality holds only if all values in the sample are the same.

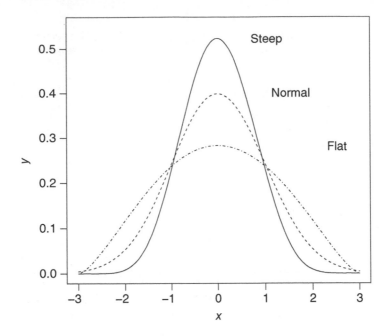

Figure 2.13   *Normal, steep, and flat distributions.*

Additional statistics to measure the dispersion are

$$\text{(i) The interquartile range IQR} = Q_3 - Q_1, \tag{2.4.8}$$

and

$$\text{(ii) The coefficient of variation } \gamma = \frac{S}{|\overline{X}|}. \tag{2.4.9}$$

The interquartile range, IQR, is a useful measure of dispersion when there are extreme values (outliers) in the sample. It is easy to compute and can yield an estimate of S, for more details see Section 2.6.4. The coefficient of variation is a dimensionless index, used to compare the variability of different data sets, when the standard deviation tends to grow with the mean. The coefficient of variation of the log-yarn strength data is $\gamma = \dfrac{0.938}{2.924} = 0.321$.

2.5   Prediction intervals

When the data $X_1, \ldots, X_n$ represents a sample of observations from some population, we can use the sample statistics discussed in the previous sections to predict how future measurements will behave. Of course, our ability to predict accurately depends on the size of the sample.

Prediction using order statistics is very simple and is valid for any type of distribution. Since the ordered measurements partition the real line into $n + 1$ subintervals,

$$(-\infty, X_{(1)}), (X_{(1)}, X_{(2)}), \ldots, (X_{(n)}, \infty),$$

we can predict that $100/(n + 1)\%$ of all future observations will fall in any one of these subintervals; hence, $100i/(n + 1)\%$ of future sample values are expected to be less than the i-th order statistic $X_{(i)}$. It is interesting to note that the sample minimum, $X_{(1)}$, is **not** the smallest possible value. Instead, we expect to see one out of every $n + 1$ future measurements to be less than $X_{(1)}$. Similarly, one out of every $n + 1$ future measurements is expected to be greater than $X_{(n)}$.

Predicting future measurements using sample skewness and kurtosis is a bit more difficult because it depends on the type of distribution that the data follow. If the distribution is symmetric (skewness ≈ 0) and somewhat "bell-shaped" or "normal"[1] (kurtosis ≈ 3) as in Figure 2.9, for the log yarn strength data, we can make the following statements:

1) Approximately **68%** of all future measurements will lie within **one standard deviation** of the mean
2) Approximately **95%** of all future measurements will lie within **two standard deviations** of the mean
3) Approximately **99.7%** of all future measurements will lie within **three standard deviations** of the mean.

The sample mean and standard deviation for the log yarn strength measurement are $\overline{X} = 2.92$ and $S = 0.94$. Hence, we predict that 68% of all future measurements will lie between $\overline{X} - S = 1.98$ and $\overline{X} + S = 3.86$; 95% of all future observations will be between $\overline{X} - 2S = 1.04$ and $\overline{X} + 2S = 4.80$; and 99.7% of all future observations will be between $\overline{X} - 3S = 0.10$ and $\overline{X} + 3S = 5.74$. For the data in Table 2.3 there are exactly 69, 97, and 99 of the 100 values in the above intervals, respectively.

When the data does not follow a normal distribution, we may use the following result: **Chebyshev's Inequality**

For any number $k > 1$ **the percentage of future measurements within k standard deviations of the mean will be at least** $100(1 - 1/k^2)\%$.

This means that at least 75% of all future measurements will fall within 2 standard deviations ($k = 2$). Similarly, at least 89% will fall within 3 standard deviations ($k = 3$). These statements are true for any distribution; however, the actual percentages may be considerably larger. Notice that for data which is normally distributed, 95% of the values fall in the interval $[\overline{X} - 2S, \overline{X} + 2S]$. The Chebyshev inequality gives only the lower bound of 75%, and is therefore very conservative.

Any prediction statements, using the order statistics or the sample mean and standard deviation, can only be made with the understanding that they are based on a sample of data. They are accurate only to the degree that the sample is representative of the entire population. When the sample size is small, we cannot be very confident in our prediction. For example, if based on a sample of size $n = 10$, we find $\overline{X} = 20$ and $S = 0.1$, then we might make the statement that 95% of all future values will be between $19.8 = 20 - 2(0.1)$ and $20.2 = 20 + 2(0.1)$. However, it would not be too unlikely to find that a second sample produced $\overline{X} = 20.1$ and $S = 0.15$. The new prediction interval would be wider than 19.8 to 20.4; a considerable change. Also, a sample of size 10 does not provide sufficient evidence that the data has a "normal" distribution. With larger samples, say $n > 100$, we may be able to draw this conclusion with greater confidence.

In Chapter 4, we will discuss theoretical and computerized statistical inference whereby we assign a "confidence level" to such statements. This confidence level will depend on the sample size. Prediction intervals which are correct with known confidence are called **tolerance intervals**.

---

[1] The normal or Gaussian distribution will be defined in Chapter 3.

2.6 Additional techniques of exploratory data analysis

In the present section, we present additional modern graphical techniques, which are quite common today in exploratory data analysis. These techniques are the **Box and Whiskers Plot**, the **Quantile Plot**, and **Stem-and-Leaf Diagram**. We also discuss the problem of sensitivity of the sample mean and standard deviation to outlying observations, and introduce some robust statistics.

2.6.1 Box and whiskers plot

The Box and Whiskers Plot is a graphical presentation of the data, which provides an efficient display of various features, like location, dispersion, and skewness. A box is plotted, with its lower hinge at the first quartile $Q_1 = X_{(q_1)}$, and its upper hinge at the third quartile $Q_3 = X_{(q_3)}$. Inside the box a line is drawn at the median, M_e, and a cross is marked at the sample mean, $\bar{X}_n$, to mark the statistics of central location. The IQR, $Q_3 - Q_1$, which is the length of the box, is a measure of dispersion. Two whiskers are extended from the box. The lower whisker is extended toward the minimum $X_{(1)}$, but not lower than one and half of the IQR, i.e.,

$$\text{Lower whisker starts} = \max \{X_{(1)}, Q_1 - 1.5(Q_3 - Q_1)\}. \tag{2.6.1}$$

Similarly,

$$\text{Upper whisker ends} = \min \{X_{(n)}, Q_3 + 1.5(Q_3 - Q_1)\}. \tag{2.6.2}$$

Data points beyond the lower or upper whiskers are considered **outliers**. Figure 2.14 presents such a plot derived with MINITAB. The commands below generate a similar plot in R.

```
> boxplot(YARNSTRG,                # Produce box-and-whisker plot
         ylab="Log Yarn Strength")
```

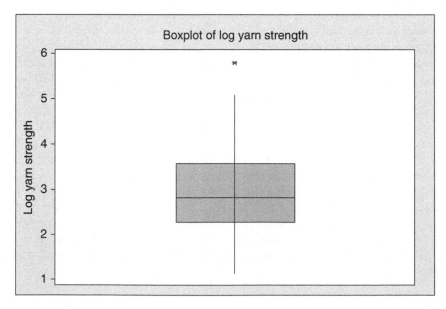

Figure 2.14 *Box–whiskers plot of log-yarn strength data (MINITAB).*

Example 2.7. In Figure 2.14, we present the box–whiskers plot of the yarn log-strength data, of Table 2.3. For this data we find the following summarizing statistic

$$X_{(1)} = 1.1514$$

$$Q_1 = 2.2790$$

$$M_e = 2.8331, \quad \overline{X}_{100} = 2.925$$

$$Q_3 = 3.5733$$

$$X_{(100)} = 5.7978$$

$$Q_3 - Q_1 = 1.2760, \quad S_{(100)} = 0.937.$$

In the box–whiskers plot, the end point of the lower whisker is at max $\{1.151, 0.367\} = X_{(1)}$. The upper whisker ends at min $\{5.798, 5.51475\} = 5.51475$. Thus, $X_{(100)}$ is an outlier. We conclude that the one measurement of yarn strength, which seems to be exceedingly large, is an outlier (could have been an error of measurement).

2.6.2 Quantile plots

The quantile plot is a plot of the sample quantiles x_p against p, $0 < p < 1$, where $x_p = X_{(p(n+1))}$. In Figure 2.15, we see the quantile plot of the log yarn-strength. From such a plot, one can obtain graphical estimates of the quantiles of the distribution. For example, from Figure 2.15, we immediately obtain the estimate 2.8 for the median, 2.23 for the first quartile, and 3.58 for the third quartile. These are close to the values presented earlier. We see also in Figure 2.15 that

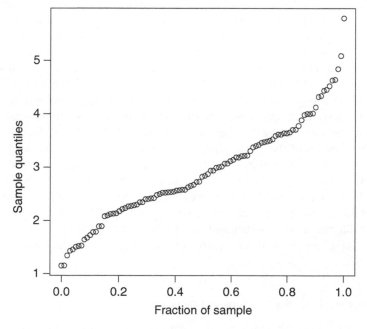

Figure 2.15 *Quantile plot of log yarn-strength data.*

the maximal point of this data set is an outlier. Tracing a straight line, beginning at the median, we can also see that from $x_{0.4}$ to $x_{0.9}$ (50% of the data points) are almost uniformly distributed, while the data between $x_{0.1}$ to $x_{0.4}$ tend to be larger (closer to the M_e) than those of a uniform distribution, while the largest 10% of the data values tend to be again larger (further away from the M_e) than those of a uniform distribution. This explains the slight positive skewness of the data, as seen in Figure 2.14.

2.6.3   Stem-and-leaf diagrams

The following is a stem-and-leaf display of the log yarn-strength data.

Character stem-and-leaf display

Stem-and-leaf of $\log y$, $N = 100$, Leaf unit $= 0.10$ (MINITAB)

5	1	11344
15	1	5556677788
34	2	0011112222233344444
(21)	2	555555555566677888999
45	3	000011112223344444
27	3	5556666677789
14	4	00013344
6	4	5668
2	5	0
1	5	7

In order to construct the stem-and-leaf diagram, the data is classified into class intervals, like in the histogram. The classes are of equal length. The 100 values in Table 2.3 start $X_{(1)} = 1.151$ and at $X_{(100)} = 5.798$. The stem-and-leaf diagram presents only the first two digits to the left, without rounding. All values between 1.0 and 1.499 are represented in the first class as 1.1, 1.1, 1.3, 1.4, 1.4. There are five such values, and this frequency is written on the left-hand side. The second class consists of all values between 1.5 and 1.999. There are 10 such values, which are represented as 1.5, 1.5, 1.5, 1.6, 1.6, 1.7, 1.7, 1.7, 1.8, 1.8. In a similar manner, all other classes are represented. The frequency of the class to which the median, M_e, belongs is written on the left in round brackets. In this way, one can immediately indicate where the median is located. The frequencies below or above the class of the median, are cumulative. Since the cumulative frequency (from above) of the class right that of the median is 45, we know that the median is located right after the 5th largest value from the top of that class, namely $M_e = 2.8$, as we have seen before. Similarly, to find Q_1, we see that $X_{(q_1)}$ is located at the third class from the top. It is the 10th value in that class, from the left. Thus, we find $Q_1 = 2.2$. Similarly, we find that $X_{(q_3)} = 4.5$. This information cannot be directly obtained from the histogram. Thus, the stem-and-leaf diagram is an important additional tool for data analysis.

In Figure 2.16, we present the stem-and-leaf diagram of the electric output data (**OELECT.csv**).

2.6.4   Robust statistics for location and dispersion

The sample mean $\overline{X}_n$ and the sample standard deviation are both sensitive statistics to extreme deviations. Let us illustrate this point. Suppose we have made three observations on the

Character stem-and-leaf display

Stem-and-leaf of Elec_Out $N = 99$ Leaf unit $= 1.0$ (MINITAB)

5	21	01111
10	21	22333
19	21	444445555
37	21	6666666677777777777
(22)	21	8888888888889999999999
40	22	0000000001111111111
21	22	22233333
13	22	44455555
5	22	6777
1	22	8

Figure 2.16   *Stem-and-leaf diagram of electric output data.*

sheer weld strength of steel and obtained the values 2350, 2400, 2500. The sample mean is $\overline{X}_3 = 2416.67$. What happens if the technician by mistake punches into the computer the value 25,000, instead of 2500. The sample mean would come out as 9916.67. If the result is checked on the spot, the mistake would likely be discovered and corrected. However, if there is no immediate checking, that absurd result would have remained and cause all kinds of difficulties later. Also, the standard deviations would have recorded wrongly as 13,063 rather than the correct value of 76.376. This simple example shows how sensitive are the mean and the standard deviation to extreme deviations (outliers) in the data.

To avoid such complexities, a more **robust** statistic can be used, instead of the sample mean, $\overline{X}_n$. This statistic is the α-**trimmed mean**. A proportion α of the data is trimmed from the lower and from the upper end of the ordered sample. The mean is then computed on the remaining $(1 - 2\alpha)$ proportion of the data. Let us denote by $\overline{T}_\alpha$ the α-trimmed mean. The formula of this statistic is

$$\overline{T}_\alpha = \frac{1}{N_\alpha} \sum_{j=[n\alpha]+1}^{[n(1-\alpha)]} X_j, \tag{2.6.3}$$

where $[\cdot]$ denotes the integer part of the number in brackets, e.g., $[7.3] = 7$, and $N_\alpha = [n(1 - \alpha)] - [n\alpha]$. For example, if $n = 100$ and $\alpha = 0.05$, we compute the mean of the 90 ordered values $X_{(6)}, \ldots, X_{(95)}$.

Example 2.8.   Let us now examine the **robustness** of the trimmed mean. We import data file **OELECT.csv** in package `mistat`, DatFiles directory. Function `mean` yields different results given specific values of the trim parameter. We use this example to show how to setup a function in R.

```
> File <- paste(              # Compose a string with
    path.package("mistat"),   # mistat package path and
    "/csvFiles/OELECT.csv",   # /DatFiles/OELECT.csv
    sep = "")                 # separate the terms with ""
>                             #
> Oelect <- read.csv(file=File) # Read a csv file and assign to
>                             # Oelect
```

```
>                                # 
> rm(File)                       # 
>                                # 
> mySummary <- function(         # Define a new function
    x, trim=0, type=6 )          # with arguments x, trim and type
                                 # 
  {                              # The new function does:
                                 # 
    qq <-quantile(x, type=type)  # Calculate quantiles
                                 # 
    qq <- c(qq[1L:3L],           # Concatenate quantiles and mean
            mean(x, trim=trim),
            qq[4L:5L])
                                 # 
    names(qq) <- c("Min.", "1st Qu.",   # Assign names to values
                   "Median", "Mean",
                   "3rd Qu.", "Max.")
                                 # 
    qq <- signif(qq)             # Round to significant values
                                 # 
    return(qq)                   # Return results
                                 # 
  }                              # Function end
>                                # 
> mySummary(Oelect$OELECT)       # Apply mySummary to Oelect data

   Min. 1st Qu.  Median    Mean 3rd Qu.    Max.
210.896 216.796 219.096 219.248 221.706 228.986

>                                # 
> mySummary(Oelect$OELECT,       # Apply mySummary to Oelect data
           trim=0.05)            # set trim to 5%

   Min. 1st Qu.  Median    Mean 3rd Qu.    Max.
210.896 216.796 219.096 219.218 221.706 228.986

>                                # 
> sd(Oelect$OELECT)              # Computes the standard deviation

[1] 4.003992
```

We see that $\bar{X}_{99} = 219.25$ and $\bar{T}_{0.05} = 219.22$. Let us order the sample value by using the function sort() and reassign values to column V with assignment operator <-. The largest value in $C2$ is $C2(99) = 228.986$. Let us now change this value to be $V(99) = 2289.86$ (an error in punching the data), and look at results when we apply the same commands

```
> OutVolt <- sort(Oelect$OELECT)  # Assign a vector
>                                 # of sorted values
>                                 # 
```

```
> OutVolt[99] <- 2289.86          # Assign a specific value
>                                 # at position 99
>                                 #
> mySummary(OutVolt)     # Apply function mySummary

    Min.   1st Qu.   Median       Mean   3rd Qu.       Max.
 210.896   216.796  219.096    240.065   221.706   2289.860

>                        #
> mySummary(OutVolt,     # Apply mySummary with
         trim=0.05)      # trim = 5%

    Min.   1st Qu.   Median       Mean   3rd Qu.       Max.
 210.896   216.796  219.096    219.218   221.706   2289.860

>                        #
> sd(OutVolt)            # Computes the standard deviation

[1] 208.1505
```

We see by comparing the two outputs that $\overline{X}_{99}$ changed from 219.25 to 240.1, S_{99} (STDEV) changed dramatically from 4.00 to 208.2 (and correspondingly SEMEAN $= S/\sqrt{n}$ changed).

On the other hand, M_e, $\overline{T}_\alpha$, Q_1, and Q_3 did not change at all. These statistics are called **robust** (nonsensitive) against extreme deviations (outliers).

We have seen that the standard deviation S is very sensitive to deviations in the extremes. A robust statistic for dispersion is

$$\tilde{\sigma} = \frac{Q_3 - Q_1}{1.3490}. \tag{2.6.4}$$

The denominator 1.3490 is the distance between Q_3 and Q_1 in the theoretical normal distribution (see Chapter 4). Indeed, Q_3 and Q_1 are robust against outliers. Hence, $\tilde{\sigma}$, which is about 3/4 of the IQR, is often a good statistic to replace S.

Another statistic is the α-**trimmed standard deviation**

$$S_\alpha = \left(\frac{1}{N_\alpha - 1} \sum_{j=[n\alpha]+1}^{[n(1-\alpha)]} (X_j - \overline{T}_\alpha)^2 \right)^{1/2}. \tag{2.6.5}$$

For the OELECT data S_α equals 3.5969. The command below calculates a robust statistic for dispersion $\tilde{\sigma}$ from the OELECT data.

```
> IQR(OutVolt)/1.349  # Robust estimate of S

[1] 3.587843
```

We see these two robust statistics, $\tilde{\sigma}$ and S_α yield close results. The sample standard deviation of OELECT is $S = 4.00399$.

2.7   Chapter highlights

The chapter focuses on statistical variability and on various methods of analyzing random data. Random results of experiments are illustrated with distinction between deterministic and random components of variability. The difference between accuracy and precision is explained. Frequency distributions are defined to represent random phenomena. Various characteristics of location and dispersion of frequency distributions are defined. The elements of exploratory data analysis are presented.

The main concepts and definitions introduced in this chapter include

- Random Variable
- Fixed and Random Components
- Accuracy and Precision
- The Population and the Sample
- Random Sampling With Replacement (RSWR)
- Random Sampling Without Replacement (RSWOR)
- Frequency Distributions
- Discrete and Continuous Random Variables
- Quantiles
- Sample Mean and Sample Variance
- Skewness
- Kurtosis
- Prediction Intervals
- Box and Whiskers Plots
- Quantile Plots
- Stem and Leaf Diagrams
- Robust Statistics

2.8   Exercises

2.1 In the present problem, we are required to generate at random 50 integers from the set $\{1, 2, 3, 4, 5, 6\}$. To do this, we can use the MINITAB command

MTB> RANDOM 50 $C1$;

SUBC> INTEGER 1 6.

Use this method of simulation and count the number of times the different integers have been repeated. This counting can be done by using the MINITAB command

MTB> TABLE $C1$

How many times you expect each integer to appear if the process generates the numbers at random?

2.2 Construct a sequence of 50 numbers having a linear trend for deterministic components with random deviations around it. This can be done by using the MINITAB commands

MTB> Set $C1$

DATA> 1(1 : 50/1)1

DATA> End.

MTB> Let $C2 = 5 + 2.5 * C1$

MTB> Random 50 $C3$;

SUBC> Uniform −10 10.

MTB> Let $C4 = C2 + C3$

MTB> Plot $C4 * C1$

By plotting $C4$ versus $C1$, one sees the random variability around the linear trend.

2.3 Generate a sequence of 50 random binary numbers $(0, 1)$, when the likelihood of 1 is p, by using the command

MTB> RANDOM 50 $C1$;

SUBC> Bernoulli p.

Do this for the values $p = 0.1$, 0.3, 0.7, 0.9. Count the number of 1's in these random sequences, by the command

SUM($C1$)

2.4 The following are two sets of measurements of the weight of an object, which correspond to two different weighing instruments. The object has a true weight of 10 kg.

Instrument 1:

[1] 9.490950	10.436813	9.681357	10.996083	10.226101	10.253741
[7] 10.458926	9.247097	8.287045	10.145414	11.373981	10.144389
[13] 11.265351	7.956107	10.166610	10.800805	9.372905	10.199018
[19] 9.742579	10.428091				

Instrument 2:

[1] 11.771486	10.697693	10.687212	11.097567	11.676099	10.583907
[7] 10.505690	9.958557	10.938350	11.718334	11.308556	10.957640
[13] 11.250546	10.195894	11.804038	11.825099	10.677206	10.249831
[19] 10.729174	11.027622				

Which instrument seems to be more accurate? Which instrument seems to be more precise?

2.5 The quality control department of a candy factory uses a scale to verify compliance of the weight of packages. What could be the consequences of problems with the scale accuracy, precision, and stability.

2.6 Draw a random sample with replacement (RSWR) of size $n = 20$ from the set of integers $\{1, 2, \ldots, 100\}$.

2.7 Draw a random sample without replacement (RSWOR) of size $n = 10$ from the set of integers $\{11, 12, \ldots, 30\}$.

2.8 (i) How many words of five letters can be composed ($N = 26$, $n = 5$)?

(ii) How many words of five letters can be composed, if all letters are different?

(iii) How many words of five letters can be written if the first and the last letters are x?

(iv) An electronic signal is a binary sequence of 10 zeros or ones. How many different signals are available?

(v) How many electronic signals in a binary sequence of size 10 are there in which the number 1 appears exactly 5 times?

2.9 For each of the following variables, state whether it is discrete or continuous:

(i) The number of "heads" among the results of 10 flippings of a coin;

(ii) The number of blemishes on a ceramic plate;

(iii) The thickness of ceramic plates;

(iv) The weight of an object.

2.10 Data file **FILMSP.csv** contains data gathered from 217 rolls of film. The data consists of the film speed as measured in a special lab.

(i) Prepare a histogram of the data.

2.11 Data file **COAL.csv** contains data on the number of yearly disasters in coal mines in England. Prepare a table of frequency distributions of the number of coalmine disasters. [You can use the MINITAB command TABLE $C1$.]

2.12 Data file **CAR.csv** contains information on 109 different car models. For each car, there are values of five variables

1. Number of cylinders (4, 6, 8)
2. Origin (1, 2, 3)
3. Turn Diameter [m]
4. Horsepower [HP]
5. Number of miles/gallon in city driving [mpg].

   Prepare frequency distributions of variables 1, 2, 3, 4, 5.

2.13 Compute the following five quantities for the data in file **FILMSP.csv**

   (i) Sample minimum, $X_{(1)}$;
   (ii) Sample first quartile, Q_1;
   (iii) Sample median, M_e;
   (iv) Sample third quartile, Q_3;
   (v) Sample maximum, $X_{(217)}$.
   (vi) The 0.8-quantile.
   (vii) The 0.9-quantile.
   (viii) The 0.99-quantile.

   Show how you get these statistics by using the formulae. [The order statistics of the sample can be obtained by first ordering the values of the sample. For this, use the MINITAB command

   MTB> SORT $C1$ $C2$.

   Certain order statistics can be put into constants by the commands, e.g.,

   MTB> Let $k1 = 1$

   MTB> Let $k2 = C2(k1)$

   The sample minimum is $C2(1)$, the sample maximum is $C2(217)$, etc.].

2.14 Compute with MINITAB the indices of skewness and kurtosis of the **FILMSP.csv**, by defining the constants: For skewness:

   MTB> Let $k1$ =mean$((C1- \text{mean}(C1)) ** 3)/(\text{mean}((C1- \text{Mean}(C1)) ** 2)) ** 1.5$

   For kurtosis:

   MTB> Let $k2 = \text{mean}((C1- \text{mean}(C1)) ** 4)/(\text{mean}((C1- \text{mean}(C1)) ** 2)) ** 2$

   Interpret the skewness and kurtosis of this sample in terms of the shape of the distribution of film speed.

2.15 Compare the means and standard deviations of the number of miles per gallon/city of cars by origin (1 = US; 2 = Europe; 3 = Asia) according to the data of file **CAR.csv**.

2.16 Compute the coefficient of variation of the Turn Diameter of US made cars (Origin = 1) in file **CAR.csv**.

2.17 Compare the mean $\overline{X}$ and the geometric mean G of the Turn Diameter of US made and Japanese cars in **CAR.csv**.

2.18 Compare the prediction proportions to the actual frequencies of the intervals

$$\overline{X} \pm kS, \quad k = 1, 2, 3$$

for the film speed data, given in **FILMSP.csv** file.

2.19 Present side by side the box plots of Miles per Gallon/City for cars by origin. Use data file **CAR.csv**.

2.20 Prepare a stem-leaf diagram of the piston cycle time in file **OTURB.csv**. Compute the five summary statistics $(X_{(1)}, Q_1, M_e, Q_3, X_{(n)})$ from the stem-leaf.

2.21 Compute the trimmed mean $\overline{T}_{0.10}$ and trimmed standard-deviation, $S_{0.10}$ of the piston cycle time of file **OTURB.csv**.

2.22 The following data is the time (in seconds) to get from 0 to 60 mph for a sample of 15 German made cars and 20 Japanese cars

German Made Cars			Japanese Made Cars			
10.0	10.9	4.8	9.4	9.5	7.1	8.0
6.4	7.9	8.9	8.9	7.7	10.5	6.5
8.5	6.9	7.1	6.7	9.3	5.7	12.5
5.5	6.4	8.7	7.2	9.1	8.3	8.2
5.1	6.0	7.5	8.5	6.8	9.5	9.7

Compare and contrast the acceleration times of German and Japanese made cars, in terms of their five summary statistics.

2.23 Summarize variables Res 3 and Res 7 in data set HADPAS.csv by computing sample statistics, histograms and stem and leaf diagrams.

2.24 Are there outliers in the Res 3 data of **HADPAS.csv**? Show your calculations.

3

Probability Models and Distribution Functions

3.1 Basic probability

3.1.1 Events and sample spaces: formal presentation of random measurements

Experiments, or trials of interest, are those which may yield different results with outcomes that are not known ahead of time with certainty. We have seen in Chapter 2 a large number of examples in which outcomes of measurements vary. It is of interest to find, before conducting a particular experiment, what are the chances of obtaining results in a certain range. In order to provide a quantitative answer to such a question, we have to formalize the framework of the discussion so that no ambiguity is left.

When we say a "trial" or "experiment," in the general sense, we mean a well-defined process of measuring certain characteristic(s), or variable(s). For example, if the experiment is to measure the compressive strength of concrete cubes, we must specify exactly how the concrete mixture was prepared, i.e., proportions of cement, sand, aggregates, and water in the batch. Length of mixing time, dimensions of mold, number of days during which the concrete has hardened. The temperature and humidity during preparation and storage of the concrete cubes, etc. All these factors influence the resulting compressive strength. Well-documented protocol of an experiment enables us to replicate it as many times as needed. In a well-controlled experiment, we can assume that the variability in the measured variables is due to **randomness**. We can think of the random experimental results as sample values from a hypothetical population. The set of all possible sample values is called the **sample space**. In other words, the sample space is the set of all possible outcomes of a specified experiment. The outcomes do not have to be numerical. They could be names, categorical values, functions, or collection of items. The individual outcome of an experiment will be called an **elementary event** or a **sample point** (element). We provide a few examples.

Modern Industrial Statistics: With Applications in R, MINITAB and JMP, Third Edition.
Ron S. Kenett and Shelemyahu Zacks.
© 2021 John Wiley & Sons, Ltd. Published 2021 by John Wiley & Sons, Ltd.

Example 3.1.   The experiment consists of choosing 10 names (without replacement) from a list of 400 undergraduate students at a given university. The outcome of such an experiment is a list of 10 names. The sample space is the collection of **all** possible such sublists that can be drawn from the original list of 400 students

Example 3.2.   The experiment is to produce 20 concrete cubes, under identical manufacturing conditions, and count the number of cubes with compressive strength above 200 [kg/cm^2]. The sample space is the set $S = \{0, 1, 2, \ldots, 20\}$. The elementary events, or sample points, are the elements of S

Example 3.3.   The experiment is to choose a steel bar from a specific production process, and measure its weight. The sample space S is the interval (ω_0, ω_1) of possible weights. The weight of a particular bar is a sample point

Thus, sample spaces could be finite sets of sample points, or countable or noncountable infinite sets.

Any subset of the sample space, S, is called an **event**. S itself is called the **sure event**. The empty set, $\emptyset$, is called the **null event**. We will denote events by the letters $A, B, C, \ldots$ or $E_1, E_2, \ldots$. All events under consideration are subsets of the same sample space S. Thus, events are sets of sample points.

For any event $A \subseteq S$, we denote by A^c the **complementary event**, i.e., the set of all points of S which are not in A.

An event A is said to **imply** an event B, if all elements of A are elements of B. We denote this **inclusion relationship** by $A \subset B$. If $A \subset B$ and $B \subset A$, then the two events are **equivalent**, $A \equiv B$.

Example 3.4.   The experiment is to select a sequence of five letters for transmission of a code in a money transfer operation. Let $A_1, A_2, \ldots, A_5$ denote the first, second, $\ldots$, fifth letter chosen. The sample space is the set of all possible sequences of five letters. Formally,

$$S = \{(A_1 A_2 A_3 A_4 A_5) : A_i \in \{a, b, c, \ldots, z\}, \ i = 1, \ldots, 5\}$$

This is a finite sample space containing 26^5 possible sequences of five letters. Any such sequence is a sample point.
Let E be the event that all the five letters in the sequence are the same. Thus,

$$E = \{aaaaa, bbbbb, \ldots, zzzzz\}.$$

This event contains 26 sample points. The complement of E, E^c, is the event that at least one letter in the sequence is different from the other ones.

3.1.2   Basic rules of operations with events: unions, intersections

Given events $A, B, \ldots$ of a sample space S, we can generate new events, by the operations of union, intersection and complementation.

The **union** of two events A and B, denoted $A \cup B$, is an event having elements which belong **either** to A **or** to B.

The intersection of two events, $A \cap B$, is an event whose elements belong both to A **and** to B. By pairwise union or intersection, we immediately extend the definition to finite number of events $A_1, A_2, \ldots, A_n$, i.e.,

$$\bigcup_{i=1}^{n} A_i = A_1 \cup A_2 \cup \cdots \cup A_n$$

and

$$\bigcap_{i=1}^{n} A_i = A_1 \cap A_2 \cap \cdots \cap A_n.$$

The finite union $\bigcup_{i=1}^{n} A_i$ is an event whose elements belong to **at least one** of the n events. The finite intersection $\bigcap_{i=1}^{n} A_i$ is an event whose elements belong to **all** the n events.

Any two events, A and B, are said to be mutually **exclusive** or **disjoint** if $A \cap B = \emptyset$, i.e., they do not contain common elements. Obviously, by definition, any event is disjoint of its complement, i.e., $A \cap A^c = \emptyset$. The operations of union and intersection are

1. **Commutative**:

$$A \cup B = B \cup A,$$

$$A \cap B = B \cap A;$$

2. **Associative**:

$$(A \cup B) \cup C = A \cup (B \cup C)$$

$$= A \cup B \cup C$$

$$(A \cap B) \cap C = A \cap (B \cap C)$$

$$= A \cap B \cap C \qquad (3.1.1)$$

3. **Distributive**:

$$A \cap (B \cup C) = (A \cap B) \cup (A \cap C)$$

$$A \cup (B \cap C) = (A \cup B) \cap (A \cup C). \qquad (3.1.2)$$

The intersection of events is sometimes denoted as a product, i.e.,

$$A_1 \cap A_2 \cap \cdots \cap A_n \equiv A_1 A_2 A_3 \cdots A_n.$$

The following law, called **De-Morgan Rule**, is fundamental to the algebra of events and yields the complement of the union, or intersection, of two events, namely:

1. $(A \cup B)^c = A^c \cap B^c$ (3.1.3)

2. $(A \cap B)^c = A^c \cup B^c$.

Finally, we define the notion of **partition**. A collection of n events $E_1, \ldots, E_n$ is called a **partition** of the sample space S, if

(i) $\bigcup_{i=1}^{n} E_i = S$,

(ii) $E_i \cap E_j = \emptyset$ for all $i \neq j$ $(i, j = 1, \ldots, n)$. That is, the events in any partition are mutually **disjoint**, and their union exhaust all the sample space.

Example 3.5. The experiment is to generate on the computer a random number, U, in the interval $(0, 1)$. A random number in $(0, 1)$ can be obtained as

$$U = \sum_{j=1}^{\infty} I_j 2^{-j},$$

where I_j is the random result of tossing a coin, i.e.,

$$I_j = \begin{cases} 1, & \text{if Head} \\ 0, & \text{if Tail.} \end{cases}$$

For generating random numbers from a set of integers, the summation index j is bounded by a finite number N. This method is, however, not practical for generating random numbers on a continuous interval. Computer programs generate "pseudo-random" numbers. Methods for generating random numbers are described in various books on simulation (see Bratley et al. 1983). Most commonly applied is the **linear congruential generator**. This method is based on the recursive equation

$$U_i = (aU_{i-1} + c) \bmod m, \quad i = 1, 2, \ldots.$$

The parameters a, c, and m depend on the computer's architecture. In many programs, $a = 65{,}539$, $c = 0$ and $m = 2^{31} - 1$. The first integer X_0 is called the "seed." Different choices of the parameters a, c and m yield "pseudo-random" sequences with different statistical properties.

The sample space of this experiment is

$$S = \{u : 0 \leq u \leq 1\}.$$

Let E_1 and E_2 be the events

$$E_1 = \{u : 0 \leq u \leq 0.5\},$$
$$E_2 = \{u : 0.35 \leq u \leq 1\}.$$

The union of these two events is

$$E_3 = E_1 \cup E_2 = \{u : 0 \le u \le 1\} = S.$$

The intersection of these events is

$$E_4 = E_1 \cap E_2 = \{u : 0.35 \le u < 0.5\}.$$

Thus, E_1 and E_2 are **not** disjoint.
The complementary events are

$$E_1^c = \{u : 0.5 \le u < 1\} \text{ and } E_2^c = \{u : u < 0.35\}$$

$E_1^c \cap E_2^c = \emptyset$; i.e., the complementary events are disjoint. By DeMorgan's law

$$(E_1 \cap E_2)^c = E_1^c \cup E_2^c$$
$$= \{u : u < 0.35 \text{ or } u \ge 0.5\}.$$

However,

$$\emptyset = S^c = (E_1 \cup E_2)^c = E_1^c \cap E_2^c.$$

Finally, the following is a partition of S:

$$B_1 = \{u : u < 0.1\}, \quad B_2 = \{u : 0.1 \le u < 0.2\},$$
$$B_3 = \{u : 0.2 \le u < 0.5\}, \quad B_4 = \{u : 0.5 \le u < 1\}.$$

Notice that $B_4 = E_1^c$

Different identities can be derived by the above rules of operations on events, a few will be given as exercises.

3.1.3 Probabilities of events

A probability function $\Pr\{\cdot\}$ assigns to events of S real numbers, following the following basic axioms.

1. $\Pr\{E\} \ge 0$.
2. $\Pr\{S\} = 1$.
3. **If $E_1, \ldots, E_n$ $(n \ge 1)$ are mutually disjoint events, then**

$$\Pr\left\{ \bigcup_{i=1}^{n} E_i \right\} = \sum_{i=1}^{n} \Pr\{E_i\}.$$

From these three basic axioms, we deduce the following results:

Result 1. If $A \subset B$ then

$$\Pr\{A\} \le \Pr\{B\}.$$

Indeed, since $A \subset B$, $B = A \cup (A^c \cap B)$. Moreover, $A \cap A^c \cap B = \emptyset$. Hence, by Axioms 1 and 3, $\Pr\{B\} = \Pr\{A\} + \Pr\{A^c \cap B\} \geq \Pr\{A\}$.

Thus, if E is any event, since $E \subset S$, $0 \leq \Pr\{E\} \leq 1$.

Result 2. For any event E, $\Pr\{E^c\} = 1 - \Pr\{E\}$.

Indeed, $S = E \cup E^c$. Since $E \cap E^c = \emptyset$,

$$1 = \Pr\{S\} = \Pr\{E\} + \Pr\{E^c\}. \tag{3.1.3}$$

This implies the result.

Result 3. For any events A, B

$$\Pr\{A \cup B\} = \Pr\{A\} + \Pr\{B\} - \Pr\{A \cap B\}. \tag{3.1.4}$$

Indeed, we can write

$$A \cup B = A \cup A^c \cap B,$$

where $A \cap (A^c \cap B) = \emptyset$. Thus, by the third axiom,

$$\Pr\{A \cup B\} = \Pr\{A\} + \Pr\{A^c \cap B\}.$$

Moreover, $B = A^c \cap B \cup A \cap B$, where again $A^c \cap B$ and $A \cap B$ are disjoint. Thus, $\Pr\{B\} = \Pr\{A^c \cap B\} + \Pr\{A \cap B\}$, or $\Pr\{A^c \cap B\} = \Pr\{B\} - \Pr\{A \cap B\}$. Substituting this above we obtain the result.

Result 4. If $B_1, \ldots, B_n$ $(n \geq 1)$ is a partition of S, then for any event E,

$$\Pr\{E\} = \sum_{i=1}^{n} \Pr\{E \cap B_i\}.$$

Indeed, by the distributive law,

$$E = E \cap S = E \cap \left(\bigcup_{i=1}^{n} B_i \right)$$

$$= \bigcup_{i=1}^{n} EB_i.$$

Finally, since $B_1, \ldots, B_n$ are mutually disjoint, $(EB_i) \cap (EB_j) = E \cap B_i \cap B_j = \emptyset$ for all $i \neq j$. Therefore, by the third axiom

$$\Pr\{E\} = \Pr\left\{ \bigcup_{i=1}^{n} EB_i \right\} = \sum_{i=1}^{n} \Pr\{EB_i\}. \tag{3.1.5}$$

Example 3.6. Fuses are used to protect electronic devices from unexpected power surges. Modern fuses are produced on glass plates through processes of metal deposition and photographic lythography. On each plate several hundred fuses are simultaneously produced. At the end of the process, the plates undergo precise cutting with special saws. A certain fuse is handled on one of three alternative cutting machines. Machine M_1 yields 200 fuses per hour. Machine M_2 yields 250 fuses per hour and machine M_3 yields 350 fuses per hour. The fuses are then mixed together. The proportions of defective parts that are typically produced on these machines are 0.01, 0.02, and 0.005, respectively. A fuse is chosen at random from the production of a given hour. What is the probability that it is compliant with the amperage requirements (nondefective)?

Let E_i be the event that the chosen fuse is from machine M_i ($i = 1, 2, 3$). Since the choice of the fuse is random, each fuse has the same probability $\frac{1}{800}$ to be chosen. Hence, $\Pr\{E_1\} = \frac{1}{4}$, $\Pr\{E_2\} = \frac{5}{16}$ and $\Pr\{E_3\} = \frac{7}{16}$.

Let G denote the event that the selected fuse is nondefective. For example for machine M_1, $\Pr\{G\} = 1 - 0.01 = 0.99$. We can assign $\Pr\{G \cap M_1\} = 0.99 \times 0.25 = 0.2475$, $\Pr\{G \cap M_2\} = 0.98 \times \frac{5}{16} = 0.3062$ and $\Pr\{G \cap M_3\} = 0.995 \times \frac{7}{16} = 0.4353$. Hence, the probability of selecting a nondefective fuse is, according to Result 4,

$$\Pr\{G\} = \Pr\{G \cap M_1\} + \Pr\{G \cap M_2\} + \Pr\{G \cap M_3\} = 0.989.$$

Example 3.7. Consider the problem of generating random numbers, discussed in Example 3.5. Suppose that the probability function assigns any interval $I(a, b) = \{u : a < u < b\}$, $0 \le a < b \le 1$, the probability
$$\Pr\{I(a, b)\} = b - a.$$

Let $E_3 = I(0.1, 0.4)$ and $E_4 = I(0.2, 0.5)$. $C = E_3 \cup E_4 = I(0.1, 0.5)$. Hence,
$$\Pr\{C\} = 0.5 - 0.1 = 0.4.$$

On the other hand, $\Pr\{E_3 \cap E_4\} = 0.4 - 0.2 = 0.2$.

$$\Pr\{E_3 \cup E_4\} = \Pr\{E_3\} + \Pr\{E_4\} - \Pr\{E_3 \cap E_4\}$$
$$= (0.4 - 0.1) + (0.5 - 0.2) - 0.2 = 0.4.$$

This illustrates Result 3

3.1.4   Probability functions for random sampling

Consider a finite population P, and suppose that the random experiment is to select a random sample from P, with or without replacement. More specifically, let $L_N = \{w_1, w_2, \ldots, w_N\}$ be a **list** of the elements of P, where N is its size. w_j ($j = 1, \ldots, N$) is an identification number of the j-th element.

Suppose that a sample of size n is drawn from L_N [respectively, P] **with** replacement. Let W_1 denote the first element selected from L_N. If j_1 is the index of this element then $W_1 = w_{j_1}$. Similarly, let W_i ($i = 1, \ldots, n$) denote the i-th element of the sample. The corresponding sample space is the collection

$$S = \{(W_1, \ldots, W_n) : W_i \in L_N, \quad i = 1, 2, \ldots, n\}$$

of all samples, with replacement from L_N. The total number of possible samples is N^n. Indeed, w_{j_1} could be any one of the elements of L_N, and so are $w_{j_2}, \ldots, w_{j_n}$. With each one of the N

possible choices of w_{j_1} we should combine the N possible choices of w_{j_2} and so on. Thus, there are N^n possible ways of selecting a sample of size n, with replacement. The sample points are the elements of S (possible samples). The sample is called **random with replacement**, RSWR, if each one of these N^n possible samples is assigned the same probability, $1/N^n$, for being selected.

Let $M(i)$ $(i = 1, \ldots, N)$ be the number of samples in S, which contain the i-th element of L_N (at least once). Since sampling is **with** replacement

$$M(i) = N^n - (N - 1)^n.$$

Indeed, $(N - 1)^n$ is the number of samples with replacement, which do not include w_i. Since all samples are equally probable, the probability that a RSWR $\mathbf{S}_n$ includes w_i $(i = 1, \ldots, N)$ is

$$\Pr\{w_i \in \mathbf{S}_n\} = \frac{N^n - (N - 1)^n}{N^n}$$

$$= 1 - \left(1 - \frac{1}{N}\right)^n.$$

If $n > 1$, then the above probability is larger than $1/N$ which is the probability of selecting the element W_i in any given trial, but smaller than n/N. Notice also that this probability does not depend on i, i.e., all elements of L_N have the same probability to be included in a RSWR. It can be shown that the probability that w_i is included in the sample exactly once is $\frac{n}{N}\left(1 - \frac{1}{N}\right)^{n-1}$. If sampling is **without** replacement, the number of sample points in S is $N(N - 1) \cdots (N - n + 1)/n!$, since the order of selection is immaterial. The number of sample points which include w_i is $M(i) = (N - 1)(N - 2) \cdots (N - n + 1)/(n - 1)!$. A sample $\mathbf{S}_n$ is called **random without replacement**, RSWOR, if all possible samples are equally probable. Thus, under RSWOR,

$$\Pr\{w_i \in \mathbf{S}_n\} = \frac{n!M(i)}{N(N - 1) \cdots (N - n + 1)} = \frac{n}{N},$$

for all $i = 1, \ldots, N$.

We consider now events, which depend on the attributes of the elements of a population. Suppose that we sample to obtain information on the number of defective (nonstandard) elements in a population. The attribute in this case is "the element complies to the requirements of the standard." Suppose that M out of N elements in L_N are nondefective (have the attribute). Let E_j be the event that j out of the n elements in the sample are nondefective. Notice that $E_0, \ldots, E_n$ is a partition of the sample space. What is the probability, under RSWR, of E_j? Let K_j^n denote the number of sample points in which j out of n are G elements (nondefective) and $(n - j)$ elements are D (defective). To determine K_j^n, we can proceed as follows:

Choose first j G's and $(n - j)$ D's from the population. This can be done in $M^j(N - M)^{n-j}$ different ways. We have now to assign the j G's into j out of n components of the vector $(w_1, \ldots, w_n)$. This can be done in $n(n - 1) \cdots (n - j + 1)/j!$ possible ways. This is known as the number of combinations of j out of n, i.e.,

$$\binom{n}{j} = \frac{n!}{j!(n-j)!}, \quad j = 0, 1, \ldots, n \tag{3.1.6}$$

where $k! = 1 \cdot 2 \cdot \cdots \cdot k$ is the product of the first k positive integers, $0! = 1$. Hence, $K_j^n = \binom{n}{j} M^j (N-M)^{n-j}$. Since every sample is equally probable, under RSWR,

$$\Pr\{E_{j:n}\} = K_j^n / N^n = \binom{n}{j} P^j (1-P)^{n-j}, \quad j = 0, \ldots, n, \tag{3.1.7}$$

where $P = M/N$. If sampling is without replacement, then

$$K_j^n = \binom{M}{j} \binom{N-M}{n-j}$$

and

$$\Pr\{E_j\} = \frac{\binom{M}{j} \binom{N-M}{n-j}}{\binom{N}{n}}. \tag{3.1.8}$$

These results are valid since the order of selection is immaterial for the event E_j.

These probabilities of E_j under RSWR and RSWOR are called, respectively, the **binomial** and **hypergeometric** probabilities.

Example 3.8. The experiment consists of randomly transmitting a sequence of binary signals, 0 or 1. What is the probability that 3 out of 6 signals are 1s? Let E_3 denote this event.
The sample space of 6 signals consists of 2^6 points. Each point is equally probable. The probability of E_3 is

$$\Pr\{E_3\} = \binom{6}{3} \frac{1}{2^6} = \frac{6 \cdot 5 \cdot 4}{1 \cdot 2 \cdot 3 \cdot 64}$$

$$= \frac{20}{64} = \frac{5}{16} = 0.3125.$$

Example 3.9. Two out of ten television sets are defective. A RSWOR of $n = 2$ sets is chosen. What is the probability that the two sets in the sample are good (nondefective). This is the hypergeometric probability of E_0 when $M = 2, N = 10, n = 2$, i.e.,

$$\Pr\{E_0\} = \frac{\binom{8}{2}}{\binom{10}{2}} = \frac{8 \cdot 7}{10 \cdot 9} = 0.622.$$

3.1.5   Conditional probabilities and independence of events

In this section, we discuss the notion of conditional probabilities. When different events are related, the realization of one event may provide us relevant information to improve our probability assessment of the other event(s). In Section 3.1.3, we gave an example with three machines which manufacture the same part but with different production rates and different proportions of defective parts in the output of those machines. The random experiment was to choose at random a part from the mixed yield of the three machines.

We saw earlier that the probability that the chosen part is nondefective is 0.989. If we can identify, before the quality test, from which machine the part came, the probabilities of nondefective would be conditional on this information.

The probability of choosing at random a nondefective part from machine M_1, is 0.99. If we are given the information that the machine is M_2, the probability is 0.98, and given machine M_3, the probability is 0.995. These probabilities are called **conditional probabilities**. The information given changes our probabilities.

We define now formally the concept of conditional probability.

Let A and B be two events such that $\Pr\{B\} > 0$. The conditional probability of A, given B, is

$$\Pr\{A \mid B\} = \frac{\Pr\{A \cap B\}}{\Pr\{B\}}. \tag{3.1.9}$$

Example 3.10.   The random experiment is to measure the length of a steel bar. The sample space is $S = (19.5, 20.5)$ [cm]. The probability function assigns any subinterval a probability equal to its length. Let $A = (19.5, 20.1)$ and $B = (19.8, 20.5)$. $\Pr\{B\} = 0.7$. Suppose that we are told that the length belongs to the interval B, and we have to guess whether it belongs to A. We compute the conditional probability

$$\Pr\{A \mid B\} = \frac{\Pr\{A \cap B\}}{\Pr\{B\}} = \frac{0.3}{0.7} = 0.4286.$$

On the other hand, if the information that the length belongs to B is not given,, then $\Pr\{A\} = 0.6$. Thus, there is difference between the conditional and nonconditional probabilities. This indicates that the two events A and B are dependent.

Definition.   Two events A, B are called **independent** if

$$\Pr\{A \mid B\} = \Pr\{A\}.$$

If A and B are independent events, then

$$\Pr\{A\} = \Pr\{A \mid B\} = \frac{\Pr\{A \cap B\}}{\Pr\{B\}}$$

or, equivalently,

$$\Pr\{A \cap B\} = \Pr\{A\} \Pr\{B\}.$$

If there are more than two events, $A_1, A_2, \ldots, A_n$, we say that the events are **pairwise independent** if

$$\Pr\{A_i \cap A_j\} = \Pr\{A_i\} \Pr\{A_j\} \text{ for all } i \neq j, \ i,j = 1, \ldots, n.$$

The n events are said to be **mutually independent** if, for any subset of k events, $k = 2, \ldots, n$, indexed by $A_{i_1}, \ldots, A_{i_k}$,

$$\Pr\{A_{i_1} \cap A_{i_2} \cdots \cap A_{i_k}\} = \Pr\{A_{i_1}\} \cdots \Pr\{A_{i_n}\}.$$

In particular, if n events are mutually independent, then

$$\Pr\left\{\bigcap_{i=1}^{n} A_i\right\} = \prod_{i=1}^{n} \Pr\{A_i\}. \tag{3.1.10}$$

One can show examples of events which are pairwise independent but **not** mutually independent.

We can further show (see exercises) that if two events are independent, then the corresponding complementary events are independent. Furthermore, if n events are mutually independent, then any pair of events is pairwise independent, every three events are triplewise independent, etc.

Example 3.11. Five identical parts are manufactured in a given production process. Let $E_1, \ldots, E_5$ be the events that these five parts comply with the quality specifications (nondefective). Under the model of mutual independence, the probability that **all** the five parts are indeed nondefective is

$$\Pr\{E_1 \cap E_2 \cap \cdots \cap E_5\} = \Pr\{E_1\}\Pr\{E_2\}\cdots\Pr\{E_5\}.$$

Since these parts come from the same production process, we can assume that $\Pr\{E_i\} = p$, all $i = 1, \ldots, 5$. Thus, the probability that **all** the five parts are nondefective is p^5.

What is the probability that one part is defective and all the other four are nondefective? Let A_1 be the event that one out of five parts is defective. In order to simplify the notation, we write the intersection of events as their product. Thus,

$$A_1 = E_1^c E_2 E_3 E_4 E_5 \cup E_1 E_2^c E_3 E_4 E_5 \cup E_1 E_2 E_3^c E_4 E_5 \cup E_1 E_2 E_3 E_4^c E_5 \cup E_1 E_2 E_3 E_4 E_5^c.$$

A_1 is the union of five **disjoint** events. Therefore,

$$\Pr\{A_1\} = \Pr\{E_1^c E_2 \cdots E_5\} + \cdots + \Pr\{E_1 E_2 \cdots E_5^c\}$$
$$= 5p^4(1-p).$$

Indeed, since $E_1, \ldots, E_5$ are **mutually** independent events

$$\Pr\{E_1^c E_2 \cdots E_5\} = \Pr\{E_1^c\}\Pr\{E_2\}\cdots\Pr\{E_5\} = (1-p)p^4.$$

Also,

$$\Pr\{E_1 E_2^c E_3 E_4 E_5\} = (1-p)p^4,$$

etc. Generally, if J_5 denotes the number of defective parts among the five ones,

$$\Pr\{J_5 = i\} = \binom{5}{i} p^{(5-i)}(1-p)^i, \quad i = 0, 1, 2, \ldots, 5.$$

3.1.6   Bayes' formula and its application

Bayes' formula, which is derived in the present section, provides us with a fundamental formula for weighing the evidence in the data concerning unknown parameters, or some unobservable events.

Suppose that the results of a random experiment depend on some event(s) which is (are) not directly observable. The observable event is related to the unobservable one(s) via the conditional probabilities. More specifically, suppose that $\{B_1, \ldots, B_m\}$ ($m \geq 2$) is a partition of the sample space. The events $B_1, \ldots, B_m$ are not directly observable, or verifiable. The random experiment results in an event A (or its complement). The conditional probabilities $\Pr\{A \mid B_i\}$, $i = 1, \ldots, m$ are known. The question is whether, after observing the event A, we can assign probabilities to the events $B_1, \ldots, B_m$? In order to weigh the evidence that A has on $B_1, \ldots, B_m$, we first assume some probabilities $\Pr\{B_i\}$, $i = 1, \ldots, m$, which are called **prior probabilities**. The prior probabilities express our degree of belief in the occurrence of the events B_i ($i = 1, \ldots, m$). After observing the event A, we convert the prior probabilities of B_i ($i = 1, \ldots, m$) to **posterior probabilities** $\Pr\{B_i \mid A\}$, $i = 1, \ldots, m$ by using **Bayes' formula**

$$\Pr\{B_i \mid A\} = \frac{\Pr\{B_i\}\Pr\{A \mid B_i\}}{\sum_{j=1}^{m}\Pr\{B_j\}\Pr\{A \mid B_j\}}, \quad i = 1, \ldots, m. \tag{3.1.11}$$

These posterior probabilities reflect the weight of evidence that the event A has concerning $B_1, \ldots, B_m$.

Bayes' formula can be obtained from the basic rules of probability. Indeed, assuming that $\Pr\{A\} > 0$,

$$\Pr\{B_i \mid A\} = \frac{\Pr\{A \cap B_i\}}{\Pr\{A\}}$$
$$= \frac{\Pr\{B_i\}\Pr\{A \mid B_i\}}{\Pr\{A\}}.$$

Furthermore, since $\{B_1, \ldots, B_m\}$ is a partition of the sample space,

$$\Pr = \{A\} = \sum_{j=1}^{m}\Pr\{B_j\}\Pr\{A \mid B_j\}.$$

Substituting this expression above, we obtain Bayes' formula.

The following example illustrates the applicability of Bayes' formula to a problem of decision-making.

Example 3.12. Two vendors B_1, B_2 produce ceramic plates for a given production process of hybrid micro circuits. The parts of vendor B_1 have probability $p_1 = 0.10$ of being defective. The parts of vendor B_2 have probability $p_2 = 0.05$ of being defective. A delivery of $n = 20$ parts arrive, but the label which identifies the vendor is missing. We wish to apply Bayes' formula to assign a probability that the package came from vendor B_1.

Suppose that it is a priori, equally likely that the package was mailed by vendor B_1 or vendor B_2. Thus, the prior probabilities are $\Pr\{B_1\} = \Pr\{B_2\} = 0.5$. We inspect the 20 parts in the package and find $J_{20} = 3$ defective items. A is the event $\{J_{20} = 3\}$. The conditional probabilities of A, given B_i ($i = 1, 2$) are

$$\Pr\{A \mid B_1\} = \binom{20}{3} p_1^3 (1 - p_1)^{17}$$

$$= 0.1901.$$

Similarly,

$$\Pr\{A \mid B_2\} = \binom{20}{3} p_2^3 (1 - p_2)^{17}$$

$$= 0.0596.$$

According to Bayes' formula

$$\Pr\{B_1 \mid A\} = \frac{0.5 \times 0.1901}{0.5 \times 0.1901 + 0.5 \times 0.0596} = 0.7613$$

$$\Pr\{B_2 \mid A\} = 1 - \Pr\{B_1 \mid A\} = 0.2387.$$

Thus, after observing three defective parts in a sample of $n = 20$ ones, we believe that the delivery came from vendor B_1. The posterior probability of B_1, given A, is more than three times higher than that of B_2 given A. The a priori odds of B_1 against B_2 were 1:1. The a posteriori odds are 19:6.

In the context of a graph representing directed links between variables, a directed acyclic graph (DAG) represents a qualitative causality model. The model parameters are derived by applying the Markov property, where the conditional probability distribution at each node depends only on its parents. For discrete random variables, this conditional probability is often represented by a table, listing the local probability that a child node takes on each of the feasible values – for each combination of values of its parents. The joint distribution of a collection of variables can be determined uniquely by these local conditional probability tables. A Bayesian Network (BN) is represented by a DAG. A BN reflects a simple **conditional independence** statement, namely that each variable is independent of its nondescendants in the graph given the state of its parents. This property is used to reduce, sometimes significantly, the number of parameters that are required to characterize the joint probability distribution of the variables. This reduction provides an efficient way to compute the posterior probabilities given the evidence present in the data. We do not cover BN here. For examples of applications of BN with a general introduction to this topic see Kenett (2012).

3.2   Random variables and their distributions

Random variables are formally defined as **real-valued functions**, $X(w)$, **over the sample space**, S, **such that, events** $\{w : X(w) \le x\}$ **can be assigned probabilities, for all** $-\infty < x < \infty$, where w are the elements of S.

Example 3.13. Suppose that S is the sample space of all RSWOR of size n, from a finite population, P, of size N. $1 \leq n < N$. The elements w of S are subsets of distinct elements of the population P. A random variable $X(w)$ is some function which assigns w a finite real number, e.g., the number of "defective" elements of w. In the present example, $X(w) = 0, 1, \ldots, n$ and

$$\Pr\{X(w) = j\} = \frac{\binom{M}{j}\binom{N-M}{n-j}}{\binom{N}{n}}, \quad j = 0, \ldots, n,$$

where M is the number of "defective" elements of P.

Example 3.14. Another example of random variable is the compressive strength of a concrete cube of a certain dimension. In this example, the random experiment is to manufacture a concrete cube according to a specified process. The sample space S is the space of all cubes, w, that can be manufactured by this process. $X(w)$ is the compressive strength of w. The probability function assigns each event $\{w : X(w) \leq \xi\}$ a probability, according to some mathematical model which satisfies the laws of probability. Any continuous nondecreasing function $F(x)$, such that $\lim_{x \to -\infty} F(x) = 0$ and $\lim_{x \to \infty} F(x) = 1$ will do the job. For example, for compressive strength of concrete cubes, the following model has been shown to fit experimental results

$$\Pr\{X(w) \leq x\} = \begin{cases} 0, & x \leq 0 \\ \dfrac{1}{\sqrt{2\pi}\sigma} \displaystyle\int_0^x \frac{1}{y} \exp\left\{ -\frac{(\ln y - \mu)^2}{2\sigma^2} \right\} dy, & 0 < x < \infty. \end{cases}$$

The constants μ and σ, $-\infty < \mu < \infty$ and $0 < \sigma < \infty$, are called **parameters** of the model. Such parameters characterize the manufacturing process.

We distinguish between **two types** of random variables: **discrete** and **continuous**.

Discrete random variables, $X(w)$, are random variables having a finite or countable range. For example, the number of "defective" elements in a random sample is a discrete random variable. The number of blemishes on a ceramic plate is a discrete random variable. A **continuous random variable** is one whose range consists of whole intervals of possible values. The weight, length, compressive strength, tensile strength, cycle time, output voltage, etc., are continuous random variables.

3.2.1 Discrete and continuous distributions

3.2.1.1 Discrete random variables

Suppose that a discrete random variable can assume the distinct values $x_0, \ldots, x_k$ (k is finite or infinite). The function

$$p(x) = \Pr\{X(w) = x\}, -\infty < x < \infty \tag{3.2.1}$$

is called the **probability distribution function** (p.d.f.) of X.

Notice that if x is not one of the values in the specified range $S_X = \{x_j; j = 0, 1, \ldots, k\}$ then $\{X(w) = x\} = \phi$ and $p(x) = 0$. Thus, $p(x)$ assumes positive values only on the specified sequence S_X (S_X is also called the sample space of X), such that

1. $p(x_j) \geq 0, j = 0, \ldots, k$ ⠀⠀⠀⠀⠀⠀⠀⠀⠀⠀⠀⠀⠀⠀⠀⠀⠀⠀⠀⠀⠀⠀⠀⠀⠀⠀⠀⠀(3.2.2)

2. $\sum_{j=0}^{k} p(x_j) = 1$.

Example 3.15. Suppose that the random experiment is to cast a die once. The sample points are six possible faces of the die, $\{w_1, \ldots, w_6\}$. Let $X(w_j) = j, j = 1, \ldots, 6$, be the random variable, representing the face number. The probability model yields

$$p(x) = \begin{cases} \dfrac{1}{6}, & \text{if } x = 1, 2, \ldots, 6 \\ 0, & \text{otherwise.} \end{cases}$$

Example 3.16. Consider the example of Section 3.1.5, of drawing independently $n = 5$ parts from a production process, and counting the number of "defective" parts in this sample. The random variable is $X(w) = J_5$. $S_X = \{0, 1, \ldots, 5\}$ and the p.d.f. is

$$p(x) = \begin{cases} \binom{5}{x} p^{5-x}(1-p)^x, & x = 0, 1, \ldots, 5 \\ 0, & \text{otherwise.} \end{cases}$$

The probability of the event $\{X(w) \leq x\}$, for any $-\infty < x < \infty$ can be computed by summing the probabilities of the values in S_X, which belong to the interval $(-\infty, x]$. This sum is called the **cumulative distribution function** (c.d.f.) of X, and denoted by

$$P(x) = \Pr\{X(w) \leq x\}$$

$$= \sum_{\{x_j \leq x\}} p(x_j), \qquad (3.2.3)$$

where $x_j \in S_X$.

The c.d.f. corresponding to Example 3.16 is

$$P(x) = \begin{cases} 0, & x < 0 \\ \sum_{j=0}^{[x]} \binom{5}{j} p^{5-j}(1-p)^j, & 0 \leq x < 5 \\ 1, & 5 \leq x, \end{cases}$$

where $[x]$ denotes the **integer part of** x, i.e., the **largest integer** smaller or equal to x.

Generally, the graph of the p.d.f. of a discrete variable is a bar-chart (see Figure 3.1). The corresponding c.d.f. is a step function, as shown in Figure 3.2.

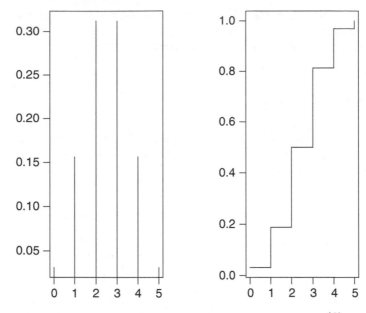

Figure 3.1 (left)–3.2 (right)   *The graph of the p.d.f. and c.d.f. $P(x) = \sum_{j=0}^{[x]} \frac{\binom{5}{j}}{2^5}$ random variable.*

3.2.1.2   *Continuous random variables*

In the case of continuous random variables, the model assigns the variable under consideration a function $F(x)$ which is

(i) continuous;
(ii) Nondecreasing, i.e., if $x_1 < x_2$ then $F(x_1) \le F(x_2)$ and
(iii) $\lim_{x \to -\infty} F(x) = 0$ and $\lim_{x \to \infty} F(x) = 1$.

Such a function can serve as a **c.d.f.**, for X.

An example of a c.d.f. for a continuous random variable which assumes nonnegative values, e.g., the operation total time until a part fails, is

$$F(x) = \begin{cases} 0, & \text{if } x \le 0 \\ 1 - e^{-x}, & \text{if } x > 0. \end{cases}$$

This function (see Figure 3.3) is continuous, monotonically increasing, and $\lim_{x \to \infty} F(x) = 1 - \lim_{x \to \infty} e^{-x} = 1$. If the c.d.f. of a continuous random variable, can be represented as

$$F(x) = \int_{-\infty}^{x} f(y)dy, \tag{3.2.4}$$

for some $f(y) \ge 0$, then we say that $F(x)$ is **absolutely continuous** and $f(x) = \frac{d}{dx}F(x)$. (The derivative $f(x)$ may not exist on a finite number of x values, in any finite interval.) The function $f(x)$ is called the **p.d.f.** of X.

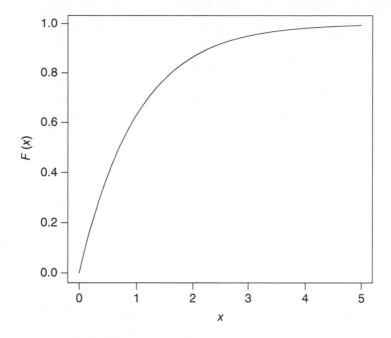

Figure 3.3   c.d.f. of $F(x) = 1 - e^{-x}$.

In the above example of total operational time, the p.d.f. is

$$f(x) = \begin{cases} 0, & \text{if } x < 0 \\ e^{-x}, & \text{if } x \geq 0. \end{cases}$$

Thus, as in the discrete case, we have $F(x) = \Pr\{X \leq x\}$. It is now possible to write

$$\Pr\{a \leq X < b\} = \int_a^b f(t)dt = F(b) - F(a) \tag{3.2.5}$$

or

$$\Pr\{X \geq b\} = \int_b^\infty f(t)dt = 1 - F(b). \tag{3.2.6}$$

Thus, if X has the exponential c.d.f.

$$\Pr\{1 \leq X \leq 2\} = F(2) - F(1) = e^{-1} - e^{-2} = 0.2325.$$

There are certain phenomena which require more complicated modeling. The random variables under consideration may not have purely discrete or purely absolutely continuous distribution. There are many random variables with c.d.f.s which are absolutely continuous within certain intervals and have jump points (points of discontinuity) at the end points of the intervals (Figure 3.4). Distributions of such random variables can be expressed as mixtures of purely discrete c.d.f., $F_d(x)$, and of absolutely continuous c.d.f., $F_{ac}(x)$ i.e.,

$$F(x) = pF_d(x) + (1-p)F_{ac}(x), -\infty < x < \infty, \tag{3.2.7}$$

where $0 \leq p \leq 1$.

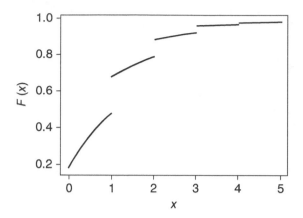

Figure 3.4   *The c.d.f. of the mixture distribution $F(x) = 0.5\,(1 - e^{-x}) + 0.5 \times e^{-1} \sum_{j=0}^{[x]} \frac{1}{j!}$.*

Example 3.17.   A distribution which is a mixture of discrete and continuous distributions is obtained for example, when a measuring instrument is not sensitive enough to measure small quantities or large quantities which are outside its range. This could be the case for a weighing instrument which assigns the value 0 [mg] to any weight smaller than 1 [mg], the value 1 [g] to any weight greater than 1 gram, and the correct weight to values in between. Another example is the total number of minutes, within a given working hour, that a service station is busy serving customers. In this case, the c.d.f. has a jump at 0, of height p, which is the probability that the service station is idle at the beginning of the hour, and no customer arrives during that hour. In this case,

$$F(x) = p + (1 - p)G(x), \quad 0 \le x < \infty,$$

where $G(x)$ is the c.d.f. of the total service time, $G(0) = 0$.

3.2.2   Expected values and moments of distributions

The **expected value** of a function $g(X)$, under the distribution $F(x)$, is

$$E_F\{g(X)\} = \begin{cases} \displaystyle\int_{-\infty}^{\infty} g(x)f(x)dx, & \text{if } X \text{ is continuous} \\[2ex] \displaystyle\sum_{j=0}^{k} g(x_j)p(x_j), & \text{if } X \text{ is discrete.} \end{cases}$$

In particular,

$$\mu_l(F) = E_F\{X^l\}, \quad l = 1, 2, \ldots \tag{3.2.8}$$

is called the *l*th **moment** of $F(x)$. $\mu_1(F) = E_F\{X\}$ is the expected value of X, or the **population mean**, according to the model $F(x)$.

Moments around $\mu_1(F)$ are called **central moments**, which are

$$\mu_l^*(F) = E\{(X - \mu_1(F))^l\}, \quad l = 1, 2, 3, \ldots. \tag{3.2.9}$$

Obviously, $\mu_1^*(F) = 0$. The **second central moment** is called the **variance** of $F(x)$, $V_F\{X\}$.

In the following, the notation $\mu_l(F)$ will be simplified to μ_l, if there is no room for confusion.

Expected values of a function $g(X)$, and in particular the moments, may not exist, since an integral $\int_{-\infty}^{\infty} x^l f(x) dx$ may not be well defined. Example of such a case is the distribution, called the **Cauchy distribution**, with p.d.f.

$$f(x) = \frac{1}{\pi} \cdot \frac{1}{1 + x^2}, -\infty < x < \infty.$$

Notice that under this model, moments do not exist for any $l = 1, 2, \ldots$. Indeed, the integral

$$\frac{1}{\pi} \int_{-\infty}^{\infty} \frac{x}{1 + x^2} dx$$

does not exist. If the second moment exists, then

$$V\{X\} = \mu_2 - \mu_1^2.$$

Example 3.18. Consider the random experiment of casting a die once. The random variable, X, is the face number. Thus, $p(x) = \frac{1}{6}, x = 1, \ldots, 6$ and

$$\mu_1 = E\{X\} = \frac{1}{6} \sum_{j=1}^{6} j = \frac{6(6+1)}{2 \times 6} = \frac{7}{2} = 3.5$$

$$\mu_2 = \frac{1}{6} \sum_{j=1}^{6} j^2 = \frac{6(6+1)(2 \times 6 + 1)}{6 \times 6} = \frac{7 \times 13}{6} = \frac{91}{6} = 15.167.$$

The variance is

$$V\{X\} = \frac{91}{6} - \left(\frac{7}{2}\right)^2 = \frac{182 - 147}{12}$$

$$= \frac{35}{12}.$$

Example 3.19. X has a continuous distribution with p.d.f.

$$f(x) = \begin{cases} 0, & \text{otherwise} \\ 1, & \text{if } 1 \le x \le 2. \end{cases}$$

Thus,

$$\mu_1 = \int_1^2 x dx = \frac{1}{2}\left(x^2\big|_1^2\right) = \frac{1}{2}(4 - 1) = 1.5$$

$$\mu_2 = \int_1^2 x^2 dx = \frac{1}{3}\left(x^3|_1^2\right) = \frac{7}{3}$$

$$V\{X\} = \mu_2 - \mu_1 = \frac{7}{3} - \frac{9}{4} = \frac{28 - 27}{12} = \frac{1}{12}.$$

The following is a useful formula when X assumes only positive values, i.e., $F(x) = 0$ for all $x \le 0$,

$$\mu_1 = \int_0^\infty (1 - F(x))dx, \tag{3.2.10}$$

for continuous c.d.f. $F(x)$. Indeed,

$$\mu_1 = \int_0^\infty xf(x)dx$$

$$= \int_0^\infty \left(\int_0^x dy\right) f(x)dx$$

$$= \int_0^\infty \left(\int_y^\infty f(x)dx\right) dy$$

$$= \int_0^\infty (1 - F(y))dy.$$

For example, suppose that $f(x) = \mu e^{-\mu x}$, for $x \ge 0$. Then $F(x) = 1 - e^{-\mu x}$ and

$$\int_0^\infty (1 - F(x))dx = \int_0^\infty e^{-\mu x}dx = \frac{1}{\mu}.$$

When X is discrete, assuming the values $\{1, 2, 3, \dots\}$, then we have a similar formula

$$E\{X\} = 1 + \sum_{i=1}^\infty (1 - F(i)).$$

3.2.3 The standard deviation, quantiles, measures of skewness and kurtosis

The **standard deviation** of a distribution $F(x)$ is $\sigma = (V\{X\})^{1/2}$. The standard deviation is used as a measure of dispersion of a distribution. An important theorem in probability theory, called the **Chebychev Theorem**, relates the standard deviation to the probability of deviation from the mean. More formally, the theorem states that, if σ exists then

$$\Pr\{|X - \mu_1| > \lambda\sigma\} \le \frac{1}{\lambda^2}. \tag{3.2.11}$$

Thus, by this theorem, the probability that a random variable will deviate from its expected value by more than three standard deviations is less than 1/9, whatever the distribution is. This theorem has important implications, which will be highlighted later.

The *p*th quantile of a distribution $F(x)$ is the smallest value of x, ξ_p such that $F(x) \ge p$. We also write $\xi_p = F^{-1}(p)$.

For example, if $F(x) = 1 - e^{-\lambda x}$, $0 \le x < \infty$, where $0 < \lambda < \infty$ then ξ_p is such that

$$F(\xi_p) = 1 - e^{-\lambda \xi_p} = p.$$

Solving for ξ_p we get

$$\xi_p = -\frac{1}{\lambda} \cdot \ln(1 - p).$$

The **median** of $F(x)$ is $f^{-1}(0.5) = \xi_{0.5}$. Similarly $\xi_{0.25}$ and $\xi_{0.75}$ are the first and third quartiles of F.

A distribution $F(x)$ is **symmetric** about the mean $\mu_1(F)$ if

$$F(\mu_1 + \delta) = 1 - F(\mu_1 - \delta)$$

for all $\delta \ge 0$.

In particular, if F is symmetric, then $F(\mu_1) = 1 - F(\mu_1)$ or $\mu_1 = F^{-1}(0.5) = \xi_{0.5}$. Accordingly, the mean and median of a symmetric distribution coincide. In terms of the p.d.f., a distribution is symmetric about its mean if

$$f(\mu_1 + \delta) = f(\mu_1 - \delta), \text{ for all } \delta \ge 0.$$

A commonly used index of **skewness** (asymmetry) is

$$\beta_3 = \frac{\mu_3^*}{\sigma^3}, \tag{3.2.12}$$

where μ_3^* is the third central moment of F. One can prove that **if $F(x)$ is symmetric then $\beta_3 = 0$**. If $\beta_3 > 0$ we say that $F(x)$ is positively skewed, otherwise, it is negatively skewed.

Example 3.20. Consider the **binomial** distribution, with p.d.f.

$$p(x) = \binom{n}{x} p^x (1 - p)^{n-x}, \quad x = 0, 1, \ldots, n.$$

In this case,

$$\mu_1 = \sum_{x=0}^{n} x \binom{n}{x} p^x (1 - p)^{n-x}$$

$$= np \sum_{x=1}^{n} \binom{n-1}{x-1} p^{x-1} (1 - p)^{n-1-(x-1)}$$

$$= np \sum_{j=0}^{n-1} \binom{n-1}{j} p^j (1 - p)^{n-1-j}$$

$$= np.$$

Indeed,

$$x \binom{n}{x} = x \frac{n!}{x!(n-x)!} = \frac{n!}{(x-1)!((n-1)-(x-1))!}$$

$$= n \binom{n-1}{x-1}.$$

Similarly, we can show that

$$\mu_2 = n^2 p^2 + np(1-p),$$

and

$$\mu_3 = np[n(n-3)p^2 + 3(n-1)p + 1 + 2p^2].$$

The third central moment is

$$\mu_3^* = \mu_3 - 3\mu_2\mu_1 + 2\mu_1^3$$
$$= np(1-p)(1-2p).$$

Furthermore,

$$V\{X\} = \mu_2 - \mu_1^2$$
$$= np(1-p).$$

Hence,

$$\sigma = \sqrt{np(1-p)}$$

and the index of asymmetry is

$$\beta_3 = \frac{\mu_3^*}{\sigma_3} = \frac{np(1-p)(1-2p)}{(np(1-p))^{3/2}}$$
$$= \frac{1-2p}{\sqrt{np(1-p)}}.$$

Thus, if $p = \frac{1}{2}$, then $\beta_3 = 0$, and the distribution is symmetric. If $p < \frac{1}{2}$, the distribution is positively skewed, and it is negatively skewed if $p > \frac{1}{2}$.

In Chapter 2, we mentioned also the index of **kurtosis** (steepness). This is given by

$$\beta_4 = \frac{\mu_4^*}{\sigma^4}. \tag{3.2.13}$$

Example 3.21.   Consider the exponential c.d.f.

$$F(x) = \begin{cases} 0, & \text{if } x < 0 \\ 1 - e^{-x}, & \text{if } x \geq 0. \end{cases}$$

The p.d.f. is $f(x) = e^{-x}$, $x \geq 0$. Thus, for this distribution

$$\mu_1 = \int_0^\infty x e^{-x} dx = 1$$

$$\mu_2 = \int_0^\infty x^2 e^{-x} dx = 2$$

$$\mu_3 = \int_0^\infty x^3 e^{-x} dx = 6$$

$$\mu_4 = \int_0^\infty x^4 e^{-x} dx = 24.$$

Therefore,

$$V\{X\} = \mu_2 - \mu_1^2 = 1,$$

$$\sigma = 1$$

$$\mu_4^* = \mu_4 - 4\mu_3 \cdot \mu_1 + 6\mu_2 \mu_1^2 - 3\mu_1^4$$

$$= 24 - 4 \times 6 \times 1 + 6 \times 2 \times 1 - 3 = 9.$$

Finally, the index of kurtosis is

$$\beta_4 = 9.$$

3.2.4 Moment generating functions

The **moment generating function** (m.g.f.) of a distribution of X, is defined as a function of a real variable t,

$$M(t) = E\{e^{tX}\}. \tag{3.2.14}$$

$M(0) = 1$ for all distributions. $M(t)$, however, may not exist for some $t \neq 0$. To be useful, it is sufficient that $M(t)$ will exist in some interval containing $t = 0$.

For example, if X has a continuous distribution with p.d.f.

$$f(x) = \begin{cases} \dfrac{1}{b-a}, & \text{if } a \leq x \leq b, \ a < b \\ 0, & \text{otherwise} \end{cases}$$

then

$$M(t) = \frac{1}{b-a} \int_a^b e^{tx} dx = \frac{1}{t(b-a)} (e^{tb} - e^{ta}).$$

This is a differentiable function of t, for all t, $-\infty < t < \infty$.

On the other hand, if for $0 < \lambda < \infty$,

$$f(x) = \begin{cases} \lambda e^{-\lambda x}, & 0 \leq x < \infty \\ 0, & x < 0 \end{cases}$$

then

$$M(t) = \lambda \int_0^\infty e^{tx - \lambda x} dx$$

$$= \frac{\lambda}{\lambda - t}, \quad t < \lambda.$$

This m.g.f. exists only for $t < \lambda$. The m.g.f. $M(t)$ is a transform of the distribution $F(x)$, and the correspondence between $M(t)$ and $F(x)$ is one-to-one. In the above example, $M(t)$ is the Laplace transform of the p.d.f. $\lambda e^{-\lambda x}$. This correspondence is often useful in identifying the distributions of some statistics, as will be shown later.

Another useful property of the m.g.f. $M(t)$ is that often we can obtain the moments of $F(x)$ by differentiating $M(t)$. More specifically, consider the rth order derivative of $M(t)$. Assuming that this derivative exists, and differentiation can be interchanged with integration (or summation), then

$$M^{(r)}(t) = \frac{d^r}{dt^r} \int e^{tx} f(x) dx = \int \left(\frac{d^r}{dt^r} e^{tx} \right) f(x) dx$$

$$= \int x^r e^{tx} f(x) dx.$$

Thus, if these operations are justified, then

$$M^{(r)}(t)|_{t=0} = \int x^r f(x) dx = \mu_r. \tag{3.2.15}$$

In the following sections, we will illustrate the usefulness of the m.g.f.

3.3   Families of discrete distribution

In the present section, we discuss several families of discrete distributions and illustrate possible application in modeling industrial phenomena.

3.3.1   The binomial distribution

Consider n **identical** independent trials. In each trial the probability of "success" is fixed at some value p, and successive events of "success" or "failure" are **independent**. Such trials are called **Bernoulli trials**. The distribution of the number of "successes," J_n, is binomial with p.d.f.

$$b(j; n, p) = \binom{n}{j} p^j (1 - p)^{n-j}, \quad j = 0, 1, \ldots, n. \tag{3.3.1}$$

This p.d.f. was derived in Example 3.11 as a special case.

A binomial random variable, with parameters (n, p) will be designated as $B(n, p)$. n is a given integer and p belongs to the interval $(0, 1)$. The collection of all such binomial distributions is called the **binomial family**.

The binomial distribution is a proper model whenever we have a sequence of independent binary events ($0 - 1$, or "Success" and "Failure") with the same probability of "Success."

Example 3.22.   We draw a random sample of $n = 10$ items from a mass production line of light bulbs. Each light bulb undergoes an inspection and if it complies with the production specifications, we say that the bulb is compliant (successful event). Let $X_i = 1$ if the ith bulb is compliant and $X_i = 0$ otherwise. If we can assume that the probability of $\{X_i = 1\}$ is the same, p, for all bulbs and if the n events are mutually independent, then the number of bulbs in the sample which comply with the specifications, i.e., $J_n = \sum_{i=1}^{n} X_i$, has the binomial p.d.f. $b(i; n, p)$. Notice that if we draw a sample at random **with** replacement, RSWR, from a lot of size N, which contains M compliant units then J_n is $B\left(n, \frac{M}{N}\right)$.

Indeed, if sampling is with replacement, the probability that the ith item selected is compliant is $p = \frac{M}{N}$ for all $i = 1, \ldots, n$. Furthermore, selections are **independent** of each other.

The binomial c.d.f. will be denoted by $B(i; n, p)$. Recall that

$$B(i; n, p) = \sum_{j=0}^{i} b(j; n, p), \tag{3.3.2}$$

$i = 0, 1, \ldots, n$. The m.g.f. of $B(n, p)$ is

$$M(t) = E\{e^{tX}\}$$

$$= \sum_{j=0}^{n} \binom{n}{j} (pe^t)^j (1 - p)^{n-j} \tag{3.3.3}$$

$$= (pe^t + (1 - p))^n, -\infty < t < \infty.$$

Notice that

$$M'(t) = n(pe^t + (1 - p))^{n-1} pe^t$$

and

$$M''(t) = n(n - 1)p^2 e^{2t}(pe^t + (1 - p))^{n-2} + npe^t(pe^t + (1 - p))^{n-1}.$$

The expected value and variance of $B(n, p)$ are

$$E\{J_n\} = np, \tag{3.3.4}$$

and

$$V\{J_n\} = np(1 - p). \tag{3.3.5}$$

This was shown in Example 3.20 and can be verified directly by the above formulae of $M'(t)$ and $M''(t)$. To obtain the values of $b(i; n, p)$, we can use R or MINITAB. For example suppose we wish to tabulate the values of the p.d.f. $b(i; n, p)$ and those of the c.d.f. $B(i; n, p)$ for $n = 30$ and $p = 0.60$. Below R commands to generate a data frame with values as illustrated in Table 3.1.

```
> X <- data.frame(i=0:30,
            b=dbinom(x=0:30, size=30, prob=0.6),
            B=pbinom(q=0:30, size=30, prob = 0.6))
> rm(X)
```

In MINITAB, we first put in column C1 the integers $0, 1, \ldots, 30$ and put the value of $b(i; 30, .60)$ in column C2, and those of $B(i; 30, .60)$ in C3. To make MINITAB commands visible in the session window go to Editor > Enable Commands. We then type the commands:

Table 3.1   *Values of the p.d.f. and c.d.f. of B(30,.6)*

i	$b(i; 30, .6)$	$B(i; 30, .6)$
8	0.0002	0.0002
9	0.0006	0.0009
10	0.0020	0.0029
11	0.0054	0.0083
12	0.0129	0.0212
13	0.0269	0.0481
14	0.0489	0.0971
15	0.0783	0.1754
16	0.1101	0.2855
17	0.1360	0.4215
18	0.1474	0.5689
19	0.1396	0.7085
20	0.1152	0.8237
21	0.0823	0.9060
22	0.0505	0.9565
23	0.0263	0.9828
24	0.0115	0.9943
25	0.0041	0.9985
26	0.0012	0.9997
27	0.0003	1.0000

```
MTB> Set C1
DATA> 1(0:30/1)1
DATA> End.
MTB> PDF C1 C2;
SUBC> Binomial 30 0.60.
MTB> CDF C1 C3;
SUBC> Binomial 30 0.60.
```

In Table 3.1, we present these values

An alternative option is to use the pull down window.

After tabulating the values of the c.d.f., we can obtain the quantiles (or fractiles) of the distribution. Recall that in the discrete case, the pth quantile of a random variable X is

$$x_p = \text{smallest } x \text{ such that } F(x) \geq p.$$

Thus, from Table 3.1, we find that the lower quartile, the median, and upper quartile of $B(30, .6)$ are $Q_1 = 16$, $M_e = 18$ and $Q_3 = 20$. These values can also be obtained directly with R code

```
> qbinom(p=0.5, size=30, prob=0.6)
```

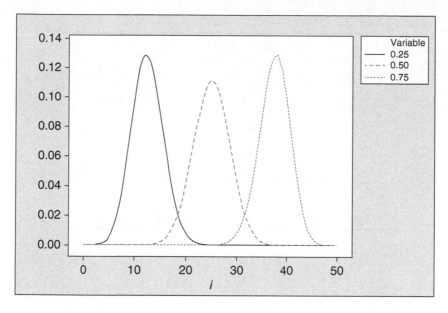

Figure 3.5 *The p.d.f. of B(50, p), p = 0.25, 0.50, 0.75 (MINITAB).*

or with MINITAB commands

```
MTB> InvCDF.5 k1;
SUBC> Binomial 30.6.
```

The value of the median is stored in the constant $k1$.

In Figure 3.5, we present the p.d.f. of three binomial distributions, with $n = 50$ and $p = .25$, 0.50 and 0.75. We see that if $p = .25$, the p.d.f. is positively skewed. When $p = .5$ it is symmetric, and when $p = .75$ it is negatively skewed. This is in accordance with the index of skewness β_3, which was presented in Example 3.20.

3.3.2 The hypergeometric distribution

Let J_n denote the number of units, in a RSWOR of size n, from a population of size N, having a certain property. The number of population units before sampling, having this property is M. The distribution of J_n is called the hypergeometric distribution. We denote a random variable having such a distribution by $H(N, M, n)$. The p.d.f. of J_n is

$$h(j; N, M, n) = \frac{\binom{M}{j}\binom{N-M}{n-j}}{\binom{N}{n}}, \quad j = 0, \ldots, n. \tag{3.3.6}$$

This formula was shown already in Section 3.1.4.

The c.d.f. of $H(N, M, n)$ will be designated by $H(j; N, M, n)$. In Table 3.2, we present the p.d.f. and c.d.f. of $H(75, 15, 10)$.

In Figure 3.6, we show the p.d.f. of $H(500, 350, 100)$.

Table 3.2   *The p.d.f. and c.d.f. of H(75, 15, 10)*

j	h(j; 75, 15, 10)	H(j; 75, 15, 10)
0	0.0910	0.0910
1	0.2675	0.3585
2	0.3241	0.6826
3	0.2120	0.8946
4	0.0824	0.9770
5	0.0198	0.9968
6	0.0029	0.9997
7	0.0003	1.0000

Table 3.3   *The p.d.f. of H(500,350, 20) and B(20, 0.7)*

i	h(i; 500,350, 20)	b(i; 20, 0.7)
5	0.00003	0.00004
6	0.00016	0.00022
7	0.00082	0.00102
8	0.00333	0.00386
9	0.01093	0.01202
10	0.02928	0.03082
11	0.06418	0.06537
12	0.11491	0.11440
13	0.16715	0.16426
14	0.19559	0.19164
15	0.18129	0.17886
16	0.12999	0.13042
17	0.06949	0.07160
18	0.02606	0.02785
19	0.00611	0.00684
20	0.00067	0.00080

The expected value and variance of $H(N, M, n)$ are

$$E\{J_n\} = n \cdot \frac{M}{N} \tag{3.3.7}$$

and

$$V\{J_n\} = n \cdot \frac{M}{N} \cdot \left(1 - \frac{M}{N}\right)\left(1 - \frac{n-1}{N-1}\right). \tag{3.3.8}$$

Notice that when $n = N$, the variance of J_n is $V\{J_N\} = 0$. Indeed, if $n = N$, $J_N = M$, which is not a random quantity. Derivation of these formulae are given in Section 5.2.2. There is no simple expression for the m.g.f.

If the sample size n is small relative to N, i.e., $n/N \ll 0.1$, the hypergeometric p.d.f. can be approximated by that of the binomial $B\left(n, \frac{M}{N}\right)$. In Table 3.3, we compare the p.d.f. of $H(500,350, 20)$ to that of $B(20, 0.7)$.

The expected value and variance of the binomial and the hypergeometric distributions are compared in Table 3.4. We see that the expected values have the same formula, but that the

Table 3.4 *The expected value and variance of the hypergeometric and binomial distribution*

	$H(a; N, M, n)$ hypergeometric	$B\left(n, \dfrac{M}{N}\right)$ binomial
Expected value	$n\dfrac{M}{N}$	$n\dfrac{M}{N}$
Variance	$n\dfrac{M}{N}\left(1 - \dfrac{M}{N}\right)\left(1 - \dfrac{n-1}{N-1}\right)$	$n\dfrac{M}{N}\left(1 - \dfrac{M}{N}\right)$

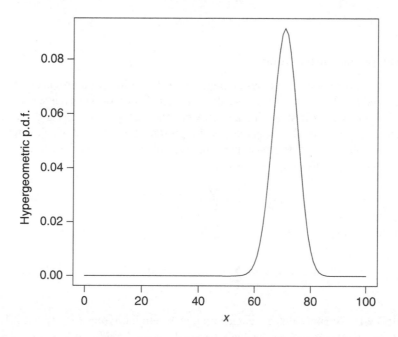

Figure 3.6 *The p.d.f. $h(i; 500, 350, 100)$.*

variance formulae differ by the correction factor $(N - n)/(N - 1)$ which becomes 1 when $n = 1$ and 0 when $n = N$.

Example 3.23. At the end of a production day, printed circuit boards (PCB) soldered by wave soldering process are subjected to sampling audit. A RSWOR of size n is drawn from the lot, which consists of all the PCBs produced on that day. If the sample has any defective PCB, another RSWOR of size $2n$ is drawn from the lot. If there are more than three defective boards in the combined sample, the lot is sent for rectification, in which every PCB is inspected. If the lot consists of $N = 100$ PCB's, and the number of defective ones is $M = 5$, what is the probability that the lot will be rectified, when $n = 10$?

Let J_1 be the number of defective items in the first sample. If $J_1 > 3$, then the lot is rectified without taking a second sample. If $J_1 = 1, 2$ or 3, a second sample is drawn. Thus, if R denotes the event "the lot is sent for rectification,"

$$\Pr\{R\} = 1 - H(3; 100, 5, 10)$$

$$+ \sum_{i=1}^{3} h(i; 100, 5, 10) \cdot [1 - H(3 - i; 90, 5 - i, 20)]$$

$$= 0.00025 + 0.33939 \times 0.03313$$

$$+ 0.07022 \times 0.12291$$

$$+ 0.00638 \times 0.397 = 0.0227.$$

3.3.3 The poisson distribution

A third discrete distribution that plays an important role in quality control is the Poisson distribution, denoted by $P(\lambda)$. It is sometimes called the distribution of rare events, since it is used as an approximation to the Binomial distribution when the sample size, n, is large and the proportion of defectives, p, is small. The parameter λ represents the "rate" at which defectives occur, i.e., the expected number of defectives per time interval or per sample. The Poisson probability distribution function is given by the formula

$$p(j; \lambda) = \frac{e^{-\lambda} \lambda^j}{j!}, \quad j = 0, 1, 2, \dots \tag{3.3.9}$$

and the corresponding c.d.f. is

$$P(j; \lambda) = \sum_{i=0}^{j} p(i; \lambda), \quad j = 0, 1, 2, \dots. \tag{3.3.10}$$

Example 3.24.   Suppose that a machine produces aluminum pins for airplanes. The probability p that a single pin emerges defective is small, say $p = 0.002$. In one hour, the machine makes $n = 1000$ pins (considered here to be a random sample of pins). The number of defective pins produced by the machine in one hour has a binomial distribution with a mean of $\mu = np = 1000(0.002) = 2$, so the rate of defective pins for the machine is $\lambda = 2$ pins per hour. In this case, the binomial probabilities are very close to the Poisson probabilities. This approximation is illustrated below in Table 3.5, by considering processes which produce defective items at a rate of $\lambda = 2$ parts per hour, based on various sample sizes. In Exercise [3.46], the student is asked to prove that the binomial p.d.f. converges to that of the Poisson with mean λ when $n \to \infty$, $p \to 0$ but $np \to \lambda$.

The m.g.f. of the Poisson distribution is

$$M(t) = e^{-\lambda} \sum_{j=0}^{\infty} e^{tj} \frac{\lambda^j}{j!}$$

$$= e^{-\lambda} \cdot e^{\lambda e^t} = e^{-\lambda(1 - e^t)}, -\infty < t < \infty. \tag{3.3.11}$$

Table 3.5 *Binomial distributions for np = 2 and the Poisson distribution with $\lambda = 2$*

			Binomial		Poisson
	$n = 20$	$n = 40$	$n = 100$	$n = 1000$	
k	$p = 0.1$	$p = 0.05$	$p = 0.02$	$p = 0.002$	$\lambda = 2$
0	0.121577	0.128512	0.132620	0.135065	0.135335
1	0.270170	0.270552	0.270652	0.270670	0.270671
2	0.285180	0.277672	0.273414	0.270942	0.270671
3	0.190120	0.185114	0.182276	0.180628	0.180447
4	0.089779	0.090122	0.090208	0.090223	0.090224
5	0.031921	0.034151	0.035347	0.036017	0.036089
6	0.008867	0.010485	0.011422	0.011970	0.012030
7	0.001970	0.002680	0.003130	0.003406	0.003437
8	0.000356	0.000582	0.000743	0.000847	0.000859
9	0.000053	0.000109	0.000155	0.000187	0.000191

Thus,

$$M'(t) = \lambda M(t)e^t$$

$$M''(t) = \lambda^2 M(t)e^{2t} + \lambda M(t)e^t$$

$$= (\lambda^2 e^{2t} + \lambda e^t)M(t).$$

Hence, the mean and variance of the Poisson distribution are

$$\mu = E\{X\} = \lambda \qquad (3.3.12)$$

and

$$\sigma^2 = V\{X\} = \lambda.$$

The Poisson distribution is used not only as an approximation to the Binomial. It is a useful model for describing the number of "events" occurring in a unit of time (or area, volume, etc.) when those events occur "at random." The rate at which these events occur is denoted by λ. An example of a Poisson random variable is the number of decaying atoms, from a radioactive substance, detected by a Geiger counter in a fixed period of time. If the rate of detection is 5 per second, then the number of atoms detected in a second has a Poisson distribution with mean $\lambda = 5$. The number detected in 5 seconds, however, will have a Poisson distribution with $\lambda = 25$. A rate of 5 per second equals a rate of 25 per 5 seconds. Other examples of Poisson random variables include

1. The number of blemishes found in a unit area of a finished surface (ceramic plate).
2. The number of customers arriving at a store in one hour.
3. The number of defective soldering points found on a circuit board.

The p.d.f., c.d.f., and quantiles of the Poisson distribution can be computed using R, MINITAB or JMP. In Figure 3.7, we illustrate the p.d.f. for three values of λ.

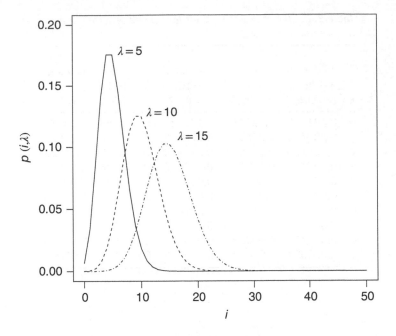

Figure 3.7   *Poisson p.d.f.* $\lambda = 5, 10, 15$.

3.3.4   The geometric and negative binomial distributions

Consider a sequence of **independent** trials, each one having the same probability for "Success," say p. Let N be a random variable which counts the number of trials until the first "Success" is realized, including the successful trial. N may assume positive integer values with probabilities

$$\Pr\{N = n\} = p(1-p)^{n-1}, \quad n = 1, 2, \dots. \tag{3.3.13}$$

This probability function is the p.d.f. of the **geometric** distribution.

Let $g(n;p)$ designate the p.d.f. The corresponding c.d.f. is

$$G(n;p) = 1 - (1-p)^n, \quad n = 1, 2, \dots.$$

From this we obtain that the α-quantile $(0 < \alpha < 1)$ is given by

$$N_\alpha = \left[\frac{\log(1-\alpha)}{\log(1-p)} \right] + 1,$$

where $[x]$ designates the integer part of x.

The expected value and variance of the geometric distribution are

$$E\{N\} = \frac{1}{p}, \tag{3.3.14}$$

and

$$V\{N\} = \frac{1-p}{p^2}.$$

Indeed, the m.g.f. of the geometric distribution is

$$M(t) = pe^t \sum_{j=0}^{\infty} (e^t(1-p))^j$$

$$= \frac{pe^t}{1 - e^t(1-p)}, \quad \text{if } t < -\log(1-p). \tag{3.3.15}$$

Thus, for $t < -\log(1-p)$,

$$M'(t) = \frac{pe^t}{(1 - e^t(1-p))^2}$$

and

$$M''(t) = \frac{pe^t}{(1 - e^t(1-p))^2} + \frac{2p(1-p)e^{2t}}{(1 - e^t(1-p))^3}.$$

Hence,

$$\mu_1 = M'(0) = \frac{1}{p}$$

$$\mu_2 = M''(0) = \frac{2-p}{p^2}, \tag{3.3.16}$$

and the above formulae of $E\{X\}$ and $V\{X\}$ are obtained.

The geometric distribution is applicable in many problems. We illustrate one such application in the following example.

Example 3.25. An insertion machine stops automatically if there is failure in handling a component during an insertion cycle. A cycle starts immediately after the insertion of a component and ends at the insertion of the next component. Suppose that the probability of stopping is $p = 10^{-3}$ per cycle. Let N be the number of cycles until the machine stops. It is assumed that events at different cycles are mutually independent. Thus, N has a geometric distribution and $E\{N\} = 1000$. We expect a run of 1000 cycles between consecutive stopping. The number of cycles, N, however, is a random variable with standard deviation of $\sigma = \left(\frac{1-p}{p^2}\right)^{1/2} = 999.5$. This high value of σ indicates that we may see very short runs and also long ones. Indeed, for $\alpha = 0.5$, the quantiles of N are, $N_{0.05} = 52$ and $N_{0.95} = 2995$.

The number of failures until the first success, $N - 1$, has a shifted geometric distribution, which is a special case of the family of **Negative-Binomial** distribution.

We say that a nonnegative integer valued random variable X has a negative-binomial distribution, with parameters (p, k), where $0 < p < 1$ and $k = 1, 2, \ldots$, if its p.d.f. is

$$g(j; p, k) = \binom{j+k-1}{k-1} p^k (1-p)^j, \tag{3.3.17}$$

$j = 0, 1, \ldots$. The shifted geometric distribution is the special case of $k = 1$.

A more general version of the negative-binomial distribution can be formulated, in which $k - 1$ is replaced by a positive real parameter. A random variable having the above

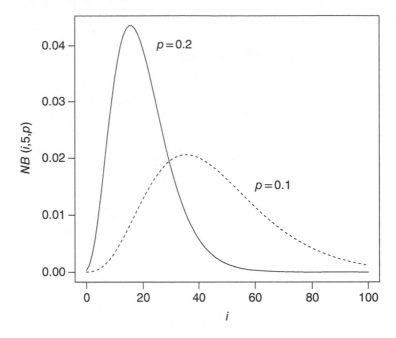

Figure 3.8 *The p.d.f. of NB(p, 5) with p = 0.10, 0.20.*

negative-binomial will be designated by $NB(p, k)$. The $NB(p, k)$ represents the number of failures observed until the kth success. The expected value and variance of $NB(p, k)$ are

$$E\{X\} = k\frac{1-p}{p}, \tag{3.3.18}$$

and

$$V\{X\} = k\frac{1-p}{p^2}.$$

In Figure 3.8, we present the p.d.f. of $NB(p, k)$. The negative binomial distributions have been applied as a model of the distribution for the periodic demand of parts in inventory theory.

3.4 Continuous distributions

3.4.1 The uniform distribution on the interval (a, b), $a < b$

We denote a random variable having this distribution by $U(a, b)$. The p.d.f. is given by

$$f(x; a, b) = \begin{cases} 1/(b-a), & a \le x \le b \\ 0, & \text{elsewhere,} \end{cases} \tag{3.4.1}$$

and the c.d.f. is

$$F(x; a, b) = \begin{cases} 0, & \text{if } x < a \\ (x-a)/(b-a), & \text{if } a \le x < b \\ 1, & \text{if } b \le x \end{cases} \tag{3.4.2}$$

The expected value and variance of $U(a, b)$ are

$$\mu = (a + b)/2, \tag{3.4.3}$$

and

$$\sigma^2 = (b - a)^2/12.$$

The pth fractile is $x_p = a + p(b - a)$.
To verify the formula for μ, we set

$$\mu = \frac{1}{b - a} \int_a^b x \, dx = \frac{1}{b - a} \left| \frac{1}{2} x^2 \right|_a^b = \frac{1}{2(b - a)}(b^2 - a^2)$$

$$= \frac{a + b}{2}.$$

Similarly,

$$\mu_2 = \frac{1}{b - a} \int_a^b x^2 \, dx = \frac{1}{b - a} \left| \frac{1}{3} x^3 \right|_a^b$$

$$= \frac{1}{3(b - a)}(b^3 - a^3) = \frac{1}{3}(a^2 + ab + b^2).$$

Thus,

$$\sigma^2 = \mu_2 - \mu_1^2 = \frac{1}{3}(a^2 + ab + b^2) - \frac{1}{4}(a^2 + 2ab + b^2)$$

$$= \frac{1}{12}(4a^2 + 4ab + 4b^2 - 3a^2 - 6ab - 3b^2)$$

$$= \frac{1}{12}(b - a)^2.$$

We can get these moments also from the m.g.f., which is

$$M(t) = \frac{1}{t(b - a)}(e^{tb} - e^{ta}), -\infty < t < \infty.$$

Moreover, for values of t close to 0

$$M(t) = 1 + \frac{1}{2}t(b + a) + \frac{1}{6}t^2(b^2 + ab + a^2) + \cdots.$$

3.4.2   The normal and log-normal distributions

3.4.2.1   *The normal distribution*

The Normal or Gaussian distribution denoted by $N(\mu, \sigma)$, occupies a central role in statistical theory. Its density function (p.d.f.) is given by the formula

$$n(x; \mu, \sigma) = \frac{1}{\sigma \sqrt{2\pi}} \exp \left\{ -\frac{1}{2\sigma^2}(x - \mu)^2 \right\}. \tag{3.4.4}$$

This p.d.f. is symmetric around the location parameter, μ. σ is a scale parameter. The m.g.f. of $N(0, 1)$ is

$$M(t) = \frac{1}{\sqrt{2\pi}} e^{tx - \frac{1}{2}x^2} dx$$

$$= \frac{e^{t^2/2}}{\sqrt{2\pi}} \int_{-\infty}^{\infty} e^{-\frac{1}{2}(x^2 - 2tx + t^2)} dx \qquad (3.4.5)$$

$$= e^{t^2/2}.$$

Indeed, $\frac{1}{\sqrt{2\pi}} \exp\left\{ -\frac{1}{2}(x - t)^2 \right\}$ is the p.d.f. of $N(t, 1)$. Furthermore,

$$M'(t) = tM(t)$$

$$M''(t) = t^2 M(t) + M(t) = (1 + t^2)M(t)$$

$$M'''(t) = (t + t^3)M(t) + 2tM(t)$$

$$= (3t + t^3)M(t)$$

$$M^{(4)}(t) = (3 + 6t^2 + t^4)M(t).$$

Thus, by substituting $t = 0$, we obtain that

$$E\{N(0, 1)\} = 0,$$

$$V\{N(0, 1)\} = 1,$$

$$\mu_3^* = 0,$$

$$\mu_4^* = 3. \qquad (3.4.6)$$

To obtain the moments in the general case of $N(\mu, \sigma^2)$, we write $X = \mu + \sigma N(0, 1)$. Then

$$E\{X\} = E\{\mu + \sigma N(0, 1)\}$$

$$= \mu + \sigma E\{N(0, 1)\} = \mu$$

$$V\{X\} = E\{(X - \mu)^2\} = \sigma^2 E\{N^2(0, 1)\} = \sigma^2$$

$$\mu_3^* = E\{(X - \mu)^3\} = \sigma^3 E\{N^3(0, 1)\} = 0$$

$$\mu_4^* = E\{(X - \mu)^4\} = \sigma^4 E\{N^4(0, 1)\} = 3\sigma^4.$$

Thus, the index of kurtosis in the normal case is $\beta_4 = 3$.

The graph of the p.d.f. $n(x; \mu, \sigma)$ is a symmetric bellshaped curve that is centered at μ (shown in Figure 3.9). The spread of the density is determined by the variance σ^2 in the sense that most of the area under the curve (in fact, 99.7% of the area) lies between $\mu - 3\sigma$ and $\mu + 3\sigma$. Thus, if X has a normal distribution with mean $\mu = 25$ and standard deviation $\sigma = 2$, the probability is 0.997 that the observed value of X will fall between 19 and 31.

Areas (that is, probabilities) under the normal p.d.f. are found in practice using a table or appropriate software like MINITAB. Since it is not practical to have a table for each pair of parameters μ and σ, we use the standardized form of the normal random variable. A random

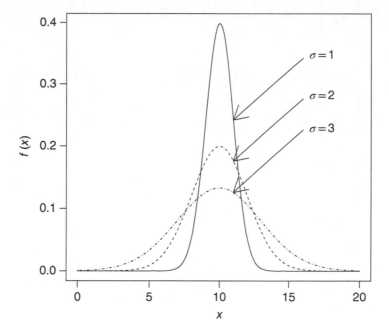

Figure 3.9   *The p.d.f. of N(μ, σ), $\mu = 10$, $\sigma = 1, 2, 3$.*

variable Z is said to have a **standard normal distribution** if it has a normal distribution with mean zero and variance one. The standard normal density function is $\phi(x) = n(x; 0, 1)$ and the standard cumulative distribution function is denoted by $\Phi(x)$. This function is also called the **standard normal** integral, i.e.,

$$\Phi(x) = \int_{-\infty}^{x} \phi(t)dt = \int_{-\infty}^{x} \frac{1}{\sqrt{2\pi}} e^{-\frac{1}{2}t^2} dt. \tag{3.4.7}$$

The c.d.f., $\Phi(x)$, represents the area over the x-axis under the standard normal p.d.f. to the left of the value x.

If we wish to determine the probability that a standard normal random variable is less than 1.5, for example, we use R code

```
> pnorm(q=1.5, mean=0, sd=1)
```

or the MINITAB commands

```
MTB> CDF 1.5;
SUBC> NORMAL 0 1.
```

We find that $\Pr\{Z \leq 1.5\} = \Phi(1.5) = 0.9332$. To obtain the probability that Z lies between 0.5 and 1.5 we first find the probability that Z is less than 1.5, then subtract from this number the probability that Z is less than 0.5. This yields

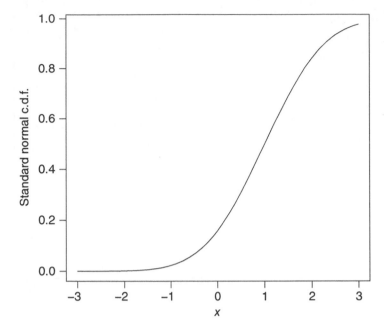

Figure 3.10   *Standard Normal c.d.f.*

$$\Pr\{0.5 < Z < 1.5\} = \Pr\{Z < 1.5\} - \Pr\{Z < .5\}$$
$$= \Phi(1.5) - \Phi(0.5) = 0.9332 - 0.6915 = 0.2417.$$

Many tables of the normal distribution do not list values of $\Phi(x)$ for $x < 0$. This is because the normal density is symmetric about $x = 0$, and we have the relation

$$\Phi(-x) = 1 - \Phi(x), \quad \text{for all } x. \tag{3.4.8}$$

Thus, to compute the probability that Z is less than -1, for example, we write (Figure 3.11)

$$\Pr\{Z < -1\} = \Phi(-1) = 1 - \Phi(1) = 1 - 0.8413 = 0.1587.$$

The **pth quantile (percentile of fractile)** of the standard normal distribution is the number z_p that satisfies the statement

$$\Phi(z_p) = \Pr\{Z \leq z_p\} = p. \tag{3.4.9}$$

If X has a normal distribution with mean μ and standard deviation σ, we denote the pth fractile of the distribution by x_p. We can show that x_p is related to the standard normal fractile by

$$x_p = \mu + z_p\sigma.$$

The pth fractile of the normal distribution can be obtained by using R code

```
> qnorm(p=0.95, mean=0, sd=1)
```

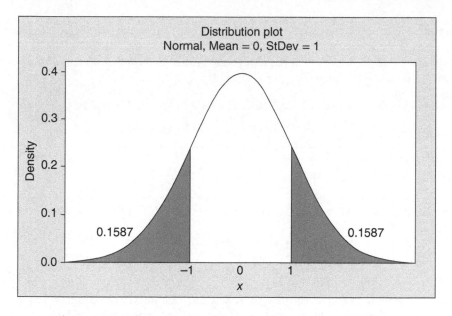

Figure 3.11   *The symmetry of the normal distribution (MINITAB).*

or the MINITAB command

```
MTB> InvCDF 0.95;
SUBC> Normal 0.0 1.0.
```

In this command, we used $p = 0.95$. The printed result is $z_{0.95} = 1.6449$. We can use any value of μ and σ in the subcommand. Thus, for $\mu = 10$ and $\sigma = 1.5$

$$x_{0.95} = 10 + z_{0.95} \times \sigma = 12.4673.$$

Now, suppose that X is a random variable having a normal distribution with mean μ and variance σ^2. That is, X has a $N(\mu, \sigma)$ distribution. We define the **standardized form** of X as

$$Z = \frac{X - \mu}{\sigma}.$$

By subtracting the mean from X and then dividing by the standard deviation, we transform X to a standard normal random variable. (That is, Z has expected value zero and standard deviation one.) This will allow us to use the standard normal table to compute probabilities involving X. Thus, to compute the probability that X is less than a, we write

$$\Pr\{X \le a\} = \Pr\left\{\frac{X - \mu}{\sigma} < \frac{a - \mu}{\sigma}\right\}$$
$$= \Pr\left\{Z < \frac{a - \mu}{\sigma}\right\} = \Phi\left(\frac{a - \mu}{\sigma}\right).$$

Example 3.26. Let X represent the length (with cap) of a randomly selected aluminum pin. Suppose we know that X has a normal distribution with mean $\mu = 60.02$ and standard deviation $\sigma = 0.048$ [mm]. What is the probability that the length with cap of a randomly selected pin will be less than 60.1[mm]? Using R

```
> pnorm(q=60.1, mean=60.02, sd=0.048, lower.tail=TRUE)
```

or the MINITAB command

```
MTB> CDF 60.1;
SUBC> Normal 60.02 0.048.
```

we obtain $\Pr\{X \leq 60.1\} = 0.9522$. If we have to use the table of $\Phi(Z)$, we write

$$\Pr\{X \leq 60.1\} = \Phi\left(\frac{60.1 - 60.02}{0.048}\right)$$

$$= \Phi(1.667) = 0.9522.$$

Continuing with the example, consider the following question: If a pin is considered "acceptable" when its length is between 59.9 and 60.1 mm, what proportion of pins is expected to be rejected? To answer this question, we first compute the probability of accepting a single pin. This is the probability that X lies between 59.9 and 60.1, i.e.,

$$\Pr\{50.9 < X < 60.1\} = \Phi\left(\frac{60.1 - 60.02}{0.048}\right) - \Phi\left(\frac{59.9 - 60.02}{0.048}\right)$$

$$= \Phi(1.667) - \Phi(-2.5)$$

$$= 0.9522 - 0.0062 = 0.946.$$

Thus, we expect that 94.6% of the pins will be accepted, and that 5.4% of them will be rejected.

3.4.2.2   *The log-normal distribution*

A random variable X is said to have a **log-normal distribution**, $LN(\mu, \sigma^2)$, if $Y = \log X$ has the normal distribution $N(\mu, \sigma^2)$.

The log-normal distribution has been applied for modeling distributions of strength variables, like the tensile strength of fibers (see Chapter 2), the compressive strength of concrete cubes, etc. It has also been used for random quantities of pollutants in water or air, and other phenomena with skewed distributions.

The p.d.f. of $LN(\mu, \sigma)$ is given by the formula

$$f(x; \mu, \sigma^2) = \begin{cases} \dfrac{1}{\sqrt{2\pi}\sigma x} \exp\left\{-\dfrac{1}{2\sigma^2}(\log x - \mu)^2\right\}, & 0 < x < \infty \\ 0, & x \leq 0. \end{cases} \tag{3.4.10}$$

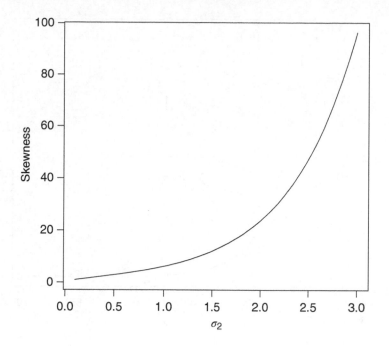

Figure 3.12   *The index of skewness of LN(μ, σ).*

The c.d.f. is expressed in terms of the standard normal integral as

$$F(x) = \begin{cases} 0, & x \leq 0 \\ \Phi\left(\dfrac{\log x - \mu}{\sigma}\right), & 0 < x < \infty. \end{cases} \tag{3.4.11}$$

The expected value and variance of LN(μ, σ) are

$$E\{X\} = e^{\mu + \sigma^2/2} \tag{3.4.12}$$

and

$$V\{X\} = e^{2\mu + \sigma^2}(e^{\sigma^2} - 1).$$

One can show that the third central moment of LN(μ, σ^2) is

$$\mu_3^* = e^{3\mu + \frac{3}{2}\sigma^2}(e^{3\sigma^2} - 3e^{\sigma^2} + 2).$$

Hence, the **index of skewness** of this distribution is

$$\beta_3 = \frac{\mu_3^*}{\sigma^3} = \frac{e^{3\sigma^2} - 3e^{\sigma^2} + 2}{(e^{\sigma^2} - 1)^{3/2}}. \tag{3.4.13}$$

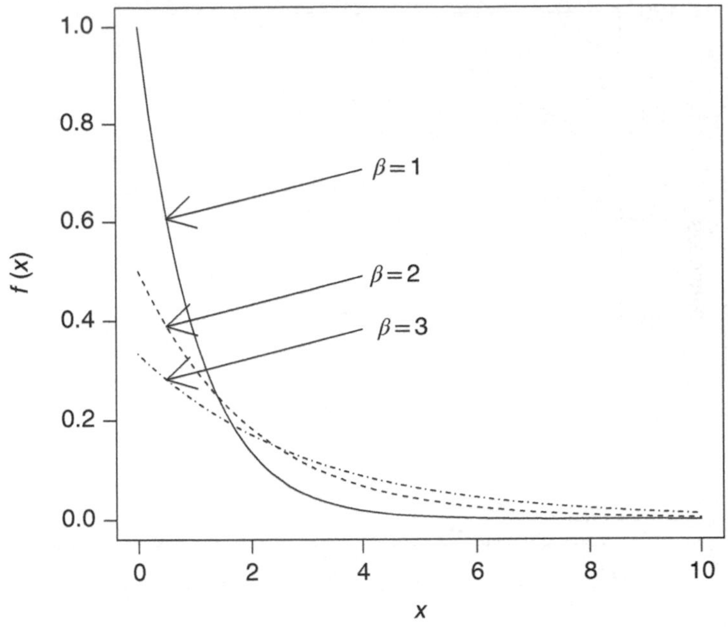

Figure 3.13   *The p.d.f. of $E(\beta)$, $\beta = 1, 2, 3$.*

It is interesting that the index of skewness does not depend on μ, and is positive for all $\sigma^2 > 0$. This index of skewness grows very fast as σ^2 increases. This is shown in Figure 3.12.

3.4.3   The exponential distribution

We designate this distribution by $E(\beta)$. The p.d.f. of $E(\beta)$ is given by the formula

$$f(x; \beta) = \begin{cases} 0, & \text{if } x < 0 \\ (1/\beta)e^{-x/\beta}, & \text{if } x \geq 0, \end{cases} \tag{3.4.14}$$

where β is a positive parameter, i.e., $0 < \beta < \infty$. In Figure 3.13, we present these p.d.f.s for various values of β.

The corresponding c.d.f. is

$$F(x; \beta) = \begin{cases} 0, & \text{if } x < 0 \\ 1 - e^{-x/\beta}, & \text{if } x \geq 0. \end{cases} \tag{3.4.15}$$

The expected value and the variance of $E(\beta)$ are

$$\mu = \beta,$$

and

$$\sigma^2 = \beta^2.$$

Indeed,

$$\mu = \frac{1}{\beta} \int_0^\infty x e^{-x/\beta} dx.$$

Making the change of variable to $y = x/\beta$, $dx = \beta dy$, we obtain

$$\mu = \beta \int_0^\infty y e^{-y} dy$$

$$= \beta.$$

Similarly,

$$\mu_2 = \frac{1}{\beta} \int_0^\infty x^2 e^{-x/\beta} dx = \beta^2 \int_0^\infty y^2 e^{-y} dy$$

$$= 2\beta^2.$$

Hence,

$$\sigma^2 = \beta^2.$$

The pth quantile is $x_p = -\beta \ln(1 - p)$.

The exponential distribution is related to the Poisson model in the following way: If the number of events occurring in a period of time follows a Poisson distribution with rate λ, then the time between occurrences of events has an exponential distribution with parameter $\beta = 1/\lambda$. The exponential model can also be used to describe the lifetime (i.e., time to failure) of certain electronic systems. For example, if the mean life of a system is 200 hours then, the probability that it will work at least 300 hours without failure is

$$\text{Pr}\{X \geq 300\} = 1 - \text{Pr}\{X < 300\}$$

$$= 1 - F(300) = 1 - (1 - e^{-300/200}) = 0.223.$$

The exponential distribution is positively skewed, and its index of skewness is

$$\beta_3 = \frac{\mu_3^*}{\sigma^3} = 2,$$

irrespective of the value of β. We have seen before that the kurtosis index is $\beta_4 = 9$.

3.4.4 The Gamma and Weibull distributions

Two important distributions for studying the reliability and failure rates of systems are the Gamma and the Weibull distributions. We will need these distributions in our study of reliability methods (Chapter 16). These distributions are discussed here as further examples of continuous distributions.

Suppose we use in a manufacturing process a machine which mass produces a particular part. In a random manner, it produces defective parts at a rate of λ per hour. The number of defective parts produced by this machine in a time period $[0, t]$ is a random variable $X(t)$ having a Poisson distribution with mean λt, i.e.,

$$\text{Pr}\{X(t) = j\} = (\lambda t)^j e^{-(\lambda t)}/j!, \quad j = 0, 1, 2, \ldots. \tag{3.4.16}$$

Suppose we wish to study the distribution of the time until the kth defective part is produced. Call this continuous random variable Y_k. We use the fact that the kth defect will occur before time t (i.e., $Y_k \le t$) if and only if at least k defects occur up to time t (i.e., $X(t) \ge k$). Thus, the c.d.f. for Y_k is

$$G(t; k, \lambda) = \Pr\{Y_k \le t\}$$

$$= \Pr\{X(t) \ge k\} \qquad (3.4.17)$$

$$= 1 - \sum_{j=0}^{k-1} (\lambda t)^j e^{-\lambda t}/j!.$$

The corresponding p.d.f. for Y_k is

$$g(t; k, \lambda) = \frac{\lambda^k}{(k-1)!} t^{k-1} e^{-\lambda t}, \text{ for } t \ge 0. \qquad (3.4.18)$$

This p.d.f. is a member of a general family of distributions which depend on two parameters, v and β, and are called the **gamma** distributions $G(v, \beta)$. The p.d.f. of a gamma distribution $G(v, \beta)$ is

$$g(x; v, \beta) = \begin{cases} \dfrac{1}{\beta^v \Gamma(v)} x^{v-1} e^{-x/\beta}, & x \ge 0, \\ 0, & x < 0. \end{cases} \qquad (3.4.19)$$

In R function, pgamma computes c.d.f. of a gamma distribution having $v =$ shape and $\beta =$ scale

```
> pgamma (q=1, shape=1, scale=1)
```

```
[1]  0.6321206
```

where $0 < v, \beta < \infty$, $\Gamma(v)$ is called the **gamma function** of v and is defined as the integral

$$\Gamma(v) = \int_0^\infty x^{v-1} e^{-x} dx, \quad v > 0. \qquad (3.4.20)$$

The gamma function satisfies the relationship

$$\Gamma(v) = (v-1)\Gamma(v-1), \quad \text{for all } v > 1. \qquad (3.4.21)$$

Hence, for every positive integer k, $\Gamma(k) = (k-1)!$. Also, $\Gamma\left(\frac{1}{2}\right) = \sqrt{\pi}$. We note also that the exponential distribution, $E(\beta)$, is a special case of the gamma distribution with $v = 1$. Some gamma p.d.f.s are presented in Figure 3.14. The value of $\Gamma(v)$ can be computed in R by the following commands which compute $\Gamma(5)$. Generally, replace 5 in line 2 by v.

```
> gamma (5)
```

```
[1]  24
```

The expected value and variance of the gamma distribution $G(v, \beta)$ are, respectively,

$$\mu = v\beta, \qquad (3.4.22)$$

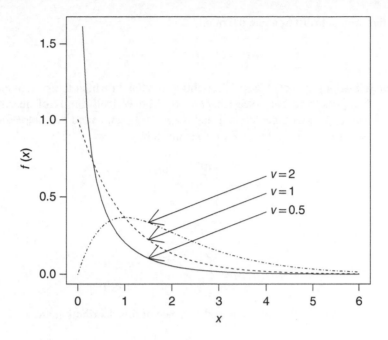

Figure 3.14   *The Gamma densities, with β = 1 and v = 0.5, 1, 2.*

and

$$\sigma^2 = v\beta^2.$$

To verify these formulae, we write

$$\mu = \frac{1}{\beta^v \Gamma(v)} \int_0^\infty x \cdot x^{v-1} e^{-x/\beta} dx$$

$$= \frac{\beta^{v+1}}{\beta^v \Gamma(v)} \int_0^\infty y^v e^{-y} dy$$

$$= \beta \frac{\Gamma(v+1)}{\Gamma(v)} = v\beta.$$

Similarly,

$$\mu_2 = \frac{1}{\beta^v \Gamma(v)} \int_0^\infty x^2 \cdot x^{v-1} e^{-x/\beta} dx$$

$$= \frac{\beta^{v+2}}{\beta^v \Gamma(v)} \int_0^\infty y^{v+1} e^{-y} dy$$

$$= \beta^2 \frac{\Gamma(v+2)}{\Gamma(v)} = (v+1)v\beta^2.$$

Hence,

$$\sigma^2 = \mu_2 - \mu_1^2 = v\beta^2.$$

An alternative way is to differentiate the m.g.f.

$$M(t) = (1 - t\beta)^{-\nu}, \quad t < \frac{1}{\beta}. \tag{3.4.23}$$

Weibull distributions are often used in reliability models in which the system either "ages" with time or becomes "younger" (see Chapter 16). The Weibull family of distributions will be denoted by $W(\alpha, \beta)$. The parameters α and β, $\alpha, \beta > 0$, are called the shape and the scale parameters, respectively. The p.d.f. of $W(\alpha, \beta)$ is given by

$$w(t; \alpha, \beta) = \begin{cases} \dfrac{\alpha t^{\alpha-1}}{\beta^\alpha} e^{-(t/\beta)^\alpha}, & t \geq 0, \\ 0, & t < 0. \end{cases} \tag{3.4.24}$$

The corresponding c.d.f. is

$$W(t; \alpha, \beta) = \begin{cases} 1 - e^{-(t/\beta)^\alpha}, & t \geq 0 \\ 0, & t < 0. \end{cases} \tag{3.4.25}$$

Notice that $W(1, \beta) = E(\beta)$. The mean and variance of this distribution are

$$\mu = \beta \cdot \Gamma\left(1 + \frac{1}{\alpha}\right) \tag{3.4.26}$$

and

$$\sigma^2 = \beta^2 \left\{ \Gamma\left(1 + \frac{2}{\alpha}\right) - \Gamma^2\left(1 + \frac{1}{\alpha}\right) \right\}, \tag{3.4.27}$$

respectively. The values of $\Gamma(1 + (1/\alpha))$ and $\Gamma(1 + (2/\alpha))$ can be computed by R, MINITAB or JMP. If, for example $\alpha = 2$, then

$$\mu = \beta\sqrt{\pi}/2 = 0.8862\beta$$
$$\sigma^2 = \beta^2(1 - \pi/4) = 0.2145\beta^2,$$

since

$$\Gamma\left(1 + \frac{1}{2}\right) = \frac{1}{2} \cdot \Gamma\left(\frac{1}{2}\right) = \frac{1}{2}\sqrt{\pi},$$

and

$$\Gamma\left(1 + \frac{2}{2}\right) = \Gamma(2) = 1.$$

In Figure 3.15, we present three p.d.f.s of $W(\alpha, \beta)$ for $\alpha = 1.5, 2.0$ and $\beta = 1$.

3.4.5   The Beta distributions

Distributions having p.d.f. of the form

$$f(x; \nu_1, \nu_2) = \begin{cases} \dfrac{1}{\text{Be}(\nu_1, \nu_2)} x^{\nu_1-1}(1 - x)^{\nu_2-1}, & 0 < x < 1, \\ 0, & \text{otherwise}, \end{cases} \tag{3.4.28}$$

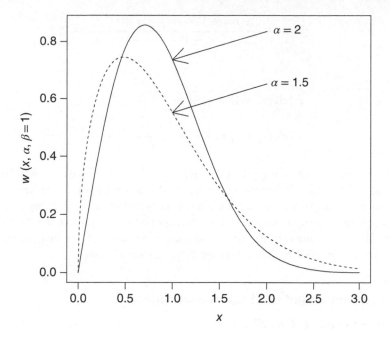

Figure 3.15　*Weibull density functions, $\alpha = 1.5, 2$.*

where for v_1, v_2 positive,

$$\mathrm{Be}(v_1, v_2) = \int_0^1 x^{v_1-1}(1-x)^{v_2-1}dx \tag{3.4.29}$$

are called Beta distributions. The function $\mathrm{Be}(v_1, v_2)$ is called the Beta integral. One can prove that

$$\mathrm{Be}(v_1, v_2) = \frac{\Gamma(v_1)\Gamma(v_2)}{\Gamma(v_1 + v_2)}. \tag{3.4.30}$$

The parameters v_1 and v_2 are shape parameters. Notice that when $v_1 = 1$ and $v_2 = 1$ the Beta reduces to $U(0, 1)$. We designate distributions of this family by $\mathrm{Be}(v_1, v_2)$. The c.d.f. of $\mathrm{Be}(v_1, v_2)$ is denoted also by $I_x(v_1, v_2)$, which is known as the **incomplete beta function ratio**, i.e.,

$$I_x(v_1, v_2) = \frac{1}{\mathrm{Be}(v_1, v_2)} \int_0^x u^{v_1-1}(1-u)^{v_2-1}du, \tag{3.4.31}$$

for $0 \leq x \leq 1$. Notice that $I_x(v_1, v_2) = 1 - I_{1-x}(v_2, v_1)$. The density functions of the p.d.f. $\mathrm{Be}(2.5, 5.0)$ and $\mathrm{Be}(2.5, 2.5)$ are plotted in Figure 3.16. Notice that if $v_1 = v_2$, then the p.d.f. is symmetric around $\mu = \frac{1}{2}$. There is no simple formula for the m.g.f. of $\mathrm{Be}(v_1, v_2)$. However, the mth moment is equal to

$$\begin{aligned}
\mu_m &= \frac{1}{\mathrm{Be}(v_1, v_2)} \int_0^1 u^{m+v_1-1}(1-u)^{v_2-1}du \\
&= \frac{\mathrm{Be}(v_1 + m, v_2)}{\mathrm{Be}(v_1, v_2)}
\end{aligned} \tag{3.4.32}$$

$$= \frac{v_1(v_1 + 1) \cdots (v_1 + m - 1)}{(v_1 + v_2)(v_1 + v_2 + 1) \cdots (v_1 + v_2 + m - 1)}.$$

Hence,

$$E\{\text{Be}(v_1, v_2)\} = \frac{v_1}{v_1 + v_2}$$

$$V\{\text{Be}(v_1, v_2)\} = \frac{v_1 v_2}{(v_1 + v_2)^2 (v_1 + v_2 + 1)}. \tag{3.4.33}$$

The beta distribution has an important role in the theory of statistics. As will be seen later, many methods of statistical inference are based on the order statistics (see Section 3.7). The distribution of the order statistics are related to the beta distribution. Moreover, since the beta distribution can get a variety of shapes, it has been applied in many cases in which the variable has a distribution on a finite domain. By introducing a location and a scale parameter, one can fit a shifted-scaled beta distribution to various frequency distributions.

3.5 Joint, marginal and conditional distributions

3.5.1 Joint and marginal distributions

Let $X_1, \ldots, X_k$ be random variables which are jointly observed at the same experiments. In Section 3.6, we present various examples of bivariate and multivariate frequency distributions. In the present section, we present only the fundamentals of the theory, mainly for future reference.

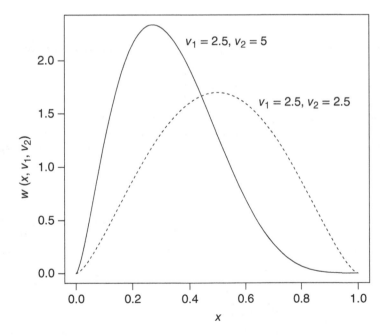

Figure 3.16   *Beta densities, $v_1 = 2.5$, $v_2 = 2.5$; $v_1 = 2.5$, $v_2 = 5.00$.*

We make the presentation here, focusing on continuous random variables. The theory holds generally for discrete or for mixture of continuous and discrete random variables.

A function $F(x_1, \ldots, x_k)$ is called the joint c.d.f. of $X_1, \ldots, X_k$ if

$$F(x_1, \ldots, x_k) = \Pr\{X_1 \leq x_1, \ldots, X_k \leq x_k\} \tag{3.5.1}$$

for all $(x_1, \ldots, x_k) \in \mathbb{R}^k$ (the Euclidean k-space). By letting one or more variables tend to infinity, we obtain the joint c.d.f. of the remaining variables. For example,

$$F(x_1, \infty) = \Pr\{X_1 \leq x_1, X_2 \leq \infty\}$$
$$= \Pr\{X_1 \leq x_1\} = F_1(x_1). \tag{3.5.2}$$

The c.d.f.s of the individual variables are called the **marginal** distributions. $F_1(x_1)$ is the marginal c.d.f. of X_1.

A nonnegative function $f(x_1, \ldots, x_k)$ is called the **joint p.d.f.** of $X_1, \ldots, X_k$, if

(i) $f(x_1, \ldots, x_k) \geq 0$ for all $(x_1, \ldots, x_k)$, where $-\infty < x_i < \infty$ $(i = 1, \ldots, k)$

(ii)
$$\int_{-\infty}^{\infty} \cdots \int_{-\infty}^{\infty} f(x_1, \ldots, x_k) dx_1, \ldots, dx_k = 1.$$

and

(iii)
$$F(x_1, \ldots, x_k) = \int_{-\infty}^{x_1} \cdots \int_{-\infty}^{x_k} f(y_1, \ldots, y_k) dy_1 \ldots dy_k.$$

The **marginal p.d.f.** of X_i $(i = 1, \ldots, k)$ can be obtained from the joint p.d.f. $f(x_1, \ldots, x_k)$, by integrating the joint p.d.f. with respect to all $x_j, j \neq i$. For example if $k = 2$, then $f(x_1, x_2)$ is the joint p.d.f. of X_1, X_2. The marginal p.d.f. of X_1 is

$$f_1(x_1) = \int_{-\infty}^{\infty} f(x_1, x_2) dx_2.$$

Similarly, the marginal p.d.f. of X_2 is

$$f_2(x_2) = \int_{-\infty}^{\infty} f(x_1, x_2) dx_1.$$

Indeed, the marginal c.d.f. of X_i is

$$F(x_1) = \int_{-\infty}^{x_1} \int_{-\infty}^{\infty} f(y_1, y_2) dy_1 dy_2.$$

Differentiating $F(x_1)$ with respect to x_1, we obtain the marginal p.d.f. of X_1, i.e.,

$$f(x_1) = \frac{d}{dx_1} \int_{-\infty}^{x_1} \int_{-\infty}^{\infty} f(y_1, y_2) dy_1 dy_2$$
$$= \int_{-\infty}^{\infty} f(x_1, y_2) dy_2.$$

If $k = 3$, we can obtain the marginal joint p.d.f. of a pair of random variables by integrating with respect to the third variable. For example the joint marginal p.d.f. of (X_1, X_2), can be obtained from that of (X_1, X_2, X_3) as

$$f_{1,2}(x_1, x_2) = \int_{-\infty}^{\infty} f(x_1, x_2, x_3) dx_3.$$

Similarly,

$$f_{1,3}(x_1, x_3) = \int_{-\infty}^{\infty} f(x_1, x_2, x_3) dx_2,$$

and

$$f_{2,3}(x_2, x_3) = \int_{-\infty}^{\infty} f(x_1, x_2, x_3) dx_1.$$

Example 3.27. The present example is theoretical and is designed to illustrate the above concepts.

Let (X, Y) be a pair of random variables having a joint uniform distribution on the region

$$T = \{(x, y) : 0 \le x, y, \ x + y \le 1\}.$$

T is a triangle in the (x, y)-plane with vertices at $(0, 0)$, $(1, 0)$, and $(0, 1)$. According to the assumption of uniform distribution, the joint p.d.f. of (X, Y) is

$$f(x, y) = \begin{cases} 2, & \text{if } (x, y) \in T \\ 0, & \text{otherwise.} \end{cases}$$

The marginal p.d.f. of X is obtained as

$$f_1(x) = 2 \int_0^{1-x} dy = 2(1 - x), \quad 0 \le x \le 1.$$

Obviously, $f_1(x) = 0$ for x outside the interval $[0, 1]$. Similarly, the marginal p.d.f. of Y is

$$f_2(y) = \begin{cases} 2(1 - y), & 0 \le y \le 1 \\ 0, & \text{otherwise.} \end{cases}$$

Both X and Y have the same marginal $B(1, 2)$ distribution. Thus,

$$E\{X\} = E\{Y\} = \frac{1}{3}$$

and

$$V\{X\} = V\{Y\} = \frac{1}{18}.$$

3.5.2 Covariance and correlation

Given any two random variables (X_1, X_2) having a joint distribution with p.d.f. $f(x_1, x_2)$, the **covariance** of X_1 and X_2 is defined as

$$\text{Cov}(X_1, X_2) = \int_{-\infty}^{\infty} \int_{-\infty}^{\infty} (x_1 - \mu_1)(x_2 - \mu_2) f(x_1, x_2) dx_1 dx_2, \tag{3.5.3}$$

where

$$\mu_i = \int_{-\infty}^{\infty} x f_i(x) dx, \quad i = 1, 2,$$

is the expected value of X_i. Notice that

$$\text{Cov}(X_1, X_2) = E\{(X_1 - \mu_1)(X_2 - \mu_2)\}$$
$$= E\{X_1 X_2\} - \mu_1 \mu_2.$$

The **correlation** between X_1 and X_2 is defined as

$$\rho_{12} = \frac{\text{Cov}(X_1, X_2)}{\sigma_1 \sigma_2}, \tag{3.5.4}$$

where σ_i $(i = 1, 2)$ is the standard deviation of X_i.

Example 3.28. In continuation of the previous example, we compute $\text{Cov}(X, Y)$. We have seen that $E\{X\} = E\{Y\} = \frac{1}{3}$. We compute now the expected value of their product

$$E\{XY\} = 2 \int_0^1 x \int_0^{1-x} y \, dy$$

$$= 2 \int_0^1 x \cdot \frac{1}{2}(1 - x)^2 dx$$

$$= B(2, 3) = \frac{\Gamma(2)\Gamma(3)}{\Gamma(5)} = \frac{1}{12}.$$

Hence,

$$\text{Cov}(X, Y) = E\{XY\} - \mu_1 \mu_2 = \frac{1}{12} - \frac{1}{9}$$

$$= -\frac{1}{36}.$$

Finally, the correlation between X, Y is

$$\rho_{XY} = -\frac{1/36}{1/18} = -\frac{1}{2}.$$

The following are some properties of the covariance:

(i) $$|\text{Cov}(X_1, X_2)| \leq \sigma_1 \sigma_2,$$

where σ_1 and σ_2 are the standard deviations of X_1 and X_2, respectively.

(ii) If c is any constant then,

$$\text{Cov}(X, c) = 0. \tag{3.5.5}$$

(iii) For any constants a_1 and a_2,

$$\text{Cov}(a_1 X_1, a_2 X_2) = a_1 a_2 \text{Cov}(X_1, X_2). \tag{3.5.6}$$

(iv) For any constants a, b, c, and d,

$$\text{Cov}(aX_1 + bX_2, cX_3 + dX_4) = ac\,\text{Cov}(X_1, X_3) + ad\,\text{Cov}(X_1, X_4)$$
$$+ bc\,\text{Cov}(X_2, X_3) + bd\,\text{Cov}(X_2, X_4).$$

Property (iv) can be generalized to be

$$\text{Cov}\left(\sum_{i=1}^{m} a_i X_i, \sum_{j=1}^{n} b_j Y_j\right) = \sum_{i=1}^{m}\sum_{j=1}^{n} a_i b_j \text{Cov}(X_i, Y_j). \tag{3.5.7}$$

From property (i) above, we deduce that $-1 \le \rho_{12} \le 1$. The correlation obtains the values ± 1 only if the two variables are linearly dependent.

3.5.2.1   *Definition of independence*

Random variables $X_1, \ldots, X_k$ are said to be **mutually independent** if, for every $(x_1, \ldots, x_k)$,

$$f(x_1, \ldots, x_k) = \prod_{i=1}^{k} f_i(x_i), \tag{3.5.8}$$

, where $f_i(x_i)$ is the marginal p.d.f. of X_i. The variables X, Y of Example 3.26 are dependent, since $f(x, y) \ne f_1(x) f_2(y)$.

If two random variables are independent, then their correlation (or covariance) is zero. The converse is generally not true. Zero correlation **does not** imply independence.

We illustrate this in the following example.

Example 3.29.   Let (X, Y) be discrete random variables having the following joint p.d.f.

$$p(x, y) = \begin{cases} \dfrac{1}{3}, & \text{if } X = -1, Y = 0 \text{ or } X = 0, Y = 0 \text{ or } X = 1, Y = 1 \\ 0, & \text{elsewhere.} \end{cases}$$

In this case, the marginal p.d.f.s are

$$p_1(x) = \begin{cases} \dfrac{1}{3}, & x = -1, 0, 1 \\ 0, & \text{otherwise} \end{cases}$$

$$
p_2(y) = \begin{cases} \dfrac{1}{3}, & y = 0 \\[2mm] \dfrac{2}{3}, & y = 1. \end{cases}
$$

$p(x, y) \neq p_1(x)p_2(y)$ if $X = 1$, $Y = 1$ for example. Thus, X and Y are dependent. On the other hand, $E\{X\} = 0$ and $E\{XY\} = 0$. Hence, $\text{COV}(X, Y) = 0$.

The following result is very important for independent random variables.

If $X_1, X_2, \ldots, X_k$ **are mutually independent then, for any integrable functions** $g_1(X_1), \ldots, g_k(X_k)$,

$$
E\left\{ \prod_{i=1}^{k} g_i(X_i) \right\} = \prod_{i=1}^{k} E\{g_i(X_i)\}. \tag{3.5.9}
$$

Indeed,

$$
E\left\{ \prod_{i=1}^{k} g_i(X_i) \right\} = \int \cdots \int g_1(x_1) \cdots g_k(x_k) \cdot
$$

$$
f(x_1, \ldots, x_k)dx_1, \ldots, dx_k = \int \cdots \int g_1(x_1) \cdots g_k(x_k) f_1(x_1) \cdots f_k(x_k) dx_1 \cdots dx_k
$$

$$
= \int g_1(x_1) f_1(x_1) dx_1 \cdot \int g_2(x_2) f_2(x_2) dx_2 \cdots \int g_k(x_k) f_k(x_k) dx_k
$$

$$
= \prod_{i=1}^{k} E\{g_i(X_i)\}.
$$

3.5.3   Conditional distributions

If (X_1, X_2) are two random variables having a joint p.d.f. $f(x_1, x_2)$ and marginals ones, $f_1(\cdot)$, and $f_2(\cdot)$, respectively, then the **conditional** p.d.f. of X_2, given $\{X_1 = x_1\}$, where $f_1(x_1) > 0$, is defined to be

$$
f_{2 \cdot 1}(x_2 \mid x_1) = \frac{f(x_1, x_2)}{f_1(x_1)}. \tag{3.5.10}
$$

Notice that $f_{2 \cdot 1}(x_2 \mid x_1)$ is a p.d.f. Indeed, $f_{2 \cdot 1}(x_2 \mid x_1) \geq 0$ for all x_2, and

$$
\int_{-\infty}^{\infty} f_{2 \cdot 1}(x_2 \mid x_1) dx_2 = \frac{\int_{-\infty}^{\infty} f(x_1, x_2) dx_2}{f_1(x_1)}
$$

$$
= \frac{f_1(x_1)}{f_1(x_1)} = 1.
$$

The **conditional expectation** of X_2, given $\{X_1 = x_1\}$ such that $f_1(x_1) > 0$, is the expected value of X_2 with respect to the conditional p.d.f. $f_{2 \cdot 1}(x_2 \mid x_1)$, i.e.,

$$
E\{X_2 \mid X_1 = x_1\} = \int_{-\infty}^{\infty} x f_{2 \cdot 1}(x \mid x_1) dx.
$$

Similarly, we can define the **conditional variance** of X_2, given $\{X_1 = x_1\}$, as the variance of X_2, with respect to the conditional p.d.f. $f_{2 \cdot 1}(x_2 \mid x_1)$. If X_1 and X_2 are independent, then by substituting $f(x_1, x_2) = f_1(x_1) f_2(x_2)$, we obtain

$$f_{2 \cdot 1}(x_2 \mid x_1) = f_2(x_2),$$

and

$$f_{1 \cdot 2}(x_1 \mid x_2) = f_1(x_1).$$

Example 3.30.   Returning to Example 3.26, we compute the conditional distribution of Y, given $\{X = x\}$, for $0 < x < 1$.

According to the above definition, the conditional p.d.f. of Y, given $\{X = x\}$, for $0 < x < 1$, is

$$f_{Y \mid X}(y \mid x) = \begin{cases} \dfrac{1}{1 - x}, & \text{if } 0 < y < (1 - x) \\ 0, & \text{otherwise.} \end{cases}$$

Notice that this is a uniform distribution over $(0, 1 - x)$, $0 < x < 1$. If $x \notin (0, 1)$, then the conditional p.d.f. does not exist. This is, however, an event of probability zero. From the above result, the conditional expectation of Y, given $\{X = x\}$, for $0 < x < 1$, is

$$E\{Y \mid X = x\} = \frac{1 - x}{2}.$$

The conditional variance is

$$V\{Y \mid X = x\} = \frac{(1 - x)^2}{12}.$$

In a similar fashion, we show that the conditional distribution of X, given $Y = y$, $0 < y < 1$, is uniform on $(0, 1 - y)$.

One can immediately prove that if X_1 and X_2 are independent, then the conditional distribution of X_1 given $\{X_2 = x_2\}$, when $f_2(x_2) > 0$, is just the marginal distribution of X_1. Thus, X_1 and X_2 are independent if, and only if,

$$f_{2 \cdot 1}(x_2 \mid x_1) = f_2(x_2) \text{ for all } x_2$$

and

$$f_{1 \cdot 2}(x_1 \mid x_2) = f_1(x_1) \text{ for all } x_1,$$

provided that the conditional p.d.f. are well defined.

Notice that for a pair of random variables (X, Y), $E\{Y \mid X = x\}$ changes with x, as shown in Example 3.30, if X and Y are dependent. Thus, we can consider $E\{Y \mid X\}$ to be a random variable, which is a function of X. It is interesting to compute the expected value of this function of X, i.e.,

$$E\{E\{Y \mid X\}\} = \int E\{Y \mid X = x\} f_1(x) dx$$

$$= \int \left\{ \int y f_{Y \cdot X}(y \mid x) dy \right\} f_1(x) dx$$

$$= \int\int y\frac{f(x,y)}{f_1(x)}f_1(x)dydx.$$

If we can interchange the order of integration (whenever $\int |y|f_2(y)dy < \infty$), then

$$E\{E\{Y \mid X\}\} = \int y\left\{\int f(x,y)dx\right\}dy$$

$$= \int yf_2(y)dy \qquad\qquad (3.5.11)$$

$$= E\{Y\}.$$

This result, known as **the law of the iterated expectation**, is often very useful. An example of the use of the law of the iterated expectation is the following.

Example 3.31. Let (J, N) be a pair of random variables. The conditional distribution of J, given $\{N = n\}$, is the binomial $B(n, p)$. The marginal distribution of N is Poisson with mean λ. What is the expected value of J?
By the law of the iterated expectation,

$$E\{J\} = E\{E\{J \mid N\}\}$$

$$= E\{Np\} = pE\{N\} = p\lambda.$$

One can show that the marginal distribution of J is Poisson, with mean $p\lambda$.

Another important result relates variances and conditional variances. That is, if (X, Y) is a pair of random variables, having finite Variances, then

$$V\{Y\} = E\{V\{Y \mid X\}\} + V\{E\{Y \mid X\}\}. \qquad\qquad (3.5.12)$$

We call this relationship the **law of total variance**.

Example 3.32. Let (X, Y) be a pair of independent random variables having finite variances σ_X^2 and σ_Y^2 and expected values μ_X, μ_Y. Determine the variance of $W = XY$. By the law of total variance,

$$V\{W\} = E\{V\{W \mid X\}\} + V\{E\{W \mid X\}\}.$$

Since X and Y are independent

$$V\{W \mid X\} = V\{XY \mid X\} = X^2 V\{Y \mid X\}$$

$$= X^2 \sigma_Y^2.$$

Similarly,

$$E\{W \mid X\} = X\mu_Y.$$

Hence,

$$V\{W\} = \sigma_Y^2 E\{X^2\} + \mu_Y^2 \sigma_X^2$$
$$= \sigma_Y^2(\sigma_X^2 + \mu_X^2) + \mu_Y^2 \sigma_X^2$$
$$= \sigma_X^2 \sigma_Y^2 + \mu_X^2 \sigma_Y^2 + \mu_Y^2 \sigma_X^2.$$

3.6   Some multivariate distributions

3.6.1   The multinomial distribution

The multinomial distribution is a generalization of the binomial distribution to cases of n **independent** trials in which the results are classified to k possible categories (e.g., Excellent, Good, Average, Poor). The random variables $(J_1, J_2, \ldots, J_k)$ are the number of trials yielding results in each one of the k categories. These random variables are dependent, since $J_1 + J_2 + \cdots + J_k = n$. Furthermore, let $p_1, p_2, \ldots, p_k; p_i \geq 0$, $\sum_{i=1}^k p_i = 1$, be the probabilities of the k categories. The binomial distribution is the special case of $k = 2$. Since $J_k = n - (J_1 + \cdots + J_{k-1})$, the joint probability function is written as a function of $k - 1$ arguments, and its formula is

$$p(j_1, \ldots, j_{k-1}) = \binom{n}{j_1, \ldots, j_{k-1}} p_1^{j_1} \cdots p_{k-1}^{j_{k-1}} p_k^{j_k} \tag{3.6.1}$$

for $j_1, \ldots, j_{k-1} \geq 0$ such that $\sum_{i=1}^{k-1} j_i \leq n$. In this formula,

$$\binom{n}{j_1, \ldots, j_{k-1}} = \frac{n!}{j_1! j_2! \cdots j_k!}, \tag{3.6.2}$$

and $j_k = n - (j_1 + \cdots + j_{k-1})$. For example, if $n = 10$, $k = 3$, $p_1 = 0.3$, $p_2 = 0.4$, $p_3 = 0.3$,

$$p(5, 2) = \frac{10!}{5!2!3!}(0.3)^5(0.4)^2(0.3)^3$$

$$= 0.02645.$$

The marginal distribution of each one of the k variables is binomial, with parameters n and p_i ($i = 1, \ldots, k$). The joint marginal distribution of (J_1, J_2) is trinomial, with parameters n, p_1, p_2 and $(1 - p_1 - p_2)$. Finally, the conditional distribution of $(J_1, \ldots, J_r)$, $1 \leq r < k$, given $\{J_{r+1} = j_{r+1}, \ldots, J_k = j_k\}$ is $(r+1)$-nomial, with parameters $n_r = n - (j_{r+1} + \cdots + j_k)$ and $p_1', \ldots, p_r', p_{r+1}'$, where

$$p_i' = \frac{p_i}{(1 - p_{r+1} - \cdots - p_k)}, \quad i = 1, \ldots, r$$

and

$$p_{r+1}' = 1 - \sum_{i=1}^r p_i'.$$

Finally, we can show that, for $i \neq j$,

$$\text{Cov}(J_i, J_j) = -np_i p_j. \tag{3.6.3}$$

Example 3.33. An insertion machine is designed to insert components into computer printed circuit boards. Every component inserted on a board is scanned optically. An insertion is either error free or its error is classified to the following two main categories: misinsertion (broken lead, off pad, etc.) or wrong component. Thus, we have altogether three general categories. Let

$$J_1 = \text{\# of error free components;}$$

$$J_2 = \text{\# of misinsertion;}$$

$$J_3 = \text{\# of wrong components.}$$

The probabilities that an insertion belongs to one of these categories is $p_1 = 0.995$, $p_2 = 0.001$ and $p_2 = 0.004$.

The insertion rate of this machine is $n = 3500$ components per hour of operation. Thus, we expect during one hour of operation $n \times (p_2 + p_3) = 175$ insertion errors.

Given that there are 16 insertion errors during a particular hour of operation, the conditional distribution of the number of misinsertions is binomial $B\left(16, \frac{0.01}{0.05}\right)$.

Thus,

$$E\{J_2 \mid J_2 + J_3 = 16\} = 16 \times 0.2 = 3.2.$$

On the other hand,

$$E\{J_2\} = 3500 \times 0.001 = 3.5.$$

We see that the information concerning the total number of insertion errors makes a difference.

Finally,

$$\text{Cov}(J_2, J_3) = -3500 \times 0.001 \times 0.004$$

$$= -0.014$$

$$V\{J_2\} = 3500 \times 0.001 \times 0.999 = 3.4965$$

and

$$V\{J_3\} = 3500 \times 0.004 \times 0.996 = 13.944.$$

Hence, the correlation between J_2 and J_3 is

$$\rho_{2,3} = \frac{-0.014}{\sqrt{3.4965 \times 13.944}} = -0.0020.$$

This correlation is quite small.

3.6.2 The multi-hypergeometric distribution

Suppose that we draw from a population of size N a RSWOR of size n. Each one of the n units in the sample is classified to one of k categories. Let $J_1, J_2, \ldots, J_k$ be the number of sample units belonging to each one of these categories. $J_1 + \cdots + J_k = n$. The distribution of $J_1, \ldots, J_k$

is k-variate hypergeometric. If $M_1, \ldots, M_k$ are the number of units in the population in these categories, before the sample is drawn, then the joint p.d.f. of $J_1, \ldots, J_k$ is

$$p(j_1, \ldots, j_{k-1}) = \frac{\binom{M_1}{j_1}\binom{M_2}{j_2}\cdots\binom{M_k}{j_k}}{\binom{N}{n}}, \tag{3.6.4}$$

where $j_k = n - (j_1 + \cdots + j_{k-1})$. This distribution is a generalization of the hypergeometric distribution $H(N, M, n)$. The hypergeometric distribution $H(N, M_i, n)$ is the marginal distribution of J_i ($i = 1, \ldots, k$). Thus,

$$E\{J_i\} = n\frac{M_i}{N}, \quad i = 1, \ldots, k$$

$$V\{J_i\} = n\frac{M_i}{N}\left(1 - \frac{M_i}{N}\right)\left(1 - \frac{n-1}{N-1}\right), \quad i = 1, \ldots, k \tag{3.6.5}$$

and for $i \neq j$

$$\mathrm{Cov}(J_i, J_j) = -n\frac{M_i}{N} \cdot \frac{M_j}{N}\left(1 - \frac{n-1}{N-1}\right).$$

Example 3.34. A lot of 100 spark plugs contains 20 plugs from vendor V_1, 50 plugs from vendor V_2 and 30 plugs from vendor V_3.
A random sample of $n = 20$ plugs is drawn from the lot without replacement.
Let J_i be the number of plugs in the sample from the vendor V_i, $i = 1, 2, 3$. Accordingly,

$$\Pr\{J_1 = 5, J_2 = 10\} = \frac{\binom{20}{5}\binom{50}{10}\binom{30}{5}}{\binom{100}{20}}$$

$$= 0.00096.$$

If we are told that 5 out of the 20 plugs in the sample are from vendor V_3, then the conditional distribution of J_1 is

$$\Pr\{J_1 = j_1 \mid J_3 = 5\} = \frac{\binom{20}{j_1}\binom{50}{15-j_1}}{\binom{70}{15}}, \quad j_1 = 0, \ldots, 15.$$

Indeed, given $J_3 = 5$, then J_1 can assume only the values $0, 1, \ldots, 15$. The conditional probability that j_1 out of the 15 remaining plugs in the sample are from vendor V_1, is the same like that of choosing a RSWOR of size 15 from a lot of size $70 = 20 + 50$, with 20 plugs from vendor V_1.

3.6.3 The bivariate normal distribution

The bivariate normal distribution is the joint distribution of two continuous random variables (X, Y) having a joint p.d.f.

$$f(x, y; \mu, \eta, \sigma_X, \sigma_Y, \rho) = \frac{1}{2\pi\sigma_X\sigma_Y\sqrt{1-\rho^2}} \cdot$$

$$\exp\left\{-\frac{1}{2(1-\rho^2)}\left[\left(\frac{x-\mu}{\sigma_x}\right)^2 - 2\rho\frac{x-\mu}{\sigma_Y}\cdot\frac{y-\eta}{\sigma_Y} + \left(\frac{y-\eta}{\sigma_Y}\right)^2\right]\right\},$$

$$-\infty < x, y < \infty. \tag{3.6.6}$$

$\mu, \eta, \sigma_X, \sigma_Y$ and ρ are parameters of this distribution.

Integration of y yields that the marginal distribution of X is $N(\mu, \sigma_x^2)$. Similarly, the marginal distribution of Y is $N(\eta, \sigma_Y^2)$. Furthermore, ρ is the correlation between X and Y. Notice that if $\rho = 0$, then the joint p.d.f. becomes the product of the two marginal ones, i.e.,

$$f(x, y); \mu, \eta, \sigma_X, \sigma_Y, 0) = \frac{1}{\sqrt{2\pi}\sigma_X}\exp\left\{-\frac{1}{2}\left(\frac{x-\mu}{\sigma_X}\right)^2\right\} \cdot$$

$$\frac{1}{\sqrt{2\pi}\sigma_Y}\exp\left\{-\frac{1}{2}\left(\frac{y-\eta}{\sigma_Y}\right)^2\right\}, \quad \text{for all} -\infty < x, y < \infty.$$

Hence, if $\rho = 0$, then X and Y are independent. On the other hand, if $\rho \neq 0$, then $f(x, y; \mu, \eta, \sigma_X, \sigma_Y, \rho) \neq f_1(x; \mu, \sigma_X)f_2(y; \eta, \sigma_Y)$, and the two random variables are dependent.

In Figure 3.17, we present the bivariate p.d.f. for $\mu = \eta = 0$, $\sigma_X = \sigma_Y = 1$ and $\rho = 0.5$.

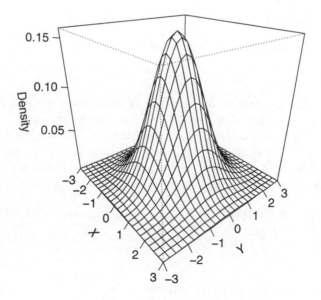

Figure 3.17 *Bivariate normal p.d.f.*

One can verify also that the conditional distribution of Y, given $\{X = x\}$ is normal with mean

$$\mu_{Y \cdot x} = \eta + \rho \frac{\sigma_Y}{\sigma_X}(x - \mu) \tag{3.6.7}$$

and variance

$$\sigma^2_{Y \cdot x} = \sigma_Y^2(1 - \rho^2). \tag{3.6.8}$$

It is interesting to see that $\mu_{Y \cdot x}$ is a linear function of x. We can say that $\mu_{Y \cdot x} = E\{Y \mid X = x\}$ is, in the bivariate normal case, the theoretical (linear) regression of Y on X (see Chapter 5). Similarly,

$$\mu_{X \cdot y} = \mu + \rho \frac{\sigma_X}{\sigma_Y}(y - \eta),$$

and

$$\sigma^2_{X \cdot y} = \sigma_X^2(1 - \rho^2).$$

If $\mu = \eta = 0$ and $\sigma_X = \sigma_Y = 1$, we have the **standard** bivariate normal distribution. The joint c.d.f. in the standard case is denoted by $\Phi_2(x, y; \rho)$ and its formula is

$$\Phi_2(x, y; \rho) = \frac{1}{2\pi\sqrt{1 - \rho^2}} \int_{-\infty}^{x} \int_{-\infty}^{y} \exp\left\{-\frac{1}{2(1 - \rho^2)}(z_1^2 - 2\rho z_1 z_2 + z^2)\right\} dz_1 dz_2$$

$$= \int_{-\infty}^{x} \phi(z_1)\Phi\left(\frac{y - \rho z_1}{\sqrt{1 - \rho^2}}\right) dz_1 \tag{3.6.9}$$

values of $\Phi_2(x, y; \rho)$ can be obtained by numerical integration. If one has to compute the bivariate c.d.f. in the general case, the following formula is useful

$$F(x, y; \mu, \eta, \sigma_X, \sigma_Y, \rho) = \Phi_2\left(\frac{x - \mu}{\sigma_X}, \frac{y - \eta}{\sigma_Y}; \rho\right).$$

For computing $\Pr\{a \leq X \leq b, c \leq Y \leq d\}$ we use the formula,

$$\Pr\{a \leq X \leq b, c \leq Y \leq d\} = F(b, d; -)$$
$$- F(a, d; -) - F(b, c; -) + F(a, c; -).$$

Example 3.35. Suppose that (X, Y) deviations in components placement on PCB by an automatic machine have a bivariate normal distribution with means $\mu = \eta = 0$, standard deviations $\sigma_X = 0.00075$, and $\sigma_Y = 0.00046$ [in.] and $\rho = 0.160$. The placement errors are within the specifications if $|X| < 0.001$ [in.] and $|Y| < 0.001$ [in.]. What proportion of components are expected to have X, Y deviations compliant with the specifications? The standardized version of the spec limits are $Z_1 = \frac{0.001}{0.00075} = 1.33$ and $Z_2 = \frac{0.001}{0.00046} = 2.174$. We compute

$$\Pr\{|X| < 0.001, |Y| < 0.001\} = \Phi_2(1.33, 2.174, .16) - \Phi_2(-1.33, 2.174, .16)$$

$$- \Phi_2(1.33, -2.174; .16) + \Phi_2(-1.33, -2.174; .16)$$

$$= 0.793.$$

This is the expected proportion of good placements.

3.7   Distribution of order statistics

As defined in Chapter 2, the order statistics of the sample are the sorted data. More specifically, let $X_1, \ldots, X_n$ be identically distributed independent (i.i.d.) random variables. The order statistics are $X_{(i)}, i = 1, \ldots, n$, where

$$X_{(1)} \le X_{(2)} \le \cdots \le X_{(n)}.$$

In the present section, we discuss the distributions of these order statistics, when $F(x)$ is (absolutely) continuous, having a p.d.f. $f(x)$.

We start with the extremal statistics $X_{(1)}$ and $X_{(n)}$.

Since the random variables X_i $(i = 1, \ldots, n)$ are i.i.d., the c.d.f. of $X_{(1)}$ is,

$$F_{(1)}(x) = \Pr\{X_{(1)} \le x\}$$

$$= 1 - \Pr\{X_{(1)} \ge x\} = 1 - \prod_{i=1}^{n} \Pr\{X_i \ge x\}$$

$$= 1 - (1 - F(x))^n.$$

By differentiation, we obtain that the p.d.f. of $X_{(1)}$ is

$$f_{(1)}(x) = nf(x)[1 - F(x)]^{n-1}. \tag{3.7.1}$$

Similarly, the c.d.f. of the sample maximum $X_{(n)}$ is

$$F_{(n)}(x) = \prod_{i=1}^{n} \Pr\{X_i \le x\}$$

$$= (F(x))^n.$$

The p.d.f. of $X_{(n)}$ is

$$f_{(n)}(x) = nf(x)(F(x))^{n-1}. \tag{3.7.2}$$

Example 3.36. (i) A switching circuit consists of n modules, which operate independently and which are connected in **series** (see Figure 3.18). Let X_i be the time till failure of the ith module. The system fails when any module fails. Thus, the time till failure of the system is $X_{(1)}$. If all X_i are exponentially distributed with mean life β, then the c.d.f. of $X_{(1)}$ is

$$F_{(1)}(x) = 1 - e^{-nx/\beta}, \quad x \ge 0.$$

Thus, $X_{(1)}$ is distributed like $E\left(\frac{\beta}{n}\right)$. It follows that the expected time till failure of the circuit is $E\{X_{(1)}\} = \frac{\beta}{n}$.

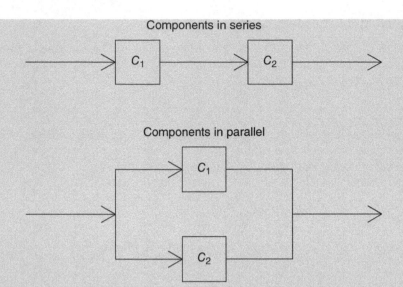

Figure 3.18 *Series and parallel systems.*

(ii) If the modules are connected in parallel, then the circuit fails at the instant the last of the n modules fail, which is $X_{(n)}$. Thus, if X_i is $E(\beta)$, the c.d.f. of $X_{(n)}$ is

$$F_{(n)}(x) = (1 - e^{-(x/\beta)})^n.$$

The expected value of $X_{(n)}$ is

$$E\{X_{(n)}\} = \frac{n}{\beta} \int_0^\infty x e^{-x/\beta} (1 - e^{-x/\beta})^{n-1} dx$$

$$= n\beta \int_0^\infty y e^{-y} (1 - e^{-y})^{n-1} dy$$

$$= n\beta \sum_{j=0}^{n-1} (-1)^j \binom{n-1}{j} \int_0^\infty y e^{-(1+j)y} dy$$

$$= n\beta \sum_{j=1}^{n} (-1)^{j-1} \binom{n-1}{j-1} \frac{1}{j^2}.$$

Furthermore, since $n\binom{n-1}{j-1} = j\binom{n}{j}$, we obtain that

$$E\{X_{(n)}\} = \beta \sum_{j=1}^{n} (-1)^{j-1} \binom{n}{j} \frac{1}{j}.$$

One can also show that this formula is equivalent to

$$E\{X_{(n)}\} = \beta \sum_{j=1}^{n} \frac{1}{j}.$$

Accordingly, if the parallel circuit consists of three modules, and the time till failure of each module is exponential with $\beta = 1000$ [hours], the expected time till failure of the system is 1833.3[hours].

Generally, the distribution of $X_{(i)}$ $(i = 1, \ldots, n)$ can be obtained by the following argument. The event $\{X_{(i)} \leq x\}$ is equivalent to the event that the number of X_i values in the random example which are smaller or equal to x is at least i.

Consider n independent and identical trials, in which "success" is that $\{X_i \leq x\}$ $(i = 1, \ldots, n)$. The probability of "success" is $F(x)$. The distribution of the number of successes is $B(n, F(x))$. Thus, the c.d.f. of $X_{(i)}$ is

$$F_{(i)}(x) = \Pr\{X_{(i)} \leq x\} = 1 - B(i - 1; n, F(x))$$

$$= \sum_{j=i}^{n} \binom{n}{j} (F(x))^j (1 - F(x))^{n-j}.$$

Differentiating this c.d.f. with respect to x yields, the p.d.f. of $X_{(i)}$, namely

$$f_{(i)}(x) = \frac{n!}{(i - 1)!(n - i)!} f(x)(F(x))^{i-1}(1 - F(x))^{n-i}. \tag{3.7.3}$$

Notice that if X has a uniform distribution on $(0, 1)$, then the distribution of $X_{(i)}$ is like that of $B(i, n - i + 1)$, $i = 1, \ldots, n$. In a similar manner, one can derive the joint p.d.f. of $(X_{(i)}, X_{(j)})$, $1 \leq i < j \leq n$, etc. This joint p.d.f. is given by

$$f_{(i),(j)}(x, y) = \frac{n!}{(i - 1)!(j - 1 - i)!(n - j)!} f(x)f(y)$$

$$\cdot (F(x))^{i-1}[F(y) - F(x)]^{j-i-1}(1 - F(y))^{n-j}, \tag{3.7.4}$$

for $-\infty < x < y < \infty$.

3.8 Linear combinations of random variables

Let $X_1, X_2, \ldots, X_n$ be random variables having a joint distribution, with joint p.d.f. $f(x_1, \ldots, x_n)$. Let $\alpha_1, \ldots, \alpha_n$ be given constants. Then

$$W = \sum_{i=1}^{n} \alpha_i X_i$$

is a linear combination of the X's. The p.d.f. of W can generally be derived, using various methods. We discuss in the present section only the formulae of the expected value and variance of W.

It is straightforward to show that

$$E\{W\} = \sum_{i=1}^{n} \alpha_i E\{X_i\}. \tag{3.8.1}$$

That is, the expected value of a linear combination is the same linear combination of the expectations.

The formula for the variance is somewhat more complicated and is given by

$$V\{W\} = \sum_{i=1}^{n} \alpha_i^2 V\{X_i\} + \sum_{i \neq j} \sum \alpha_i \alpha_j \text{COV}(X_i, X_j). \tag{3.8.2}$$

Example 3.37. Let $X_1, X_2, \ldots, X_n$ be i.i.d. random variables, with common expectations μ and common finite variances σ^2. The sample mean $\overline{X}_n = \frac{1}{n} \sum_{i=1}^{n} X_i$ is a particular linear combination, with

$$\alpha_1 = \alpha_2 = \cdots = \alpha_n = \frac{1}{n}.$$

Hence,

$$E\{\overline{X}_n\} = \frac{1}{n} \sum_{i=1}^{n} E\{X_i\} = \mu$$

and, since $X_1, X_2, \ldots, X_n$ are mutually independent, $\text{COV}(X_i, X_j) = 0$, all $i \neq j$. Hence,

$$V\{\overline{X}_n\} = \frac{1}{n^2} \sum_{i=1}^{n} V\{X_i\} = \frac{\sigma^2}{n}.$$

Thus, we have shown that in a random sample of n i.i.d. random variables, the sample mean has the same expectation as that of the individual variables, but its sample variance is reduced by a factor of $1/n$.

Moreover, from Chebychev's inequality, for any $\epsilon > 0$

$$\Pr\{|\overline{X}_n - \mu| > \epsilon\} < \frac{\sigma^2}{n\epsilon^2}.$$

Therefore, since $\lim_{n \to \infty} \frac{\sigma^2}{n\epsilon^2} = 0$,

$$\lim_{n \to \infty} \Pr\{|\overline{X}_n - \mu| > \epsilon\} = 0.$$

This property is called the **convergence in probability** of $\overline{X}_n$ to μ.

Example 3.38. Let U_1, U_2, U_3 be three i.i.d. random variables having uniform distributions on $(0, 1)$. We consider the statistic

$$W = \frac{1}{4} U_{(1)} + \frac{1}{2} U_{(2)} + \frac{1}{4} U_{(3)},$$

where $0 < U_{(1)} < U_{(2)} < U_{(3)} < 1$ are the order statistics. We have seen in Section 3.7 that the distribution of $U_{(i)}$ is like that of $B(i, n - i + 1)$. Hence,

$$E\{U_{(1)}\} = E\{B(1, 3)\} = \frac{1}{4}$$

$$E\{U_{(2)}\} = E\{B(2, 2)\} = \frac{1}{2}$$

$$E\{U_{(3)}\} = E\{B(3, 1)\} = \frac{3}{4}.$$

It follows that

$$E\{W\} = \frac{1}{4} \cdot \frac{1}{4} + \frac{1}{2} \cdot \frac{1}{2} + \frac{1}{4} \cdot \frac{1}{4} = \frac{1}{2}.$$

To find the variance of W, we need more derivations.
First

$$V\{U_{(1)}\} = V\{B(1,3)\} = \frac{3}{4^2 \times 5} = \frac{3}{80}$$

$$V\{U_{(2)}\} = V\{B(2,2)\} = \frac{4}{4^2 \times 5} = \frac{1}{20}$$

$$V\{U_{(3)}\} = V\{B(3,1)\} = \frac{3}{4^2 \times 5} = \frac{3}{80}.$$

We need to find $\mathrm{Cov}(U_{(1)}, U_{(2)})$, $\mathrm{Cov}(U_{(1)}, U_{(3)})$ and $\mathrm{Cov}(U_{(2)}, U_{(3)})$. From the joint p.d.f. formula of order statistics, the joint p.d.f. of $(U_{(1)}, U_{(2)})$ is

$$f_{(1),(2)}(x, y) = 6(1 - y), \quad 0 < x \leq y < 1.$$

Hence,

$$E\{U_{(1)}U_{(2)}\} = 6 \int_0^1 x \left(\int_0^1 y(1 - y) dy \right) dx$$

$$= \frac{6}{40}.$$

Thus,

$$\mathrm{Cov}(U_{(1)}, U_{(2)}) = \frac{6}{40} - \frac{1}{4} \cdot \frac{1}{2}$$

$$= \frac{1}{40}.$$

Similarly, the p.d.f. of $(U_{(1)}, U_{(3)})$ is

$$f_{(1),(3)}(x, y) = 6(y - x), \quad 0 < x \leq y < 1.$$

Thus,

$$E\{U_{(1)}U_{(3)}\} = 6 \int_0^1 x \left(\int_x^1 y(y - x) dy \right) dx$$

$$= 6 \int_0^1 x \left(\frac{1}{3}(1 - x^3) - \frac{x}{2}(1 - x^2) \right) dx$$

$$= \frac{1}{5},$$

and

$$\mathrm{Cov}(U_{(1)}, U_{(3)}) = \frac{1}{5} - \frac{1}{4} \cdot \frac{3}{4} = \frac{1}{80}.$$

The p.d.f. of $(U_{(2)}, U_{(3)})$ is

$$f_{(2),(3)}(x, y) = 6x, \quad 0 < x \leq y \leq 1,$$

and

$$\text{Cov}(U_{(2)}, U_{(3)}) = \frac{1}{40}.$$

Finally,

$$V\{W\} = \frac{1}{16} \cdot \frac{3}{80} + \frac{1}{4} \cdot \frac{1}{20} + \frac{1}{16} \cdot \frac{3}{80}$$
$$+ 2 \cdot \frac{1}{4} \cdot \frac{1}{2} \cdot \frac{1}{40} + 2 \cdot \frac{1}{4} \cdot \frac{1}{4} \cdot \frac{1}{80}$$
$$+ 2 \cdot \frac{1}{2} \cdot \frac{1}{4} \cdot \frac{1}{40}$$
$$= \frac{1}{32} = 0.03125.$$

The following is a useful result:

If $X_1, X_2, \ldots, X_n$ are mutually independent, then the m.g.f. of $T_n = \sum_{i=1}^{n} X_i$ is

$$M_{T_n}(t) = \prod_{i=1}^{n} M_{X_i}(t). \qquad (3.8.3)$$

Indeed, as shown in Section 4.5.2, when $X_1, \ldots, X_n$ are independent, the expected value of the product of functions is the product of their expectations. Therefore,

$$M_{T_n}(t) = E\left\{ e^{t \sum_{i=1}^{n} X_i} \right\}$$
$$= E\left\{ \prod_{i=1}^{n} e^{tX_i} \right\}$$
$$= \prod_{i=1}^{n} E\{e^{tX_i}\}$$
$$= \prod_{i=1}^{n} M_{X_i}(t).$$

The expected value of the product is equal to the product of the expectations, since $X_1, \ldots, X_n$ are mutually independent.

Example 3.39.   In the present example, we illustrate some applications of the last result.

(i) Let $X_1, X_2, \ldots, X_k$ be independent random variables having binomial distributions like $B(n_i, p)$, $i = 1, \ldots, k$, then their sum T_k has the binomial distribution. To show this,

$$M_{T_k}(t) = \prod_{i=1}^{k} M_{X_i}(t)$$
$$= [e^t p + (1 - p)]^{\sum_{i=1}^{k} n_i}.$$

That is, T_k is distributed like $B\left(\sum_{i=1}^{k} n_i, p\right)$. This result is intuitively clear.

(ii) If $X_1, \ldots, X_n$ are independent random variables, having Poisson distributions with parameters λ_i $(i = 1, \ldots, n)$, then the distribution of $T_n = \sum_{i=1}^{n} X_i$ is Poisson with parameter $\mu_n = \sum_{i=1}^{n} \lambda_i$. Indeed,

$$M_{T_n}(t) = \prod_{j=1}^{n} \exp\{-\lambda_j(1 - e^t)\}$$

$$= \exp\left\{-\sum_{j=1}^{n} \lambda_j(1 - e^t)\right\}$$

$$= \exp\{-\mu_n(1 - e^t)\}.$$

(iii) Suppose $X_1, \ldots, X_n$ are independent random variables, and the distribution of X_i is normal $N(\mu_i, \sigma_i^2)$, then the distribution of $W = \sum_{i=1}^{n} \alpha_i X_i$ is normal like that of

$$N\left(\sum_{i=1}^{n} \alpha_i \mu_i, \sum_{i=1}^{n} \alpha_i^2 \sigma_i^2\right).$$

To verify this, we recall that $X_i = \mu_i + \sigma_i Z_i$, where Z_i is $N(0, 1)$ $(i = 1, \ldots, n)$. Thus,

$$M_{\alpha_i X_i}(t) = E\{e^{t(\alpha_i \mu_i + \alpha_i \sigma_i Z_i)}\}$$

$$= e^{t\alpha_i \mu_i} M_{Z_i}(\alpha_i \sigma_i t).$$

We derived before that $M_{Z_i}(u) = e^{u^2/2}$. Hence,

$$M_{\alpha_i X_i}(t) = \exp\left\{\alpha_i \mu_i t + \frac{\alpha_i^2 \sigma_i^2}{2} t^2\right\}.$$

Finally,

$$M_W(t) = \prod_{i=1}^{n} M_{\alpha_i X_i}(t)$$

$$= \exp\left\{\left(\sum_{i=1}^{n} \alpha_i \mu_i\right) t + \frac{\sum_{i=1}^{n} \alpha_i^2 \sigma_i^2}{2} t^2\right\}.$$

This implies that the distribution of W is normal, with

$$E\{W\} = \sum_{i=1}^{n} \alpha_i \mu_i$$

and

$$V\{W\} = \sum_{i=1}^{n} \alpha_i^2 \sigma_i^2.$$

(iv) If $X_1, X_2, \ldots, X_n$ are independent random variables, having gamma distribution like $G(\nu_i, \beta)$, respectively, $i = 1, \ldots, n$, then the distribution of $T_n = \sum_{i=1}^{n} X_i$ is gamma, like that of $G\left(\sum_{i=1}^{n} \nu_i, \beta\right)$. Indeed,

$$M_{T_n}(t) = \prod_{i=1}^{n} (1 - t\beta)^{-\nu_i}$$

$$= (1 - t\beta)^{-\sum_{i=1}^{n} \nu_i}.$$

3.9   Large sample approximations

3.9.1   The law of large numbers

We have shown in Example 3.37 that the mean of a random sample, $\overline{X}_n$, converges in probability to the expected value of X, μ (the population mean). This is the **law of large numbers** (L.L.N.) which states that, if $X_1, X_2, \ldots$ are i.i.d. random variables and $E\{|X_1|\} < \infty$, then for any $\epsilon > 0$,

$$\lim_{n \to \infty} \Pr\{|\overline{X}_n - \mu| > \epsilon\} = 0.$$

We also write,

$$\lim_{n \to \infty} \overline{X}_n = \mu, \text{ in probability.}$$

This is known as the **weak** L.L.N. There is a stronger law, which states that, under the above conditions,

$$\Pr\left\{\lim_{n \to \infty} \overline{X}_n = \mu\right\} = 1.$$

It is beyond the scope of the book to discuss the meaning of the strong L.L.N.

3.9.2   The central limit theorem

The **Central Limit Theorem**, C.L.T., is one of the most important theorems in probability theory. We formulate here the simplest version of this theorem, which is often sufficient for applications. The theorem states that if $\overline{X}_n$ is the sample mean of n i.i.d. random variables then, if the population variance σ^2 is positive and finite, the sampling distribution of $\overline{X}_n$ is approximately normal, as $n \to \infty$. More precisely,

If $X_1, X_2, \ldots$ is a sequence of i.i.d. random variables, with $E\{X_1\} = \mu$ and $V\{X_1\} = \sigma^2$, $0 < \sigma^2 < \infty$, then

$$\lim_{n \to \infty} \Pr\left\{\frac{(\overline{X}_n - \mu)\sqrt{n}}{\sigma} \leq z\right\} = \Phi(z), \tag{3.9.1}$$

where $\Phi(z)$ is the c.d.f. of $N(0, 1)$.

The proof of this basic version of the C.L.T. is based on a result in probability theory, stating that if $X_1, X_2, \ldots$ is a sequence of random variables having m.g.f.s, $M_n(T)$, $n = 1, 2, \ldots$ and if

$\lim_{n\to\infty} M_n(t) = M(t)$ is the m.g.f. of a random variable X^*, having a c.d.f. $F^*(x)$, then $\lim_{n\to\infty} F_n(x) = F^*(x)$, where $F_n(x)$ is the c.d.f. of X_n.

The m.g.f. of

$$Z_n = \frac{\sqrt{n}(\bar{X}_n - \mu)}{\sigma},$$

can be written as

$$M_{Z_n}(t) = E\left\{ \exp\left\{ \frac{t}{\sqrt{n}\sigma} \sum_{i=1}^{n} (X_i - \mu) \right\} \right\}$$

$$= \left(E\left\{ \exp\left\{ \frac{t}{\sqrt{n}\sigma} (X_1 - \mu) \right\} \right\} \right)^n,$$

since the random variables are independent. Furthermore, Taylor expansion of $\exp\left\{ \frac{t}{\sqrt{n}\sigma}(X_1 - \mu) \right\}$ is

$$1 + \frac{t}{\sqrt{n}\sigma}(X_1 - \mu) + \frac{t^2}{2n\sigma^2}(X_1 - \mu)^2 + o\left(\frac{1}{n}\right),$$

for n large. Hence, as $n \to \infty$

$$E\left\{ \exp\left\{ \frac{t}{\sqrt{n}\sigma}(X_1 - \mu) \right\} \right\} = 1 + \frac{t^2}{2n} + o\left(\frac{1}{n}\right).$$

Hence,

$$\lim_{n\to\infty} M_{Z_n}(t) = \lim_{n\to\infty} \left(1 + \frac{t^2}{2n} + o\left(\frac{1}{n}\right) \right)^n$$

$$= e^{t^2/2},$$

which is the m.g.f. of $N(0, 1)$. This is a sketch of the proof. For rigorous proofs and extensions, see textbooks on probability theory.

3.9.3   Some normal approximations

The C.L.T. can be applied to provide an approximation to the distribution of the sum of n i.i.d. random variables, by a standard normal distribution, when n is large. We list below a few such useful approximations.

(i) *Binomial distribution*:
   When n is large, then the c.d.f. of $B(n, p)$ can be approximated by

$$B(k; n, p) \cong \Phi\left(\frac{k + \frac{1}{2} - np}{\sqrt{np(1 - p)}} \right). \tag{3.9.2}$$

We add $\frac{1}{2}$ to k, in the argument of $\Phi(\cdot)$ to obtain a better approximation when n is not too large. This modification is called "correction for discontinuity."

How large should n be to get a "good" approximation? A General rule is

$$n > \frac{9}{p(1-p)}. \tag{3.9.3}$$

(ii) *Poisson distribution*:

The c.d.f. of Poisson with parameter λ can be approximated by

$$P(k; \lambda) \cong \Phi \left(\frac{k + \frac{1}{2} - \lambda}{\sqrt{\lambda}} \right), \tag{3.9.4}$$

if λ is large (greater than thirty).

(iii) *Gamma distribution*:

The c.d.f. of $G(\nu, \beta)$ can be approximated by

$$G(x; \nu, \beta) \cong \Phi \left(\frac{x - \nu\beta}{\beta\sqrt{\nu}} \right), \tag{3.9.5}$$

for large values of ν.

Example 3.40.

(i) A lot consists of $n = 10,000$ screws. The probability that a screw is defective is $p = 0.01$. What is the probability that there are more than 120 defective screws in the lot?

The number of defective screws in the lot, J_n, has a distribution like $B(10,000, 0.01)$. Hence,

$$\Pr\{J_{10,000} > 120\} = 1 - B(120; 10,000, 0.01)$$

$$\cong 1 - \Phi \left(\frac{120.5 - 100}{\sqrt{99}} \right)$$

$$= 1 - \Phi(2.06) = 0.0197.$$

(ii) In the production of industrial film, we find on the average 1 defect per $100[\text{ft}]^2$ of film. What is the probability that fewer than 100 defects will be found on $12,000[\text{ft}]^2$ of film?

We assume that the number of defects per unit area of film is a Poisson random variable. Thus, our model is that the number of defects, X, per $12,000[\text{ft}]^2$ has a Poisson distribution with parameter $\lambda = 120$. Thus,

$$\Pr\{X < 100\} \cong \Phi \left(\frac{99.5 - 120}{\sqrt{120}} \right)$$

$$= 0.0306.$$

(iii) The time till failure, T, of radar equipment is exponentially distributed with mean time till failure (MTTF) of $\beta = 100$ [hours].

A sample of $n = 50$ units is put on test. Let $\overline{T}_{50}$ be the sample mean. What is the probability that $\overline{T}_{50}$ will fall in the interval $(95,105)$ [hours]?

We have seen that $\sum_{i=1}^{50} T_i$ is distributed like $G(50,100)$, since $E(\beta)$ is distributed like $G(1,\beta)$. Hence, $\overline{T}_{50}$ is distributed like $\frac{1}{50}G(50,100)$ which is $G(50,2)$. By the normal approximation

$$\Pr\{95 < \overline{T}_{50} < 105\} \cong \Phi\left(\frac{105-100}{2\sqrt{50}}\right)$$

$$- \Phi\left(\frac{95-100}{2\sqrt{50}}\right) = 2\Phi(0.3536) - 1 = 0.2763.$$

3.10 Additional distributions of statistics of normal samples

In the present section, we assume that $X_1, X_2, \ldots, X_n$ are i.i.d. $N(\mu, \sigma^2)$ random variables. In Sections 3.10.1–3.10.3, we present the chi-squared, t- and F-distributions, which play an important role in the theory of statistical inference (Chapter 4).

3.10.1 Distribution of the sample variance

Writing $X_i = \mu + \sigma Z_i$, where $Z_1, \ldots, Z_n$ are i.i.d. $N(0, 1)$, we obtain that the sample variance S^2 is distributed like

$$S^2 = \frac{1}{n-1} \sum_{i=1}^{n} (X_i - \overline{X}_n)^2$$

$$= \frac{1}{n-1} \sum_{i=1}^{n} (\mu + \sigma Z_i - (\mu + \sigma \overline{Z}_n))^2$$

$$= \frac{\sigma^2}{n-1} \sum_{i=1}^{n} (Z_i - \overline{Z}_n)^2.$$

One can show that $\sum_{i=1}^{n} (Z_i - \overline{Z}_n)^2$ is distributed like $\chi^2[n-1]$, where $\chi^2[\nu]$ is called a **chi-squared random variable with ν degrees of freedom**. Moreover, $\chi^2[\nu]$ is distributed like $G\left(\frac{\nu}{2}, 2\right)$.

The αth quantile of $\chi^2[\nu]$ is denoted by $\chi_\alpha^2[\nu]$. Accordingly, the c.d.f. of the sample variance is

$$H_{S^2}(x; \sigma^2) = \Pr\left\{\frac{\sigma^2}{n-1}\chi^2[n-1] \le x\right\}$$

$$= \Pr\left\{\chi^2[n-1] \le \frac{(n-1)x}{\sigma^2}\right\}$$

$$= \Pr\left\{G\left(\frac{n-1}{2}, 2\right) \le \frac{(n-1)x}{\sigma^2}\right\}$$

$$= G\left(\frac{(n-1)x}{2\sigma^2}; \frac{n-1}{2}, 1\right). \tag{3.10.1}$$

The probability values of the distribution of $\chi^2[\nu]$, as well as the α-quantiles, can be computed by R, MINITAB or JMP, or read from appropriate tables.

The expected value and variance of the sample variance are

$$E\{S^2\} = \frac{\sigma^2}{n-1}E\{\chi^2[n-1]\}$$

$$= \frac{\sigma^2}{n-1}E\left\{G\left(\frac{n-1}{2},2\right)\right\}$$

$$= \frac{\sigma^2}{n-1} \cdot (n-1) = \sigma^2.$$

Similarly

$$V\{S^2\} = \frac{\sigma^4}{(n-1)^2}V\{\chi^2[n-1]\}$$

$$= \frac{\sigma^4}{(n-1)^2}V\left\{G\left(\frac{n-1}{2},2\right)\right\}$$

$$= \frac{\sigma^4}{(n-1)^2} \cdot 2(n-1)$$

$$= \frac{2\sigma^4}{n-1}. \tag{3.10.2}$$

Thus, applying the Chebychev's inequality, for any given $\epsilon > 0$,

$$\Pr\{|S^2 - \sigma^2| > \epsilon\} < \frac{2\sigma^4}{(n-1)\epsilon^2}.$$

Hence, S^2 converges in probability to σ^2. Moreover,

$$\lim_{n\to\infty} \Pr\left\{\frac{(S^2 - \sigma^2)}{\sigma^2\sqrt{2}}\sqrt{n-1} \leq z\right\} = \Phi(z). \tag{3.10.3}$$

That is, the distribution of S^2 can be approximated by the normal distributions in large samples.

3.10.2   The "Student" t-statistic

We have seen that

$$Z_n = \frac{\sqrt{n}(\overline{X}_n - \mu)}{\sigma}$$

has a $N(0, 1)$ distribution. As we will see in Chapter 6, when σ is unknown, we test hypotheses concerning μ by the statistic

$$t = \frac{\sqrt{n}(\overline{X}_n - \mu_0)}{S},$$

where S is the sample standard deviation. If $X_1, \ldots, X_n$ are i.i.d. like $N(\mu_0, \sigma^2)$, then the distribution of t is called the **Student t-distribution with $\nu = n - 1$ degrees of freedom**. The corresponding random variable is denoted by $t[\nu]$.

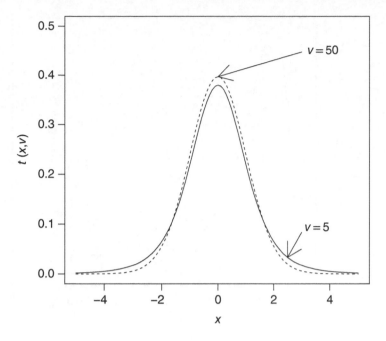

Figure 3.19 *Density functions of t[v], v = 5, 50.*

The p.d.f. of $t[v]$ is symmetric about 0 (see Figure 3.19). Thus,

$$E\{t[v]\} = 0, \text{ for } v \geq 2 \tag{3.10.4}$$

and

$$V\{t[v]\} = \frac{v}{v-2}, \quad v \geq 3. \tag{3.10.5}$$

The α-quantile of $t[v]$ is denoted by $t_\alpha[v]$. It can be read from a table, or determined by R, JMP or MINITAB.

3.10.3   Distribution of the variance ratio

$$F = \frac{S_1^2 \sigma_2^2}{S_2^2 \sigma_1^2}.$$

Consider now two independent samples of size n_1 and n_2, respectively, which have been taken from normal populations having variances σ_1^2 and σ_2^2. Let

$$S_1^2 = \frac{1}{n_1 - 1} \sum_{i=1}^{n_1} (X_{1i} - \overline{X}_1)^2$$

and

$$S_2^2 = \frac{1}{n_2 - 1} \sum_{i=1}^{n_2} (X_{2i} - \overline{X}_2)^2$$

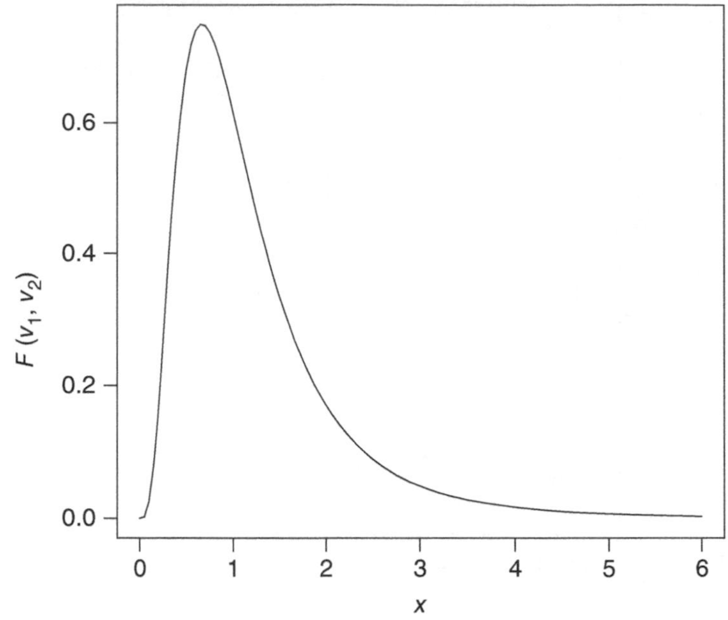

Figure 3.20   *Density function of $F(v_1, v_2)$.*

be the variances of the two samples, where $\overline{X}_1$ and $\overline{X}_2$ are the corresponding sample means. The F-ratio has a distribution denoted by $F[v_1, v_2]$, with $v_1 = n_1 - 1$ and $v_2 = n_2 - 1$. This distribution is called the F-**distribution with v_1 and v_2 degrees of freedom**. A graph of the densities of $F[v_1, v_2]$ is given in Figure 3.20.

The expected value and the variance of $F[v_1, v_2]$ are

$$E\{F[v_1, v_2]\} = v_2/(v_2 - 2), \quad v_2 > 2, \tag{3.10.6}$$

and

$$V\{F[v_1, v_2]\} = \frac{2v_2^2(v_1 + v_2 - 2)}{v_1(v_2 - 2)^2(v_2 - 4)}, \quad v_2 > 4. \tag{3.10.7}$$

The $(1 - \alpha)$th quantile of $F[v_1, v_2]$, i.e. $F_{1-\alpha}[v_1, v_2]$, can be computed by MINITAB. If we wish to obtain the αth fractile $F_\alpha[v_1, v_2]$ for values of $\alpha < 0.5$, we can apply the relationship:

$$F_{1-\alpha}[v_1, v_2] = \frac{1}{F_\alpha[v_2, v_1]}. \tag{3.10.8}$$

Thus, for example, to compute $F_{0.05}[15, 10]$, we write

$$F_{0.05}[15, 10] = 1/F_{0.95}[10, 15] = 1/2.54 = 0.3937.$$

3.11   Chapter highlights

The chapter provides the basics of probability theory and of the theory of distribution functions. The probability model for random sampling is discussed. This is fundamental for sampling procedures to be discussed in Chapter 6. Bayes theorem has important ramifications in statistical inference, as will be discussed in Chapter 4. The concepts and definitions introduced are

- Sample Space
- Elementary Events
- Operations with Events
- Disjoint Events
- Probability of Events
- Random Sampling With Replacement (RSWR)
- Random Sampling Without Replacement (RSWOR)
- Conditional Probabilities
- Independent Events
- Bayes Formula
- Prior Probability
- Posterior Probability
- Probability Distribution Function (p.d.f.)
- Discrete Random Variable
- Continuous Random Variable
- Cumulative Distribution Function
- Central Moments
- Expected Value
- Standard Deviation
- Chebychev Inequality
- Moment Generating Function
- Skewness
- Kurtosis
- Independent Trials
- pth Quantile
- Joint Distribution
- Marginal Distribution
- Conditional Distribution
- Mutual Independence
- Conditional Independence
- Law of Total Variance
- Law of Iterated Expectation
- Order Statistics
- Convergence in Probability
- Central Limit Theorem
- Law of Large Numbers

3.12   Exercises

3.1 An experiment consists of making 20 observations on the quality of chips. Each observation is recorded as G or D.
   (i) What is the sample space, S, corresponding to this experiment?
   (ii) How many elementary events in S?
   (iii) Let A_n, $n = 0, \ldots, 20$, be the event that exactly n G observations are made. Write the events A_n formally. How many elementary events belong to A_n?

3.2 An experiment consists of 10 measurements $w_1, \ldots, w_{10}$ of the weights of packages. All packages under consideration have weights between 10 and 20 lb. What is the sample space S? Let $A = \{(w_1, w_2, \ldots, w_{10}) : w_1 + w_2 = 25\}$. Let $B = \{(w_1, \ldots, w_{10}) : w_1 + w_2 \leq 25\}$. Describe the events A and B graphically. Show that $A \subset B$.

3.3 Strings of 30 binary $(0, 1)$ signals are transmitted.
   (i) Describe the sample space, S.
   (ii) Let A_{10} be the event that the first 10 signals transmitted are all 1s. How many elementary events belong to A_{10}.
   (iii) Let B_{10} be the event that exactly 10 signals, out of 30 transmitted, are 1s. How many elementary events belong to B_{10}? Does $A_{10} \subset B_{10}$?

3.4 Prove DeMorgan laws
   (i) $(A \cup B)^c = A^c \cap B^c$.
   (ii) $(A \cap B)^c = A^c \cup B^c$.

3.5 Consider Exercise [3.1] Show that the events $A_0, A_1, \ldots, A_{20}$ are a partition of the sample space S.

3.6 Let $A_1, \ldots, A_n$ be a partition of S. Let B be an event. Show that $B = \bigcup_{i=1}^{n} A_i B$, where $A_i B = A_i \cap B$, is a union of disjoint events.

3.7 Develop a formula for the probability $\Pr\{A \cup B \cup C\}$, where A, B, C are arbitrary events.

3.8 Show that if $A_1, \ldots, A_n$ is a partition, then for any event B, $P\{B\} = \sum_{i=1}^{n} P\{A_i B\}$. [Use the result of [3.6].]

3.9 An unbiased die has the numbers $1, 2, \ldots, 6$ written on its faces. The die is thrown twice. What is the probability that the two numbers shown on its upper face sum up to 10?

3.10 The time till failure, T, of electronic equipment is a random quantity. The event $A_t = \{T > t\}$ is assigned the probability $\Pr\{A_t\} = \exp\{-t/200\}, t \geq 0$. What is the probability of the event $B = \{150 < T < 280\}$?

3.11 A box contains 40 parts, 10 of type A, 10 of type B, 15 of type C and 5 of type D. A random sample of 8 parts is drawn without replacement. What is the probability of finding two parts of each type in the sample?

3.12 How many samples of size $n = 5$ can be drawn from a population of size $N = 100$,
   (i) with replacement?
   (ii) without replacement?

3.13 A lot of 1000 items contain $M = 900$ "good" ones, and 100 "defective" ones. A random sample of size $n = 10$ is drawn from the lot. What is the probability of observing in the sample at least eight good items,
   (i) when sampling is with replacement?
   (ii) when sampling is without replacement?

3.14 In continuation of the previous exercise, what is the probability of observing in an RSWR at least one defective item.

3.15 Consider the problem of Exercise [3.10]. What is the conditional probability $\Pr\{T > 300 \mid T > 200\}$.

3.16 A point (X, Y) is chosen at random within the unit square, i.e.

$$S = \{(x, y) : 0 \le x, y \le 1\}.$$

Any set A contained in S having area given by

$$\text{Area}\{A\} = \iint_A dx\, dy$$

is an event, whose probability is the area of A. Define the events

$$B = \left\{(x, y) : x > \frac{1}{2}\right\}$$
$$C = \{(x, y) : x^2 + y^2 \le 1\}$$
$$D = \{(x, y) : (x + y) \le 1\}.$$

(i) Compute the conditional probability $\Pr\{D \mid B\}$.
(ii) Compute the conditional probability $\Pr\{C \mid D\}$.

3.17 Show that if A and B are independent events, then A^c and B^c are also independent events.

3.18 Show that if A and B are disjoint events, then A and B are dependent events.

3.19 Show that if A and B are independent events, then

$$\Pr\{A \cup B\} = \Pr\{A\}(1 - \Pr\{B\}) + \Pr\{B\}$$
$$= \Pr\{A\} + \Pr\{B\}(1 - \Pr\{A\}).$$

3.20 A machine which tests whether a part is defective, D, or good, G, may err. The probabilities of errors are given by

$$\Pr\{A \mid G\} = 0.95,$$
$$\Pr\{A \mid D\} = 0.10,$$

where A is the event "the part is considered G after testing." If $\Pr\{G\} = 0.99$, what is the probability of D given A?

Additional problems in combinatorial and geometric probabilities

3.21 Assuming 365 days in a year, if there are 10 people in a party, what is the probability that their birthdays fall on different days? Show that if there are more than 22 people in the party, the probability is greater than 1/2 that at least 2 will have birthdays on the same day.

3.22 A number is constructed at random by choosing 10 digits from $\{0, \ldots, 9\}$ with replacement. We allow the digit 0 at any position. What is the probability that the number does not contain three specific digits?

3.23 A caller remembers all the seven digits of a telephone number, but is uncertain about the order of the last four. He keeps dialing the last four digits at random, without repeating the same number, until he reaches the right number. What is the probability that he will dial at least 10 wrong numbers?

3.24 One hundred lottery tickets are sold. There are 4 prizes and 10 consolation prizes. If you buy five tickets, what is the probability that you win:
  (i) one prize?
  (ii) a prize and a consolation prize?
  (iii) Something?

3.25 Ten PCBs are in a bin, two of these are defectives. The boards are chosen at random, one by one, without replacement. What is the probability that exactly five good boards will be found between the drawing of the first and second defective PCB?

3.26 A random sample of 11 integers is drawn without replacement from the set $\{1, 2, \ldots, 20\}$. What is the probability that the sample median, Me, is equal to the integer k? $6 \le k \le 15$.

3.27 A stick is broken at random into three pieces. What is the probability that these pieces can be the sides of a triangle?

3.28 A particle is moving at a uniform speed on a circle of unit radius and is released at a random point on the circumference. Draw a line segment of length $2h$ ($h < 1$) centered at a point A of distance $a > 1$ from the center of the circle, O. Moreover the line segment is perpendicular to the line connecting O with A. What is the probability that the particle will hit the line segment? [The particle flies along a straight line tangential to the cirlce.]

3.29 A block of 100 bits is transmitted over a binary channel, with probability $p = 10^{-3}$ of bit error. Errors occur independently. Find the probability that the block contains at least three errors.

3.30 A coin is tossed repeatedly until 2 "heads" occur. What is the probability that 4 tosses are required.

3.31 Consider the sample space S of all sequences of 10 binary numbers (0–1 signals). Define on this sample space two random variables and derive their probability distribution function, assuming the model that all sequences are equally probable.

3.32 The number of blemishes on a ceramic plate is a discrete random variable. Assume the probability model, with p.d.f.

$$p(x) = e^{-5}\frac{5^x}{x!}, \quad x = 0, 1, \ldots$$

  (i) Show that $\sum_{x=0}^{\infty} p(x) = 1$
  (ii) What is the probability of at most 1 blemish on a plate?
  (iii) What is the probability of no more than seven blemishes on a plate?

3.33 Consider a distribution function of a mixed type with c.d.f.

$$F_x(x) = \begin{cases} 0, & \text{if } x < -1 \\ 0.3 + 0.2(x+1), & \text{if } -1 \le x < 0 \\ 0.7 + 0.3x, & \text{if } 0 \le x < 1 \\ 1, & \text{if } 1 \le x. \end{cases}$$

  (i) What is $\Pr\{X = -1\}$?

    (ii) What is $\Pr\{-0.5 < X < 0\}$?

    (iii) What is $\Pr\{0 \le X < 0.75\}$?

    (iv) What is $\Pr\{X = 1\}$?

    (v) Compute the expected value, $E\{X\}$ and variance, $V\{X\}$.

3.34 A random variable has the Rayleigh distribution, with c.d.f.

$$F(x) = \begin{cases} 0, & x < 0 \\ 1 - e^{-x^2/2\sigma^2}, & x \ge 0 \end{cases}$$

where σ^2 is a positive parameter. Find the expected value $E\{X\}$.

3.35 A random variable X has a discrete distribution over the integers $\{1, 2, \ldots, N\}$ with equal probabilities. Find $E\{X\}$ and $V\{X\}$.

3.36 A random variable has expectation $\mu = 10$ and standard deviation $\sigma = 0.5$. Use Chebychev's inequality to find a lower bound to the probability

$$\Pr\{8 < X < 12\}.$$

3.37 Consider the random variable X with c.d.f.

$$F(x) = \frac{1}{2} + \frac{1}{\pi}\tan^{-1}(x), -\infty < x < \infty.$$

Find the 0.25th, 0.50th, and 0.75th quantiles of this distribution.

3.38 Show that the central moments μ_l^* relate to the moments μ_l around the origin, by the formula

$$\mu_l^* = \sum_{j=0}^{l-2}(-1)^j \binom{l}{j} \mu_{l-j}\mu_1^j + (-1)^{l-1}(l-1)\mu_1^l.$$

3.39 Find the expected value μ_1 and the second moment μ_2 of the random variable whose c.d.f. is given in Exercise 3.2.3.

3.40 A random variable X has a continuous uniform distribution over the interval (a, b), i.e.,

$$f(x) = \begin{cases} \dfrac{1}{b-a}, & \text{if } a \le x \le b \\ 0, & \text{otherwise.} \end{cases}$$

Find the m.g.f. of X. Find the mean and variance by differentiating the m.g.f.

3.41 Consider the moment generating function, m.g.f. of the exponential distribution, i.e.,

$$M(t) = \frac{\lambda}{\lambda - t}, \quad t < \lambda.$$

    (i) Find the first four moments of the distribution, by differentiating $M(t)$.

    (ii) Convert the moments to central moments.

    (iii) What is the index of kurtosis β_4?

3.42 Using R, MINITAB, or JMP prepare a table of the p.d.f. and c.d.f. of the binomial distribution $B(20, .17)$.

3.43 What are the first quantile, Q_1, median, Me, and third quantile, Q_3, of $B(20, .17)$?

3.44 Compute the mean $E\{X\}$ and standard deviation, σ, of $B(45, .35)$.

3.45 A PCB is populated by 50 chips which are randomly chosen from a lot. The probability that an individual chip is nondefective is p. What should be the value of p so that no defective chip is installed on the board is $\gamma = 0.99$? [The answer to this question shows why the industry standards are so stringent.]

3.46 Let $b(j; n, p)$ be the p.d.f. of the binomial distribution. Show that as $n \to \infty$, $p \to 0$ so that $np \to \lambda, 0 < \lambda < \infty$, then

$$\lim_{\substack{n \to \infty \\ p \to 0 \\ np \to \lambda}} b(j; n, p) = e^{-\lambda} \frac{\lambda^j}{j!}, \quad j = 0, 1, \ldots.$$

3.47 Use the result of the previous exercise to find the probability that a block of 1000 bits, in a binary communication channel, will have less than 4 errors, when the probability of a bit error is $p = 10^{-3}$.

3.48 Compute $E\{X\}$ and $V\{X\}$ of the hypergeometric distribution $H(500, 350, 20)$.

3.49 A lot of size $N = 500$ items contains $M = 5$ defective ones. A random sample of size $n = 50$ is drawn from the lot without replacement (RSWOR). What is the probability of observing more than one defective item in the sample?

3.50 Consider Example 3.23. What is the probability that the lot will be rectified if $M = 10$ and $n = 20$?

3.51 Use the m.g.f. to compute the third and fourth central moments of the Poisson distribution $P(10)$. What is the index of skewness and kurtosis of this distribution?

3.52 The number of blemishes on ceramic plates has a Poisson distribution with mean $\lambda = 1.5$. What is the probability of observing more than two blemishes on a plate?

3.53 The error rate of an insertion machine is 380 PPM (per 10^6 parts inserted). What is the probability of observing more than six insertion errors in 2 hours of operation, when the insertion rate is 4000 parts per hour?

3.54 In continuation of the previous Exercise, let N be the number of parts inserted until an error occurs. What is the distribution of N? Compute the expected value and the standard deviation of N.

3.55 What are Q_1, Me, and Q_3 of the negative binomial N.B. (p, k) with $p = 0.01$ and $k = 3$?

3.56 Derive the m.g.f. of N.B. (p, k).

3.57 Differentiate the m.g.f. of the geometric distribution, i.e.,

$$M(t) = \frac{pe^t}{(1 - e^t(1 - p))}, \quad t < -\log(1 - p),$$

to obtain its first four moments and derive then the indices of skewness and kurtosis.

3.58 The proportion of defective RAM chips is $p = 0.002$. You have to install 50 chips on a board. Each chip is tested before its installation. How many chips should you order so that with probability greater than $\gamma = 0.95$ you will have at least 50 good chips to install?

3.59 The random variable X assumes the values $\{1, 2, \ldots\}$ with probabilities of a geometric distribution, with parameter p, $0 < p < 1$. Prove the "memoryless" property of the geometric distribution, namely

$$P[X > n + m \mid X > m] = P[X > n],$$

for all $n, m = 1, 2, \ldots.$

3.60 Let X be a random variable having a continuous c.d.f. $F(x)$. Let $Y = F(X)$. Show that Y has a uniform distribution on $(0, 1)$. Conversely, if U has a uniform distribution on $(0, 1)$, then $X = F^{-1}(U)$ has the c.d.f. $F(x)$.

3.61 Compute the expected value and the standard deviation of a uniform distribution $U(10, 50)$.

3.62 Show that if U is uniform on $(0, 1)$, then $X = -\log(U)$ has an exponential distribution $E(1)$.

3.63 Use R, MINITAB or JMP to compute the probabilities, for $N(100, 15)$, of
 (i) $92 < X < 108$;
 (ii) $X > 105$;
 (iii) $2X + 5 < 200$.

3.64 The.9-quantile of $N(\mu, \sigma)$ is 15 and its .99-quantile is 20. Find the mean μ and standard deviation σ.

3.65 A communication channel accepts an arbitrary voltage input v and outputs a voltage $v + E$, where $E \sim N(0, 1)$. The channel is used to transmit binary information as follows:
 to transmit 0, input $-v$
 to transmit 1, input v
 The receiver decides a 0 if the voltage Y is negative, and 1 otherwise. What should be the value of v so that the receiver's probability of bit error is $\alpha = 0.01$?

3.66 Aluminum pins manufactured for an aviation industry have a random diameter, whose distribution is (approximately) normal with mean of $\mu = 10$ [mm] and standard deviation $\sigma = 0.02$ [mm]. Holes are automatically drilled on aluminum plates, with diameters having a normal distribution with mean μ_d [mm] and $\sigma = 0.02$ [mm]. What should be the value of μ_d so that the probability that a pin will not enter a hole (too wide) is $\alpha = 0.01$?

3.67 Let $X_1, \ldots, X_n$ be a random sample (i.i.d.) from a normal distribution $N(\mu, \sigma^2)$. Find the expected value and variance of $Y = \sum_{i=1}^{n} iX_i$.

3.68 Concrete cubes have compressive strength with log-normal distribution LN(5, 1). Find the probability that the compressive strength X of a random concrete cube will be greater than 300 [kg/cm^2].

3.69 Using the m.g.f. of $N(\mu, \sigma)$, derive the expected value and variance of LN(μ, σ). [Recall that $X \sim e^{N(\mu,\sigma)}$.]

3.70 What are Q_1, Me, and Q_3 of $E(\beta)$?

3.71 Show that if the life length of a chip is exponential $E(\beta)$, then only 36.7% of the chips will function longer than the mean time till failure β.

3.72 Show that the m.g.f. of $E(\beta)$ is $M(t) = (1 - \beta t)^{-1}$, for $t < \frac{1}{\beta}$.

3.73 Let X_1, X_2, X_3 be independent random variables having an identical exponential distribution $E(\beta)$. Compute $\Pr\{X_1 + X_2 + X_3 \geq 3\beta\}$.

3.74 Establish the formula

$$G(t; k, \frac{1}{\lambda}) = 1 - e^{-\lambda t} \sum_{j=0}^{k-1} \frac{(\lambda t)^j}{j!},$$

by integrating in parts the p.d.f. of

$$G\left(k; \frac{1}{\lambda}\right).$$

3.75 Use **R** to compute $\Gamma(1.17)$, $\Gamma\left(\frac{1}{2}\right)$, $\Gamma\left(\frac{3}{2}\right)$.

3.76 Using m.g.f., show that the sum of k independent exponential random variables, $E(\beta)$, has the gamma distribution $G(k, \beta)$.

3.77 What is the expected value and variance of the Weibull distribution $W(2, 3.5)$?

3.78 The time till failure (days) of an electronic equipment has the Weibull distribution $W(1.5, 500)$. What is the probability that the failure time will not be before 600 days?

3.79 Compute the expected value and standard deviation of a random variable having the beta distribution $B\left(\frac{1}{2}, \frac{3}{2}\right)$.

3.80 Show that the index of kurtosis of $B(v, v)$ is $\beta_2 = \frac{3(1+2v)}{3+2v}$.

3.81 The joint p.d.f. of two random variables (X, Y) is

$$f(x, y) = \begin{cases} \frac{1}{2}, & \text{if } (x, y) \in S \\ 0, & \text{otherwise}, \end{cases}$$

where S is a square of area 2, whose vertices are $(1, 0)$, $(0, 1)$, $(-1, 0)$, $(0, -1)$.
 (i) Find the marginal p.d.f. of X and of Y.
 (ii) Find $E\{X\}$, $E\{Y\}$, $V\{X\}$, $V\{Y\}$.

3.82 Let (X, Y) have a joint p.d.f.

$$f(x, y) = \begin{cases} \frac{1}{y} \exp\left\{-y - \frac{x}{y}\right\}, & \text{if } 0 < x, y < \infty \\ 0, & \text{otherwise}. \end{cases}$$

Find $\text{COV}(X, Y)$ and the coefficient of correlation ρ_{XY}.

3.83 Show that the random variables (X, Y) whose joint distribution is defined in Example 3.27 are dependent. Find $\text{COV}(X, Y)$.

3.84 Find the correlation coefficient of N and J of Example 3.31.

3.85 Let X and Y be independent random variables, $X \sim G(2, 100)$ and $W(1.5, 500)$. Find the variance of XY.

3.86 Consider the trinomial distribution of Example 3.33.
 (i) What is the probability that during one hour of operation there will be no more than 20 errors.
 (ii) What is the conditional distribution of wrong components, given that there are 15 misinsertions in a given hour of operation?
 (iii) Approximating the conditional distribution of (ii) by a Poisson distribution, compute the conditional probability of no more than 15 wrong components?

3.87 In continuation of Example 3.34, compute the correlation between J_1 and J_2.

3.88 In a bivariate normal distribution, the conditional variance of Y given X is 150 and the variance of Y is 200. What is the correlation ρ_{XY}?

3.89 $n = 10$ electronic devices start to operate at the same time. The times till failure of these devices are independent random variables having an identical $E(100)$ distribution.
 (i) What is the expected value of the first failure?
 (ii) What is the expected value of the last failure?

3.90 A factory has $n = 10$ machines of a certain type. At each given day, the probability is $p = 0.95$ that a machine will be working. Let J denote the number of machines that work on a given day. The time it takes to produce an item on a given machine is $E(10)$, i.e., exponentially distributed with mean $\mu = 10$ [minutes]. The machines operate independently of each other. Let $X_{(1)}$ denote the minimal time for the first item to be produced. Determine

(i) $P[J = k, X_{(1)} \leq x], k = 1, 2, \ldots$

(ii) $P[X_{(1)} \leq x \mid J \geq 1]$.

(iii) Notice that when $J = 0$ no machine is working. The probability of this event is $(0.05)^{10}$.

3.91 Let $X_1, X_2, \ldots, X_{11}$ be a random sample of exponentially distributed random variables with p.d.f. $f(x) = \lambda e^{-\lambda x}, x \geq 0$.

(i) What is the p.d.f. of the median Me $= X_{(6)}$?

(ii) What is the expected value of Me?

3.92 Let X and Y be independent random variables having an $E(\beta)$ distribution. Let $T = X + Y$ and $W = X - Y$. Compute the variance of $T + \frac{1}{2}W$.

3.93 Let X and Y be independent random variables having a common variance σ^2. What is the covariance $\text{COV}(X, X + Y)$?

3.94 Let (X, Y) have a bivariate normal distribution. What is the variance of $\alpha X + \beta Y$?

3.95 Let X have a normal distribution $N(\mu, \sigma)$. Let $\Phi(z)$ be the standard normal c.d.f. Verify that $E\{\Phi(X)\} = P\{U < X\}$, where U is independent of X and $U \sim N(0, 1)$. Show that

$$E\{\Phi(X)\} = \Phi\left(\frac{\eta}{\sqrt{1 + \sigma^2}}\right).$$

3.96 Let X have a normal distribution $N(\mu, \sigma)$. Show that

$$E\{\Phi^2(X)\} = \Phi_2\left(\frac{\mu}{\sqrt{1 + \sigma^2}}, \frac{\mu}{\sqrt{1 + \sigma^2}}; \frac{\sigma^2}{1 + \sigma^2}\right).$$

3.97 X and Y are independent random variables having Poisson distributions, with means $\lambda_1 = 5$ and $\lambda_2 = 7$, respectively. Compute the probability that $X + Y$ is greater than 15.

3.98 Let X_1 and X_2 be independent random variables having continuous distributions with p.d.f. $f_1(x)$ and $f_2(x)$, respectively. Let $Y = X_1 + X_2$. Show that the p.d.f. of Y is

$$g(y) = \int_{-\infty}^{\infty} f_1(x) f_2(y - x) dx.$$

[This integral transform is called the convolution of $f_1(x)$ with $f_2(x)$. The convolution operation is denoted by $f_1 * f_2$.]

3.99 Let X_1 and X_2 be independent random variables having the uniform distributions on $(0, 1)$. Apply the convolution operation to find the p.d.f. of $Y = X_1 + X_2$.

3.100 Let X_1 and X_2 be independent random variables having a common exponential distribution $E(1)$. Determine the p.d.f. of $U = X_1 - X_2$. [The distribution of U is called bi-exponential or Laplace and its p.d.f. is $f(u) = \frac{1}{2}e^{-|u|}$.]

3.101 Apply the central limit theorem to approximate $P\{X_1 + \cdots + X_{20} \leq 50\}$, where $X_1, \ldots, X_{20}$ are independent random variables having a common mean $\mu = 2$ and a common standard deviation $\sigma = 10$.

3.102 Let X have a binomial distribution $B(200, .15)$. Find the normal approximation to $\Pr\{25 < X < 35\}$.

3.103 Let X have a Poisson distribution with mean $\lambda = 200$. Find, approximately, $\Pr\{190 < X < 210\}$.

3.104 $X_1, X_2, \ldots, X_{200}$ are 200 independent random variables have a common beta distribution $Be(3, 5)$. Approximate the probability $\Pr\{|\overline{X}_{200} - 0.375| < 0.2282\}$, where

$$\overline{X}_n = \frac{1}{n}\sum_{i=1}^{n} X_i, \quad n = 200.$$

3.105 Use R, MINITAB, or JMP to compute the 0.95-quantiles of $t[10]$, $t[15]$, $t[20]$.

3.106 Use R, MINITAB or JMP to compute the 0.95-quantiles of $F[10, 30]$, $F[15, 30]$, $F[20, 30]$.

3.107 Show that, for each $0 < \alpha < 1$, $t_{1-\alpha/2}^2[n] = F_{1-\alpha}[1, n]$.

3.108 Verify the relationship

$$F_{1-\alpha}[\nu_1, \nu_2] = \frac{1}{F_\alpha[\nu_2, \nu_1]}, \quad 0 < \alpha < 1,$$

$\nu_1, \nu_2 = 1, 2, \ldots$.

3.109 Verify the formula

$$V\{t[\nu]\} = \frac{\nu}{\nu - 2}, \quad \nu > 2.$$

3.110 Find the expected value and variance of $F[3, 10]$.

4

Statistical Inference and Bootstrapping

In this chapter, we introduce basic concepts and methods of statistical inference. The focus is on estimating the parameters of statistical distributions and of testing hypotheses about them. Problems of testing whether or not certain distributions fit observed data are considered too. We begin with some basic problems of estimation theory.

A statistical population is represented by the distribution function(s) of the observable random variable(s) associated with its elements. The actual distributions representing the population under consideration are generally unspecified or only partially specified. Based on some theoretical considerations, and/or practical experience, we often assume that a distribution belongs to a particular family such as normal, Poisson, Weibull. Such assumptions are called the **statistical model**. If the model assumes a specific distribution with known parameters, there is no need to estimate the parameters. We may, however, use sample data to test whether the hypothesis concerning the specific distribution in the model is valid. This is a "**goodness of fit**" testing problem. If the model assumes only the family to which the distribution belongs, while the specific values of the parameters are unknown, the problem is that of estimating the unknown parameters. This chapter presents the basic principles and methods of statistical estimation and testing hypotheses for infinite population models.

4.1 Sampling characteristics of estimators

The means and the variances of random samples vary randomly around the true values of the parameters. In practice, we usually take one sample of data, then, construct a single estimate for each population parameter. To illustrate the concept of error in estimation, consider what happens if we take many samples from the same population. The collection of estimates (one from each sample) can itself be thought of as a sample taken from a hypothetical population of all possible estimates. The distribution of all possible estimates is called the **sampling distribution**. The sampling distributions of the estimates may be of a different type than the distribution of the original observations. In Figures 4.1 and 4.2, we present the frequency distributions of $\bar{X}_{10}$

Modern Industrial Statistics: With Applications in R, MINITAB and JMP, Third Edition.
Ron S. Kenett and Shelemyahu Zacks.
© 2021 John Wiley & Sons, Ltd. Published 2021 by John Wiley & Sons, Ltd.

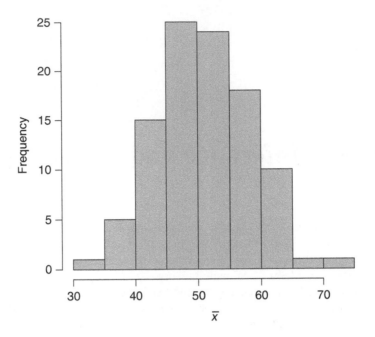

Figure 4.1   *Histogram of 100 sample means.*

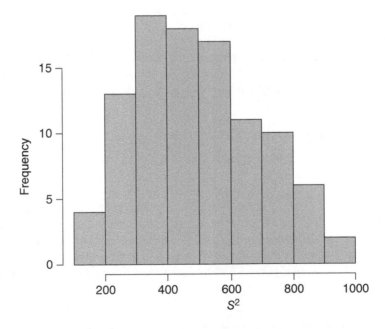

Figure 4.2   *Histogram of 100 sample variances.*

and of S_{10}^2 for 100 random samples of size $n = 10$, drawn from the uniform distribution over the integers $\{1, \ldots, 100\}$.

We see in Figure 4.1 that the frequency distribution of sample means does not resemble a uniform distribution but seems to be close to normal. Moreover, the spread of the sample means is from 35 to 72, rather than the original spread from 1 to 100. We have discussed in Chapter 3 the central limit theorem (CLT) which states that **when the sample size is large the sampling distribution of the sample mean of a simple random sample, $\bar{X}_n$, for any population having a finite positive variance σ^2, is approximately normal with mean**

$$E\{\bar{X}\} = \mu \tag{4.1.1}$$

and variance

$$V\{\bar{X}_n\} = \frac{\sigma^2}{n}. \tag{4.1.2}$$

Notice that

$$\lim_{n \to \infty} V\{\bar{X}_n\} = 0.$$

This means that the precision of the sample mean, as an estimator of the population mean μ, grows with the sample size.

Generally, if a function of the sample values $X_1, \ldots, X_n$, $\hat{\theta}(X_1, \ldots, X_n)$, is an estimator of a parameter θ of a distribution, then $\hat{\theta}_n$ is called an **unbiased estimator** if

$$E\{\hat{\theta}_n\} = \theta \quad \text{for all } \theta. \tag{4.1.3}$$

Furthermore, $\hat{\theta}_n$ is called a **consistent estimator** of θ, if for any $\epsilon > 0$, $\lim_{n \to \infty} \Pr\{|\hat{\theta}_n - \theta| > \epsilon\} = 0$. Applying the Chebyshev inequality, we see that a sufficient condition for consistency is that $\lim_{n \to \infty} V\{\hat{\theta}_n\} = 0$. The sample mean is generally a consistent estimator. The standard deviation of the sampling distribution of $\hat{\theta}_n$ is called the **standard error** of $\hat{\theta}_n$, i.e., SE $\{\hat{\theta}_n\} = (V\{\hat{\theta}_n\})^{1/2}$.

4.2 Some methods of point estimation

Consider a statistical model, which specifies the family F of the possible distributions of the observed random variable. The family F is called a **parametric family** if the distributions in F are of the same functional type and differ only by the values of their parameters. For example, the family of all exponential distributions $E(\beta)$, when $0 < \beta < \infty$, is a parametric family. In this case, we can write

$$\mathcal{F} = \{E(\beta) : 0 < \beta < \infty\}.$$

Another example of a parametric family is $\mathcal{F} = \{N(\mu, \sigma); -\infty < \mu < \infty, 0 < \sigma < \infty\}$, which is the family of all normal distributions. The range Θ of the parameter(s) θ, is called the **parameter space**. Thus, a parametric statistical model specifies the parametric family $\mathcal{F}$. This specification gives both the functional form of the distribution and its parameter(s) space Θ.

We observe a random sample from the infinite population, which consists of the values of independent and identically distributed (i.i.d.) random variables $X_1, X_2, \ldots, X_n$, whose common distribution $F(x; \theta)$ is an element of $\mathcal{F}$.

A function of the observable random variables is called a **statistic**. A statistic cannot depend on unknown parameters. A statistic is thus a random variable, whose value can be determined from the sample values $(X_1, \ldots, X_n)$. In particular, a statistic $\hat{\theta}(X_1, \ldots, X_n)$, which yields values in the parameter space problem is called a **point estimator** of θ. If the distributions in F depend on several parameters, we have to determine point estimators for each parameter, or for a function of the parameters. For example, the p-th quantile of a normal distribution is $\xi_p = \mu + z_p\sigma$, where μ and σ are the parameters and $z_p = \Phi^{-1}(p)$. This is a function of two parameters. An important problem in quality control is to estimate such quantiles. In this section, we discuss a few methods of deriving point estimators.

4.2.1   Moment equation estimators

If $X_1, X_2, \ldots, X_n$ are i.i.d. random variables (a random sample), then the sample l-th moment $(l = 1, 2, \ldots)$ is

$$M_l = \frac{1}{n}\sum_{i=1}^{n} X_i^l. \tag{4.2.1}$$

The law of large numbers (strong) says that if $E\{|X|^l\} < \infty$, then M_l converges with probability one to the population l-th moment $\mu_l(F)$. Accordingly, we know that if the sample size n is large, then, with probability close to 1, M_l is close to $\mu_l(F)$. The method of moments, for parametric models, equates M_l to μ_l, which is a function of θ, and solves for θ. Generally, if $F(x; \theta)$ depends on k parameters $\theta_1, \theta_2, \ldots, \theta_k$, then we set up k equations

$$M_1 = \mu_1(\theta_1, \ldots, \theta_k),$$
$$M_2 = \mu_2(\theta_1, \ldots, \theta_k),$$
$$\vdots$$
$$M_k = \mu_k(\theta_1, \ldots, \theta_k), \tag{4.2.2}$$

and solve for $\theta_1, \ldots, \theta_k$. The solutions are functions of the sample statistics $M_1, \ldots, M_k$, and are therefore estimators. This method does not always yield simple or good estimators. We give now a few examples in which the estimators obtained by this method are reasonable.

Example 4.1.   Consider the family F of Poisson distributions, i.e.,

$$F = \{P(x; \theta); 0 < \theta < \infty\}.$$

The parameter space is $\Theta = (0, \infty)$. The distributions depend on one parameter, and

$$\mu_1(\theta) = E_\theta\{X\} = \theta.$$

Thus, the method of moments yields the estimator

$$\hat{\theta}_n = \bar{X}_n.$$

This is an unbiased estimator with $V\{\hat{\theta}_n\} = \dfrac{\theta}{n}$.

Example 4.2.   Consider a random sample of $X_1, X_2, \ldots, X_n$ from a log normal distribution $LN(\mu, \sigma)$. The distributions depend on $k = 2$ parameters.
We have seen that

$$\mu_1(\mu, \sigma^2) = \exp\{\mu + \sigma^2/2),$$

$$\mu_2(\mu, \sigma^2) = \exp\{2\mu + \sigma^2\}(e^{\sigma^2} - 1).$$

Thus, let $\theta_1 = \mu$, $\theta_2 = \sigma^2$ and set the equations

$$\exp\{\theta_1 + \theta_2/2\} = M_1$$

$$\exp\{2\theta_1 + \theta_2\}(e^{\theta_2} - 1) = M_2.$$

The solutions $\hat{\theta}_1$ and $\hat{\theta}_2$ of this system of equations is

$$\hat{\theta}_1 = \log M_1 - \frac{1}{2} \log \left(1 + \frac{M_2}{M_1^2} \right),$$

and

$$\hat{\theta}_2 = \log \left(1 + \frac{M_2}{M_1^2} \right).$$

The estimators obtained are **biased**, but we can show that they are **consistent**. Simple formulae for $V\{\hat{\theta}_1\}$, $V\{\hat{\theta}_2\}$ and $\text{cov}(\hat{\theta}_1, \hat{\theta}_2)$ do not exist. We can derive large sample approximations to these characteristics, or approximate them by a method of resampling, called bootstrapping, which is discussed later.

4.2.2   The method of least squares

If $\mu = E\{X\}$, then the method of least squares, chooses the estimator $\hat{\mu}$, which minimizes

$$Q(\mu) = \sum_{i=1}^{n} (X_i - \mu)^2. \tag{4.2.3}$$

It is immediate to show that the **least squares estimator** (LSE) is the sample mean, i.e.,

$$\hat{\mu} = \bar{X}_n.$$

Indeed, write

$$Q(\mu) = \sum_{i=1}^{n} (X_i - \bar{X}_n + \bar{X}_n - \mu)^2$$

$$= \sum_{i=1}^{n} (X_i - \bar{X}_n)^2 + n(\bar{X}_n - \mu)^2.$$

Thus, $Q(\hat{\mu}) \geq Q(\bar{X}_n)$ for all μ and $Q(\hat{\mu})$ is minimized only if $\hat{\mu} = \bar{X}_n$. This estimator is in a sense nonparametric. It is unbiased, and consistent. Indeed,

$$V\{\hat{\mu}\} = \frac{\sigma^2}{n},$$

provided that $\sigma^2 < \infty$.

The LSE is more interesting in the case of linear regression (see Chapter 5).

In the simple linear regression case, we have n independent random variables $Y_1, \ldots, Y_n$, with equal variances, σ^2, but expected values which depend linearly on known regressors (predictors) $x_1, \ldots, x_n$. That is,

$$E\{Y_i\} = \beta_0 + \beta_1 x_i, \quad i = 1, \ldots, n. \tag{4.2.4}$$

The least squares estimators of the regression coefficients β_0 and β_1, are the values which minimize

$$Q(\beta_0, \beta_1) = \sum_{i=1}^{n} (Y_i - \beta_0 - \beta_1 x_i)^2. \tag{4.2.5}$$

These LSEs are

$$\hat{\beta}_0 = \bar{Y}_n - \hat{\beta}_1 \bar{x}_n, \tag{4.2.6}$$

and

$$\hat{\beta}_1 = \frac{\sum_{i=1}^{n} Y_i (x_i - \bar{x}_n)}{\sum_{i=1}^{n} (x_i - \bar{x}_n)^2}, \tag{4.2.7}$$

where $\bar{x}_n$ and $\bar{Y}_n$ are the sample means of the x's and the Y's, respectively. Thus, $\hat{\beta}_0$ and $\hat{\beta}_1$ are linear combinations of the Y's, with known coefficients. From the results of Section 3.8

$$E\{\hat{\beta}_1\} = \sum_{i=1}^{n} \frac{(x_i - \bar{x}_n)}{SS_x} E\{Y_i\}$$

$$= \sum_{i=1}^{n} \frac{(x_i - \bar{x}_n)}{SS_x} (\beta_0 + \beta_1 x_i)$$

$$= \beta_0 \sum_{i=1}^{n} \frac{(x_i - \bar{x}_n)}{SS_x} + \beta_1 \sum_{i=1}^{n} \frac{(x_i - \bar{x}_n) x_i}{SS_x},$$

where $SS_x = \sum_{i=1}^{n} (x_i - \bar{x}_n)^2$. Furthermore,

$$\sum_{i=1}^{n} \frac{x_i - \bar{x}_n}{SS_x} = 0$$

and

$$\sum_{i=1}^{n} \frac{(x_i - \bar{x}_n) x_i}{SS_x} = 1.$$

Hence, $E\{\hat{\beta}_1\} = \beta_1$. Also,

$$E\{\hat{\beta}_0\} = E\{\bar{Y}_n\} - \bar{x}_n E\{\hat{\beta}_1\}$$

$$= (\beta_0 + \beta_1 \bar{x}_n) - \beta_1 \bar{x}_n$$

$$= \beta_0.$$

Thus, $\hat{\beta}_0$ and $\hat{\beta}_1$ are both **unbiased**. The variances of these LSE are given by

$$V\{\hat{\beta}_1\} = \frac{\sigma^2}{SS_x},$$

$$V\{\hat{\beta}_0\} = \frac{\sigma^2}{n} + \frac{\sigma^2 \bar{x}_n^2}{SS_x}, \qquad (4.2.8)$$

and

$$\mathrm{cov}(\hat{\beta}_0, \hat{\beta}_1) = -\frac{\sigma^2 \bar{x}_n}{SS_x}. \qquad (4.2.9)$$

Thus, $\hat{\beta}_0$ and $\hat{\beta}_1$ are **not** independent. A hint for deriving these formulae is given in Exercise 4.7. The correlation between $\hat{\beta}_0$ and $\hat{\beta}_1$ is

$$\rho = -\frac{\bar{x}_n}{\left(\frac{1}{n}\sum_{i=1}^{n} x_i^2\right)^{1/2}}. \qquad (4.2.10)$$

4.2.3   Maximum likelihood estimators

Let $X_1, X_2, \ldots, X_n$ be i.i.d. random variables having a common distribution belonging to a parametric family $\mathcal{F}$. Let $f(x; \theta)$ be the probability density function (p.d.f.) of X, $\theta \in \Theta$. This is either a density function or a probability distribution function of a discrete random variable. Since $X_1, \ldots, X_n$ are independent, their joint p.d.f. is

$$f(x_1, \ldots, x_n; \theta) = \prod_{i=1}^{n} f(x_i; \theta).$$

The **likelihood function** of θ over Θ is defined as

$$L(\theta; x_1, \ldots, x_n) = \prod_{i=1}^{n} f(x_i; \theta). \qquad (4.2.11)$$

The likelihood of θ is thus the probability in the discrete case, or the joint density in the continuous case, of the observed sample values under θ. In the likelihood function $L(\theta; x_1, \ldots, x_n)$, the sample values $(x_1, \ldots, x_n)$ are playing the role of parameters. A **maximum likelihood estimator** (MLE) of θ is a point in the parameter space, $\hat{\theta}_n$, for which $L(\theta; X_1, \ldots, X_n)$ is maximized. The notion of maximum is taken in a general sense. For example, the function

$$f(x; \lambda) = \begin{cases} \lambda e^{-\lambda x}, & x \geq 0 \\ 0, & x < 0 \end{cases}$$

as a function of λ, $0 < \lambda < \infty$, attains a maximum at $\lambda = \frac{1}{x}$.

On the other hand, the function

$$f(x; \theta) = \begin{cases} \dfrac{1}{\theta}, & 0 \leq x \leq \theta \\ 0, & \text{otherwise} \end{cases}$$

as a function of θ, over $(0, \infty)$ attains a lowest upper bound (supremum) at $\theta = x$, which is $\frac{1}{x}$. We say that it is maximized at $\theta = x$. Notice that it is equal to zero for $\theta < x$. We give a few examples.

Example 4.3. Suppose that $X_1, X_2, \ldots, X_n$ is a random sample from a normal distribution. Then, the likelihood function of (μ, σ^2) is

$$L(\mu, \sigma^2; X_1, \ldots, X_n) = \frac{1}{(2\pi)^{n/2}\sigma^n} \exp\left\{ -\frac{1}{2\sigma^2} \sum_{i=1}^{n} (X_i - \mu)^2 \right\}$$

$$= \frac{1}{(2\pi)^{n/2}(\sigma^2)^{n/2}} \exp\left\{ -\frac{1}{2\sigma^2} \sum_{i=1}^{n} (X_i - \bar{X}_n)^2 - \frac{n}{2\sigma^2}(\bar{X}_n - \mu)^2 \right\}.$$

Notice that the likelihood function of (μ, σ^2) depends on the sample variables only through the statistics $(\bar{X}_n, Q_n)$, where $Q_n = \sum_{i=1}^{n} (X_i - \bar{X}_n)^2$. These statistics are called the **likelihood statistics** or **sufficient statistics**. To maximize the likelihood, we can maximize the log-likelihood

$$l(\mu, \sigma^2; \bar{X}_n, Q_n) = -\frac{n}{2}\log(2\pi) - \frac{n}{2}\log(\sigma^2) - \frac{Q_n}{2\sigma^2} - \frac{n(\bar{X}_n - \mu)^2}{2\sigma^2}.$$

With respect to μ, we maximize by $\hat{\mu}_n = \bar{X}_n$. With respect to σ^2, differentiate

$$l(\hat{\mu}_n, \sigma^2; \bar{X}_n, Q_n) = -\frac{n}{2}\log(2\pi) - \frac{n}{2}\log(\sigma^2) - \frac{Q_n}{2\sigma^2}.$$

This is

$$\frac{\partial}{\partial \sigma^2} \log(\hat{\mu}, \sigma^2; \bar{X}_n, Q_n) = -\frac{n}{2\sigma^2} + \frac{Q_n}{2\sigma^4}.$$

Equating the derivative to zero and solving yields the MLE

$$\hat{\sigma}_n^2 = \frac{Q_n}{n}.$$

Thus, the MLEs are $\hat{\mu}_n = \bar{X}_n$ and

$$\hat{\sigma}_n^2 = \frac{n-1}{n} S_n^2.$$

$\hat{\sigma}_n^2$ is biased, but the bias goes to zero as $n \to \infty$.

Example 4.4. Let X have a negative binomial distribution $NB(k, p)$. Suppose that k is known, and $0 < p < 1$. The likelihood function of p is

$$L(p; X, k) = \binom{X + k - 1}{k - 1} p^k (1 - p)^X.$$

Thus, the log-likelihood is

$$l(p; X, k) = \log\binom{X + k - 1}{k - 1} + k \log p + X \log(1 - p).$$

The MLE of p is

$$\hat{p} = \frac{k}{X + k}.$$

We can show that $\hat{p}$ has a positive bias, i.e., $E\{\hat{p}\} > p$. For large values of k, the bias is approximately

$$\text{Bias}(\hat{p}; k) = E\{\hat{p}; k\} - p$$
$$\cong \frac{3p(1 - p)}{2k}, \qquad \text{large } k.$$

The variance of $\hat{p}$ for large k is approximately $V\{\hat{p}; k\} \cong \frac{p^2(1-p)}{k}$.

4.3   Comparison of sample estimates

4.3.1   Basic concepts

Statistical hypotheses are statements concerning the parameters, or some characteristics, of the distribution representing a certain random variable (or variables) in a population. For example, consider a manufacturing process. The parameter of interest may be the proportion, p, of non-conforming items. If $p \leq p_0$, the process is considered to be acceptable. If $p > p_0$, the process should be corrected.

Suppose that 20 items are randomly selected from the process and inspected. Let X be the number of nonconforming items in the sample. Then X has a binomial distribution $B(20, p)$. On the basis of the observed value of X, we have to decide whether the process should be stopped for adjustment. In the statistical formulation of the problem, we are testing the hypothesis

$$H_0 : p \leq p_0,$$

against the hypothesis

$$H_1 : p > p_0.$$

The hypothesis H_0 is called the **null hypothesis**, while H_1 is called the **alternative hypothesis**. Only when the data provides significant evidence that the null hypothesis is wrong do we reject it in favor of the alternative. It may not be justifiable to disrupt a production process unless we have ample evidence that the proportion of nonconforming items is too high. It is important to distinguish between **statistical significance** and **practical or technological significance**. The statistical level of significance is the probability of rejecting H_0 when it is true. If we reject H_0 at a low level of significance, the probability of committing an error is small, and we are confident that our conclusion is correct. Rejecting H_0 might not be technologically significant, if the true value of p is not greater than $p_0 + \delta$, where δ is some acceptable level of indifference. If $p_0 < p < p_0 + \delta$, H_1 is true, but there is no technological significance to the difference $p - p_0$.

To construct a statistical test procedure based on a **test statistic**, X, consider first all possible values that could be observed. In our example, X can assume the values $0, 1, 2, \ldots, 20$. Determine a **critical region** or **rejection region**, so that, whenever the observed value of X belongs to this region the null hypothesis H_0 is rejected. For example if we were testing $H_0 : P \leq 0.10$ against $H_1 : P > 0.10$, we might reject H_0 if $X > 4$. The complement of this region, $X \leq 3$, is called the **acceptance region**.

There are two possible errors that can be committed. If the true proportion of nonconforming items, for example, were only 0.05 (unknown to us) and our sample happened to produce four items, we would incorrectly decide to reject H_0 and shut down the process that was performing acceptably. This is called a **type I error**. On the other hand, if the true proportion were 0.15 and only 3 nonconforming items were found in the sample, we would incorrectly allow the process to continue with more than 10% defectives (**a type II error**).

We denote the probability of committing a type I error by $\alpha(p)$, for $p \leq p_0$, and the probability of committing a type II error by $\beta(p)$, for $p > p_0$.

In most problems, the critical region is constructed in such a way that the probability of committing a type I error will not exceed a preassigned value called the **significance level** of the test. Let α denote the significance level. In our example, the significance level is

$$\alpha = \Pr\{X \geq 4; p = 0.1\} = 1 - B(3; 20, 0.1) = 0.133.$$

Notice that the significance level is computed with $p = 0.10$, which is the largest p value for which the null hypothesis is true.

To further evaluate the test procedure, we would like to know the probability of accepting the null hypothesis for various values of p. Such a function is called the **operating characteristic function** and is denoted by $OC(p)$. The graph of $OC(p)$ versus p is called the **OC curve**. Ideally, we would like $OC(p) = 1$ whenever $H_0 : p \leq p_0$ is true, and $OC(p) = 0$ when $H_1 : p > p_0$ is true. This, however, cannot be obtained when the decision is based on a random sample of items.

In our example, we can compute the OC function as

$$OC(p) = \Pr\{X \leq 3; p\} = B(3; 20, p).$$

From Table 4.1 we find that

$$OC(0.10) = 0.8670$$

$$OC(0.15) = 0.6477$$

$$OC(0.20) = 0.4114$$

$$OC(0.25) = 0.2252.$$

Table 4.1   *The binomial c.d.f. $B(x; n, p)$, for $n = 20$, $p = 0.10(0.05)0.25$*

x	$p = 0.10$	$p = 0.15$	$p = 0.20$	$p = 0.25$
0	0.1216	0.0388	0.0115	0.0032
1	0.3917	0.1756	0.0692	0.0243
2	0.6769	0.4049	0.2061	0.0913
3	0.8670	0.6477	0.4114	0.2252
4	0.9568	0.8298	0.6296	0.4148
5	0.9887	0.9327	0.8042	0.6172
6	0.9976	0.9781	0.9133	0.7858
7	0.9996	0.9941	0.9679	0.8982
8	0.9999	0.9987	0.9900	0.9591
9	1.0000	0.9998	0.9974	0.9861
10	1.0000	1.0000	0.9994	0.9961

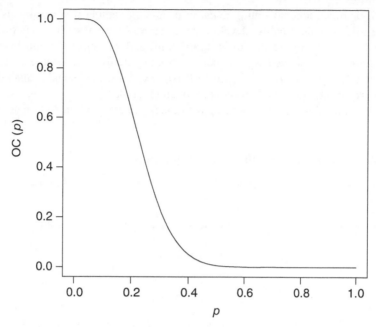

Figure 4.3 *The OC curve for testing $H_0 : p \leq 0.1$ against $H_1 : p > 0.1$ with a Sample of Size $n = 20$ and Rejection Region $X \geq 4$.*

Notice that the significance level α is the maximum probability of rejecting H_0 when it is true. Accordingly, $OC(p_0) = 1 - \alpha$. The OC curve for this example is shown in Figure 4.3.

We see that as p grows the value of $OC(p)$ decreases, since the probability of observing at least four nonconforming items out of 20 is growing with p.

Suppose that the significance level of the test is decreased in order to reduce the probability of incorrectly interfering with a good process. We may choose the critical region to be $X \geq 5$. The OC function for this new critical region is

$$OC(0.10) = 0.9568,$$

$$OC(0.15) = 0.8298,$$

$$OC(0.20) = 0.6296,$$

$$OC(0.25) = 0.4148.$$

The new significance level is $\alpha = 1 - OC(0.1) = 0.0432$. Notice that, although we reduced the risk of committing a type I error, we increased the risk of committing a type II error. Only with a larger sample size we can reduce simultaneously the risks of both type I and type II errors.

Instead of the OC function, one may consider the **power function**, for evaluating the sensitivity of a test procedure. The power function, denoted by $\psi(p)$, is the probability of **rejecting** the null hypothesis when the alternative is true. Thus, $\psi(p) = 1 - OC(p)$.

Finally, we consider an alternative method of performing a test. Rather than specifying in advance the desired significance level, say $\alpha = 0.05$, we can compute the probability of observing X_0 or more nonconforming items in a random sample if $p = p_0$. This probability is called the **attained significance level** or the *P*-**value** of the test. If the *P*-value is small, say ≤ 0.05, we consider the results to be **significant**, and we reject the null hypothesis. For example, suppose we observed $X_0 = 6$ nonconforming items in a sample of size 20. The *P*-value is $\Pr\{X \geq 6; p = 0.10\} = 1 - B(5; 20, 0.10) = 0.0113$. This small probability suggests that we could reject H_0 in favor of H_1 without much of a risk.

The term *P*-value should not be confused with the parameter p of the binomial distribution.

4.3.2   Some common one-sample tests of hypotheses

4.3.2.1   *The Z-test: testing the mean of a normal distribution, σ^2 known*

One-sided test
The hypothesis for a one-sided test on the mean of a normal distribution are

$$H_0 : \mu \leq \mu_0,$$

against

$$H_1 : \mu > \mu_0,$$

where μ_0 is a specified value. Given a sample $X_1, \ldots, X_n$, we first compute the sample mean $\bar{X}_n$. Since large values of $\bar{X}_n$, relative to μ_0, would indicate that H_0 is possibly not true, the critical region should be of the form $\bar{X} \geq C$, where C is chosen so that the probability of committing a type I error is equal to α. (In many problems, we use $\alpha = 0.01$ or 0.05 depending on the consequences of a type I error.) For convenience, we use a modified form of the test statistic, given by the Z-statistic

$$Z = \sqrt{n}(\bar{X}_n - \mu_0)/\sigma. \tag{4.3.1}$$

The critical region, in terms of Z, is given by

$$\{Z : Z \geq z_{1-\alpha}\},$$

where $z_{1-\alpha}$ is the $1 - \alpha$ quantile of the standard normal distribution. This critical region is equivalent to the region

$$\{\bar{X}_n : \bar{X}_n \geq \mu_0 + z_{1-\alpha}\sigma/\sqrt{n}\}.$$

These regions are illustrated in Figure 4.4.

The operating characteristic function of this test is given by

$$\text{OC}(\mu) = \Phi(z_{1-\alpha} - \delta\sqrt{n}), \tag{4.3.2}$$

where

$$\delta = (\mu - \mu_0)/\sigma. \tag{4.3.3}$$

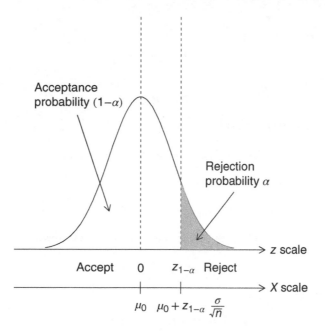

Figure 4.4   *Critical regions for the one-sided Z-test.*

Example 4.5.   Suppose we are testing the hypothesis $H_0 : \mu \le 5$, against $H_1 : \mu > 5$, with a sample of size $n = 100$ from a normal distribution with known standard deviation $\sigma = 0.2$. With a significance level of size $\alpha = 0.05$, we reject H_0 if

$$Z \ge z_{0.95} = 1.645.$$

The values of the OC function are computed in Table 4.2. In this table, $z = z_{1-\alpha} - \delta\sqrt{n}$ and $OC(\mu) = \Phi(z)$.

Table 4.2   *OC values in the normal case*

μ	$\delta\sqrt{n}$	z	$OC(\mu)$
5.	0	1.645	0.9500
5.01	0.5	1.145	0.8739
5.02	1.0	0.645	0.7405
5.03	1.5	0.045	0.5179
5.04	2.0	−0.355	0.3613
5.05	3.0	−1.355	0.0877

If the null hypothesis is $H_0 : \mu \ge \mu_0$ against the alternative $H_1 : \mu < \mu_0$, we reverse the direction of the test and reject H_0 if $Z \le -z_{1-\alpha}$.

Two-sided test

The two-sided test has the form

$$H_0 : \mu = \mu_0$$

against

$$H_1 : \mu \neq \mu_0.$$

The corresponding critical region is given by

$$\{Z : Z \geq z_{1-\alpha/2}\} \cup \{Z : Z \leq -z_{1-\alpha/2}\}.$$

The operating characteristic function is

$$OC(\mu) = \Phi(z_{1-\alpha/2} + \delta\sqrt{n}) - \Phi(-z_{1-\alpha/2} - \delta\sqrt{n}). \tag{4.3.4}$$

The *P*-value of the two-sided test can be determined in the following manner. First compute

$$|Z_0| = \sqrt{n}|\bar{X}_n - \mu_0|/\sigma,$$

and then compute the *P*-value

$$P = \Pr\{Z \geq |Z_0|\} + P\{Z \leq -|Z_0|\}$$

$$= 2(1 - \Phi(|Z_0|)). \tag{4.3.5}$$

4.3.2.2   *The t-test: testing the mean of a normal distribution, σ^2 unknown*

In this case, we replace σ in the above Z-test with the sample standard deviation, S, and $z_{1-\alpha}$ (or $z_{1-\alpha/2}$) with $t_{1-\alpha}[n-1]$ (or $t_{1-\alpha/2}[n-1]$). Thus, the critical region for the two-sided test becomes

$$\{t : |t| \geq t_{1-\alpha/2}[n-1]\},$$

where

$$t = (\bar{X}_n - \mu_0)\sqrt{n}/S. \tag{4.3.6}$$

The operating characteristic function of the one-sided test is given approximately by

$$OC(\mu) \cong 1 - \Phi\left(\frac{\delta\sqrt{n} - t_{1-\alpha}[n-1](1 - 1/8(n-1))}{(1 + t_{1-\alpha}^2[n-1]/2(n-1))^{1/2}}\right), \tag{4.3.7}$$

where $\delta = |\mu - \mu_0|/\sigma$. (This is a good approximation to the exact formula, which is based on the complicated noncentral *t*-distribution.)

In Table 4.3, we present some numerical comparisons of the **power** of the one-sided test for the cases of σ^2 known and σ^2 unknown, when $n = 20$ and $\alpha = 0.05$. Notice that when σ is unknown, the power of the test is somewhat smaller than when it is known.

Table 4.3 *Power functions of Z- and t-tests*

δ	σ Known	σ Unknown
0.0	0.050	0.050
0.1	0.116	0.111
0.2	0.226	0.214
0.3	0.381	0.359
0.4	0.557	0.527
0.5	0.723	0.691

Example 4.6. The cooling system of a large computer consists of metal plates that are attached together, so as to create an internal cavity, allowing for the circulation of special purpose cooling liquids. The metal plates are attached with steel pins that are designed to measure 0.5 mm in diameter. Experience with the process of manufacturing similar steel pins, has shown that the diameters of the pins are normally distributed, with mean μ and standard deviation σ. The process is aimed at maintaining a mean of $\mu_0 = 0.5$ [mm]. For controlling this process, we want to test $H_0 : \mu = 0.5$ against $H_1 : \mu \neq 0.5$. If we have prior information that the process standard deviation is constant at $\sigma = 0.02$, we can use the Z-test to test the above hypotheses. If we apply a significance level of $\alpha = 0.05$, then we will reject H_0 if $|Z| \geq z_{1-\alpha/2} = 1.96$.

Suppose that the following data were observed:

$$.53, 0.54, 0.48, 0.50, 0.50, 0.49, 0.52.$$

The sample size is $n = 7$ with a sample mean of $\bar{X} = 0.509$. Therefore,

$$Z = |0.509 - 0.5| \sqrt{7}/0.02 = 1.191.$$

Since this value of Z does not exceed the critical value of 1.96, do not reject the null hypothesis.

If there is no prior information about σ, use the sample standard deviation S and perform a t-test, and reject H_0 if $|t| > t_{1-\alpha/2}[6]$. In the present example $S = 0.022$, and $t = 1.082$. Since $|t| < t_{0.975}[6] = 2.447$, we reach the same conclusion.

4.3.2.3 *The chi-squared test: testing the variance of a normal distribution*

Consider a one-sided test of the hypothesis:

$$H_0 : \sigma^2 \leq \sigma_0^2,$$

against

$$H_1 : \sigma^2 > \sigma_0^2.$$

The test statistic corresponding to this hypothesis is

$$Q^2 = (n-1)S^2/\sigma^2, \tag{4.3.8}$$

with a critical region

$$\{Q^2 : Q^2 \geq \chi^2_{1-\alpha}[n-1]\}.$$

The operating characteristic function for this test is given by

$$OC(\sigma^2) = \Pr\{\chi^2[n-1] \leq \frac{\sigma_0^2}{\sigma^2}\chi^2_{1-\alpha}[n-1]\}, \tag{4.3.9}$$

where $\chi^2[n-1]$ is a Chi-Squared random variable with $n-1$ degrees of freedom.
    Continuing the Example 4.6 let us test the hypothesis

$$H_0 : \sigma^2 \leq 0.0004,$$

against

$$H_1 : \sigma^2 > 0.0004.$$

Since the sample standard deviation is $S = 0.022$, we find

$$Q^2 = (7-1)(0.022)^2/0.0004 = 7.26.$$

H_0 is rejected at level $\alpha = 0.05$ if

$$Q^2 \geq \chi^2_{0.95}[6] = 12.59.$$

Since $Q^2 < \chi^2_{0.95}[6]$, H_0 is not rejected. Whenever n is odd, that is $n = 2m+1$ $(m = 0, 1, \ldots)$, the cumulative distribution function (c.d.f.) of $\chi^2[n-1]$ can be computed according to the formula:

$$\Pr\{\chi^2[2m] \leq x\} = 1 - P\left(m-1; \frac{x}{2}\right),$$

where $P(a; \lambda)$ is the c.d.f. of the Poisson distribution with mean λ. For example, if $n = 21, m = 10$ and $\chi^2_{0.95}[20] = 31.41$. Thus, the value of the OC function at $\sigma^2 = 1.5\,\sigma_0^2$ is

$$OC(1.5\sigma_0^2) = \Pr\left\{\chi^2[20] \leq \frac{31.41}{1.5}\right\}$$

$$= 1 - P(9; 10.47) = 1 - 0.4007$$

$$= 0.5993.$$

If n is even, i.e., $n = 2m$, we can compute the OC values for $n = 2m-1$ and for $n = 2m+1$ and take the average of these OC values. This will yield a good approximation.
    The power function of the test is obtained by subtracting the OC function from 1.
    In Table 4.4, we present a few numerical values of the **power function** for $n = 20, 30, 40$ and for $\alpha = 0.05$. Here, we have let $\rho = \sigma^2/\sigma_0^2$ and have used the values $\chi^2_{0.95}[19] = 30.1$, $\chi^2_{0.95}[29] = 42.6$, and $\chi^2_{0.95}[39] = 54.6$.
    As illustrated in Table 4.4, the power function changes more rapidly as n grows.

Table 4.4   *Power of the χ^2-Test, $\alpha = 0.05$, $\rho = \sigma^2/\sigma_0^2$*

		n	
ρ	20	30	40
1.00	0.050	0.050	0.050
1.25	0.193	0.236	0.279
1.50	0.391	0.497	0.589
1.75	0.576	0.712	0.809
2.00	0.719	0.848	0.920

4.3.2.4   *Testing hypotheses about the success probability, p, in binomial trials*

Consider one-sided tests, for which

The null hypothesis is $H_0 : p \leq p_0$.
The alternative hypothesis is $H_1 : p > p_0$.
The critical region is $\{X : X > c_\alpha(n, p_0)\}$,

where X is the number of successes among n trials and $c_\alpha(n, p_0)$ is the first value of k for which the Binomial c.d.f., $B(k; n, p_0)$, exceeds $1 - \alpha$.

The operating characteristic function:

$$OC(p) = B(c_\alpha(n, p_0); n, p). \tag{4.3.10}$$

Notice that $c_\alpha(n, p_0) = B^{-1}(1 - \alpha; n, p_0)$ is the $(1 - \alpha)$ quantile of the binomial distribution $B(n, p_0)$. In order to determine $c(n, p_0)$, one can use R or MINITAB commands which, for $\alpha = 0.05$, $n = 20$, and $p_0 = 0.20$ are

```
> qbinom(p=0.95, size=20, prob=0.2)
```

```
MTB> INVCDF 0.95;
SUBC> BINOM 20 0.2.
```

Table 4.5 is an output for the binomial distribution with $n = 20$ and $p = 0.2$.

The smallest value of k for which $B(k; 20, 0.2) = \Pr\{X \leq k\} \geq 0.95$ is 7. Thus, we set $c_{0.05}(20, 0.20) = 7$. H_0 is rejected whenever $X > 7$. The level of significance of this test is actually 0.032, which is due to the discrete nature of the binomial distribution. The OC function of the test for $n = 20$ can be easily determined from the corresponding distribution of $B(20, p)$. For example, the $B(n, p)$ distribution for $n = 20$ and $p = 0.25$ is presented in Table 4.6.

We see that $B(7; 20, 0.25) = 0.8982$. Hence, the probability of accepting H_0 when $p = 0.25$ is $OC(0.25) = 0.8982$.

A large sample test in the Binomial case can be based on the normal approximation to the Binomial distribution. If the sample is indeed large, we can use the test statistic

$$Z = \frac{\hat{p} - p_0}{\sqrt{p_0 q_0}} \sqrt{n}, \tag{4.3.11}$$

Table 4.5   The p.d.f. and c.d.f. of $B(20, 0.2)$

Binomial Distribution: $n = 20$ $p = 0.2$

a	$\Pr(X = a)$	$\Pr(X \leq a)$
0	0.0115	0.0115
1	0.0576	0.0692
2	0.1369	0.2061
3	0.2054	0.4114
4	0.2182	0.6296
5	0.1746	0.8042
6	0.1091	0.9133
7	0.0546	0.9679
8	0.0222	0.9900
9	0.0074	0.9974
10	0.0020	0.9994
11	0.0005	0.9999
12	0.0001	1.0000

Table 4.6   The p.d.f. and c.d.f. of $B(20, 0.25)$

Binomial Distribution: $n = 20$ $p = 0.25$

a	$\Pr(X = a)$	$\Pr(X \leq a)$
0	0.0032	0.0032
1	0.0211	0.0243
2	0.0669	0.0913
3	0.1339	0.2252
4	0.1897	0.4148
5	0.2023	0.6172
6	0.1686	0.7858
7	0.1124	0.8982
8	0.0609	0.9591
9	0.0271	0.9861
10	0.0099	0.9961
11	0.0030	0.9991
12	0.0008	0.9998
13	0.0002	1.0000

with the critical region

$$\{Z : Z \geq z_{1-\alpha}\},$$

where $q_0 = 1 - p_0$. Here $\hat{p}$ is the sample proportion of successes. The operating characteristic function takes the form:

$$OC(p) = 1 - \Phi\left(\frac{(p - p_0)\sqrt{n}}{\sqrt{pq}} - z_{1-\alpha}\sqrt{\frac{p_0 q_0}{pq}}\right), \tag{4.3.12}$$

where $q = 1 - p$, and $q_0 = 1 - p_0$.

For example, suppose that $n = 450$ and the hypotheses are $H_0 : p \leq 0.1$ against $H_1 : p > 0.1$. The critical region, for $\alpha = 0.05$, is

$$\{\hat{p} : \hat{p} \geq 0.10 + 1.645\sqrt{(0.1)(0.9)/450}\} = \{\hat{p} : \hat{p} \geq 0.1233\}.$$

Thus, H_0 is rejected whenever $\hat{p} \geq 0.1233$. The OC value of this test, at $p = 0.15$ is approximately

$$OC(0.15) \cong 1 - \Phi\left(\frac{0.05\sqrt{450}}{\sqrt{(0.15)(0.85)}} - 1.645\sqrt{\frac{(0.1)(0.9)}{(0.15)(0.85)}}\right)$$

$$= 1 - \Phi(2.970 - 1.382)$$

$$= 1 - 0.944 = 0.056.$$

The corresponding value of the power function is 0.949. Notice that the power of rejecting H_0 for H_1 when $p = 0.15$ is so high because of the large sample size.

4.4 Confidence intervals

Confidence intervals for unknown parameters are intervals, determined around the sample estimates of the parameters, having the property that, whatever the true value of the parameter is, in repetitive sampling a prescribed proportion of the intervals, say $1 - \alpha$, will contain the true value of the parameter. The prescribed proportion, $1 - \alpha$, is called the **confidence level** of the interval. In Figure 4.5, we illustrate 50 simulated confidence intervals, which correspond to independent samples. All of these intervals are designed to estimate the mean of the population

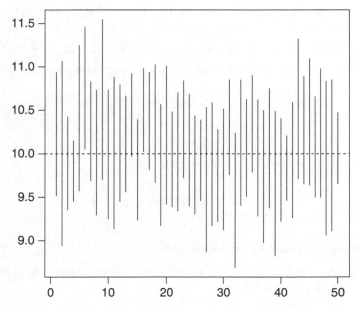

Figure 4.5 *Simulated confidence intervals for the mean of a normal distribution, samples of size n = 10 from N(10, 1).*

from which the samples were drawn. In this particular simulation, the population was normally distributed with mean $\mu = 10$. We see from the figure that 48 out of these 50 random intervals cover the true value of μ.

If the sampling distribution of the estimator $\hat{\theta}_n$ is approximately normal, one can use, as a rule of thumb, the interval estimator with limits

$$\hat{\theta}_n \pm 2 \, \mathrm{SE}\{\hat{\theta}_n\}.$$

The confidence level of such an interval will be close to 0.95 for all θ.

Generally, if one has a powerful test procedure for testing the hypothesis $H_0 : \theta = \theta_0$ versus $H_1 : \theta \neq \theta_0$, one can obtain good confidence intervals for θ by the following method:

Let $T = T(\mathbf{X})$ be a test statistic for testing $H_0 : \theta = \theta_0$. Suppose that H_0 is rejected if $T \geq \bar{K}_\alpha(\theta_0)$ or if $T \leq \underline{K}_\alpha(\theta_0)$, where α is the significance level. The interval $(\underline{K}_\alpha(\theta_0), \bar{K}_\alpha(\theta_0))$ is the acceptance region for H_0. We can now consider the family of acceptance regions $\Theta = \{(\underline{K}_\alpha(\theta), \bar{K}_\alpha(\theta)), \theta \in \Theta\}$, where Θ is the parameter space. The interval $(L_\alpha(T), U_\alpha(T))$ defined as

$$L_\alpha(T) = \inf\{\theta : T \leq \bar{K}_\alpha(\theta)\}$$

$$U_\alpha(T) = \sup\{\theta : T \geq \underline{K}_\alpha(\theta)\}, \tag{4.4.1}$$

is a confidence interval for θ at level of confidence $1 - \alpha$. Indeed, any hypothesis H_0 with $L_\alpha(T) < \theta_0 < U_\alpha(T)$ is accepted with the observed value of the test statistic. By construction, the probability of accepting such hypothesis is $1 - \alpha$. That is, if θ_0 is the true value of θ, the probability that H_0 is accepted is $(1 - \alpha)$. But H_0 is accepted if, and only if, θ_0 is covered by the interval $(L_\alpha(T), U_\alpha(T))$.

4.4.1   Confidence intervals for μ; σ known

For this case, the sample mean $\bar{X}$ is used as an estimator of μ, or as a test statistic for the hypothesis $H_0 : \mu = \mu_0$. H_0 is accepted, at level of significance α if $\bar{X} \geq \mu_0 - z_{1-\alpha/2}\frac{\sigma}{\sqrt{n}}$ or $\bar{X} \leq \mu_0 + z_{1-\alpha/2}\frac{\sigma}{\sqrt{n}}$, where $z_{1-\alpha/2} = \Phi^{-1}(1 - \alpha/2)$. Thus, $\bar{K}_\alpha(\mu) = \mu + z_{1-\alpha/2}\frac{\sigma}{\sqrt{n}}$ and $\underline{K}_\alpha(\mu) = \mu - z_{1-\alpha/2}\frac{\sigma}{\sqrt{n}}$. The limits of the confidence interval are, accordingly the roots μ of the equation

$$\bar{K}_\alpha(\mu) = \bar{X}$$

and

$$\underline{K}_\alpha(\mu) = \bar{X}.$$

These equations yield the confidence interval for μ,

$$\left(\bar{X} - z_{1-\alpha/2}\frac{\sigma}{\sqrt{n}}, \bar{X} + z_{1-\alpha/2}\frac{\sigma}{\sqrt{n}} \right). \tag{4.4.2}$$

4.4.2   Confidence intervals for μ; σ unknown

A confidence interval for μ, at level $1 - \alpha$, when σ is unknown is obtained from the corresponding t-test. The confidence interval is

$$\left(\bar{X} - t_{1-\alpha/2}[n-1]\frac{S}{\sqrt{n}}, \bar{X} + t_{1-\alpha/2}[n-1]\frac{S}{\sqrt{n}} \right), \tag{4.4.3}$$

where $\bar{X}$ and S are the sample mean and standard deviation, respectively. $t_{1-\alpha/2}[n-1]$ is the $(1-\alpha/2)$th quantile of the t-distribution with $n-1$ degrees of freedom.

4.4.3   Confidence intervals for σ^2

We have seen that, in the normal case, the hypothesis $H_0 : \sigma = \sigma_0$, is rejected at level of significance α if

$$S^2 \geq \frac{\sigma_0^2}{n-1} \chi_{1-\alpha/2}^2[n-1]$$

or

$$S^2 \leq \frac{\sigma_0^2}{n-1} \chi_{\alpha/2}^2[n-1],$$

where S^2 is the sample variance, and $\chi_{\alpha/2}^2[n-1]$ and $\chi_{1-\alpha/2}^2[n-1]$ are the $\alpha/2$-th and $(1-\alpha/2)$th quantiles of χ^2, with $(n-1)$ degrees of freedom. The corresponding confidence interval for σ^2, at confidence level $(1-\alpha)$ is

$$\left(\frac{(n-1)S^2}{\chi_{1-\alpha/2}^2[n-1]}, \frac{(n-1)S^2}{\chi_{\alpha/2}^2[n-1]} \right). \tag{4.4.4}$$

Example 4.7.   Consider a normal distribution with unknown mean μ and unknown standard deviation σ. Suppose that we draw a random sample of size $n = 16$ from this population, and the sample values are the following:

$$\begin{array}{llll} 16.16, & 9.33, & 12.96, & 11.49, \\ 12.31, & 8.93, & 6.02, & 10.66, \\ 7.75, & 15.55, & 3.58, & 11.34, \\ 11.38, & 6.53, & 9.75, & 9.47. \end{array}$$

The mean and variance of this sample are $\bar{X} = 10.20$, and $S^2 = 10.977$. The sample standard deviation is $S = 3.313$. For a confidence level of $1 - \alpha = 0.95$, we find

$$t_{0.975}[15] = 2.131,$$

$$\chi_{0.975}^2[15] = 27.50,$$

$$\chi_{0.025}^2[15] = 6.26.$$

Thus, the confidence interval for μ is $(8.435, 11.965)$. The confidence interval for σ^2 is $(5.987, 26.303)$.

4.4.4   Confidence intervals for p

Let X be the number of "success" in n independent trials, with unknown probability of "success," p. The sample proportion, $\hat{p} = X/n$, is an unbiased estimator of p. To construct a confidence interval for p, using $\hat{p}$, we must find limits $p_L(\hat{p})$ and $p_U(\hat{p})$ that satisfy

$$\Pr\{p_L(\hat{p}) < p < p_U(\hat{p})\} = 1 - \alpha.$$

The null hypothesis $H_0 : p = p_0$ is rejected if $\hat{p} \geq \bar{K}_\alpha(p_0)$ or $\hat{p} \leq \underline{K}_\alpha(p_0)$, where

$$\bar{K}_\alpha(p_0) = \frac{1}{n}B^{-1}(1 - \alpha/2; n, p_0)$$

and

$$\underline{K}_\alpha(p_0) = \frac{1}{n}B^{-1}(\alpha/2; n, p_0). \tag{4.4.5}$$

$B^{-1}(\gamma; n, p)$ is the γ-th quantile of the binomial distribution $B(n, p)$. Thus, if $X = n\hat{p}$, the upper confidence limit for p, $p_U(\hat{p})$, is the largest value of p satisfying the equation

$$B(X; n, p) \geq \alpha/2.$$

The lower confidence limit for p is the smallest value of p satisfying

$$B(X; n, p) \leq 1 - \alpha/2.$$

Exact solutions to this equation can be obtained using tables of the binomial distribution. This method of searching for the solution in binomial tables is tedious. However, from the relationship between the F-distribution, the beta distribution, and the binomial distribution, the lower and upper limits are given by the formulae:

$$p_L = \frac{X}{X + (n - X + 1)F_1} \tag{4.4.6}$$

and

$$p_U = \frac{(X + 1)F_2}{n - X + (X + 1)F_2}, \tag{4.4.7}$$

where

$$F_1 = F_{1-\alpha/2}[2(n - X + 1), 2X] \tag{4.4.8}$$

and

$$F_2 = F_{1-\alpha/2}[2(X + 1), 2(n - X)] \tag{4.4.9}$$

are the $(1 - \alpha/2)$th quantiles of the F-distribution with the indicated degrees of freedom.

Example 4.8.   Suppose that among $n = 30$ Bernoulli trials, we find $X = 8$ successes. For level of confidence $1 - \alpha = 0.95$, the confidence limits are $p_L = 0.123$ and $p_U = 0.459$. Indeed,

$$B(7; 30, 0.123) = 0.975,$$

and

$$B(8; 30, 0.459) = 0.025.$$

Moreover,

$$F_1 = F_{0.975}[46, 16] = 2.49$$

and

$$F_2 = F_{0.975}[18, 44] = 2.07.$$

Hence,

$$p_L = 8/(8 + 23(2.49)) = 0.123$$

and

$$p_U = 9(2.07)/(22 + 9(2.07)) = 0.459.$$

When the sample size n is large, we may use the normal approximation to the binomial distribution. This approximation yields the following formula for a $(1 - \alpha)$ confidence interval

$$\left(\hat{p} - z_{1-\alpha/2}\sqrt{\hat{p}\hat{q}/n}, \hat{p} + z_{1-\alpha/2}\sqrt{\hat{p}\hat{q}/n} \right), \tag{4.4.10}$$

where $\hat{q} = 1 - \hat{p}$. Applying this large sample approximation to our previous example, in which $n = 30$, we obtain the approximate 0.95-confidence interval $(0.108, 0.425)$. This interval is slightly different from the interval obtained with the exact formulae. This difference is due to the inaccuracy of the normal approximation.

It is sometimes reasonable to use only a one-sided confidence interval, for example, if $\hat{p}$ is the estimated proportion of nonconforming items in a population. Obviously, the true value of p is always greater than 0, and we may wish to determine only an upper confidence limit. In this case, we apply the formula given earlier but replace $\alpha/2$ by α. For example, in the case of $n = 30$ and $X = 8$, the upper confidence limit for p, in a one-sided confidence interval, is

$$p_U = \frac{(X + 1)F_2}{n - X + (X + 1)F_2},$$

where $F_2 = F_{1-\alpha}[2(X + 1), 2(n - X)] = F_{0.95}[18, 44] = 1.855$. Thus, the upper confidence limit of a 0.95 one-sided interval is $P_U = 0.431$. This limit is smaller than the upper limit of the two-sided interval.

4.5 Tolerance intervals

Technological specifications for a given characteristic X may require that a specified proportion of elements of a statistical population satisfy certain constraints. For example, in the production of concrete, we may have the requirement that at least 90% of all concrete cubes, of a certain size, will have a compressive strength of at least $240\,\text{kg/cm}^2$. As another example, suppose that in the production of washers, it is required that at least 99% of the washers produced will have a thickness between 0.121 and 0.129 inches. In both examples, we want to be able to determine whether or not the requirements are satisfied. If the distributions of strength and thickness were completely known, we could determine if the requirements are met without data. However, if the distributions are not completely known, we can make these determinations only with a certain level of confidence and not with certainty.

4.5.1 Tolerance intervals for the normal distributions

In order to construct tolerance intervals, we first consider what happens when the distribution of the characteristic X is completely known. Suppose for example, that the compressive strength X of the concrete cubes is such that $Y = \ln X$ has a normal distribution with mean $\mu = 5.75$

and standard deviation $\sigma = 0.2$. The proportion of concrete cubes exceeding the specification of $240 \, \text{kg/cm}^2$ is

$$\Pr\{X \geq 240\} = \Pr\{Y \geq \log 240\}$$
$$= 1 - \Phi((5.481 - 5.75)/0.2)$$
$$= \Phi(1.345) = 0.911$$

Since this probability is greater than the specified proportion of 90%, the requirement is satisfied.

We can also solve this problem by determining the compressive strength that is exceeded by 90% of the concrete cubes. Since 90% of the Y values are greater than the 0.1th fractile of the $N(5.75, 0.04)$ distribution,

$$Y_{0.1} = \mu + z_{0.1}\sigma$$
$$= 5.75 - 1.28(0.2)$$
$$= 5.494.$$

Accordingly, 90% of the compressive strength values should exceed $e^{5.494} = 243.2 \, \text{kg/cm}^2$. Once again, we see that the requirement is satisfied, since more than 90% of the cubes have strength values that exceed the specification of $240 \, \text{kg/cm}^2$. Notice that no sample values are required, since the distribution of X is known. Furthermore, we are **certain** that the requirement is met.

Consider the situation in which we have only partial information on the distribution of Y. Suppose we know that Y is normally distributed with standard deviation $\sigma = 0.2$, but the mean μ is unknown. The 0.1th fractile of the distribution, $y_{0.1} = \mu + z_{0.1}\sigma$, cannot be determined exactly. Let $Y_1, \ldots, Y_n$ be a random sample from this distribution and let $\bar{Y}_n$ represent the sample mean. From the previous section, we know that

$$L(\bar{Y}_n) = \bar{Y}_n - z_{1-\alpha}\sigma/\sqrt{n}$$

is a $1 - \alpha$ lower confidence limit for the population mean. That is

$$\Pr\{\bar{Y}_n - z_{1-\alpha}\sigma/\sqrt{n} < \mu\} = 1 - \alpha.$$

Substituting this lower bound for μ in the expression for the 0.1th fractile, we obtain a **lower tolerance limit** for 90% of the log-compressive strengths, with confidence level $1 - \alpha$. More specifically, the lower tolerance limit at level of confidence $1 - \alpha$ is

$$L_{\alpha,0.1}(\bar{Y}_n) = \bar{Y}_n - (z_{1-\alpha}/\sqrt{n} + z_{0.9})\sigma.$$

In general we say that, **with confidence level of $1 - \alpha$, the proportion of population values exceeding the lower tolerance limit is at least $1 - \beta$**. This lower tolerance limit is

$$L_{\alpha,\beta}(\bar{Y}_n) = (\bar{Y}_n - (z_{1-\alpha}/\sqrt{n} + z_{1-\beta})\sigma. \tag{4.5.1}$$

It can also be shown that the **upper tolerance limit** for a proportion $1 - \beta$ of the values, with confidence level $1 - \alpha$, is

$$U_{\alpha,\beta}(\bar{Y}_n) = \bar{Y}_n + (z_{1-\alpha}/\sqrt{n} + z_{1-\beta})\sigma \tag{4.5.2}$$

and a **tolerance interval** containing a proportion $1 - \beta$ of the values, with confidence $1 - \alpha$, is

$$(\bar{Y}_n - (z_{1-\alpha/2}/\sqrt{n} + z_{1-\beta/2})\sigma, \bar{Y}_n + (z_{1-\alpha/2}/\sqrt{n} + z_{1-\beta/2})\sigma).$$

When the standard deviation σ is unknown, we should use the sample standard deviation S to construct the tolerance limits and interval. The lower tolerance limits will be of the form $\bar{Y}_n - kS_n$, where the factor $k = k(\alpha, \beta, n)$ is determined so that with confidence level $1 - \alpha$ we can state that a proportion $1 - \beta$ of the population values will exceed this limit. The corresponding upper limit is given by $\bar{Y}_n + kS_n$ and the tolerance interval is given by

$$(\bar{Y}_n - k'S_n, \bar{Y}_n + k'S_n).$$

The "two-sided" factor $k' = k'(\alpha, \beta, n)$ is determined so that the interval will contain a proportion $1 - \beta$ of the population with confidence $1 - \alpha$. Approximate solutions, for large values of n, are given by

$$k(\alpha, \beta, n) \doteq t(\alpha, \beta, n) \qquad (4.5.3)$$

and

$$k'(\alpha, \beta, n) \doteq t(\alpha/2, \beta/2, n), \qquad (4.5.4)$$

where

$$t(a, b, n) = \frac{z_{1-b}}{1 - z_{1-a}^2/2n} + \frac{z_{1-a}(1 + z_b^2/2 - z_{1-a}^2/2n)^{1/2}}{\sqrt{n}(1 - z_{1-a}^2/2n)}. \qquad (4.5.5)$$

Example 4.9. The following data represent a sample of 20 compressive strength measurements (kg/cm^2) of concrete cubes at age of 7 days.

349.09	308.88
238.45	196.20
385.59	318.99
330.00	257.63
388.63	299.04
348.43	321.47
339.85	297.10
348.20	218.23
361.45	286.23
357.33	316.69

Applying the transformation $Y = \ln X$, we find that $\bar{Y}_{20} = 5.732$ and $S_{20} = 0.184$. To obtain a lower tolerance limit for 90% of the log-compressive strengths with 95% confidence, we use the factor $k(0.05, 0.10, 20) = 2.548$. Thus, the lower tolerance limit for the transformed data is

$$\bar{Y}_{20} - kS_{20} = 5.732 - 2.548 \times 0.184 = 5.263,$$

and the corresponding lower tolerance limit for the compressive strength is

$$e^{5.263} = 193.09 \ [\text{kg/cm}^2].$$

If the tolerance limits are within the specification range, we have a satisfactory production.

4.6   Testing for normality with probability plots

It is often assumed that a sample is drawn from a population which has a normal distribution. It is, therefore, important to test the assumption of normality. We present here a simple test based on the **normal-scores** (NSCORES) of the sample values. The normal-scores corresponding to a sample $x_1, x_2, \ldots, x_n$ are obtained in the following manner. First, we let

$$r_i = \text{rank of } x_i, \quad i = 1, \ldots, n. \tag{4.6.1}$$

Here the rank of x_i is the position of x_i in a listing of the sample when it is arranged in increasing order. Thus, the rank of the smallest value is 1, that of the second smallest is 2, etc. We then let

$$p_i = (r_i - 3/8)/(n + 1/4), \quad i = 1, \ldots, n. \tag{4.6.2}$$

Then the normal-score of x_i is

$$z_i = \Phi^{-1}(p_i),$$

i.e., the p_i-th fractile of the standard normal distribution. If the sample is drawn at random from a normal distribution $N(\mu, \sigma^2)$, the relationship between the normal-scores, NSCORES, and x_i should be approximately linear. Accordingly, the correlation between $x_1, \ldots, x_n$ and their NSCORES should be close to 1 in large samples. The graphical display of the sample values versus their NSCORES is called a **Normal Q–Q Plot**.

In the following example, we provide a normal probability plotting of $n = 50$ values simulated from $N(10, 1)$, given in the previous section. If the simulation is good, and the sample is indeed generated from $N(10, 1)$ the X versus NSCORES should be scattered randomly around the line $X = 10 + $ NSCORES. We see in Figure 4.6 that this is indeed the case. Also, the correlation between the x-values and their NSCORES is 0.976.

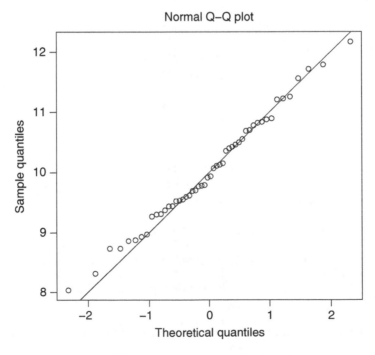

Figure 4.6   *Normal Q–Q plot of simulated values from $N(10, 1)$.*

Table 4.7   *Critical Values for the correlation between sample values and their NSCORES*

n\α	0.10	0.05	0.01
10	0.9347	0.9180	0.8804
15	0.9506	0.9383	0.9110
20	0.9600	0.9503	0.9290
30	0.9707	0.9639	0.9490
50	0.9807	0.9764	0.9664

(Adapted from Ryan et al. (1976).)

The linear regression of the x values on the NSCORES is

$$X = 10.043 + 0.953 * \text{NSCORES}.$$

We see that both the intercept and slope of the regression equation are close to the nominal values of μ and σ. Chapter 5 provides more details on linear regression, including testing statistical hypothesis on these coefficients.

In Table 4.7, we provide some critical values for testing whether the correlation between the sample values and their NSCORES is sufficiently close to 1. If the correlation is smaller than the critical value, an indication of nonnormality has been established. In the example of Figure 4.6, the correlation is $R = 0.976$. This value is almost equal to the critical value for $\alpha = 0.05$ given in the following table. The hypothesis of normality is accepted.

MINITAB provides also a graph of the Normal Probability Plot. In this graph, the probabilities corresponding to the NSCORES are plotted against x. The Normal Probability Plot of

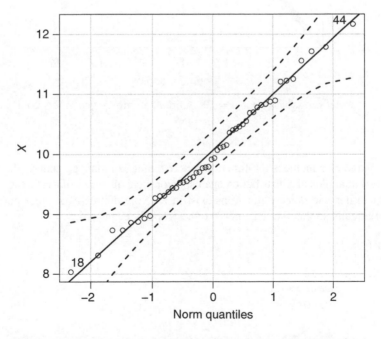

Figure 4.7   *Normal probability plot of 50 simulated N(10, 1) values.*

the above sample is given in Figure 4.7. Normal probability plot is obtained in MINITAB by the command

MTB> % NormPlot *C*1.

To demonstrate the relationship between the sample values and their normal scores, when the sample is drawn from a nonnormal distribution, consider the following two examples.

> **Example 4.10.**   Consider a sample of $n = 100$ observations from a log-normal distribution. The normal $Q-Q$ plot of this sample is shown in Figure 4.8. The correlation here is 0.73. It is apparent that the relation between the NSCORES and the sample values is not linear. We reject the hypothesis that the sample has been generated from a normal distribution.

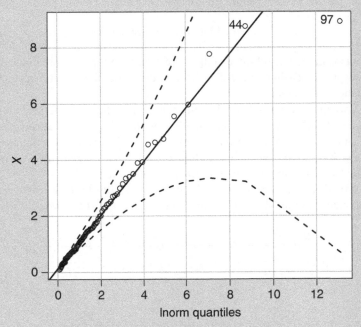

Figure 4.8   *Normal probability plot n = 100 random numbers generated from a log-normal distribution.*

In R, the package car includes the function qqPlot that can manage general distributions. It plots empirical quantiles of a distribution against theoretical quantiles of a reference distribution for which quantiles and density functions exist in R. Note that qqPlot does not standardize inputs automatically so that the scale needs to be properly evaluated.

```
> library(car)
> set.seed(123)
> X <- rlnorm(n=100,
               meanlog=2,
               sdlog=0.1)
> invisible(
```

```
qqPlot(X,
      distribution="lnorm",
      meanlog=2,
      sdlog=0.1)
)
```

Example 4.11.   We consider here a sample of $n = 100$ values, with 50 of the values generated from $N(10, 1)$ and 50 from $N(15, 1)$. Thus, the sample represents a mixture of two normal distributions. The histogram is given in Figure 4.9 and a normal probability plot in Figure 4.10. The normal probability plot is definitely not linear. Although the correlation is 0.962, the hypothesis of a normal distribution is rejected.

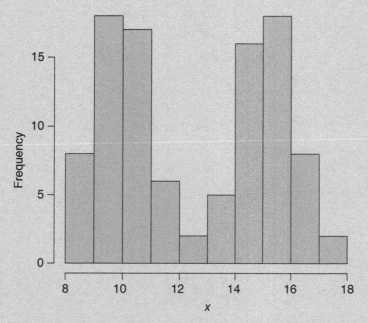

Figure 4.9   *Histogram of 100 random numbers, 50 generated from a $N(10, 1)$ and 50 from $N(15, 1)$.*

4.7   Tests of goodness of fit

4.7.1   The chi-square test (large samples)

The chi-square test is applied by comparing the observed frequency distribution of the sample to the expected one under the assumption of the model. More specifically, consider a (large) sample of size N. Let $\xi_0 < \xi_1 < \cdots < \xi_k$ be the limit points of k subintervals of the frequency distribution, and let f_i be the observed frequency in the i-th subinterval. If, according to the model, the c.d.f. is specified by the distribution function $F(x)$, then the expected frequency e_i in the i-th subinterval is

$$e_i = N(F(\xi_i) - F(\xi_{i-1})), \quad i = 1, \ldots, k.$$

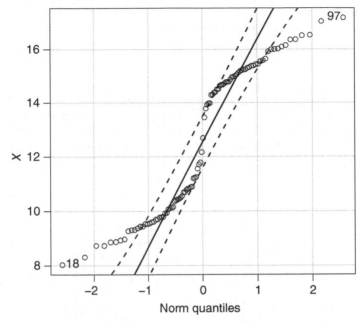

Figure 4.10   *Normal probability plot of 100 random numbers generated from a mixture of two normal distributions.*

The chi-square statistic is defined as

$$\chi^2 = \sum_{i=1}^{k} \frac{(f_i - e_i)^2}{e_i}.$$

We notice that

$$\sum_{i=1}^{k} f_i = \sum_{i=1}^{k} e_i = N,$$

and hence

$$\chi^2 = \sum_{i=1}^{k} \frac{f_i^2}{e_i} - N.$$

The value of χ^2 is distributed approximately like $\chi^2[k-1]$. Thus, if $\chi^2 \geq \chi^2_{1-\alpha}[k-1]$, the distribution $F(x)$ does not fit the observed data.

Often, the c.d.f. $F(x)$ is specified by its family, e.g., normal or Poisson, but the values of the parameters have to be estimated from the sample. In this case, we reduce the number of degrees of freedom of χ^2 by the number of estimated parameters. For example, if $F(x)$ is $N(\mu, \sigma^2)$, where both μ and σ^2 are unknown, we use $N(\bar{X}, S^2)$ and compare χ^2 to $\chi^2_{1-\alpha}[k-3]$.

Example 4.12.   In Section 4.1, we considered the sampling distribution of sample means from the uniform distribution over the integers $\{1, \ldots, 100\}$. The frequency distribution of the means of samples of size $n = 10$ is given in Figure 4.1. We test here whether the model $N(50.5, 83.325)$ fits this data.

The observed and expected frequencies (for $N = 100$) are summarized in Table 4.8.

Table 4.8 *Observed and expected frequencies of 100 sample means*

Interval	f_i	e_i
27.5–32.5	3	1.84
32.5–37.5	11	5.28
37.5–42.5	12	11.32
42.5–47.5	11	18.08
47.4–52.5	19	21.55
52.5–57.5	24	19.7
57.5–62.5	14	12.73
62.5–67.5	4	6.30
67.5–72.5	2	2.33
TOTAL	100	99.13

The sum of e_i here is 99.13 due to truncation of the tails of the normal distribution. The value of χ^2 is 12.86. The value of $\chi^2_{0.95}[8]$ is 15.5. Thus, the deviation of the observed frequency distribution from the expected one is not significant at the $\alpha = 0.05$ level.

Example 4.13.   We consider here a sample of 100 cycle times of a piston, which is described in detail in Chapter 10. We make a chi-squared test whether the distribution of cycle times is normal. The estimated values of μ and σ are $\hat{\mu} = 0.1219$ and $\hat{\sigma} = 0.0109$.
In Table 4.9, we provide the observed and expected frequencies over $k = 8$ intervals.

Table 4.9   *Observed and expected frequencies of 100 cycle times*

	Lower Limit	Upper Limit	Observed Frequency	Expected Frequency
	At or below	0.1050	7	6.1
	0.1050	0.1100	9	7.7
	0.1100	0.1150	17	12.6
	0.1150	0.1200	12	16.8
	0.1200	0.1250	18	18.1
	0.1250	0.1300	11	15.9
	0.1300	0.1350	12	11.4
Above	0.1350		14	11.4

The calculated value of χ^2 is 5.4036. We should consider the distribution of χ^2 with $k - 3 = 5$ degrees of freedom. The P value of the test is 0.37. The hypothesis of normality is not rejected.

4.7.2   The Kolmogorov–Smirnov test

The Kolmogorov–Smirnov (KS) test is a more accurate test of goodness of fit than the chi-squared test of the previous section.

Suppose that the hypothesis is that the sample comes from a specified distribution with c.d.f. $F_0(x)$. The test statistic compares the empirical distribution of the sample, $\hat{F}_n(x)$, to $F_0(x)$, and considers the maximal value, overall x values, that the distance $|\hat{F}_n(x) - F_0(x)|$ may assume. Let $x_{(1)} \leq x_{(2)} \leq \cdots \leq x_{(n)}$ be the ordered sample values. Notice that $\hat{F}_n(x_{(i)}) = \frac{i}{n}$. The KS test statistic can be computed according to the formula

$$D_n = \max_{1 \leq i \leq n} \left\{ \max \left\{ \frac{i}{n} - F_0(x_{(i)}), F_0(x_{(i)}) - \frac{i-1}{n} \right\} \right\} \tag{4.7.1}$$

We have shown earlier that $U = F(X)$ **has a uniform distribution on** $(0, 1)$.

Accordingly, if the null hypothesis is correct, $F_0(X_{(i)})$ is distributed like the i-th order statistic $U_{(i)}$ from a uniform distribution on $(0, 1)$, irrespective of the particular functional form of $F_0(x)$. The distribution of the KS test statistic, D_n, is, therefore, independent of $F_0(x)$, if the hypothesis, H, is correct. Tables of the critical values k_α and D_n are available. One can also estimate the value of k_α by the bootstrap method, discussed later.

If $F_0(x)$ is a normal distribution, i.e., $F_0(x) = \Phi\left(\frac{x-\mu}{\sigma}\right)$, and if the mean μ and the standard-deviation, σ, are unknown, one can consider the test statistic

$$D_n^* = \max_{1 \leq i \leq n} \left\{ \max \left\{ \frac{i}{n} - \Phi\left(\frac{X_{(i)} - \bar{X}_n}{S_n}\right), \Phi\left(\frac{X_{(i)} - \bar{X}_n}{S_n}\right) - \frac{i-1}{n} \right\} \right\}, \tag{4.7.2}$$

where $\bar{X}_n$ and S_n are substituted for the unknown μ and σ. The critical values k_α^* for D_n^* are given approximately by

$$k_\alpha^* = \delta_\alpha^* / \left(\sqrt{n} - 0.01 + \frac{0.85}{\sqrt{n}} \right), \tag{4.7.3}$$

where δ_α^* is given in the following table:

Table 4.10   *Some critical values δ_α^*.*

α	0.15	0.10	0.05	0.025	0.01
δ_α^*	0.775	0.819	0.895	0.995	1.035

To compute the KS statistics, in R use the function `ks.test`, in MINITAB use Stat > Basic Statistics > Normality Test and check Kolmogorov Smirnov.

For the data in Example 4.13 (file name **OTURB.csv**), we obtain $D_n^* = 0.1107$. According to Table 4.10, the critical value for $\alpha = 0.05$ is $k_{0.05}^* = 0.895/(10 - 0.01 + 0.085) = 0.089$. Thus, the hypothesis of normality for the piston cycle time data is rejected at $\alpha = 0.05$.

4.8   Bayesian decision procedures

It is often the case that optimal decision depends on unknown parameters of statistical distributions. The Bayesian decision framework provides us the tools to integrate information that one may have on the unknown parameters with the information obtained from the observed sample in such a way that the expected loss due to erroneous decisions will be minimized. In order to illustrate an industrial decision problem of such nature consider the following example.

Example 4.14.   Inventory Management. The following is the simplest inventory problem that is handled daily by organizations of all sizes worldwide. One such organization is Starbread Express that supplies bread to a large community in the Midwest. Every night, the shift manager has to decide how many loafs of bread, s, to bake for the next day consumption. Let X (a random variable) be the number of units demanded during the day. If a manufactured unit is left at the end of the day, we lose \$ c_1 on that unit. On the other hand, if a unit is demanded and is not available, due to shortage, the loss is \$ c_2. How many units, s, should be manufactured so that the total expected loss due to overproduction or to shortages will be minimized?

The loss at the end of the day is

$$L(s,X) = c_1(s-X)^+ + c_2(X-s)^+, \tag{4.8.1}$$

where $a^+ = \max(a,0)$. The loss function $L(s,X)$ is a random variable. If the p.d.f. of X is $f(x)$, $x = 0,1,\ldots$, then the **expected loss**, is a function of the quantity s, is

$$\begin{aligned}
R(s) &= c_1 \sum_{x=0}^{s} f(x)(s-x) + c_2 \sum_{x=s+1}^{\infty} f(x)(x-s) \\
&= c_2 E\{X\} - (c_1+c_2) \sum_{x=0}^{s} xf(x) \\
&\quad + s(c_1+c_2)F(s) - c_2 s,
\end{aligned} \tag{4.8.2}$$

where $F(s)$ is the c.d.f. of X, at $X = s$, and $E\{X\}$ is the expected demand.

The **optimal** value of s, s^0, is the smallest integer s for which $R(s+1) - R(s) \geq 0$. Since, for $s = 0,1,\ldots$

$$R(s+1) - R(s) = (c_1+c_2)F(s) - c_2,$$

we find that

$$s^0 = \textbf{smallest nonnegative integer } s, \textbf{ such that } F(s) \geq \frac{c_2}{c_1+c_2}. \tag{4.8.3}$$

In other words, s^0 is the $c_2/(c_1+c_2)$th quantile of $F(x)$. We have seen that the optimal decision is a function of $F(x)$. If this distribution is unknown, or only partially known, one cannot determine the optimal value s^0.

After observing a large number, N, of X values one can consider the empirical distribution, $F_N(x)$, of the demand and determine the level $S^0(F_N) = $ smallest s value such that $F_N(s) \geq \frac{c_2}{c_1+c_2}$. The question is what to do when N is small.

4.8.1   Prior and posterior distributions

We will focus attention here on parametric models. Let $f(x;\theta)$ denote the p.d.f. of some random variable X, which depends on a parameter θ. θ could be a vector of several real parameters, like in the case of a normal distribution. Let Θ denote the set of all possible parameters θ. Θ is called the **parameter space**. For example, the parameter space Θ of the family of normal distribution is the set $\Theta = \{(\mu, \sigma); -\infty < \mu < \infty, 0 < \sigma < \infty\}$. In the case of Poisson distributions,

$$\Theta = \{\lambda; 0 < \lambda < \infty\}.$$

In a Bayesian framework, we express our prior belief (based on prior information) on which θ values are plausible, by a p.d.f. on Θ, which is called the **prior** p.d.f. Let $h(\theta)$ denote the prior p.d.f. of θ. For example, suppose that X is a discrete random variable having a binomial distribution $B(n, \theta)$. n is known, but θ is unknown. The parameter space is $\Theta = \{\theta; 0 < \theta < 1\}$. Suppose we believe that θ is close to 0.8, with small dispersion around this value. In Figure 4.11, we illustrate the p.d.f. of a Beta distribution Beta(80,20), whose functional form is

$$h(\theta; 80, 20) = \frac{99!}{79!19!}\theta^{79}(1-\theta)^{19}, \quad 0 < \theta < 1.$$

If we wish, however, to give more weight to small values of θ, we can choose the Beta(8,2) as a prior density, i.e.,

$$h(\theta; 8, 2) = 72\theta^{7}(1-\theta), \quad 0 < \theta < 1$$

(see Figure 4.12).

The average p.d.f. of X, with respect to the prior p.d.f. $h(\theta)$ is called the **predictive p.d.f.** of X. This is given by

$$f_h(x) = \int_{\Theta} f(x; \theta)h(\theta) \, d\theta. \tag{4.8.4}$$

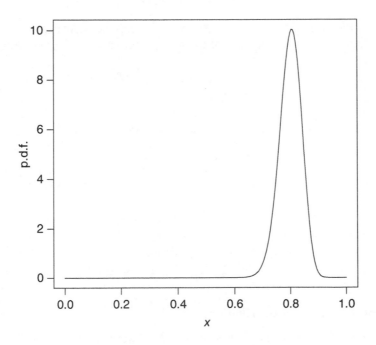

Figure 4.11 *The p.d.f. of Beta(80, 20).*

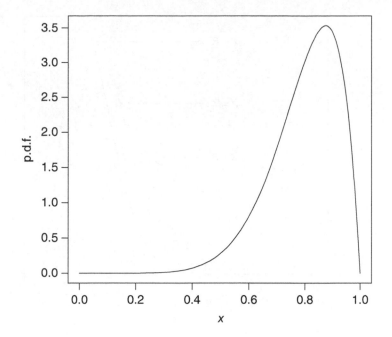

Figure 4.12   *The p.d.f. of Beta(8, 2).*

For the example above, the predictive p.d.f. is

$$f_h(x) = 72 \binom{n}{x} \int_0^1 \theta^{7+x}(1-\theta)^{n-x+1} \, d\theta$$

$$= 72 \binom{n}{x} \frac{(7+x)!(n+1-x)!}{(n+9-x)!}, \quad x = 0, 1, \ldots, n.$$

Before taking observations on X, we use the predictive p.d.f. $f_h(x)$, to predict the possible outcomes of observations on X. After observing the outcome of X, say x, we convert the prior p.d.f. to a **posterior** p.d.f., by employing **Bayes formula**. If $h(\theta \mid x)$ denotes the posterior p.d.f. of θ, given that $\{X = x\}$, Bayes formula yields

$$h(\theta \mid x) = \frac{f(x \mid \theta)h(\theta)}{f_h(x)}. \tag{4.8.5}$$

In the example above,

$$f(x \mid \theta) = \binom{n}{x} \theta^x (1-\theta)^{n-x}, \quad x = 0, 1, \ldots, n,$$

$$h(\theta) = 72\theta^7(1-\theta), \quad 0 < \theta < 1,$$

and hence,

$$h(\theta \mid x) = \frac{(n+9)!}{(7+x)!(n+1-x)!}\theta^{7+x}(1-\theta)^{n+1-x}, \quad 0 < \theta < 1.$$

This is again the p.d.f. of a Beta distribution Beta$(8+x, n-x+2)$.

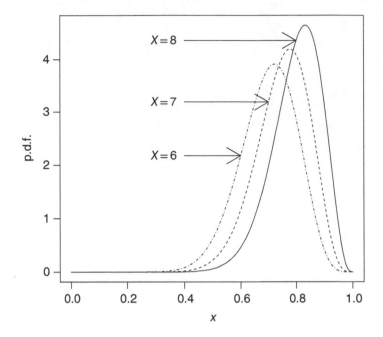

Figure 4.13   *The posterior p.d.f. of θ, n = 10, X = 6, 7, 8.*

In Figure 4.13, we present some of these posterior p.d.f. for the case of $n = 10$, $x = 6$, 7, 8. Notice that the posterior p.d.f. $h(\theta \mid x)$ is the **conditional** p.d.f. of θ, given $\{X = x\}$. If we observe a random sample of n i.i.d. random variables, and the observed values of $X_1, \ldots, X_n$ are $x_1, \ldots, x_n$, then the posterior p.d.f. of θ is

$$h(\theta \mid x_1, \ldots, x_n) = \frac{\prod\limits_{i=1}^{n} f(x_i, \theta)h(\theta)}{f_h(x_1, \ldots, x_n)}, \tag{4.8.6}$$

where

$$f_h(x_1, \ldots, x_n) = \int_\Theta \prod_{i=1}^{n} f(x_i, \theta)h(\theta)\, d\theta \tag{4.8.7}$$

is the **joint predictive** p.d.f. of $X_1, \ldots, X_n$. If the i.i.d. random variables $X_1, X_2, \ldots$ are observed sequentially (time wise), then the posterior p.d.f. of θ, given $x_1, \ldots, x_n, n \geq 2$ can be determined recursively, by the formula

$$H(\theta \mid x_1, \ldots, x_n) = \frac{f(x_n; \theta)h(\theta \mid x_1, \ldots, x_{n-1})}{\int_\Theta f(x_n; \theta')h(\theta' \mid x_1, \ldots, x_{n-1})\, d\theta'}.$$

The function

$$f_h(x_n \mid x_1, \ldots, x_{n-1}) = \int_\Theta f(x_n; \theta)h(\theta \mid x_1, \ldots, x_{n-1})\, d\theta$$

is called the **conditional predictive** p.d.f. of X_n, given $X_1 = x_1, \ldots, X_{n-1} = x_{n-1}$. Notice that

$$f_h(x_n \mid x_1, \ldots, x_{n-1}) = \frac{f_h(x_1, \ldots, x_n)}{f_h(x_1, \ldots, x_{n-1})}. \tag{4.8.8}$$

4.8.2 Bayesian testing and estimation

4.8.2.1 Bayesian testing

We discuss here the problem of testing hypotheses as a Bayesian decision problem. Suppose that we consider a null hypothesis H_0 concerning a parameter θ of the p.d.f. of X. Suppose also that the parameter space Θ is partitioned to two sets Θ_0 and Θ_1. Θ_0 is the set of θ values corresponding to H_0, and Θ_1 is the complementary set of elements of Θ which are not in Θ_0. If $h(\theta)$ is a prior p.d.f. of θ, then the **prior probability** that H_0 is true is $\pi = \int_{\Theta_0} h(\theta) \, d\theta$. The prior probability that H_1 is true is $\bar{\pi} = 1 - \pi$.

The statistician has to make a decision whether H_0 is true or H_1 is true. Let $d(\pi)$ be a decision function, assuming the values 0 and 1, i.e.,

$$d(\pi) = \begin{cases} 0, & \text{decision to accept } H_0 \ (H_0 \text{ is true}) \\ 1, & \text{decision to reject } H_0 \ (H_1 \text{ is true}). \end{cases}$$

Let w be an indicator of the true situation, i.e.,

$$w = \begin{cases} 0, & \text{if } H_0 \text{ is true.} \\ 1, & \text{if } H_1 \text{ is true.} \end{cases}$$

We also impose a **loss function** for erroneous decision

$$L(d(\pi), w) = \begin{cases} 0, & \text{if } d(\pi) = w \\ r_0, & \text{if } d(\pi) = 0, w = 1 \\ r_1, & \text{if } d(\pi) = 1, w = 0, \end{cases} \tag{4.8.9}$$

where r_0 and r_1 are finite positive constants. The **prior risk** associated with the decision function $d(\pi)$ is

$$R(d(\pi), \pi) = d(\pi) r_1 \pi + (1 - d(\pi)) r_0 (1 - \pi)$$
$$= r_0 (1 - \pi) + d(\pi)[\pi(r_0 + r_1) - r_0]. \tag{4.8.10}$$

We wish to choose a decision function which **minimizes** the prior risk $R(d(\pi), \pi)$. Such a decision function is called the **Bayes decision function**, and the prior risk associated with the Bayes decision function is called the **Bayes risk**. According to the above formula of $R(d(\pi), \pi)$, we should choose $d(\pi)$ to be 1 if, and only if, $\pi(r_0 + r_1) - r_0 < 0$. Accordingly, the Bayes decision function is

$$d^0(\pi) = \begin{cases} 0, & \text{if } \pi \geq \dfrac{r_0}{r_0 + r_1} \\[3mm] 1, & \text{if } \pi < \dfrac{r_0}{r_0 + r_1}. \end{cases} \qquad (4.8.11)$$

Let $\pi^* = r_0/(r_0 + r_1)$ and define the indicator function

$$I(\pi; \pi^*) = \begin{cases} 1, & \text{if } \pi \geq \pi^* \\[2mm] 0, & \text{if } \pi < \pi^* \end{cases}$$

then, the Bayes risk is

$$R^0(\pi) = r_0(1 - \pi)I(\pi; \pi^*) + \pi r_1(1 - I(\pi; \pi^*)). \qquad (4.8.12)$$

In Figure 4.14, we present the graph of the Bayes risk function $R^0(\pi)$, for $r_0 = 1$ and $r_1 = 5$. We see that the function $R^0(\pi)$ attains its maximum at $\pi = \pi^*$. The maximal Bayes risk is $R^0(\pi^*) = r_0 r_1/(r_0 + r_1) = 5/6$. If the value of π is close to π^*, the Bayes risk is close to $R^0(\pi^*)$.

The analysis above can be performed even before observations commenced. If π is close to 0 or to 1, the Bayes risk $R^0(\pi)$ is small, and we may reach a decision concerning the hypotheses without even making observations. Recall that observations cost money, and it might not

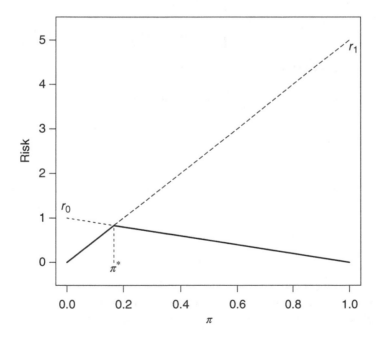

Figure 4.14   *The Bayes risk function.*

be justifiable to spend this money. On the other hand, if the cost of observations is negligible compared to the loss due to erroneous decision, it might be prudent to take as many observations as required to reduce the Bayes risk.

After observing a random sample, $x_1, \ldots, x_n$, we convert the prior p.d.f. of θ to posterior and determine the posterior probability of H_0, namely

$$\pi_n = \int_\Theta h(\theta \mid x_1, \ldots, x_n) \, d\theta.$$

The analysis then proceeds as before, replacing π with the posterior probability π_n.

Accordingly, the Bayes decision function is

$$d^0(x_1, \ldots, x_n) = \begin{cases} 0, & \text{if } \pi_n \geq \pi^* \\ 1, & \text{if } \pi_n < \pi^* \end{cases}$$

and the Bayes posterior risk, is

$$R^0(\pi_n) = r_0(1 - \pi_n)I(\pi_n; \pi^*) + \pi_n r_1(1 - I(\pi_n; \pi^*)).$$

Under certain regularity conditions, $\lim\limits_{n\to\infty} \pi_n = 1$ or 0, according to whether H_0 is true or false. We illustrate this with a simple example.

Example 4.15.   Suppose that X has a normal distribution, with known $\sigma^2 = 1$. The mean μ is unknown. We wish to test $H_0 : \mu \leq \mu_0$ against $H_1 : \mu > \mu_0$. Suppose that the prior distribution of μ is also normal, $N(\mu^*, \tau^2)$. The posterior distribution of μ, given $X_1, \ldots, X_n$, is normal with mean

$$E\{\mu \mid X_1, \ldots, X_n\} = \mu^* \frac{1}{(1 + n\tau^2)} + \frac{n\tau^2}{1 + n\tau^2} \bar{X}_n$$

and posterior variance

$$V\{\mu \mid X_1, \ldots, X_n\} = \frac{\tau^2}{1 + n\tau^2}.$$

Accordingly, the posterior probability of H_0 is

$$\pi_n = \Phi\left(\frac{\mu_0 - \dfrac{\mu^*}{1 + n\tau^2} - \dfrac{n\tau^2}{1 + n\tau^2} \bar{X}_n}{\sqrt{\dfrac{\tau^2}{1 + n\tau^2}}} \right).$$

According to the Law of Large Numbers, $\bar{X}_n \to \mu$ (the true mean), as $n \to \infty$, with probability one. Hence,

$$\lim_{n\to\infty} \pi_n = \begin{cases} 1, & \text{if } \mu < \mu_0 \\ \dfrac{1}{2}, & \text{if } \mu = \mu_0 \\ 0, & \text{if } \mu > \mu_0. \end{cases}$$

Notice that the prior probability that $\mu = \mu_0$ is zero. Thus, if $\mu < \mu_0$ or $\mu > \mu_0$, $\lim_{n\to\infty} R^0(\pi_n) = 0$, with probability one. That is, if n is sufficiently large, the Bayes risk is, with probability close to one, smaller than some threshold r^*. This suggests to continue, stepwise or sequentially, collecting observations, until the Bayes risk $R^0(\pi_n)$ is, for the first time, smaller than r^*. At stopping, $\pi_n \geq 1 - \dfrac{r^*}{r_0}$ or $\pi_n \leq \dfrac{r^*}{r_1}$. We obviously choose $r^* < \dfrac{r_0 r_1}{r_0 + r_1}$.

4.8.2.2   Bayesian estimation

In an estimation problem, the decision function is an estimator $\hat{\theta}(x_1, \ldots, x_n)$, which yields a point in the parameter space Θ. Let $L(\hat{\theta}(x_1, \ldots, x_n), \theta)$ be a **loss function** which is nonnegative, and $L(\theta, \theta) = 0$. The **posterior risk** of an estimator $\hat{\theta}(x_1, \ldots, x_n)$ is the expected loss, with respect to the posterior distribution of θ, given $(x_1, \ldots, x_n)$, i.e.,

$$R_h(\hat{\theta}, \mathbf{x}_n) = \int_\Theta L(\hat{\theta}(\mathbf{x}_n), \theta) h(\theta \mid \mathbf{x}_n) \, d\theta, \tag{4.8.13}$$

where $\mathbf{x}_n = (x_1, \ldots, x_n)$. We choose an estimator which **minimizes the posterior risk**. Such an estimator is called a **Bayes estimator** and designated by $\hat{\theta}_B(\mathbf{x}_n)$. We present here a few cases of importance.

Case A. θ real, $L(\hat{\theta}, \theta) = (\hat{\theta} - \theta)^2$.
  In this case, the Bayes estimator of θ, is posterior expectation of θ, i.e.,

$$\hat{\theta}_B(\mathbf{x}_n) = E_h\{\theta \mid \mathbf{x}_n\}. \tag{4.8.14}$$

The Bayes risk is the expected posterior variance, i.e.,

$$R_h^0 = \int V_h\{\theta \mid \mathbf{x}_n\} f_h(x_1, \ldots, x_n) \, dx_1, \ldots, dx_n.$$

Case B. θ real, $L(\hat{\theta}, \theta) = c_1(\hat{\theta} - \theta)^+ + c_2(\theta - \hat{\theta})^+$, with $c_1, c_2 > 0$, and $(a)^+ = \max(a, 0)$.
  As shown in the inventory example, at the beginning of the section, the Bayes estimator is

$$\hat{\theta}_B(\mathbf{x}_n) = \frac{c_2}{c_1 + c_2}\text{-th quantile of the posterior distribution of } \theta, \text{ given } \mathbf{x}_n.$$

When $c_1 = c_2$, we obtain the posterior median.

4.8.3   Credibility intervals for real parameters

We restrict attention here to the case of a real parameter, θ. Given the values $x_1, \ldots, x_n$ of a random sample, let $h(\theta \mid \mathbf{x}_n)$ be the posterior p.d.f. of θ. An interval $C_{1-\alpha}(\mathbf{x}_n)$ such that

$$\int_{C_{1-\alpha}(\mathbf{x}_n)} h(\theta \mid \mathbf{x}_n) \, d\theta \geq 1 - \alpha \tag{4.8.15}$$

is called a **credibility interval** for θ. A credibility interval $C_{1-\alpha}(\mathbf{x}_n)$ is called a **highest posterior density (HPD) interval** if for any $\theta \in C_{1-\alpha}(\mathbf{x}_n)$ and $\theta' \notin C_{1-\alpha}(\mathbf{x}_n)$, $h(\theta \mid \mathbf{x}_n) > h(\theta' \mid \mathbf{x}_n)$.

Example 4.16.   Let $x_1, \ldots, x_n$ be the values of a random sample from a Poisson distribution $P(\lambda)$, $0 < \lambda < \infty$. We assign λ a Gamma distribution $G(\nu, \tau)$. The posterior p.d.f. of λ, given $\mathbf{x}_n = (x_1, \ldots, x_n)$ is

$$h(\lambda \mid \mathbf{x}_n) = \frac{(1 + n\tau)^{\nu + \Sigma x_i}}{\Gamma(\nu + \Sigma x_i)\tau^{\nu + \Sigma x_i}} \cdot \lambda^{\nu + \Sigma x_i - 1} e^{-\lambda \frac{1 + n\tau}{\tau}}.$$

In other words, the posterior distribution is a Gamma distribution $G\left(\nu + \sum_{i=1}^{n} x_i, \frac{\tau}{1 + n\tau}\right)$. From the relationship between the Gamma and the χ^2-distributions, we can express the limits of a credibility interval for λ, at level $(1 - \alpha)$ as

$$\frac{\tau}{2(1 + n\tau)} \chi^2_{\alpha/2}[\phi] \quad \text{and} \quad \frac{\tau}{2(1 + n\tau)} \chi^2_{1-\alpha/2}[\phi]$$

where $\phi = 2\nu + 2\sum_{i=1}^{n} x_i$. This interval is called an **equal tail credibility interval**. However, it is not an HPD credibility interval. In Figure 4.15, we present the posterior density for the special case of $n = 10$, $\nu = 2$, $\tau = 1$, and $\sum_{i=1}^{10} x_i = 15$. For these values, the limits of the credibility interval for λ, at level 0.95, are 0.9 and 2.364. As we see in Figure 4.15, $h(0.9 \mid \mathbf{x}_n) > h(2.364 \mid \mathbf{x}_n)$. Thus, the equal-tail credibility interval is **not** an HPD interval. The limits of the HPD interval can be determined by trial and error. In the present case, they are approximately 0.86 and 2.29, as shown in Figure 4.15.

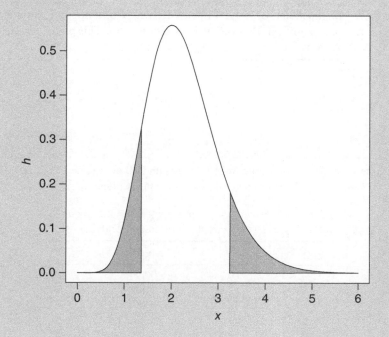

Figure 4.15   *The posterior p.d.f. and credibility intervals.*

4.9 Random sampling from reference distributions

We have seen in Section 2.4.1 an example of blemishes on ceramic plates. In that example (Table 2.2), the proportion of plates having **more** than one blemish is 0.23. Suppose that we decide to improve the manufacturing process and reduce this proportion. How can we test whether an alternative production process with new operating procedures and machine settings is indeed better so that the proportion of plates with more than one blemish is significantly smaller. The objective is to operate a process with a proportion of defective units (i.e., with more than one blemish) which is smaller than 0.10. After various technological modifications, we are ready to test whether the modified process conforms with the new requirement. Suppose that a random sample of ceramic plates is drawn from the modified manufacturing process. One has to test whether the proportion of defective plates in the sample is not significantly larger than 0.10. In the parametric model, it was assumed that the number of plates having more than one defect, in a random sample of n plates, has a binomial distribution $B(n, p)$. For testing $H_0 : p \leq 0.1$, a test was constructed based on the reference distribution $B(n, 0.1)$.

One can create, artificially on a computer, a population having 90 zeros and 10 ones. In this population, the proportion of ones is $p_0 = 0.10$. From this population, one can draw a large number, M, of random samples with replacement (RSWR) of a given size n. In each sample, the sample mean $\bar{X}_n$ is the proportion of 1's in the sample. The sampling distribution of the M sample means is our **empirical reference distribution** for the hypothesis that the proportion of defective plates is $p \leq p_0$. We pick a value α close to zero, and determine the $(1 - \alpha)$th quantile of the empirical reference distribution. If the observed proportion in the real sample is greater than this quantile, the hypothesis $H : p \leq p_0$ is rejected.

Example 4.17. To illustrate, we created, using MINITAB, an empirical reference distribution of $M = 100$ proportions of 1's in RSWR of size $n = 50$. This was done by executing 1000 times the MACRO:

```
Sample 50 C1 C2;
Replace.
Let k1 = mean (C2)
stack C3 k1 C3
end
```

It was assumed that column $C1$ contained 90 zeros and 10 ones. The frequency distribution of column $C3$ represents the reference distribution. This is given in Table 4.11.

Table 4.11 Frequency distribution of M = 100 means of RSWR from a set with 90 zeros and 10 ones

$\bar{x}$	f	$\bar{x}$	f
0.03	10	0.11	110
0.04	17	0.12	100
0.05	32	0.13	71
0.06	61	0.14	50
0.07	93	0.15	28
0.08	128	0.16	24
0.09	124	0.17	9
0.10	133	>0.17	10

For $\alpha = 0.05$, the 0.95-quantile of the empirical reference distribution is 0.15, since less than 50 out of 1000 observations are greater than 0.15. Thus, if in a real sample, of size $n = 50$, the proportion defectives is greater than 0.15, the null hypothesis is rejected.

Example 4.18. Consider a hypothesis on the length of aluminum pins (with cap), $H_0 : \mu \geq$ 60.1 [mm]. We create now an empirical reference distribution for this hypothesis. In the data set **ALMPIN.csv**, we have the actual sample values. The mean of the variable lenWcp is $\bar{X}_{70} = 60.028$. Since the hypothesis states that the process mean is $\mu \geq 60.1$, we transform the sample values to $Y = X - 60.028 + 60.1$. This transformed sample has mean of 60.1. We now create a reference distribution of sample means by drawing M RSWR of size $n = 70$ from the transformed sample. We can perform this using the following MINITAB's MACRO:

```
Sample 70 C7 C8;
Replace.
let k1 = mean(C8)
stack k1 C9 C9
end
```

Executing this MACRO $M = 1000$ times, we obtain an empirical reference distribution whose frequency distribution is given in Table 4.12.

In the data set **ALMPIN.csv**, there are six variables, measuring various dimensions of aluminum pins. The data are stored in columns $C1-C6$. Column $C7$ contains the values of the transformed variable Y.

Table 4.12 *Frequency distribution of $\bar{X}_{70}$ from 1000 RSWR from lengthwcp*

Midpoint	Count
60.080	3
60.085	9
60.090	91
60.095	268
60.100	320
60.105	217
60.110	80
60.115	11
60.120	1

Since $\bar{X}_{70} = 60.028$ is smaller than 60.1, we consider as a test criterion the α-quantile of the reference distribution. If $\bar{X}_{70}$ is smaller than this quantile, we reject the hypothesis. For $\alpha = 0.01$, the 0.01-quantile in the above reference distribution is 60.0869. Accordingly we reject the hypothesis, since it is very implausible (less than one chance in a hundred) that $\mu \geq 60.1$. The estimated P-value is less than 10^{-3}, since the smallest value in the reference distribution is 60.0803.

4.10 Bootstrap sampling

4.10.1 The bootstrap method

The bootstrap methodology was introduced in 1979 by B. Efron, as an elegant method of performing statistical inference by harnessing the power of the computer, and without the need for extensive assumptions and intricate theory. Some of the ideas of statistical inference with the aid of computer sampling, were presented in the previous sections. In the present section, we introduce the bootstrap method in more detail.

Given a sample of size n, $S_n = \{x_1, \ldots, x_n\}$, let t_n denote the value of some specified sample statistic T. The bootstrap method draws M random samples **with** replacement (RSWR) of size n from S_n. For each such sample, the statistic T is computed. Let $\{t_1^*, t_2^*, \ldots, t_M^*\}$ be the collection of these sample statistics. The distribution of these M values of T, is called the **Empirical Bootstrap Distribution** (EBD). It provides an approximation, if M is large, to the **Bootstrap Distribution**, of all possible values of the statistic T, that can be generated by repeatedly sampling from S_n.

General Properties of the (EBD)

1. The EBD is centered at the sample statistic t_n.
2. The mean of the EBD is an estimate of the mean of the sampling distribution of the statistic T, over all possible samples.
3. The standard deviation of the EBD, is the **bootstrap estimate of the standard-error of T**.
4. The $\alpha/2$-th and $(1 - \alpha/2)$th quantiles of the EBD are **bootstrap confidence limits** for the parameter which is estimated by t_n, at level of confidence $(1 - \alpha)$.

Example 4.19. We illustrate the bootstrap method with data set **ETCHRATE.csv** in which we want to test if the sample is derived from a population with a specific mean. After installing

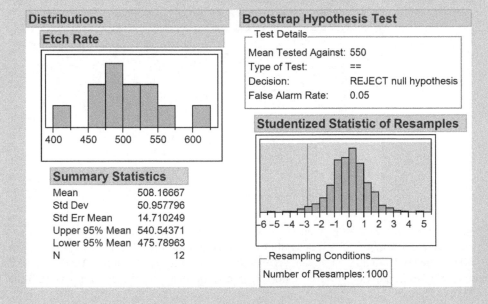

Figure 4.16 Output of a bootstrap studentized test for the mean (JMP).

JMP download, the addin file **com.jmp.cox.ian.bootstrapCI.jmpaddin** from the book web-site and click on it. This will open and "Add-Ins" window within JMP. Opening "Bootstrap Confidence Intervals" will make available to you the etchrate data set and options to run bootstrap studentized tests for one and two samples. If you open other data sets, the same procedures will also work. As an example consider running a *Studentized Test for the Mean (One Sample)* for testing if the mean ETCHRATE has dropped below $\mu = 550$ and with a confidence level of $(1 - \alpha = 0.95)$. As shown in Figure 4.16, we reject the null hypothesis. The line in the figure indicates the position of the studentized tested mean.

The JMP PRO version of JMP includes a bootstrap analysis of most statistical reports. It is activated by right clicking on the JMP report itself.

Using the MINITAB **BOOTMEAN.MTB** MACRO, we can perform the same calculation by drawing **from the original sample** 1000 RSWR samples. For each such sample, we compute the bootstrap means.

```
Sample 1000 C1 C2;
Replace.
let k1 = mean(C2)
stack C3 k1 C3
end
```

The column $C3$ contains the 1000 bootstrap means. The standard deviation of $C3$ is the bootstrap estimate of the standard-error of $\bar{X}_{1000}$. We denote it by $\text{SE}^*\{\bar{X}_{1000}\}$. To obtain the bootstrap confidence limits, at confidence level $(1 - \alpha) = 0.95$, we sort the values in $C3$. The 0.025-quantile of $\bar{X}^*_{1000}$ is 475.817 the 0.975-quantile of $\bar{X}^*_{1000}$ is 539.139. The bootstrap interval $(475.817, 539.139)$ is called a bootstrap confidence interval for μ. We see that this interval does not cover the tested mean of 550.

4.10.2 Examining the bootstrap method

In the previous section, we introduced the bootstrap method as a computer intensive technique for making statistical inference. In this section, some of the properties of the bootstrap methods are examined in light of the theory of sampling from finite populations. As we recall, the bootstrap method is based on drawing repeatedly M simple RSWR from the original sample.

Let $S_X = \{x_1, \ldots, x_n\}$ be the values of the n original observations on X. We can consider S_X as a finite population P of size n. Thus, the mean of this population μ_n, is the sample mean $\bar{X}_n$, and the variance of this population is $\sigma_n^2 = \frac{n-1}{n}S_n^2$, where S_n^2 is the sample variance, $S_n^2 = \frac{1}{n-1}\sum_{i=1}^n (x_i - \bar{X}_n)^2$. Let $S_X^* = \{X_1^*, \ldots, X_n^*\}$ denote a simple RSWR from S_X. S_X^* is the bootstrap sample. Let $\bar{X}_n^*$ denote the mean of the bootstrap sample.

We have shown in Chapter 3 that the mean of a simple RSWR is an unbiased estimator of the corresponding sample mean. Thus,

$$E^*\{\bar{X}_n^*\} = \bar{X}_n, \tag{4.10.1}$$

where $E^*\{\cdot\}$ is the expected value with respect to the bootstrap sampling. Moreover, the bootstrap variance of $\bar{X}_n^*$ is

$$V^*\{\bar{X}_n^*\} = \frac{\dfrac{n-1}{n}S_n^2}{n}$$

$$= \frac{S_n^2}{n}\left(1 - \frac{1}{n}\right). \qquad (4.10.2)$$

Thus, in large sample

$$V^*\{\bar{X}_n^*\} \cong \frac{S_n^2}{n}. \qquad (4.10.3)$$

If the original sample S_X is a realization of n i.i.d. random variables, having a c.d.f. $F(x)$, with finite expected value μ_F and a finite variance σ_F^2, then, as shown in Section 4.8, the variance of $\bar{X}_n$ is σ_F^2/n. The sample variance S_n^2 is an unbiased estimator of σ_F^2. Thus, $\dfrac{S_n^2}{n}$ is an unbiased estimator of σ_F^2/n. Finally, the variance of the EBD of $\bar{X}_1^*, \ldots, \bar{X}_M^*$ obtained by repeating the bootstrap sampling M times independently, is an unbiased estimator of $\dfrac{S_n^2}{n}\left(1 - \dfrac{1}{n}\right)$. Thus, the variance of the EBD is an approximation to the variance of $\bar{X}_n$.

We remark that this estimation problem is a simple one, and there is no need for bootstrapping in order to estimate the variance, or standard error of the estimator $\bar{X}_n$.

4.10.3   Harnessing the bootstrap method

The effectiveness of the bootstrap method manifests itself when formula for the variance of an estimator are hard to obtain. In Section 3.2.3, we provided a formula for the variance of the estimator S_n^2, in simple RSWR. By bootstrapping from the sample S_X, we obtain an EBD of S_n^{*2}. The variance of this EBD is an approximation to the true variance of S_n^2. Thus, for example, when $P = \{1, 2, \ldots, 100\}$, the true variance of S_n^2 is 31,131.2, while the bootstrap approximation, for a particular sample is 33,642.9. Another sample will yield a different approximation. The approximation obtained by the bootstrap method becomes more precise as the sample size grows. For the above problem, if $n = 100$, $V\{S_n\} = 5693.47$, then the bootstrap approximation is distributed around this value.

The following are values of four approximations of $V\{S_n\}$ for $n = 100$, when $M = 100$. Each approximation is based on different random samples from P:

$$6293.28, \quad 5592.07, \quad 5511.71, \quad 5965.89.$$

Each bootstrap approximation is an estimate of the true value of $V\{S_n\}$.

4.11   Bootstrap testing of hypotheses

In this section, we present some of the theory and the methods of testing hypotheses by bootstrapping. Given a test statistic $\mathbf{T} = T(X_1, \ldots, X_n)$, the critical level for the test, k_α, is determined according to the distribution of $\mathbf{T}$ under the null hypothesis, which is the **reference distribution**.

The bootstrapping method, as explained before, is a randomization method which resamples the sample values, and thus constructs a reference distribution for $\mathbf{T}$, independently of the unknown distribution F of $\mathbf{X}$. For each bootstrap sample, we compute the value of the test statistic $\mathbf{T}^* = T(x_1^*, \ldots, x_n^*)$. Let $\mathbf{T}_1^*, \ldots, \mathbf{T}_M^*$ be the M values of the test statistic obtained from the M samples from bootstrap population (BP). Let $F_M^*(t)$ denote the empirical c.d.f. of these values. $F_M^*(t)$ is an estimator of the bootstrap distribution $F^*(t)$, from which we can estimate the critical value k^*. Specific procedures are given in the following sections.

4.11.1   Bootstrap testing and confidence intervals for the mean

Suppose that $\{x_1, \ldots, x_n\}$ is a random sample from a parent population, having an unknown distribution, F, with mean μ and a finite variance σ^2.

We wish to test the hypothesis

$$H_0 : \mu \le \mu_0 \text{ against } H_1 : \mu > \mu_0.$$

Let $\bar{X}_n$ and S_n be the sample mean and sample standard-deviation. Suppose that we draw from the original sample M bootstrap samples. Let $\bar{X}_1^*, \ldots, \bar{X}_M^*$ be the means of the bootstrap samples. Recall that, since the bootstrap samples are RSWR, $E^*\{\bar{X}_j^*\} = \bar{X}_n$ for $j = 1, \ldots, M$, where $E^*\{\cdot\}$ designates the expected value, with respect to the bootstrap sampling. Moreover, for large n,

$$\text{SE}^*\{\bar{X}_j^*\} \cong \frac{S_n}{\sqrt{n}}, \quad j = 1, \ldots, M.$$

Thus, if n is not too small, the CLT implies that $F_M^*(\bar{X}^*)$ is approximately $\Phi\left(\dfrac{\bar{X}^* - \bar{X}_n}{S_n/\sqrt{n}}\right)$, i.e., the bootstrap means $\bar{X}_1^*, \ldots, \bar{X}_m^*$ are distributed approximately normally around $\mu^* = \bar{X}_n$. We wish to reject H_0 if $\bar{X}_n$ is significantly larger than μ_0. According to this normal approximation to $F_M^*(\bar{X}^*)$, we should reject H_0, at level of significance α, if $\dfrac{\mu_0 - \bar{X}_n}{S_n/\sqrt{n}} \le z_\alpha$ or $\bar{X}_n \ge \mu_0 + z_{1-\alpha}\dfrac{S_n}{\sqrt{n}}$.

This is approximately the t-test of Section 4.3.2.2.

Notice that the reference distribution can be obtained from the EBD by subtracting $\Delta = \bar{X}_n - \mu_0$ from $\bar{X}_j^*$ ($j = 1, \ldots, M$). The reference distribution is centered at μ_0. The $(1 - \alpha/2)$th quantile of the reference distribution is $\mu_0 + z_{1-\alpha/2}\dfrac{S_n}{\sqrt{n}}$. Thus, if $\bar{X}_n \ge \mu_0 + z_{1-\alpha/2}\dfrac{S_n}{\sqrt{n}}$, we reject the null hypothesis $H_0 : \mu \le \mu_0$.

If the sample size n is not large, it might not be justified to use the normal approximation. We use bootstrap procedures in the following sections.

4.11.2   Studentized test for the mean

A **studentized test statistic**, for testing the hypothesis $H_0 : \mu \le \mu_0$, is

$$t_n = \frac{\bar{X}_n - \mu_0}{S_n/\sqrt{n}}. \tag{4.11.1}$$

H is rejected if t_n is significantly greater than zero. To determine what is the rejection criterion, we construct an EBD by following procedure:

1. Draw a RSWR, of size n, from the original sample.
2. Compute $\bar{X}_n^*$ and S_n^* of the bootstrap sample.
3. Compute the studentized statistic

$$t_n^* = \frac{\bar{X}_n^* - \bar{X}_n}{S_n^*/\sqrt{n}}.$$ (4.11.2)

4. Repeat this procedure M times.

Let t_p^* denote the p-th quantile of the EBD.

Case I. $H : \mu \leq \mu_0$.
The hypothesis H is rejected if

$$t_n \geq t_{1-\alpha}^*.$$

Case II. $H : \mu \geq \mu_0$.
We reject H if

$$t_n \leq t_\alpha^*.$$

Case III. $H : \mu = \mu_0$.
We reject H if

$$|t_n| \geq t_{1-\alpha/2}^*.$$

The corresponding P^*-**levels** are the following:

For Case I: The proportions of t_n^* values greater than t_n.
For Case II: The proportions of t_n^* values smaller than t_n.
For Case III: The proportion of t_n^* values greater than $|t_n|$ or smaller than $-|t_n|$. H is rejected if P^* is small.

Notice the difference in definition between t_n and t_n^*. t_n is centered around μ_0 while t_n^* around $\bar{X}_n$.

Example 4.20. In data file **HYBRID1.csv**, we find the resistance (in ohms) of Res3 in a hybrid microcircuit labeled hybrid 1 on $n = 32$ boards. The mean of Res 3 in hybrid 1 is $\bar{X}_{32} = 2143.4$. The question is whether Res 3 in hybrid 1 is significantly different from $\mu_0 = 2150$. We consider the hypothesis

$$H : \mu = 2150 \quad \text{(Case III)}.$$

With $M = 500$, we obtain with the R commands below the following 0.95-confidence level bootstrap interval (2109,2178). We see that $\mu_0 = 2150$ is covered by this interval. We, therefore, infer that $\bar{X}_{32}$ is not significantly different than μ_0. The hypothesis H is **not** rejected. With R, JMP PRO or the JMP bootstrap studentized test add-in, we see that the studentized difference between the sample mean $\bar{X}_{32}$ and μ_0 is $t_n = -0.374$. $M = 500$ bootstrap replicas yield the value $P^* = 0.708$. The hypothesis is **not** rejected.
The commands in R are the following:

```
> library(boot)
> data(HYBRID1)
> set.seed(123)
> boot.ci(boot(data=HYBRID1,
              statistic=function(x, i)
```

```
                      mean(x[i]),
                 R=500),
             type = "perc")

BOOTSTRAP CONFIDENCE INTERVAL CALCULATIONS
Based on 500 bootstrap replicates

CALL :
boot.ci(boot.out = boot(data = HYBRID1, statistic =
    function(x, i) mean(x[i]), R = 500), type = "perc")

Intervals :
Level      Percentile
95%    (2112, 2176 )
Calculations and Intervals on Original Scale

> t.test(HYBRID1, mu=2150)


        One Sample t-test

data:  HYBRID1
t = -0.37432, df = 31, p-value = 0.7107
alternative hypothesis: true mean is not equal to 2150
95 percent confidence interval:
 2107.480 2179.333
sample estimates:
mean of x
 2143.406

> set.seed(123)
> B <- boot(data=HYBRID1,
            statistic=function(x, i, mu)
               t.test(x[i],
                        mu=mu)$p.value,
            R=500,
            mu=2150)
> sum(B$t <
        t.test(HYBRID1,
               mu=2150)$p.value) /
    nrow(B$t)

[1] 0.724
```

4.11.3   Studentized test for the difference of two means

The problem is whether two population means μ_1 and μ_2 are the same. This problem is important in many branches of science and engineering, when two "treatments" are compared.

Suppose that one observes a random sample $X_1, \ldots, X_{n_1}$ from population 1 and another random sample $Y_1, \ldots, Y_{n_2}$ from population 2. Let $\bar{X}_{n_1}$, $\bar{Y}_{n_2}$, S_{n_1}, and S_{n_2} be the means and standard deviations of these two samples, respectively. Compute the studentized difference of the two sample means as

$$t = \frac{\bar{X}_{n_1} - \bar{Y}_{n_2} - \delta_0}{\left(\frac{S_{n_1}^2}{n_1} + \frac{S_{n_2}^2}{n_2} \right)^{1/2}}, \tag{4.11.3}$$

where $\delta = \mu_1 - \mu_2$. The question is whether this value is significantly different from zero. The hypothesis under consideration is

$$H : \mu_1 = \mu_2, \text{ or } \delta_0 = 0.$$

By the bootstrap method, we draw RSWR of size n_1 from the x-sample, and an RSWR of size n_2 from the y-sample. Let $X_1^*, \ldots, X_{n_1}^*$ and $Y_1^*, \ldots, Y_{n_2}^*$ be these two bootstrap samples, with means and standard deviations $\bar{X}_{n_1}^*$, $\bar{Y}_{n_2}^*$ and $S_{n_1}^*$, $S_{n_2}^*$. We compute then the studentized difference

$$t^* = \frac{\bar{X}_{n_1}^* - \bar{Y}_{n_2}^* - (\bar{X}_{n_1} - \bar{Y}_{n_2})}{\left(\frac{S_{n_1}^{*2}}{n_1} + \frac{S_{n_2}^{*2}}{n_2} \right)^{1/2}}. \tag{4.11.4}$$

This procedure is repeated independently M times, to generate an EBD of $t_1^*, \ldots, t_M^*$.

Let $(D_{\alpha/2}^*, D_{1-\alpha/2}^*)$ be a $(1 - \alpha)$ level confidence interval for δ, based on the EBD. If $t_{\alpha/2}^*$ is the $\alpha/2$-quantile of t^* and $t_{1-\alpha/2}^*$ is its $(1 - \alpha/2)$-quantile, then

$$D_{\alpha/2}^* = (\bar{X}_{n_1} - \bar{Y}_{n_2}) + t_{\alpha/2}^* \left(\frac{S_{n_1}^2}{n_1} + \frac{S_{n_2}^2}{n_2} \right)^{1/2}$$

$$D_{1-\alpha/2}^* = (\bar{X}_{n_1} - \bar{Y}_{n_2}) + t_{1-\alpha/2}^* \left(\frac{S_{n_1}^2}{n_1} + \frac{S_{n_2}^2}{n_2} \right)^{1/2}. \tag{4.11.5}$$

If this interval does **not** cover the value $\delta_0 = 0$, we **reject** the hypothesis $H : \mu_1 = \mu_2$. The P^*-value of the test is the proportion of t_i^* values which are either smaller than $-|t|$ or greater than $|t|$.

Example 4.21.   We compare the resistance coverage of Res 3 in hybrid 1 and in hybrid 2. The data file **HYBRID2.csv** consists of two columns. The first represents the sample of $n_1 = 32$ observations on hybrid 1 and the second column consists of $n_2 = 32$ observations on hybrid 2. The output file consists of $M = 500$ values of t_i^* ($i = 1, \ldots, M$).

We see that $\bar{X}_{n_1} = 2143.41$, $\bar{Y}_{n_2} = 1902.81$, $S_{n_1} = 99.647$ and $S_{n_2} = 129.028$. The studentized difference between the means is $t = 8.348$. The bootstrap $(1 - \alpha)$-level confidence interval for $\delta / \left(\frac{S_{n_1}^2}{n_1} + \frac{S_{n_2}^2}{n_2} \right)^{1/2}$ is $(6.326, 10.297)$. The hypothesis that $\mu_1 = \mu_2$ or $\delta = 0$ is **rejected** with $P^* \approx 0$. In Figure 4.17, we present the histogram of the EBD of t^*.

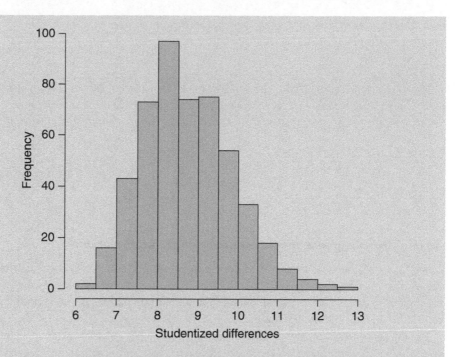

Figure 4.17   *Histogram of the EBD of M = 500 studentized differences.*

In R

```
> data(HYBRID2)
> t.test(HYBRID2$hyb1, HYBRID2$hyb2)

        Welch Two Sample t-test

data:  HYBRID2$hyb1 and HYBRID2$hyb2
t = 8.3483, df = 58.276, p-value = 1.546e-11
alternative hypothesis: true difference in means is not
    equal to 0
95 percent confidence interval:
 182.9112 298.2763
sample estimates:
mean of x mean of y
 2143.406  1902.812

> set.seed(123)
> boot(data=HYBRID2,
       statistic=function(x, i)
         t.test(x=x[i,1],
                y=x[i,2])$p.value,
       R=500)
```

```
ORDINARY NONPARAMETRIC BOOTSTRAP


Call:
boot(data = HYBRID2, statistic = function(x, i) t.test(x =
    x[i, 1], y = x[i, 2])$p.value, R = 500)


Bootstrap Statistics :
        original                bias              std. error
t1* 1.546468e-11 0.0000000005176439 0.000000003024552
```

4.11.4   Bootstrap tests and confidence intervals for the variance

Let $S_1^{*2}, \ldots, S_M^{*2}$ be the variances of M bootstrap samples. These statistics are distributed around the sample variance S_n^2. Consider the problem of testing the hypotheses $H_0 : \sigma^2 \le \sigma_0^2$ against $H_1 : \sigma^2 > \sigma_0^2$, where σ^2 is the variance of the parent population. As in Section 4.3.2.3, H_0 is rejected if S^2/σ_0^2 is sufficiently large.

Let $G_M^*(x)$ be the bootstrap empirical c.d.f. of $S_1^{*2}, \ldots, S_M^{*2}$. The bootstrap P^* value for testing H_0 is

$$P^* = 1 - G_M^*\left(\frac{S^2}{\sigma_0^2}\right). \tag{4.11.6}$$

If P^* is sufficiently small, we reject H_0. For example in a random sample of size $n = 20$, the sample standard deviation is $S_{20} = 24.812$. Suppose that we wish to test whether it is significantly larger than $\sigma_0 = 20$. We can run $M = 1000$ bootstrapped samples with the JMP Bootstrap add-in. The P^* value for testing the hypothesis H_0 is the proportion of bootstrap standard deviations greater than $S_{20}^2/\sigma_0^2 = 1:5391$. Running the program we obtain $P^* = 0.559$. The hypothesis H^0 is not rejected. S_{20} is not determined significantly greater than $\sigma_0 = 20$. In a similar manner, we test the hypotheses $H_0 : \sigma^2 \ge \sigma_0^2$ against $H_1 : \sigma^2 < \sigma_0^2$, or the two-sided hypothesis $H_0 : \sigma^2 = \sigma_0^2$ against $H_0 : \sigma^2 \ne \sigma_0^2$. Percentile bootstrap confidence limits for σ^2, at level $1 - \alpha$, are given by $\frac{\alpha}{2}$th and $\left(1 - \frac{\alpha}{2}\right)$th quantiles of $G_M^*(x)$, or

$$S_{(j_{\alpha/2})}^{*2} \text{ and } S_{(1+j_{1-\alpha/2})}^{*2}.$$

These bootstrap confidence limits for σ^2 at level 0.95, are 210.914 and 1024.315. The corresponding chi-squared confidence limits (see Section 4.4.3) are 355.53 and 1312.80. Another type of bootstrap confidence interval is given by the limits

$$\frac{S_n^4}{S_{(j_{1-\alpha/2})}^{*2}}, \frac{S_n^4}{S_{(j_{\alpha/2})}^{*2}}.$$

These limits are similar to the chi-squared confidence interval limits, but use the quantiles of S_n^{*2}/S_n instead of those of $\chi^2[n-1]$. For the sample of size $n = 20$, with $S_{20} = 24.812$ the above confidence interval for σ^2 is (370.01, 1066.033).

4.11.5 Comparing statistics of several samples

It is often the case that we have to test whether the means or the variances of three or more populations are equal. In Chapters 13–15, we discuss the design and analysis of experiments, where we study the effect of changing levels of different factors. Typically, we perform observations at different experimental conditions. The question is whether the observed differences between the means and variances of the samples observed under different factor-level combinations are significant. The test statistic which we will introduce to test differences between means, might be effected also by differences between variances. It is, therefore, prudent to test first whether the population variances are the same. If this hypothesis is rejected one should not use the test for means, which is discussed below, but refer to a different type of analysis.

4.11.5.1 Comparing variances of several samples

Suppose we have k samples, $k \geq 2$. Let $S_{n_1}^2, S_{n_2}^2, \ldots, S_{n_k}^2$ denote the variances of these samples. Let $S_{\max}^2 = \max \{S_{n_1}^2, \ldots, S_{n_k}^2\}$ and $S_{\min}^2 = \min \{S_{n_1}^2, \ldots, S_{n_k}^2\}$. The test statistic which we consider is the ratio of the maximal to the minimal variances, i.e.,

$$\tilde{F} = S_{\max}^2 / S_{\min}^2. \tag{4.11.7}$$

The hypothesis under consideration is

$$H : \sigma_1^2 = \sigma_2^2 = \cdots = \sigma_k^2.$$

To test this hypothesis, we construct the following EBD:

- *Step 1*: Sample independently RSWR of sizes $n_1, \ldots, n_k$ respectively, from the given samples. Let $S_{n_1}^{*2}, \ldots, S_{n_k}^{*2}$ be the sample variances of these bootstrap samples.
- *Step 2*: Compute $W_i^{*2} = \dfrac{S_{n_i}^{*2}}{S_{n_i}^2}, i = 1, \ldots, k.$
- *Step 3*: Compute $\tilde{F}^* = \max\limits_{1 \leq i \leq k} \{W_i^{*2}\} / \min\limits_{1 \leq i \leq k} \{W_i^{*2}\}.$

Repeat these steps M times to obtain the EBD of $\tilde{F}_1^*, \ldots, \tilde{F}_M^*$.

Let $\tilde{F}_{1-\alpha}^*$ denote the $(1 - \alpha)$th quantile of this EBD distribution. The hypothesis H is **rejected** with level of significance α, if $\tilde{F} > \tilde{F}_{1-\alpha}^*$. The corresponding P^* level is the proportion of $\tilde{F}^*$ values which are greater than $\tilde{F}$.

Example 4.22. We compare now the variances of the resistance Res 3 in three hybrids. The data file is **HYBRID.csv**. In the present example, $n_1 = n_2 = n_3 = 32$. We find that $S_{n_1}^2 = 9929.54$, $S_{n_2}^2 = 16648.35$ and $S_{n_3}^2 = 21001.01$. The ratio of the maximal to minimal variance is $\tilde{F} = 2.11$. With $M = 500$ bootstrap samples, we find that $P^* = 0.582$. For $\alpha = 0.05$, we find that $\tilde{F}_{0.95}^* = 2.515$. The sample $\tilde{F}$ is smaller than $\tilde{F}_{0.95}^*$. The hypothesis of equal variances cannot be rejected at a level of significance of $\alpha = 0.05$.

In R

```
> data(HYBRID)
> set.seed(123)
> B <- apply(HYBRID, MARGIN=2,
```

```
                    FUN=boot,
                    statistic=function(x, i){
                       var(x[i])
                    },
                    R = 500)
> Bt0 <- sapply(B,
                    FUN=function(x) x$t0)
> Bt <-   sapply(B,
                    FUN=function(x) x$t)
> Bf <- max(Bt0)/min(Bt0)
> FBoot <- apply(Bt, MARGIN=1,
                    FUN=function(x){
                       max(x)/min(x)
                    })
> Bf

[1] 2.115003

> quantile(FBoot, 0.95)

      95%
4.214902

> sum(FBoot <= Bf)/length(FBoot)

[1] 0.606

> rm(Bt0, Bt, Bf, FBoot)
```

4.11.5.2   Comparing several means: the one-way analysis of variance

The one-way analysis of variance, ANOVA, is a procedure of testing the equality of means, assuming that the variances of the populations are all equal. The hypothesis under test is

$$H : \mu_1 = \mu_2 \cdots = \mu_k.$$

Let $\bar{X}_{n_1}, S^2_{n_1}, \ldots, \bar{X}_{n_k}, S^2_{n_k}$ be the means and variances of the k samples. We compute the test statistic

$$F = \frac{\sum_{i=1}^{k} n_i (\bar{X}_{n_i} - \bar{\bar{X}})^2 / (k-1)}{\sum_{i=1}^{k} (n_i - 1) S^2_{n_i} / (N - k)}, \tag{4.11.8}$$

where

$$\bar{\bar{X}} = \frac{1}{N} \sum_{i=1}^{k} n_i \bar{X}_{n_i} \tag{4.11.9}$$

is the weighted average of the sample means, called the **grand mean**, and $N = \sum_{i=1}^{k} n_i$ is the total number of observations.

According to the bootstrap method for testing equality of means, we repeat the following procedure M times:

- *Step 1*: Draw k RSWR of sizes $n_1, \ldots, n_k$ from the k given samples.
- *Step 2*: For each bootstrap sample, compute the mean and variance $\bar{X}_{n_i}^*$ and $S_{n_i}^{*2}$, $i = 1, \ldots, k$.
- *Step 3*: For each $i = 1, \ldots, k$ compute

$$\bar{Y}_i^* = \bar{X}_{n_i}^* - (\bar{X}_{n_i} - \bar{\bar{X}}).$$

[Notice that $\bar{\bar{Y}}^* = \frac{1}{N} \sum_{i=1}^{k} n_i \bar{Y}_i^* = \bar{\bar{X}}^*$, which is the grand mean of the k bootstrap samples.]

- *Step 4*: Compute

$$
F^* = \frac{\sum_{i=1}^{k} n_i (\bar{Y}_i^* - \bar{\bar{Y}}^*)^2 / (k-1)}{\sum_{i=1}^{k} (n_i - 1) S_{n_i}^{*2} / (N-k)}
$$

$$
= \frac{\left[\sum_{i=1}^{k} n_i (\bar{X}_{n_i}^* - \bar{X}_{n_i})^2 - N(\bar{\bar{X}} - \bar{\bar{X}}^*)^2 \right] / (k-1)}{\sum_{i=1}^{k} (n_i - 1) S_{n_i}^{*2} / (N-k)}. \tag{4.11.10}
$$

After M repetitions, we obtain the EBD of $F_1^*, \ldots, F_M^*$.

Let $F_{1-\alpha}^*$ be the $(1-\alpha)$th quantile of this EBD. The hypothesis H_0 is **rejected**, at level of significance α, if $F > F_{1-\alpha}^*$. Alternatively, H is rejected if the P^*-level is small, where

$$P^* = \text{proportion of } F^* \text{ values greater than } F$$

Example 4.23. Testing post hoc power for stating nonequality of the means in the **HYBRID.csv** file, we obtain, using $M = 500$ bootstrap replicates, the following statistics:

$$\text{Hybrid1: } \bar{X}_{32} = 2143.406, \quad S_{32}^2 = 9929.539.$$

$$\text{Hybrid2: } \bar{X}_{32} = 1902.813, \quad S_{32}^2 = 16648.351.$$

$$\text{Hybrid3: } \bar{X}_{32} = 1850.344, \quad S_{32}^2 = 21001.007.$$

If we skip step 3 above, i.e. we compute the F statistic from bootstrapping each sample separately without the adjustment in step 3, the test statistic is $F = 49.274$. The P^* level for this F is 0.0000. Thus, the hypothesis H is **rejected** with posthoc power of 1. This represents the probability of rejecting the hypothesis of equal populations for the difference in the samples observed in the data, or in more extreme differences. JMP has a feature for computing posthoc power. The histogram of the EBD of the F^* values is presented in Figure 4.18.

In R

```
> onewayTestBoot <- function(x, i){
    x <- x[i,]
    y <- stack(x)
    names(y) <- c("v", "g")
    oneway.test(v ~ g,
                data=y,
                var.equal=TRUE)$statistic
}
> set.seed(123)
```

```
> B <- boot(data=HYBRID,
            statistic=onewayTestBoot,
            R=500)
> B$t0

        F
49.27359

> sum(B$t > B$t0)/nrow(B$t)

[1] 0.616.
```

We can state that with a difference in population means, such as in the data, we will observe these and more extreme values with a probability of 0.616. Such differences are therefore highly identified as significant.

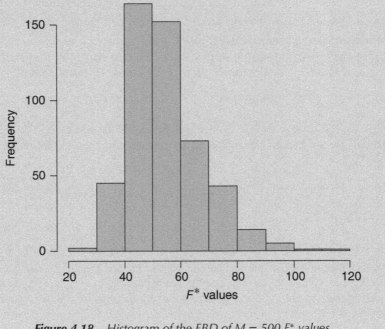

Figure 4.18   *Histogram of the EBD of M = 500 F* values.*

4.12   Bootstrap tolerance intervals

4.12.1   Bootstrap tolerance intervals for Bernoulli samples

Trials (experiments) are called **Bernoulli** trials, if the results of the trials are either 0 or 1 (Head or Tail; Good or Defective, etc.); the trials are independently performed, and the probability for 1 in a trial is a fixed constant p, $0 < p < 1$. A random sample (RSWR) of size n, from a population of 0's and 1's, whose mean is p (proportion of 1's) will be called a **Bernoulli Sample**.

The number of 1's in such a sample has a binomial distribution. This is the sampling distribution of the number of 1's in all possible Bernoulli sample of size n, and population mean p. p is the probability that in a random drawing of an element from the population, the outcome is 1.

Let X be the number of 1's in a RSWR of size n from such a population. If p is known, we can determine two integers $I_{\beta/2}(p)$ and $I_{1-\beta/2}(p)$ such that, the proportion of Bernoulli samples for which $I_{\beta/2}(p) \leq X \leq I_{1-\beta/2}(p)$ is $(1 - \beta)$. Using R

```
> qbinom(p=c(0.025, 0.975), size=50, prob=0.1)
```

```
[1]  1  9
```

Using MINITAB, we obtain these integers with the command

```
MTB> let k1 = p
MTB> Inv_CDF β/2 k₂;
SUBC> Binomial n p.
```

For example, if $n = 50$, $p = 0.1$, $\beta = 0.05$, we obtain $I_{0.025}(0.1) = 1 \ (= k_2)$, and $I_{0.975}(0.1) = 9$.

If p is unknown and has to be estimated from a given Bernoulli sample of size n, we determine first the bootstrap $(1 - \alpha)$ level confidence interval for p. If the limits for this interval are $(p_{\alpha/2}^*, p_{1-\alpha/2}^*)$, then the prediction interval $(I_{\beta/2}(p_{\alpha/2}^*), I_{1-\beta/2}(p_{1-\alpha/2}^*))$ is a Bootstrap **tolerance interval** of confidence $(1 - \alpha)$ and content $(1 - \beta)$. The MACRO **BINOPRED.MTB** can be executed to obtain tolerance intervals. In this macro, $k1$ is the future Bernoulli sample size, n, $k3$ is $\beta/2$ and $k4$ is $1 - \beta/2$. $k7$ is the size of the given Bernoulli sample, which is stored in column $C1$. Initiate $C3(1) = k1$, $C(4)(1) = k1$.

Example 4.24.   Consider the $n = 99$ electric voltage outputs of circuits, which is in data file **OELECT.csv**. Suppose that it is required that the output X will be between 216 and 224 V. We create a Bernoulli sample in which, we give a circuit the value 1 if its electric output is in the interval $(216, 224)$ and the value 0 otherwise. This Bernoulli sample is stored in file **ELECINDX.csv**.

The objective is to determine a $(0.95, 0.95)$ tolerance interval for a future batch of $n = 100$ circuits from this production process. Using MINITAB, we import file **ELECINDX.csv** to column $C1$. We set $k7 = 99$, $k1 = 100$, $k3 = 0.025$ and $k4 = 0.975$, and then apply MACRO **BINOPRED.MTB** $M = 500$ times. The next step is to order the columns $C3$ and $C4$, by the commands

```
MTB> SORT C3 C5
MTB> SORT C4 C6
```

Since $M\beta/2 = 500 \times 0.025 = 12.5$,

$$I_{0.025}(p_{0.025}^*) = (C5(12) + C5(13))/2 = 48$$

and

$$I_{0.975}(p_{0.975}^*) = (C6(487) + C6(488))/2 = 84.$$

```
> data(OELECT)
> ELECINDX <- ifelse(
    test=OELECT >= 216 &
      OELECT <= 224,
    yes=1, no=0)
> qbinomBoot <- function(x, i,
                        size,
                        probs=c(0.025,
                                0.975)){

    qbinom(p=probs,
          size=size,
          prob=mean(x[i]))
  }
> set.seed(123)
> B <- boot(data=ELECINDX,
            statistic=qbinomBoot,
            R = 500, size = 100)
> quantile(x=B$t[,1],
          probs=c(0.025, 0.975))


 2.5% 97.5%
   49    68
```

The Bootstrap tolerance interval is (48,84). In other words, with confidence level of 0.95, we predict that 95% of future batches of $n = 100$ circuits, will have between 48 and 84 circuits which comply to the standard. The exact tolerance intervals are given by

$$\text{Lower} = B^{-1}\left(\frac{\beta}{2}; n, \underline{p}_\alpha\right)$$

$$\text{Upper} = B^{-1}\left(1 - \frac{\beta}{2}; n, \bar{p}_\alpha\right),$$

(4.12.1)

where $(\underline{p}_\alpha, \bar{p}_\alpha)$ is a $(1 - \alpha)$ confidence interval for p. In the present data, the 0.95-confidence interval for p is (0.585,0.769). Thus, the (0.95,0.95) tolerance interval is (48,84), which is equal to the bootstrap interval.

4.12.2   Tolerance interval for continuous variables

In a RSWR of size n, the p-th quantile, i.e., $X_{(np)}$, is an estimator of the pth quantile of the distribution. Thus, we expect that the proportion of X-values in the population, falling in the interval $(X_{(n\beta/2)}, X_{(n(1-\beta/2))})$ is approximately $(1 - \beta)$ in large samples. As was explained in Chapter 2, $X_{(j)}$, $(j = 1, \ldots, n)$ is the j-th order statistic of the sample, and for $0 < p < 1$, $X_{(j,p)} = X_{(j)} + p(X_{(j+1)} - X_{(j)})$. By the bootstrap method, we generate M replicas of the statistics $X^*_{(n\beta/2)}$ and $X^*_{(n(1-\beta/2))}$. The $(1 - \alpha, 1 - \beta)$-tolerance interval is given by $(Y^*_{(M\alpha/2)}, Y^{**}_{(M(1-\alpha/2))})$, where $Y^*_{(M\alpha/2)}$ is the $\alpha/2$-quantile of the EBD of $X^*_{(n\beta/2)}$ and $Y^{**}_{(M(1-\alpha/2))}$ is the $(1 - \alpha/2)$-quantile of the EBD of $X^*_{(n(1-\beta/2))}$. Macro **CONTPRED.MTB** provides M bootstrap copies of $X^*_{(n\beta/2)}$ and $X^*_{(n(1-\beta/2))}$, from which we determine the tolerance limits.

Example 4.25. Let us determine (0.95,0.95)-tolerance interval for samples of size $n = 100$, of piston cycle times. Use the sample in the data file **CYCLT.csv**. The original sample is of size $n_0 = 50$. Since future samples are of size $n = 100$, we draw from the original sample RSWR of size $n = 100$. This bootstrap sample is put into column $C2$, and the ordered bootstrap sample is put in column $C3$. Since $n\beta/2 = 2.5$, $X^*_{(2.5)} = (C3(2) + C2(3))/2$. Similarly, $X^*_{(97.5)} = (C3(97) + C3(98))/2$. M replicas of $X^*_{(2.5)}$ and $X^*_{(97.5)}$ are put, respectively, in columns $C4$ and $C5$, $M = 500$. Finally, we sort column $C4$ and put it in $C6$, and sort $C5$ and put it in $C7$. $Y^*_{(M\alpha/2)} = (C6(12) + C6(13))/2$ and $Y^{**}_{(M(1-\alpha/2))} = (C7(487) + C7(488))/2$. A copy of this MINITAB session is given in the following window where we first opened a folder called MISTAT.

```
MTB> store 'C:\MISTAT\CONTPRED.MTB'
Storing in file: C:\MISTAT\CONTPRED.MTB
STOR> sample 100 C1 C2;
STOR> replace.
STOR> sort C2 C3
STOR> let k1 = (C3(2) + C3(3))/2
STOR> let k2 = (C3(97) + C3(98))/2
STOR> stack C4 k1 C4
STOR> stack C5 k2 C5
STOR> end
MTB> exec 'C:\MISTAT\CONTPRED.MTB' 500
Executing from file: C:\MISTAT\CONTPRED.MTB
MTB> sort C4 C6
MTB> sort C5 C7
MTB> let k3 = (C6(12) + C6(13))/2
MTB> let k4 = (C7(487) + C7(488))/2
```

To use MACRO **CONTPRED.MTB** on samples of size $n \neq 100$ and for $\beta \neq 0.05$, we need to edit it first. Change the sample size 100 in row 1 to the new n, and correspondingly modify $k1$ and $k2$. The bootstrap (0.95,0.95)-tolerance interval for $n = 100$ piston cycle times was estimated as (0.175,1.141).

In R, the calculations are straightforward:

```
> data(CYCLT)
> set.seed(123)
> B <- boot(CYCLT,
            statistic=function(x, i){
              quantile(x[i],
                       probs=c(0.025, 0.975))},
            R=500)
> quantile(x=B$t[,1], probs=0.025)

 2.5%
0.175

> quantile(x=B$t[,2], probs=0.975)

97.5%
1.141
```

4.12.3   Distribution free tolerance intervals

The tolerance limits described above are based on the model of normal distribution. **Distribution free** tolerance limits for $(1 - \beta)$ proportion of the population, at confidence level $(1 - \alpha)$, can be obtained for any model of continuous c.d.f. $F(x)$. As we will show below, if the sample size n is large enough, so that the following inequality is satisfied, i.e.,

$$\left(1 - \frac{\beta}{2}\right)^n - \frac{1}{2}(1 - \beta)^n \leq \frac{\alpha}{2} \qquad (4.12.2)$$

then the order statistics $X_{(1)}$ and $X_{(n)}$ are lower and upper tolerance limits. This is based on the following important property:

If X is a random variable having a continuous c.d.f. $F(x)$, then $U = F(x)$ has a uniform distribution on $(0, 1)$.

Indeed,

$$\Pr\{F(X) \leq \eta\} = \Pr\{X \leq F^{-1}(\eta)\}$$
$$= F(F^{-1}(\eta)) = \eta, \quad 0 < \eta < 1.$$

If $X_{(i)}$ is the i-th order statistic of a sample of n i.i.d. random variables having a common c.d.f. $F(x)$, then $U_{(i)} = F(X_{(i)})$ is the i-th order statistic of n i.i.d. random having a uniform distribution. Now, the interval $(X_{(1)}, X_{(n)})$ contains at least a proportion $(1 - \beta)$ of the population if $X_{(1)} \leq \xi_{\beta/2}$ and $X_{(n)} \geq \xi_{1-\beta/2}$, where $\xi_{\beta/2}$ and $\xi_{1-\beta/2}$ are the $\beta/2$ and $\left(1 - \frac{\beta}{2}\right)$ quantiles of $F(x)$.

Equivalently, $(X_{(1)}, X_{(n)})$ contains at least a proportion $(1 - \beta)$ if

$$U_{(1)} \leq F(\xi_{\beta/2}) = \frac{\beta}{2}$$
$$U_{(n)} \geq F(\xi_{1-\beta/2}) = 1 - \beta/2.$$

By using the joint p.d.f. of $(U_{(1)}, U_{(n)})$, we show that

$$\Pr\left\{U_{(1)} \leq \frac{\beta}{2}, U_{(n)} \geq 1 - \frac{\beta}{2}\right\} = 1 - 2\left(1 - \frac{\beta}{2}\right)^n + (1 - \beta)^n. \qquad (4.12.3)$$

This probability is the confidence that the interval $(X_{(1)}, X_{(n)})$ covers the interval $(\xi_{\beta/2}, \xi_{1-\beta/2})$. By finding n which satisfies

$$1 - 2\left(1 - \frac{\beta}{2}\right)^n + (1 - \beta)^n \geq 1 - \alpha, \qquad (4.12.4)$$

we can assure that the confidence level is at least $(1 - \alpha)$.

In Table 4.13, we give the values of n for some α and β values.

Table 4.13 can also be used to obtain the confidence level associated with fixed values of β and n. We see that with a sample of size 104, $(X_{(1)}, X_{(n)})$ is a tolerance interval for at least 90% of the population with approximately 99% confidence level or a tolerance interval for at least 95% of the population with slightly less than 90% confidence.

Other order statistics can be used to construct distribution-free tolerance intervals. That is, we can choose any integers j and k, where $1 \leq j, k \leq n/2$ and form the interval $(X_{(j)}, X_{(n-k+1)})$.

Table 4.13 *Sample size required for $(X_{(1)}, X_{(n)})$ to be a $(1 - \alpha, 1 - \beta)$ level tolerance interval*

β	α	n
0.10	0.10	58
	0.05	72
	0.01	104
0.05	0.10	118
	0.05	146
	0.01	210
0.01	0.10	593
	0.05	734
	0.01	1057

When $j > 1$ and $k > 1$, the interval will be shorter than the interval $(X_{(1)}, X_{(n)})$, but its confidence level will be reduced.

4.13 Nonparametric tests

Testing methods like the Z-tests, t-tests, etc. presented in this chapter, were designed for specific distributions. The Z- and t-tests are based on the assumption that the parent population is normally distributed. What would be the effect on the characteristics of the test if this basic assumption is wrong? This is an important question, which deserves special investigation. We remark that if the population variance σ^2 is finite and the sample is large, then the t-test for the mean has approximately the required properties even if the parent population is not normal. In small samples, if it is doubtful whether the distribution of the parent population, we should perform a distribution free test, or compute the P-value of the test statistic by the bootstrapping method. In the present section, we present three nonparametric tests, the so-called **sign test** the **randomization test** and the **Wilcoxon signed-rank test**.

4.13.1 The sign test

Suppose that $X_1, \ldots, X_n$ is a random sample from some **continuous** distribution, F, and has a positive p.d.f. throughout the range of X. Let ξ_p, for some $0 < p < 1$, be the p-th quantile of F. We wish to test the hypothesis that ξ_p does not exceed a specified value ξ^*, i.e.,

$$H_0 : \xi_p \leq \xi^*$$

against the alternative

$$H_1 : \xi_p > \xi^*.$$

If the null hypothesis H_0 is true, the probability of observing an X-value smaller than ξ^* is greater or equal to p; and if H_1 is true, then this probability is smaller than p. The sign test of H_0 versus H_1 reduces the problem to a test for p in a binomial model. The test statistic is $K_n = \#\{X_i \leq \xi^*\}$, i.e., the number of observed X-values in the sample which do not exceed ξ^*. K_n has a binomial distribution $B(n, \theta)$, irrespective of the parent distribution F. According to H_0, $\theta \geq p$, and according to H_1, $\theta < p$. The test proceeds then as in Section 4.3.2.4.

Example 4.26.   In continuation of Example 4.25, we wish to test whether the median, $\xi_{0.5}$, of the distribution of piston cycle times, is greater than 0.50 [minutes]. The sample data is in file **CYCLT.csv**. The sample size is $n = 50$. Let $K_{50} = \sum_{i=1}^{50} I\{X_i \leq 50\}$. The null hypothesis is $H_0 : p \leq \frac{1}{2}$ versus $H_1 : p > \frac{1}{2}$. From the sample values, we find $K_{50} = 24$. The P-value is $1 - B(23; 50, 0.5) = 0.664$. The null hypothesis H_0 is not rejected. The sample median is $M_e = 0.546$. This is however not significantly greater than 0.5.

The sign test can be applied also to test whether tolerance specifications hold. Suppose that the standard specifications require that at least $(1 - \beta)$ proportion of products will have an X value in the interval (ξ^*, ξ^{**}). If we wish to test this, with level of significance α, we can determine the $(1 - \alpha, 1 - \beta)$ tolerance interval for X, based on the observed random sample, and accept the hypothesis

$$H_0 : \xi^* \leq \xi_{\beta/2} \quad \text{and} \quad \xi_{1-\beta/2} \leq \xi^{**}$$

if the tolerance interval is included in (ξ^*, ξ^{**}).

We can also use the sign test. Given the random sample $X_1, \ldots, X_n$, we compute

$$K_n = \sum_{i=1}^{n} I\{\xi^* \leq X_i \leq \xi^{**}\}.$$

The null hypothesis H_0 above is equivalent to the hypothesis

$$H_0^* : p \geq 1 - \beta$$

in the binomial test. H_0^* is rejected, with level of significance α, if

$$K_n < B^{-1}(\alpha; n, 1 - \beta),$$

where $B^{-1}(\alpha; n, 1 - \beta)$ is the α-quantile of the binomial distribution $B(n, 1 - \beta)$.

Example 4.27.   In Example 4.25, we have found that the bootstrap $(0.95, 0.95)$ tolerance interval for the **CYCLT.csv** sample is $(0.175, 1.141)$. Suppose that the specification requires that the piston cycle time in 95% of the cases will be in the interval $(0.2, 1.1)$ [minutes]. Can we accept the hypothesis

$$H_0^* : 0.2 \leq \xi_{0.025} \quad \text{and} \quad \xi_{0.975} \leq 1.1$$

with level of significance $\alpha = 0.05$? For the data **CYCLT.csv**, we find

$$K_{50} = \sum_{i=1}^{50} I\{0.2 \leq X_i \leq 1.1\} = 41.$$

Also $B^{-1}(0.05, 50, 0.95) = 45$. Thus, since $K_{50} < 45$ H_0^* is rejected. This is in accord with the bootstrap tolerance interval, since $(0.175, 1.141)$ contains the interval $(0.2, 1.1)$.

4.13.2   The randomization test

The randomization test described here can be applied to test whether two random samples come from the same distribution, F, without specifying the distribution F.

The null hypothesis, H_0, is that the two distributions, from which the samples are generated, are the same. The randomization test constructs a reference distribution for a specified test statistic, by randomly assigning to the observations the labels of the samples. For example, let us consider two samples, which are denoted by A_1 and A_2. Each sample is of size $n = 3$. Suppose that we observed

$$A_2 \quad A_2 \quad A_2 \quad A_1 \quad A_1 \quad A_1$$
$$1.5 \quad 1.1 \quad 1.8 \quad 0.75 \quad 0.60 \quad 0.80.$$

The sum of the values in A_2 is $T_2 = 4.4$ and that of A_1 is $T_1 = 2.15$. Is there an indication that the two samples are generated from different distributions? Let us consider the test statistic $D = (T_2 - T_1)/3$ and reject H_0 if D is sufficiently large. For the given samples, $D = 0.75$. We construct now the reference distribution for D under H_0.

There are $\binom{6}{3} = 20$ possible assignments of the letters A_1 and A_2 to the six values. Each such assignment yields a value for D. The reference distribution assigns each such value of D an equal probability of 1/20. The 20 assignments of letters and the corresponding D values are given in Table 4.14.

Under the reference distribution, each one of these values of D is equally probable, and the P-value of the observed value of the observed D is $P = \dfrac{1}{20} = 0.05$. The null hypothesis is

Table 4.14   *Assignments for the randomized test*

Y_{ij}	Assignments									
0.75	1	1	1	1	1	1	1	1	1	1
0.60	1	1	1	1	2	2	2	2	2	2
0.80	1	2	2	2	1	1	1	2	2	2
1.5	2	1	2	2	1	2	2	1	1	2
1.1	2	2	1	2	2	1	2	1	2	1
1.8	2	2	2	1	2	2	1	2	1	1
D	0.750	0.283	0.550	0.083	0.150	0.417	−0.05	−0.050	−0.517	−0.250

Y_{ij}	Assignments									
0.75	2	2	2	2	2	2	2	2	2	2
0.60	1	1	1	1	1	1	2	2	2	2
0.80	1	1	1	2	2	2	1	1	1	2
1.5	1	2	2	1	1	2	1	1	2	1
1.1	2	1	2	1	2	1	1	2	1	1
1.8	2	2	1	2	1	1	2	1	1	1
D	0.250	0.517	0.050	0.050	−0.417	−0.150	−0.083	−0.550	−0.283	−0.750

rejected at the $\alpha = 0.05$ level. If n is large, it becomes impractical to construct the reference distribution in this manner. For example, if $t = 2$ and $n_1 = n_2 = 10$, we have $\binom{20}{10} = 184,756$ assignments.

We can, however, estimate the P-value, by sampling, without replacement, from this reference distribution. This can be attained by MINITAB, using the macro **RANDTES2.MTB**. Before executing **RANDTES2.MTB**, we make the following preparation. We import the data file containing the two samples into column $C1$. In this column sample, A occupies the first n_1 rows, and sample B the last n_2 rows. After this, we perform the following MINITAB commands:

```
MTB> LET K1 = (n1).
MTB> LET K2 = (n2)
MTB> LET K3 = K1 + 1
MTB> LET K4 = K1 + K2
MTB> COPY C1 C3;
SUBC> USE(1 : K1).
MTB> COPY C1 C4;
SUBC> USE(K3 : K4).
MTB> LET K5 = MEAN(C3) − MEAN(C4)
MTB> LET C5(1) = K5.
```

After this, we can execute macro **RANDTES2.MTB**. The first value in column $C5$ is the actual observed value of the test statistic.

Example 4.28. File **OELECT.csv** contains $n_1 = 99$ random values of the output in volts of a rectifying circuit. File **OELECT1.csv** contains $n_2 = 25$ values of outputs of another rectifying circuit. The question is whether the differences between the means of these two samples is significant. Let $\bar{X}$ be the mean of **OELECT** and $\bar{Y}$ be that of **OELECT1**. We find that $D = \bar{X} - \bar{Y} = -10.7219$. Executing macro **RANDTES2.MTB** 500 times yields results which, together with the original D are described by the following sample statistics.

Descriptive Statistics

Variable	N	Mean	Median	TrMEAN	StDev	SEMean
C5	501	−0.816	−0.0192	−0.0450	1.6734	0.0748
Variable	Min	Max	$Q1$	$Q3$		
C5	−10.7219	4.3893	−1.1883	1.0578		

Thus, the original mean -10.721 is the minimum, and the test rejects the hypothesis of equal means with a P-value $P = \frac{1}{501}$.

In R, we use the function randomizationTest that is included in mistat package.

```
> data(OELECT1)
> randomizationTest(list(a=OELECT, b=OELECT1),
                    R=500, calc=mean,
```

```
                  fun=function(x) x[1]-x[2],
                  seed=123)

Original stat is at quantile 1 over 501   ( 0.2 %)
Original stat is -10.72198
```

4.13.3   The Wilcoxon signed rank test

In Section 7.5.1, we discussed the sign test. The Wilcoxon signed rank (WSR) test is a modification of the sign test, which brings into consideration not only the signs of the sample values but also their magnitudes. We construct the test statistic in two steps. First, we rank the magnitudes (absolute values) of the sample values, giving the rank 1 to the value with smallest magnitude, and the rank n to that with the maximal magnitude. In the second step, we sum the ranks multiplied by the signs of the values. For example, suppose that a sample of $n = 5$ is -1.22, -0.53, 0.27, 2.25, 0.89. The ranks of the magnitudes of these values are, respectively, 4, 2, 1, 5, 3. The signed rank statistic is

$$W_5 = 0 \times 4 + 0 \times 2 + 1 + 5 + 3 = 9.$$

Here we assigned each negative value the weight 0 and each positive value the weight 1.

The WSR test can be used for a variety of testing problems. If we wish to test whether the distribution median, $\xi_{0.5}$, is smaller or greater than some specified value ξ^*, we can use the statistics

$$W_n = \sum_{i=1}^{n} I\{X_i > \xi^*\} R_i, \qquad (4.13.1)$$

where

$$I\{X_i > \xi^*\} = \begin{cases} 1, & \text{if } X_i > \xi^* \\ 0, & \text{otherwise.} \end{cases}$$

$R_i = \text{rank}(|X_i|)$.

The WSR test can be applied to test whether two random samples are generated from the same distribution against the alternative that one comes from a distribution having a larger location parameter (median) than the other. In this case, we can give the weight 1 to elements of sample 1 and the weight 0 to the elements of sample 2. The ranks of the values are determined by combining the two samples. For example, consider two random samples $X_1, \ldots, X_5$ and $Y_1, \ldots, Y_5$ generated from $N(0, 1)$ and $N(2, 1)$. These are

X	0.188	0.353	−0.257	0.220	0.168
Y	1.240	1.821	2.500	2.319	2.190

The ranks of the magnitudes of these values are

X	2	5	4	3	1
Y	6	7	10	9	8

The value of the WSR statistic is

$$W_{10} = 6 + 7 + 10 + 9 + 8 = 40.$$

Notice that all the ranks of the Y values are greater than those of the X values. This yields a relatively large value of W_{10}. Under the null hypothesis that the two samples are from the same distribution, the probability that the sign of a given rank is 1 is 1/2. Thus, the reference distribution, for testing the significance of W_n, is like that of

$$W_n^0 = \sum_{j=1}^{n} jB_j \left(1, \frac{1}{2}\right), \tag{4.13.2}$$

where $B_1\left(1, \frac{1}{2}\right), \ldots, B_n\left(1, \frac{1}{2}\right)$ are mutually independent $B\left(1, \frac{1}{2}\right)$ random variables. The distribution of W_n^0 can be determined exactly. W_n^0 can assume the values $0, 1, \ldots, \frac{n(n+1)}{2}$ with probabilities which are the coefficients of the polynomial in t

$$P(t) = \frac{1}{2^n} \prod_{j=1}^{n} \left(1 + t^j\right).$$

These probabilities can be computed exactly. For large values of n, W_n^0 is approximately normal with mean

$$E\{W_n^0\} = \frac{1}{2} \sum_{j=1}^{n} j = \frac{n(n+1)}{4} \tag{4.13.3}$$

and variance

$$V\{W_n^0\} = \frac{1}{4} \sum_{j=1}^{n} j^2 = \frac{n(n+1)(2n+1)}{24}. \tag{4.13.4}$$

This can yield a large sample approximation to the P-value of the test. The WSR test, to test whether the median of a symmetric continuous distribution F is equal to ξ^* or not can be performed in R

```
> X <- c(0.188, 0.353, -0.257, 0.220, 0.168)
> Y <- c(1.240, 1.821, 2.500, 2.319, 2.190)
> wilcox.test(x=X, y=Y,
            conf.int = TRUE)

        Wilcoxon rank sum exact test

data:  X and Y
W = 0, p-value = 0.007937
alternative hypothesis: true location shift is not equal to 0
95 percent confidence interval:
 -2.447 -1.052
sample estimates:
difference in location
                -2.002

> rm(X, Y)
```

and in MINITAB, by using the command

MTB> WTest $k1$ $C1$

where $k1$ is the value of ξ^* and $C1$ is the column in which the sample resides.

4.14   Description of MINITAB macros

CONFINT.MTB: Computes 2-sigma confidence intervals for a sample of size $k1$ to demonstrate the coverage probability of a confidence interval.

BOOT1SMP.MTB: Computes the mean and standard deviation of bootstrap distributions from a single sample. (Note: Columns where stacking occurs have to be initiated.)

BOOTPERC.MTB: Computes the first quartile, the mean and the third quartile of bootstrap distributions from a single sample.

BOOTREGR.MTB: Computes the least squares coefficients of a simple linear regression and their standard errors from bootstrap distributions from a single sample with two variables measured simultaneously.

BINOPRED.MTB: Computes bootstrapped tolerance intervals from Bernoulli samples.

CONTPRED.MTB: Computes bootstrapped 95% tolerance intervals for means from one sample.

RANDTEST.MTB: Computer randomization distribution for one sample means.

RANDTES2.MTB: Computer randomization distribution for comparison of two sample means.

RANDTES3.MTB: Computer randomization distribution for comparison of three sample means.

4.15   Chapter highlights

This chapter provides theoretical foundations for statistical inference. Inference on parameters of infinite populations is discussed using classical point estimation, confidence intervals, tolerance intervals, and hypothesis testing. Properties of point estimators such as moment equation estimators and maximum likelihood estimators are discussed in detail. Formulas for parametric confidence intervals and distribution-free tolerance intervals are provided. Statistical tests of hypothesis are presented with examples, including tests for normality with probability plots and the chi-square and KS tests of goodness of fit. The chapter includes a section on Bayesian testing and estimation.

Statistical inference is introduced by exploiting the power of the personal computer. Reference distributions are constructed through bootstrapping methods. Testing for statistical significance and the significance of least square methods in simple linear regression using bootstrapping is demonstrated. Industrial applications are used throughout with specially written software simulations. Through this analysis, confidence intervals and reference distributions are derived and used to test statistical hypothesis. Bootstrap analysis of variance is developed for testing the equality of several population means. Construction of tolerance intervals with bootstrapping is also presented. Three nonparametric procedures for testing are given: The sign test, randomization test, and the WSR test.

The main concepts and definitions introduced in this chapter include

- Statistical Inference
- Sampling Distribution

- Unbiased Estimators
- Consistent Estimators
- Standard Error
- Parameter Space
- Statistic
- Point Estimator
- Least Squares Estimators
- Maximum Likelihood Estimators
- Likelihood Function
- Confidence Intervals
- Tolerance Intervals
- Testing Statistical Hypotheses
- Operating Characteristic Function
- Rejection Region
- Acceptance Region
- Type I Error
- Type II Error
- Power Function
- OC Curve
- Significance Level
- P-Value
- Normal Scores
- Normal Probability Plot
- Chi-Squared Test
- Kolmogorov–Smirnov Test
- Bayesian Decision Procedures
- Statistical Inference
- The Bootstrap Method
- Sampling Distribution of an Estimate
- Reference Distribution
- Bootstrap Confidence Intervals
- Bootstrap Tolerance Interval
- Bootstrap ANOVA
- Nonparametric Tests

4.16   Exercises

4.1 The consistency of the sample mean, $\bar{X}_n$, in RSWR, is guaranteed by the weak law of large numbers (WLLN), whenever the mean exists. Let $M_l = \frac{1}{n} \sum_{i=1}^{n} X_i^l$ be the sample estimate of the l-th moment, which is assumed to exist ($l = 1, 2, \ldots$). Show that M_r is a consistent estimator of μ_r.

4.2 Consider a population with mean μ and standard deviation $\sigma = 10.5$. Use the CLT to find, approximately how large should the sample size, n, be so that $\Pr\{|\bar{X}_n - \mu| < 1\} = 0.95$.

4.3 Let $X_1, \ldots, X_n$ be a random sample from a normal distribution $N(\mu, \sigma)$. What is the moments equation estimator of the p-th quantile $\xi_p = \mu + z_p \sigma$?

4.4 Let $(X_1, Y_1), \ldots, (X_n, Y_n)$ be a random sample from a bivariate normal distribution. What is the moments equations estimator of the correlation ρ?

4.5 Let $X_1, X_2, \ldots, X_n$ be a sample from a beta distribution Beta(v_1, v_2); $0 < v_1, v_2 < \infty$. Find the moment-equation estimators of v_1 and v_2.

4.6 Let $\bar{Y}_1, \ldots, \bar{Y}_k$ be the means of k independent RSWR from normal distributions, $N(\mu, \sigma_i)$, $i = 1, \ldots, k$, with common means and variances σ_i^2 **known**. Let $n_1, \ldots, n_k$ be the sizes of these samples. Consider a weighted average $\bar{Y}_w = \frac{\sum_{i=1}^{k} w_i \bar{Y}_i}{\sum_{i=1}^{k} w_i}$, with $w_i > 0$. Show that for the estimator $\bar{Y}_w$ having smallest variance, the required weights are $w_i = \frac{n_i}{\sigma_i^2}$.

4.7 Using the formula

$$\hat{\beta}_1 = \sum_{i=1}^{n} w_i Y_i,$$

with $w_i = \frac{x_i - \bar{x}_n}{SS_x}$, $i = 1, \ldots, n$, for the LSE of the slope β in a simple linear regression, derive the formula for $V\{\hat{\beta}_1\}$. We assume that $V\{Y_i\} = \sigma^2$ for all $i = 1, \ldots, n$. You can refer to Chapter 5 for a detailed exposition of linear regression.

4.8 In continuation of Exercise 4.7, derive the formula for the variance of the LSE of the intercept β_0 and Cov$(\hat{\beta}_0, \hat{\beta}_1)$.

4.9 Show that the correlation between the LSE's, $\hat{\beta}_0$ and $\hat{\beta}_1$ in the simple linear regression is

$$\rho = -\frac{\bar{x}_n}{\left(\frac{1}{n} \sum x_i^2\right)^{1/2}}.$$

4.10 Let $X_1, \ldots, X_n$ be i.i.d. random variables having a Poisson distribution $P(\lambda)$, $0 < \lambda < \infty$. Show that the MLE of λ is the sample mean $\bar{X}_n$.

4.11 Let $X_1, \ldots, X_n$ be i.i.d. random variables from a gamma distribution, $G(v, \beta)$, with **known** v. Show that the MLE of β is $\hat{\beta}_n = \frac{1}{v} \bar{X}_n$, where $\bar{X}_n$ is the sample mean. What is the variance of $\hat{\beta}_n$?

4.12 Consider Example 4.4. Let $X_1, \ldots, X_n$ be a random sample from a negative-binomial distribution, NB$(2, p)$. Show that the MLE of p is

$$\hat{p}_n = \frac{2}{\bar{X}_n + 2},$$

where $\bar{X}_n$ is the sample mean.

 (i) On the basis of the WLLN, show that $\hat{p}_n$ is a consistent estimator of p [Hint: $\bar{X}_n \to E\{X\} = (2 - p)/p$ in probability as $n \to \infty$].

 (ii) Using the fact that if $X_1, \ldots, X_n$ are i.i.d. like NB(k, p) then $T_n = \sum_{i=1}^{n} X_i$ is distributed like NB(nk, p), and the results of Example 4.4, show that for large values of n,

$$\text{Bias}(\hat{p}_n) \cong \frac{3p(1-p)}{4n} \text{ and}$$

$$V\{\hat{p}_n\} \cong \frac{p^2(1-p)}{2n}.$$

4.13 Let $X_1, \ldots, X_n$ be a random sample from a shifted exponential distribution

$$f(x; \mu, \beta) = \frac{1}{\beta} \exp\left\{-\frac{x-\mu}{\beta}\right\}, \quad x \geq \mu,$$

where $0 < \mu, \beta < \infty$.
   (i)   Show that the sample minimum $X_{(1)}$ is an MLE of μ.
   (ii) Find the MLE of β.
   (iii) What are the variances of these MLEs?

4.14 We wish to test that the proportion of defective items in a given lot is smaller than $P_0 = 0.03$. The alternative is that $P > P_0$. A random sample of $n = 20$ is drawn from the lot **with replacement** (RSWR). The number of observed defective items in the sample is $X = 2$. Is there sufficient evidence to reject the null hypothesis that $P \leq P_0$?

4.15 Compute and plot the operating characteristic curve OC(p), for binomial testing of $H_0 : P \leq P_0$ versus $H_1 : P > P_0$, when the hypothesis is accepted if 2 or less defective items are found in a RSWR of size $n = 30$.

4.16 For testing the hypothesis $H_0 : P = 0.01$ versus $H_1 : P = 0.03$, concerning the parameter P of a binomial distribution, how large should the sample be, n, and what should be the critical value, k, if we wish that error probabilities will be $\alpha = 0.05$ and $\beta = 0.05$? [Use the normal approximation to the binomial.]

4.17 As will be discussed in Chapter 10, the Shewhart $3 - \sigma$ control charts, for statistical process control provide repeated tests of the hypothesis that the process mean is equal to the nominal one, μ_0. If a sample mean $\bar{X}_n$ falls outside the limits $\mu_0 \pm 3\frac{\sigma}{\sqrt{n}}$, the hypothesis is rejected.
   (i)   What is the probability that $\bar{X}_n$ will fall outside the control limits when $\mu = \mu_0$?
   (ii) What is the probability that when the process is in control, $\mu = \mu_0$, all sample means of 20 consecutive independent samples, will be within the control limits?
   (iii) What is the probability that a sample mean will fall outside the control limits when μ changes from μ_0 to $\mu_1 = \mu_0 + 2\frac{\sigma}{\sqrt{n}}$?
   (iv) What is the probability that, a change from μ_0 to $\mu_1 = \mu_0 + 2\frac{\sigma}{\sqrt{n}}$, will not be detected by the next 10 sample means?

4.18 Consider the data in file **SOCELL.csv**. Use R, MINITAB, or JMP to test whether the mean short circuit current of solar cells (ISC) at time t_1 is significantly smaller than 4 (Amp). [Use 1-sample t-test.]

4.19 Is the mean of ISC for time t_2 significantly larger than 4 (Amp)?

4.20 Consider a one-sided t-test based on a sample of size $n = 30$, with $\alpha = 0.01$. Compute the OC(δ) as a function of $\delta = (\mu - \mu_0)/\sigma$, $\mu > \mu_0$.

4.21 Compute the OC function for testing the hypothesis $H_0 : \sigma^2 \leq \sigma_0^2$ versus $H_1 : \sigma^2 > \sigma_0^2$, when $n = 31$ and $\alpha = 0.10$.

4.22 Compute the OC function in testing $H_0 : p \leq p_0$ versus $H_1 : p > p_0$ in the binomial case, when $n = 100$ and $\alpha = 0.05$.

4.23 Let $X_1, \ldots, X_n$ be a random sample from a normal distribution $N(\mu, \sigma)$. For testing $H_0 : \sigma^2 \leq \sigma_0^2$ against $H_1 : \sigma^2 > \sigma_0^2$, we use the test which reject H_0 if $S_n^2 \geq \frac{\sigma_0^2}{n-1}\chi_{1-\alpha}^2[n-1]$, where S_n^2 is the sample variance. What is the power function of this test?

4.24 Let $S_{n_1}^2$ and $S_{n_2}^2$ be the variances of two independent samples from normal distributions $N(\mu_i, \sigma_i)$, $i = 1, 2$. For testing $H_0 : \frac{\sigma_1^2}{\sigma_2^2} \leq 1$ against $H_1 : \frac{\sigma_1^2}{\sigma_2^2} > 1$, we use the F-test, which rejects H_0 when $F = \dfrac{S_{n_1}^2}{S_{n_2}^2} > F_{1-\alpha}[n_1 - 1, n_2 - 1]$. What is the power of this test, as a function of $\rho = \sigma_1^2/\sigma_2^2$?

4.25 A random sample of size $n = 20$ from a normal distribution gave the following values: 20.74, 20.85, 20.54, 20.05, 20.08, 22.55, 19.61, 19.72, 20.34, 20.37, 22.69, 20.79, 21.76, 21.94, 20.31, 21.38, 20.42, 20.86, 18.80, 21.41. Compute

  (i) Confidence interval for the mean μ, at level of confidence $1 - \alpha = 0.99$.

  (ii) Confidence interval for the variance σ^2, at confidence level $1 - \alpha = 0.99$.

  (iii) A confidence interval for σ, at level of confidence $1 - \alpha = 0.99$.

4.26 Let C_1 be the event that a confidence interval for the mean, μ, covers it. Let C_2 be the event that a confidence interval for the standard deviation σ covers it. The probability that both μ **and** σ are simultaneously covered is

$$\Pr\{C_1 \cap C_2\} = 1 - \Pr\{\overline{C_1 \cap C_2}\}$$
$$= 1 - \Pr\{\bar{C}_1 \cup \bar{C}_2\} \geq 1 - \Pr\{\bar{C}_1\} - \Pr\{\bar{C}_2\}.$$

This inequality is called the **Bonferroni inequality**. Apply this inequality and the results of the previous exercise to determine the confidence interval for $\mu + 2\sigma$, at level of confidence not smaller than 0.98.

4.27 A total of 20 independent trials yielded $X = 17$ successes. Assuming that the probability for success in each trial is the same, θ, determine the confidence interval for θ at level of confidence 0.95.

4.28 Let $X_1, \ldots, X_n$ be a random sample from a Poisson distribution with mean λ. Let $T_n = \sum_{i=1}^{n} X_i$. Using the relationship between the Poisson and the gamma c.d.f., we can show that a confidence interval for the mean λ, at level $1 - \alpha$, has lower and upper limits, λ_L and λ_U, where

$$\lambda_L = \frac{1}{2n}\chi_{\alpha/2}^2[2T_n + 2], \text{ and}$$

$$\lambda_U = \frac{1}{2n}\chi_{1-\alpha/2}^2[2T_n + 2].$$

The following is a random sample of size $n = 10$ from a Poisson distribution 14, 16, 11, 19, 11, 9, 12, 15, 14, 13. Determine a confidence interval for λ at level of confidence 0.95. [Hint: for large number of degrees of freedom $\chi_p^2[v] \approx v + z_p\sqrt{2v}$, where z_p is the p-th quantile of the standard normal distribution.]

4.29 The mean of a random sample of size $n = 20$, from a normal distribution with $\sigma = 5$, is $\bar{Y}_{20} = 13.75$. Determine a $1 - \beta = 0.90$ content tolerance interval with confidence level $1 - \alpha = 0.95$.

4.30 Use the **YARNSTRG.csv** data file to determine a $(0.95, 0.95)$ tolerance interval for log-yarn strength. [Hint: Notice that the interval is $\bar{Y}_{100} \pm kS_{100}$, where $k = t(0.025, 0.025, 100)$.]

4.31 Use the minimum and maximum of the log-yarn strength (see previous problem) to determine a distribution free tolerance interval. What are the values of α and β for your interval. How does it compare with the interval of the previous problem?

4.32 Make a normal Q–Q plot to test, graphically, whether the variable-t_1 in data file **SOCELL.csv**, is normally distributed.

4.33 Using R, MINITAB, or JMP and data file **CAR.csv**.

  (i) Test graphically whether the turn diameter is normally distributed.

  (ii) Test graphically whether the log (horse-power) is normally distributed.

4.34 Use the **CAR.csv** file. Make a frequency distribution of turn-diameter, with $k = 11$ intervals. Fit a normal distribution to the data and make a chi-squared test of the goodness of fit.

4.35 Using R, MINITAB, or JMP and the **CAR.csv** data file, compute the KS test statistic D_n^* for the turn-diameter variable, testing for normality. Compute k_α^* for $\alpha = 0.05$. Is D_n^* significant?

4.36 The daily demand (loaves) for whole wheat bread at a certain bakery has a Poisson distribution with mean $\lambda = 100$. The loss to the bakery for undemanded unit at the end of the day is $C_1 = \$0.10$. On the other hand, the penalty for a shortage of a unit is $C_2 = \$0.20$. How many loaves of whole wheat bread should be baked every day?

4.37 A random variable X has the binomial distribution $B(10, p)$. The parameter p has a beta prior distribution Beta$(3, 7)$. What is the posterior distribution of p, given $X = 6$?

4.38 In continuation to the previous exercise, find the posterior expectation and posterior standard deviation of p.

4.39 A random variable X has a Poisson distribution with mean λ. The parameter λ has a gamma, $G(2, 50)$, prior distribution.

  (i) Find the posterior distribution of λ given $X = 82$.

  (ii) Find the 0.025-th and 0.975th quantiles of this posterior distribution.

4.40 A random variable X has a Poisson distribution with mean which is either $\lambda_0 = 70$ or $\lambda_1 = 90$. The prior probability of λ_0 is $1/3$. The losses due to wrong actions are $r_1 = \$100$ and $r_2 = \$150$. Observing $X = 72$, which decision would you take?

4.41 A random variable X is normally distributed, with mean μ and standard deviation $\sigma = 10$. The mean μ is assigned a prior normal distribution with mean $\mu_0 = 50$ and standard deviation $\tau = 5$. Determine a credibility interval for μ, at level 0.95. Is this credibility interval also a HPD interval?

4.42 Read file **CAR.csv** into MINITAB. There are five variables stored in columns $C1$–$C5$. Write a macro which samples 64 values from column $C5$ (MPG/City), with replacement, and puts the sample in column $C6$. Let $k1$ be the mean of $C6$, and stack $k1$ in column $C7$. Execute this macro $M = 200$ times to obtain a sampling distribution of the sample means. Check graphically whether this sampling distribution is approximately normal. Also check whether the standard deviation of the sampling distribution is approximately $S/8$, where S is the standard deviation of $C5$.

4.43 Read file **YARNSTRG.csv** into column $C1$ of MINITAB. Execute macro **CONFINT.MTB** $M = 500$ times, to obtain confidence intervals for the mean of $C1$. Use samples of size $n = 30$. Check in what proportion of samples the confidence intervals cover the mean of $C1$.

4.44 The average turn diameter of 58 US made cars, in data file **CAR.csv**, is $\bar{X} = 37.203$ [m]. Is this mean significantly larger than 37 [m]? In order to check this, use MINITAB with the following commands (assuming you stored the CSV data files in a folder called MISTAT):

```
MTB> READ 'C: MISTAT\CAR.csv' C1–C5
MTB> SORT C2 C3 C6 C7;
SUBC> BY C2.
MTB> COPY C7 C8;
SUBC> USE (1:58).
```

Column $C8$ contains the turn diameter of the 58 US made cars. Write a macro which samples with replacement from $C8$ 58 values, and put them in $C9$. Stack the means of $C9$ in $C10$. Execute this macro 100 times. An estimate of the P-value is the proportion of means in $C10$ smaller than 36, greater than $2 \times 37.203 - 37 = 37.406$. What is your estimate of the P-value?

4.45 You have to test whether the proportion of nonconforming units in a sample of size $n = 50$ from a production process is significantly greater than $p = 0.03$. Use R, MINITAB, or JMP to determine when should we reject the hypothesis that $p \leq 0.03$ with $\alpha = 0.05$.

4.46 Generate 1000 bootstrap samples of the sample mean and sample standard deviation of the data in **CYCLT.csv** on 50 piston cycle times.
  (i) Compute 95% confidence intervals for the sample mean and sample standard deviation.
  (ii) Draw histograms of the EBD of the sample mean and sample standard deviation.

4.47 Use **BOOTPERC.MTB** to generate 1000 bootstrapped quartiles of the data in **CYCLT.csv**.
  (i) Compute 95% confidence intervals for the 1st quartile, the median and the third quartile.
  (ii) Draw histograms of the bootstrap quartiles.

4.48 Generate the EBD of size $M = 1000$, for the sample correlation ρ_{XY} between $t1$ and $t2$ in data file **SOCELL.csv**. [For running the program, you have to prepare a temporary data file containing only the first two columns of **SOCELL.csv**.] Compute the bootstrap confidence interval for ρ_{XY}, at confidence level of 0.95.

4.49 Generate the EBD of the regression coefficients (a, b) of Miles per Gallon/City, Y, versus Horsepower, X, in data file **CAR.csv**. See that X is in column $C1$, Y is in column $C2$. Run a simple regression with the command
MTB> Regr $C2$ 1 $C1$ $C3$ $C4$
The residuals are stored in $C3$. Let $k9$ be the sample size (109). Let $k1$ be the intercept a
  and $k2$ the slope b. Use the commands

```
MTB> Let k9 = 109
MTB> Let k1 = 30.7
MTB> Let k2 = −0.0736
MTB> Let C7(1) = k1
MTB> Let C8(1) = k2
```

Execute macro **BOOTREGR.MTB** $M = 100$ times.
  (i) Determine a bootstrap confidence interval for the intercept a, at level 0.95.
  (ii) Determine a bootstrap confidence interval for the slope b, at level 0.95.
  (iii) Compare the bootstrap standard errors of a and b to those obtained from the formulae of Section 4.2.2.

4.50 Test the hypothesis that the data in **CYCLT.csv** comes from a distribution with mean $\mu_0 = 0.55$ seconds.
  (i) What is the P-value?
  (ii) Does the confidence interval derived in Exercise 4.11.1 include $\mu_0 = 0.55$?
  (iii) Could we have guessed the answer of part (ii) after completing part (i)?

4.51 Compare the variances of the two measurements recorded in data file **ALMPIN2.csv**

   (i) What is the *P*-value?

   (ii) Draw box plots of the two measurements.

4.52 Compare the means of the two measurements on the two variables diam1 and diam2 in **ALMPIN2.csv**.

   (i) What is the bootstrap estimate of the *P*-values for the means and variances?

4.53 Compare the variances of the gasoline consumption (MPG/City) of cars by origin. The data is saved in file **MPG.csv**. There are $k = 3$ samples of sizes $n_1 = 58$, $n_2 = 14$, and $n_3 = 37$. Do you accept the null hypothesis of equal variances?

4.54 Test the equality of mean gas consumption (MPG/City) of cars by origin. The data file to use is **MPG.csv**. The sample sizes are $n_1 = 58$, $n_2 = 14$, and $n_3 = 37$. The number of samples is $k = 3$. Do you accept the null hypothesis of equal means?

4.55 Use MINITAB to generate 50 random Bernoulli numbers, with $p = 0.2$ into $C1$. Use macro **BINOPRED.MTB** to obtain tolerance limits with $\alpha = 0.05$ and $\beta = 0.05$, for the number of nonconforming items in future batches of 50 items, when the process proportion defectives is $p = 0.2$. Repeat this for $p = 0.1$ and $p = 0.05$.

4.56 Use macro **CONTPRED.MTB** to construct a $(0.95, 0.95)$ tolerance interval for the piston cycle time from the data in **OTURB.csv**.

4.57 Using macro **CONTPRED.MTB** and data file **OTURB.csv**, determine $(0.95, 0.95)$ tolerance interval for the piston cycle times.

4.58 Using the sign test, test the hypothesis that the median, $\xi_{0.5}$, of the distribution of cycle time of the piston, is not exceeding $\xi^* = 0.7$ [minutes]. The sample data is in file **CYCLT.csv**. Use $\alpha = 0.10$ for level of significance.

4.59 Use the WSR Test on the data of file **OELECT.csv** to test whether the median of the distribution $\xi_{0.5} = 220$ [V].

4.60 Apply the randomization test on the **CAR.csv** file to test whether the turn diameter of foreign cars, having four cylinders, is different from that of US made cars with four cylinders.

5

Variability in Several Dimensions and Regression Models

When surveys or experiments are performed, measurements are usually taken on several characteristics of the observation elements in the sample. In such cases, we have multivariate observations, and the statistical methods which are used to analyze the relationships between the values observed on different variables are called multivariate methods. In this chapter, we introduce some of these methods. In particular, we focus attention on graphical methods, linear regression methods and the analysis of contingency tables. The linear regression methods explore the linear relationship between a variable of interest and a set of variables, by which we try to predict the values of the variable of interest. Contingency tables analysis studies the association between qualitative (categorical) variables, on which we cannot apply the usual regression methods. We start the chapter with multivariate graphical analysis, using methods available in modern statistical software packages. We then introduce the concepts of multivariate frequency distributions and marginal and conditional frequency distributions. Following this, we present the most common methods of correlation and regression analysis, and end with the analysis of contingency tables. Several industrial data sets are analyzed.

5.1   Graphical display and analysis

5.1.1   Scatterplots

Suppose we are given a data set consisting of N records (elements). Each record contains observed values on k variables. Some of these variables might be qualitative (categorical) and some quantitative. Scatterplots display the values of pairwise quantitative variables, in two-dimensional plots.

Modern Industrial Statistics: With Applications in R, MINITAB and JMP, Third Edition.
Ron S. Kenett and Shelemyahu Zacks.

Example 5.1. Consider the data set **PLACE.csv**. The observations are the displacements (position errors) of electronic components on printed circuit boards. The data was collected by a large US manufacturer of automatic insertion machines used in mass production of electronic devices. The components are fed to the machine on reals. A robot arm picks the components and places them in a prescribed location on a printed circuit board. The placement of the component is controlled by a computer built into the insertion machine. There are 26 boards. A total of 16 components are placed on each board. Each component has to be placed at a specific location (x, y) on a board and with correct orientation θ (theta). Due to mechanical and other design or environmental factors some errors are committed in placement. It is interesting to analyze whether these errors are within the specified tolerances. There are $k = 4$ variables in the data set. The first one is categorical and gives the board number. The three other variables are continuous. The variable 'x-dev' provides the error in placement along the x-axis of the system. The variable "y-dev" presents the error in placement along the y-axis. The variable "theta-dev" is the error in angular orientation.

In Figure 5.1, we present a scatterplot of y-dev versus x-dev of each record. The picture reveals immediately certain unexpected clustering of the data points. The y-dev of placements should not depend on their x-dev. The scatterplot of Figure 5.1 shows three distinct clusters of points, which will be investigated later.

In a similar manner, we can plot the values of theta-dev against those of x-dev or y-dev. This can be accomplished by performing what is called a **multiple scatterplot**, or a **scatterplot matrix**. In Figure 5.2, we present the scatterplot matrix of x-dev, y-dev, and theta-dev. The multiple (matrix) scatterplot gives us a general picture of the relationships between the three variables. Figure 5.1 is the middle left box in Figure 5.2. Figure 5.2 directs us into

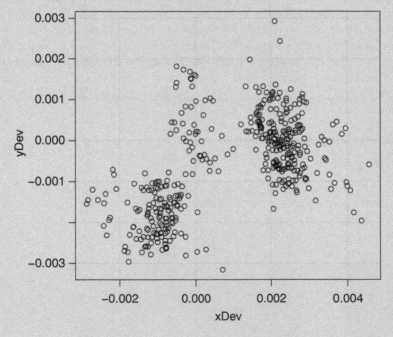

Figure 5.1   *Scatterplots of y-dev versus x-dev.*

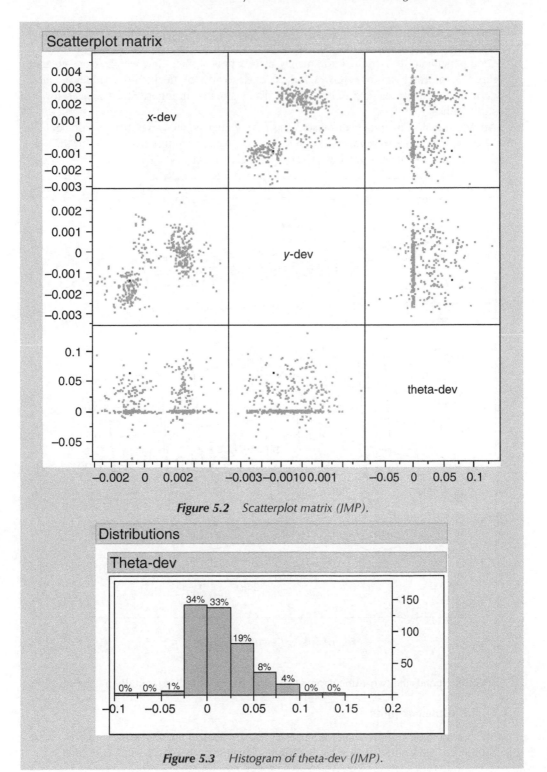

Figure 5.2   *Scatterplot matrix (JMP).*

Figure 5.3   *Histogram of theta-dev (JMP).*

further investigations. For example, we see in Figure 5.2, that the variable theta-dev has high concentration around zero with many observations to bigger than zero indicating a tilting of the components to the right. The frequency distribution of theta-dev, which is presented in Figure 5.3, reinforces this conclusion. Indeed, close to 50% of the theta-dev values are close to zero. The other values tend to be positive. The histogram in Figure 5.3 is skewed toward positive values.

An additional scatterplot can present the three-dimensional variability simultaneously. This graph is called a **3D-scatterplot**. In Figure 5.4, we present this scatterplot for the three variables x-dev (*X* direction), y-dev (*Y* direction), and "theta-dev" (*Z* direction).

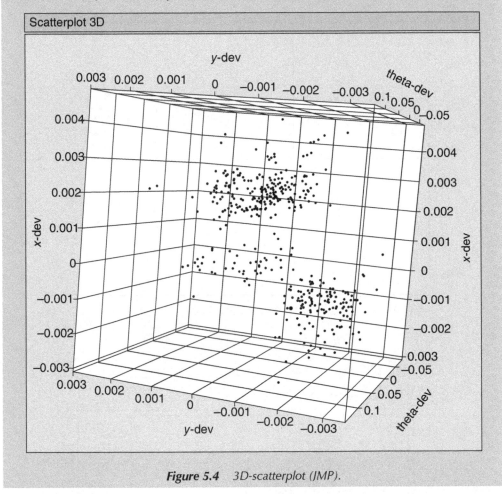

Figure 5.4   *3D-scatterplot (JMP).*

This plot expands the two-dimensional scatter plot by adding horizontally a third variable.

5.1.2   Multiple box-plots

Multiple boxplots or side-by-side boxplot is another graphical technique by which we present distributions of a quantitative variable at different categories of a categorical variable.

Example 5.2. Returning to the data set **PLACE.csv**, we wish to further investigate the apparent clusters, indicated in Figure 5.1. As mentioned before, the data was collected in an experiment in which components were placed on 26 boards in a successive manner. The board number "board n" is in the first column of the data set. We would like to examine whether the deviations in x, y, or θ tend to change with time. We can, for this purpose, plot the x-dev, y-dev, or theta-dev against board n. A more concise presentation is to graph multiple boxplots, by board number. In Figure 5.5, we present these multiple boxplots of the x-dev against board #. We see in this figure an interesting picture. Boards 1–9 yield similar boxplots, while those of boards 10–12 are significantly above those of the first group, and those of boards 13–26 constitute a third group. These groups seem to be connected with the three clusters seen in Figure 5.1. To verify it, we introduce a **code variable** to the data set, which assumes the value 1 if board # ≤ 9, the value 2 if $10 \leq$ board# ≤ 12 and the value 3 if board # ≥ 13. We then plot again y-dev against x-dev, denoting the points in the scatterplot by the code variable symbols 0, +, ×.

In Figure 5.6, we see this coded scatterplot. It is clear now that the three clusters are formed by these three groups of boards. The differences between these groups might be due to some deficiency in the placement machine, which caused the apparent time- related drift in the errors. Other possible reasons could be the printed circuit board composition or different batches of raw material, such as the glue used for placing the components.

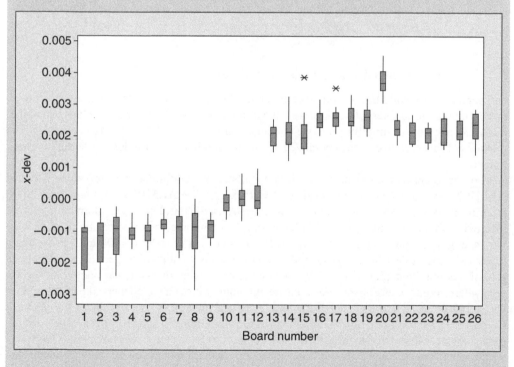

Figure 5.5   *Multiple box-plots of x-dev versus board number (MINITAB).*

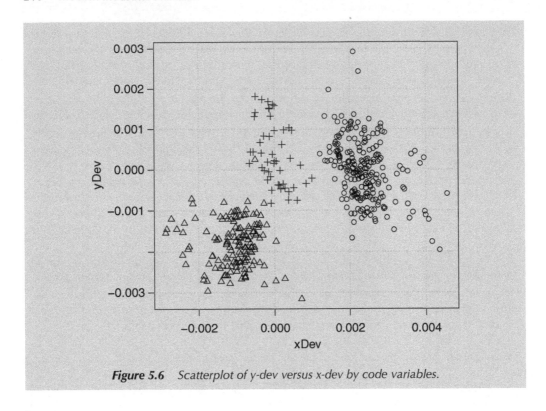

Figure 5.6   *Scatterplot of y-dev versus x-dev by code variables.*

5.2   Frequency distributions in several dimensions

In Chapter 2, we studied how to construct frequency distributions of single variables, categorical, or continuous. In the present section, we extend those concepts to several variables simultaneously. For the sake of simplification, we restrict the discussion to the case of two variables. The methods of this section can be generalized to a larger numbers of variables in a straightforward manner.

In order to enrich the examples, we introduce here two additional data sets. One is called **ALMPIN.csv** and the other one is called **HADPAS.csv**. The **ALMPIN.csv** set consists of 70 records on six variables measured on aluminum pins used in airplanes. The aluminum pins are inserted with air-guns in predrilled holes in order to combine critical airplane parts such as wings, engine supports, and doors. Typical lot sizes consist of at least 1000 units providing a prime example of the discrete mass production operations mentioned in Section 1.1. The main role of the aluminum pins is to reliably secure the connection of two metal parts. The surface area where contact is established between the aluminum pins and the connected part determines the strength required to disconnect the part. A critical feature of the aluminum pin is that it fits perfectly the predrilled holes. Parallelism of the aluminum pin is therefore essential, and the parts diameter is measured in three different locations producing three measurements of the parts width. Diameters 1, 2, and 3 should be all equal. Any deviation indicates lack of parallelism and therefore potential reliability problems since the surface area with actual contact is not uniform. The measurements were taken in a computerized numerically controlled (CNC) metal cutting operation. The six variables are Diameter 1, Diameter 2, Diameter 3, Cap Diameter, Lengthncp, and Lengthwcp. All the measurements are in millimeters. The first three variables give the pin diameter at three specified locations. Cap Diameter is the diameter of the cap on top of the pin. The last two variables are the length of the pin, without and with the cap, respectively.

Data set **HADPAS.csv** provides several resistance measurements (ohms) of five types of resistances (Res 3, Res 18, Res 14, Res 7, and Res 20), which are located in six hybrid microcircuits (three rows and two columns) simultaneously manufactured on ceramic substrates. There are altogether 192 records for 32 ceramic plates.

5.2.1 Bivariate joint frequency distributions

A joint frequency distribution is a function which provides the frequencies in the data set of elements (records) having values in specified intervals. More specifically, consider two variables X and Y. We assume that both variables are continuous. We partition the X-axis to k subintervals (ξ_{i-1}, ξ_i), $i = 1, \ldots, k_1$. We then partition the Y-axis to k_2 subintervals (η_{j-1}, η_j), $j = 1, \ldots, k_2$. We denote by f_{ij} the number (count) of elements in the data set (sample) having X values in (ξ_{i-1}, ξ_i) and Y values in (η_{j-1}, η_j), simultaneously. f_{ij} is called the **joint frequency** of the rectangle $(\xi_{i-1}, \xi_i) \times (\eta_{j-1}, \eta_j)$. If N denotes the total number of elements in the data set, then obviously

$$\sum_i \sum_j f_{ij} = N. \tag{5.2.1}$$

The frequencies f_{ij} can be represented in a table called a table of the frequency distribution. The column totals provide the frequency distribution of the variable Lengthwcp. These row and column totals are called **marginal frequencies**. Generally, the marginal frequencies are

$$f_{i.} = \sum_{j=1}^{k_2} f_{ij}, \quad i = 1, \ldots, k_1 \tag{5.2.2}$$

and

$$f_{.j} = \sum_{i=1}^{k_1} f_{ij}, \quad j = 1, \ldots, k_2. \tag{5.2.3}$$

These are the sums of the frequencies in a given row or in a given column.

Example 5.3. In Table 5.1, we present the joint frequency distribution of Lengthncp and Lengthwcp of the data set **ALMPIN.csv**

Table 5.1 *Joint frequency distribution*

Lengthncp \ Lenthwcp	59.9–60.0	60.0–60.1	60.1–60.2	Row total
49.8–49.9	16	17	0	33
49.9–50.0	5	27	2	34
50.0–50.1	0	0	3	3
Column total	21	44	5	70

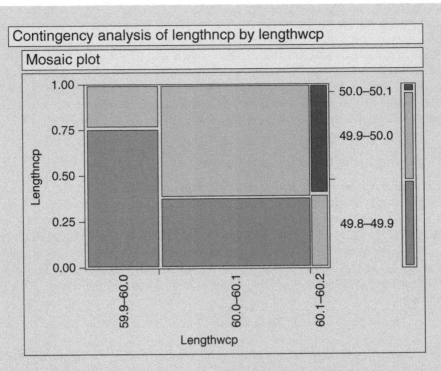

Figure 5.7   *Mosaic plot of data in Table 5.1 (JMP).*

The row totals provide the frequency distribution of Lengthncp. We can visualize this table using a mosaic plot where the table entries are proportional to the size of the rectangles in the plot. Figure 5.7 presents a mosaic plot of the data in Table 5.1. Figure 5.8 is a MINITAB output of the crosstabulation feature with the cell values also represented as a percent of rows, columns, and total. Similar tabulation can be done of the frequency distributions of resistances, in data set **HADPAS.csv**. In Table 5.2, we provide the joint frequency distribution of Res 3 and Res 7.

Table 5.2   *Joint frequency distribution of Res 3 and Res 7 (in ohms)*

Res 3 \ Res 7	1300–1500	1500–1700	1700–1900	1900–2100	2100–2300	Row totals
1500–1700	1	13	1	0	0	15
1700–1900	0	15	31	1	0	47
1900–2100	0	1	44	40	2	87
2100–2300	0	0	5	31	6	42
2300–2500	0	0	0	0	1	1
Column total	1	29	81	72	9	192

Minitab project report

Tabulated statistics: Lengthncp, Lengthwcp

Rows: Lengthncp   Columns: Lengthwcp

	59.9-60.0	60.0-60.1	60.1-60.2	All
49.8-49.9	16	17	0	33
	48.48	51.52	0.00	100.00
	76.19	38.64	0.00	47.14
	22.86	24.29	0.00	47.14
49.9-50.0	5	27	2	34
	14.71	79.41	5.88	100.00
	23.81	61.36	40.00	48.57
	7.14	38.57	2.86	48.57
50.0-50.1	0	0	3	3
	0.00	0.00	100.00	100.00
	0.00	0.00	60.00	4.29
	0.00	0.00	4.29	4.29
All	21	44	5	70
	30.00	62.86	7.14	100.00
	100.00	100.00	100.00	100.00
	30.00	62.86	7.14	100.00

Cell contents: Count
                      % of Row
                      % of Column
                      % of Total

Figure 5.8   *Crosstabulation of data in Table 5.1 (MINITAB).*

The bivariate frequency distribution also provides us information on the association, or dependence between the two variables. In Table 5.2, we see that resistance values of Res 3 tend to be similar to those of Res 7. For example, if the resistance value of Res 3 is in the interval (1500, 1700), 13 out of 15 resistance values of Res 7 are in the same interval. This association can be illustrated by plotting the box and whiskers plots of the variable Res 3 by the categories (intervals) of the variable Res 7. In order to obtain these plots, we partition first the 192 cases to five subgroups, according to the resistance values of Res 7. The single case having Res 7 in the interval (1300, 1500) belongs to subgroup 1. The 29 cases having Res 7 values in (1500, 1700) belong to subgroup 2, and so on. We can then perform an analysis by subgroups. Such an analysis yields the following table.

Table 5.3   *Means and standard deviations of Res 3*

Subgroup	Interval of Res 7	Sample size	Mean	Standard deviation
1	1300–1500	1	1600.0	–
2	1500–1700	29	1718.9	80.27
3	1700–1900	77	1932.9	101.17
4	1900–2100	76	2068.5	99.73
5	2100–2300	9	2204.0	115.49

We see in Table 5.3 that the subgroup means grow steadily with the values of Res 7. The standard deviations do not change much. (There is no estimate of the standard deviation of subgroup 1.) A better picture of the dependence of Res 3 on the intervals of Res 7 is given by Figure 5.9, in which the boxplots of the Res 3 values are presented by subgroup.

5.2.2   Conditional distributions

Consider a population (or a sample) of elements. Each element assumes random values of two (or more) variables $X, Y, Z, \ldots$. The distribution of X, over elements whose Y value is restricted to a given interval (or set) A, is called the **conditional distribution of X, given Y is in** A. If the conditional distributions of X given Y are different from the marginal distribution of X, we say that the variables X and Y are **statistically dependent**. We will learn later how to test whether the differences between the conditional distributions and the marginal ones are significant, and not due just to randomness in small samples.

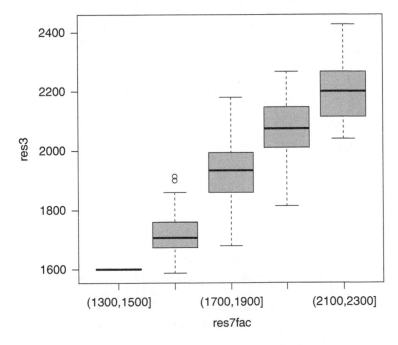

Figure 5.9   *Boxplots of Res 3 by intervals of Res 7.*

Example 5.4.   If we divide the frequencies in Table 5.2 by their column Sums, we obtain the proportional frequency distributions of Res 3, given the intervals of Res 7. In Table 5.4, we compare these conditional frequency distributions, with the marginal frequency distribution of Res 3. We see in Table 5.4 that the proportional frequencies of the conditional distributions of Res 3 depend strongly on the intervals of Res 7 to which they are restricted.

Table 5.4   *Conditional and marginal frequency distributions of Res 3*

Res 3 \ Res 7	1300–1500	1500–1700	1700–1900	1900–2100	2100–2300	Marginal distrib.
1500–1700	100.0	44.8	1.2	0	0	7.8
1700–1900	0	51.7	38.3	1.4	0	24.5
1900–2100	0	3.4	54.3	55.6	22.2	45.3
2100–2300	0	0	6.2	43.0	66.7	21.9
2300–2500	0	0	0	0	11.1	0.5
Column sums	100.0	100.0	100.0	100.0	100.0	100.0

5.3   Correlation and regression analysis

In the previous sections, we presented various graphical procedures for analyzing multivariate data. In particular, we showed the multivariate scatterplots, three-dimensional histograms, conditional boxplots, etc. In the present section, we start with numerical analysis of multivariate data.

5.3.1   Covariances and correlations

We introduce now a statistic which summarize the simultaneous variability of two variables. The statistic is called the **sample covariance**. It is a generalization of the sample variance statistics, S_x^2, of one variable, X. We will denote the sample covariance of two variables, X and Y by S_{xy}. The formula of S_{xy} is

$$S_{xy} = \frac{1}{n-1} \sum_{i=1}^{n} (X_i - \overline{X})(Y_i - \overline{Y}), \tag{5.3.1}$$

where $\overline{X}$ and $\overline{Y}$ are the sample means of X and Y, respectively. Notice that S_{xx} is the sample variance S_x^2 and S_{yy} is S_y^2. The sample covariance can assume positive or negative values. If one of the variables, say X, assumes a constant value c, for all X_i ($i = 1, \ldots, n$) then $S_{xy} = 0$. This can be immediately verified, since $\overline{X} = c$ and $X_i - \overline{X} = 0$ for all $i = 1, \ldots, n$.

It can be proven that, for any variables X and Y,

$$S_{xy}^2 \leq S_x^2 \cdot S_y^2. \tag{5.3.2}$$

This inequality is the celebrated **Schwarz inequality**. By dividing S_{xy} by $S_x \cdot S_y$, we obtain a standardized index of dependence, which is called the **sample correlation** (Pearson's product-moment correlation), namely

$$R_{xy} = \frac{S_{xy}}{S_x \cdot S_y}. \tag{5.3.3}$$

From the Schwarz inequality, the sample correlation always assumes values between -1 and $+1$. In Table 5.5, we present the sample covariances of the six variables measured on the aluminum pins. Since $S_{xy} = S_{yx}$ (Covariances and correlations are symmetric statistics), it is sufficient to present the values at the bottom half of the table (on and below the diagonal).

Example 5.5. In Tables 5.5 and 5.6, we present the sample covariances and sample correlations in the data file **ALMPIN.csv**.

Table 5.5 *Sample covariances of aluminum pins variables*

$_x \backslash^Y$	Diameter 1	Diameter 2	Diameter 3	Cap Diameter	Length nocp	Length wcp
Diameter 1	0.0270					
Diameter 2	0.0285	0.0329				
Diameter 3	0.0255	0.0286	0.0276			
Cap Diameter	0.0290	0.0314	0.0285	0.0358		
Lengthnocp	−0.0139	−0.0177	−0.0120	−0.0110	0.1962	
Lengthwcp	−0.0326	−0.0418	−0.0333	−0.0319	0.1503	0.2307

Table 5.6 *Sample correlations of aluminum pins variables*

$_x \backslash^Y$	Diameter 1	Diameter 2	Diameter 3	Cap Diameter	Length nocp	Length wcp
Diameter 1	1.000					
Diameter 2	0.958	1.000				
Diameter 3	0.935	0.949	1.000			
Cap Diameter	0.933	0.914	0.908	1.000		
Lengthnocp	−0.191	−0.220	−0.163	−0.132	1.000	
Lengthwcp	−0.413	−0.480	−0.417	−0.351	0.707	1.000

We see in Table 5.6 that the sample correlations between Diameter 1, Diameter 2, and Diameter 3 and Cap Diameter are all greater than 0.9. As we see in Figure 5.10 (the multivariate scatterplots), the points of these variables are scattered close to straight lines. On the other hand, no clear relationship is evident between the first four variables and the length of the pin (with or without the cap). The negative correlations, usually indicate that the points are scattered around a straight line having a negative slope. In the present case, it seems that the magnitude of these negative correlations are due to the one outlier (pin # 66). If we delete it from the data sets, the correlations are reduced in magnitude, as shown in Table 5.7.

We see that the correlations between the four diameter variables and the lengthnocp are much closer to zero after excluding the outlier. Moreover, the correlation with the Cap

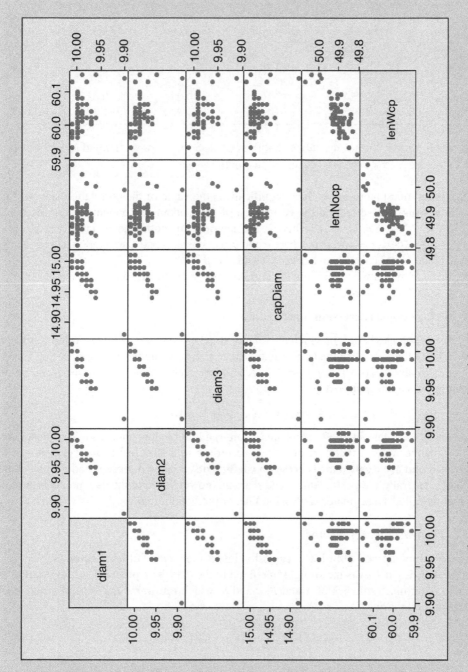

Figure 5.10 *Multiple scatterplots of the aluminum pins measurements (MINITAB).*

Table 5.7 *Sample correlations of aluminum pins variables, after excluding outlying observation #66*

$_x\backslash^Y$	Diameter 1	Diameter 2	Diameter 3	Cap Diameter	Length nocp
Diameter 2	0.925				
Diameter 3	0.922	0.936			
Cap Diameter	0.876	0.848	0.876		
Lengthnocp	−0.056	−0.103	−0.054	0.022	
Lengthwcp	−0.313	−0.407	−0.328	−0.227	0.689

Diameter changed its sign. This shows that the sample correlation, as defined above, is sensitive to the influence of extreme observations (outliers).

An important question to ask is, how **significant** is the value of the correlation statistic? In other words, what is the effect on the correlation of the random components of the measurements? If $X_i = \xi_i + e_i$, $i = 1, \ldots, n$, where ξ_i are deterministic components and e_i are random, and if $Y_i = \alpha + \beta \xi_i + f_i$, $i = 1, \ldots, n$, where α and β are constants and f_i are random components, how large could be the correlation between X and Y if $\beta = 0$?

Questions which deal with assessing the **significance** of the results will be discussed later.

5.3.2   Fitting simple regression lines to data

We have seen examples before in which the relationship between two variables X and Y is close to linear. This is the case when the (x, y) points scatter along a straight line. Suppose that we are given n pairs of observations $\{(x_i, y_i), i = 1, \ldots, n\}$. If the Y observations are related to those on X, according to the **linear model**

$$y_i = \alpha + \beta x_i + e_i, \quad i = 1, \ldots, n, \tag{5.3.4}$$

where α and β are constant coefficients, and e_i are random components, with zero mean and constant variance, we say that Y relates to X according to a **simple linear regression**. The coefficients α and β, are called the **regression coefficients**. α is the **intercept** and β is the **slope coefficient**. Generally, the coefficients α and β are unknown. We fit to the data points a straight line, which is called the estimated regression line, or prediction line.

5.3.2.1   The least squares method

The most common method of fitting a regression line is the method of **least squares**.

Suppose that $\hat{y} = a + bx$ is the straight line fitted to the data. The **principle** of **least squares** requires to determine estimates of α and β, a and b, which **minimize** the **sum of squares of residuals** around the line, i.e.,

$$\text{SSE} = \sum_{i=1}^{n}(y_i - a - bx_i)^2. \tag{5.3.5}$$

If we require that the regression line will pass through the point $(\bar{x}, \bar{y})$, where $\bar{x}, \bar{y}$ are the sample means of the x's and y's, then

$$\bar{y} = a + b\bar{x},$$

or the coefficient a should be determined by the equation

$$a = \bar{y} - b\bar{x}. \tag{5.3.6}$$

Substituting this equation above, we obtain that

$$SSE = \sum_{i=1}^{n}(y_i - \bar{y} - b(x_i - \bar{x}))^2$$

$$= \sum_{i=1}^{n}(y_i - \bar{y})^2 - 2b\sum_{i=1}^{n}(x_i - \bar{x})(y_i - \bar{y}) + b^2\sum_{i=1}^{n}(x_i - \bar{x})^2.$$

Dividing the two sides of the equation by $(n-1)$, we obtain

$$\frac{SSE}{n-1} = S_y^2 - 2bS_{xy} + b^2 S_x^2.$$

The coefficient b should be determined to minimize this quantity. One can write

$$\frac{SSE}{n-1} = S_y^2 + S_x^2\left(b^2 - 2b\frac{S_{xy}}{S_x^2} + \frac{S_{xy}^2}{S_x^4}\right) - \frac{S_{xy}^2}{S_x^2}$$

$$= S_y^2(1 - R_{xy}^2) + S_x^2\left(b - \frac{S_{xy}}{S_x^2}\right)^2.$$

It is now clear that the least squares estimate of β is

$$b = \frac{S_{xy}}{S_x^2} = R_{xy}\frac{S_y}{S_x}. \tag{5.3.7}$$

The value of $SSE/(n-1)$, corresponding to the least squares estimate is

$$S_{y|x}^2 = S_y^2(1 - R_{xy}^2). \tag{5.3.8}$$

$S_{y|x}^2$ is the sample variance of the residuals around the least squares regression line. By definition, $S_{y|x}^2 \geq 0$, and hence $R_{xy}^2 \leq 1$, or $-1 \leq R_{xy} \leq 1$. $R_{xy} = \pm 1$ only if $S_{y|x}^2 = 0$. This is the case when all the points (x_i, y_i), $i = 1, \ldots, n$, lie on a straight line. If $R_{xy} = 0$, then the slope of the regression line is $b = 0$ and $S_{y|x}^2 = S_y^2$.

Notice that

$$R_{xy}^2 = \left(1 - \frac{S_{y|x}^2}{S_y^2}\right). \tag{5.3.9}$$

Thus, R_{xy}^2 is the proportion of variability in Y, which is explainable by the linear relationship $\hat{y} = a + bx$. For this reason, R_{xy}^2 is also called the **coefficient of determination**. The coefficient of correlation (squared) measures the extent of linear relationship in the data. The linear regression line, or prediction line, could be used to predict the values of Y corresponding to X values, when R_{xy}^2 is not too small. To interpret the coefficient of determination – particularly when dealing with multiple regression models (see Section 5.4) – it is sometimes useful to consider an "adjusted" R^2. The adjustment accounts for the number of predictor or explanatory variables in the model and the sample size. In simple linear regression, we define

$$R_{xy}^2(\text{adjusted}) = 1 - \left[(1 - R_{xy}^2)\frac{n-1}{n-2}\right]. \tag{5.3.10}$$

Example 5.6. Telecommunication satellites are powered while in orbit by solar cells. Tadicell, a solar cells producer that supplies several satellite manufacturers, was requested to provide data on the degradation of its solar cells over time. Tadicell engineers performed a simulated experiment in which solar cells were subjected to temperature and illumination changes similar to those in orbit and measured the short-circuit current ISC (amperes) of solar cells at three different time periods, in order to determine their rate of degradation. In Table 5.8, we present the ISC values of $n = 16$ solar cells, measured at three time epochs, one month apart. The data is given in file **SOCELL.csv**. In Figure 5.11, we see the scatter of the ISC values at t_1, and at t_2

Table 5.8   *ISC values of solar-cells at three time epochs*

Cell\time	t_1	t_2	t_3
1	4.18	4.42	4.55
2	3.48	3.70	3.86
3	4.08	4.39	4.45
4	4.03	4.19	4.28
5	3.77	4.15	4.22
6	4.01	4.12	4.16
7	4.49	4.56	4.52
8	3.70	3.89	3.99
9	5.11	5.37	5.44
10	3.51	3.81	3.76
11	3.92	4.23	4.14
12	3.33	3.62	3.66
13	4.06	4.36	4.43
14	4.22	4.47	4.45
15	3.91	4.17	4.14
16	3.49	3.90	3.81

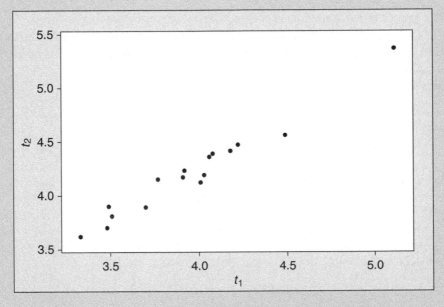

Figure 5.11   *Relationship of ISC values at t_1 and t_2 (MINITAB).*

We now make a regression analysis of ISC at time t_2, Y, versus ISC at time t_1, X. The computations can be easily performed by R.

```
> data(SOCELL)
> LmISC <- lm(t2 ~ 1 + t1,
            data=SOCELL)
> summary(LmISC)

Call:
lm(formula = t2 ~ 1 + t1, data = SOCELL)

Residuals:
      Min        1Q     Median        3Q        Max
-0.145649 -0.071240  0.008737  0.056562  0.123051

Coefficients:
            Estimate Std. Error t value Pr(>|t|)
(Intercept)  0.53578    0.20314   2.638   0.0195 *
t1           0.92870    0.05106  18.189 3.88e-11 ***
---
Signif. codes:
0 '***' 0.001 '**' 0.01 '*' 0.05 '.' 0.1 ' ' 1

Residual standard error: 0.08709 on 14 degrees of freedom
Multiple R-squared: 0.9594,        Adjusted R-squared: 0.9565
F-statistic: 330.8 on 1 and 14 DF,  p-value: 3.877e-11
```

And in MINITAB, as shown in the following exhibit.

MTB > regress c2 on 1 pred in c1
The regression equation is C2 = 0.536 + 0.929 C1

Predictor	Coef	Stdev	t-ratio	P
Constant	0.5358	0.2031	2.64	0.019
C1	0.92870	0.05106	18.19	0.0000

$s = 0.08709$ R-sq $= 95.9\%$ R-sq(adj) $= 95.7\%$

We see in the exhibit of the MINITAB analysis that the least squares regression (prediction) line is $\hat{y} = 0.536 + 0.929x$. We read also that the coefficient of determination is $R_{xy}^2 = 0.959$. This means that only 4% of the variability in the ISC values, at time period t_2, are not explained by the linear regression on the ISC values at time t_1. Observation #9 is an "unusual observation." It has relatively much influence on the regression line, as can be seen in Figure 5.11.

The MINITAB output provides also additional analysis. The Stdev corresponding to the least squares regression coefficients are the square-roots of the variances of these estimates, which are given by the formulae:

$$S_a^2 = S_e^2 \left[\frac{1}{n} + \frac{\bar{x}^2}{\sum_{i=1}^{n}(x_i - \bar{x})^2} \right] \qquad (5.3.11)$$

and

$$S_b^2 = S_e^2 \Big/ \sum_{i=1}^{n}(x_i - \bar{x})^2, \qquad (5.3.12)$$

where

$$S_e^2 = \frac{(1 - R_{xy}^2)}{n - 2} \sum_{i=1}^{n}(y_i - \bar{y})^2. \qquad (5.3.13)$$

We see here that $S_e^2 = \frac{n-1}{n-2} S_{y|x}^2$. The reason for this modification is for testing purposes. The value of S_e^2 in the above analysis is 0.0076. The standard deviation of y is $S_y = 0.4175$. The standard deviation of the residuals around the regression line is $S_e = 0.08709$. This explains the high value of $R_{y|x}^2$.

In Table 5.9, we present the values of ISC at time t_2, y, and their predicted values, according to those at time t_1, $\hat{y}$. We present also a graph (Figure 5.12) of the residuals, $\hat{e} = y - \hat{y}$, versus the predicted values $\hat{y}$. If the simple linear regression explains the variability adequately, the residuals should be randomly distributed around zero, without any additional relationship to the regression x.

In Figure 5.12, we plot the residuals $\hat{e} = y - \hat{y}$, versus the predicted values $\hat{y}$, of the ISC values for time t_2. It seems that the residuals are randomly dispersed around zero. Later, we will learn how to test whether this dispersion is indeed random.

Table 5.9 *Observed and predicted values of ISC at time t_2*

i	y_i	$\hat{y}_i$	$\hat{e}_i$
1	4.42	4.419	0.0008
2	3.70	3.769	−0.0689
3	4.39	4.326	0.0637
4	4.19	4.280	−0.0899
5	4.15	4.038	0.1117
6	4.12	4.261	−0.1413
7	4.56	4.707	−0.1472
8	3.89	3.973	−0.0833
9	5.37	5.283	0.0868
10	3.81	3.797	0.0132
11	4.23	4.178	0.0523
12	3.62	3.630	−0.0096
13	4.36	4.308	0.0523
14	4.47	4.456	0.0136
15	4.17	4.168	0.0016
16	3.90	3.778	0.1218

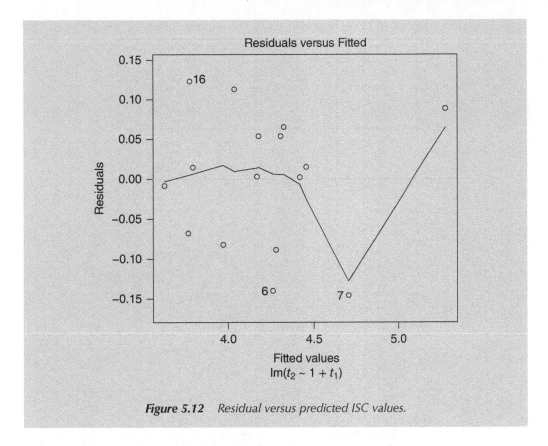

Figure 5.12 *Residual versus predicted ISC values.*

5.3.2.2 *Regression and prediction intervals*

Suppose that we wish to predict the possible outcomes of the Y for some specific value of X, say x_0. If the true regression coefficients α and β are known, then the predicted value of Y is $\alpha + \beta x_0$. However, when α and β are unknown, we predict the outcome at x_0 to be $\hat{y}(x_0) = a + bx_0$. We know, however, that the actual value of Y to be observed will not be exactly equal to $\hat{y}(x_0)$. We can determine a prediction interval around $\hat{y}(x_0)$ such that, the likelihood of obtaining a Y value within this interval will be high. Generally, the prediction interval limits, given by the formula

$$\hat{y}(x_0) \pm 3S_e^2 \cdot \left[1 + \frac{1}{n} + \frac{(x_0 - \bar{x})^2}{\sum_i (x_i - \bar{x})^2} \right]^{1/2}, \tag{5.3.14}$$

will yield good predictions. In Table 5.10, we present the prediction intervals for the ISC values at time t_2, for selected ISC values at time t_1. In Figure 5.13, we present the scatterplot, regression line, and prediction limits for Res 3 versus Res 7, of the **HADPAS.csv** set.

5.4 Multiple regression

In the present section, we generalize the regression to cases where the variability of a variable Y of interest can be explained, to a large extent, by the linear relationship between Y and k predicting or explaining variables $X_1, \ldots, X_k$. The number of explaining variables is $k \geq 2$.

Table 5.10   *Prediction intervals for ISC values at time t_2*

x_0	$\hat{y}(x_0)$	Lower limit	Upper limit
4.0	4.251	3.987	4.514
4.4	4.622	4.350	4.893
4.8	4.993	4.701	5.285
5.2	5.364	5.042	5.687

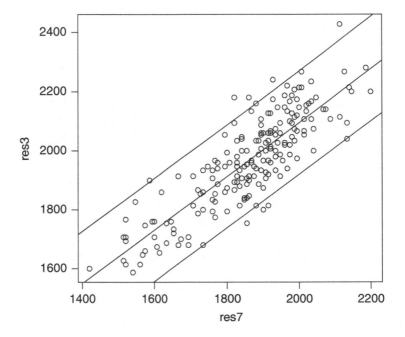

Figure 5.13   *Prediction intervals for Res 3 values, given the Res 7 values.*

All the k variables $X_1, \ldots, X_k$ are continuous ones. The regression analysis of Y on several predictors is called **multiple regression**; multiple regression analysis is an important statistical tool for exploring the relationship between the dependence of one variable Y on a set of other variables. Applications of multiple regression analysis can be found in all areas of science and engineering. This method plays an important role in the statistical planning and control of industrial processes.

The statistical linear model for multiple regression is

$$y_i = \beta_0 + \sum_{j=1}^{k} \beta_j x_{ij} + e_i, \quad i = 1, \ldots, n,$$

where $\beta_0, \beta_1, \ldots, \beta_k$ are the linear regression coefficients, and e_i are random components. The commonly used method of estimating the regression coefficients and testing their significance is called **multiple regression analysis**. The method is based on the **principle of least squares**, according to which the regression coefficients are estimated by choosing $b_0, b_1, \ldots, b_k$ to minimize the sum of residuals

$$\text{SSE} = \sum_{i=1}^{n} (y_i - (b_0 + b_1 x_{i1} + \cdots + b_k x_{ik}))^2.$$

The first subsections of the present chapter are devoted to the methods of regression analysis, when both the regressant Y and the regressors $x_1, \ldots, x_k$ are quantitative variables. In Section 5.10, we present quantal response regression, in which the regressant is qualitative (binary) variable and the regressors $x_1, \ldots, x_k$ are quantitative. In particular, we present the logistic model and the logistic regression. In Section 5.11, we discuss the analysis of variance (ANOVA), for the comparison of sample means, when the regressant is quantitative but the regressors are categorical variables. It is important to emphasize that the regression coefficients represent an association and not a causal effect. In fact, they indicate the difference of responses between groups of the input variables. For example, b_1 corresponds to the change in response between two groups with 1 unit difference in x_1.

5.4.1   Regression on two variables

The multiple regression linear model, in the case of two predictors, assumes the form

$$y_i = \beta_0 + \beta_1 x_{1i} + \beta_2 x_{2i} + e_i, \quad i = 1, \ldots, n. \tag{5.4.1}$$

$e_1, \ldots, e_n$ are independent r.v.s, with $E\{e_i\} = 0$ and $V\{e_i\} = \sigma^2$, $i = 1, \ldots, n$. The principle of least-squares calls for the minimization of SSE. One can differentiate SSE with respect to the unknown parameters. This yields the least squares estimators (LSEs), b_0, b_1, and b_2 of the regression coefficients, β_0, β_1, β_2. The formula for these estimators are the following:

$$b_0 = \overline{Y} - b_1 \overline{X}_1 - b_2 \overline{X}_2; \tag{5.4.2}$$

and b_1 and b_2 are obtained by solving the set of linear equations

$$\left. \begin{array}{l} S_{x_1}^2 b_1 + S_{x_1 x_2} b_2 = S_{x_1 y} \\ S_{x_1 x_2} b_1 + S_{x_2}^2 b_2 = S_{x_2 y}. \end{array} \right\} \tag{5.4.3}$$

As before, $S_{x_1}^2$, $S_{x_1 x_2}$, $S_{x_2}^2$, $S_{x_1 y}$, and $S_{x_2 y}$ denote the sample variances and covariances of x_1, x_2, and y.

By simple substitution, we obtain for b_1 and b_2 the explicit formulae:

$$b_1 = \frac{S_{x_2}^2 S_{x_1 y} - S_{x_1 x_2} S_{x_2 y}}{S_{x_1}^2 S_{x_2}^2 - S_{x_1 x_2}^2} \tag{5.4.4}$$

and

$$b_2 = \frac{S_{x_1}^2 S_{x_2 y} - S_{x_1 x_2} S_{x_1 y}}{S_{x_1}^2 S_{x_2}^2 - S_{x_1 x_2}^2}. \tag{5.4.5}$$

The values $\hat{y}_i = b_0 + b_1 x_{1i} + b_2 x_{2i}$ $(i = 1, \ldots, n)$ are called the **predicted values** of the regression, and the residuals around the regression plane are

$$\hat{e}_i = y_i - \hat{y}_i$$
$$= y_i - (b_0 + b_1 x_{1i} + b_2 x_{2i}), \quad i = 1, \ldots, n.$$

The mean-square of the residuals around the regression plane is

$$S^2_{y|(x_1,x_2)} = S^2_y(1 - R^2_{y|(x_1,x_2)}),$$ (5.4.6)

where

$$R^2_{y|(x_1,x_2)} = \frac{1}{S^2_y}(b_1 S_{x_1 y} + b_2 S_{x_2 y}),$$ (5.4.7)

is the **multiple squared-correlation** (multiple-R^2), and S^2_y is the sample variance of y. The interpretation of the multiple-R^2 is as before, i.e., the proportion of the variability of y which is **explainable** by the predictors (regressors) x_1 and x_2.

Example 5.7. We illustrate the fitting of a multiple regression on the following data, labeled **GASOL.csv**. The data set consists of 32 measurements of distillation properties of crude oils (see Daniel and Wood 1971, p. 165). There are five variables, $x_1, \ldots, x_4$ and y. These are

x_1 : crude oil **gravity**, °API;

x_2 : crude oil **vapor** pressure, psi;

x_3 : crude oil **ASTM** 10% point, °F;

x_4 : gasoline ASTM **endpoint**, °F;

y : **yield** of gasoline (in percentage of crude oil).

The measurements of crude oil, and gasoline volatility, measure the temperatures at which a given amount of liquid has been evaporized.

The sample correlations between these five variables are

	x_2	x_3	x_4	y
x_1	0.621	−0.700	−0.322	0.246
x_2		−0.906	−0.298	0.384
x_3			0.412	−0.315
x_4				0.712

We see that the yield y is highly correlated with x_4 and with x_2 (or x_3). The following is an R output of the regression of y on x_3 and x_4:

```
> data(GASOL)
> LmYield <- lm(yield ~ 1 + astm + endPt,
               data=GASOL)
> summary(LmYield)

Call:
lm(formula = yield ~ 1 + astm + endPt, data = GASOL)

Residuals:
    Min       1Q   Median       3Q      Max
-3.9593  -1.9063  -0.3711   1.6242   4.3802
```

```
Coefficients:
             Estimate Std. Error t value Pr(>|t|)
(Intercept) 18.467633   3.009009    6.137 1.09e-06 ***
astm        -0.209329   0.012737  -16.435 3.11e-16 ***
endPt        0.155813   0.006855   22.731  < 2e-16 ***
---
Signif. codes:
0 '***' 0.001 '**' 0.01 '*' 0.05 '.' 0.1 ' ' 1

Residual standard error: 2.426 on 29 degrees of freedom
Multiple R-squared:  0.9521,       Adjusted R-squared: 0.9488
F-statistic: 288.4 on 2 and 29 DF,  p-value: < 2.2e-16
```

The following is a MINITAB output of the regression of y on x_3 and x_4:

The regression equation is			
Yield = 18.5 − 0.209 ASTM + 0.156 End_pt			
Predictor	Coef	Stdev	*t*-ratio
Constant	18.468	3.009	6.14
ASTM	−0.20933	0.01274	−16.43
End_pt	0.155813	0.006855	22.73
$s = 2.426$	#R#-sq = 95.2%		

We compute now these estimates of the regression coefficients using the above formulae. The variances and covariances of x_3, x_4, and y are (as derived from MINITAB)

	ASTM	End_pt	Yield
ASTM	1409.355		
End_pt	1079.565	4865.894	
Yield	−126.808	532.188	114.970

The means of these variables are $\overline{X}_3 = 241.500$, $\overline{X}_4 = 332.094$, $\overline{Y} = 19.6594$. Thus, the least-squares estimators of b_1 and b_2 are obtained by solving the equations:

$$1409.355b_1 + 1079.565b_2 = -126.808$$
$$1079.565b_1 + 4865.894b_2 = 532.188.$$

The solution is

$$b_1 = -0.20933,$$

and

$$b_2 = 0.15581.$$

Finally, the estimate of β_0 is

$$b_0 = 19.6594 + 0.20933 \times 241.5 - 0.15581 \times 332.094$$

$$= 18.469.$$

These are the same results as in the MINITAB output. Moreover, the multiple R^2 is

$$R^2_{y|(x_3,x_4)} = \frac{1}{114.970}[0.20932 \times 126.808 + 0.15581 \times 532.88]$$

$$= 0.9530.$$

In addition,

$$S^2_{y|(x_1,x_2)} = S^2_y(1 - R^2_{y|(x_1,x_2)})$$

$$= 114.97(1 - 0.9530)$$

$$= 5.4036.$$

In Figure 5.14, we present a scatterplot of the residuals $\hat{e}_i$ $(i = 1, \ldots, n)$ against the predicted values $\hat{y}_i$ $(i = 1, \ldots, n)$. This scatterplot does not reveal any pattern different than random. It can be concluded that the regression of y on x_3 and x_4 accounts for all the systematic variability in the yield, y. Indeed, $R^2 = 0.952$, and no more than 4.8% of the variability in y is unaccounted by the regression.

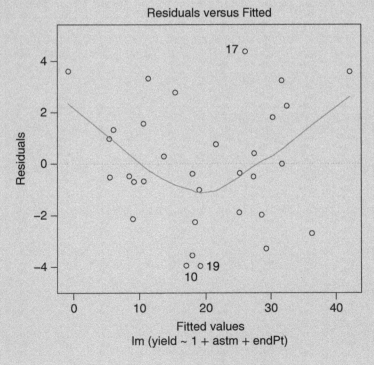

Figure 5.14   *Scatterplot of $\hat{e}$ versus $\hat{Y}$.*

The following are formulae for the variances of the least squares coefficients. First, we convert $S^2_{y|(x_1,x_2)}$ to S^2_e, i.e.,

$$S^2_e = \frac{n-1}{n-3} S^2_{y|x}. \tag{5.4.8}$$

S^2_e is an unbiased estimator of σ^2. The variance formulae are

$$S^2_{b_0} = \frac{S^2_e}{n} + \bar{x}_1^2 S^2_{b_1} + \bar{x}_2^2 S^2_{b_2} + 2\bar{x}_1 \bar{x}_2 S_{b_1 b_2},$$

$$S^2_{b_1} = \frac{S^2_e}{n-1} \cdot \frac{S^2_{x_2}}{D},$$

$$S^2_{b_2} = \frac{S^2_e}{n-1} \cdot \frac{S^2_{x_1}}{D},$$

$$S_{b_1 b_2} = -\frac{S^2_e}{n-1} \cdot \frac{S_{x_1 x_2}}{D}, \tag{5.4.9}$$

where

$$D = S^2_{x_1} S^2_{x_2} - (S_{x_1 x_2})^2.$$

Example 5.8.   Using the numerical Example 5.7 on the **GASOL.csv** data, we find that

$$S^2_e = 5.8869,$$

$$D = 5692311.4,$$

$$S^2_{b_1} = 0.0001624,$$

$$S^2_{b_2} = 0.0000470,$$

$$S_{b_1,b_2} = -0.0000332,$$

and

$$S^2_{b_0} = 9.056295$$

The squared roots of these variance estimates are the "Stdev" values printed in the MINITAB output, and $s = S_e$.

5.5   Partial regression and correlation

In performing the multiple least squares regression, one can study the effect of the predictors on the response in stages. This more pedestrian approach does not simultaneously provide all regression coefficients, but studies the effect of predictors in more detail.

In Stage I, we perform a simple linear regression of the yield y on one of the predictors, x_1 say. Let $a_0^{(1)}$ and $a_1^{(1)}$ be the intercept and slope coefficients of this simple linear regression. Let $\hat{e}^{(1)}$ be the vector of residuals

$$\hat{e}_i^{(1)} = y_i - \left(a_0^{(1)} + a_1^{(1)} x_{1i} \right), \quad i = 1, \ldots, n. \tag{5.5.1}$$

In Stage II, we perform a simple linear regression of the second predictor, x_2, on the first predictor x_1. Let $c_0^{(2)}$ and $c_1^{(2)}$ be the intercept and slope coefficients of this regression. Let $\hat{e}^{(2)}$ be the vector of residuals,

$$\hat{e}_i^{(2)} = x_{2i} - \left(c_0^{(2)} + c_1^{(2)} x_{1i}\right), \quad i = 1, \ldots, n. \tag{5.5.2}$$

In Stage III, we perform a simple linear regression of $\hat{e}^{(1)}$ on $\hat{e}^{(2)}$. It can be shown that this linear regression must pass through the origin, i.e., it has a zero intercept. Let $d^{(3)}$ be the slope coefficient.

The simple linear regression of $\hat{e}^{(1)}$ on $\hat{e}^{(2)}$ is called the **partial regression**. The correlation between $\hat{e}^{(1)}$ and $\hat{e}^{(2)}$ is called the **partial correlation** of y and x_2, given x_1 and is denoted by $r_{yx_2 \cdot x_1}$.

From the regression coefficients obtained in the three stages, one can determine the multiple regression coefficients of y on x_1 and x_2, according to the formulae:

$$b_0 = a_0^{(1)} - d^{(3)} c_0^{(2)},$$
$$b_1 = a_1^{(1)} - d^{(3)} c_1^{(2)},$$
$$b_2 = d^{(3)}. \tag{5.5.3}$$

Example 5.9. For the **GASOL** data, let us determine the multiple regression of the yield (y) on the ASTM (x_3) and the End_pt (x_4) in stages.
In Stage I, the simple linear regression of y on ASTM is

$$\hat{y} = 41.4 - 0.08998 \cdot x_3.$$

The residuals of this regression are $\hat{e}^{(1)}$. Also $R_{yx_3}^2 = 0.099$. In Stage II, the simple linear regression of x_4 on x_3 is

$$\hat{x}_4 = 147 + 0.766 \cdot x_3.$$

The residuals of this regression are $\hat{e}^{(2)}$. In Figure 5.15, we see the scatterplot of $\hat{e}^{(1)}$ versus $\hat{e}^{(2)}$. The partial correlation is $r_{yx_4 \cdot x_3} = 0.973$. This high partial correlation means that, after adjusting the variability of y for the variability of x_3, and the variability of x_4 for that of x_3, the adjusted x_4 values, namely $\hat{e}_i^{(2)}$ ($i = 1, \ldots, n$), are still good predictors for the adjusted y values, namely $\hat{e}_i^{(1)}$ ($i = 1, \ldots, n$). The regression of $\hat{e}^{(1)}$ on $\hat{e}^{(2)}$, determined in Stage III, is

$$\hat{e}^{(1)} = 0.156 \cdot \hat{e}^{(2)}.$$

We have found the following estimates:

$$a_0^{(1)} = 41.4, \quad a_1^{(1)} = -0.08998$$
$$c^{(2)} = 147.0, \quad c_1^{(2)} = 0.766$$
$$d^{(3)} = 0.156.$$

From the above formulae, we get

$$b_0 = 41.4 - 0.156 \times 147.0 = 18.468,$$
$$b_1 = -0.0900 - 0.156 \times 0.766 = -0.2095$$
$$b_2 = 0.156.$$

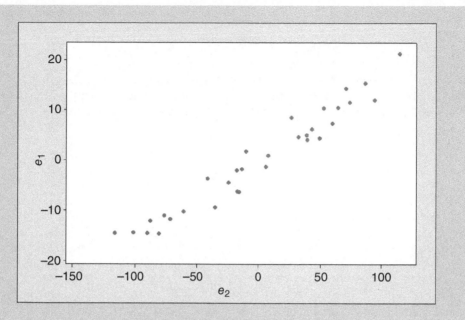

Figure 5.15   *Scatterplot of $\hat{e}_1$ versus $\hat{e}_2$ (MINITAB).*

These values coincide with the previously determined coefficients. Finally, the relationship between the multiple and the partial correlations is

$$R^2_{y|(x_1,x_2)} = 1 - (1 - R^2_{yx_1})(1 - r^2_{yx_2 \cdot x_1}).\qquad(5.5.4)$$

In the present example,

$$R^2_{y|(x_3,x_4)} = 1 - (1 - 0.099)(1 - 0.94673) = 0.9520.$$

In R:

```
> summary(LmYield <-
          update(object=LmYield,
                 formula.=. ~ 1 + astm))

Call:
lm(formula = yield ~ astm, data = GASOL)

Residuals:
     Min       1Q   Median       3Q      Max
-14.6939  -9.5838  -0.2001   7.6575  21.4069

Coefficients:
            Estimate Std. Error t value Pr(>|t|)
(Intercept) 41.38857   12.09117   3.423  0.00181 **
astm        -0.08998    0.04949  -1.818  0.07906 .
```

```
---
Signif. codes:
0 '***' 0.001 '**' 0.01 '*' 0.05 '.' 0.1 ' ' 1

Residual standard error: 10.34 on 30 degrees of freedom
Multiple R-squared:  0.09924,        Adjusted R-squared: 0.06921
F-statistic: 3.305 on 1 and 30 DF, p-value: 0.07906


> summary(LmYield2 <-
          lm(endPt ~ 1 + astm,
             data=GASOL))

Call:
lm(formula = endPt ~ 1 + astm, data = GASOL)


Residuals:
    Min        1Q    Median        3Q       Max
-116.627   -46.215   -1.894    50.120   114.355


Coefficients:
            Estimate Std. Error t value Pr(>|t|)
(Intercept) 147.1050    75.5101   1.948   0.0608.
astm          0.7660     0.3091   2.478   0.0190 *
---
Signif. codes:
0 '***' 0.001 '**' 0.01 '*' 0.05 '.' 0.1 ' ' 1

Residual standard error: 64.6 on 30 degrees of freedom
Multiple R-squared:  0.1699,         Adjusted R-squared: 0.1423
F-statistic: 6.142 on 1 and 30 DF,  p-value: 0.01905
```

5.6   Multiple linear regression

In the general case, we have k predictors ($k \geq 1$). Let (X) denote an array of n rows and $(k + 1)$ columns, in which, the first column consists of the value 1 in all entries, and the second to $(k + 1)$st columns consist of the values of the predictors $x_1, \ldots, x_k$. (X) is called the predictors matrix. Let Y be an array of n rows and one column, consisting of the values of the regressant. The linear regression model can be written in matrix notation as

$$Y = (X)\beta + e,   \tag{5.6.1}$$

where $\beta' = (\beta_0, \beta_1, \ldots, \beta_k)$ is the vector of regression coefficients, and e is a vector of random residuals.

The sum of squares of residuals, can be written as

$$SSE = (Y - (X)\beta)'(Y - (X)\beta).   \tag{5.6.2}$$

( )$'$ denotes the transpose of the vector of residuals. Differentiating SSE partially with respect to the components of β, and equating the partial derivatives to zero, we obtain a set of linear equations in the LSE **b**, namely

$$(X)'(X)\mathbf{b} = (X)'Y. \tag{5.6.3}$$

$(X)'$ is the transpose of the matrix (X). These linear equations are called the **normal equations**.
If we define the matrix

$$B = [(X)'(X)]^{-1}(X)', \tag{5.6.4}$$

where $[\ ]^{-1}$ is the inverse of $[\ ]$, then the general formula of the least-squares regression coefficients vector $\mathbf{b}' = (b_0, \ldots, b_k)$, is given in matrix notation as

$$\mathbf{b} = (B)\mathbf{Y}. \tag{5.6.5}$$

The vector of predicted y values, or FITS, is given by $\hat{\mathbf{y}} = (H)\mathbf{y}$, where $(H) = (X)(B)$. The vector of residuals $\hat{e} = \mathbf{y} - \hat{\mathbf{y}}$ is given by

$$\hat{e} = (I - H)\mathbf{y}, \tag{5.6.6}$$

where (I) is the $n \times n$ identity matrix. The variance of $\hat{e}$, around the regression surface, is

$$S_e^2 = \frac{1}{n-k-1} \sum_{i=1}^{n} \hat{e}_i^2$$

$$= \frac{1}{n-k-1} \mathbf{Y}'(I - H)\mathbf{Y}.$$

The sum of squares of $\hat{e}_i$ ($i = 1, \ldots, n$) is divided by $(n-k-1)$ to attain an unbiased estimator of σ^2. The multiple-R^2 is given by

$$R_{y|(x)}^2 = \frac{1}{(n-1)S_y^2}(\mathbf{b}'(X)'\mathbf{Y} - n\bar{y}^2), \tag{5.6.7}$$

where $x_{i0} = 1$ for all $i = 1, \ldots, n$, and S_y^2 is the sample variance of y. Finally, an estimate of the variance–covariance matrix of the regression coefficients $b_0, \ldots, b_k$ is

$$(S_b) = S_e^2[(X)'(X)]^{-1}. \tag{5.6.8}$$

Example 5.10.   We use again the **ALMPIN** data set, and regress the Cap Diameter (y) on Diameter 1 (x_1), Diameter 2 (x_2), and Diameter 3 (x_3). The "stdev" of the regression coefficients are the squared-roots of the diagonal elements of the (S_b) matrix. To see this, we present first the inverse of the $(X)'(X)$ matrix, which is given by the following symmetric matrix

$$[(X)'(X)]^{-1} = \begin{bmatrix} 5907.11 & -658.57 & 557.99 & -490.80 \\ \cdot & 695.56 & -448.14 & -181.94 \\ \cdot & \cdot & 739.76 & -347.37 \\ \cdot & \cdot & \cdot & 578.75 \end{bmatrix}$$

The value of S_e^2 is the square of the printed s value, i.e., $S_e^2 = 0.0000457$. Thus, the variances of the regression coefficients are

$$S_{b_0}^2 = 0.0000457 \times 5907.11 = 0.2700206$$

$$S_{b_1}^2 = 0.0000457 \times 695.56 = 0.0317871$$

$$S_{b_2}^2 = 0.0000457 \times 739.76 = 0.033807$$

$$S_{b_3}^2 = 0.0000457 \times 578.75 = 0.0264489.$$

Thus, S_{b_i} ($i = 0, \ldots, 3$) are the "Stdev" in the printout. The t-ratios are given by

$$t_i = \frac{b_i}{S_{b_i}}, \quad i = 0, \ldots, 3.$$

The t-ratios should be large to be considered significant. The significance criterion is given by the P-value. Large value of P indicates that the regression coefficient is not significantly different from zero. In the above table, see that b_2 is not significant. Notice that Diameter 2 by itself, as the sole predictor of Cap Diameter, is very significant. This can be verified by running a simple regression of y on x_2. However, in the presence of x_1 and x_3, x_2 loses its significance. This analysis can be done in R as shown below.

```
Call:
lm(formula = capDiam ~ 1 + diam1 + diam2 + diam3, data
          = ALMPIN)

Residuals:
      Min         1Q     Median         3Q        Max
-0.013216  -0.004185  -0.002089   0.007543   0.015466

Coefficients:
            Estimate Std. Error t value Pr(>|t|)
(Intercept)  4.04111    0.51961   7.777 6.63e-11 ***
diam1        0.75549    0.17830   4.237 7.18e-05 ***
diam2        0.01727    0.18388   0.094   0.9255
diam3        0.32269    0.16265   1.984   0.0514 .
---
Signif. codes:
0 '***' 0.001 '**' 0.01 '*' 0.05 '.' 0.1 ' ' 1

Residual standard error: 0.006761 on 66 degrees of freedom
Multiple R-squared:  0.879,          Adjusted R-squared:  0.8735
F-statistic: 159.9 on 3 and 66 DF,  p-value: < 2.2e-16

            Df   Sum Sq  Mean Sq F value Pr(>F)
diam1        1 0.021657 0.021657 473.823 <2e-16 ***
diam2        1 0.000084 0.000084   1.832 0.1805
diam3        1 0.000180 0.000180   3.936 0.0514 .
Residuals   66 0.003017 0.000046
---
```

```
Signif. codes:
0 '***' 0.001 '**' 0.01 '*' 0.05 '.' 0.1 ' ' 1
```

If we perform this analysis with MINITAB, we get the following output:MINITAB output of the multiple regression

---

The regression equation is
CapDiam = 4.04 + 0.755 Diam 1 + 0.017 Diam2 + 0.323 Diam3

Predictor	Coef	Stdev	t-ratio	P
Constant	4.0411	0.5196	7.78	0.000
Diam1	0.7555	0.1783	4.24	0.000
Diam2	0.0172	0.1839	0.09	0.926
Diam3	0.3227	0.1626	1.98	0.051

$s = 0.006761$ R-sq = 87.9% R-sq(adj) = 87.4%

Analysis of Variance

SOURCE	DF	SS	MS	F	P
Regression	3	0.0219204	0.0073068	159.86	0.000
Error	66	0.0030167	0.0000457		
Total	69	0.0249371			

The "Analysis of Variance" table provides a global summary of the contribution of the various factors to the variability of y. The total sum of squares of y, around its mean is

$$\text{SST} = (n-1)S_y^2 = \sum_{i=1}^{n}(y_i - \bar{y})^2 = 0.0249371.$$

This value of SST is partitioned into the sum of the variability explainable by the regression (SSR) and that due to the residuals around the regression (Error, SSE). These are given by

$$\text{SSE} = (n - k - 1)S_e^2$$

$$\text{SSR} = \text{SST} - \text{SSE}$$

$$= \mathbf{b}'(X)'\mathbf{y} - n\bar{\mathbf{Y}}_n^2$$

$$= \sum_{j=0}^{k} b_j \cdot \sum_{i=1}^{n} X_{ij} y_i - \left(\sum_{i=1}^{n} y(i)\right)^2 / n.$$

We present in Figure 5.16, the scatterplot of the residuals, $\hat{e}_i$, versus the predicted values (FITS), $\hat{y}_i$. We see in this figure one point, corresponding to element # 66 in the data set, whose x-values have strong influence on the regression.

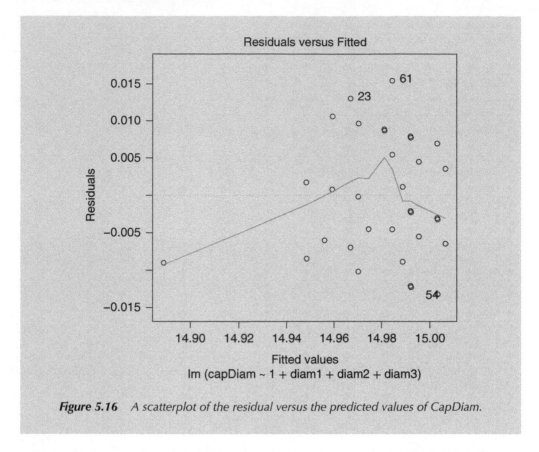

Figure 5.16   *A scatterplot of the residual versus the predicted values of CapDiam.*

The multiple regression can be used to test whether two or more simple linear regressions are parallel (same slopes) or have the same intercepts. We will show this by comparing two simple linear regressions.

Let $(x_i^{(1)}, Y_i^{(1)})$, $i = 1, \ldots, n$, be data set of one simple linear regression of Y on x, and let $(x_j^{(2)}, Y_j^{(2)})$, $j = 1, \ldots, n_2$ be that of the second regression. By combining the data on the regression x from the two sets, we get the $\mathbf{x}$ vector

$$\mathbf{x} = (x_1^{(1)}, \ldots, x_{n_1}^{(1)}, x_1^{(2)}, \ldots, x_{n_2}^{(2)})'.$$

In a similar fashion, we combine the Y values and set

$$\mathbf{Y} = (Y_1^{(1)}, \ldots, Y_{n_1}^{(1)}, Y_1^{(2)}, \ldots, Y_{n_2}^{(2)})'.$$

Introduce a dummy variable z. The vector $\mathbf{z}$ has n_1 zeros at the beginning followed by n_2 ones. Consider now the multiple regression

$$\mathbf{Y} = b_0 \mathbf{1} + b_1 \mathbf{x} + b_2 \mathbf{z} + b_3 \mathbf{w} + \mathbf{e}, \tag{5.6.9}$$

where $\mathbf{1}$ is a vector of $(n_1 + n_2)$ ones, and $\mathbf{w}$ is a vector of length $(n_1 + n_2)$, whose i-th component is the product of the corresponding components of $\mathbf{x}$ and $\mathbf{z}$, i.e., $w_i = x_i z_i$ ($i = 1, \ldots, n_1 + n_2$).

Perform the regression analysis of $\mathbf{Y}$ on $(\mathbf{x}, \mathbf{z}, \mathbf{w})$. If b_2 is significantly different than 0 we conclude that the two simple regression lines have different intercepts. If b_3 is significantly different from zero we conclude that the two lines have different slopes.

Example 5.11. In the present example, we compare the simple linear regressions of Turn-diameter (Y) on MPG/City (x) of US made cars and of Japanese cars. The data is in the file **CAR.csv**. The simple linear regression for US cars is

$$\hat{Y} = 49.0769 - 0.7565x$$

with $R^2 = 0.432$, $S_e = 2.735$ [56 degrees of freedom]. The simple linear regression for Japanese cars is

$$\hat{Y} = 42.0860 - 0.5743x,$$

with $R^2 = 0.0854$, $S_e = 3.268$ [35 degrees of freedom]. The combined multiple regression of Y on $\mathbf{x}, \mathbf{z}, \mathbf{w}$ yields the following table of P-values of the coefficients

Coefficients:

	Value	Std. Error	t value	$\Pr(> |t|)$
(Intercept)	49.0769	5.3023	9.2557	0.0000
mpgc	−0.7565	0.1420	−5.3266	0.0000
z	−6.9909	10.0122	−0.6982	0.4868
w	0.1823	0.2932	0.6217	0.5357

We see in this table that the P-values corresponding to $\mathbf{z}$ and $\mathbf{w}$ are 0.4868 and 0.5357, respectively. Accordingly, both b_2 and b_3 are **not** significantly different than zero. We can conclude that the two regression lines are not significantly different. We can combine the data and have one regression line for both US and Japanese cars, namely:

$$\hat{Y} = 44.8152 - 0.6474x$$

with $R^2 = 0.3115$, $S_e = 3.337$ [93 degrees of freedom].

5.7 Partial F-tests and the sequential SS

In the MINITAB output for multiple regression, a column entitled SEQ SS provides a partition of the regression sum of squares, SSR, to additive components of variance, each one with 1 degree of freedom. We have seen that the multiple R^2, $R^2_{y|(x_1,\ldots,x_k)} = $ SSR/SST, is the proportion of the total variability which is explainable by the linear dependence of Y on all the k regressors. A simple linear regression on the first variable x_1 yields a smaller $R^2_{y|x_1}$. The first component of the SEQ SS is $\text{SSR}_{y|(x_1)} = \text{SST} \cdot R^2_{y|(x_1)}$. If we determine the multiple regression of Y on x_1 and x_2, then $\text{SSR}_{y|(x_1,x_2)} = \text{SST} R^2_{y|(x_1,x_2)}$ is the amount of variability explained by the linear relationship with the two variables. The difference

$$\text{DSS}_{x_2|x_1} = \text{SST} \left(R^2_{y|(x_1x_2)} - R^2_{y|(x_1)} \right) \tag{5.7.1}$$

is the additional amount of variability explainable by x_2, after accounting for x_1. Generally, for $i = 2, \ldots, k$

$$\text{DSS}_{x_i | x_1 \ldots x_{i-1}} = \text{SST} \left(R^2_{y|(x_1, \ldots, x_i)} - R^2_{y|(x_1, \ldots, x_{i-1})} \right) \qquad (5.7.2)$$

is the additional contribution of the i-th variable after controlling for the first $(i - 1)$ variables.

Let

$$s^2_{e(i)} = \frac{\text{SST}}{n - i - 1} \left(1 - R^2_{y|(x_1, \ldots, x_i)} \right), \quad i = 1, \ldots, k, \qquad (5.7.3)$$

then

$$F^{(i)} = \frac{\text{DSS}_{x_i | x_1, \ldots, x_{i-1}}}{s^2_{e(i)}}, \quad i = 1, \ldots, k \qquad (5.7.4)$$

is called the **partial-F** for testing the significance of the contribution of the variable x_i, after controlling for $x_1, \ldots, x_{i-1}$. If $F^{(i)}$ is greater than the $(1 - \alpha)$th quantile $F_{1-\alpha}[1, n - i - 1]$ of the F distribution, the additional contribution of X_i is significant. The partial F-test is used to assess whether the addition of the i-th regression significantly improves the prediction of Y, given that the first $(i - 1)$ regressors have already been included.

Example 5.12.   In the previous example, we have examined the multiple regression of Cap-Diam, Y, on Diam1, Diam2, and Diam3 in the **ALMPIN.csv** file. We compute here the partial F statistics corresponding to the SEQ SS values.

Variable	SEQ SS	SSE	d.f.	Partial-F	P-value
Diam1	0.0216567	0.003280	68	448.98	0
Diam2	0.0000837	0.003197	67	1.75	0.190
Diam3	0.0001799	0.003167	66	3.93	0.052

We see from these partial-F values, and their corresponding P-values, that after using Diam1 as a predictor, the additional contribution of Diam2 is insignificant. Diam3, however, in addition to the regressor Diam1, significantly decreases the variability which is left unexplained.

The partial-F test is called sometimes **sequential-F** test (Draper and Smith 1981, p. 612). We use the terminology Partial-F statistic because of the following relationship between the partial-F and the partial correlation. In Section 5.5, we defined the partial correlation $r_{yx_2 \cdot x_1}$, as the correlation between $\hat{e}^{(1)}$, which is the vector of residuals around the regression of Y on x_1, and the vector of residuals $\hat{e}^{(2)}$, of x_2 around its regression on x_1. Generally, suppose that we have determined the multiple regression of Y on $(x_1, \ldots, x_{i-1})$. Let $\hat{e}(y \mid x_1, \ldots, x_{i-1})$ be the vector of residuals around this regression. Let $\hat{e}(x_i \mid x_1, \ldots, x_{i-1})$ be the vector of residuals around the multiple regression of x_i on $x_1, \ldots, x_{i-1}$ ($i \geq 2$). **The correlation between** $\hat{e}(y \mid x_1, \ldots, x_{i-1})$ **and** $\hat{e}(x_i \mid x_1, \ldots, x_{i-1})$ **is the partial correlation between** Y **and** x_i, **given** $x_1, \ldots, x_{i-1}$. We denote this partial correlation by $r_{yx_i \cdot x_1, \ldots, x_{i-1}}$. The following relationship holds between the partial-F, $F^{(i)}$, and the partial correlation

$$F^{(i)} = (n - i - 1) \frac{r^2_{yx_i \cdot x_1, \ldots, x_{i-1}}}{1 - r^2_{yx_i \cdot x_1, \ldots, x_{i-1}}}, \quad i \geq 2. \qquad (5.7.5)$$

This relationship is used to test whether $r_{yx_i \cdot x_1, \ldots, x_{i-1}}$ is significantly different than zero. $F^{(i)}$ should be larger than $F_{1-\alpha}[1, n - i - 1]$.

5.8   Model construction: stepwise regression

It is often the case that data can be collected on a large number of regressors, which might help us predict the outcomes of a certain variable, Y. However, the different regressors vary generally with respect to the amount of variability in Y which they can explain. Moreover, different regressors or predictors are sometimes highly correlated and therefore not all of them might be needed to explain the variability in Y and to be used as predictors.

The following example is given by Draper and Smith (1981, p. 615). The amount of steam [Pds] which is used monthly, Y, in a plant may depend on nine regressors:

x_1 = Pounds of real fatty acid in storage per month;

x_2 = Pounds of crude glycerin made in a month;

x_3 = Monthly average wind velocity [Miles/hour]

x_4 = Plant operating days per month;

x_5 = Number of days per month with temperature below $32°F$;

x_6 = Monthly average atmospheric temperature [F],

etc. Are all these six regressors required to be able to predict Y? If not, which variables should be used? This is the problem of model construction.

There are several techniques for constructing a regression model. R, MINITAB and JMP use a stepwise method which is based on forward selection, backward elimination, and user intervention, which can force certain variables to be included. We present only the **forward** selection procedure.

In the first step, we select the variable x_j ($j = 1, \ldots, k$) whose correlation with Y has maximal magnitude, provided it is significantly different than zero.

At each step, the procedure computes a partial-F, or partial correlation, for each variable, x_l, which has not been selected in the previous steps. A variable having the **largest** significant partial-F is selected. The procedure stops when no additional variables can be selected. We illustrate the forward stepwise regression in the following example.

Example 5.13.   In Example 5.7, we introduced the data file **GASOL.csv** and performed a multiple regression of Y on x_3 and x_4. In the present example, we apply the MINITAB stepwise regression procedure to arrive at a linear model, which includes all variables which contributes significantly to the prediction. In R.

```
> LmYield <- lm(yield ~ 1, data=GASOL)
> Step <- step(LmYield, direction="both",
               scope=list(
                 lower= ~ 1,
                 upper= ~ endPt + astm + x1 + x2),
               trace=FALSE)
> Step$anova
```

	Step	Df	Deviance	Resid. Df	Resid. Dev	AIC
1		NA	NA	31	3564.0772	152.81359
2	+ endPt	-1	1804.38359	30	1759.6936	132.22909
3	+ astm	-1	1589.08205	29	170.6115	59.55691

```
4     + x1 -1    24.61041         28    146.0011   56.57211
5     + x2 -1    11.19717         27    134.8040   56.01874
```

The MINITAB output is given in the following
MINITAB output for stepwise regression

Stepwise Regression
F-to-Enter: 4.00   *F*-to-Remove: 4.00
Response is gasy on four predictors, with $N = 32$

Step	1	2	3
Constant	−16.662	18.468	4.032
gasx4	0.1094	0.1558	0.1565
T-Ratio	5.55	22.73	24.22
gasx3		−0.209	−0.187
T-Ratio		−16.43	−11.72
gasx1			0.22
T-Ratio			2.17
S	7.66	2.43	2.28
R-Sq(%)	50.63	95.21	95.90

More? (Yes, No, Subcommand, or Help)
SUBC> No.

The MINITAB procedure includes at each stage the variable whose partial-*F* value is maximal, but greater than "*F*-to-enter," which is 4.00. Since each partial-*F* statistic has 1 degree of freedom in the denominator, and since

$$(F_{1-\alpha}[1, v])^{1/2} = t_{1-\alpha/2}[v],$$

the output prints the corresponding values of *t*. Thus, in Step 1, variable x_4 is selected (gasx4). The fitted regression equation is

$$\hat{Y} = -16.662 + 0.1094x_4$$

with $R^2_{y|(x_4)} = 0.5063$. The partial-*F* for x_4 is $F = (5.55)^2 = 30.8025$. Since this value is greater than "*F*-to-remove," which is 4.00, x_4 remains in the model. In Step 2 the maximal partial correlation of *Y* and x_1, x_2, x_3 given x_4, is that of x_3, with a partial-$F = 269.9449$. Variable x_3 is selected, and the new regression equation is

$$\hat{Y} = 18.468 + 0.1558x_4 - 0.2090x_3,$$

with $R^2_{y|(x_4,x_3)} = 0.9521$. Since the partial-*F* of x_4 is $(22.73)^2 = 516.6529$, the two variables remain in the model. In Step 3 the variable x_1 is chosen. Since its partial-*F* is $(2.17)^2 = 4.7089$, it is included too. The final regression equation is

$$\hat{Y} = 4.032 + 0.1565x_4 - 0.1870x_3 + 0.2200x_1$$

with $R^2_{y|(x_4,x_3,x_1)} = 0.959$. Only 4.1% of the variability in *Y* is left unexplained.

To conclude the example, we provide the three regression analyses, from the MINITAB. One can compute the partial-F values from these tables.

MTB> Regress "gasy" 1 "gasx4";
SUBC> Constant.

Regression Analysis

The regression equation is gasy $= -16.7 + 0.109$ gasx4

Predictor	Coef	Stdev	t-ratio	p
Constant	-16.662	6.687	-2.49	0.018
gasx4	0.10937	0.01972	5.55	0.000

$s = 7.659$ $\quad R - \text{sq} = 50.6\%$ $\quad R - \text{sq(adj)} = 49.0\%$

Analysis of Variance

SOURCE	DF	SS	MS	F	p
Regression	1	1804.4	1804.4	30.76	0.000
Error	30	1759.7	58.7		
Total	31	3564.1			

Unusual observations

Obs.	gasx4	gasy	Fit	Stdev.Fit	Residual	St.Resid
4	407	45.70	27.85	2.00	17.85	2.41R

R denotes an obs. with a large st. resid.

MTB> Regress "gasy" 2 "gasx4" "gasx3";
SUBC> Constant.

Regression Analysis

The regression equation is gasy $= 18.5 + 0.156$ gasx4 $- 0.209$ gasx3

Predictor	Coef	Stdev	t-ratio	p
Constant	18.468	3.009	6.14	0.000
gasx4	0.155813	0.006855	22.73	0.000
gasx3	-0.20933	0.01274	-16.43	0.000

$s = 2.426$ $\quad R\text{-sq} = 95.2\%$ $\quad R\text{-sq(adj)} = 94.9\%$

Analysis of Variance

SOURCE	DF	SS	MS	F	p
Regression	2	3393.5	1696.7	288.41	0.000
Error	29	170.6	5.9		
Total	31	3564.1			

SOURCE	DF	SEQ SS
gasx4	1	1804.4
gasx3	1	1589.1

```
MTB> Regress "gasy" 3 "gasx4" "gasx3" "gasx1";
SUBC> Constant.
```

Regression Analysis

The regression equation is gasy = 4.03 + 0.157 gasx4 − 0.187 gasx3 + 0.222 gasx1

Predictor	Coef	Stdev	t-ratio	p
Constant	4.032	7.223	0.56	0.581
gasx4	0.156527	0.006462	24.22	0.000
gasx3	−0.18657	0.01592	−11.72	0.000
gasx1	0.2217	0.1021	2.17	0.038

$s = 2.283$ $R\text{-sq} = 95.9\%$ $R\text{-sq(adj)} = 95.5\%$

Analysis of Variance

SOURCE	DF	SS	MS	F	p
Regression	3	3418.1	1139.4	218.51	0.000
Error	28	146.0	5.2		
Total	31	3564.1			

SOURCE	DF	SEQ SS
gasx4	1	1804.4
gasx3	1	1589.1
gasx1	1	24.6

Unusual Observations

Obs.	gasx4	gasy	Fit	Stdev.Fit	Residual	St.Resid
17	340	30.400	25.634	0.544	4.766	2.15R

R denotes an obs. with a large st. resid.
```
MTB>
```

5.9   Regression diagnostics

As mentioned earlier, the least-squares regression line is sensitive to extreme x or y values of the sample elements. Sometimes even one point may change the characteristics of the regression line substantially. We illustrate this in the following example.

Example 5.14.   Consider again the SOCELL data. We have seen earlier that the regression line (L1) of ISC at time $t2$ on ISC at time $t1$ is $\hat{y} = 0.536 + 0.929x$, with $R^2 = 0.959$.

The point having the largest x-value has a y-value of 5.37. If the y-value of this point is changed to 4.37, we obtain a different regression line (L2), given by $\hat{y} = 2.04 + 0.532x$, with $R^2 = 0.668$. In R

```
> print(influence.measures(LmISC), digits=2)

Influence measures of
        lm(formula = t2 ~ 1 + t1, data = SOCELL):
```

	dfb.1_	dfb.tl	dffit	cov.r	cook.d	hat	inf
1	-0.0028	0.00356	0.0076	1.26	0.000031	0.080	
2	-0.2716	0.24914	-0.3346	1.22	0.057266	0.140	
3	-0.0361	0.05761	0.2057	1.14	0.021816	0.068	
4	0.0180	-0.04752	-0.2766	1.05	0.037942	0.064	
5	0.1955	-0.15761	0.3949	0.95	0.073055	0.074	
6	0.0090	-0.05878	-0.4647	0.80	0.093409	0.064	
7	0.6262	-0.68909	-0.8816	0.80	0.318895	0.161	
8	-0.1806	0.15398	-0.2995	1.10	0.044951	0.085	
9	-1.4355	1.50371	1.6030	1.73	1.170543	0.521	*
10	0.0529	-0.04820	0.0667	1.33	0.002391	0.131	
11	0.0306	-0.01344	0.1615	1.17	0.013630	0.063	
12	-0.0451	0.04225	-0.0511	1.44	0.001407	0.197	*
13	-0.0220	0.03951	0.1662	1.17	0.014439	0.066	
14	-0.0233	0.02838	0.0539	1.26	0.001559	0.087	
15	0.0019	-0.00095	0.0089	1.24	0.000043	0.063	
16	0.5149	-0.47139	0.6392	0.94	0.183649	0.137	

The MINITAB diagnostics singles this point as an unusual point, as seen in the following box:

Unusual Observations						
Obs.	ISC1	ISC2	Fit	Stdev.Fit	Residual	St.Resid
9	5.11	4.3700	4.7609	0.1232	−0.3909	−3.30RX

In the present section, we present the diagnostic tools which are commonly used. The objective is to measure the degree of influence the points have on the regression line.

We start with the notion of the x-**leverage** of a point.

Consider the matrix (H) defined in Section 5.6. The vector of predicted values, $\hat{\mathbf{y}}$, is obtained as $(H)\mathbf{y}$. The x-**leverage** of the i-th point is measured by the i-th diagonal element of (H), which is

$$h_i = \mathbf{x}_i'((X)'(X))^{-1}\mathbf{x}_i, \quad i = 1, \ldots, n. \tag{5.9.1}$$

Here $\mathbf{x}_i'$ denotes the i-th row of the predictors matrix (X), i.e.,

$$\mathbf{x}_i' = (1, x_{i1}, \ldots, x_{ik}).$$

In the special case of simple linear regression ($k = 1$), we obtain the formula

$$h_i = \frac{1}{n} + \frac{(x_i - \bar{x})^2}{\sum_{j=1}^{n} (x_j - \bar{x})^2}, \quad i = 1, \ldots, n. \tag{5.9.2}$$

Notice that $S_e \sqrt{h_{ii}}$ is the **standard-error** (squared root of variance) **of the predicted value** $\hat{y}_i$. This interpretation holds also in the multiple regression case ($k > 1$). In Figure 5.17, we present the x-leverage values of the various points in the **SOCELL** example.

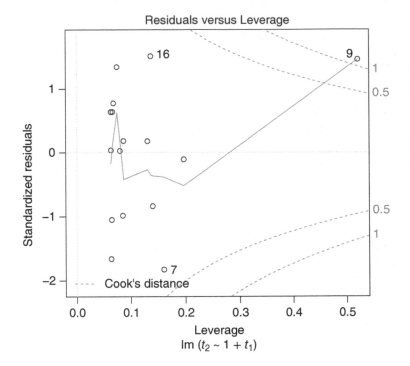

Figure 5.17   *x-leverage of ISC values.*

From the above formula we deduce that, when $k = 1$, $\sum_{i=1}^{n} h_i = 2$. Generally, for any k, $\sum_{i=1}^{n} h_i = k + 1$. Thus, the average x-leverage is $\bar{h} = \frac{k+1}{n}$. In the above solar cells example, the average x-leverage of the 16 points is $\frac{2}{16} = 0.125$. Point #9, (5.21,4.37), has a leverage value of $h_9 = 0.521$. This is indeed a high x-leverage.

The standard-error of the i-th residual, $\hat{e}_i$, is given by

$$S\{\hat{e}_i\} = S_e \sqrt{1 - h_i}. \tag{5.9.3}$$

The **standardized residuals** are therefore given by

$$\hat{e}_i^* = \frac{\hat{e}_i}{S\{\hat{e}_i\}} = \frac{\hat{e}_i}{S_e \sqrt{1 - h_i}}, \tag{5.9.4}$$

$i = 1, \ldots, n$. There are several additional indices, which measure the effects of the points on the regression. We mention here two such measures, the **Cook distance** and the **fits distance**.

If we delete the i-th point from the data set and recompute the regression, we obtain a vector of regression coefficients $\mathbf{b}^{(i)}$ and standard deviation of residuals $S_e^{(i)}$. The standardized difference

$$D_i = \frac{(\mathbf{b}^{(i)} - \mathbf{b})'((X)'(X))(\mathbf{b}^{(i)} - \mathbf{b})}{(k + 1)S_e} \tag{5.9.5}$$

is the so-called **Cook's distance**.

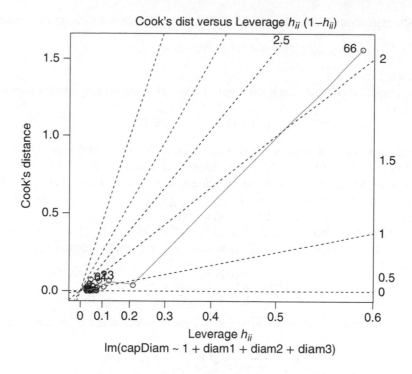

Figure 5.18   *Cook's distance for aluminum pins data.*

The influence of the fitted values, denoted by difference in fits distance (DFIT), is defined as

$$\text{DFIT}_i = \frac{\hat{Y}_i - \hat{Y}_i^{(i)}}{S_e^{(i)} \sqrt{h_i}}, \quad i = 1, \ldots, n, \tag{5.9.6}$$

where $\hat{Y}_i^{(i)} = b_0^{(i)} + \sum_{j=1}^{k} b_j^{(i)} x_{ij}$ are the predicted values of Y, at $(1, x_{i1}, \ldots, x_{ik})$, when the regression coefficients are $\mathbf{b}^{(i)}$.

In Figure 5.18, we present the Cook Distance, for the **ALMPIN** data set.

5.10 Quantal response analysis: logistic regression

We consider the case where the regressant Y is a binary random variable, and the regressors are quantitative. The distribution of Y at a given combination of x values $\mathbf{x} = (x_1, \ldots, x_k)$ is binomial $B(n, p(\mathbf{x}))$, where n is the number of identical and independent repetitions of the experiment at $\mathbf{x}$. $p(\mathbf{x}) = P\{Y = 1 \mid \mathbf{x}\}$. The question is how to model the function $p(\mathbf{x})$. An important class of models, is the so-called **quantal response models**, according to which

$$p(\mathbf{x}) = F(\boldsymbol{\beta}'\mathbf{x}), \tag{5.10.1}$$

where $F(\cdot)$ is a cumulative distribution function (c.d.f.), and

$$\boldsymbol{\beta}'\mathbf{x} = \beta_0 + \beta_1 x_1 + \cdots + \beta_k x_k. \tag{5.10.2}$$

The **logistic regression** is a method of estimating the regression coefficients β, in which

$$F(z) = e^z/(1 + e^z), -\infty < z < \infty, \tag{5.10.3}$$

is the logistic c.d.f.

The experiment is conducted at m different, and linearly independent, combinations of $\mathbf{x}$ values. Thus, let

$$(X) = (1, \mathbf{x}_1, \mathbf{x}_2, \ldots, \mathbf{x}_k)$$

be the predictors matrix of m rows and $(k + 1)$ columns. We assumed that $m > (k + 1)$ and the rank of (X) is $(k + 1)$. Let $\mathbf{x}^{(i)}$, $i = 1, \ldots, m$, denote the i-th row vector of (X).

As mentioned above, we replicate the experiment at each $\mathbf{x}^{(i)}$ n times. Let $\hat{p}_i$ $(i = 1, \ldots, m)$ be the proportion of 1's observed at $\mathbf{x}^{(i)}$, i.e., $\hat{p}_{i,n} = \frac{1}{n}\sum_{j=1}^{n} Y_{ij}$, $i = 1, \ldots, m$, where $Y_{ij} = 0, 1$, is the observed value of the regressant at the j-th replication ($j = 1, \ldots, n$).

We have proven before that $E\{\hat{p}_{i,n}\} = p(\mathbf{x}^{(i)})$, and $V\{\hat{p}_{i,n}\} = \frac{1}{n}p(\mathbf{x}^{(i)})(1 - p(\mathbf{x}^{(i)}))$, $i = 1, \ldots, m$. Also, the estimators $\hat{p}_i$ $(i = 1, \ldots, m)$ are independent. According to the logistic model,

$$p(\mathbf{x}^{(i)}) = \frac{e^{\beta'\mathbf{x}^{(i)}}}{1 + e^{\beta'\mathbf{x}^{(i)}}}, \quad i = 1, \ldots, m. \tag{5.10.4}$$

The problem is to estimate the regression coefficients β. Notice that the log-odds at $\mathbf{x}^{(i)}$ is

$$\log \frac{p(\mathbf{x}^{(i)})}{1 - p(\mathbf{x}^{(i)})} = \beta'\mathbf{x}^{(i)}, \quad i = 1, \ldots, m. \tag{5.10.5}$$

Define $Y_{i,n} = \log \frac{\hat{p}_{i,n}}{1-\hat{p}_{i,n}}$, $i = 1, \ldots, m$. $Y_{i,n}$ is finite if n is sufficiently large. Since $\hat{p}_{i,n} \to p(\mathbf{x}^{(i)})$ in probability, as $n \to \infty$ (WLLN), and since $\log \frac{x}{1-x}$ is a continuous function of x on $(0, 1)$, $Y_{i,n}$ is a consistent estimator of $\beta'\mathbf{x}^{(i)}$. For large values of n, we can write the regression model

$$Y_{i,n} = \beta'\mathbf{x}^{(i)} + e_{i,n} + e_{i,n}^*, \quad i = 1, \ldots, m, \tag{5.10.6}$$

where

$$e_{i,n} = (\hat{p}_{i,n} - p(\mathbf{x}^{(i)}))/[p(\mathbf{x}^{(i)})(1 - p(\mathbf{x}^{(i)}))]. \tag{5.10.7}$$

$e_{i,n}^*$ is a negligible remainder term if n is large. $e_{i,n}^* \to 0$ in probability at the rate of $\frac{1}{n}$. If we omit the remainder term $e_{i,n}^*$, we have the approximate regression model

$$Y_{i,n} \cong \beta'\mathbf{x}^{(i)} + e_{i,n}, \quad i = 1, \ldots, m, \tag{5.10.8}$$

where

$$E\{e_{i,n}\} = 0,$$

$$V\{e_{i,n}\} = \frac{1}{n} \cdot \frac{1}{p(\mathbf{x}^{(i)})(1 - p(\mathbf{x}^{(i)}))}$$

$$= \frac{(1 + e^{\beta'\mathbf{x}^{(i)}})^2}{n \cdot e^{\beta'\mathbf{x}^{(i)}}}, \tag{5.10.9}$$

$i = 1, \ldots, m$. The problem here is that $V\{e_{i,n}\}$ depends on the unknown β and varies from one $\mathbf{x}^{(i)}$ to another. An ordinary LSE of β is given by $\hat{\beta} = [(X)'(X)]^{-1}(X)'Y$, where

$\mathbf{Y}' = (Y_{1,n}, \ldots, Y_{m,n})$. Since the variances of $e_{i,n}$ are different, an estimator having smaller variances is the weighted LSE

$$\hat{\boldsymbol{\beta}}_w = [(X)'W(\boldsymbol{\beta})(X)]^{-1}(X)'W(\boldsymbol{\beta})Y, \qquad (5.10.10)$$

where $W(\boldsymbol{\beta})$ is a diagonal matrix, whose i-th term is

$$W_i(\boldsymbol{\beta}) = \frac{ne^{\boldsymbol{\beta}'\mathbf{x}^{(i)}}}{(1 + e^{\boldsymbol{\beta}'\mathbf{x}^{(i)}})^2}, \qquad i = 1, \ldots, m. \qquad (5.10.11)$$

The problem is that the weights $W_i(\boldsymbol{\beta})$ depend on the unknown vector $\boldsymbol{\beta}$. An iterative approach to obtain $\hat{\boldsymbol{\beta}}_w$ is to substitute on the r.h.s. the value of $\hat{\boldsymbol{\beta}}$ obtained in the previous iteration, starting with the ordinary LSE, $\hat{\boldsymbol{\beta}}$. Other methods of estimating the coefficients $\boldsymbol{\beta}$ of the logistic regression are based on the maximum likelihood method. For additional information, see Kotz and Johnson (1985) and Ruggeri et al. (2007).

5.11 The analysis of variance: the comparison of means

5.11.1 The statistical model

When the regressors $x_1, x_2, \ldots, x_k$ are qualitative (categorical) variables and the variable of interest Y is quantitative, the previously discussed methods of multiple regression are invalid. The different values that the regressors obtain are different categories of the variables. For example, suppose that we study the relationship between film speed (Y) and the type of gelatin x used in the preparation of the chemical emulsion for coating the film, the regressor is a categorical variable. The values it obtains are the various types of gelatin, as classified according to manufacturers.

When we have $k, k \geq 1$, such qualitative variables, the combination of categorical levels of the k variables are called **treatment combinations** (a term introduced by experimentalists). Several observations, n_i, can be performed at the i-th treatment combination. These observations are considered a random sample from the (infinite) population of all possible observations under the specified treatment combination. The statistical model for the j-th observation is

$$Y_{ij} = \mu_i + e_{ij}, \quad i = 1, \ldots t \ j = 1, \ldots, n_i,$$

where μ_i is the population mean for the i-th treatment combination, t is the number of treatment combinations, e_{ij} ($i = 1, \ldots, t; j = 1, \ldots, n_i$) are assumed to be independent random variables (experimental errors) with $E\{e_{ij}\} = 0$ for all (i,j) and $v\{e_{ij}\} = \sigma^2$ for all (i,j). The comparison of the means μ_i ($i = 1, \ldots, t$) provides information on the various effects of the different treatment combinations. The method used to do this analysis is called **analysis of variance** (ANOVA).

5.11.2 The one-way analysis of variance (ANOVA)

In Section 4.11.5.1, we introduced the ANOVA F-test statistics, and presented the algorithm for bootstrap ANOVA for comparing the means of k populations. In the present section, we develop the rationale for the ANOVA. We assume here that the errors e_{ij} are independent and normally distributed. For the i-th treatment combination (sample), let

$$\overline{Y}_i = \frac{1}{n_i} \sum_{j=1}^{n_i} Y_{ij}, \quad i = 1, \ldots, t \qquad (5.11.1)$$

and

$$\mathrm{SSD}_i = \sum_{j=1}^{n_i} (Y_{ij} - \overline{Y}_i)^2, \quad i = 1, \ldots, t. \tag{5.11.2}$$

Let $\overline{\overline{Y}} = \frac{1}{N} \sum_{i=1}^{t} n_i \overline{Y}_i$ be the grand mean of all the observations.

The one-way ANOVA is based on the following partition of the total sum of squares of deviations around $\overline{\overline{Y}}$,

$$\sum_{i=1}^{t} \sum_{j=1}^{n_i} (Y_{ij} - \overline{\overline{Y}})^2 = \sum_{i=1}^{t} \mathrm{SSD}_i + \sum_{i=1}^{t} n_i (\overline{Y}_i - \overline{\overline{Y}})^2. \tag{5.11.3}$$

We denote the l.h.s. by SST and the r.h.s. by SSW and SSB, i.e.,

$$\mathrm{SST} = \mathrm{SSW} + \mathrm{SSB}. \tag{5.11.4}$$

SST, SSW, and SSB are symmetric quadratic form in deviations like $Y_{ij} - \overline{\overline{Y}}$, $Y_{ij} - \overline{Y}_i$ and $\overline{Y}_i - \overline{\overline{Y}}$. Since $\sum_i \sum_j (Y_{ij} - \overline{\overline{Y}}) = 0$, only $N - 1$, linear functions $Y_{ij} - \overline{\overline{Y}} = \sum_{i'} \sum_{j'} c_{i'j'} Y_{i'j'}$, with

$$c_{i'j'} = \begin{cases} 1 - \frac{1}{N}, & i' = i, \ j' = j \\ -\frac{1}{N}, & \text{otherwise} \end{cases}$$

are linearly independent, where $N = \sum_{i=1}^{t} n_i$. For this reason, we say that the quadratic form SST has $(N - 1)$ degrees of freedom (d.f.). Similarly, SSW has $(N - t)$ degrees of freedom, since $\mathrm{SSW} = \sum_{i=1}^{t} \mathrm{SSD}_i$, and the number of degrees of freedom of SSD_i is $(n_i - 1)$. Finally, SSB has $(t - 1)$ degrees of freedom. Notice that SSW is the total sum of squares of deviations **within** the t samples, and SSB is the sum of squares of deviations **between** the t sample means.

Dividing a quadratic form by its number of degrees of freedom, we obtain the mean-squared statistic. We summarize all these statistics in a table called the **ANOVA table**. The ANOVA table for comparing t treatments is given in Table 5.11.

Generally, in an ANOVA table, DF designates degrees of freedom, SS designates the sum of squares of deviations, and MS designates the mean-squared. In all tables,

$$\mathrm{MS} = \frac{\mathrm{SS}}{\mathrm{DF}} \tag{5.11.5}$$

We show now that

$$E\{\mathrm{MSW}\} = \sigma^2. \tag{5.11.6}$$

Table 5.11   *ANOVA table for one-way layout*

Source of variation	DF	SS	MS
Between treatments	$t - 1$	SSB	MSB
Within treatments	$N - t$	SSW	MSW
Total (adjusted for mean)	$N - 1$	SST	–

Indeed, according to the model, and since $\{Y_{ij}, j = 1, \ldots, n_i\}$ is a RSWR from the population corresponding to the i-th treatment,

$$E\left\{\frac{\text{SSD}_i}{n_i - 1}\right\} = \sigma^2, \quad i = 1, \ldots, t.$$

Since $\text{MSW} = \sum_{i=1}^{t} v_i \left(\frac{\text{SSD}_i}{n_i - 1}\right)$, where $v_i = \frac{n_i - 1}{N - t}, i = 1, \ldots, t,$

$$E\{\text{MSW}\} = \sum_{i=1}^{t} v_i E\left\{\frac{\text{SSD}_i}{n_i - 1}\right\} = \sigma^2 \sum_{i=1}^{t} v_i$$

$$= \sigma^2.$$

Another important result is

$$E\{\text{MSB}\} = \sigma^2 + \frac{1}{t - 1} \sum_{i=1}^{t} n_i \tau_i^2, \tag{5.11.7}$$

where $\tau_i = \mu_i - \overline{\mu}$ ($i = 1, \ldots, t$) and $\overline{\mu} = \frac{1}{N} \sum_{i=1}^{t} n_i \mu_i$. Thus, under the null hypothesis H_0 : $\mu_1 = \ldots = \mu_t$, $E\{\text{MSB}\} = \sigma^2$. This motivates us to use, for testing H_0, the F-statistic

$$F = \frac{\text{MSB}}{\text{MSW}}. \tag{5.11.8}$$

H_0 is rejected, at level of significance α, if

$$F > F_{1-\alpha}[t - 1, N - t].$$

Example 5.15. Three different vendors are considered for supplying cases for floppy disk drives. The question is whether the latch mechanism that opens and closes the disk loading slot is sufficiently reliable. In order to test the reliability of this latch, three independent samples of cases, each of size $n = 10$, were randomly selected from the production lots of these vendors. The testing was performed on a special apparatus that opens and closes a latch, until it breaks. The number of cycles required until latch failure was recorded. In order to avoid uncontrollable environmental factors to bias the results, the order of testing of cases of different vendors was completely randomized. In file **VENDOR.csv**, we can find the results of this experiment, arranged in three columns. Column 1 represents the sample from vendor A_1; column 2 that of vendor A_2, and column 3 of vendor A_3. An ANOVA was performed, using MINITAB. The analysis was done on $Y = (\text{Number of Cycles})^{1/2}$, in order to have data which is approximately normally distributed. The original data are expected to have positively skewed distribution, since it reflects the life length of the latch. The following is the one-way ANOVA, as performed by R

```
> data(VENDOR)
> VENDOR <- stack(VENDOR)
> VENDOR$ind <- as.factor(VENDOR$ind)
> VENDOR$values <- sqrt(VENDOR$values)
> oneway.test(values ~ ind,
```

```
            data = VENDOR,
            var.equal=T)

    One-way analysis of means

data:  values and ind
F = 14.024, num df = 2, denom df = 27, p-value =
0.00006658

> confint(lm(values ~ -1 + ind,
            data=VENDOR))

                2.5 %      97.5 %
indvendor1    58.00093    84.12333
indvendor2    91.54669   117.66910
indvendor3   104.11425   130.23666
```

The following and Figure 5.19 is the one-way ANOVA, as performed by MINITAB, following the command

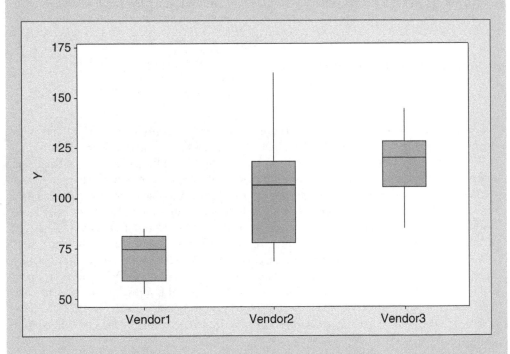

Figure 5.19 *Boxplots of Y by vendor (MINITAB).*

```
MTB> Oneway C6 C4
ANALYSIS OF VARIANCE ON C6
SOURCE      DF        SS       MS       F        p
C4           2     11366     5683    14.02    0.000
ERROR       27     10941      405
TOTAL       29     22306
                              INDIVIDUAL 95 PCT CI'S FOR MEAN
                              BASED ON POOLED STDEV
LEVEL  N          MEAN  STDEV  --+---------+---------+---------+----
    1  10        71.06  11.42     (------*-----)
    2  10       104.61  27.87                (-----*------)
    3  10       117.18  17.56                    (------*-----)
                                 --+---------+---------+---------+---
  POOLED STDEV = 20.13          60        80       100       120
```

In columns $C1-C3$, we stored the cycles data (importing file **VENDOR.csv**). Column $C4$ has indices of the samples. Column $C5$ contains the data of $C1-C3$ in a stacked form. Column $C6 = sqrt(C5)$.

The ANOVA table shows that the F statistic is significantly large, having a P-value close to 0. The null hypothesis H_0 is rejected. The reliability of the latches from the three vendors is not the same. The 0.95-confidence intervals for the means show that vendors A_2 and A_3 manufacture latches with similar reliability. That of vendor A_1 is significantly lower.

5.12 Simultaneous confidence intervals: multiple comparisons

Whenever the hypothesis of no difference between the treatment means are rejected, the question arises, which of the treatments have similar effects, and which ones differ significantly? In Example 5.15, we analyzed data on the strength of latches supplied by three different vendors. It was shown that the differences are very significant. We also saw that the latches from vendor A_1 were weaker from those of vendors A_2 and A_3, which were of similar strength. Generally, if there are t treatments, and the ANOVA shows that the differences between the treatment means are significant, we may have to perform up to $\binom{t}{2}$ comparisons, to rank the different treatments in term of their effects.

If we compare the means of all pairs of treatments, we wish to determine $\binom{t}{2} = \frac{t(t-1)}{2}$ confidence intervals to the true differences between the treatment means. If each confidence interval has confidence level $(1 - \alpha)$, the probability that **all** $\binom{t}{2}$ confidence intervals cover the true differences simultaneously is smaller than $(1 - \alpha)$. The simultaneous confidence level might be as low as $(1 - t\alpha)$.

There are different types of simultaneous confidence intervals. We present here the method of Scheffé, for simultaneous confidence intervals for any number of contrasts (Scheffé 1959, p. 66). A **contrast** between t means $\overline{Y}_1, \ldots, \overline{Y}_t$, is a linear combination $\sum_{i=1}^{t} c_i \overline{Y}_i$, such that $\sum_{i=1}^{t} c_i = 0$. Thus, any difference between two means is a contrast, e.g., $\overline{Y}_2 - \overline{Y}_1$. Any second-order difference, e.g.,

$$(\overline{Y}_3 - \overline{Y}_2) - (\overline{Y}_2 - \overline{Y}_1) = \overline{Y}_3 - 2\overline{Y}_2 + \overline{Y}_1,$$

is a contrast. The space of all possible linear contrasts has dimension $(t-1)$. For this reason, the coefficient we use, according to Scheffé's method, to obtain simultaneous confidence intervals of level $(1-\alpha)$ is

$$S_\alpha = ((t-1)F_{1-\alpha}[t-1, t(n-1)])^{1/2}, \tag{5.12.1}$$

where $F_{1-\alpha}[t-1, t(n-1)]$ is the $(1-\alpha)$th quantile of the F-distribution. It is assumed that all the t samples are of equal size n. Let $\hat{\sigma}_p^2$ denote the pooled estimator of σ^2, i.e.,

$$\hat{\sigma}_p^2 = \frac{1}{t(n-1)}\sum_{i=1}^{t}\text{SSD}_i, \tag{5.12.2}$$

then the simultaneous confidence intervals for all contrasts of the form $\sum_{i=1}^{t}c_i\mu_i$, have limits

$$\sum_{i=1}^{t}c_i\overline{Y}_i \pm S_\alpha\frac{\hat{\sigma}_p}{\sqrt{n}}\left(\sum_{i=1}^{t}c_i^2\right)^{1/2}. \tag{5.12.3}$$

Example 5.16.   In data file **HADPAS.csv**, we have the resistance values (ohms) of several resistors on 6 different hybrids at 32 cards. We analyze here the differences between the means of the $n = 32$ resistance values of resistor RES3, where the treatments are the $t = 6$ hybrids. The boxplots of the samples corresponding to the 6 hybrids are presented in Figure 5.20.

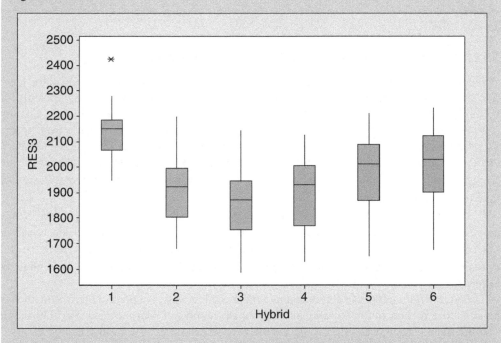

Figure 5.20   *Boxplots of six hybrids (MINITAB).*

In Table 5.12, we present the means and standard deviations of these six samples (treatments).

Table 5.12 *Means and Std. of resistance RES3 by hybrid*

Hybrid	$\bar{Y}$	S_y
1	2143.41	99.647
2	1902.81	129.028
3	1850.34	144.917
4	1900.41	136.490
5	1980.56	146.839
6	2013.91	139.816

The pooled estimator of σ is

$$\hat{\sigma}_p = 133.74.$$

The Scheffé coefficient, for $\alpha = 0.05$ is

$$S_{0.05} = (5F_{0.95}[5,186])^{1/2} = 3.332.$$

Upper and lower simultaneous confidence limits, with 0.95 level of significance, are obtained by adding to the differences between means $\pm S_\alpha \frac{\hat{\sigma}_p}{\sqrt{16}} = \pm 111.405$.
Differences which are smaller in magnitude than 111.405 are considered insignificant. Thus, if we order the sample means, we obtain

Hybrid	Mean	Group mean
1	2143.41	2143.41
6	2013.91	1997.235
5	1980.56	
2	1902.81	1884.52
4	1900.41	
3	1850.34	

Thus, the difference between the means of Hybrid 1 and all the others are significant. The mean of Hybrid 6 is significantly different than those of 2, 4, and 3. The mean of Hybrid 5 is significantly larger than that of Hybrid 3. We suggest therefore the following **homogeneous group** of treatments (all treatments within the same homogeneous group have means which are not significantly different):

Homog group	Means of groups
{1}	2143.41
{5,6}	1997.24
{2,3,4}	1884.52

The difference between the means of $\{5,6\}$ and $\{2,3,4\}$ is the contrast

$$-\frac{1}{3}\overline{Y}_2 - \frac{1}{3}\overline{Y}_3 - \frac{1}{3}\overline{Y}_4 + \frac{1}{2}\overline{Y}_5 + \frac{1}{2}\overline{Y}_6.$$

This contrast is significant, if it is greater than

$$S_\alpha \frac{\hat{\sigma}_p}{\sqrt{32}} \sqrt{\left(\frac{1}{2}\right)^2 + \left(\frac{1}{2}\right)^2 + \left(\frac{1}{3}\right)^2 + \left(\frac{1}{3}\right)^2 + \left(\frac{1}{3}\right)^2} = 71.912.$$

The above difference is thus significant.

```
> HADPAS$hyb <- factor(HADPAS$hyb)
> TukeyHSD(aov(res3 ~ hyb, data=HADPAS))

  Tukey multiple comparisons of means
    95% family-wise confidence level

Fit: aov(formula = res3 ~ hyb, data = HADPAS)

$hyb
          diff         lwr         upr      p adj
2-1 -240.59375 -336.87544 -144.31206 0.0000000
3-1 -293.06250 -389.34419 -196.78081 0.0000000
4-1 -243.00000 -339.28169 -146.71831 0.0000000
5-1 -162.84375 -259.12544  -66.56206 0.0000347
6-1 -129.50000 -225.78169  -33.21831 0.0020359
3-2  -52.46875 -148.75044   43.81294 0.6197939
4-2   -2.40625  -98.68794   93.87544 0.9999997
5-2   77.75000  -18.53169  174.03169 0.1891788
6-2  111.09375   14.81206  207.37544 0.0135101
4-3   50.06250  -46.21919  146.34419 0.6664332
5-3  130.21875   33.93706  226.50044 0.0018805
6-3  163.56250   67.28081  259.84419 0.0000315
5-4   80.15625  -16.12544  176.43794 0.1625016
6-4  113.50000   17.21831  209.78169 0.0107198
6-5   33.34375  -62.93794  129.62544 0.9183436
```

5.13   Contingency tables

5.13.1   The structure of contingency tables

When the data is categorical, we generally summarize it in a table which presents the frequency of each category, by variable, in the data. Such a table is called a **contingency table**.

Example 5.17. Consider a test of a machine which inserts components into a board. The displacement errors of such a machine were analyzed in Example 5.1. In this test, we perform a large number of insertions with $k = 9$ different components. The result of each trial (insertion) is either Success (no insertion error) or Failure (insertion error). In the present test, there are two categorical variables: Component type and Insertion Result. The first variable has nine categories:

$C1$: Diode
$C2$: 1/2 Watt Canister
$C3$: Jump Wire
$C4$: Small Corning
$C5$: Large Corning
$C6$: Small Bullet
$C7$: 1/8 Watt Dogbone
$C8$: 1/4 Watt Dogbone
$C9$: 1/2 Watt Dogbone

The second variable, Insertion Result, has two categories only (Success, Failure). The contingency table below (Table 5.13) presents the frequencies of the various insertion results by component type.

Table 5.13   *Contingency table of insertion results by component type*

| Component | Insertion result | | Row |
Type	Failure	Success	Total
C1	61	108 058	108 119
C2	34	136 606	136 640
C3	10	107 328	107 338
C4	23	105 042	105 065
C5	25	108 829	108 854
C6	9	96 864	96 873
C7	12	107 379	107 391
C8	3	105 851	105 854
C9	13	180 617	180 630
Column total	190	1 056 574	1 056 764

Table 5.13 shows that the proportional frequency of errors in insertions is very small ($190/1056764 = 0.0001798$), which is about 180 FPM (failures per million). This may be judged to be in conformity with the industry standard. We see, however, that there are apparent differences between the failure proportions, by component types. In Figure 5.21, we present the FPMs of the insertion failures, by component type. The largest one is that of $C1$ (Diode), followed by components $\{C2, C4, C5\}$. Smaller proportions are those of $\{C3, C6, C9\}$. The smallest error rate is that of $C8$.

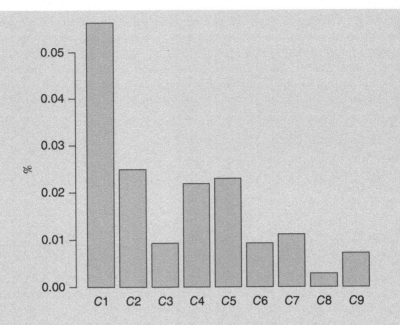

Figure 5.21   *Bar-chart of components error rates.*

The differences in the components error rates can be shown to be very significant. The structure of the contingency table might be considerably more complicated than that of Table 5.13. We illustrate here a contingency table with three variables.

Example 5.18.   The data are the placement errors of an OMNI 4621 automatic insertion machine. The variables are the following:

(i) *Machine structure*: Basic, EMI1, EMI2, EMI4;
(ii) *Components*: 805, 1206, SOT_23;
(iii) *Placement result*: Error, No_Error.

The contingency table is given in Table 5.14, and summarizes the results of 436 431 placements.

Table 5.14   *Contingency table of placement errors*

Comp. Structure	805 Err	805 N_Err	1206 Er	1206 N_Err	SOT_23 Er	SOT_23 N_Err	Total comp Er	Total comp N_Err	Total rows
Basic	11	40 279	7	40 283	16	40 274	34	120 836	120 870
EMI1	11	25 423	8	25 426	2	25 432	21	76 281	76 302
EMI2	19	54 526	15	54 530	12	54 533	46	163 589	163 635
EMI4	14	25 194	4	25 204	5	25 203	23	75 601	75 624
Total	55	145 422	34	145 443	35	145 442	124	436 307	436 431

We see in Table 5.14 that the total failure rate of this machine type is $124/436307 = 284$ (FPM). The failure rates, by machine structure, in FPMs, are 281, 275, 281, and 304, respectively. The first three structural types have almost the same FPMs, while the fourth one is slightly larger. The components failure rates are 378, 234, and 241 FPM, respectively. It remains to check the failure rates according to Structure × Component. These are given in Table 5.15:

Table 5.15   *Failure rates (FPM) by structure and component type*

	Component		
Structure	805	1206	SOT_23
Basic	273	174	397
EMI1	433	315	79
EMI2	348	275	220
EMI4	555	159	198

We see that the effect of the structure is different on different components. Again, one should test whether the observed differences are statistically significant or due only to chance variability. Methods for testing this will be discussed in Chapter 14.

The construction of contingency tables can be done in R or by using MINITAB or JMP. We illustrate this on the data in file **CAR.csv**. This file consists of information on 109 car models from 1989. The file contains 109 records on 5 variables: Number of cylinders (4,6,8), origin (US = 1, Europe = 2, ASIA = 3), turn diameter [meters], horsepower, and number of miles per gallon in city driving. One variable, Origin, is categorical, while the other four are interval scaled variables. One discrete (number of cylinders) and the other three are continuous. In R

```
> data(CAR)
> with(data=CAR,
       expr=table(cyl, origin))

    origin
cyl  1  2  3
  4 33  7 26
  6 13  7 10
  8 12  0  1
```

and in MINITAB by using the command

```
MTB> Table C1-C2;
SUBC> Counts.
```

we obtain the contingency table, which is illustrated in Table 5.16.

One can prepare a contingency table also from continuous data, by selecting the number and length of intervals for each variable, and counting the frequencies of each cell in the table.

Table 5.16   *Contingency table of number of cylinders and origin*

Num. cyc.	Origin 1	Origin 2	Origin 3	Total
4	33	7	26	66
6	13	7	10	30
8	12	0	1	13
Total	58	14	37	109

Table 5.17   *Contingency table of turn diameter versus miles/gallon city*

Turn diameter	Miles/gallon city 12–18	Miles/gallon city 19–24	Miles/gallon city 25–	Total
27–30.6	2	0	4	6
30.7–34.2	4	12	15	31
34.3–37.8	10	26	6	42
37.9–	15	15	0	30
Total	31	53	25	109

For example, for the car data, if we wish to construct a contingency table of turn diameter versus miles/gallon, we obtain the contingency table presented in Table 5.17.

5.13.2   Indices of association for contingency tables

In the present section, we construct several indices of association, which reflect the degree of dependence, or association between variables. For the sake of simplicity, we consider here indices for two-way tables, i.e., association between two variables.

5.13.2.1   *Two interval scaled variables*

If the two variables are continuous ones, measured on an interval scale, or some transformation of it, we can use some of the dependence indices discussed earlier. For example, we can represent each interval by its midpoint, and compute the correlation coefficient between these midpoints. As in Section 5.2, if variable X is classified into k intervals,

$$(\xi_0, \xi_1), (\xi_1, \xi_2), \ldots, (\xi_{k-1}, \xi_k)$$

and variable Y is classified into m intervals $(\eta_0, \eta_1), \ldots, (\eta_{m-1}, \eta_m)$, let

$$\tilde{\xi}_i = \tfrac{1}{2}(\xi_{i-1} + \xi_i), \quad i = 1, \ldots, k$$

$$\tilde{\eta}_j = \tfrac{1}{2}(\eta_{j-1} + \eta_j), \quad j = 1, \ldots, m.$$

Let $p_{ij} = f_{ij}/N$ denote the proportional frequency of the (i,j)-th cell, i.e., X values in (ξ_{i-1}, ξ_i) and Y values in (η_{j-1}, η_j). Then, an estimate of the coefficient of correlation obtained from the contingency table is

$$\hat{\rho}_{XY} = \frac{\sum_{i=1}^k \sum_{j=1}^m p_{ij}(\tilde{\xi}_i - \overline{\tilde{\xi}})(\tilde{\eta}_j - \overline{\tilde{\eta}})}{\left[\sum_{i=1}^k p_{i.}(\tilde{\xi}_i - \overline{\tilde{\xi}})^2\right]^{1/2}\left[\sum_{j=1}^m p_{.j}(\tilde{\eta}_j - \overline{\tilde{\eta}})^2\right]^{1/2}}, \tag{5.13.1}$$

where

$$p_{i.} = \sum_{j=1}^{m} p_{ij}, \quad i = 1, \ldots, k,$$

$$p_{.j} = \sum_{i=1}^{k} p_{ij}, \quad j = 1, \ldots, m,$$

$$\overline{\xi} = \sum_{i=1}^{k} p_{i.} \tilde{\xi}_{i},$$

and

$$\overline{\eta} = \sum_{j=1}^{m} p_{.j} \tilde{\eta}_{j}.$$

Notice that the sample correlation r_{XY}, obtained from the sample data, is different from $\hat{\rho}_{XY}$, due to the reduced information that is given by the contingency table. We illustrate this in the following example:

Example 5.19. Consider the data in file **CAR.csv**. The sample correlation between the turn diameter, X, and the gas consumption (Miles/Gal) in a city, is $r_{XY} = -0.539$. If we compute this correlation on the basis of the data in Table 5.17, we obtain $\hat{\rho}_{XY} = -0.478$. The approximation given by $\hat{\rho}_{XY}$ depends on the number of intervals, k and m, on the length of the intervals, and the sample size N.

5.13.2.2 *Indices of association for categorical variables*

If one of the variables or both are categorical, there is no meaning to the correlation coefficient. We should devise another index of association. Such an index should not depend on the labeling or ordering of the categories. Common indices of association are based on comparison of the observed frequencies f_{ij} of the cells ($i = 1, \ldots, k; j = 1, \ldots, m$) to the expected ones if the events associated with the categories are independent. The concept of independence, in a probability sense, is defined in Chapter 4. We have seen earlier conditional frequency distributions. If $N_{i.} = \sum_{j=1}^{m} f_{ij}$, the conditional **proportional** frequency of the j-th category of Y, given the i-th category of X, is

$$p_{j|i} = \frac{f_{ij}}{N_{i.}}, \quad j = 1, \ldots, m.$$

We say that X and Y are **not associated** if

$$p_{j|i} = p_{.j} \quad \text{for all } i = 1, \ldots, k,$$

where

$$p_{.j} = \frac{N_{.j}}{N} \quad j = 1, \ldots, m$$

and

$$N_{.j} = \sum_{i=1}^{k} f_{ij}.$$

Accordingly, the expected frequency of cell (i,j), if there is **no association**, is

$$\tilde{f}_{ij} = \frac{N_{i.}N_{.j}}{N}, \quad i = 1, \ldots, k, j = 1, \ldots, m.$$

A common index of discrepancy between f_{ij} and $\tilde{f}_{ij}$ is

$$X^2 = \sum_{i=1}^{k} \sum_{j=1}^{m} \frac{(f_{ij} - \tilde{f}_{ij})^2}{\tilde{f}_{ij}}. \tag{5.13.2}$$

This index is called the **chi-squared statistic**. One can compute this statistic using MINITAB, with the commands

```
MTB> Table C1 C2;
SUBC> Counts;
SUBC> ChiSquare.
```

Example 5.20.   For the **CAR.csv** data, the chi-squared statistic for the association between Origin and Num Cycl is $X^2 = 12.13$. In Chapter 14, we will study how to assess the statistical significance of such a magnitude of X^2.
In R

```
> chisq.test(x=CAR$origin, y=CAR$cyl)
```

Another option in MINITAB is to set the observed frequencies of the contingency table into columns, and use the command

MTB> ChiSquare $C_- - C_-$.

For example, let us set the frequencies of Table 5.17 into three columns, say $C6$–$C8$. The above command yields Table 5.18, in which the expected frequencies $\tilde{f}_{ij}$ are printed below the observed ones.

Table 5.18   *Observed and expected frequencies of turn diameter by miles/gallon, **CAR.csv** ($\tilde{f}_{ij}$ under $\tilde{f}_{ij}$)*

Turn diameter	Miles/gallon city			Total
	12–18	18–24	24–	
27–30.6	2	0	4	6
	1.71	2.92	1.38	
30.6–34.2	4	12	15	31
	8.82	15.07	7.11	
34.2–37.8	10	26	6	42
	11.94	20.42	9.63	
37.8–	15	15	0	30
	8.53	14.59	6.88	
Total	31	53	25	109

The chi-squared statistic is

$$X^2 = \frac{(2 - 1.71)^2}{1.71} + \cdots + \frac{6.88^2}{6.88} = 34.99.$$

There are several association indices in the literature, based on the X^2. Three popular indices are the following:

Mean Squared Contingency:

$$\Phi^2 = \frac{X^2}{N}. \tag{5.13.3}$$

Tschuprow's Index:

$$T = \Phi/\sqrt{(k-1)(m-1)}. \tag{5.13.4}$$

Cramér's Index:

$$C = \Phi/\sqrt{\min(k-1, m-1)}. \tag{5.13.5}$$

No association corresponds to $\Phi^2 = T = C = 0$. The larger the index the stronger the association. For the data of Table 5.16,

$$\Phi^2 = \frac{34.99}{109} = 0.321$$

$$T = 0.283$$

$$C = 0.401.$$

We provide an additional example of contingency tables analysis, using the Cramér Index.

Example 5.21.   Compu Star, a service company providing technical support and sales of personal computers and printers, decided to investigate the various components of customer satisfaction that are specific to the company. A special questionnaire with 13 questions was designed and, after a pilot run, was mailed to a large sample of customers with a self addressed stamped envelope and a prize incentive. The prize was to be awarded by lottery among the customers who returned the questionnaire.

The customers were asked to rate, on a 1–6 ranking order, various aspects of the service. The rating of 1 corresponding to VERY POOR and the rating of 6 to VERY GOOD. These questions include the following:

Q1: First impression of service representative.
Q2: Friendliness of service representative.
Q3: Speed in responding to service request.
Q4: Technical knowledge of service representative.
Q5: Professional level of service provided.
Q6: Helpfulness of service representative.
Q7: Additional information provided by service representative.
Q8: Clarity of questions asked by service representative.
Q9: Clarity of answers provided by service representative.
Q10: Efficient use of time by service representative.
Q11: Overall satisfaction with service.
Q12: Overall satisfaction with product.
Q13: Overall satisfaction with company.

The response ranks are

1. Very poor,
2. poor,
3. below average,
4. above average,
5. good
6. very good.

The responses were tallied and contingency tables were computed linking the questions on overall satisfaction with questions on specific service dimensions. For example Table 5.19 is a contingency table of responses to Q13 versus Q3.

Table 5.19 *Two by two contingency table of customer responses, for Q3 and Q13*

Q3\Q13	1	2	3	4	5	6
1	0	1	0	0	3	1
2	0	2	0	1	0	0
3	0	0	4	2	3	0
4	0	1	1	10	7	5
5	0	0	0	10	71	38
6	0	0	0	1	30	134

Cramer's Index for Table 5.19 is:

$$C = \frac{1.07}{2.23} = 0.478.$$

There were 10 detailed questions (Q1–Q10) and 3 questions on overall customer satisfaction (Q11–Q13). A table was constructed for every combination of the 3 overall customer satisfaction questions and the 10 specific questions. For each of these 30 tables, Cramer's Index was computed and using a code of graphical symbols, we present these indices in Table 5.20.

Table 5.20 *Cramer's indices of Q1–Q10 by Q11–Q13*

		Q1	Q2	Q3	Q4	Q5	Q6	Q7	Q8	Q9	Q10
Q11:	Overall satisfaction with Service		•	++	•			+	•		
Q12:	Overall satisfaction with Product	+	•	•			++	•	•		
Q13:	Overall satisfaction with Company		•	++			+	++	•		•

The indices are coded according to the following key:

Cramer's Index	Code
0–0.2	
0.2–0.3	•
0.3–0.4	+
0.4–0.5	++
0.5–	+++

We can see from Table 5.20 that "Overall satisfaction with company" (Q13) is highly correlated with "Speed in responding to service requests" (Q3). However, the "Efficient use of time" (Q10) was not associated with overall satisfaction.

On the other hand, we also notice that questions Q1, Q5, Q10 show no correlation with overall satisfaction. Many models have been proposed in the literature for the analysis of customer satisfaction surveys. For a comprehensive review with applications using R see Kenett and Salini (2011). For more examples of indices of association and graphical analysis of contingency tables see Kenett (1983). Contingency tables are closely related to the data mining techniques of Association Rules. For more on this, see Kenett and Salini (2008).

5.14   Categorical data analysis

If all variables $x_1, \ldots, x_k$ and Y are categorical, we cannot perform the ANOVA without special modifications. In the present section, we discuss the analysis appropriate for such cases.

5.14.1   Comparison of binomial experiments

Suppose that we have performed t independent Binomial experiments, each one corresponding to a treatment combination. In the i-th experiment, we ran n_i independent trials. The yield variable, J_i, is the number of successes among the n_i trials ($i = 1, \ldots, t$). We further assume that in each experiment, the n_i trials are independent and have the same, unknown, probability for success, θ_i; i.e., J_i has a binomial distribution $B(n_i, \theta_i)$, $i = 1, \ldots, t$. We wish to compare the probabilities of success, θ_i ($i = 1, \ldots, k$). Accordingly, the null hypothesis is of equal success probabilities, i.e.,

$$H_0 : \theta_1 = \theta_2 = \cdots = \theta_k.$$

We describe here a test, which is good for large samples. Since by the Central Limit Theorem (CLT), $\hat{p}_i = \frac{J_i}{n_i}$ has a distribution which is approximately normal for large n_i, with mean θ_i and variance $\frac{\theta_i(1-\theta_i)}{n_i}$, one can show that, the large sample distribution of

$$Y_i = 2 \arcsin\left(\sqrt{\frac{J_i + 3/8}{n_i + 3/4}} \right) \tag{5.14.1}$$

(in radians) is approximately normal, with mean $\eta_i = 2 \arcsin(\sqrt{\theta_i})$ and variance $V\{Y_i\} = \frac{1}{n_i}$, $i = 1, \ldots, t$.

Using this result, we obtain that under the assumption of H_0, the sampling distribution of the test statistic

$$Q = \sum_{i=1}^{k} n_i (Y_i - \overline{Y})^2, \tag{5.14.2}$$

where

$$\overline{Y} = \frac{\sum_{i=1}^{k} n_i Y_i}{\sum_{i=1}^{k} n_i}, \tag{5.14.3}$$

is approximately chi-squared with $k - 1$ DF, $\chi^2[k - 1]$. In this test, we reject H_0, at level of significance α, if $Q > \chi^2_{1-\alpha}[k - 1]$.

Another test statistic for general use in contingency tables, will be given in the following section.

Example 5.22. In Table 5.13, we presented the frequency of failures of nine different components in inserting a large number of components automatically. In the present example, we test the hypothesis that the failure probabilities, θ_i, are the same for all components. In the following table, we present the values of J_i (# of failures), n_i and $Y_i = 2 \arcsin\left(\sqrt{\frac{J_i + 3/8}{n_i + 3/4}}\right)$, for each component. Using MINITAB, if the values of J_i are stored in C1 and those of n_i in C2, we compute Y_i (stored in C3) with the command

MTB> let $C3 = 2$ * asin(sqrt((C1 + 0.375)/(C2 + 0.75))) (see Table 5.21).

Table 5.21   *The arcsin transformation*

i	J_i	n_i	Y_i
1	61	108 119	0.0476556
2	34	136 640	0.0317234
3	10	107 338	0.0196631
4	23	105 065	0.0298326
5	25	108 854	0.0305370
6	9	96 873	0.0196752
7	12	107 391	0.0214697
8	3	105 854	0.0112931
9	13	180 630	0.0172102

The test statistic Q can be computed by the MINITAB command (the constant $k1$ stands for Q)

MTB> let $k1 = \text{sum}(C2 * C3 ** 2) - \text{sum}(C2 * C3) ** 2/\text{sum}(C2)$.

The value of Q is 105.43. The P-value of this statistic is 0. The null hypothesis is rejected. To determine this P value using MINITAB, since the distribution of Q under H_0 is $\chi^2[8]$, we use the commands:

MTB> CDF 105.43;
SUBC> Chisquare 8.

We find that $\Pr\{\chi^2[8] \leq 105.43\} \doteq 1$. This implies that $P = 0$.

5.15   Chapter highlights

Several techniques for graphical analysis of data in several dimensions are introduced and demonstrated using case studies. These include matrix scatterplots, 3D-scatterplots, and multiple box-plots. Topics covered also include simple linear regression, multiple regression models, and contingency tables. Prediction intervals are constructed for currents of solar cells and resistances on hybrid circuits. Robust regression is used to analyze data on placement errors of components on circuit boards. A special section on indices of association for categorical variables includes an analysis of a customer satisfaction survey designed to identify the main components of customer satisfaction and dissatisfaction. The material covered by this chapter can be best studied in front of a personal computer so that the reader can reproduce and even expand the data analysis offered in the text.

The chapter provides an introduction to multiple regression methods, in which the relationship of k explanatory (predicting) quantitative variables to a variable of interest is explored. In particular, the least squares estimation procedure is presented in detail for regression on two variables. Partial regression and correlation are discussed. The least squares estimation of the regression coefficients for multiple regressions ($k > 2$) is presented with matrix formulae. The contributions of the individual regressors is tested by the partial-F test. The sequential SS partition of the total sum of squares due to the departure on the regressors is defined and explained. The partial correlation, given a set of predictors, is defined and its relationship to the partial-F statistic is given.

The ANOVA for testing the significance of differences between several sample means is introduced, as well as the method of multiple comparisons, which protects the overall level of significance. The comparisons of proportions for categorical data (binomial or multinomial) is also discussed. The chapter contains also a section on regression diagnostics, in which the influence of individual points on the regression is studied. In particular, one wishes to measure the effects of points which seem to deviate considerably from the rest of the sample.

The main concepts and tools introduced in this chapter include

- Matrix Scatterplots
- 3D-Scatterplots
- Multiple Box-Plots
- Code Variables
- Joint, Marginal, and Conditional Frequency Distributions
- Sample Correlation
- Coefficient of Determination
- Simple Linear Regression
- Multiple Regression
- Predicted Values, FITS
- Residuals Around the Regression
- Multiple Squared Correlation

- Partial Regression
- Partial Correlation
- Partial-F Test
- Sequential SS
- Stepwise Regression
- Regression Diagnostics
- x-Leverage of a Point
- Standard Error of Predicted Value
- Standardized Residual
- Cook Distance
- Fits Distance, DFIT
- Analysis of Variance
- Treatment Combinations
- Simultaneous Confidence Intervals
- Multiple Comparisons
- Contrasts
- Scheffé's Method
- Contingency Tables Analysis
- Categorical Data Analysis
- Arcsin Transformation
- Chi-Squared Test for Contingency Tables

5.16   Exercises

5.1 Use file **CAR.csv** to prepare multiple or matrix scatter plots of Turn Diameter versus Horsepower versus Miles per Gallon. What can you learn from these plots?

5.2 Make a multiple (side by side) box plots of the Turn Diameter by Car Origin, for the data in file **CAR.csv**. Can you infer that Turn Diameter depends on the Car Origin?

5.3 Data file **HADPAS.csv** contains the resistance values (Ohms) of five resistors placed in six hybrids on 32 ceramic substrates. The file contains eight columns. The variables in these columns are the following:
  1. Record Number
  2. Substrate Number
  3. Hybrid Number
  4. Res 3.
  5. Res 18.
  6. Res 14.
  7. Res 7.
  8. Res 20.
   (i) Make a multiple box plot of the resistance in Res 3 by hybrid.
   (ii) Make a matrix plot of all the Res variables. What can you learn from the plots?

5.4 Construct a joint frequency distribution of the variables Horsepower and MPG/City for the data in file **CAR.csv**.

5.5 Construct a joint frequency distribution for the resistance values of RES 3 and RES 14, in data file **HADPAS.csv**. [Code the variables first, see instructions in Exercise 5.7.]

5.6 Construct the conditional frequency distribution of RES 3, given that the resistance values of RES 14 is between 1300 and 1500 (Ω).

5.7 In the present exercise, we compute the **conditional** means and standard deviations of one variable given another one. Use file **HADPAS.csv**. We classify the data according to the values of Res 14 (Column *C*5) to five subgroups. This is done by using the CODE command in MINITAB, which is

MTB> CODE(900:1200)1 (1201:1500)2 (1501:1800)3 (1801:2100)4
(2101:2700)5 *C*5 *C*8

We use then the command

MTB> DESC *C*3;

SUBC> By *C*8.

In this way, we can obtain the conditional means and stand-deviations of Res 3 given the subgroups of Res 7. Use these commands and write a report on the obtained results.

5.8 Given below are four data sets of (X, Y) observations
  (i) Compute the least squares regression coefficients of Y on X, for the four data sets.
  (ii) Compute the coefficient of determination, R^2, for each set.

Data set 1		Data set 2		Data set 3		Data set 4	
$X^{(1)}$	$Y^{(1)}$	$X^{(2)}$	$Y^{(2)}$	$X^{(3)}$	$Y^{(3)}$	$X^{(4)}$	$Y^{(4)}$
10.0	8.04	10.0	9.14	10.0	7.46	8.0	6.68
8.0	6.95	8.0	8.14	8.0	6.67	8.0	5.76
13.0	7.58	13.0	8.74	13.0	12.74	8.0	7.71
9.0	8.81	9.0	8.77	9.0	7.11	8.0	8.84
11.0	8.33	11.0	9.26	11.0	7.81	8.0	8.47
14.0	9.96	14.0	8.1	14.0	8.84	8.0	7.04
6.0	7.24	6.0	6.13	6.0	6.08	8.0	5.25
4.0	4.26	4.0	3.1	4.0	5.39	19.0	12.5
12.0	10.84	11.0	9.13	12.0	8.16	8.0	5.56
7.0	4.82	7.0	7.26	7.0	6.42	8.0	7.91
5.0	5.68	5.0	4.74	5.0	5.73	8.0	6.89

5.9 Compute the correlation matrix of the variables Turn Diameter, Horsepower, and Miles per Gallon/City for the data in file **CAR.csv**.

5.10 (i) Differentiate partially the quadratic function

$$\text{SSE} = \sum_{i=1}^{n}(Y_i - \beta_0 - \beta_1 X_{i1} - \beta_2 X_{i2})^2$$

with respect to β_0, β_1, and β_2 to obtain the linear equations in the least squares estimates b_0, b_1, b_2. These linear equations are called **the normal equations**.
  (ii) Obtain the formulae for b_0, b_1, and b_2 from the normal equations.

5.11 Consider the variables Miles per Gallon, Horsepower, and Turn Diameter in the data set **CAR.csv**. Find the least squares regression line of MPG (y) on Horsepower (x_1) and Turn

Diameter (x_2). For this purpose, use first the equations in Section 5.4 and then verify your computations by using the MINITAB command "regress."

5.12 Compute the partial correlation between Miles per Gallon and Horsepower, give the Number of Cylinders, in data file **CAR.csv**.

5.13 Compute the partial regression of Miles per Gallon and Turn Diameter, Given Horsepower, in data file **CAR.csv**.

5.14 Use the three-stage algorithm of Section 5.5 to obtain the multiple regression of Exercise 5.2 from the results of 5.5.

5.15 Consider Example 5.4. From the MINITAB output, we see that, when regression Cap Diam on Diam1, Diam2, and Diam3, the regression coefficient of Diam2 is not significant (P value $= 0.926$), and this variable can be omitted. Perform a regression of Cap Diam on Diam2 and Diam3. Is the regression coefficient for Diam2 significant? How can you explain the difference between the results of the two regressions?

5.16 Regress the yield in **GASOL.csv** on all the four variables x_1, x_2, x_3, x_4.
   (i) What is the regression equation?
   (ii) What is the value of R^2?
   (iii) Which regression coefficient(s) is (are) nonsignificant?
   (iv) Which factors are important to control the yield?
   (v) Are the residuals from the regression distributed normally? Make a graphical test.

5.17 (i) Show that the matrix $(H) = (X)(B)$ is idempotent, i.e., $(H)^2 = (H)$.
   (ii) Show that the matrix $(Q) = (I - H)$ is idempotent, and therefore, $s_e^2 = \mathbf{y}'(Q)\mathbf{y}/(n - k - 1)$.

5.18 Show that the vectors of fitted values, $\hat{\mathbf{y}}$, and of the residuals, $\hat{e}$, are orthogonal, i.e., $\hat{\mathbf{y}}'\hat{e} = 0$.

5.19 Show that the $1 - R^2_{y|(x)}$ is proportional to $||\hat{e}||^2$, which is the squared Euclidean norm of $\hat{e}$.

5.20 In Section 4.5, we presented properties of the $\text{cov}(X, Y)$ operator. Prove the following generalization of property (iv). Let $\mathbf{X}' = (X_1, \ldots, X_n)$ be a vector of n random variables. Let (Σ) be an $n \times n$ matrix whose (i,j)th element is $\Sigma_{ij} = \text{cov}(X_i, X_j), i, j = 1, \ldots, n$. Notice that the diagonal elements of (Σ) are the variances of the components of $\mathbf{X}$. Let β and γ be two n-dimensional vectors. Prove that $\text{cov}(\beta'\mathbf{X}, \gamma'\mathbf{X}) = \beta'(\Sigma)\gamma$. [The matrix (Σ) is called the variance–covariance matrix of $\mathbf{X}$.]

5.21 Let $\mathbf{X}$ be an n-dimensional random vector, having a variance–covariance matrix (Σ). Let $\mathbf{W} = (B)\mathbf{X}$, where (B) is an $m \times n$ matrix. Show that the variance–covariance matrix of $\mathbf{W}$ is $(B)(\Sigma)(B)'$.

5.22 Consider the linear regression model $\mathbf{y} = (X)\beta + \mathbf{e}$. $\mathbf{e}$ is a vector of random variables, such that $E\{e_i\} = 0$ for all $i = 1, \ldots, n$ and

$$\text{cov}(e_i, e_j) = \begin{cases} \sigma^2, & \text{if } i = j \\ \\ 0, & \text{if } i \neq j \end{cases}$$

$i, j = 1, \ldots, n$. Show that the variance–covariance matrix of the LSE $\mathbf{b} = (B)\mathbf{y}$ is $\sigma^2[(\mathbf{X})'(\mathbf{X})]^{-1}$.

5.23 Consider **SOCELL.csv** data file. Compare the slopes and intercepts of the two simple regressions of ISC at time t_3 on that at time t_1, and ISC at t_3 on that at t_2.

5.24 The following data (see Draper and Smith 1981, p. 629) gives the amount of heat evolved in hardening of element (in calories per gram of cement), and the percentage of four various chemicals in the cement (relative to the weight of clinkers from which the cement was made). The four regressors are

x_1 :   amount of tricalcium aluminate;

x_2 :   amount of tricalcium silicate;

x_3 : amount of tetracalcium alumino ferrite;

x_4 :   amount of dicalcium silicate.

The regressant Y is the amount of heat evolved. The data are given in the following table:

X_1	X_2	X_3	X_4	Y
7	26	6	60	78
1	29	15	52	74
11	56	8	20	104
11	31	8	47	87
7	52	6	33	95
11	55	9	22	109
3	71	17	6	102
1	31	22	44	72
2	54	18	22	93
21	47	4	26	115
1	40	23	34	83
11	66	9	12	113
10	68	8	12	109

Compute in a sequence the regressions of Y on X_1; of Y on X_1, X_2; of Y on X_1, X_2, X_3; of Y on X_1, X_2, X_3, X_4. For each regression, compute the partial-F of the new regression added, the corresponding partial correlation with Y, and the sequential SS.

5.25 For the data of Exercise 5.24, construct a linear model of the relationship between Y and $X_1, \ldots, X_4$, by the forward stepwise regression method.

5.26 Consider the linear regression of Miles per Gallon on Horsepower for the cars in data file **CAR.csv**, with Origin = 3. Compute for each car the residuals, RESI, the standardized residuals, SRES, the leverage HI, and the Cook distance, D.

5.27 A simulation of the operation of a piston is available as the R piston simulator function *pistonSimulation* and with a JMP addin. In order to test whether changing the piston weight from 30 to 60 [k effects the cycle time significantly.Run the simulation program four times at weight 30, 40, 50, 60 [k, keeping all other factors at their low level. In each run, make $n = 5$ observations. Perform a one-way ANOVA of the results, and state your conclusions [You can use R, MINITAB or JMP.].

5.28 In experiments performed for studying the effects of some factors on the integrated circuits fabrication process, the following results were obtained, on the preetch line width (μ_m)

Exp. 1	Exp. 2	Exp. 3
2.58	2.62	2.22
2.48	2.77	1.73
2.52	2.69	2.00
2.50	2.80	1.86
2.53	2.87	2.04
2.46	2.67	2.15
2.52	2.71	2.18
2.49	2.77	1.86
2.58	2.87	1.84
2.51	2.97	1.86

Perform an ANOVA to find whether the results of the three experiments are significantly different by using MINITAB and R. Do the two test procedures yield similar results?

5.29 In manufacturing film for industrial use, samples from two different batches gave the following film speed:

Batch A: 103, 107, 104, 102, 95, 91, 107, 99, 105, 105

Batch B: 104, 103, 106, 103, 107, 108, 104, 105, 105, 97

Test whether the differences between the two batches are significant, by using (i) a randomization test; (ii) an ANOVA.

5.30 Use the MINITAB macro **RANDTES3.MTB** to test the significance of the differences between the results of the three experiments in Exercise 5.28.

5.31 In data file **PLACE.csv**, we have 26 samples, each one of size $n = 16$, of x-, y-, θ-deviations of components placements. Make an ANOVA, to test the significance of the sample means in the x-deviation. Classify the samples into homogeneous groups such that the differences between sample means in the same group are not significant, and those in different groups are significant. Use the Scheffé coefficient S_α for $\alpha = 0.05$.

5.32 The frequency distribution of cars by origin and number of cylinders is given in the following table.

Num. cylinders	US	Europe	Asia	Total
4	33	7	26	66
6 or more	25	7	11	43
Total	58	14	37	109

Perform a chi-square test of the dependence of number of cylinders and the origin of car.

5.33 Perform a chi-squared test of the association between turn diameter and miles/gallon based on Table 5.17.

5.34 In a customer satisfaction survey several questions were asked regarding specific services and products provided to customers. The answers were on a 1–5 scale, where 5 means "very satisfied with the service or product" and 1 means "very dissatisfied." Compute the Mean Squared Contingency, Tschuprow's Index and Cramer's Index for both contingency tables.

Question 3	Question			1	
	1	2	3	4	5
1	0	0	0	1	0
2	1	0	2	0	0
3	1	2	6	5	1
4	2	1	10	23	13
5	0	1	1	15	100

Question 3	Question			2	
	1	2	3	4	5
1	1	0	0	3	1
2	2	0	1	0	0
3	0	4	2	3	0
4	1	1	10	7	5
5	0	0	1	30	134

6

Sampling for Estimation of Finite Population Quantities

6.1 Sampling and the estimation problem

6.1.1 Basic definitions

In the present chapter, we consider the problem of estimating quantities (parameters) of a finite population. The problem of testing hypotheses concerning such quantities, in the context of sampling inspection of product quality, will be studied in Chapter 18. Estimation and testing the parameters of statistical models for infinite populations were discussed in Chapter 4.

Let P designate a finite population of N units. It is assumed that the population size, N, is **known**. Also assume that a list (or a frame) of the population units $L_N = \{u_1, \ldots, u_N\}$ is available.

Let X be a variable of interest and $x_i = X(u_i)$, $i = 1, \ldots, N$ the value ascribed by X to the ith unit, u_i, of P.

The population **mean** and population variance, for the variable X, i.e.,

$$\mu_N = \frac{1}{N} \sum_{i=1}^{N} x_i$$

and

(6.1.1)

$$\sigma_N^2 = \frac{1}{N} \sum_{i=1}^{N} (x_i - \mu_N)^2,$$

are called **population quantities**. In some books (Cochran 1977), these quantities are called "population parameters." We distinguish between population quantities and parameters of distributions, which represent variables in infinite populations. Parameters are not directly observable and can only be estimated, while finite population quantities can be determined exactly if the whole population is observed.

Modern Industrial Statistics: With Applications in R, MINITAB and JMP, Third Edition.
Ron S. Kenett and Shelemyahu Zacks.
© 2021 John Wiley & Sons, Ltd. Published 2021 by John Wiley & Sons, Ltd.

The population quantity μ_N is the expected value of the distribution of X in the population, whose c.d.f. is

$$\hat{F}_N(x) = \frac{1}{N}\sum_{i=1}^{N} I(x;x_i),$$

where   (6.1.2)

$$I(x;x_i) = \begin{cases} 1, & \text{if } x_i \leq x \\ \\ 0, & \text{if } x_i > x. \end{cases}$$

σ_N^2 is the variance of $F_N(x)$.

In this chapter, we focus attention on estimating the population mean, μ_N, when a sample of size n, $n < N$, is observed. The problem of estimating the population variance σ_N^2 will be discussed in the context of estimating the standard errors of estimators of μ_N.

Two types of sampling strategies will be considered. One type consists of random samples (with or without replacement) from the whole population. Such samples are called **simple random samples**. The other type of sampling strategy is that of **stratified random sampling**. In stratified random sampling, the population is first partitioned to strata (blocks) and then a simple random sample is drawn from each stratum independently. If the strata are determined so that the variability within strata is smaller relative to the general variability in the population, the precision in estimating the population mean μ_N, using a stratified random sampling will generally be higher than that in simple random sampling. This will be shown in Section 6.3.

As an example of a case where stratification could be helpful consider the following. At the end of each production day, we draw a random sample from the lot of products of that day to estimate the proportion of defective item. Suppose that several machines operate in parallel and manufacture the same item. Stratification by machine will provide higher precision for the global estimate, as well as information on the level of quality of each machine. Similarly, if we can stratify by shift, by vendor or by other factors that may contribute to the variability we may increase the precision of our estimates.

6.1.2   Drawing a random sample from a finite population

Given a finite population consisting of N distinct elements, we first make a **list** of all the elements of the population, which are all labeled for identification purposes. Suppose we wish to draw a random sample of size n from this population, where $1 \leq n \leq N$. We distinguish between two methods of random sampling: (1) **sampling with replacement** and (2) **sampling without replacement**. A sample drawn with replacement is obtained by returning the selected element, after each choice, to the population before the next item is selected. In this method of sampling, there are altogether N^n possible samples. A sample is called **random sample with replacement** (RSWR) if it is drawn by a method which gives every possible sample the same probability to be drawn. A sample is **without replacement** if an element drawn is not replaced and hence cannot be drawn again. There are $N(N-1)\cdots(N-n+1)$ such possible samples of size n from a population of size N. If each of these has the same probability of being drawn, the sample is called **random sample without replacement** (RSWOR). Bootstrapping discussed in Chapter 4 is an application of RSWR.

Practically speaking, the choice of a particular random sample is accomplished with the aid of **random numbers**. Random numbers can be generated by various methods. For example, an integer has to be drawn at random from the set $0, 1, \ldots, 99$. If we had a ten-faced die, we could label its faces with the numbers $0, \ldots, 9$ and cast it twice. The results of these two drawings would yield a two-digit integer, e.g., 13. Since in general we do not have such a die, we could, instead, use a coin and, flipping it seven times, generate a random number between 0 and 99 in the following way. Let X_j ($j = 1, \ldots, 7$) be 0 or 1, corresponding to whether a head or tail appeared on the jth flip of the coin. We then compute the integer I, which can assume one of the values $0, 1, \ldots, 127$, according to the formula

$$I = X_1 + 2X_2 + 4X_3 + 8X_4 + 16X_5 + 32X_6 + 64X_7.$$

If we obtain a value greater than 99, we disregard this number and flip the coin again seven times. Adding 1 to the outcome produces random numbers between 1 and 100. In a similar manner, a roulette wheel could also be used in constructing random numbers. A computer algorithm for generating pseudo random numbers was described in Example 3.5. In actual applications, we use readymade **tables of random numbers** or computer routines for generating random numbers.

Example 6.1. The following 10 numbers were drawn by using a random number generator on a computer: 76, 49, 95, 23, 31, 52, 65, 16, 61, 24. These numbers form a random sample of size 10 from the set $1, \ldots, 100$. If by chance two or more numbers are the same, the sample would be acceptable if the method is RSWR. If the method is RSWOR any number that was already selected would be discarded. In R to for draw a RSWOR of 10 integers from the set $\{1, \ldots, 100\}$ use

```
> set.seed(123)
> sample(x=100, size=10)
```

MINITAB commands for drawing a RSWOR of 10 integers from the set $\{1, \ldots, 100\}$ and storing them in column C1 are

```
MTB> Random 10 C1;
SUBC> Integer 1 100.
```

6.1.3 Sample estimates of population quantities and their sampling distribution

So far we have discussed the nature of variable phenomena and presented some methods of exploring and presenting the results of experiments. More specifically, the methods of analysis described in Chapter 2 explore the given data, but do not provide an assessment of what might happen in future experiments.

If we draw from the same population several different random samples, of the same size, we will find generally that statistics of interest assume different values at the different samples.

This can be illustrated in R we draw samples, with or without replacement, from a collection of numbers (population) which is stored in a vector. To show it, let us store in X the integers $1, 2, \ldots, 100$. To sample at random with replacement (RSWR), a sample of size $n = 20$ from X and put the random sample in a vector X *Sample*, we use

```
> X <- 1:100
> XSample <- sample(X, size=20, replace=TRUE)
```

This can be repeated four times, each time we put the sample in a new column with cbind. In Table 6.1, we present the results of this sampling, and for each sample, we present its mean and standard deviation (std.).

Notice that the "population" mean (that of column $C1$) is 50.5 and its standard deviation is 29.011. The sample means and standard deviations are estimates of these population parameters and as seen above, they vary around the parameters. The distribution of sample estimates of a parameter is called the **sampling distribution of an estimate**.

Theoretically (hypothetically), the number of possible different random samples, with replacement, is either infinite, if the population is infinite, or of magnitude N^n, if the population is finite (n is the sample size and N is the population size). This number is practically too large even if the population is finite (100^{20} in the above example). We can, however, approximate this distribution by drawing a large number, M, of such samples. In Figure 6.1, we present the histogram of the sampling distribution of $\overline{X}_n$, for $M = 1000$ random samples with replacement of size $n = 20$, from the population $\{1, 2, \ldots, 100\}$ of the previous example.

Table 6.1 *Four random samples with replacement of size 20, from* $\{1, 2, \ldots, 100\}$

		Sample		
1	2		3	4
26	54		4	15
56	59		81	52
63	73		87	46
46	62		85	98
1	57		5	44
4	2		52	1
31	33		6	27
79	54		47	9
21	97		68	28
5	6		50	52
94	62		89	39
52	70		18	34
79	40		4	30
33	70		53	58
6	45		70	18
33	74		7	14
67	29		68	14
33	40		49	32
21	21		70	10
8	43		15	52
		Means		
37.9	49.6		46.4	33.6
	Stand.		*Dev.*	
28.0	23.7		31.3	22.6

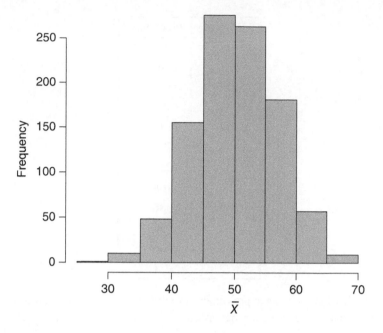

Figure 6.1 *Histogram of 1000 sample means.*

This can be effectively done in R with the function `boot`, defining a simple function to calculate a statistic over a sample of size n and looking at the returning object, component t:

```
> library(boot)
> set.seed(123)
> B <- boot(data=X,
            statistic=function(x, i, n){
              mean(x[i[1:n]])
              },
            R=1000, n=20)
> head(B$t, 3)

      [,1]
[1,] 47.85
[2,] 58.40
[3,] 34.20

> table(cut(B$t, 12))

(29.1,32.4] (32.4,35.8] (35.8,39.1] (39.1,42.4]
          8           6          28          70
(42.4,45.7]  (45.7,49]  (49,52.4] (52.4,55.7]
        133         191        200         154
  (55.7,59]  (59,62.4] (62.4,65.7]  (65.7,69]
        126          53          25           6
```

Mid-point	Frequency
28	1
32	10
36	19
40	55
44	153
48	224
52	251
56	168
60	86
64	26
68	6
72	1

This frequency distribution is an approximation to the sampling distribution of $\overline{X}_{20}$. It is interesting to notice that this distribution has mean $\overline{\overline{X}} = 50.42$ and standard deviation $\overline{S} = 6.412$. $\overline{\overline{X}}$ is quite close to the population mean 50.5, and $\overline{S}$ is approximately $\sigma/\sqrt{20}$, where σ is the population standard deviation. A proof of this is given in the following section.

Our computer sampling procedure provided a very close estimate of this standard error. Very often, we are interested in properties of statistics for which it is difficult to derive formulae for their standard errors. Computer sampling techniques, like bootstrapping discussed in Chapter 4, provide good approximations to the standard errors of sample statistics.

6.2   Estimation with simple random samples

In the present section, we investigate the properties of estimators of the population quantities when sampling is simple random.

The probability structure for simple random samples with or without replacements, RSWR and RSWOR, was studied in Section 3.1.4.

Let $X_1, \ldots, X_n$ denote the values of the variable $X(u)$ of the n elements in the random samples. The marginal distributions of X_i ($i = 1, \ldots, n$) if the sample is random, with or without replacement, is the distribution $\hat{F}_N(x)$. If the sample is random **with** replacement, then $X_1, \ldots, X_n$ are **independent**. If the sample is random **without** replacement, then $X_1, \ldots, X_n$ are **correlated** (dependent).

For an estimator of μ_N, we use the sample mean

$$\overline{X}_n = \frac{1}{n} \sum_{j=1}^{n} X_j.$$

For an estimator of σ_N^2, we use the sample variance

$$S_n^2 = \frac{1}{n-1} \sum_{j=1}^{n} (X_j - \overline{X}_n)^2.$$

Both estimators are random variables, which may change their values from one sample to another.

An estimator is called **unbiased** if its expected value is equal to the population value of the quantity it estimates. The **precision** of an estimator is the inverse of its sampling variance.

Example 6.2. We illustrate the above with the following numerical example. The population is of size $N = 100$. For simplicity, we take $X(u_i) = i$ $(i = 1, \ldots, 100)$. For this simple population, $\mu_{100} = 50.5$ and $\sigma^2_{100} = 833.25$.

Draw from this population 100 independent samples, of size $n = 10$, **with** and **without** replacement.

This can be done in R by using the function boot as in the previous section and modifying the function arguments. The means and standard deviations (Std.) of the 100 sample estimates are (Table 6.2)

Table 6.2 *Statistics of sampling distributions*

	RSWR		RSWOR	
Estimate	Mean	Std.	Mean	Std.
$\overline{X}_{10}$	49.85	9.2325	50.12	9.0919
S^2_{10}	774.47	257.96	782.90	244.36

As will be shown in the following section, the theoretical expected value of $\overline{X}_{10}$, both in RSWR and RSWOR, is $\mu = 50.5$. We see above that the means of the sample estimates are close to the value of μ. The theoretical standard deviation of $\overline{X}_{10}$ is 9.128 for RSWR and 8.703 for RSWOR. The empirical standard deviations are also close to these values. The empirical means of S^2_{10} are somewhat lower than their expected values of 833.25 and 841.67, for RSWR and RSWOR, respectively. But, as will be shown later, they are not significantly smaller than σ^2.

6.2.1 Properties of $\overline{X}_n$ and S^2_n under RSWR

If the sampling is RSWR, the random variables $X_1, \ldots, X_n$ are independent, having the same c.d.f. $\hat{F}_N(x)$. The corresponding p.d.f. is

$$p_N(x) = \begin{cases} \dfrac{1}{N}, & \text{if } x = x_j, \quad j = 1, \ldots, N \\ 0, & \text{otherwise.} \end{cases} \tag{6.2.1}$$

Accordingly,

$$E\{X_j\} = \frac{1}{N} \sum_{j=1}^{N} x_j = \mu_N, \quad \text{all } j = 1, \ldots, N. \tag{6.2.2}$$

It follows from the results of Section 3.8 that

$$E\{\overline{X}_n\} = \frac{1}{n}\sum_{j=1}^{n} E\{X_j\}$$

$$= \mu_N. \tag{6.2.3}$$

Thus, **the sample mean is an unbiased estimator of the population mean**.

The variance of X_j, is the variance associated with $F_N(x)$, i.e.,

$$V\{X_j\} = \frac{1}{N}\sum_{j=1}^{N} x_j^2 - \mu_N^2$$

$$= \frac{1}{N}\sum_{j=1}^{N} (x_j - \mu_N)^2$$

$$= \sigma_N^2.$$

Moreover, since $X_1, X_2, \ldots, X_n$ are i.i.d,

$$V\{\overline{X}_n\} = \frac{\sigma_N^2}{n}. \tag{6.2.4}$$

Thus, as explained in Section 3.8, the sample mean converges in probability to the population mean, as $n \to \infty$. An estimator having such a property is called **consistent**.

We show now that S_n^2 **is an unbiased estimator of** σ_N^2.

Indeed, if we write

$$S_n^2 = \frac{1}{n-1}\sum_{j=1}^{n} (X_j - \overline{X}_n)^2$$

$$= \frac{1}{n-1}\left(\sum_{j=1}^{n} X_j^2 - n\overline{X}_n^2\right),$$

we obtain

$$E\{S_n^2\} = \frac{1}{n-1}\left(\sum_{j=1}^{n} E\{X_j^2\} - nE\{\overline{X}_n^2\}\right).$$

Moreover, since $X_1, \ldots, X_n$ are i.i.d.,

$$E\{X_j^2\} = \sigma_N^2 + \mu_n^2, \quad j = 1, \ldots, n$$

and

$$E\{\overline{X}_n^2\} = \frac{\sigma_N^2}{n} + \mu_N^2.$$

Substituting these in the expression for $E\{S_n^2\}$ we obtain

$$E\{S_n^2\} = \frac{1}{n-1}\left(n(\sigma_N^2 + \mu_N^2) - n\left(\frac{\sigma_N^2}{n} + \mu_N^2\right)\right)$$

$$= \sigma_N^2. \tag{6.2.5}$$

An estimator of the standard error of $\overline{X}_n$ is $\frac{S_n}{\sqrt{n}}$. This estimator is slightly biased.

In large samples, the distribution of $\overline{X}_n$ is approximately normal, like $N\left(\mu_N, \frac{\sigma_N^2}{n}\right)$, as implied by the CLT. Therefore, the interval

$$\left(\overline{X}_n - z_{1-\alpha/2}\frac{S_n}{\sqrt{n}}, \overline{X}_n + z_{1-\alpha/2}\frac{S_n}{\sqrt{n}}\right)$$

has, in large samples the property that $\Pr\left\{\overline{X}_n - z_{1-\alpha/2}\frac{S_n}{\sqrt{n}} < \mu_N < \overline{X}_n + z_{1-\alpha/2}\frac{S_n}{\sqrt{n}}\right\} \cong 1 - \alpha$. An interval having this property is called a **confidence interval** for μ_N, with an approximate confidence level $(1 - \alpha)$. In the above formula, $z_{1-\alpha/2} = \Phi^{-1}\left(1 - \frac{\alpha}{2}\right)$.

It is considerably more complicated to derive the formula for $V\{S_n^2\}$. An approximation for large samples is

$$V\{S_n^2\} \cong \frac{\mu_{4,N} - (\sigma_N^2)^2}{n} + \frac{2(\sigma_N)^2 - \mu_{3,N}}{n^2} + \frac{\mu_{4,N} - 3(\sigma_N^2)^2}{n^3}, \tag{6.2.6}$$

where

$$\mu_{3,N} = \frac{1}{N}\sum_{j=1}^{N}(x_j - \mu_N)^3, \tag{6.2.7}$$

and

$$\mu_{4,N} = \frac{1}{N}\sum_{j=1}^{N}(x_j - \mu_N)^4. \tag{6.2.8}$$

Example 6.3.   In file **PLACE.csv**, we have data on x, y, and θ deviations of $N = 416$ placements of components by automatic insertion in 26 PCBs.

Let us consider this record as a finite population. Suppose that we are interested in the population quantities of the variable x-dev. Using R or MINITAB we find that the population mean, variance, third, and fourth central moments are

$$\mu_N = 0.9124,$$

$$\sigma_N^2 = 2.91999,$$

$$\mu_{3,N} = -0.98326,$$

$$\mu_{4,N} = 14.655.$$

The unit of measurements of the x-dev is 10^{-3} [inch].

Thus, if we draw a simple RSWR, of size $n = 50$, the variance of $\overline{X}_n$ will be $V\{\overline{X}_{50}\} = \frac{\sigma_N^2}{50} = 0.0584$. The variance of S_{50}^2 will be

$$V\{S_{50}^2\} \cong \frac{14.655 - (2.9199)^2}{50} + \frac{2(2.9199)^2 + 0.9833}{2500} + \frac{14.655 - 3(2.9199)^2}{125000}$$

$$= 0.1297.$$

6.2.2   Properties of $\overline{X}_n$ and S_n^2 under RSWOR

We show first that $\overline{X}_n$ is an unbiased estimator of μ_N, under RSWOR.

Let I_j be an indicator variable, which assumes the value 1 if u_j belongs to the selected sample, s_n, and equal to zero otherwise. Then we can write

$$\overline{X}_n = \frac{1}{n}\sum_{j=1}^{N} I_j x_j. \tag{6.2.9}$$

Accordingly,

$$E\{\overline{X}_n\} = \frac{1}{n}\sum_{j=1}^{N} x_j E\{I_j\}$$

$$= \frac{1}{n}\sum_{j=1}^{N} x_j \Pr\{I_j = 1\}.$$

As shown in Section 3.1.4,

$$\Pr\{I_j = 1\} = \frac{n}{N}, \quad \text{all } j = 1, \ldots, N.$$

Substituting this above yields that

$$E\{\overline{X}_n\} = \mu_N. \tag{6.2.10}$$

It is shown below that

$$V\{\overline{X}_n\} = \frac{\sigma_N^2}{n}\left(1 - \frac{n-1}{N-1}\right). \tag{6.2.11}$$

To derive the formula for the variance of $\overline{X}_n$, under RSWOR, we use the result of Section 4.8 on the variance of linear combinations of random variables. Write first,

$$V\{\overline{X}_n\} = V\left\{\frac{1}{n}\sum_{i=1}^{N} x_i I_i\right\}$$

$$= \frac{1}{n^2}V\left\{\sum_{i=1}^{N} x_i I_i\right\}.$$

$\sum_{i=1}^{N} x_i I_i$ is a linear combination of the random variables $I_1, \ldots, I_N$.

First, we show that

$$V\{I_i\} = \frac{n}{N}\left(1 - \frac{n}{N}\right), \quad i = 1, \ldots, N.$$

Indeed, since $I_i^2 = I_i$,

$$V\{I_i\} = E\{I_i^2\} - (E\{I_i\})^2$$

$$= E\{I_i\}(1 - E\{I_i\})$$

$$= \frac{n}{N}\left(1 - \frac{n}{N}\right), \quad i = 1, \ldots, N.$$

Moreover, for $i \neq j$,

$$\mathrm{Cov}(I_i, I_j) = E\{I_i I_j\} - E\{I_i\} E\{I_j\}.$$

But

$$E\{I_i I_j\} = \Pr\{I_i = 1, I_j = 1\}$$

$$= \frac{n(n-1)}{N(N-1)}.$$

Hence, for $i \neq j$,

$$\mathrm{Cov}(I_i, I_j) = -\frac{n}{N^2} \cdot \frac{N-n}{N-1}.$$

Finally,

$$V\left\{ \sum_{i=1}^{N} x_i I_i \right\} = \sum_{i=1}^{N} x_i^2 V\{I_i\} + \sum_i \sum_{i \neq j} x_i x_j \mathrm{cov}(X_i, X_j).$$

Substituting these expressions in

$$V\{\overline{X}_n\} = \frac{1}{n^2} V\left\{ \sum_{i=1}^{N} x_i I_i \right\},$$

we obtain

$$V\{\overline{X}_n\} = \frac{1}{n^2} \left\{ \frac{n}{N}\left(1 - \frac{n}{N}\right) \sum_{i=1}^{N} x_i^2 - \frac{n(N-n)}{N^2(N-1)} \sum_i \sum_{i \neq j} x_i x_j \right\}.$$

But, $\sum_i \sum_{i \neq j} x_i x_j = \left(\sum_{i=1}^{N} x_i \right)^2 - \sum_{i=1}^{N} x_i^2$. Hence,

$$V\{\overline{X}_n\} = \frac{N-n}{nN^2} \left\{ \frac{N}{N-1} \sum_{i=1}^{N} x_i^2 - \frac{1}{N-1} \left(\sum_{i=1}^{N} x_i \right)^2 \right\}$$

$$= \frac{N-n}{n \cdot (N-1) \cdot N} \sum_{i=1}^{N} (x_i - \mu_N)^2$$

$$= \frac{\sigma_N^2}{n} \left(1 - \frac{n-1}{N-1} \right).$$

We see that the variance of $\overline{X}_n$ is smaller under RSWOR than under RSWR, by a factor of $\left(1 - \frac{n-1}{N-1}\right)$. This factor is called the **finite population multiplier**.

The formula we have in Section 3.3.2 for the variance of the hypergeometric distribution can be obtained from the above formula. In the hypergeometric model, we have a finite population of size N. M elements have a certain attribute. Let

$$x_i = \begin{cases} 1, & \text{if } w_i \text{ has the attribute} \\ \\ 0, & \text{if } w_i \text{ does not have it.} \end{cases}$$

Since $\sum_{i=1}^{N} x_i = M$ and $x_i^2 = x_i$,

$$\sigma_N^2 = \frac{M}{N}\left(1 - \frac{M}{N}\right).$$

If $J_n = \sum_{i=1}^{n} X_i$, we have

$$V\{J_n\} = n^2 V\{\overline{X}_n\}$$

$$= n\frac{M}{N}\left(1 - \frac{M}{N}\right)\left(1 - \frac{n-1}{N-1}\right). \tag{6.2.12}$$

To estimate σ_N^2, we can again use the sample variance S_n^2. The sample variance has, however, a slight positive bias. Indeed,

$$E\{S_n^2\} = \frac{1}{n-1}E\left\{\sum_{j=1}^{n} X_j^2 - n\overline{X}_n^2\right\}$$

$$= \frac{1}{n-1}\left(n\left(\sigma_N^2 + \mu_N^2\right) - n\left(\mu_N^2 + \frac{\sigma^2}{n}\left(1 - \frac{n-1}{N-1}\right)\right)\right)$$

$$= \sigma_N^2\left(1 + \frac{1}{N-1}\right).$$

This bias is negligible if σ_N^2/N is small. Thus, the standard-error of $\overline{X}_n$ can be estimated by

$$SE\{\overline{X}_n\} = \frac{S_n}{\sqrt{n}}\left(1 - \frac{n-1}{N-1}\right)^{1/2}. \tag{6.2.13}$$

When sampling is RSWOR, the random variables $X_1, \ldots, X_n$ are **not independent**, and we cannot justify theoretically the usage of the normal approximation to the sampling distribution of $\overline{X}_n$. However, if n/N is small the normal approximation is expected to yield good results. Thus, if $\frac{n}{N} < 0.1$, we can approximate the confidence interval, of level $(1 - \alpha)$, for μ_N, by the interval with limits

$$\overline{X}_n \pm z_{1-\alpha/2} \cdot SE\{\overline{X}_n\}.$$

In order to estimate the coverage probability of this interval estimator, when $\frac{n}{N} = 0.3$, we perform the following simulation example.

Example 6.4.   We can use MINITAB or R to select RSWOR of size $n = 30$ from the population $P = \{1, 2, \ldots, 100\}$ of $N = 100$ units, whose values are $x_i = i$.
For this purpose, set the integers $1, \ldots, 100$ into object X. Notice that, when $n = 30, N = 100$, $\alpha = 0.05$, $z_{1-\alpha/2} = 1.96$, and $\frac{1.96}{\sqrt{n}}\left(1 - \frac{n-1}{N-1}\right)^{1/2} = 0.301$.

```
Sample 30 C1 C4
let k1 = mean(C4) − 0.301 ∗ stan(C4)
let k2 = mean(C4) + 0.301 ∗ stan(C4)
stack C2 k1 C2
stack C3 k2 C3
end
```

```
> X <- 1:100
> set.seed(123)
> XSmp <- replicate(1000, sample(X,
                                  size=30,
                                  replace=FALSE))
> Confint <- function(x, p, n=length(x), N){
    p <- if(p >= 0.5)
      1-((1-p)/2)
    else
      1-(p/2)
    m <- mean(x)
    z <- qnorm(p=p)/sqrt(n)*(1-((n-1)/(N-1)))^(1/2)
    s <- sd(x)
    res <- m - z*s
    res <- c(res, m + z*s)
    names(res) <- c("lower", "upper")
    return(res)
  }
> XSmpCnf <- t(apply(XSmp, MARGIN=2,
                     FUN=Confint,
                     p=0.95,
                     N=100))
> head(XSmpCnf, 3)

        lower     upper
[1,]  44.61377  61.91956
[2,]  44.23539  62.96461
[3,]  42.95316  58.91351

> sum(apply(XSmpCnf, MARGIN=1,
            FUN=function(x, m){
              x[1]< m && x[2] > m
            },
          m =50.5))/nrow(XSmpCnf)

[1]  0.959
```

The true population mean is $\mu_N = 50.5$. The estimated coverage probability is the proportion of cases for which $k_1 \le \mu_N \le k_2$. In the present simulation, the proportion of coverage is 0.947. The nominal confidence level is $1 - \alpha = 0.95$. The estimated coverage probability is 0.947. Thus, the present example shows that even in cases, where $n/N > 0.1$ the approximate confidence limits are quite effective.

6.3   Estimating the mean with stratified RSWOR

We consider now the problem of estimating the population mean, μ_N, with stratified RSWOR. Thus, suppose that the population P is partitioned into k strata (subpopulations) $P_1, P_2, \ldots, P_k$, $k \ge 2$.

Let $N_1, N_2, \ldots, N_k$ denote the sizes; $\mu_{N_1}, \ldots, \mu_{N_k}$ the means and $\sigma_{N_1}^2, \ldots, \sigma_{N_k}^2$ the variances of these strata, respectively. Notice that the population mean is

$$\mu_N = \frac{1}{N} \sum_{i=1}^{k} N_i \mu_{N_i} \tag{6.3.1}$$

and according to the formula of total variance (see Section 4.8), the population variance is

$$\sigma_N^2 = \frac{1}{N} \sum_{i=1}^{k} N_i \sigma_{N_i}^2 + \frac{1}{N} \sum_{i=1}^{k} N_i (\mu_{N_i} - \mu_N)^2. \tag{6.3.2}$$

We see that if the means of the strata are not the same, the population variance is greater than the weighted average of the within strata variances, $\sigma_{N_i}^2$ $(i = 1, \ldots, k)$.

A stratified RSWOR is a sampling procedure in which k independent random samples without replacement are drawn from the strata. Let n_i, $\overline{X}_{n_i}$ and $S_{n_i}^2$ be the size, mean, and variance of the RSWOR from the ith stratum, P_i $(i = 1, \ldots, k)$.

We have shown in the previous section that $\overline{X}_{n_i}$ is an unbiased estimator of μ_{N_i}. Thus, an unbiased estimator of μ_N is the weighted average

$$\hat{\mu}_N = \sum_{i=1}^{k} W_i \overline{X}_{n_i}, \tag{6.3.3}$$

where $W_i = \frac{N_i}{N}$, $i = 1, \ldots, k$. Indeed,

$$E\{\hat{\mu}_N\} = \sum_{i=1}^{k} W_i E\{\overline{X}_{n_i}\}$$

$$= \sum_{i=1}^{k} W_i \mu_{N_i} \tag{6.3.4}$$

$$= \mu_N.$$

Since $\overline{X}_{n_1}, \overline{X}_{n_2}, \ldots, \overline{X}_{n_k}$ are independent random variables, the variance of $\hat{\mu}_N$ is

$$V\{\hat{\mu}_N\} = \sum_{i=1}^{k} W_i^2 V\{\overline{X}_{n_i}\}$$

$$= \sum_{i=1}^{k} W_i^2 \frac{\sigma_{n_i}^2}{n_i} \left(1 - \frac{n_i - 1}{N_i - 1}\right) \tag{6.3.5}$$

$$= \sum_{i=1}^{k} W_i^2 \frac{\tilde{\sigma}_{N_i}^2}{n_i} \left(1 - \frac{n_i}{N_i}\right),$$

where

$$\tilde{\sigma}_{N_i}^2 = \frac{N_i}{N_i - 1} \sigma_{N_i}^2.$$

Example 6.5. Returning to the data of Example 6.3, on deviations in the x-direction of automatically inserted components, the units are partitioned to $k = 3$ strata. Boards $1-10$ in stratum 1, boards $11-13$ in stratum 2 and boards $14-26$ in stratum 3. The population characteristics of these strata are the following:

Stratum	Size	Mean	Variance
1	160	-0.966	0.4189
2	48	0.714	1.0161
3	208	2.403	0.3483.

The relative sizes of the strata are $W_1 = 0.385$, $W_2 = 0.115$, and $W_3 = 0.5$. If we select a stratified RSWOR of sizes $n_1 = 19$, $n_2 = 6$, and $n_3 = 25$, the variance of $\hat{\mu}_N$ will be

$$V\{\hat{\mu}_N\} = (0.385)^2 \frac{0.4189}{19}\left(1 - \frac{18}{159}\right) + (0.115)^2 \frac{1.0161}{6}\left(1 - \frac{5}{47}\right)$$
$$+ (0.5)^2 \frac{0.3483}{25}\left(1 - \frac{24}{207}\right)$$
$$= 0.00798.$$

This variance is considerably smaller than the variance of $\overline{X}_{50}$ in a simple RSWOR, which is

$$V\{\overline{X}_{50}\} = \frac{2.9199}{50}\left(1 - \frac{49}{415}\right)$$
$$= 0.0515.$$

6.4   Proportional and optimal allocation

An important question in designing the stratified RSWOR is how to allocate the total number of observations, n, to the different strata, i.e., the determination of $n_i \geq 0$ ($i = 1, \ldots, k$) so that $\sum_{i=1}^{k} n_i = n$, for a given n. This is called the **sample allocation**. One type of sample allocation is the so-called **proportional allocation**, i.e.,

$$n_i = nW_i, \quad i = 1, \ldots, k. \tag{6.4.1}$$

The variance of the estimator $\hat{\mu}_N$ under proportional allocation is

$$V_{\text{prop}}\{\hat{\mu}_N\} = \frac{1}{n}\sum_{i=1}^{k} W_i \tilde{\sigma}_{N_i}^2 \left(1 - \frac{n}{N}\right)$$
$$= \frac{\overline{\sigma}_N^2}{n}\left(1 - \frac{n}{N}\right), \tag{6.4.2}$$

where

$$\overline{\sigma}_N^2 = \sum_{i=1}^{k} W_i \tilde{\sigma}_{N_i}^2$$

is the weighted average of the within strata variances.

We have shown in the previous section that if we take a simple RSWOR, the variance of $\overline{X}_n$ is

$$V_{\text{simple}}\{\overline{X}_n\} = \frac{\sigma_N^2}{n}\left(1 - \frac{n-1}{N-1}\right)$$

$$= \frac{\tilde{\sigma}_N^2}{n}\left(1 - \frac{n}{N}\right),$$

where

$$\tilde{\sigma}_N^2 = \frac{N}{N-1}\sigma_N^2.$$

In large-sized populations, σ_N^2 and $\tilde{\sigma}_N^2$ are very close, and we can write

$$V_{\text{simple}}\{\overline{X}_n\} \cong \frac{\sigma_N^2}{N}\left(1 - \frac{n}{N}\right)$$

$$= \frac{1}{n}\left(1 - \frac{n}{N}\right)\left\{\sum_{i=1}^{k}W_i\sigma_{N_i}^2 + \sum_{i=1}^{k}W_i(\mu_{N_i} - \mu_N)^2\right\}$$

$$\cong V_{\text{prop}}\{\hat{\mu}_N\} + \frac{1}{n}\left(1 - \frac{n}{N}\right)\sum_{i=1}^{k}W_i(\mu_{N_i} - \mu_N)^2.$$

This shows that $V_{\text{simple}}\{\overline{X}_n\} > V_{\text{prop}}\{\hat{\mu}_N\}$; i.e., the estimator of the population mean, μ_N, under stratified RSWOR, with proportional allocation, generally has smaller variance (more precise) than the estimator under a simple RSWOR. The difference grows with the variance between the strata means, $\sum_{i=1}^{k}W_i(\mu_{N_i} - \mu_N)^2$. Thus, effective stratification is one which partitions the population to strata which are homogeneous within (small values of $\sigma_{N_i}^2$) and heterogeneous between (large value of $\sum_{i=1}^{k}W_i(\mu_{N_i} - \mu_N)^2$). If sampling is stratified RSWR, then the variance $\hat{\mu}_N$, under proportional allocation is

$$V_{\text{prop}}\{\hat{\mu}_N\} = \frac{1}{n}\sum_{i=1}^{k}W_i\sigma_{N_i}^2. \qquad (6.4.3)$$

This is strictly smaller than the variance of $\overline{X}_n$ in a simple RSWR. Indeed,

$$V_{\text{simple}}\{\overline{X}_n\} = \frac{\sigma_N^2}{n}$$

$$= V_{\text{prop}}\{\hat{\mu}_N\} + \frac{1}{n}\sum_{i=1}^{k}W_i(\mu_{N_i} - \mu_N)^2.$$

Example 6.6. Defective circuit breakers are a serious hazard since their function is to protect electronic systems from power surges or power drops. Variability in power supply voltage levels can cause major damage to electronic systems. Circuit breakers are used to shield electronic systems from such events. The proportion of potentially defective circuit breakers is a key parameter in designing redundancy levels of protection devices and preventive maintenance programs. A lot of $N = 10,000$ circuit breakers was put together by purchasing the products from $k = 3$ different vendors. We want to estimate the proportion of defective breakers, by sampling and testing $n = 500$ breakers. Stratifying the lot by vendor, we have three strata of sizes $N_1 = 3000$, $N_2 = 5000$, and $N_3 = 2000$. Before installing the circuit breakers, we drew from the lot a stratified RSWOR, with proportional allocation, i.e., $n_1 = 150$, $n_2 = 250$, and $n_3 = 100$. After testing, we found in the first sample $J_1 = 3$ defective circuit breakers. In the second sample $J_2 = 10$ and in the third sample $J_3 = 2$ defectives. Testing is done with a special purpose device, simulating intensive usage of the product.

In the present case, we set $X = 1$ if the item is defective and $X = 0$ otherwise. Then μ_N is the proportion of defective items in the lot. μ_{N_i} $(i = 1, 2, 3)$ is the proportion defectives in the ith stratum.

The unbiased estimator of μ_N is

$$\hat{\mu}_N = 0.3 \times \frac{J_1}{150} + 0.5 \times \frac{J_2}{250} + 0.2 \times \frac{J_3}{100}$$
$$= 0.03.$$

The variance within each stratum is $\sigma_{N_i}^2 = P_{N_i}(1 - P_{N_i})$, $i = 1, 2, 3$, where P_{N_i} is the proportion in the ith stratum. Thus, the variance of $\hat{\mu}_N$ is

$$V_{\text{prop}}\{\hat{\mu}_N\} = \frac{1}{500}\bar{\sigma}_N^2\left(1 - \frac{500}{10,000}\right),$$

where

$$\bar{\sigma}_N^2 = 0.3\tilde{\sigma}_{N_1}^2 + 0.5\tilde{\sigma}_{N_2}^2 + 0.2\tilde{\sigma}_{N_3}^2,$$

or

$$\bar{\sigma}_N^2 = 0.3 \times \frac{3000}{2999}P_{N_1}(1 - P_{N_1}) + 0.5\frac{5000}{4999}P_{N_2}(1 - P_{N_2}) + 0.2\frac{2000}{1999}P_{N_3}(1 - P_{N_3}).$$

Substituting $\frac{3}{150}$ for an estimate of P_{N_1}, $\frac{10}{250}$ for that of P_{N_2} and $\frac{2}{100}$ for P_{N_3}, we obtain the estimate of $\bar{\sigma}_N^2$,

$$\bar{\sigma}_N^2 = 0.029008.$$

Finally, an estimate of $V_{\text{prop}}\{\hat{\mu}_N\}$ is

$$\hat{V}_{\text{prop}}\{\hat{\mu}_N\} = \frac{0.029008}{500}\left(1 - \frac{500}{10,000}\right)$$

$$= 0.00005511.$$

The standard error of the estimator is 0.00742.
Confidence limits for μ_N, at level $1 - \alpha = 0.95$, are given by

$$\hat{\mu}_N \pm 1.96 \times \text{SE}\{\hat{\mu}_N\} = \begin{cases} 0.0446 \\ \\ 0.0154. \end{cases}$$

These limits can be used for spare parts policy.

When the variances $\tilde{\sigma}_{N_i}^2$ within strata are known, we can further reduce the variance of μ_N by an allocation, which is called **optimal allocation**.

We wish to minimize

$$\sum_{i=1}^{k} W_i^2 \frac{\tilde{\sigma}_{N_i}^2}{n_i} \left(1 - \frac{n_i}{N_i}\right)$$

subject to the constraint:

$$n_1 + n_2 + \cdots + n_k = n.$$

This can be done by minimizing

$$L(n_1, \ldots, n_k, \lambda) = \sum_{i=1}^{k} W_i^2 \frac{\tilde{\sigma}_{N_i}^2}{n_i} - \lambda \left(n - \sum_{i=1}^{k} n_i\right),$$

with respect to $n_1, \ldots, n_k$ and λ. This function is called the **Lagrangian** and λ is called the **Lagrange multiplier**.

The result is

$$n_i^0 = n \frac{W_i \tilde{\sigma}_{N_i}}{\sum_{j=1}^{k} W_j \tilde{\sigma}_j}, \quad i = 1, \ldots, k. \tag{6.4.4}$$

We see that the proportional allocation is optimal when all $\tilde{\sigma}_{N_i}^2$ are equal.

The variance of $\hat{\mu}_N$, corresponding to the optimal allocation is

$$V_{\text{opt}}\{\hat{\mu}_N\} = \frac{1}{N} \left(\sum_{i=1}^{k} W_i \tilde{\sigma}_{N_i}\right)^2 - \frac{1}{N} \sum_{i=1}^{k} W_i \tilde{\sigma}_{N_i}^2. \tag{6.4.5}$$

6.5   Prediction models with known covariates

In some problems of estimating, the mean μ_N of a variable Y in a finite population, we may have information on variables $X_1, X_2, \ldots, X_k$ which are related to Y. The variables $X_1, \ldots, X_k$ are called **covariates**. The model relating Y to $X_1, \ldots, X_k$ is called a **prediction model**. If the values of Y are known only for the units in the sample, while the values of the covariates are known for all the units of the population, we can utilize the prediction model to improve the precision of the estimator. The method can be useful, for example, when the measurements of Y are destructive, while the covariates can be measured without destroying the units. There are many such examples, like the case of measuring the compressive strength of a concrete cube. The measurement is destructive. The compressive strength Y is related to the ratio of cement to water in the mix, which is a covariate that can be known for all units. We will develop the ideas with a simple prediction model.

Let $\{u_1, u_2, \ldots, u_N\}$ be a finite population, P. The values of $x_i = X(u_i)$, $i = 1, \ldots, N$ are known for all the units of P. Suppose that $Y(u_i)$ is related linearly to $X(u_i)$ according to the prediction model

$$y_i = \beta x_i + e_i, \quad i = 1, \ldots, N, \tag{6.5.1}$$

where β is an unknown regression coefficient, and $e_1, \ldots, e_N$ are i.i.d. random variables such that

$$\begin{aligned} E\{e_i\} &= 0, \quad i = 1, \ldots, N, \\ V\{e_i\} &= \sigma^2, \quad i = 1, \ldots, N. \end{aligned}$$

The random variable e_i in the prediction model is due to the fact that the linear relationship between Y and X is not perfect, but subject to random deviations.

We are interested in the population quantity $\bar{y}_N = \frac{1}{N} \sum_{i=1}^{N} y_i$. We cannot, however, measure all the Y values. Even if we know the regression coefficient β, we can only predict $\bar{y}_N$ by $\beta \bar{x}_N$, where $\bar{x}_N = \frac{1}{N} \sum_{j=1}^{N} x_j$. Indeed, according to the prediction model, $\bar{y}_N = \beta \bar{x}_N + \bar{e}_N$, and $\bar{e}_N$ is a random variable with

$$E\{\bar{e}_N\} = 0, \quad V\{\bar{e}_N\} = \frac{\sigma^2}{N}. \tag{6.5.2}$$

Thus, since $\bar{y}_N$ has a random component, and since $E\{\bar{y}_N\} = \beta \bar{x}_N$, we say that a predictor of $\bar{y}_N$, say $\hat{y}_N$, is **unbiased**, if $E\{\hat{Y}_N\} = \beta \bar{x}_N$. Generally, β is unknown. Thus, we draw a sample of units from P and measure their Y values, in order to estimate β. For estimating β, we draw a simple RSWOR from P of size n, $1 < n < N$.

Let $(X_1, Y_1), \ldots, (X_n, Y_n)$ be the values of X and Y in the random sample. A predictor of $\bar{y}_N$ is some function of the observed sample values. Notice that after drawing a random sample, we have **two** sources of variability. One due to the random error components $e_1, \ldots, e_n$, associated with the sample values, and the other one is due to the random sampling of the n units of P. Notice that the error variables $e_1, \ldots, e_n$ are independent of the X values, and thus, independent of $X_1, X_2, \ldots, X_n$, randomly chosen to the sample. In the following, expectation and variances are taken with respect to the errors model and with respect to the sampling procedure. We will examine now a few alternative predictors of $\bar{y}_N$.

(i) **The sample mean, $\bar{Y}_n$.**

  Since

$$\bar{Y}_n = \beta \bar{X}_n + \bar{e}_n,$$

  we obtain that

$$E\{\bar{Y}_n\} = \beta E\{\bar{X}_n\} + E\{\bar{e}_n\}$$

$E\{\bar{e}_n\} = 0$ and since the sampling is RSWOR, $E\{\bar{X}_n\} = \bar{x}_N$. Thus, $E\{\bar{Y}_n\} = \beta \bar{x}_N$, and the predictor is **unbiased**. The variance of the predictor is, since $\bar{e}_n$ is independent of $\bar{X}_n$,

$$V\{\bar{Y}_n\} = \beta^2 V\{\bar{X}_n\} + \frac{\sigma^2}{n}$$

$$= \frac{\sigma^2}{n} + \frac{\beta^2 \sigma_x^2}{n} \left(1 - \frac{n-1}{N-1}\right), \tag{6.5.3}$$

  where

$$\sigma_x^2 = \frac{1}{N} \sum_{j=1}^{N} (x_j - \bar{x}_N)^2.$$

(ii) **The ratio predictor,**

$$\hat{Y}_R = \bar{x}_N \frac{\bar{Y}_n}{\bar{X}_n}. \tag{6.5.4}$$

The ratio predictor will be used when all $x_i > 0$. In this case, $\overline{X}_n > 0$ in every possible sample. Substituting $\overline{Y}_n = \beta \overline{X}_n + \bar{e}_n$ we obtain

$$E\{\hat{Y}_R\} = \beta \bar{x}_N + \bar{x}_N E\left\{\frac{\bar{e}_n}{\overline{X}_n}\right\}.$$

Again, since $\bar{e}_n$ and $\overline{X}_n$ are independent, $E\left\{\frac{\bar{e}_n}{\overline{X}_n}\right\} = 0$ and $\hat{Y}_R$ is an **unbiased** predictor. The variance of $\hat{Y}_R$ is

$$V\{\hat{Y}_R\} = (\bar{x}_N)^2 V\left\{\frac{\bar{e}_n}{\overline{X}_n}\right\}.$$

Since $\bar{e}_n$ and $\overline{X}_n$ are independent, and $E\{\bar{e}_n\} = 0$, the law of the total variance implies that

$$
\begin{aligned}
V\{\hat{Y}_R\} &= \frac{\sigma^2}{n} \bar{x}_N^2 E\left\{\frac{1}{\overline{X}_n^2}\right\} \\
&= \frac{\sigma^2}{n} E\left\{\left(1 + \frac{(\overline{X}_n - \bar{x}_N)}{\bar{x}_N}\right)^{-2}\right\} \\
&= \frac{\sigma^2}{n} E\left\{1 - \frac{2}{\bar{x}_N}(\overline{X}_n - \bar{x}_N) + \frac{3}{\bar{x}_N^2}(\overline{X}_n - \bar{x}_N)^2 + \cdots\right\} \\
&\cong \frac{\sigma^2}{n}\left(1 + \frac{3\gamma_x^2}{n}\left(1 - \frac{n-1}{N-1}\right)\right),
\end{aligned}
\tag{6.5.5}
$$

where $\gamma_x = \sigma_x/\bar{x}_N$ is the coefficient of variation of X. The above approximation is effective in large samples.

Using the large sample approximation, we see that the ratio predictor $\hat{Y}_R$ has a smaller variance than $\overline{Y}_n$ if

$$\frac{3\sigma^2 \gamma_x^2}{n^2}\left(1 - \frac{n-1}{N-1}\right) < \frac{\beta^2 \sigma_x^2}{n}\left(1 - \frac{n-1}{N-1}\right)$$

or, if

$$n > \frac{3\sigma^2}{(\beta \bar{x}_N)^2}.$$

Other possible predictors for this model are

$$\hat{Y}_{RA} = \bar{x}_N \cdot \frac{1}{N}\sum_{i=1}^{n}\frac{Y_i}{X_i} \tag{6.5.6}$$

and

$$\hat{Y}_{RG} = \bar{x}_N \cdot \frac{\sum_{i=1}^{n} Y_i X_i}{\sum_{i=1}^{N} X_i^2}. \tag{6.5.7}$$

We leave it as an exercise to prove that both $\hat{Y}_{RA}$ and $\hat{Y}_{RG}$ are unbiased predictors and to derive their variances.

What happens, under the above prediction model, if the sample drawn is not random, but the units are chosen to the sample by some nonrandom fashion?

Suppose that a nonrandom sample $(x_1, y_1), \ldots, (x_n, y_n)$ is chosen. Then

$$E\{\bar{y}_n\} = \beta \bar{x}_n$$

and

$$V\{\bar{y}_n\} = \frac{\sigma^2}{n}.$$

The predictor $\bar{y}_n$ is biased, unless $\bar{x}_n = \bar{x}_N$. A sample which satisfies this property is called a **balanced sample** with respect to X. Generally, the **mean squared error** (MSE) of $\bar{y}_n$, under nonrandom sampling is

$$\text{MSE}\{\bar{y}_n\} = E\{(\bar{y}_n - \beta \bar{x}_N)^2\}$$

$$= \frac{\sigma^2}{n} + \beta^2 (\bar{x}_n - \bar{x}_N)^2. \qquad (6.5.8)$$

Thus, if the sample is balanced with respect to X then $\bar{y}_n$ is a more precise predictor than all the above, which are based on simple random samples.

Example 6.7. Electronic systems such as television sets, radios, or computers contain printed circuit boards with electronic components positioned in patterns determined by design engineers. After assembly (either by automatic insertion machines or manually), the components are soldered to the board. In the relatively new Surface Mount Technology, minute components are simultaneously positioned and soldered to the boards. The occurrence of defective soldering points impacts the assembly plant productivity and is therefore closely monitored. In file **PRED.csv**, we find 1000 records on variable X and Y. X is the number of soldering points on a board, and Y is the number of defective soldering points. The mean of Y is $\bar{y}_{1000} = 7.495$ and that of X is $\bar{x}_{1000} = 148.58$. Moreover, $\sigma_x^2 = 824.562$ and the coefficient of variation is $\gamma_x = 0.19326$. The relationship between X and Y is $y_i = \beta x_i + e_i$, where $E\{e_i\} = 0$ and $V\{e_i\} = 7.5$, $\beta = 0.05$. Thus, if we have to predict $\bar{y}_{1000}$ by a predictor based on a RSWR, of size $n = 100$, the variances of $\bar{Y}_{100}$ and $\hat{Y}_R = \bar{x}_{1000} \frac{\bar{Y}_{100}}{\bar{x}_{100}}$ are

$$V\{\bar{Y}_{100}\} = \frac{7.5}{100} + \frac{0.0025 \times 824.562}{100} = 0.0956.$$

On the other hand, the large sample approximation yields,

$$V\{\hat{Y}_R\} = \frac{7.5}{100} \left(1 + \frac{3 \times 0.037351}{100}\right)$$

$$= 0.07508.$$

We see that, if we have to predict $\bar{y}_{1000}$ on the basis of an RSWR of size $n = 100$, the ratio predictor, $\hat{Y}_R$, is more precise.

In Figures 6.2 and 6.3, we present the histograms of 500 predictors $\bar{Y}_{100}$ and 500 $\hat{Y}_R$ based on RSWR of size 100 from this population.

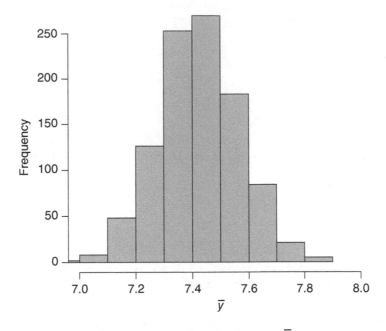

Figure 6.2   *Sampling distribution of* $\overline{Y}$.

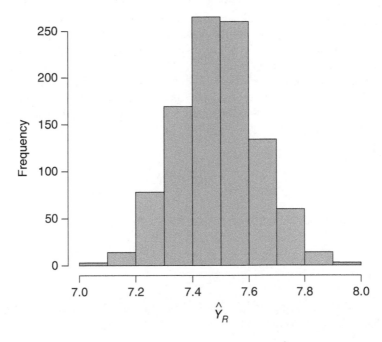

Figure 6.3   *Sampling distribution of* $\hat{Y}_R$.

6.6   Chapter highlights

Techniques for sampling finite populations and estimating population parameters are presented. Formulas are given for the expected value and variance of the sample mean and sample variance of simple random samples with and without replacement. Stratification is studied as a method to increase the precision of estimators. Formulas for proportional and optimal allocation are provided and demonstrated with case studies. The chapter is concluded with a section on prediction models with known covariates.

The main concepts and definitions introduced in this chapter include the following:

- Population Quantiles
- Simple Random Samples
- Stratified Random Samples
- Unbiased Estimators
- Precision of an estimator
- Finite Population Multiplier
- Sample Allocation
- Proportional Allocation
- Optimal Allocation
- Prediction Models
- Covariates
- Ratio Predictor
- Prediction Unbiasedness
- Prediction MSE

6.7   Exercises

6.1 Consider a finite population of size N, whose elements have values $x_1, \ldots, x_N$. Let $\hat{F}_N(x)$ be the c.d.f., i.e.,

$$\hat{F}_N(x) = \frac{1}{N} \sum_{i=1}^{N} I\{x_i \leq x\}.$$

Let $X_1, \ldots, X_n$ be the values of a RSWR. Show that $X_1, \ldots, X_n$ are independent having a common distribution $\hat{F}_N(x)$.

6.2 Show that if $\overline{X}_n$ is the mean of a RSWR, then $\overline{X}_n \to \mu_N$ as $n \to \infty$ in probability (weak law of large numbers [WLLN]).

6.3 What is the large sample approximation to $\Pr\{\sqrt{n} \mid \overline{X}_n - \mu_N \mid < \delta\}$ in RSWR?

6.4 Use MINITAB to draw random samples with or without replacement from data file **PLACE.csv**. Write a MACRO which computes the sample correlation between the x-dev and y-dev in the sample values. Execute this MACRO 100 times and make a histogram of the sample correlations.

6.5 Use file **CAR.csv** and MINITAB. Construct a MACRO which samples at random, without replacement (RSWOR), 50 records. Stack the medians of the variables turn-diameter, horsepower, and mpg (3, 4, 5). Execute the MACRO 200 times and present the histograms of the sampling distributions of the medians.

6.6 In continuation of Example 6.3, how large should the sample be from the three strata so that the SE $\{\overline{X}_i\}$ ($i = 1, \ldots, 3$) will be smaller than $\delta = 0.005$?

6.7 The proportion of defective chips in a lot of $N = 10,000$ chips is $\mathbf{P} = 5 \times 10^{-4}$. How large should a RSWOR be so that, the width of the confidence interval for P, with coverage probability $1 - \alpha = 0.95$, will be 0.002?

6.8 Use MINITAB to perform stratified random samples from the three strata of the data file **PLACE.csv** (see Example 6.3). Allocate 500 observations to the three samples proportionally. Estimate the population mean (of x-dev). Repeat this 100 times and estimate the standard-error or your estimates. Compare the estimated standard error to the exact one.

6.9 Derive the formula for n_i^0 $(i = 1, \ldots, k)$ in the optimal allocation, by differentiating $L(n_1, \ldots, n_k, \lambda)$ and solving the equations.

6.10 Consider the prediction model

$$y_i = \beta + e_i, \quad i = 1, \ldots, N,$$

where $E\{e_i\} = 0$, $V\{e_i\} = \sigma^2$ and $\mathrm{COV}(e_i, e_j) = 0$ for $i \neq j$. We wish to predict the population mean $\mu_N = \frac{1}{N} \sum_{i=1}^{N} y_i$. Show that the sample mean $\overline{Y}_n$ is prediction unbiased. What is the prediction MSE of $\overline{Y}_n$?

6.11 Consider the prediction model

$$y_i = \beta_0 + \beta_1 x_i + e_i, \quad i = 1, \ldots, N,$$

where $e_1, \ldots, e_N$ are independent r.v.s with $E\{e_i\} = 0$, $V\{e_i\} = \sigma^2 x_i$ $(i = 1, \ldots, n)$. We wish to predict $\mu_N = \frac{1}{N} \sum_{i=1}^{N} y_i$. What should be a good predictor for μ_N?

6.12 Prove that $\hat{Y}_{RA}$ and $\hat{Y}_{RG}$ are unbiased predictors and derive their prediction variances.

7

Time Series Analysis and Prediction

7.1 The components of a time series

A time series $\{X_t, t = 1, 2, \ldots\}$ is a sequence of random variables ordered according to the observation time. The analysis of the fluctuation of a time series assists us in analyzing the current behavior and forecasting the future behavior of the series. In the following sections, we introduce elementary concepts. There are three important components of a time series: the trend, the correlation structure among the observations, and the stochastic nature of the random deviations around the trend (the noise). If these three components are known, a reasonably good prediction, or forecasting can be made. However, there are many types of time series in which the future behavior of the series is not necessarily following the past behavior. In the present chapter, we discuss these two types of forecasting situations. For more details see, Box and Jenkins (1970), Zacks (2009), and Shumway and Stoffer (2017).

7.1.1 The trend and covariances

The function $f(t) : t \mapsto E\{X_t\}$ is called the ***trend*** of the time Series. A smooth trend can often be fitted to the time series data. Such a trend could be locally described as a polynomial of certain order plus a trigonometric Fourier sequence of orthogonal periodic functions.

Example 7.1. In Figure 7.1, we present the time series **DOW1941.csv**, with the Dow Jones index in 302 working days of the year 1941. The smooth curve traced among the points is the fitted local trend given by the function:

$$f(t) = 123.34 + 27.73((t-151)/302) - 15.83((t-151)/302)^2 - 237.00((t-151)/302)^{3\cdot}$$
$$+ 0.1512\cos(4\pi t/302) + 1.738\sin(4\pi t/302) + 1.770\cos(8\pi t/302)$$
$$- 0.208\sin(8\pi t/302) - 0.729\cos(12\pi t/302) + 0.748\sin(12\pi t/302).$$

Modern Industrial Statistics: With Applications in R, MINITAB and JMP, Third Edition.
Ron S. Kenett and Shelemyahu Zacks.
© 2021 John Wiley & Sons, Ltd. Published 2021 by John Wiley & Sons, Ltd.

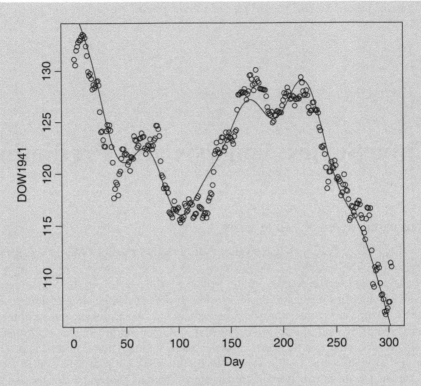

Figure 7.1   *Dow Jones values in 1941.*

To fit such a trend by the method of least squares, we use the multiple linear regression technique, in which the dependent variable Y is the vector of the time series, while the X vectors are the corresponding polynomial and trigonometric variables $((t - 151)/302)^j, j = 0, 1, 2, 3$ and $\cos(j4\pi t/302), \sin(j4\pi t/302), j = 1, 2, 3$.

7.1.2   Applications with MINITAB and JMP

In this subsection, we apply MINITAB and JMP to **DOW1941.csv** in order to demonstrate how to apply these software platforms in time series analysis.

In MINITAB, one can apply the command **Stat> Time Series> Decomposition** using an additive model and a linear trend and monthly seasonal adjustment to produces Figure 7.2. The top two graphs in the figure show the original data and the linearly detrended data using $Y_t = 125.987 - 0.02654t$. Monthly adjustments are derived from subtracting the monthly averages. The bottom right chart accounts for both the linear trend and the monthly average. It shows the residual plot of the original data after correcting for both the linear trend and the monthly adjustment. The residual plot shows that correcting for these effects is not producing white noise and, therefore, there is more structure in the data to account for.

Applying the fit model option in JMP (**Analyze>Fit Model**), we fit a model consisting of a cubic polynomial in time and a seasonal effect accounting for a month specific effect. The original data and the fitted model are presented in Figure 7.3. The adjusted R square for this

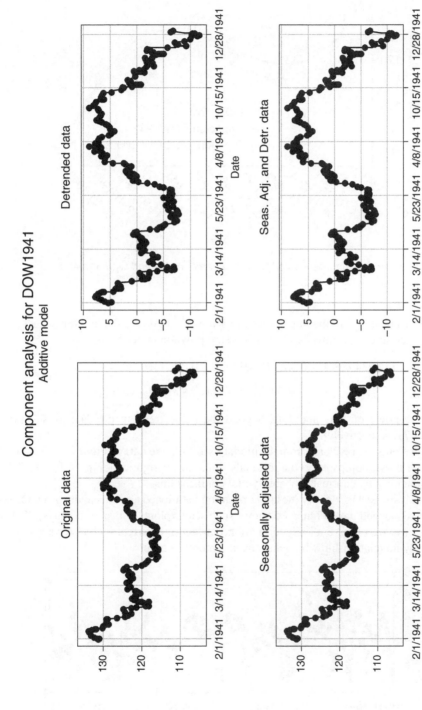

Figure 7.2 *Decomposition of the Dow 1941 time series using an additive model with a linear trend and monthly seasonality adjustment (MINITAB).*

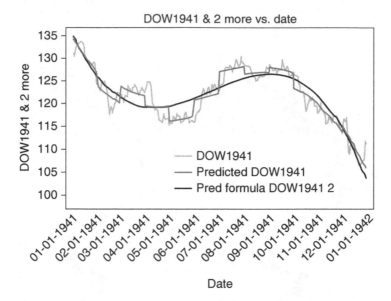

Figure 7.3   *The Dow1941 time series with a fit to a cubic model and with a fit to a cubic model with monthly seasonal effects (JMP).*

model is 90.3%. The jumps of the fitted curve in Figure 7.3 are due to the monthly effects. The model, with the coefficient estimates and noncentered polynomial terms, is

$$134.23492675 - 0.359762701\, \text{Day} + 0.0029216242\, \text{Day}^2$$

$$- 0.00000679154\, \text{Day}^3 + \text{Match (Month)} \ldots$$

We see that the effect of May and July, beyond the cubic trend in the data, is substantial (−4.65 and +3.71, respectively).

Figure 7.4 presents the residuals from the model used to fit the data in Figure 7.3. This looks more like white noise, compared to the residuals in the lower right quadrant of Figure 7.2. A formal assessment using a normal probability plot confirms this.

A simple analysis used to identify the lag-correlation mentioned above is to draw scatterplots of the data versus lagged data. Figure 7.5 shows such scatterplots for lags of 60 days, 15 days, 5 days, and 1 day. We observe a high correlation between the current and yesterday's Dow's index and barely no relationship with values from 2 months earlier.

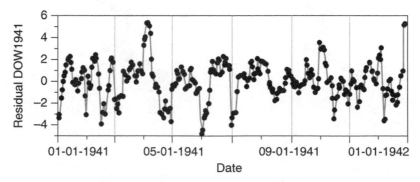

Figure 7.4   *Residuals from model used in Figure 7.3 to fit the Dow1941 data (JMP).*

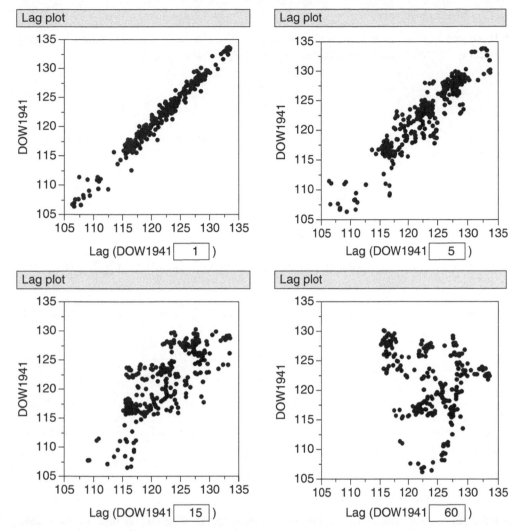

Figure 7.5   *Scatter plots of Dow1941 time series with a lagged series of lag 60, 15, 5 and 1 day (JMP).*

This correlation structure is not accounted for by the least squares' regression models used in Figure 7.3 to model the DOW1941 time series. This lack of independence between successive observations is affecting our ability to properly predict future observations. In the next sections, we show how to account for such auto-correlations.

7.2   Covariance stationary time series

Let $X_t = f(t) + U_t$. It is assumed that $E\{U_t\} = 0$ for all $t = 1, 2, \ldots$ Furthermore, the sequence of residuals $\{U_t, t = 1, 2, \ldots\}$ is called covariance stationary if $K(h) = \text{cov}(U_t, U_{t+h})$ is independent of t for all $h = 1, 2, \ldots$ Notice that in this case, the variance of each U_t is $K(0)$. The lag-correlation of order, h, is $\rho(h) = K(h)/K(0), h = 1, 2, \ldots$ The simplest covariance stationary time series is $\{e_t, t = 0, \pm 1, \pm 2, \ldots\}$, where $E\{e_t\} = 0$, the variance $V\{e_t\} = \sigma^2$ for all t, and $\rho(h) = 0$, for all $|h| > 0$. Such a time series is called a **white-noise**. We denote it by $\text{WN}(0, \sigma^2)$.

7.2.1   Moving averages

A linear combination of WN random variables is called a ***moving average***. A moving average of order q, MA(q), is the linear combination of q WN Variables, i.e.

$$X_t = \sum_{j=0}^{q} \beta_j e_{t-j}, \tag{7.1}$$

where the coefficients β_j are the same for all t. The covariance function of an MA(q) is stationary and is given by

$$K(h) = \sigma^2 \sum_{j=0}^{q-|h|} \beta_j \beta_{j+|h|}, |h| = 0, \dots, q$$

$$= 0, |h| > q. \tag{7.2}$$

Notice that $K(h) = K(-h)$, and a moving average of infinite order exists, if $\sum_{j=-\infty}^{\infty} |\beta_j| < \infty$.

Example 7.2.   Consider an MA(3) in which $X = 3e_t + 2.5e_{t-1} - 1.5e_{t-2} + e_{t-3}$ and $\sigma^2 = 1$. This covariance stationary time series has

$$K(0) = 9 + 6.25 + 2.25 + 1 = 18.50;$$

$$K(1) = 7.5 - 3.75 - 1.5 = 2.25;$$

$$K(2) = -4.5 + 2.5 = -2; \text{ and } K(3) = 3.$$

The lag-correlations are $\rho(0) = 1$, $\rho(1) = 0.1216$, $\rho(2) = -0.1081$, and $\rho(3) = 0.1622$. All lag-correlations for $|h| > 3$ are zero. □

7.2.2   Auto-regressive time series

Another important class of time series is the ***auto-regressive*** model. A time series is called auto-regressive of order p, AR(p), if $E\{X_t\} = \mu$, for all t, and $X_t = \sum_{j=1}^{p} \gamma_j X_{t-j} + e_t$, for all t, where e_t is a WN($0, \sigma^2$). Equivalently, we can specify an AR(p) time series as

$$X_t + a_1 X_{t-1} + \cdots + a_p X_{t-p} = e_t. \tag{7.3}$$

This time series can be converted to a moving average time series by applying the **Z-transform**

$$A_p(z) = 1 + a_1 z^{-1} + \cdots + a_p z^{-p},$$

p is an integer, $A_0(z) = 1$ and $z^{-j} X_t = X_{t-j}$. Accordingly, $A_p(z) X_t = X_t + a_1 X_{t-1} + \cdots + a_p X_{t-p} = e_t$. From this, we obtain that

$$X_t = (A_p(z))^{-1} e_t = \sum_{j=0}^{\infty} \beta_j e_{t-j}, \tag{7.4}$$

where

$$(A_p(z))^{-1} = 1/(1 + a_1 z^{-1} + \cdots + a_p z^{-p}) = \sum_{j=0}^{\infty} \beta_j z^{-j}. \tag{7.5}$$

This inverse transform can be computed by the algebra of power series. The inverse power series always exists since $\beta_0 \neq 0$. We can obtain the coefficients β_j as illustrated in the following example. Notice that an infinite power series obtained in this way might not converge. If it does not converge, the inversion is not useful. The transform $(A_p(z))^{-1}$ is called a **Transfer Function**.

The polynomial $A_p^*(z) = z^p A_p(z)$ is called the **characteristic polynomial** of the AR(p). The auto-regressive time series AR(p) is covariance stationary only if all its characteristic roots belong to the interior of the unit circle or the roots of $A_p(z)$ are all outside the unit circle.

The covariance function $K(h)$ can be determined by the following equations called the **Yule–Walker equations**

$$K(0) + a_1 K(1) + \cdots + a_p K(p) = \sigma^2,$$

$$K(h) + a_1 K(h - 1) + \cdots + a_p K(h - p) = 0, h > 0. \tag{7.6}$$

Example 7.3. Consider the AR(2)

$$X_t - X_{t-1} + 0.89 X_{t-2} = e_t, t = 0, \pm 1, \pm 2, \ldots$$

In this case, the characteristic polynomial is $A_2^*(z) = 0.89 - z + z^2$. The two characteristic roots are the complex numbers $\zeta_1 = 0.5 + 0.8i$, and $\zeta_2 = 0.5 - 0.8i$. These two roots are inside the unit circle, and thus this AR(2) is covariance stationary.
Using series expansion, we obtain

$$1/(1 - z^{-1} + 0.89 z^{-2}) = 1 + z^{-1} + 0.11 z^{-2} - 0.78 z^{-3} - 0.8779 z^{-4} - 0.1837 z^{-5} + \cdots$$

The corresponding MA(5), which is $X_t^* = e_t + e_{t-1} + 0.11 e_{t-2} - 0.78 e_{t-3} - 0.8779 e_{t-4} - 0.1837 e_{t-5}$ is a finite approximation to the infinite order MA representing X_t. This approximation is not necessarily good. To obtain a good approximation to the variance and covariances, we need a longer moving average.
As for the above AR(2), the $K(h)$ for $h = 0, 1, 2$ are determined by solving the Yule–Walker linear equations:

$$\begin{pmatrix} 1 & a_1 & a_2 \\ a_1 & 1 + a_2 & 0 \\ a_2 & a_1 & 1 \end{pmatrix} \begin{pmatrix} K(0) \\ K(1) \\ K(2) \end{pmatrix} = \begin{pmatrix} \sigma^2 \\ 0 \\ 0 \end{pmatrix}$$

We obtain for $\sigma = 1, K(0) = 6.6801, K(1) = 3.5344$, and $K(2) = -2.4108$. Correspondingly, the lag-correlations are

$$\rho(0) = 1, \rho(1) = 0.5291, \text{ and } \rho(2) = -0.3609.$$

For $h \geq 3$, we use the recursive equation:

$$K(h) = -a_1 K(h - 1) - a_2 K(h - 2).$$

Accordingly, $K(3) = 2.4108 - 0.89 * 3.5344 = -0.7348$, and so on. $\square$

An important tool for determining the order p of an auto-regressive time series is the ***partial lag correlation***, *denoted as* $\rho^*(h)$. This index is based on the lag-correlations in the following manner:

Let R_k denote a symmetric $k \times k$ matrix, called the ***Toeplitz matrix***, which is

$$
\begin{pmatrix}
1 & \rho(1) & \rho(2) & \rho(3) & \cdots & \rho(k-1) \\
\rho(1) & 1 & \rho(1) & \rho(2) & \cdots & \rho(k-2) \\
\rho(2) & \rho(1) & 1 & \rho(1) & \cdots & \rho(k-3) \\
\rho(3) & \rho(2) & \rho(1) & 1 & \cdots & \rho(k-4) \\
\cdots & \cdots & \cdots & \cdots & \cdots & \cdots \\
\rho(k-1) & \rho(k-2) & \rho(k-3) & \rho(k-4) & \cdots & 1
\end{pmatrix}
$$

The solution $\phi^{(k)}$ of the normal equations:

$$
R_k \phi^{(k)} = \rho_k \tag{7.7}
$$

yields least-squares estimators of X_t and of X_{t+k+1}, based on the values of $X_{t+1}, ..., X_{t+k}$. These are

$$
\hat{X}_t = \sum_{j=1}^{k} \phi_j^{(k)} X_{t+j} \tag{7.8}
$$

and

$$
\hat{X}_{t+k+1} = \sum_{j=1}^{k} \phi_j^{(k)} X_{t+k+1-j}. \tag{7.9}
$$

One obtains the following formula for the partial correlation of lag $k+1$,

$$
\rho^*(k+1) = [\rho(k+1) - \rho_k' R_k^{-1} \rho_k^*]/(1 - \rho_k' R_k^{-1} \rho_k), \tag{7.10}
$$

where $\rho_k = (\rho(1), ..., \rho(k))'$ and $\rho_k^* = (\rho(k), ..., \rho(1))'$.

In package `mistat`, the function `toeplitz` forms the above matrix.

Example 7.4. Using R, MINITAB, or JMP, we obtain Table 7.1:

Table 7.1 *Lag-correlations and partial lag correlations for the DOW1941 data*

k	$\rho(k)$	$\rho^*(k)$
1	0.9789	0.9789
2	0.9540	−0.1035
3	0.9241	−0.1228
4	0.8936	−0.0122
5	0.8622	−0.0254
6	0.8328	0.0357
7	0.8051	0.0220
8	0.7774	−0.0323
9	0.7486	−0.0504
10	0.7204	−0.0005
11	0.6919	−0.0139
12	0.6627	−0.0357
13	0.6324	−0.0335
14	0.6027	0.0012
15	0.5730	−0.0197

The above 15 lag-correlations are all significant since the DOW1941 series is not stationary. Only the first partial lag correlation is significantly different than zero. On the other hand, if we consider the residuals around the trend curve $f(t)$, we get the following significant lag-correlations (Table 7.2):

Table 7.2 *Lag-correlations of deviations from the trend of DOW1941*

1	0.842
2	0.641
3	0.499
4	0.375
5	0.278
6	0.232
7	0.252
8	0.256
9	0.272
10	0.301

The lag-correlations and the partial ones can be obtained in R by the functions `acf` and `pacf`.

7.2.3 Auto-regressive moving averages time series

A time series of the form

$$X_t + a_1 X_{t-1} + \cdots + a_p X_{t-p} = e_t + b_1 e_{t-1} + \cdots + b_q e_{t-q} \tag{7.11}$$

is called an ARMA(p, q) time series If the characteristic roots of $A_p^*(z)$ are within the unit disk, then this time series is covariance stationary. We could write

$$X_t = \sum_{j=0}^{\infty} \beta_j z^{-j} (1 + b_1 z^{-1} + \cdots + b_q z^{-q}) e_t.$$

$$= \sum_{k=0}^{\infty} v_k z^{-k} e_t. \tag{7.12}$$

Here $v_k = \sum_{l=0}^{k} \beta_l b_{k-l}$, where $b_{k-l} = 0$ when $k - l > q$. one can obtain this by series expansion of $B_q(x)/A_p(x)$

Example 7.5. We consider here the ARMA(2,2), where, as in Example 7.2, $A_2(z) = 1 - z^{-1} + 0.89\, z^{-2}$, and $B_2(z) = 1 + z^{-1} - 1.5z^{-2}$. In this case

$$(1 + x - 1.5x^2)/(1 - x + 0.89x^2) = 1 + 2x - 0.39x^2 - 2.17x^3 - 1.8229x^4 + 0.1084x^5 + \cdots$$

Thus, we can approximate the ARMA(2,2) by an MA(5), namely

$$X_t = e_t + 2e_{t-1} - 0.39e_{t-2} - 2.17e_{t-3} - 1.8229e_{t-4} + 0.1084e_{t-5}. \quad \square$$

7.2.4   Integrated auto-regressive moving average time series

A first-order difference of a time series is

$$\Delta\{X_t\} = (1 - z^{-1})\{X_t\} = \{X_t - X_{t-1}\}.$$

Similarly, a kth order difference of $\{X_t\}$ is

$$\Delta^k\{X_t\} = (1 - z^{-1})^k\{X_t\} = \sum_{j=0}^{k} \binom{k}{j}(-1)^j z^{-j}\{X_t\}.$$

If $\Delta^k\{X_t\}$ is an ARMA(p, q) we call the time series an **Integrated**
ARMA(p, q) of order k, or in short ARIMA(p, k, q). This time series has the structure

$$A_p(z)(1 - z^{-1})^k X_t = B_q(z)e_t \tag{7.13}$$

where $e_t \sim i.i.d.(0, \sigma^2)$. Accordingly, we can express

$$[A_p(z)/B_q(z))](1 - z^{-1})^k X_t = e_t. \tag{7.14}$$

Furthermore,

$$[A_p(z)/B_q(z)](1 - z^{-1})^k X_t = \sum_{j=0}^{\infty} \varphi_j z^{-j} X_t, \tag{7.15}$$

It follows that

$$X_t = \sum_{j=1}^{\infty} \pi_j X_{t-j} + e_t, \tag{7.16}$$

where $\pi_j = -\varphi_j$, for all $j = 1, 2, \ldots$ This shows that the ARIMA(p, k, q) time series can be approximated by a linear combination of its previous values. This can be utilized, if the coefficients of $A_p(z^1)$ and those of $B_q(z)$ are known, for prediction of future values. We illustrate it in the following example:

Example 7.6.   As in Example 7.5, consider the time series ARIMA$(2, 2, 2)$, where $A_2(z) = 1 - z^{-1} + 0.89z^{-2}$ and $B_2(z^1) = 1 + z^{-1} - 1.5z^{-2}$.
We have

$$[1 - x + 0.89x^2]/[1 + x - 1.5x^2] = 1 - 2x + 4.39x^2 - 7.39x^3 + 13.975x^4 - 25.060x^5\ldots$$

Multiplying by $(1 - x)^2$ we obtain that

$$X_t = 4X_{t-1} - 9.39X_{t-2} + 18.17X_{t-3} - 33.24X_{t-4} + 60.30X_{t-5}$$
$$- 73.625X_{t-6} + 25.06X_{t-7} - + \cdots + e_t$$

The predictors of X_{t+m} are obtained recursively from the above, as follows:

$$\hat{X}_{t+1} = 4X_t - 9.39X_{t-1} + 18.17X_{t-2} - + \cdots$$
$$\hat{X}_{t+2} = 4\hat{X}_{t+1} - 9.39X_t + 18.17X_{t-1} - 33.24X_{t-2} + - \cdots$$
$$\hat{X}_{t+3} = 4\hat{X}_{t+2} - 9.39\hat{X}_{t+1} + 18.17X_t - 33.24X_{t-1} + - \cdots \quad \square$$

7.2.5   Applications with JMP and R

To conclude this section, we revisit the Dow1941 data using an ARMA model with JMP. This additional modeling effort aims at picking up the auto-correlation in the data. We apply **JMP>Specialized Modeling>Time series** to the residuals from the cubic and monthly effect model used in Figure 7.3. In the red triangle, we invoke the **ARIMA Model Group** option that fits a range of models. We set the auto-regressive and moving-average order range to be 0–4. The best fitting model is ARMA(2,2). The estimates of the model parameters are shown in Figure 7.6.

The ARMA(2,2) model, with 95% confidence intervals and 30 predictions are shown in Figure 7.7. To fit an ARMA model in R, the package `forecast` provides the function `auto.arima`.

To generate predictions for January 1942 values of the DOW, we use the prediction of residuals shown in Figure 7.7 and add these to the cubic and month effect model shown in Figure 7.3.

Term	Lag	Estimate	Std Error	t Ratio	Prob>|t|
AR1	1	1.766255	0.0433265	40.77	<.0001*
AR2	2	−0.792413	0.0432359	−18.33	<.0001*
MA1	1	0.803897	0.0704196	11.42	<.0001*
MA2	2	0.196092	0.0697972	2.81	0.0053*
Intercept	0	0.003855	0.0237583	0.16	0.8712

Figure 7.6   *Estimates of ARMA(2,2) model applied to residuals of DOW1941 shown in Figure 7.4 (JMP).*

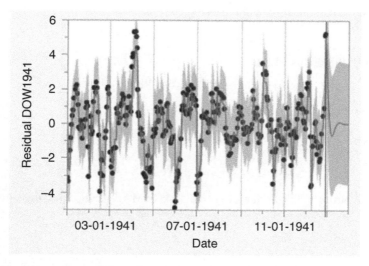

Figure 7.7   *Confidence intervals based on ARMA(2,2) model and prediction errors of future 30 observations (JMP).*

7.3   Linear predictors for covariance stationary time series

Let $\{X_t\}$ be a covariance stationary time series, such that $X_t = \sum_{j=0}^{\infty} w_j e_{t-j}$ and $e_t \sim \mathrm{WN}(0, \sigma^2)$. Notice that $E\{X_t\} = 0$ for all t. The covariance function is $K(h)$.

7.3.1   Optimal linear predictors

A linear predictor of X_{t+s} based on the n data points $\mathbf{X}_t^{(n)} = (X_t, X_{t-1}, \ldots, X_{t-n+1})'$ is

$$\hat{X}_{t+s}^{(n)} = (\boldsymbol{b}_s^{(n)})' \mathbf{X}_t^{(n)}, \tag{7.17}$$

where

$$\boldsymbol{b}_s^{(n)} = (b_{0,s}^{(n)}, b_{1,s}^{(n)}, \ldots, b_{n-1,s}^{(n)})'.$$

The prediction mean squared error (PMSE) of this linear predictor is

$$E\left\{ (\hat{X}_{t+s}^{(n)} - X_{t+s})^2 \right\} = V\left\{ \left(\boldsymbol{b}_s^{(n)} \right)' \mathbf{X}_t^{(n)} \right\} - 2\mathrm{cov}(\hat{X}_{t+s}, \left(\boldsymbol{b}_s^{(n)} \right)' \mathbf{X}_t^{(n)}) + V\{X_{t+s}\}. \tag{7.18}$$

The covariance matrix of $\mathbf{X}_t^{(n)}$ is the Toeplitz matrix $K(0)R_n$ and the covariance of X_{t+s} and $\mathbf{X}_t^{(n)}$ is

$$\gamma_s^{(n)} = K(0) \begin{pmatrix} \rho(s) \\ \rho(s+1) \\ \rho(s+2) \\ \cdots \\ \rho(n+s-1) \end{pmatrix} = K(0)\rho_s^{(n)}. \tag{7.19}$$

Hence, we can write the prediction PMSE as

$$\mathrm{PMSE}(\hat{X}_{t+s}^{(n)}) = K(0)((\boldsymbol{b}_s^{(n)})' R_n \boldsymbol{b}_s^{(n)} - 2(\boldsymbol{b}_s^{(n)})' \rho_s^{(n)} + 1) \tag{7.20}$$

It follows that the best linear predictor based on $\mathbf{X}_t^{(n)}$ is

$$\mathrm{BLP}(X_{t+s}|\mathbf{X}_t^{(n)}) = (\rho_s^{(n)})' R_n^{-1} \mathbf{X}_t^{(n)}. \tag{7.21}$$

The minimal PMSE of this BLP is

$$\mathrm{PMSE}* = K(0)(1 - (\rho_s^{(n)})' R_n^{-1} \rho_s^{(n)}). \tag{7.22}$$

Notice that the minimal PMSE is an increasing function of s.

Example 7.7.   Let

$$X_t = a_{-1} e_{t-1} + a_0 e_t + a_1 e_{t+1}, t = 0, \pm 1, \ldots$$

where $e_t \sim \mathrm{WN}(0, \sigma^2)$. This series is a special kind of MA(2), called **moving smoother**, with

$$K(0) = \sigma^2(a_{-1}^2 + a_o^2 + a_1^2) = B_0\sigma^2,$$

$$K(1) = \sigma^2(a_{-1}a_0 + a_0 a_1) = B_1\sigma^2,$$

$$K(2) = \sigma^2 a_{-1}a_1 = \sigma^2 B_2.$$

and $K(h) = 0$, for all $|h| > 2$. Moreover, $\gamma_n^{(s)} = 0$ if $s > 2$. This implies that $\hat{X}_{t+s}^{(n)} = 0$, for all $s > 2$. If $s = 2$, then

$$\gamma_2^{(n)} = \sigma^2(B_2, 0'_{n-1})$$

and

$$\hat{b}_2^{(n)} = \sigma^2 R_n^{-1}(B_2, 0'_{n-1})'$$

In the special case of $a_{-1} = 0.25, a_0 = 0.5, a_1 = 0.25, \sigma = 1$ the Toeplitz matrix is

$$R_3 = \frac{1}{16}\begin{pmatrix} 6 & 4 & 1 \\ 4 & 6 & 4 \\ 1 & 4 & 6 \end{pmatrix}$$

Finally, the best linear predictors are

$$\hat{X}_{t+1}^{(3)} = 1.2X_t - 0.9X_{t-1} + 0.4X_{t-2},$$

and

$$\hat{X}_{t+2}^{(3)} = 0.4X_t - 0.4X_{t-1} + 0.2X_{t-2}.$$

The PMSE are, correspondingly, 0.1492 and 0.3410. □

Example 7.8. In the previous example, we applied the optimal linear predictor to a simple case of an MA(2) series. Now, we examine the optimal linear predictor in a more elaborate case, where the deviations around the trend are covariance stationary. We use the **DOW1941.csv** time series stationary ARMA series.

In Figure 7.3, we present the one-day ahead predictors for the DOW1941 data. This was obtained by applying the R function `pred1` to the series of the deviations around the trend function f(t) for the DOW1941 data, with a window of size 10, and adding the results to the trend data resT1941. The corresponding total prediction risk is 0.6429.

7.4   Predictors for nonstationary time series

The linear predictor discussed in the previous section was based on the covariance stationarity of the deviations from the trend function. Such predictors are valid if we can assume that the future behavior of the time series is similar to the observed part of the past. This however is seldom the case. In the present section, we develop an adaptive procedure, which extrapolates the observed trend in a small window of the past. Such predictors would generally be good ones only for small values of time units in the future (Figure 7.8).

7.4.1   Quadratic LSE predictors

For a specified window size $n, n > 5$, we fit a polynomial of degree $p = 2$ (quadratic) to the last n observations. We then extrapolate to estimate $f(t + s), s \geq 1$. This approach is based on the assumption that in a close neighborhood of t, say $(t - n, t + n), f(t)$ can be approximated by a quadratic whose parameters may change with t, i.e.

$$f_2(t) = \beta_0(t) + \beta_1(t)t + \beta_2(t)t^2. \tag{7.23}$$

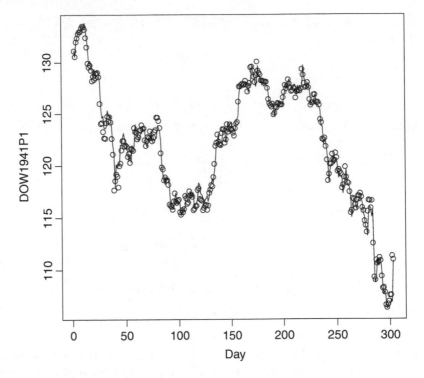

Figure 7.8   *One-step ahead predictors for the DOW1941 data.*

The quadratic moving LSE algorithm applies the method of ordinary least-squares to estimate $\beta_j(t), j = 0, 1, 2$, based on the data in the moving window $\{X_{t-n+1}, \ldots, X_t\}$. With these estimates, it predicts X_{t+s} with $f_2(t + s)$.

We provide here some technical details. In order to avoid the problem of unbalanced matrices when t is large, we shift in each cycle of the algorithm the origin to t. Thus, let $\mathbf{X}_t^{(n)} = (X_t, X_{t-1}, \ldots, X_{t-n+1})'$ be a vector consisting of the values in the window. Define the matrix

$$A_{(n)} = \begin{pmatrix} 1 & 0 & 0 \\ 1 & -1 & 1 \\ . & . & . \\ 1 & -(n-1) & (n-1)^2 \end{pmatrix}$$

The LSE of $\beta_j(t)$, $j = 0, 1, 2$ are given in the vector

$$\hat{\beta}^{(n)}(t) = (A'_{(n)}A_{(n)})^{-1}A'_{(n)}\mathbf{X}_t^{(n)}. \tag{7.24}$$

With these LSEs, the predictor of X_{t+s} is

$$\hat{X}_{t+s}^{(n)}(t) = \hat{\beta}_0^{(n)}(t) + \hat{\beta}_1^{(n)}(t)s + \hat{\beta}_2^{(n)}(t)s^2. \tag{7.25}$$

Example 7.9.   In the present example, we illustrate the quadratic predictor on a nonstationary time series. We apply the R function `predPoly` according to which we can predict the outcomes of an auto-regressive series, step by step, after the first n observations. In Figure 7.9, we see the one-step ahead prediction, $s = 1$, with a window of size $n = 20$ for the data of DOW1941.

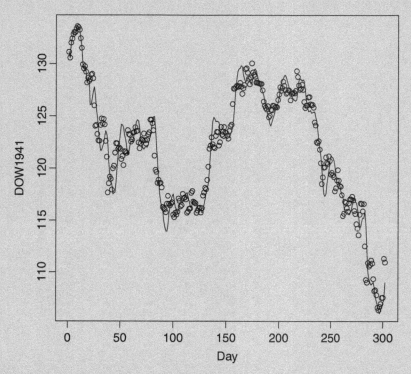

Figure 7.9   *One-step ahead prediction, s = 1, with a window of size n = 20 for the data of DOW1941.*

As we see in this figure, that the quadratic LSE predictor is quite satisfactory. The results depend strongly on the size of the window,n. In the present case, the PMSE is 2.0959.

7.4.2   Moving average smoothing predictors

A moving average smoother, MAS(m), is a sequence which replaces X_t by the a fitted polynomial based on the window of size $n = 2m + 1$, around X_t. The simplest smoother is the linear one. That is, we fit by LSE a linear function to a given window.

Let $S_{(m)} = (A_{(m)})'(A_{(m)})$. In the linear case

$$S_{(m)} = \begin{pmatrix} 2m + 1 & 0 \\ 0 & m(m + 1)(2m + 1)/3 \end{pmatrix} \tag{7.26}$$

Then, the vector of coefficients is

$$\hat{\beta}^{(m)}(t) = S_{(m)}(A_{(m)})' \mathbf{X}_t^{(m)}. \tag{7.27}$$

The components of this vector are

$$\beta_0^{(m)}(t) = \frac{1}{2m+1} \sum_{j=-m}^{m} X_{t+j} \tag{7.28}$$

$$\beta_1^{(m)}(t) = (3/(m(m+1)(2m+1))) \sum_{j=1}^{m} j(X_{t+j} - X_{t-j}). \tag{7.29}$$

Example 7.10. In Figure 7.10, we present this linear smoother predictor for the DOW1941 data, with $m = 3$ and $s = 1$. See also the R function `masPred1`. The observed PMSE is 2.1059.

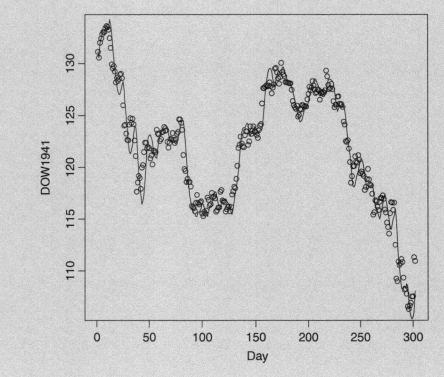

Figure 7.10 *Linear smoother predictor for the DOW1941 data, with m = 3 and s = 1.*

7.5 Dynamic linear models

The dynamic linear model (DLM) relates recursively the current observation, possibly vector of several dimensions, to a linear function of parameters, possibly random. Formally, we consider the random linear model

$$\mathbf{X_t} = \mathbf{A_t}\theta_t + \epsilon_t, \tag{7.30}$$

where

$$\theta_t = \mathbf{G_t}\theta_{t-1} + \eta_t. \tag{7.31}$$

In this model, $\mathbf{X_t}$ and ϵ_t are q dimensional vectors. θ_t and η_t are p dimensional vectors. $\mathbf{A}$ is a qxp matrix of known constants, and $\mathbf{G}$ is a pxp matrix of known constants. $\{\epsilon_t\}$ and $\{\eta_t\}$ are mutually independent vectors. Furthermore, for each t, $\epsilon_t \sim N(0, \mathbf{V}_t)$ and $\eta_t \sim N(0, \mathbf{W}_t)$

7.5.1 Some special cases

Different kinds of time series can be formulated as DLMs. For example, a time series with a polynomial trend can be expressed as Eq. (7.5.1) with $q = 1, \mathbf{A}_t = (1, t, t^2, \ldots, , t^{p-1})$, and θ_t is the random vector of the coefficients of the polynomial trend, which might change with t.

The normal random walk

The case of $p = 1, q = 1, A = 1x1$ and $G = 1x1$ is called a Normal Random Walk. For this model, $V_0 = v$ and $W_0 = w$ are prior variances, and the prior distribution of θ is $N(m, c)$. The posterior distribution of θ given X_t is $N(m_t, c_t)$, where

$$c_t = (c_{t-1} + w)v/(c_{t-1} + v + w), \tag{7.32}$$

and

$$m_t = (1 - c_{t-1}/v)m_{t-1} + (c_{t-1}/v)X_t. \tag{7.33}$$

Example 7.11. Figure 7.6 represents a Normal Random Walk with a Bayesian prediction. The prediction line in Figure 7.11 is the posterior expectation m_t. This figure was computed according to R function `nrwm`. Each application of this function yields another random graph.

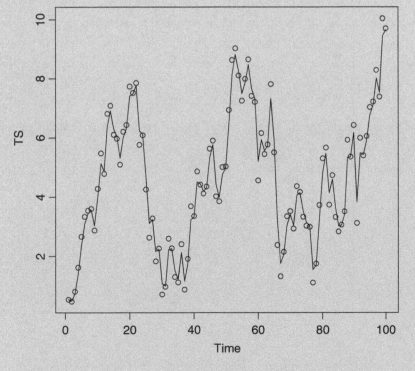

Figure 7.11 *Normal random walk with a Bayesian prediction.*

Dynamic linear model with linear growth

Another application is the DLM with linear growth. In this case, $\mathbf{X_t} = \theta_0 + \theta_1 \mathbf{t} + \epsilon_t$, with random coefficient θ. As a special case consider the following coefficients, $\mathbf{A}_t = (1, t)'$, $\mathbf{G}_t = \mathbf{I}_2$. Here $\mathbf{m}_t$ and $\mathbf{C}_t$ are the posterior mean and covariance matrix, given recursively by m

$$\mathbf{m}_t = \mathbf{m}_{t-1} + (1/r_t)(X_t - \mathbf{A}_t'\mathbf{m}_{t-1})(\mathbf{C}_{t-1} + \mathbf{W})\mathbf{A}_t, \tag{7.34}$$

$$r_t = v + \mathbf{A_t'}(\mathbf{C_{t-1}} + \mathbf{W})\mathbf{A_t} \tag{7.35}$$

and

$$\mathbf{C_t} = \mathbf{C_{t-1}} + \mathbf{W} - (1/r_t)(\mathbf{C_{t-1}} + \mathbf{W})\mathbf{a_t}\mathbf{a_t'}(\mathbf{C_{t-1}} + \mathbf{W}). \tag{7.36}$$

The predictor of X_{t+1} at time t is $\hat{X}_{t+1}(t) = \mathbf{A_{t+1}'}\mathbf{m_t}$.

Example 7.12. In Figure 7.12, we present the one-day ahead prediction ($s = 1$) for the DOW1941 data. We applied the R function `dlmLg` with parameters X, C_0, v, W, M_0. For M_0 and C_0, we used the LSE of a regression line fitted to the first 50 data points. These are $M_0 = (134.234, -0.3115)'$, with covariance matrix

$$C_0 = \begin{pmatrix} 0.22325 & -0.00668 \\ -0.00668 & 0.00032 \end{pmatrix}$$

and the random vector η has the covariance matrix

$$W = \begin{pmatrix} 0.3191 & -0.0095 \\ -0.0095 & 0.0004 \end{pmatrix}$$

As seen in Figure 7.12, the prediction using this method is very good.

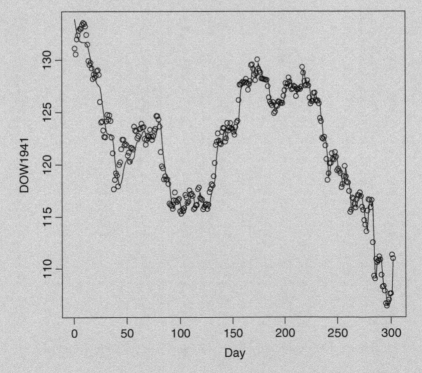

Figure 7.12   *One-step ahead prediction (s = 1) for the DOW1941 data.*

Dynamic linear model for ARMA(p,q)

In this model, for a time series stationary around a mean zero, we can write

$$X_t = \sum_{j=1}^{p} a_j X_{t-j} + \sum_{j=0}^{q} b_j \epsilon_{t-j}, \tag{7.37}$$

where $b_0 = 1, a_p \neq 0, b_q \neq 0$. Let $n = \max(1+q,p)$. If $p < n$, we insert the extra coefficients $a_{p+1} = \cdots = a_n = 0$, and if $q < n-1$, we insert $b_{q+1} = \cdots = b_{n-1} = 0$. We let $\mathbf{A} = (1,0,\ldots,,0)$, and

$$\mathbf{G} = \begin{pmatrix} a_1 & a_2 & \ldots & a_n \\ 1 & 0 & \ldots & 0 \\ . & 1 & \ldots & 0 \\ . & & . & . \\ . & & . & 1 & 0 \end{pmatrix}.$$

Furthermore, let $\mathbf{h}' = (1, b_1, \ldots, b_{n-1})$, $\theta_t' = (X_t, \ldots, X_{t-n+1})$, and $\eta_t' = (\mathbf{h}'\epsilon_t, 0, \ldots,, 0)$. Then, we can write the ARMA(p,q) time series as $\{Y_t\}$, where

$$Y_t = \mathbf{A}\theta_t, \tag{7.38}$$

$$\theta_t = \mathbf{G}\theta_{t-1} + \eta_t, \tag{7.39}$$

with $V = 0$ and $\mathbf{W} = (W_{i,j} : i,j = 1, \ldots,, n)$, in which $W_{i,j} = I\{i = j = 1\}\sigma^2(1 + \sum_{j=1}^{n-1} b_j^2)$. The recursive formulas for $\mathbf{m}_t$ and $\mathbf{C}_t$ are

$$\mathbf{m}_t = \left(\sum_{i=1}^{n} a_i m_{t-1,i}, m_{t-1,1}, \ldots,, m_{t-1,n-1}\right)' + (1/r_t)\left(Y_t - \sum_{i=1}^{n} a_i m_{t-1,i}\right)(\mathbf{G}\mathbf{C}_{t-1}\mathbf{G}' + \mathbf{W})(1, 0')', \tag{7.40}$$

where

$$r_t = W_{11} + \mathbf{a}'\mathbf{C}_{t-1}\mathbf{a}, \quad \text{and} \quad \mathbf{a}' = (a_1, \ldots,, a_n). \tag{7.41}$$

We start the recursion with $\mathbf{m}_0' = (X_p, \ldots,, X_1)$ and $\mathbf{C}_0 = K_X(0)Toeplitz(1, \rho_X(1), \ldots,, \rho_X(n-1))$. The predictor of X_{t+1} at time t is

$$\hat{X}_{t+1} = \mathbf{A}\mathbf{G}\mathbf{m}_t = \mathbf{a}'\mathbf{X}_t. \tag{7.42}$$

We illustrate it in the following example:

Example 7.13.   Consider the stationary ARMA(3,2) given by

$$X_t = 0.5X_{t-1} + 0.3X_{t-2} + 0.1X_{t-3} + \epsilon_t + 0.3\epsilon_{t-1} + 0.5\epsilon_{t-2}.$$

in which $\{\epsilon_t\}$ is an i.i.d. sequence of N(0,1) random variables. The initial random variables are $X_1 \sim N(0,1), X_2 \sim X_1 + (0,1)$ and $X_3 \sim X_1 + X_2 + N(0,1)$. Here $a = (0.5, 0.3, 0.1)$, and $b_1 = 0.3, b_2 = 0.5$. The matrix $\mathbf{G}$ is

$$\mathbf{G} = \begin{pmatrix} 0.5 & 0.3 & 0.1 \\ 1 & 0 & 0 \\ 0 & 1 & 0 \end{pmatrix},$$

$$V = 0, \quad \text{and}$$

$$\mathbf{W} = \begin{pmatrix} 1.34 & 0 & 0 \\ 0 & 0 & 0 \\ 0 & 0 & 0 \end{pmatrix}.$$

We start with $\mathbf{m}_0 = (X_3, X_2, X_1)$. All the characteristic roots are in the unit circle. Thus, this time series is covariance stationary. The Yule–Walker equations yield the covariances:

$K_X(0) = 7.69, K_X(1) = 7.1495, K_X(2) = 7.0967, K_X(3) = 6.4622$. Thus, we start the recursion with

$$\mathbf{C}_0 = \begin{pmatrix} 7.6900 & 7.1495 & 7.0967 \\ 7.1495 & 7.6900 & 7.1495 \\ 7.0967 & 7.1495 & 7.6900 \end{pmatrix}.$$

In Figure 7.8, we see a random realization of this ARMA(3,2) and the corresponding prediction line for $s = 1$. The time series was generated by the R function ARMA. Its random realization is given by the dots in the figure. The one-day ahead prediction was computed by the R function predARMA and is given by the solid line. The empirical PMSE is 0.01124.

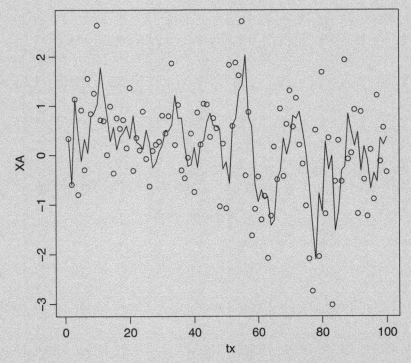

Figure 7.13 *Random realization of ARMA(3,2) and the corresponding prediction line for s = 1.*

7.6   Chapter highlights

In the present chapter, we presented essential parts of time series analysis, with the objective of predicting or forecasting its future development. Predicting future behavior is generally more successful for stationary series, which do not change their stochastic characteristics as time proceeds. We developed and illustrated time series which were of both types, namely covariance stationary, and nonstationary.

We started by fitting a smooth function, showing the trend to a complex time series consisting of the Dow Jones Industrial Average index in the 302 trading days of 1941. We defined then the notions of covariance stationarity of the deviations from the trend function, and their lag-correlation and partial correlation. Starting with the simple White Noise, we studied the properties of Moving Averages of white noise, which are covariance stationary. After this, we introduced the auto-regressive time series, in which the value of an observed variable at a given time, is a linear function of several past values in the series plus a White Noise error. A criterion for covariance stationarity of auto-regressive series was given in terms of the roots of its characteristic polynomial. We showed also how to express these stationary series as an infinite linear combination of white noise variables. More complex covariance stationary time series, which are combinations of auto-regressive and moving averages, called ARMA series, and Integrated ARMA series, called ARIMA, were discussed too.

The second part of the chapter deals with prediction of future values. We started with the optimal linear predictor of covariance stationary time series. These optimal predictors are based on moving windows of the last n observations in the series. We demonstrated that even for the **DOW1941.csv** series, if we apply the optimal linear predictor on windows of size $n = 20$, of the deviations from the trend, and then added the forecasts to the trend values, we obtain very good predictors for the next day index.

For cases where it cannot be assumed that the deviations are covariance stationary, we developed a prediction algorithm which is based on the values of the original series. Again, based on moving windows of size n, we fit by least-squares a quadratic polynomial to the last n values, and extrapolates s time units forward. As expected, this predictor is less accurate than the optimal linear for covariance stationary series, but can be useful for predicting one unit ahead, $s = 1$.

The third part of the chapter deals with and introduces DLMs, which can incorporate Bayesian analysis and vector valued observations.

The main concepts and definitions introduced in this chapter include

- Trend function
- Covariance Stationary
- White Noise
- Lag-Correlation
- Partial Lag-Correlation
- Moving Averages
- Auto-Regressive
- z-Transform
- Characteristic Polynomials
- Yule–Walker Equations
- Toeplitz Matrix
- MA Model

- ARMA Model
- ARIMA Model
- Linear Predictors
- Polynomial Predictors
- Moving Average Smoother
- Dynamic Linear Models

7.7   Exercises

7.1 Evaluate trends and peaks in the data on COVID19 related mortality available in https://www.euromomo.eu/graphs-and-maps/. Evaluate the impact of the time window on the line chart pattern. Identify periods with changes in mortality and periods with stability in mortality.

7.2 The data set "SeasCom" provides the monthly demand for a seasonal commodity during 102 months.
   (i) Plot the data to see the general growth of the demand ;
   (ii) Fit to the data the trend function
$$f(t) = \beta_1 + \beta_2((t-51)/102) + \beta_3 \cos(\pi t/6) + \beta_4 \sin(\pi t/6);$$
   (iii) Plot the deviations of the data from the fitted trend, i.e., $\hat{U}_t = X_t - \hat{f}(t)$;
   (iv) Compute the correlations between $(\hat{U}_t, \hat{U}_{t+1})$ and $(\hat{U}_t, \hat{U}_{t+2})$;
   (v) What can you infer from these results?

7.3 Write the formula for the lag-correlation $\rho(h)$ for the case of stationary MA(q).

7.4 For a stationary MA(5), with coefficients $\beta' = (1, 1.05, 0.76, -0.35, 0.45, 55)$ make a table of the covariances, $K(h)$ and lag correlations $\rho(h)$.

7.5 Consider the infinite moving average $X_t = \sum_{j=0}^{\infty} q^j e_{t-j}$, where $e_t \sim \text{WN}(0, \sigma^2)$, . where $0 < q < 1$. Compute
   (i) $E\{X_t\}$,
   (ii) $V\{X_t\}$.

7.6 Consider the AR(1) given by $X_t = 0.75X_{t-1} + e_t$, where $e_t \sim \text{WN}(0, \sigma^2)$, and $\sigma^2 = 5$. Answer the following,
   (i)   Is this sequence covariance stationary?
   (ii) Find $E\{X_t\}$,
   (iii) Determine $K(0)$,
   (iv) Determine $K(1)$.

7.7 Consider the auto-regressive series AR(2) namely,
$X_t = 0.5X_{t-1} - 0.3X_{t-2} + e_t$, where $e_t \sim \text{WN}(0, \sigma^2)$.
   1. Is this series covariance stationary?
   2. Express this AR(2) in the form $(1 - \phi_1 z^{-1})(1 - \phi_2 z^{-1})X_t = e_t$, and find the values of ϕ_1 and ϕ_2.
   3. Write this AR(2) as an MA(∞). (Hint: write $(1 - \phi z^{-1})^{-1} = \sum_{j=0}^{\infty} \phi^j z^{-j}$)

7.8 Consider the AR(3) given by $X_t - 0.5X_{t-1} + 0.3X_{t-2} - 0.2X_{t-3} = e_t$, where $e_t \sim \text{WN}(0, 1)$.
   Use the Yule–Walker equations to determine $K(h)$, $|h| = 0, 1, 2, 3$.

7.9 Write the Toeplitz matrix R_4 corresponding to the series in EX. 7.7.

7.10 Consider the series $X_t - X_{t-1} + 0.25X_{t-2} = e_t + .4e_{t-1} - .45e_{t-2}$,

  (i) Is this an ARMA(2,2) series?

  (ii) Write the process as an MA(∞) series.

7.11 Consider the second-order difference, $\Delta^2 X_t$ of the DOW1941 series.

  (i) Plot the acf and the pacf of these differences.

  (ii) Can we infer that the DOW1941 series is an integrated ARIMA(9,2,2) ?

7.12 Consider again the data set SeasCom and the trend function $f(t)$, which was determined in Exercise 7.1. Apply the R function "pred1" to the deviations of the data from its trend function. Add the results to the trend function to obtain a one-day ahead prediction of the demand.

Appendix

The following functions are required for implementing the code listed below:

```
1.  std<-function(x){
        sqrt(var(x)}
2.  inv<-function(x){
        solve(x)}
  3.    jay<-function(k,l){
                n<-k*l
                a<-c(1:n)/c(1:n)
                res<-matrix(a,nrow=k)
                res}
```

R code for generating figures

Figure 7.1

The algorithm is based on two steps.

```
Step 1:
Create a file "DOW1941" consisting of n=302 values.
Step 2:
Create a file "Trend1941" based on the function
TR(t)<-function(t){
123.34+27.73((t-151)/302)-15.83((t-151)/302)2-237.00((t-151)/302)3.
+0.1512cos(4πt/302) + 1.738sin(4πt/302) + 1.770cos(8πt/302)
- 0.208sin(8πt/302) -0.729cos(12πt/302) + 0.748sin(12πt/302).
    }
        Trend1941<-c(1:302)
            for(i in 1:302){
                Trend1941[i]<-TR(i)
                }
Step 3:
            Set
            Tx<-c(1:302)
                Plot( Tx,DOW1941,xlab="day",ylab="DOW1941")_
                    Lines(tx,Trend1941)
```

Table 7.1

Use the R function

$$Acf(DOW1941,20)$$

Figure 7.8

The R program pred1 below is applied on the deviations of the points of DOW1941 around their trend. Therefore, we create the file AR1941<-DOW1941-Trend1941. Furthermore, the function pred1 uses the function Toeplitz, which depends on a sequence of lag-correlations

```
Toeplitz<-function(a){
        n<-length(a)
        res<-jay(n,n)
        res[1,]<-a
        for(i in 2:n){
                res[i,(i:n)]<-a[1:(n-i+1)]
                        for(j in i:(i-i)){
                                res[i,j]<-res[j,i]}
        }
          res
        }
```

```
pred1<-function(x,n){
        ns<-length(x)
        res<-c(1:ns)
        res[1:n]<-x[1:n]
             for(i in (n+1):ns){
                   xs<-x[1:i]
                   ak<-acf(xs,10,plot=F)
             a<-ak$acf[1:9]
             R<-toeplitz(a)
             r<-ak$acf[2:10]
             b<-inv(R)%*%r
             res[i]<-sum(x[(i-1):(i-9)]*b)
             }
           res
           }
```

The dots in Figure 7.8 are the DOW1941 values, the prediction line among the points is the sum of the trend and the output of the pred1 function.

Figure 7.9

Figure 7.9 is produced by the following function

```
pred.poly<-function(X,n,s){
        ns<-length(X)
        res<-c(1:ns)
        res[1:n]<-X[1:n]
            for(i in n:(ns-s)){
                   xs<-X[i:(i-n+1)]
                   nx<-c(0:-(n-1))
                   A<-cbind(nx^0,nx,nx^2)
                   b<-inv(t(A)%*%A)%*%t(A)%*%xs
                   ts<-c(1,s,s^2)
                   pred<-b[1,1]+s*b[2,1]+(s^2)*b[3,1]
```

```
                                   res[i+s]<-pred
                                }
                                res
                                }
```

Figure 7.10

Figure 7.10 is based on the function

```
mas.pred1<-function(x,m,s){
            n<-length(x)
            k<-2*m+s
            res<-c(1:n)
            res[1:k]<-x[1:k]
            for(i in (m+1):(n-m-s)){
                        xtm<-x[(i-m):(i+m)]
                b0m<-mean(xtm)
                am<-c(1:m)
                amm<-c(-m:-1)
                wm<-c(amm,0,am)
                b1m<-(3*sum(wm*xtm))/(m*(m+1)*(2*m+1))
                res[i+m+s]<-b0m+(m+s)*b1m
                }
            Res
            }
```

Figure 7.11

Figure 7.11 is based on the following function

```
nrwm<-function(n,v,w,c){
            s1<-sqrt(v)
            s2<-sqrt(w)
            X<-c(1:n)
            Tet<-c(1:n)
            cf<-c(1:n)
            Mt<-c(1:n)
            cf[1]<-(c+w)*v/(c+w+v)
            Tet[1]<-s2*rnorm(1)
            X[1]<-Tet[1]+s1*rnorm(1)
            Mt[1]<-X[1]
                for(i in 2:n){
                        Tet[i]<-Tet[i-1]+s2*rnorm(1)
                        X[i]<-Tet[i]+s1*rnorm(1)
                        cf[i]<-(cf[i-1]+w)*v/(cf[i-1]+w+v)
```

```
          at<-(cf[i]+w)/(cf[i]+w+v)
          Mt[i]<-(1-at)*Mt[i-1]+at*X[i]
       }
        res<-list(X,Mt)
        res
     }
```

Figure 7.12

Figure 7.12 is based on the function

```
            dlm.lg<-function(X,C0,v,W,M0){
                 n<-length(X)
                 pred<-c(1:n)
                 for(i in 1:n){
                    at<-jay(2,1)
                    at[2,1]<-i
                    pred[i]<-t(at)%*%M0
                    ut<-v+t(at)%*%(C0+W)%*%at
                    ei<-(X[i]-t(at)%*%M0)
                    ev<-jay(2,1)
                    ev[1,1]<-ei/ut
                  ev[2,1]<-(i*ei)/ut
                  M0<-M0+(C0+W)%*%ev
                  B<-jay(2,2)
                  B[1,1]<-1/ut
                  B[1,2]<-i/ut
                  B[2,1]<-i/ut
                  B[2,2]<-(i^2)/ut
                  C0<-(C0+W)-(C0+W)%*%B%*%(C0+W)
                    }
                pred
            }
```

Figure 7.13

Figure 7.13 is composed of two sequences. The first one is the random realization of the ARMA(3,2) series, and the second one is the prediction line. The function for the first component is

```
         ARMA<-function(n,a,b){
             p<-length(a)
             q<-length(b)
             ns<-n+max(p,q)
```

```
              X<-c(1:ns)
              E<-c(1:ns)
              pm<-max(p,q)
               X[1]<-rnorm(1)
                  for(i in 2:pm){
                      X[i]<-X[i-1]+rnorm(1)}
                          for(i in 1:ns){
                           E[i]<-rnorm(1)
                            }
                          for(i in (pm+1):ns){
                              for(j in 1:p){
                              ai<-a[j]*X[i-j]
                               }
                            for(j in 1:q){
                           ei<-b[j]*E[i-j]
                            }
                          X[i]<-ai+ei+E[i]
                        }
                      out<-X
                      out
        }
```

The prediction line is obtained from the function

```
        ARM.pred<-function(X,a){
                n<-length(X)
                pred<-c(1:n)
                pred[1:3]<-X[1:3]
                    for(i in 4:n){
                    pred[i]<-a[1]*X[i-1]+a[2]*X[i-2]+a[3]*X[i-3]
                    }
                out<-pred
                out
                }
```

8

Modern Analytic Methods

8.1 Introduction to computer age statistics

Big data and data science applications have been facilitated by hardware developments in computer science. As data storage began to increase, more advanced software was required to process it. This led to the development of cloud computing and distributed computing. Parallel machine processing was enhanced by the development of Hadoop, based off the shelf Google File System (GFS), and Google MapReduce, for performing distributed computing, Facebook (Friedman 2016).

New analytic methods were developed to handle very large data sets that are being processed through distributed computing. These methods are typically referred to as machine learning, statistical learning, data mining, big data analytics, data science or AI (artificial intelligence). Breiman (2001) noted that models used in big data analytics are developed with a different purpose than traditional statistical models. Computer age models do not assume a probability-based structure for the data, such as: $y = X\beta + \epsilon$, where $\epsilon \sim \text{NID}(0, \sigma^2)$. In general, they make no assumptions as to a "true" model producing the data. The advantage of these computer age methods is that no assumptions are being made about model form or error so that standard goodness of fit assessment is not necessary. However, there are still assumptions being made, and overfitting is assessed with hold out sets. Conditions of nonstationarity and strong data stratification of the data pose complex challenges in assessing predictive capabilities of such models (Efron and Hastie 2016).

Without making any assumptions about the "true" form of the relationship between the x's and the y, there is no need to estimate population parameters. Rather, the emphasis of predictive analytics, and its ultimate measure of success, is prediction accuracy. This is computed by first fitting a training set and then calculating measures such as root mean square error or mean absolute deviation. The next step is moving on to such computations on hold-out data sets, on out-of-sample data or with cross- validation. Ultimately, the prediction error is assessed on new data collected under new circumstances. In contrast to classical statistical methods, this approach leads to a totally different mindset in developing models. Traditional statistical research is focused on understanding the process that generated the observed data and modeling

---

Modern Industrial Statistics: With Applications in R, MINITAB and JMP, Third Edition.
Ron S. Kenett and Shelemyahu Zacks.
© 2021 John Wiley & Sons, Ltd. Published 2021 by John Wiley & Sons, Ltd.

the process using the data using methods such as least squares or maximum likelihood. Computer age statistics is modeling the data per se and focuses on the algorithmic properties of the proposed methods. This chapter is about such algorithms. We begin with decision trees.

8.2   Decision trees

Partition models, also called decision trees, are nonparametric tools used in supervised learning in the context of classification and regression. In supervised learning, you observe multiple covariate and one or more target variables. The goal is to predict or classify the target using the values of covariates. Decision trees are based on splits in covariates or predictors that create separate but homogeneous groups. Splits are not sensitive to outliers but are based on a "greedy" one step look ahead, without accounting for overall performance. Breiman et al. (1984) implement a decision tree procedure called Classification and Regression Trees (CART). Other procedures are C4.5 and Chi-square Automatic Interaction Detector (CHAID). Trees can handle missing data without the need to perform imputation. Moreover, they produce rules on the predators that can be effectively communicated and implemented. Single trees, sometimes called exploratory decision trees, are however poor predictors. This can be improved with random forests, bootstrap forests, and boosted trees that we discuss later. To evaluate the performance of a decision tree, it is important to understand the notion of class confusion and the confusion matrix. A confusion matrix for a target variable we want to predict involving n classes is an $n \times n$ matrix with the columns labeled with actual classes and the rows labeled with predicted classes. Each data point in a training or validation set has an actual class label as well as the class predicted by the decision tree (the predicted class). This combination determines the confusion matrix. For example, consider a two-class problem with a target response being "Pass" or "Fail." This will produce a 2×2 confusion matrix, with actual Pass predicted as Pass and actual Fail predicted as Fail, these are on the diagonal. The lower left and top right values are actual Pass predicted as Fail and actual Fail predicted as Pass. These off diagonal values correspond to misclassifications. When one class is rare, for example, in large data sets with most data points being "Pass" and a relatively small number of "Fail." Because the "Fail" class is rare among the general population, the distribution of the target variable is highly unbalanced. As the target distribution becomes more skewed, evaluation based on misclassification (off diagonal in the confusion matrix) breaks down. For example, consider a domain where the unusual class appears in a 0.01% of the cases. A simple rule that would work is to always choose the most prevalent class, this gives 99.9% accuracy but is useless. In quality control, it would never detect a Fail item or, in fraud detection, it would never detect rare cases of fraud, and misclassification can be greatly misleading. With this background, let us see a decision tree in action.

Example 8.1.   Data set **SENSORS.csv** consists of 174 measurements from 63 sensors tracking performance of a system under test. Each test generates values for these 63 sensors and a status determined by the automatic test equipment. The distribution of the test results is presented in Figure 8.1. Our goal is to predict the outcome recorded by the testing equipment, using sensor data. The test results are coded as Pass (corresponding to "Good," 47% of the observations) and Fail (all other categories, marked in black). The CART based decision tree of the Pass–Fail target variable using the 67 sensors is presented in Figure 8.2. Our goal is to predict the outcome recorded by the testing equipment, using sensor data.

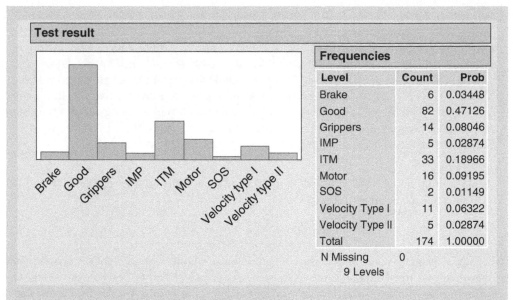

Figure 8.1   *Distribution of test results (JMP).*

The test results are coded as Pass (corresponding to "Good," 47% of the observations) and Fail (all other categories, marked in black). The column **Status** is therefore a dichotomized version of the column **Test result**. The CART-based decision tree of the Pass–Fail target variable using the 67 sensors is presented in Figure 8.2. To conduct this analysis in R use package `rpart`, in JMP go to Analyze > Predictive modeling > Partition and in MINITAB, the feature is available through Stat > Predictive Analytics > Cart Classification. In JMP,

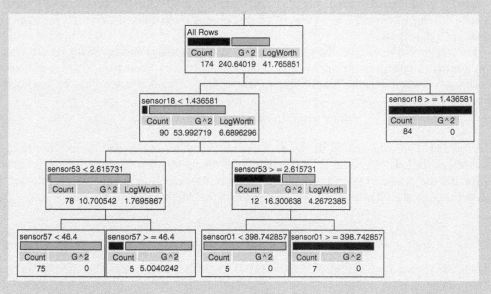

Figure 8.2   *Decision tree of sensor data with four splits (JMP).*

the method is listed by default as **Decision Tree** which is what we now use. The first split is on sensor 34 with cut-off point 1.44. All 84 observations with senor 34 > 1.44 are classified as Fail. Most observations with sensor 34 < 1.44 are classified as Pass (82 out of 90). In considering 78 observations with sensor 69 less than 2.61, we find one Fail and 77 recorded as Pass. The next split is with sensor 73 and threshold 46.4. The block below the threshold consists of 73 Pass tests. With three sensors (out of 67), we have an effective way to classify tests as Pass–Fail. These splits are based on vertical splits by determining cut off values for the predictor variables, the 67 sensors. We present next how such splits are determined in the **SENSORS.csv** data and how the performance of a decision tree is evaluated.

Node splitting in the JMP Partition platform that builds decision trees is based on the Log-Worth statistic calculated as follows:

$$\text{LogWorth} = -\log 10(p - \text{value}), \tag{8.1}$$

where the adjusted p-value takes into account the number of different ways splits can occur. This calculation is an improvement on the unadjusted p-value, which favors Xs with many levels, and the Bonferroni adjusted p-value, which favors Xs with small numbers of levels. Details about the method are discussed in Sall (2002). The LogWorth is a measure of "surprise" representing splits that generate blocks with differences that cannot be explained by random assignments. Such splits are determined as significant.

For continuous responses, the Sum of Squares (SS) is reported in node reports. This is the change in the error SS due to the split. A candidate SS used in JMP is

$$\text{SStest} = \text{SSparent} - (\text{SSright} + \text{SSleft}), \tag{8.2}$$

where SS in a node is just $s2(n - 1)$. Also reported in the JMP outputs for continuous responses is the Difference statistic. This is the difference between the predicted values for the two child nodes of a parent node.

For categorical responses, the G2 (likelihood ratio chi-square) appears in the report. This is twice the (natural log) entropy or twice the change in the entropy. Entropy is $\sum -p\log(p)$ for each observation, where p is the probability of that response. A split is determined by considering candidate G2 with

$$\text{G2test} = \text{G2parent} - (\text{G2left} + \text{G2right}) \tag{8.3}$$

The predicted probabilities for the decision tree are calculated as described below. For categorical responses, one can compute the proportion of observations at the node for each response level and the predicted probability for that node of the tree. The method for calculating the probability for the ith response level at a given node in JMP is calculated as follows:

$$\text{Prob}_i = \frac{n_i + \text{prior}_i}{\sum(n_i + \text{prior}_i)}, \tag{8.4}$$

where the summation is across all response levels; n_i is the number of observations at the node for the ith response level; and *priori* is the prior probability for the ith response level, calculated as follows:

$$\text{prior}_i = \lambda p_i + (1 - \lambda P_i), \tag{8.5}$$

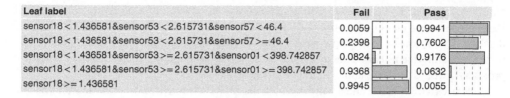

Leaf label	Fail		Pass	
sensor18 < 1.436581&sensor53 < 2.615731&sensor57 < 46.4	0.0059		0.9941	
sensor18 < 1.436581&sensor53 < 2.615731&sensor57 >= 46.4	0.2398		0.7602	
sensor18 < 1.436581&sensor53 >= 2.615731&sensor01 < 398.742857	0.0824		0.9176	
sensor18 < 1.436581&sensor53 >= 2.615731&sensor01 >= 398.742857	0.9368		0.0632	
sensor18 >= 1.436581	0.9945		0.0055	

Figure 8.3   *Classification probabilities for the five leaves in the tree of Figure 8.2 (JMP).*

where p_i is the prior$_i$ from the parent node, P_i is *Prob$_i$* of the parent node, and λ is a weighting factor set by default at 0.9. The method for calculating Prob$_i$ assures that the predicted probabilities are always nonzero. Figure 8.3 presents the probabilities for the five leaves in the tree of Figure 8.2

Assessing the performance of a decision tree is based on an evaluation of its predictive ability. The observations in each leaf are classified, as a group, according to the leaf probability and a cut of threshold. The default cutoff is typically 50% implying that all types of misclassification carry the same cost. The observations in the three leaves described by the top three rules in Figure 8.3 are all classified as Pass. The other two leaves are labeled as Fail. These leaves are determined by values in three sensors: 34, 69, 73, and 1.

Based on these probabilities, a classification of the observations in the leaves is conducted using the recorded values and the predicted values. In the case of the Pass–Fail data of sensors data, with four slits, this generates a 2×2 confusion matrix displayed in Figure 8.4. Only one observation is misclassified. It corresponds to observation 100 which was a Fail in the test equipment and was classified as Pass by the decision tree. The probability of Pass being 76%, the lowest probability in the data set. To prevent this from happening, we can change the 50%

Fit Details

Measure	Training	Definition
Entropy RSquare	0.9642	1-Loglike(model)/Loglike(0)
Generalized RSquare	0.9830	(1-(L(0)/L(model))^(2/n))/(1-L(0)^(2/n))
Mean-Log p	0.0248	Σ-Log(ρ[j])/n
RASE	0.0709	$\sqrt{\Sigma(y[j]-\rho[j])^2/n}$
Mean Abs Dev	0.0200	Σ |y[j]-ρ[j]|/n
Misclassification Rate	0.0057	Σ (ρ[j]$\neq\rho$Max)/n
N	174	n

Confusion Matrix

Training

Actual status	Predicted Count	
	Fail	Pass
Fail	91	1
Pass	0	82

Figure 8.4   *Fit details and confusion matrix for the decision tree in Figure 8.2 (JMP).*

cutoff to, say 80%. By this, we would classify as Pass-only cases with a high probability of being Pass. Under these conditions, observation 100 would be classified as Fail, as it appears in the data set. The downside is that now we get four observations recorded as Pass that get classified by the decision tree as Fail. This might not be as severe as shipping to the customer a defective product, it would however add cost to our testing effort.

If one uses a training set and a validation set, the JMP Partition has two misclassification rates: one for the training set as count to total, and another for the validation set that is biased away from zero by not having attributed probabilities of zero. With this approach, logs of probabilities can be calculated and used in calculating **Entropy R-Squar**e.

The decision tree analysis can be conducted on the original data set with a target consisting of nine values. When the target is a continuous variable, the same approach produces a regression tree where leaves are characterized not by counts, but by average and standard deviation values. We do not expand here on such cases.

Two main properties of decision trees are

1. Decision trees use decision boundaries that are perpendicular to the data set space axes. This is a direct consequence of the fact that trees select a single attribute at a time.
2. Decision trees are "piecewise" classifiers that segment the data set space recursively using a divide-and-conquer approach. In principle, a classification tree can cut up the data set space arbitrarily finely into very small regions.

It is difficult to determine, in advance, if these properties are a good match to a given dataset. A decision tree is understandable to someone without a statistics or mathematics background. If the dataset does not have two instances with exactly the same covariates, but different target values, and we continue to split the data we are left with a single observation at each leaf node. This essentially corresponds to a lookup table. The accuracy of this tree is perfect, predicting correctly the class for every training instance. However, such a tree does not generalize and will not work as well on a validation set or new data. When providing a look up table, unseen instances do not get classified. On the other hand, a decision tree will give a nontrivial classification even for data not seen before.

Tree-structured models are very flexible in what they can represent and, if allowed to grow without bound, can fit up to an arbitrary precision. But the trees may need to include a large number of splits in order to do so. The complexity of the tree lies in the number of splits. Using a training set and a validation set, we can balance accuracy and complexity.

Some strategies for obtaining a proper balance are (1) Stop growing the tree before it gets too complex, and (2) Grow the tree until it is too large, then "prune" it back, reducing its size (and thereby its complexity). There are various methods for accomplishing both. The simplest method to limit tree size is to specify a minimum number of observations that must be present in a leaf. The data at the leaf is used to derive statistical estimates of the value of the target variable for future cases that would fall to that leaf. If we make predictions of the target based on a very small subset of data, they can be inaccurate. A further option is to derive a training set, a validation set and a testing set. The training set is used to build a tree, the validation set is used to prune the tree to get a balance between accuracy and complexity, and the testing set is used to evaluate the performance of the tree with fresh data.

Following this overview of decision trees, we move on to present a variation based on computer-intensive methods that enhances the stability of the decision tree predictions. A well-known approach to reduce variability in predictions is to generate several predictions and compute a prediction based on majority votes or averages. This "ensemble" method will work best if the combined estimates are independent. In this case, the ensemble-based estimate will have a smaller variability by a factor of square root of n, the number of combined estimates.

Bootstrap Forests and Boosted Trees are ensembles derived from several trees fit to versions of the original data set. Both are available as alternatives to the Decision Tree method in JMP Pro versions and in R using packages `randomForest`, `C50` and `xgboost`. In these algorithms, we fit a number of trees, say 100 trees, to the data, and use "majority vote" among the 100 trees to classify a new observation (Breiman 2001). For prediction, we average the 100 trees. An ensemble of 100 trees makes up a "forest." However, fitting 100 trees to the same data produces redundantly the same tree, 100 times. To address this redundancy, we randomly pick subsets of the data. Technically, we use bootstrapping to create a new data set for each of the 100 trees by sampling the original data, with replacement (see Section 4.10). This creates new data sets that are based on the original data but are not identical to it. These 100 alternative data sets produce 100 different trees. The integration of multiple models, created through bootstrapping, is known as "bootstrapped aggregation," or "bagging." This is the approach in random forests. In Bootstrap Forests, besides picking the data at random, also the set of predictors used in the tree are picked at random. When determining the variable upon which to split, Bootstrap Forests consider a randomly selected subset of the original independent variables. Typically, the subset is of size around the square root of the number of predictors. This gets the algorithm to consider variables not considered in a standard decision tree. Looking at the ensemble of 100 trees produces more robust and unbiased predictions. Reporting the number of bootstrap forest splits, on each predictor variable, provides useful information on the relative importance of the predictor variables.

Chapter 15 by Galimberti and Soffritti in Kenett and Salini (2011) presents an application of decision trees to the analysis of customer surveys. The chapter invokes R code from the `party` and `rpart` applications available from CRAN. The `rpart` package implements the CART methodology for unordered categorical and numerical dependent variables. The code and the data used in Kenett and Salini (2011) are available for download from the book's website (https://www.wiley.com/en-us/Modern+Analysis+of+Customer+Surveys%3A+with+Applications+using+R-p-9781119961383).

The next section is about a competing method to decision trees, the Naive Bayes Classifier.

8.3 Naïve Bayes classifier

The basic idea of the Naïve Bayes classifier is a simple algorithm. For a given new record to be classified, $x_1, x_2, \ldots, x_n$, find other records like it (i.e. y with same values for the predictors $x_1, x_2, \ldots, x_n$). Following that, identify the prevalent class among those records (the ys) and assign that class to the new record. This is applied to categorical variables and continuous variables must be discretized, binned, and converted to categorical variables. The approach can be used efficiently with very large data sets and relies on finding other records that share same predictor values as the record-to-be-classified. We want to find the "probability of y belonging to class A, given specified values of predictors, $x_1, x_2, \ldots, x_n$." However, even with large data sets it may be hard to find other records that exactly match the record to be classified, in terms of predictor values. The Naive Bayes classifier algorithm assumes independence of predictor variables (within each class) and using the multiplication rule computes the probability that the record to be classified belongs to class A, given predictor values $x_1, x_2, \ldots x_n$, without limiting calculation only to records that exactly share these same values. From Bayes theorem (Chapter 3), we know that

$$P(y|x_1, \ldots, x_n) = \frac{P(y)P(x_1, \ldots, x_n|y)}{P(x_1, \ldots, x_n)}, \tag{8.6}$$

where y is the value to be classified and $x_1, \ldots, x_n$ are the predictors.

In the Naive Bayes classifier, we move from conditioning the predictors $x_1, x_2, \ldots, x_n$ on the target y, to conditioning the target y on the predictors $x_1, x_2, \ldots, x_n$.

To calculate the expression on the right, we rely on the marginal distribution of the predictors and assume their independence, hence (8.3.2)

$$P(y|x_1, \ldots, x_n) = \frac{P(y) \prod P(x_i|y)}{P(x_1, \ldots, x_n)} \tag{8.7}$$

This is the basis of the Naive Bayes classifier. It classifies a new observation by estimating the probability that the observation belongs to each class and reports the class with highest probability. The Naive Bayes classifier is very efficient in terms of storage space and computation time. Training consists only of storing counts of classes and feature occurrences, as each observation is recorded. However, in spite of its simplicity and the independence assumption, the Naive Bayes classifier performs surprisingly well. This is because the violation of the independence assumption tends not to hurt classification performance. Consider the following intuitive reasoning. Assume that two observations are strongly dependent so that when one sees one, we are also likely to see the other. If we treat them as independent, observing one enhances the evidence for the observed class and seeing the other also enhances the evidence for its class. To some extent, this double-counts the evidence. As long as the evidence is pointing in the right direction, classification with this double-counting will not be harmful. In fact, the probability estimates are expanded in the correct direction. The class probabilities will be therefore overestimated for the correct class and underestimated for the incorrect classes. Since for classification, we pick the class with the highest estimated probability, making these probabilities more extreme in the correct direction is not a problem. It can, however, become a problem if we use the probability estimates themselves. Naive Bayes is therefore safely used for ranking where only the relative values in the different classes are relevant. Another advantage of the Naive Bayes classifier is an "incremental learner." An incremental learner is an induction technique that updates its model, one observation at a time, and does not require to reprocess all past training data when new training data become available. Incremental learning is especially advantageous in applications where training labels are revealed in the course of the application, and the classifier needs to reflect this new information as quickly as possible. The Naive Bayes classifier is included in nearly every machine-learning toolkit and serves as a common baseline classifier against which more sophisticated methods are compared.

Example 8.2. To demonstrate the application of a Naïve Bayes classifier, we invoke the results of a customer satisfaction survey, **ABC.csv**. The data consists of 266 responses to a questionnaire with a question on overall satisfaction (q1) and responses to 125 other questions. Figure 8.5 shows the distribution of q1 and five other questions.

We can see that the response "5" in q1 corresponds to top-level responses in q4, q5, q6, and q7. Based on such responses in q4–q7, we can therefor confidently predict a response "5" to q1.

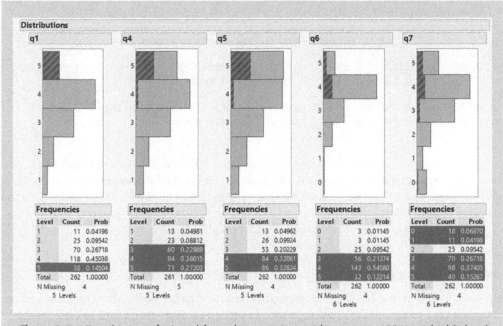

Figure 8.5 *Distribution of q1 and five other questions with response "5" in q1 highlighted (JMP).*

The Naive Bayes classifier can be applied to all 125 responses to the questionnaire. The outputs from this analysis are presented in Figure 8.6. We observe 53 misclassifications, mostly to respondents who answered "4" to q1.

Training		
Count	Misclassification Rate	Misclassifications
262	0.20229	53

Confusion Matrix

Training					
Actual	Predicated Count				
q1	1	2	3	4	5
1	11	0	0	0	0
2	0	23	2	0	0
3	0	2	60	8	0
4	0	2	17	79	20
5	0	0	0	2	36

Figure 8.6 *Confusion matrix from the Naïve Bayes classifier with q1 as target and 125 questions as predictors (JMP).*

Measures of Fit fir q1								
Creator	.2 .4 .6 .8	Entropy RSquare	Generalized RSquare	Mean-Log P	RMSE	Mean Abs Dev	Misclassification Rate	N
Naive Bayes		.	.	.	.	.	0.2023	262
Partition		0.6112	0.8661	0.5245	0.4231	0.3446	0.2443	262
Bootstrap Forest		0.6003	0.8600	0.5393	0.4211	0.4019	0.0534	262

Figure 8.7   *Comparison of performance of the Naive Bayes Classifier with q1 as target and 125 questions as predictors to the Decision Tree and Bootstrap Forrest (JMP).*

The Naive Bayes classifier's misclassification rate of 20% was obtained with an easy to compute and incremental learning algorithm. Its performance is actually better than the decision tree with 22 splits which is 24%. The Bootstrap Forrest was much better with a 5% misclassification rate (see Figure 8.7).

The next section is on methods where all variables play an equal role.

8.4   Clustering methods

Clustering methods are unsupervised methods where the data has no variable labeled as a target. Our goal is to group similar items together, in clusters. In Sections 8.2 and 8.3, we discussed supervised methods where one of the variables is labeled as a target response and the other variables as predictors that are used to predict the target. In clustering methods, all variables have an equal role. Hierarchical clustering is generated by starting with each node as its own cluster. Then, clusters are merged iteratively until only a single cluster remains. The clusters are merged in function of a distance function. The closest clusters are merged into a new cluster. The end result of an hierarchical clustering method is a dendrogram, where the j-cluster set is obtained by merging clusters from the $(j + 1)$ cluster set.

Example 8.3.   To demonstrate clustering methods, we use the **ALMPIN.csv** data set that consists of 6 measurements on 70 aluminum pins introduced in Chapter 5 as Example 5.3. To conduct this analysis in R use package `cluster`, in JMP we go to Analyze > Clustering > Hierarchical Cluster. In MINITAB, the feature is available through Stat > Multivariate > Cluster Observations. The default method in JMP is set as Ward with standardized data. Figure 8.8 is a dendrogram of the aluminum pin data created with JMP. It starts at the bottom with 70 clusters of individual observations and ends up, on the top as one cluster.

The diagram in Figure 8.8 can be cut across at any level to give any desired number of clusters. Moreover, once two clusters are joined, they remain joined in all higher levels of the hierarchy. The merging of clusters is based on computation of a distance between clusters with a merge on the closest one. There are several possible distance measures described next.

Ward's minimum variance method minimizes the total within-cluster variance. With this method, the distance between two clusters is the ANOVA sum of squares between the two clusters summed over all the variables. At each clustering step, the within-cluster sum of squares is minimized over all partitions obtainable by merging two clusters from the previous generation. Ward's method tends to join clusters with a small number of observations and is strongly biased toward producing clusters with approximately the same number of observations. It is also

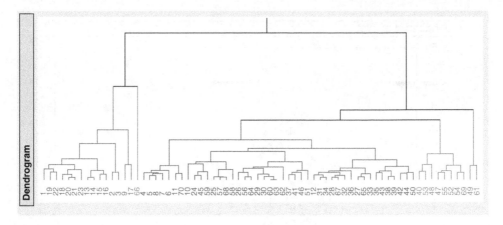

Figure 8.8   *Dendrogram of the six variables in the ALMPIN data with 70 observations (JMP).*

sensitive to outliers. The distance for Ward's method is

$$D_{KL} = \frac{\left\| \overline{x_K} - \overline{x_L} \right\|^2}{\frac{1}{N_K} + \frac{1}{N_L}} \tag{8.8}$$

Where
C_K is the Kth cluster, subset of $1, 2, \ldots, n$
N_K is the number of observations in C_K
$\overline{x_K}$ is the mean vector for cluster C_K
$\|x\|$ is the square root of the sum of the squares of the elements of x (the Euclidean length of the vector x)
$d(x_i, x_j) = \left\| x_i - x_j \right\|^2$ where x_i is the ith observation.
Other methods include single linkage, complete linkage, and average linkage.
Single Linkage. The distance for the single linkage cluster method is

$$D_{KL} = \min_{i \in C_K} \min_{j \in C_L} d(x_i, x_j) \tag{8.9}$$

Complete Linkage. The distance for the complete linkage cluster method is

$$D_{KL} = \max_{i \in C_K} \max_{j \in C_L} d(x_i, x_j) \tag{8.10}$$

Average Linkage. For average linkage, the distance between two clusters is found by computing the average dissimilarity of each item in the first cluster to each item in the second cluster. The distance for the average linkage cluster method is

$$D_{KL} = \sum_{i \in C_K} \sum_{j \in C_L} \frac{d(x_i, x_j)}{N_K N_L} \tag{8.11}$$

These distances perform differently on different clustering problems. The dendrograms from single-linkage and complete-linkage methods are invariant under monotone transformations of the pairwise distances. This does not hold for the average-linkage method. Single-linkage

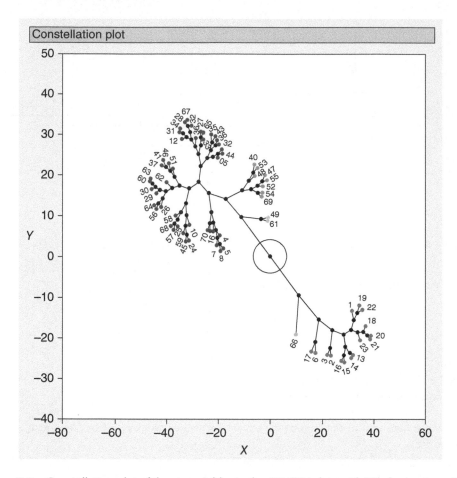

Figure 8.9   *Constellation plot of the six variables in the ALMPIN data with 70 observations (JMP).*

often leads to long "chains" of clusters, joined by individual points located near each other. Complete-linkage tends to produce many small, compact clusters. Average linkage is dependent upon the size of the clusters. Single and complete linkage depends only on the smallest or largest distance, respectively, and not on the size of the clusters.

In JMP, an alternative to the hierarchical clustering dendrogram. is the constellation plot (Figure 8.9). Each observation (row) is represented by an endpoint and each cluster join is represented by a new point. The lines that are drawn represent cluster membership. The lengths of the lines represent the distance between clusters. Longer lines represent greater distances between clusters. The length values are meaningful with respect to each other. The axis scaling, orientation of points, and angles of the lines are arbitrary. They are determined such that the ends of the nodes are spaced out and the plot does not appear cluttered.

Another clustering method is K-means. The K-means clustering is formed by an iterative fitting process. The K-means algorithm first selects a set of K points, called cluster seeds, as an initial set up for the means of the clusters. Each observation is assigned to the nearest cluster seed, to form a set of temporary clusters. The seeds are then replaced by the actual cluster means and the points are reassigned. The process continues until no further changes occur in the clusters.

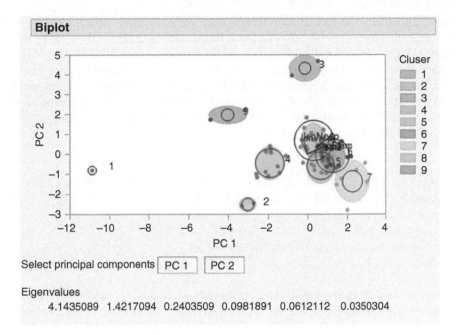

Figure 8.10 *K-Means clustering of the six variables in the ALMPIN data with 70 observations (JMP).*

The K-means algorithm is a special case of the EM algorithm, where E stands for Expectation, and M stands for maximization. In the case of the K-means algorithm, the calculation of temporary cluster means represents the Expectation step, and the assignment of points to the closest clusters represents the Maximization step. K-Means clustering supports only numeric columns. K-Means clustering ignores nominal and ordinal data characteristics and treats all variables as continuous.

In K-means, you must specify in advance the number of clusters, K. However, you can compare the results of different values of K in order to select an optimal number of clusters for your data. For background on K-means clustering, see Hastie et al. (2009).

Figure 8.10 is a graphical representation of a K-means analysis of the ALMPIN data. To derive this analysis in R, use package `cluster`. In JMP, we go to Analyze > Clustering > K Means Cluster, JMP gives you an option to apply, automatically, a range of values for K. In MINITAB, the feature is available through Stat > Multivariate > Cluster K-means. When you input a range of K, JMP indicates by CCC the optimal value of K. For the ALMPIN data, it is $K - 9$, which is what was used to generate Figure 8.10. Cluster 1 has 1 observation; clusters 2, 3; and 9 have 2 observations each. These clusters include unusual observations that can be characterized by further investigations.

8.5 Functional data analysis

When you collect data from tests or measurements over time or other dimensions, we might want to focus on the functional structure of the data. Examples can be chromatograms from high-performance liquid chromatography (HPLC) systems, dissolution profiles of drug tables over time, distribution of particle sizes or measurement of sensors. Functional data is different and individual measurements recorded at different sets of time points. It views functional observations as continuously defined so that an observation is the entire function. With functional data consisting of a set of curves representing repeated measurements, we characterize the main features of the data, for example with a functional version of principal component analysis (FPCA). The regular version of principal component analysis (PCA) is presented in detail in Chapter 12 on Multivariate Statistical Process Control. With this background, let us see an example of functional data analysis (FDA).

Example 8.4.   Data set **DISSOLUTION.csv** consists of 12 test and reference tablets measured under dissolution conditions at 5, 10, 15, 20, 30, and 45 seconds. The level of dissolution recorded at these time instances is the basis for the dissolution functions we will analyze. The test tablets behavior us compared to the reference paths. Ideally, the tested generic product is identical to the brand reference.

Functional data analysis extends the capabilities of traditional statistical techniques in a number of ways. Functional data analysis methods are considering change over time (or space, or some other dimension). Because we are observing curves rather than individual values, the vector-valued observations $X_1, \ldots, X_n$ are replaced by the univariate functions $X_1(t), \ldots, X_n(t)$, where t is a continuous index varying within a closed interval [0, T]. In functional PCA, each sample curve is considered to be an independent realization of a univariate stochastic process $X(t)$ with smooth mean function $EX(t) = \mu(t)$ and covariance function $\mathrm{cov}\{X(s), X(t)\} = \sigma(s, t)$.

In analyzing such data, JMP first smooths each individual sample curve (e.g. using spline methods or local-linear smoothers), and then applied functional PCA, assuming that the smooth curves are the completely observed curves. This gives a set of eigenvalues $\{< U + 03BB > j\}$ and (smooth) eigenfunctions $\{Vj(t)\}$ extracted from the sample covariance matrix of the smoothed data. The first and second estimated eigenfunctions are then examined to exhibit location of individual curve variation. Other approach to functional PCA have been proposed, including the use of roughness penalties and regularization, which optimize the selection of smoothing parameter and choice of the number of principal components simultaneously rather than separately in two stages. Figure 8.12–8.14 are reports from JMP when FDA is applied to the dissolution data displayed in Figure 8.11. To derive this analysis in JMP, we go to Analyze > Specialized Modeling > Functional Data Explorer. To conduct this analysis in R, use package `fdapace`.

Figure 8.12 is what you get by fitting splines to the dissolution data. Shown here is the set of reference tablets. One observes the unusual straight-line type path of T5R which was highlighted in Figure 8.11. In Figure 8.13, we display a scatterplot of the top two functional principal components of the reference paths. Here T5R clearly stands out. Could be that the dissolution at 30 seconds was misreported as too low. A double check of the record should help clarify this.

A graphical comparison of the mean functional data of the reference and test paths is shown in Figure 8.14. A formal analysis, comparing reference and test dissolution data, is presented in Chapter 12, Section 12.5.

The final topic, in this review chapter of computer age statistics, is text analytics.

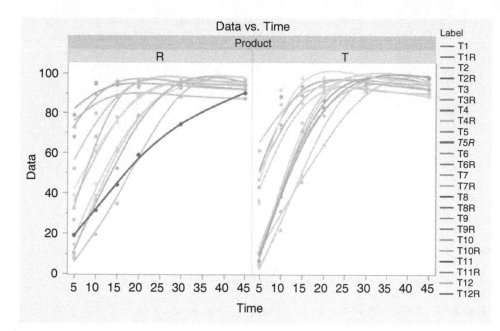

Figure 8.11 *Dissolution paths of reference and tested paths with smoother. T5R is highlighted (JMP).*

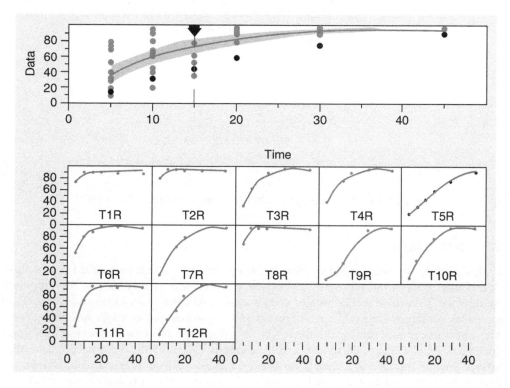

Figure 8.12 *Functional form of reference paths (JMP).*

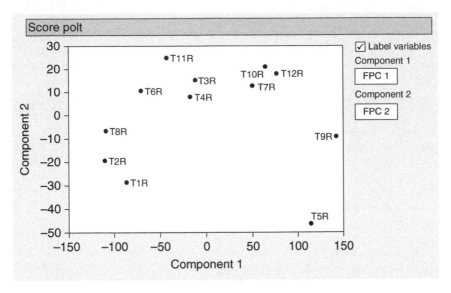

Figure 8.13   *Scatter plot of top two functional principal components of reference paths (JMP).*

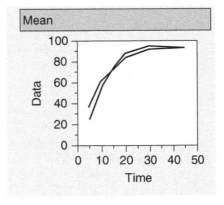

Figure 8.14   *Mean functional data of reference and test data (JMP).*

8.6   Text analytics

In this section, we discuss methods for analyzing text data, sometimes called unstructured data. Other types of unstructured data include voice recordings and images. The approach we describe is based on a collection of documents or text items that can consist of individual sentences, paragraphs, or a collection of paragraphs. As an example, consider Amazon reviews. You can consider each sentence in the review as a document or the whole review as a single document. A collection of document is called a corpus. We will look at the words, also called tokens, that are included in a document. The analysis we perform is based on a list of terms consisting of tokens included in each document. This approach to text analytics is called "bag of words" where every item has just a collection of individual words. It ignores grammar, word order, sentence structure, and punctuation. Although apparently simplistic, it performs surprisingly well. Before conducting the bag of words analysis, some text preparation is required. This involves tokenizing, phrasing, and terming. The tokenizing stage converts text to lowercase, applies tokenizing method to group characters into tokens and recodes tokens based on specified recode definitions. For example, to identify dates or other standard formats we use regular expressions (Regex). Noninformative terms are labeled as stop words and omitted. The phrasing stage collects phrases that occur in the corpus and enables you to specify that individual phrases be treated as terms. The terming stage creates the term list from the tokens and phrases that result from the previous tokenizing and phrasing. For each token, the terming stage checks that minimum and maximum length requirements are met. Tokens that contain only numbers are excluded from this operation. It also checks that a token is qualified and contains at least one alphabetical character. Stemming removes differences such as singular or plural endings. For each phrase added, the terming stage adds the phrase to the term list.

Term "frequency" shows how frequent a term is in a single document. We also look at how common the term is in the entire corpus. Text processing imposes a small lower limit on the number of items in which a term must occur so that rare terms are ignored. However, terms should also not be too common. A term occurring in every document carries no information. Overly common terms are therefore also eliminated with an upper limit on the number of documents in which a word appears.

In addition to imposing upper and lower limits on term frequency, many systems take into account the distribution of the term over items in a corpus. The fewer documents in which a term occurs, the more significant it is in the documents where it does occur. This sparseness of a term t is measured commonly by an equation called inverse document frequency (IDF). For a given document, d, and term, t, the term "frequency" is the number of times term t appears in document d: $TF(d; t) = \#$ times term t appears in document d. To account for terms that appear frequently in the domain of interest, we compute the Inverse Document Frequency of term t, calculated over the entire corpus and defined as

$$\text{IDF}(t) = \log \frac{\text{total number of documents}}{\# \text{ documents containing term} t} \qquad (8.12)$$

To demonstrate the data preparation phase of text documents, we use text describing aircraft accidents listed in the National Transportation Board data base: https://www.ntsb.gov/_layouts/ntsb.aviation/Index.aspx.

> **Example 8.5.** The Aircraft Incidents.txt data was downloaded from http://app.ntsb.gov/aviationquery/Download.ashx?type=csv. A JMP file with this data is available from the Help>Sample Data Library. It consists of 1,906 incidents in the USA. We will analyze the incident description text.

The data preparation resulted in 8787 distinct terms, a total of 313,244 terms, an average of 164.36 terms per document. This is translated to a document term matrix (DTM) with 1906 rows, one for each document and 8787 columns, one for each term (see Figure 8.15). This matrix can be binary, with entries of 1 or 0, depending on the occurrence of a term in a document, or not, respectively. It can also include the IDF score and give differential weight of terms depending on their prevalence in the corpus. This huge DTM matrix is very sparse. To conduct its analysis, we employ a basic dimension reduction procedure called partial singular value decomposition (SVD).

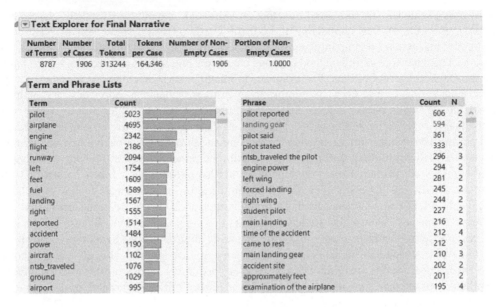

Figure 8.15   *Term and phrase list of Aircraft Incident data following data preparation (JMP).*

Partial SVD approximates the DTM using three matrices: U, S, and V'. The relationship between these matrices is defined as follows:

$$\text{DTM} \approx U * S * V \tag{8.13}$$

If k is the number of documents (rows) in the DTM, l is the number of terms (columns) in the DTM and n as a specified number of singular vectors. To achieve data reduction, n must be less than or equal to min (k, l). It follows that U is an kxn matrix that contains the left singular vectors of the DTM. S is a diagonal matrix of dimension n. The diagonal entries in S are the singular values of the DTM, V' is an n by l matrix. The rows in V' (or columns in V) are the right singular vectors.

The right singular vectors capture connections among different terms with similar meanings or topic areas. If three terms tend to appear in the same documents, the SVD is likely to produce a singular vector in V' with large values for those three terms. The U singular vectors represent the documents projected into this new term space.

Principal components, mentioned in Section 8.5, is a linear combinations of variables and a subset of them can replace the original variables. An analogous dimension reduction method can be applied to text data and is called latent semantic by applying partial SVD of the DTM. This decomposition reduces the text data into a manageable number of dimensions for analysis. For example, we can now perform topic analysis. The Rotated SVD option performs a rotation on the partial SVD of the DTM. In JMP, you must specify the number of rotated singular vectors, which corresponds to the number of topics that you want to retain from the DTM. After you specify a number of topics, the Topic Analysis report appears. Figure 8.16 is the JMP outputs for 10 topics.

Topic analysis is equivalent to a rotated PCA. The rotation takes a set of singular vectors and rotates them to make them point more directly in the coordinate directions. This rotation makes

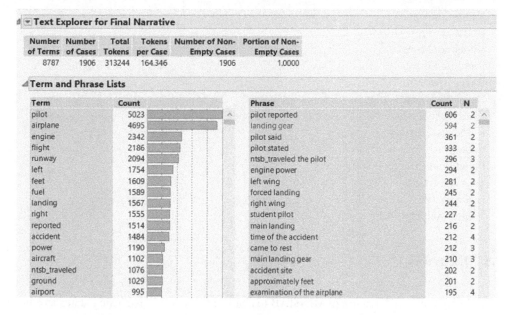

Figure 8.16   *Topic analysis of the Aircraft Incident data (JMP).*

the vectors help explain the text as each rotated vector orients toward a set of terms. Negative values indicate a repulsion force. The terms with negative values occur in a topic less frequently compared to the terms with positive values.

Looking at Figure 8.16, we identify in Topic 1 incidents related to weather conditions. Topic 2 is about failures and fatigue, Topic 3 is related to fuel and engine issues. Topic 4 is mentioning collision, Topic 5 seems related to aircraft operation, etc.

If we now link the documents to supplementary data such as incident impact, one can link label reports by topic and derive a predictive model that can drive accident prevention initiatives.

8.7   Chapter highlights

This chapter is a door opener to computer age statistics. It concludes Part A which presents the fundamental concepts and methods used in statistical and analytics. Besides basic descriptive methods, probability models, and statistical inference, Part A covers modern methods such as bootstrapping, Bayesian inference, multivariate methods, small sample estimation, and time series forecasting. Chapter 8 is designed as a review and includes tip of the iceberg examples with what we thought were interesting insights, not always available in standard texts. The chapter covers supervised and unsupervised learning methods, functional data analysis, and text analytics. The reader will be able to follow the examples with the data sets available on the book's website. More in depth study of these methods will require access to specialized books listed as references

The topics covered in the chapter include the following:

- Decision Trees
- Confusion Matrix

- Validation Data
- Boosted Tree
- Bootstrap Forest
- Random Forest
- Bayes Theorem
- Naïve Bayes Classifier
- Cluster Analysis
- *K*-Means Clusters
- Functional Data Analysis
- Functional Principal Components
- Text Analytics
- Bag of Words
- Topic Analysis

8.8   Exercises

8.1   Apply the CART feature in MINITAB to analyze the sensors data using status as a target variable. Compare the results to the outputs from JMP in Figure 8.1.

8.2   Apply the JMP Pro Boosted tree to the sensors data using status as a target variable. Compare the derived variable importance to the decision tree partition option used in Section 8.1.

8.3   Apply the JMP Pro Bootstrap forest to the sensors data using status as a target variable. Compare the derived variable importance to the decision tree partition option used in Section 8.1.

8.4   Apply the JMP Pro Partition, Boosted tree, and Bootstrap forest options to the sensors data using status as a target variable. Save the predicted values. Compare the models using the compare models option in JMP.

8.5   Apply the JMP Pro Partition, Boosted tree, and Bootstrap forest options to the sensors data using test result as a target variable.

8.6   Apply the JMP Naïve Bayes classifier to the sensors data using status as a target. Compare the confusion matrix to Figure 8.4.

8.7   Nutritional data from 961 different food items is given in the file **FOOD.csv**. For each food item, there are seven variables: fat (grams), food energy (calories), carbohydrates (grams), protein (grams), cholesterol (milligrams), weight (grams), and saturatedfat (grams). Use Ward's distance to construct 10 clusters of food items with similarity in the seven recorded variables with Minitab cluster analysis of variables.

8.8   Conduct the analysis in Exercise 8.8 with JMP and compare the results.

8.9   Apply the JMP *K*-means cluster feature to the sensor variables in SENSORS.cvs and interpret the clusters using the test result and status label.

8.10   Develop a procedure based on *K*-means for quality control using the SENSORS.cvs data. Derive its confusion matrix.

8.11   Make up a list of supervised and unsupervised applications mentioned in COVID19 related applications. See for example https://www.washington.edu/news/2020/10/29/models-show-how-covid-19-cuts-a-neighborhood-path/.

8.12 Use the JMP Functional Data Explorer to analyze the Dissolution data of reference and test tablets. Use P-splines to model dissolution data. Based on this analysis, review Figure 8.14 and identify which path is from reference tablets and which path is from test tablets.

8.13 Pick articles on a similar topic from two journals on the web. Use the same procedure for identifying stop words, phrases and other data preparation steps. Compare the topics in these two articles using five topics. Repeat the analysis using 10 topics. Report on the differences.

8.14 Pick two articles on a similar topic from the same author. Use the same procedure for identifying stop words, phrases, and other data preparation steps. Compare the topics in these two articles using five topics. Repeat the analysis using 10 topics. Report on the differences.

Part II

Modern Industrial Statistics: Design and Control of Quality and Reliability

9

The Role of Statistical Methods in Modern Industry and Services

9.1 The different functional areas in industry and services

Industrial Statistics plays a key role in establishing the level of competitiveness of industrial organizations, services, health care, government, and educational systems. The tools and methods of industrial statistics have to be viewed in the context of their applications. Their implementations are greatly affected by management style and organizational culture. In order to provide a background to Part II of the book, we begin by describing key aspects of the industrial setting.

Industrial organizations include units dedicated to product development, manufacturing, marketing, finance, human resources, purchasing, sales, quality assurance, and after-sale support. Industrial statistics is used to resolve problems in each one of these functional units. Marketing personnel typically determine customer requirements and levels of satisfaction using surveys and focus groups. Sales are usually responsible for providing forecasts to purchasing and manufacturing units. Purchasing specialists, analyze world trends in quality and prices of raw materials so that they can provide critical inputs to product developers. Budgets are prepared by the finance department using forecasts that are validated periodically. Accounting experts rely on auditing and sampling methods to ascertain inventory levels and integrity of data bases. Human resources personnel track data on absenteeism, turnover, overtime, and training needs. The quality departments commonly perform audits and special purpose tests to determine the quality and reliability of products and services. Research and development engineers perform experiments to solve problems and improve products and processes. Finally, process controls of production operations are typically developed and maintained by manufacturing personnel and process engineers.

These are only a few general examples of problem areas, where the tools of industrial statistics are used, within modern industrial and service organizations. In order to provide more examples, we first take a closer look at specific industries. Later we discuss examples from these types of industries.

Modern Industrial Statistics: With Applications in R, MINITAB and JMP, Third Edition.
Ron S. Kenett and Shelemyahu Zacks.
© 2021 John Wiley & Sons, Ltd. Published 2021 by John Wiley & Sons, Ltd.

There are basically three types of production systems: (1) continuous flow production, (2) job shops, and (3) discrete mass production. Examples of continuous flow production include paper making and chemical transformations. Such processes, typically involve expensive equipment that is very large in size, operates around the clock, and requires very rigid manufacturing steps. Continuous flow industries are both capital intensive and highly dependent on the quality of the purchased raw materials. Rapid customizing of products in a continuous flow process is virtually impossible and new products are introduced using complex scale-up procedures.

Job shops are, in many respects, exactly the opposite. Examples of job shops include metal-working of parts or call centers where customers who call in are given individual attention by nonspecialized attendants. Such operations permit production of custom-made products and are very labor-intensive. Job shops can be supported by general-purpose machinery which remains idle in periods of low demand.

Discrete mass production systems can be similar to continuous flow production, if a standard product is produced in large quantities. When flexible manufacturing is achieved, mass production can handle batches of size 1, and, in that sense, appears similar to a job shop operation. This is what Industry 4.0 mentioned in Chapter 1 is all about. Service centers, with call routing for screening calls by areas of specialization, are such an example.

Machine tool automation began in the 1950s with the development of numerical control (NC) operations. In these automatic or semiautomatic machines, tools are positioned for a desired cutting effect through computer commands. Industry 4.0 hardware and software capabilities can make a job-shop manufacturing facility as much automated as a continuous-flow enterprise. Computer-integrated manufacturing (CIM) is the integration of computer-aided design (CAD) with computer-aided manufacturing (CAM). The development of CAD has its origins in the evolution of computer graphics and computer-aided drawing and drafting often called (CADD). As an example of how these systems are used, we follow the development of an automobile suspension system designed on a computer using CAD. The new system must meet testing requirements under a battery of specific road conditions. After coming up with an initial design concept, design engineers use computer animation to show the damping effects of the new suspension design on various road conditions. The design is then iteratively improved on the basis of simulation results and established customer requirements. In parallel to the suspension system, design purchasing specialists and industrial engineers proceed with specifying and ordering the necessary raw materials, setting up the manufacturing processes, and scheduling production quantities. Throughout the manufacturing of the suspension system, several tests provide the necessary production controls. Ultimately, the objective is to minimize the costly impact of failures in a product after delivery to the customer. Statistical methods are employed throughout the design, manufacturing and servicing stages of the product. Methods for the statistical design of experiments are used to optimize the design of the suspension system (see Chapters 13–15). Control of the manufacturing process is required to assure that the product is manufactured according to specifications. Methods of statistical process control (SPC) are employed at various stages of manufacturing to identify and correct deviations from process capabilities (see Chapters 10–12). The incoming raw materials have often to be inspected for adherence to quality level (see Chapter 18). Finally, tracking and analyzing field failures of the product is carried out to assess the reliability of the suspension system and provide early warnings of product deterioration (see Chapters 16 and 17).

CAD systems provide an inexpensive environment to test and improve design concepts. CIM systems typically capture data necessary for process control. Web technology offers opportunities to set up such systems without the deployment of costly computer infrastructures. Computerized field failures tracking systems and sales forecasting are very common.

Predictive analytics and operational Business Intelligence systems tag customers that are likely to drop and allow for churn prevention initiatives. The application of industrial statistics within such computerized environments allows the practitioner to concentrate on statistical analysis as opposed to repetitive numerical computations.

Service organization can be either independent or complementary to manufacturing type operations. For example, a provider of communication systems typically also supports installation and after sale services to its customers. The service takes the form of installing the communication system, programming the system's data base with an appropriate numbering system, and responding to service calls. The delivery of services differs from manufacturing in many ways. The output of a service system is generally intangible. In many cases, the service is delivered directly to the customer without an opportunity to store or fix "defective" transactions. Some services involve very large number of transactions. Federal Express, for example, handles 1.5 million shipments per day, to 127 countries, at 1650 sites. The opportunities for error are many, and process error levels must be of only a few defective parts per million. Operating at such low-defect levels might appear at first highly expensive to maintain and therefore economically unsound. In the next section, we deal with the apparent contradiction between maintaining low-error levels and reducing costs and operating expenses.

9.2   The quality-productivity dilemma

In order to reach World War II production goals for ammunitions, airplanes, tanks, ships, and other military materiel, American industry had to restructure and raise its productivity while adhering to strict quality standards. This was partially achieved through large-scale applications of statistical methods following the pioneering work of a group of industrial scientists at Bell Laboratories. Two prominent members of this group were Walter A. Shewhart, who developed the tools and concepts of SPC, and Harold F. Dodge, who laid the foundations for statistical sampling techniques. Their ideas and methods were instrumental in the transformation of American industry in the 1940s, which had to deliver high quality and high productivity. During those years, many engineers were trained in industrial statistics throughout the United States.

After the war, a number of Americans were asked to help Japan rebuild its devastated industrial infrastructure. Two of these consultants, W. Edwards Deming and Joseph M. Juran, distinguished themselves as effective and influential teachers. Both Drs. Deming and Juran witnessed the impact of Walter Shewhart's new concepts. In the 1950s, they taught the Japanese the ideas of process control and process improvements, emphasizing the role of management and employee involvement.

The Japanese were quick to learn the basic quantitative tools for identifying and realizing improvement opportunities and for controlling processes. By improving blue-collar and white-collar processes throughout their organizations, the Japanese were able to reduce waste and rework, thus produce better products at a lower price. These changes occurred over a period of several years leading eventually to significant increases in market share for Japanese products.

In contrast, American industry had no need for improvements in quality after World War II. There was an infinite market demand for American goods, and the emphasis shifted to high productivity, without necessarily assuring high quality. This was reinforced by the Taylor approach splitting the responsibility for quality and productivity between the quality and production departments. Many managers in the US industry did not believe that high quality and high productivity can be achieved simultaneously. The Quality-Productivity Dilemma was born and managers apparently had to make a choice. By focusing attention on productivity, managers often sacrificed quality, which in turn had a negative effect on productivity. Increasing emphasis on meeting schedules and quotas made the situation even worse.

On the other hand, Japanese industrialists proved to themselves that by implementing industrial statistics tools, managers can improve process quality and, simultaneously, increase productivity. This was shown to apply in every industrial organization and thus universally resolve the Quality-Productivity Dilemma.

In the 1970s, several American companies began applying the methods taught by Deming and Juran, and by the mid-1980s, there were many companies in the US reporting outstanding successes. Quality improvements generate higher productivity since they permit to ship higher-quality products, faster. The result was better products at lower costs – an unbeatable formula for success. The key to this achievement was the implementation of Quality Management and the application of industrial statistics, which includes analyzing data, understanding variability, controlling processes, designing experiments, and making forecasts. The approach was further developed in the 1980s by Motorola who launched its famous Six Sigma initiative. A striking testimonial of such achievements is provided by Robert W. Galvin, the former chairman of the executive committee of Motorola Inc.: "At Motorola we use statistical methods daily throughout all of our disciplines to synthesize an abundance of data to derive concrete actions... How has the use of statistical methods within Motorola Six Sigma initiative, across disciplines, contributed to our growth? Over the past decade we have reduced in-process defects by over 300 fold, which has resulted in a cumulative manufacturing cost savings of over 11 billion dollars. Employee productivity measured in sales dollars per employee has increased threefold or an average 12.2 percent per year over the same period. Our product reliability as seen by our customers has increased between 5 and 10 times." Quoted from the foreword to Kenett and Zacks (1998).

As mentioned in Chapter 1, since 2011, we are seeing a deployment of the fourth industrial revolution with increased digitalization, ubiquitous sensor technology that can measure vibrations, heat and humidity, flexible manufacturing capabilities, and mostly cloud hosted high-powered analytics.

The effective implementation of industrial statistics depends on the management approach practiced in the organization. We characterize different styles of management, by a **Quality Ladder**, which is presented in Figure 9.1. Management's response to the rhetorical question: "How do you handle the inconvenience of customer complaints?" determines the position of an organization on the ladder. Some managers respond by describing an approach based on reactively waiting for complaints to be filed before initiating any corrective actions. Some try to achieve quality by extensive inspections and implement strict supervision of every activity in the organization, having several signatures of approval on every document. Others take a more proactive approach and invest in process improvement and quality by design.

The four management styles we identify are (1) reactive firefighting, (2) inspection and traditional supervision, (3) processes control and improvement, and (4) quality by design. Industrial statistics tools can have an impact only on the top three steps in the quality ladder. Levels (3) and (4) are more proactive than (2). When management's style consists exclusively of firefighting, there is typically no use for methods of industrial statistics and data is simply accumulated.

Figure 9.1   *The quality ladder.*

The Quality Ladder is matching management maturity level with appropriate statistical tools. Kenett et al. (2008) formulated and tested with 21 case studies, the **Statistical Efficiency Conjecture** which states that organizations higher up on the Quality Ladder are more efficient at solving problems with increased returns on investments. This provides an economic incentive for investments in efforts to increase the management maturity of organizations.

9.3   Firefighting

Firefighters specialize in putting down fires. Their main goal is to get to the scene of a fire as quickly as possible. In order to meet this goal, they activate sirens and flashing lights and have their equipment organized for immediate use, at a moment's notice. Firefighting is also characteristic of a particular management approach that focuses on heroic efforts to resolve problems and unexpected crisis. The seeds of these problems are often planted by the same managers, who work the extra hours required to fix them. Firefighting has been characterized as an approach where there is never enough time to do things right the first time, but always enough time for rework and fixing problems once customers are complaining and threaten to leave. This reactive approach of management is rarely conducive to serious improvements which rely on data and teamwork. Industrial statistics tools are rarely used under firefighting management. Therefore, decisions are often made without investigation of the causes for failures.

In Part I, we studied the structure of random phenomena and present the basic statistical tools used to describe and analyze such structures. The basic philosophy of statistical thinking is the realization that variation occurs in all work processes, and the recognition that reducing variation is essential to quality improvement. Failure to recognize the impact of randomness leads to unnecessary and harmful decisions. One example of the failure to understand randomness is the common practice of adjusting production quotas for the following month by relying on the current month's sales. Without appropriate tools, managers have no way of knowing whether the current month's sales are within the common variation range or not. Common variation implies that nothing significant has happened since last month and therefore no quota adjustments should be made. Under such circumstances, changes in production quotas create unnecessary, self-inflicted problems. Firefighting management, in many cases, is responsible for

avoidable costs and quick temporary fixes with negative effects on the future of the organization. Moreover, in such case, data is usually accumulated and archived without lessons learned
or proactive initiatives. Some managers in such an environment will attempt to prevent "fires"
from occurring. One approach to prevent such fires is to rely on massive inspection and traditional supervision as opposed to leadership and personal example. The next section provides
some historical background on the methods of inspection.

9.4   Inspection of products

In medieval Europe, most families and social groups made their own goods such as cloth, uten-
sils, and other household items. Then only salable cloth was woven by peasants who paid their
taxes in kind to their feudal lords. The ownership of barons or monasteries was identified through
marks put on the fabric which were also an indication of quality. Since no feudal lord would
accept payment in shoddy goods, the products were carefully inspected prior to the inscribing
of the mark. Surviving late medieval documents indicate that bales of cloth frequently changed
hands repeatedly without being opened, simply because the marks they bore were regarded
everywhere as guarantees of quality. In the new towns, fabrics were made by craftsmen who
went in for specialization and division of labor. Chinese records of the same period indicate that
silks made for export were also subjected to official quality inspections. In Ypres, the center of
the Flemish wool cloth industry, weaver's regulations were put in writing as early as 1281. These
regulations stipulated the length and width as well as the number of warp ends and the quality of
the wool to be used in each cloth. A fabric had to be of the same thickness throughout. All fabric
was inspected in the draper's hall by municipal officials. Heavy fines were levied for defective
workmanship, and the quality of fabrics which passed inspection was guaranteed by affixing
the town seal. Similar regulations existed elsewhere in France, Italy, Germany, England, and
Eastern Europe. Trademarks as a guarantee of quality used by the modern textile industry originated in Britain. They first found general acceptance in the wholesale trade and then, from the
end of the nineteenth century onward, among consumers. For a time, manufacturers still relied
on in-plant inspections of their products by technicians and merchants, but eventually technological advances introduced machines and processes which ensured the maintenance of certain
standards independently of human inspectors and their know-how. Industrial statistics played an
important role in the textile industry. In fact, it was the first large industry which analyzed its data
statistically. Simple production figures including percentages of defective products were already
compiled in British cotton mills early in the nineteenth century. The basic approach, during the
preindustrial and postindustrial period, was to guarantee quality by proper inspection of the cloth
(Juran 1995). In the early nineteen hundreds, researchers at Bell Laboratories in New Jersey
developed statistical sampling methods that provided an effective alternative to 100% inspection
(Dodge and Romig 1929). Their techniques labeled "Sampling Inspection," eventually led to the
famous MIL-STD-105 system of sampling procedures used throughout the defense industry and
elsewhere. These techniques implement statistical tests of hypotheses, in order to determine if a
certain production lot or manufacturing batch is meeting Acceptable Quality Levels. Such sampling techniques are focused on the product, as opposed to the process that makes the product.
Details on the implementation and theory of sampling inspection are provided in Chapter 18
dedicated to acceptance sampling topics. The next section introduces the approach of process
control that focuses on the performance of processes throughout the organization.

9.5   Process control

In a memorandum to his superior at Bell Laboratories Walter Shewhart documented a new approach to SPC (Godfrey 1986; Godfrey and Kenett 2007). The document dated May 16, 1924 describes a technique designed to track process quality levels over time, which Shewhart labeled a "Control Chart." The technique was further developed, and more publications followed 2 years later (Shewhart 1926). Shewhart realized that any manufacturing process can be controlled using basic engineering ideas. Control charts are a straightforward application of engineering feedback loops to the control of work-processes. The successful implementation of control charts requires management to focus on process performance, with emphasis on process control and process improvements. When a process is found capable of producing products that meet customer requirements and a system of process controls is subsequently employed, one does not need to enforce product inspection anymore. Industry can deliver its products without time-consuming and costly inspection thereby providing higher quality at reduced costs. These are prerequisites for Just-In-Time deliveries and increased customer satisfaction. Achieving quality by no relying inspection implies quicker deliveries, less testing and therefore reduced costs. Shewhart ideas are therefore essential for organizations seeking improvements in their competitive position. As mentioned earlier, W. Edwards Deming and Joseph M. Juran were instrumental in bringing this approach to Japan in the 1950s. Deming emphasized the use of statistical methods, and Juran created a comprehensive management system including the concepts of management breakthroughs, the quality trilogy of planning, improvement and control, and the strategic planning of quality. Both were awarded a medal by the Japanese emperor for their contributions to the rebuilding of Japan's industrial infrastructure. Japan's national industrial award, called the Deming Prize, has been awarded every year since the early 1950s. The US National Quality Award and the European Quality Award are awarded since the early 1990s to recognize companies in the United States and Europe that can serve as role models to others. Notable winners include Motorola, Xerox, Milliken, Globe Metallurgical, AT&T Universal Cards, and the Ritz Carlton. Similar awards exist in Australia, Israel, Mexico, and many other countries.

Part II starts with three chapters on SPC. Chapter 10 covers basic issues in SPC. After establishing the motivation for SPC, the chapter introduces the reader to process capability studies, process capability indices, the seven tools for process improvement, and the basic Shewhart charts.

Chapter 11 includes more advanced topics such as the economic design of Shewhart control charts and CUSUM procedures. The chapter concludes with special sections on Bayesian detection, process tracking, and automatic process control.

Chapter 12 is dedicated to multivariate SPC techniques including multivariate analogues to process capability indices.

9.6   Quality by design

The design of a manufactured product or a service begins with an idea and continues through a series of development and testing phases until production begins and the product is made available to the customer. Process design involves the planning and design of the physical facilities, and the information and control systems required to manufacture a good or deliver a service. The design of the product, and the associated manufacturing process, determines its ultimate performance and value. Design decisions influence the sensitivity of a product to variation in raw materials and work conditions, which in turn affect manufacturing costs. General Electric, for example, has found that 75% of failure costs in its products are determined by the design. In a series of bold design decisions in the late 1990s, IBM developed the Proprinter so that all parts and subassemblies were to snap together during final assembly without the use of fasteners. Such initiatives resulted in major cost reductions and quality improvements. These are only a few examples demonstrating how design decisions affect manufacturing capabilities with an eventual positive impact on the cost and quality of the product. Reducing the number of parts is also formulated as a statistical problem that involves clustering and grouping of similar parts. Take, for example a basic mechanical parts such as aluminum bolts. Many organizations find themselves purchasing hundreds of different types of bolts for very similar applications. Multivariate statistical techniques can be used to group together similar bolts thereby reducing the number of different purchased parts eliminating potential mistakes and lowering costs.

In the design of manufactured products, technical specifications can be precisely defined. In the design of a service process, quantitative standards may be difficult to determine. In service processes, the physical facilities, procedures, people's behavior, and professional judgment affect the quality of service. Quantitative measures in the service industry typically consist of data from periodical customer surveys and information from internal feedback loops such as waiting time in hotels front desks or supermarket cash registers. The design of products and processes, both in service and manufacturing, involves quantitative performance measurements.

A major contributor to modern quality engineering has been Genichi Taguchi, formerly from the Japanese Electronic Communications Laboratories. Since the 1950s, Taguchi has advocated the use of statistically designed experiments in industry. Already in 1959, the Japanese company NEC ran 402 planned experiments. In 1976, Nippon Denso, which is a 20 000 employee company producing electronic parts for automobiles, reported to have run 2700 designed experiments. In the summer of 1980, Taguchi came to the United States to "repay the debt" of the Japanese to Shewhart, Deming, and Juran and delivered a series of workshops at Bell Laboratories in Holmdel New Jersey. His methods slowly gained acceptance in the United States. Companies such as ITT, Ford, and Xerox have been using Taguchi methods since the mid-1980s with impressive results. For example, an ITT electrical cable and wire plant reported reduced product variability by a factor of 10. ITT Avionic Division developed, over a period of 30 years, a comprehensive approach to quality engineering, including an economic model for optimization of products and processes. Another application domain which has seen a dramatic improvement in the maturity of management is the area of system and software development. The Software Engineering Institute (SEI) was established in 1987 to improve the methods used by industry in the development of systems and software. SEI, among other things, designed a five-level capability maturity model integrated (CMMI) which represents various levels of implementation of Process Areas. The tools and techniques of Quality by Design are applied by level five organizations which are, in fact, at the top of the quality ladder. For more on CMMI and systems and software development, see Kenett and Baker (2010).

A particular industry where such initiatives are driven by regulators and industrial best practices is the pharmaceutical industry. In August 2002, the Food and Drug Administration (FDA) launched the pharmaceutical current good manufacturing practices (cGMP) for the twenty-first century initiative. In that announcement, the FDA explained the agency's intent to integrate quality systems and risk management approaches into existing quality programs with the goal of encouraging industry to adopt modern and innovative manufacturing technologies. The cGMP initiative was spurred by the fact that since 1978, when the last major revision of the cGMP regulations was published, there have been many advances in design and manufacturing technologies and in the understanding of quality systems. This initiative created several international guidance documents that operationalized this new vision of ensuring product quality through "a harmonized pharmaceutical quality system applicable across the life cycle of the product emphasizing an integrated approach to quality risk management and science." This new approach is encouraging the implementation of Quality by Design and hence, de facto, encouraging the pharmaceutical industry to move up the quality ladder. Chapter 14 covers several examples of Quality by Design initiatives in the Pharmaceutical industry using statistically designed experiments. For a broad treatment of statistical methods in health care see Faltin et al. (2012).

Chapters 13–17 present a comprehensive treatment of the principal methods of design and analysis of experiments and reliability analysis used in Quality by Design. An essential component of Quality by Design is Quality Planning. Planning, in general, is a basic engineering and management activity. It involves deciding, in advance, what to do, how to do it, when to do it, and who is to do it. Quality Planning is the process used in the design of any new product or process. In 1987, General Motors cars averaged 130 assembly defects per 100 cars. In fact, this was planned that way. A cause and effect analysis of car assembly defects pointed out causes for this poor quality that ranged from production facilities, suppliers of purchased material, manufacturing equipment, engineering tools, etc. Better choices of suppliers, different manufacturing facilities, and alternative engineering tools produced a lower number of assembly defects. Planning usually requires careful analysis, experience, imagination, foresight, and creativity. Planning for quality has been formalized by Joseph M. Juran as a series of steps (Juran 1988). These are the following:

1. Identify who are the customers of the new product or process
2. Determine the needs of those customers
3. Translate the needs into technical terms
4. Develop a product or process that can respond to those needs
5. Optimize the product of process so that it meets the needs of the customers including economic and performance goals
6. Develop the process required to actually produce the new product or to install the new process
7. Optimize that process
8. Begin production or implement the new process.

Chapter 13 presents the classical approach to the design and analysis of experiments including factorial and fractional factorial designs and blocking and randomization principles.

Chapter 14 introduces the concepts of quality by design developed by Genichi Taguchi including parameter and tolerance industrial designs. A special section covers the implementation of quality by design in the pharmaceutical industry which has had very large practical statistical efficiency (PSE) in terms of impact on drug development and drug manufacturing industries.

Chapter 15 introduces methods of designing and analyzing computer experiments including Kriging and DACE models.

The next two chapters are dedicated to reliability. Reliability is "the probability that a unit will perform its intended function until a given point in time under specified use conditions." For reliability analysis to have PSE, the specified use conditions should overlap encountered use conditions. Alternatively, one could define reliability as quality over time. Design for reliability requires detailed consideration of product (process) failure modes. Manufacturers typically have formal or informal reliability goals for their products. Such goals are generally derived from past experience with similar products, industry standards, customer requirements, or a desire to improve an existing reliability of a product.

Chapter 16 deals with classical reliability methods including system reliability, availability models, reliability demonstration tests, and accelerated life testing.

Chapter 17 presents state-of-the-art Bayesian reliability estimation and prediction methods that are implemented using applications in R.

The final chapter is on acceptance sampling and sequential methods, including a treatment of sequential Bayesian inference. Chapter 18 deals with sampling plans for product inspection using attribute or variable data. This includes simple acceptance sampling plans, double sampling plans, and sequential sampling. A two-arm bandit Bayesian application of sequential testing is A/B testing that is used to optimize web page designs. Section 18.5 is dedicated to this. The chapter takes the reader an extra step and discusses skip-lot sampling, where consistent good performance leads to skipping some lots that are accepted without inspection. A next step is to qualify the supply chain so that it can rely on process control and does not need acceptance sampling.

This completes a listing of the 10 chapters in Part II. The eight chapters in Part I where described in Chapter 1. Part I introduces Bayesian decision-making in Chapter 4. In Part II, chapters on Bayesian inference are Chapter 17 on Bayesian reliability and Chapter 18 with a section on Bayesian optimization of sequential tests. Those interested in Bayesian applications in industrial statistics will find this material a good starting point.

The next final section presents an approach for assessing the efficiency of a project driven by statistical analysis called practical statistical efficiency (PSE).

9.7   Practical statistical efficiency

The idea of adding a practical perspective to the classical mathematical definition of statistical efficiency is based on a suggestion by Churchill Eisenhart who, in a 1978 informal "Beer and Statistics" seminar at the house of George Box in Madison Wisconsin, proposed a new definition of statistical efficiency. Later, Bruce Hoadley from Bell Laboratories, picked up where Eisenhart left off and added his version nicknamed "Vador." Blan Godfrey, former CEO of the Juran Institute, used this concept in his 1988 Youden Address on "Statistics, Quality and the Bottom Line" at the Fall Technical Conference of the American Society for Quality Control. Later, Kenett et al. (2003) further expanded this idea adding an additional component, the value of the data actually collected, and defined practical statistical efficiency (PSE). The PSE formula accounts for eight components and is computed as follows:

$PSE = V\{D\} \cdot V\{M\} \cdot V\{P\} \cdot V\{PS\} \cdot P\{S\} \cdot P\{I\} \cdot T\{I\} \cdot E\{R\}$,

where

- $V\{D\}$ = value of the data actually collected.
- $V\{M\}$ = value of the statistical method employed.
- $V\{P\}$ = value of the problem to be solved.
- $V\{PS\}$ = value of the problem actually solved.
- $P\{S\}$ = probability level that the problem actually gets solved.
- $P\{I\}$ = probability level the solution is actually implemented.
- $T\{I\}$ = time the solution stays implemented.
- $E\{R\}$ = expected number of replications.

These components can be assessed qualitatively, using expert opinions, or quantitatively, if the relevant data exists. A straightforward approach to evaluate PSE is to use a scale of "1" for not very good to "5" for excellent or very high. This method of scoring can be applied uniformly for all PSE components. Some of the PSE components can be also assessed quantitatively. $P(S)$ and $P(I)$ are probability levels, $T(I)$ can be measured in months, $V(P)$ and $V(PS)$ can be evaluated in Euros, Dollars, or Pounds. $V(PS)$ is the value of the problem actually solved, as a fraction of the problem to be solved. If this is evaluated qualitatively, a large portion would be "4" or "5," a small portion "1" or "2." $V(D)$ is the value of the data actually collected for the goal to be considered. Whether PSE terms are evaluated quantitatively or qualitatively, PSE is a conceptual measure rather than a numerically precise one. A more elaborated approach to PSE evaluation can include differential weighing of the PSE components and/or nonlinear assessments.

9.8   Chapter highlights

The effective use of industrial statistics tools requires organizations to climb up the quality ladder presented in Figure 1.1. As the use of data gets gradually integrated in the decision process, both at the short-term operational level, and at the long-term strategic level, different tools are needed. The ability to plan and forecast successfully is a result of accumulated experience and proper techniques. Modern industrial organizations are described and classified in this chapter that provides background to the chapters in Part II. Different functional areas of a typical business are presented with typical problems for each area. The potential benefits of industrial statistics methods are then introduced in the context of these problems. The main theme here is that the apparent conflict between high productivity and high quality can be resolved through improvements in work processes, by introducing statistical methods and concepts. The contributions of Shewhart, Deming, and Juran's to industries seeking a more competitive position are outlined. Different approaches to the management of industrial organizations are summarized and classified using a Quality Ladder. Industrial statistics methods are then categorized according to the steps of the ladder. These consist of Firefighting, Inspection, Process Control, and Quality by Design. The chapter discusses how to match a specific set of statistical methods to the management approach and how to assess PSE to ensure that industrial statistics methods are used efficiently in organization and application areas. It is designed to provide a general background to the application of Industrial Statistics.

The chapters in Part II provide a comprehensive exposition of the tools and methods of modern industrial statistics. In this introductory chapter, we refer to the need to develop the maturity of the management approach and plan for high PSE in order to achieve high impact. This can be implemented with the methods and tools presented in the following chapters.

The main terms and concepts introduced in this first chapter include the following:

- Continuous Flow Production
- Job Shops
- Mass Production Systems
- The Quality-Productivity Dilemma
- Quality Management
- The Quality Ladder
- Firefighting as a Management Approach
- Inspection as a Management Approach
- Process Control and Improvement as a Management Approach

- Quality by Design as a Management Approach
- Practical Statistical Efficiency (PSE)

9.9   Exercises

9.1 Describe three work environments where quality is assured by 100% inspection.

9.2 Search periodicals, such as Business Week, Fortune, Time, and Newsweek and newspapers such as the *New York Times* and *Wall Street Journal* for information on quality initiatives in service, healthcare, governmental, and industrial organizations.

9.3 Provide examples of the three types of production systems:
- Continuous flow production.
- Job shops.
- Discrete mass production.

9.4 What management approach cannot work with continuous flow production?

9.5 What management approach characterizes
- a school system?
- a military unit
- a football team?

9.6 Provide examples of how you, personally, apply the four management approaches
- as a student.
- in your parents' house.
- with your friends.

9.7 Evaluate the practical statistical efficiency (PSE) of a case study provided by your instructor.

10

Basic Tools and Principles of Process Control

10.1 Basic concepts of statistical process control

In this chapter, we present the basics of **statistical process control** (SPC). The general approach is prescriptive and descriptive rather than analytical. With SPC, we do not aim at modeling the distribution of data collected from a given process. Our goal is to control the process with the aid of decision rules for signaling significant discrepancies between the observed data and the standards of a process under control. We demonstrate the application of SPC to various processes by referring to the examples of piston cycle time and strength of fibers, which have been discussed in Chapter 2. Other examples used include data on power failures in a computer center and office procedures for scheduling appointments of a university dean. The data on the piston cycle time is generated by the R piston simulator function `pistonSimulation` or the JMP addin available for download on the book's website. In order to study the causes for variability in the piston cycle time, we present, in Figure 10.1, a sketch of a piston, and, in Table 10.1, seven factors which can be controlled to change the cycle time of a piston.

Figure 10.2 is a run chart, (also called "connected line plot"), and Figure 10.3 is a histogram, of 50 piston cycle times (seconds) measured under stable operating conditions. Throughout the measurement time frame the piston operating factors remained fixed at their maximum levels. The data can be found in file **OTURB1.csv**.

The average cycle time of the 50 cycles is 0.392 [seconds] with a standard deviation of 0.114 [seconds].

Even though no changes occurred in the operating conditions of the piston, we observe variability in the cycle times. From Figure 10.2, we note that cycle times vary between 0.22 and 0.69 seconds. The histogram in Figure 10.3 indicates some skewness in the data. The normal probability plot of the 50 cycle times (Figure 10.4) also leads to the conclusion that the cycle time distribution is not normal, but skewed.

Another example of variability is provided by the yarn strength data presented in Chapter 2. The yarn strength test results indicate that there is variability in the properties of the product. High yarn strength indicates good spinning and weaving performance. Yarn strength is considered a function of the fiber length, fiber fineness, and fiber tensile strength. As a general rule,

Modern Industrial Statistics: With Applications in R, MINITAB and JMP, Third Edition.
Ron S. Kenett and Shelemyahu Zacks.
© 2021 John Wiley & Sons, Ltd. Published 2021 by John Wiley & Sons, Ltd.

Table 10.1 *Operating factors of the piston simulator and their operational levels*

Factor	Units	Minimum	Maximum
Piston weight	M [Kg]	30	60
Piston surface area	S [m^2]	0.005	0.020
Initial gas volume	V_0 [m^3]	0.002	0.010
Spring coeff.	K [N/m]	1000	5000
Atmosph. pressure	P_0 [N/m^2]	9×10^4	11×10^4
Ambient temperat.	T [^{0}K]	290	296
Filling gas temperat.	T_0 [^{0}K]	340	360

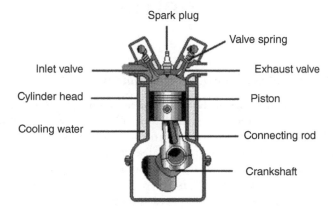

Figure 10.1   *A sketch of the piston.*

longer cottons are fine-fibered and shorter cottons are coarse-fibered. Very fine fibers, however, tend to reduce the rate of processing, so that the degree of fiber fineness depends upon the specific end-product use. Variability in fiber fineness is a major cause of variability in yarn strength and processing time.

In general, a production process has many sources or causes of variation. These can be further subdivided as process inputs and process operational characteristics including equipment, procedures, and environmental conditions. Environmental conditions consist of factors such as temperature and humidity or work-tools. Visual guides for instance, might not allow operators to precisely position parts on fixtures. The complex interactions between material, tools, machine, work methods, operators, and the environment combine to create variability in the process. Factors that are permanent, as a natural part of the process, are causing **chronic problems** and are called **common causes** of variation. The combined effect of common causes can be described using probability distributions. Such distributions were introduced in Chapter 2 and their theoretical properties presented in Chapter 3. It is important to recognize that recurring causes of variability affect every work process and that even under a stable process there are differences in performance over time. Failure to recognize variation leads to wasteful actions such as those described in Section 9.3. The only way to reduce the negative effects of chronic, common causes of variability is to modify the process. This modification can occur at the level of the process inputs, the process technology, the process controls or the process design. Some of these changes

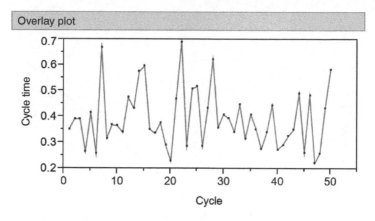

Figure 10.2   *Run chart or connected line plot (JMP) of 50 piston cycle times, [second].*

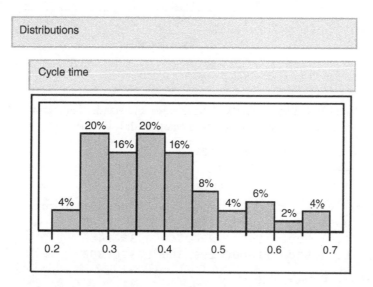

Figure 10.3   *Histogram of 50 piston cycle times (JMP).*

are technical (e.g., different process settings), some are strategic (e.g., different product specifications), and some are related to human resources management (e.g., training of operators).
Special causes, assignable causes, or **sporadic spikes** arise from external temporary sources
that are not inherent to the process. These terms are used here interchangeably. For example, an
increase in temperature can potentially affect the piston's performance. The impact can be both
in terms of changes in the average cycle times and/or the variability in cycle times.

In order to signal the occurrence of special causes, we need a control mechanism. Specifically
in the case of the piston, such a mechanism can consist of taking samples or subgroups of five
consecutive piston cycle times. Within each subgroup, we compute the subgroup average and
standard deviation.

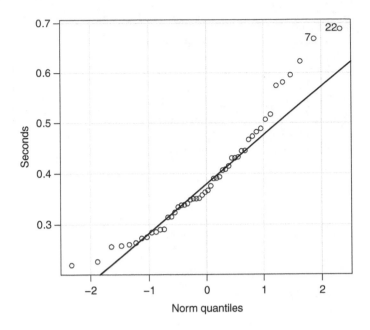

Figure 10.4   *Normal probability plot of 50 piston cycle times.*

Figures 10.5 and 10.6 display charts of the average and standard deviations of 20 samples of 5 cycle time measurements. To generate these charts with R we use:

```
> Ps <- pistonSimulation(seed=123)
> Ps <- simulationGroup(Ps, 5)
> head(Ps, 3)

    m    s   v0    k     p0  t  t0   seconds group
1  60 0.02 0.01 5000 110000 296 360 0.3503785     1
2  60 0.02 0.01 5000 110000 296 360 0.3901446     1
3  60 0.02 0.01 5000 110000 296 360 0.3907803     1

> aggregate(x=Ps["seconds"],
            by=Ps["group"],
            FUN=mean)

   group   seconds
1      1 0.3620424
2      2 0.3938172
3      3 0.4824221
4      4 0.3155764
5      5 0.4928929
6      6 0.4204112
7      7 0.3806072
8      8 0.3366084
9      9 0.3425293
10    10 0.3949534
```

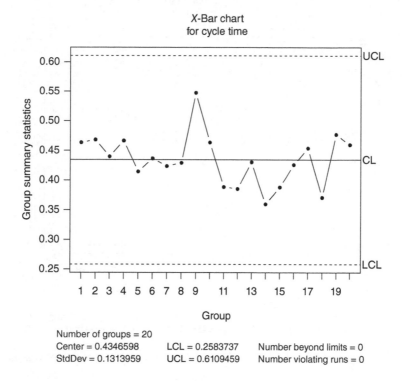

Number of groups = 20
Center = 0.4346598   LCL = 0.2583737   Number beyond limits = 0
StdDev = 0.1313959   UCL = 0.6109459   Number violating runs = 0

Figure 10.5   *X-Bar chart of cycle times under stable operating conditions.*

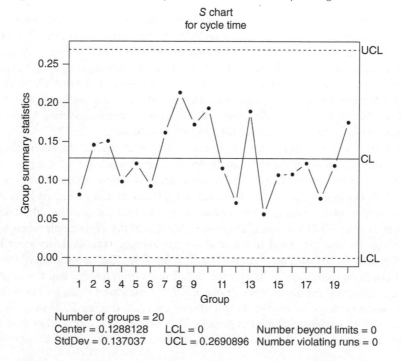

Number of groups = 20
Center = 0.1288128   LCL = 0   Number beyond limits = 0
StdDev = 0.137037   UCL = 0.2690896   Number violating runs = 0

Figure 10.6   *S Chart of cycle times under stable operating conditions.*

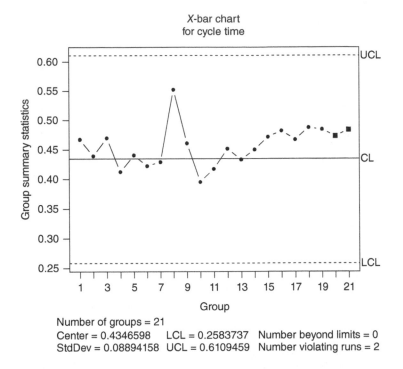

Number of groups = 21
Center = 0.4346598   LCL = 0.2583737   Number beyond limits = 0
StdDev = 0.08894158   UCL = 0.6109459   Number violating runs = 2

Figure 10.7   *X-Bar chart of cycle times with a trend in ambient temperature.*

The chart of averages is called an *X*-bar chart, the chart of standard deviations is called an *S* chart. All 100 measurements were taken under fixed operating conditions of the piston (all factors set at the maximum levels). We note that the average of cycle time averages is 0.414 seconds and that the average of the standard deviations of the 20 subgroups is 0.12 seconds. All these numbers were generated by the piston computer simulation model that allows us to change the factors affecting the operating conditions of the piston. Again we know that no changes were made to the control factors. The observed variability is due to common causes only such as variability in atmospheric pressure or filling gas temperature.

We now rerun the piston simulator introducing a forced change in the piston ambient temperature. At the beginning of the 8th sample, temperature begins to rise at a rate of 10% per cycle. Can we flag this special cause? The *X*-bar chart of this new simulated data is presented in Figure 10.7. Up to the 7th sample, the chart is identical to that of Figure 10.5. At the 8th sample, we note a small increase in cycle time. As of the 11th sample, the subgroup averages are consistently above 0.414 seconds. This run persists until the 21st sample when we stopped the simulation. To have 10 points in a row above the average is unlikely to occur by chance alone. The probability of such an event is $(1/2)^{10} = 0.00098$. The implication of the 10 points run is that common causes are no longer the only causes of variation, and that a special factor has began affecting the piston's performance. In this particular case, we know that it is an increase in ambient temperature. The *S* chart of the same data (Figure 10.8) shows a run below the average of 0.12 beginning at the 8th sample. This indication occurs earlier than that in the *X*-bar chart. The information obtained from both charts indicates that a special cause has been in

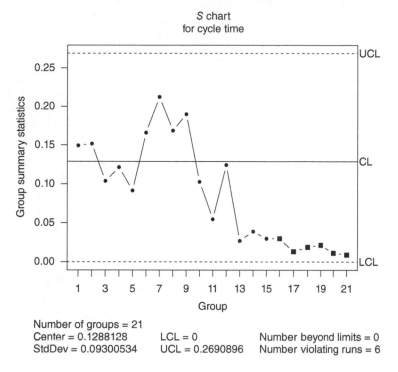

Figure 10.8   *S Chart of cycle times with a trend in ambient temperature.*

effect from the 8th sample onward. Its effect has been to increase cycle times and reduce variability. The new average cycle time appears to be around 0.49 seconds. The piston simulator allows us to try other types of changes in the operational parameters of the piston. For example, we can change the spring that controls the intake-valve in the piston gas chamber. In the next simulation, the standard deviation of the spring coefficient is increasing at a 5% rate past the 8th sample. Figures 10.9 and 10.10 are X-bar and S charts corresponding to this scenario. Until the 8th sample, these charts are identical to those in Figures 10.5 and 10.6. After the 8th sample changes appear in the chart. Points 14 and 18 fall outside the control limits, and it seems that the variability has increased. This is seen also in Figure 10.10. We see that after the 9th sample, there is a run upward of six points, and points 18 and 19 are way above the upper control limit.

Control charts have wide applicability throughout an organization. Top managers can use a control chart to study variation in sales and decide on new marketing strategies. Operators can use the same tool to determine if and when to adjust a manufacturing process. An example with universal applicability comes from the scheduling process of daily appointments in a university dean's office. At the end of each working day, the various meetings and appointment coordinated by the office of the dean were classified as being "on time" or with a problem such as "late beginning," "did not end on time," "was interrupted," etc. The ratio of problem appointments to the total number of daily appointments was tracked and control limits computed. Figure 10.11 is the dean's control chart (see Kelly et al. 1991).

Another example of a special cause is the miscalibration of spinning equipment. Miscalibration can be identified by ongoing monitoring of yarn strength. Process operators analyzing X-bar

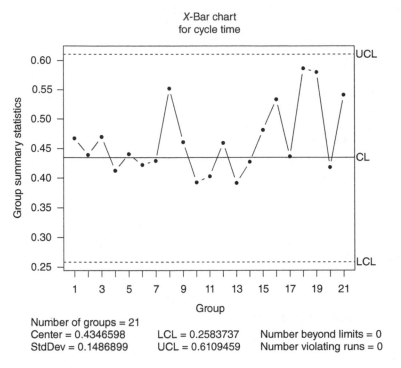

Figure 10.9   *X-Bar chart of cycle times with a trend in spring coefficient precision.*

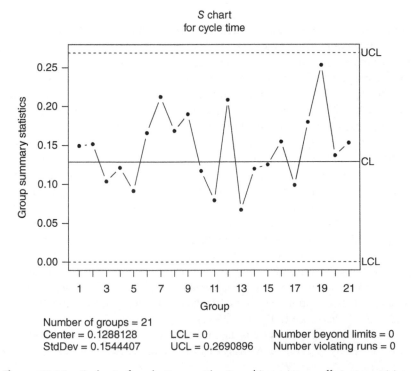

Figure 10.10   *S chart of cycle times with a trend in spring coefficient precision.*

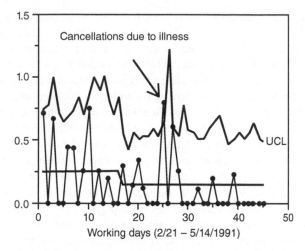

Figure 10.11   *Control chart for proportion of appointments with scheduling problems (Based on a chart prepared by Dean of the School of Management at SUNY Binghamton).*

and S charts can stop and adjust the process as trends develop or sporadic spikes appear. Timely indication of a sporadic spike is crucial to the effectiveness of process control mechanisms. **Ongoing chronic problems, however, cannot be resolved by using local operator adjustments**. The statistical approach to process control allows us to distinguish between chronic problems and sporadic spikes. This is crucial since these two different types of problems require different approaches. Process control ensures that a process performs at a level determined "doable" by a process capability study. Section 10.3 discusses how to conduct such studies and how to set control limits.

So far we focused on the analysis of data for process control. Another essential component of process control is the generation and routing of relevant and timely data through proper feedback loops. We distinguish between two types of feedback loops: **External feedback loops** and **internal feedback loops**. An external feedback loop consists of information gathered at a subsequent downstream process or by direct inspection of the process outputs.

To illustrate these concepts and ideas, let us look at the process of driving to work. The time it takes you to get to work is a variable that depends on various factors such as, how many other cars are on the road, how you happen to catch the traffic lights, your mood that morning, etc. These are factors that are part of the process, and you have little or no control over them. Such common causes create variation in the time it takes you to reach work. One day it may take you 15 minutes and the next day 12 minutes. If you are particularly unlucky and had to stop at all the red lights it might take you 18 minutes. Suppose, however that on one particular day it took you 45 minutes to reach work. Such a long trip is outside the normal range of variation and is probably associated with a special cause such as a flat tire, a traffic jam, or road constructions.

External feedback loops rely on measurements of the process outcome. They provide information like looking at a rear view mirror. The previous example consisted of monitoring time after you reached work. In most cases, identifying a special cause at that point in time is too late. Suppose that we had a local radio station that provided its listener live coverage of the traffic conditions. If we monitor, on a daily basis, the volume of traffic reported by the radio, we can avoid traffic jams, road constructions, and other unexpected delays. Such information will help us eliminate certain special causes of variation. Moreover, if we institute a preventive maintenance program for our car we can eliminate many types of engine problems, further reducing the impact of special causes. To eliminate the occasional flat tire would involve improvements in road maintenance – a much larger task. The radio station is a source of internal feedback that provides information that can be used to correct your route, and thus arrive at work on time almost every day. This is equivalent to driving the process while looking ahead. Most drivers are able to avoid getting off the road, even when obstacles present themselves unexpectedly. We now proceed to describe how control charts are used for "staying on course."

Manufacturing examples consist of physical dimensions of holes drilled by a numerically controlled CNC machine, piston cycle times, or yarn strength. The finished part leaving a CNC machine can be inspected immediately after the drilling operation or later, when the part is assembled into another part. Piston cycle times can be recorded online or stored for off-line analysis. Another example is the testing of electrical parameters at final assembly of an electronic product. The test data reflects, among other things, the performance of the components' assembly process. Information on defects such as missing components, wrong or misaligned components should be fed back, through an external feedback loop, to the assembly operators. Data collected on process variables, measured internally to the process, are the basis of an internal feedback loop information flow. An example of such data is the air pressure in the hydraulic system of a CNC machine. Air pressure can be measured so that trends or deviations in pressure are detected early enough to allow for corrective action to take place. Another example consists of the tracking of temperature in the surroundings of a piston. Such information will directly point out the trend in temperature which was indirectly observed in Figure 10.7 and 10.8. Moreover, routine direct measurements of the precision of the spring coefficient will exhibit the trend that went unnoticed in Figures 10.9 and 10.10. The relationship between a process, its suppliers, and its customers is presented in Figure 10.12. Internal and external feedback loops depend on a coherent structure of suppliers, processes, and customers. It is in this context that one can achieve effective SPC.

We discussed in this section the concepts of feedback loops, chronic problems (common causes), and sporadic spikes (special causes). Data funneled through feedback loops are used to indicate what are the type of forces affecting the measured process. SPC is "a rule of behavior that will strike a balance for the net economic loss from two sources of mistake: (1) looking for special causes too often, or overadjusting; (2) not looking often enough." (excerpt from Deming (1967)). In the implementation of SPC one distinguishes between two phases: (1) achieving control and (2) maintaining control. Achieving control consists of a study of the causes of variation

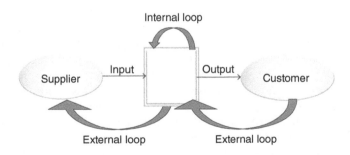

Figure 10.12   *The Supplier-process-customer structure and its feedback loops.*

followed by an effort to eliminate the special causes and a thorough understanding of the remaining permanent factors affecting the process, the common causes. Tools such as graphic displays (Chapters 2 and 5), correlation and regression analysis (Section 5.3), control charts (Chapter 10–12), and designed experiments (Chapters 13 and 14) are typically used in a process capability study whose objective is to achieve control. Section 10.3 will discuss the major steps of a process capability study and the determination of control limits on the control charts. Once control is achieved, one has to maintain it. The next section describes how control is maintained with the help of control limits.

10.2 Driving a process with control charts

Control charts allow us to determine when to take action in order to adjust a process that has been affected by a special cause. Control charts also tell us when to leave a process alone and not misinterpret variations due to common causes. Special causes need to be addressed by corrective action. Common causes are the focus of ongoing efforts aimed at improving the process.

We distinguish between **control charts for variable data** and **control charts for attribute data**. Attribute data requires an operational definition of what constitutes a problem or defect. When the observation unit is classified in one of two categories (e.g., "pass" versus "fail" or conforming versus nonconforming) we can track the proportion of nonconforming units in the observation sample. Such a chart is called a **p-chart**. If the size of the observation sample is fixed, we can simply track the number of nonconforming units and derive an **np-chart**. When an observation consists of the number of nonconformities per unit of observation, we track either number of nonconformities (**c-charts**) or rates of nonconformities (**u-charts**). Rates are computed by dividing the number of nonconformities by the number of opportunities for errors or problems. For variable data, we distinguish between processes that can be repeatedly sampled under uniform conditions, and processes were measurements are derived one at a time (e.g., monthly sales). In the latter case, we will use control charts for individual data also called moving range charts. When data can be grouped, we can use a variety of charts such as the X-bar chart or the median chart discussed in detail in Chapter 11. We proceed to demonstrate how X-bar control charts actually work using the piston cycle times discussed earlier. An X-bar control chart for the piston's cycle time is constructed by first grouping observations by time period, and then summarize the location and variability in these subgroups. An example of this was provided in Figures 10.5 and 10.6 where the average and standard deviations of 5 consecutive cycle times where tracked over 20 such subgroups. The three lines that are added to the simple run charts are the **center line**, positioned at the grand average, the **Lower Control Limits** (LCL) and the **Upper Control Limits** (UCL). The UCL and LCL indicate the range of variability we expect to observe around the center line, under stable operating conditions. Figure 10.5 shows averages of 20 subgroups of five consecutive cycle times each. The center line and control limits are computed from the average of the 20 subgroup averages and the estimated standard deviation for averages of samples of size 5. The center line is at 0.414 seconds. When using the classical 3-sigma charts developed by Shewhart, the control limits are positioned at three standard deviations of $\overline{X}$, namely $3\sigma/\sqrt{n}$, away from the center line. Using R, MINITAB, or JMP, we find that UCL = 0.585 seconds and LCL = 0.243 seconds. Under stable operating conditions, with only common causes affecting performance, the chart will typically have all points within the control limits. Specifically with 3-sigma control limits we expect to have, on the average, only one out of 370 points (1/0.0027), outside these limits, a rather rare event. Therefore, when a point falls beyond the control limits, we can safely question the stability of the process. The risk that such an alarm will turn to be false is 0.0027. A false alarm occurs when the sample mean falls outside the control limits and we suspect an assignable cause, but only common causes are operating. Moreover, stable random variation does not exhibit patterns such as upward or downwards trends, or consecutive runs of points above or below the center line. We saw earlier how a control chart was used to detect an increase in ambient temperature of a piston from the cycle times. The X-bar chart (Figure 10.7) indicates a run of six or more points above the center line. Figure 10.13 shows several patterns that indicate nonrandomness. These are:

a) A single point outside the control limits;
b) A run of nine or more points in a row above (or below) the centerline;
c) Six consecutive points increasing (trend up) or decreasing (trend down);
d) Two out of three points in a region between $\mu \pm 2\sigma/\sqrt{n}$ and $\mu \pm 3\sigma/\sqrt{n}$.

A comprehensive discussion of detection rules and properties of the classical 3-sigma control charts and of other modern control chart techniques is presented in Chapter 11.

As we saw earlier, there are many types of control charts. Selection of the control chart to use in a particular application primarily depends on the type of data that will flow through the feedback loops. The piston provided us with an example of variable data, and we used an X-bar and S chart to monitor the piston's performance. Examples of attribute data are blemishes on a given surface, wave solder defects, friendly service at the bank, and missed shipping dates. Each type of data leads to a different type of control chart. All control charts have a center line, upper and lower control limits (UCL and LCL). In general, the rule for flagging special causes is the same in every type of control chart. Figure 10.14 presents a classification of the various control charts. Properties of the different types of charts, including the more advanced EWMA and CUSUM charts are presented in Chapter 11.

We discussed earlier several examples of control charts and introduced different types of control charts. The block diagram in Figure 10.14 organizes control charts by the type of data flowing through feedback loops. External feedback loops typically rely on properties of the process' products and lead to control charts based on counts or classification. If products are classified using "pass" versus "fail" criteria, one will use np-charts or p-charts depending on whether the products are tested in fixed or variable subgroups. The advantage of such charts is that several criteria can be combined to produce a definition of what constitutes a "fail" or defective product. When counting nonconformities or incidences of a certain event or phenomenon, one is directed to use c-charts or u-charts. These charts provide more information than p-charts or np-charts since the actual number of nonconformities in a product is accounted for. The drawback is that several criteria cannot be combined without weighing the different types of nonconformities. c-Charts assume a fixed likelihood of incidence, u-charts are used in cases of varying likelihood levels. For large subgroups (subgroup sizes larger than 1000), the number of incidences, incidences per unit, number of defectives or percent defectives can be considered as individual measurements and an X chart for subgroups of size 1 can be used. Internal feedback loops and, in some cases, also external feedback loops rely on variable data derived from measuring product or process characteristics. If measurements are grouped in samples, one can combine X-bar charts with R-charts or S-charts. Such combinations provide a mechanism to control stability of a process with respect to both location and variability. X-Bar charts track the sample averages, R-charts track sample ranges (Maximum–minimum), and S-charts

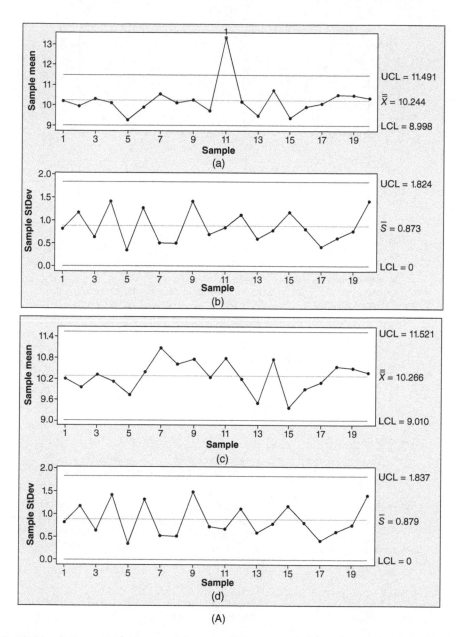

Figure 10.13   *Patterns to detect special causes (from top to bottom the patterns are (a), (b), (c) and (d)): top chart with special cause at 11th observation Bottom chart representing stable process (MINITAB).*

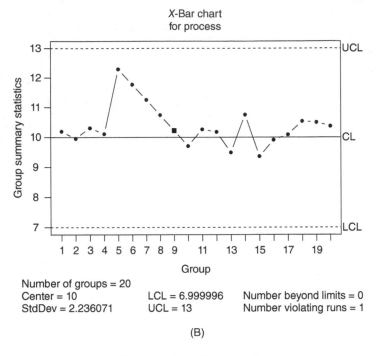

Number of groups = 20
Center = 10
StdDev = 2.236071

LCL = 6.999996
UCL = 13

Number beyond limits = 0
Number violating runs = 1

(B)

Figure 10.13   *(Continued)*

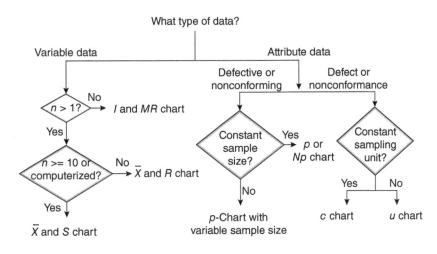

Figure 10.14   *Classification of control charts.*

are based on sample standard deviations. For samples larger than 10, S-charts are recommended over R-charts. For small samples and manual maintenance of control charts, R-charts are preferred. When sample sizes vary, only S-charts should be used to track variability.

10.3   Setting up a control chart: process capability studies

Setting up control limits of a control chart requires a detailed study of process variability and of the causes creating this variability. Control charts are used to detect occurrence of special, sporadic causes while minimizing the risk of misinterpreting special causes as common causes. In order to achieve this objective, one needs to assess the effect of chronic, common causes, and then set up control limits that reflect the variability resulting from such common causes. The study of process variability that precedes the setting up of control charts is called a **Process Capability Study**. We distinguish between attribute process capability studies and variable process capability studies.

Attribute process capability studies determine a process capability in terms of fraction of defective or nonconforming output. Such studies begin with data collected over several time periods. A rule of thumb is to use three time periods with 20–25 samples of size 50–100 units each. For each sample, the control chart statistic is computed and a control chart is drawn. This will lead to *ap, np, c* or *u*-chart and investigation patterns flagging special causes such as those in Figure 10.14. Special causes are then investigated and possibly removed. This requires changes to the process that justify removal of the measurements corresponding to the time periods when those special causes were active. The new control charts, computed without these points, indicates the capability of the process. Its center line is typically used as a measure of process capability. For example in Figure 10.13, one can see that the process capability of the scheduling of appointments at the dean's office improved from 25% of appointments with problems to 15% after introducing a change in the process. The change consisted of acknowledging appointments with a confirmation note spelling out, time, date, and topic of appointment, a brief agenda, and a scheduled ending time. On the 25th working day, there was one sporadic spike caused by illness. The Dean had to end early that day and several appointments got cancelled. When sample sizes are large (over 1000 units), control charts for attribute data become ineffective because of very narrow control limits and X-charts for individual measurements are used.

Variable process capability studies determine a process capability in terms of the distribution of measurements on product or process characteristics. Setting up of control charts for variable data requires far less data than attribute data control charts. Data is collected in samples, called **rational subgroups**, selected from a time frame so that relatively homogeneous conditions exist within each subgroup. The design strategy of rational subgroups is aimed at measuring variability due to common causes only. Control limits are then determined from measures of location and variability in each rational subgroup. The control limits are set to account for variability due to these common causes. Any deviation from stable patterns relative to the control limits (see Figure 10.13) indicates a special cause. For example, in the piston case study, a rational subgroup consists of five consecutive cycle times. The statistics used are the average and standard deviation of the subgroups. The 3-sigma control limits are computed to be UCL = 0.585 and LCL = 0.243. From an analysis of Figure 10.5 we conclude that the X-bar chart, based on a connected time plot of 20 consecutive averages, exhibits a pattern that is consistent with a stable process. We can now determine the process capability of the piston movement within the cylinder.

Process capability for variable data is a characteristic which reflects the probability of the individual outcomes of a process to be within the engineering specification limits. Assume that the piston engineering specifications stipulate a nominal value of 0.3 seconds and maximum and minimum values of 0.5 and 0.1 seconds, respectively. Table 10.2 shows the output from the

Table 10.2 *MINITAB output from storage option of process capability analysis of piston cycle time*

Process	Capability	Analysis for	OTRUB1.Cyclet
Sample size = 100 Sample mean = 0.392186 Sample standard deviation = 0.11368			

Specification	Observed Beyond Spec.	z-Score	Estimated Beyond Spec.
Upper limit: 0.5	16.000%	0.96	17.02%
Nominal: 0.3		−0.82	
Lower limit: 0.1	0.00%	−2.61	0.49%
	16.000%		17.51%
			Goodness-of-
Capability indices			fit tests

CP: 0.597	Shapiro-Wilk's W: 0.908685
CPK: 0.322	*P* value: 0.000000
(Upper): 0.21	Chi-square test: 40.160000
(Lower): 0.43	*P* value: 0.007100

process capability analysis storage option included in the SPC window of MINITAB. With R, we apply the following commands:

```
> Ps <- pistonSimulation(seed=123)
> Ps <- simulationGroup(Ps, 5)
> CycleTime <- qcc.groups(data=Ps$seconds,
                          sample=Ps$group)
> PsXbar <- qcc(CycleTime,
               type="xbar",
               nsigmas=3,
               plot=FALSE)
> process.capability(PsXbar,
                 spec.limits=c(0.1, 0.5))

Process Capability Analysis

Call:
process.capability(object = PsXbar, spec.limits = c(0.1, 0.5))

Number of obs = 50        Target = 0.3
       Center = 0.3922       LSL = 0.1
       StdDev = 0.1117       USL = 0.5

Capability indices:
```

```
          Value      2.5%     97.5%
Cp       0.5967    0.4788    0.7143
Cp_l     0.8717    0.7074    1.0360
Cp_u     0.3216    0.2275    0.4158
Cp_k     0.3216    0.2094    0.4339
Cpm      0.4602    0.3537    0.5666

Exp<LSL 0.45%           Obs<LSL 0%
Exp>USL 17%            Obs>USL 16%
```

The 50 measurements that were produced under stable conditions have a mean (average) of 0.414 seconds and a standard deviation of 0.127 seconds. The predicted proportion of cycle times beyond the specification limits is computed using the normal distribution as an approximation. The computations yield that, under stable operating conditions, an estimated 25% of future cycle times will be above 0.5 seconds, and that 0.6% will be below 0.1 seconds. We clearly see that the nominal value of 0.3 seconds is slightly lower than the process average, having a z-score of -0.90 and that the upper limit, or maximum specification limit, is 0.67 standard deviations above the average. The probability that a standard normal random variable is larger than 0.67 is 0.251. This is an estimate of the future percentage of cycle times above the upper limit of 0.5 seconds, provided stable conditions prevail. It is obvious from this analysis that the piston process is incapable of complying with the engineering specifications. The lower right-hand side of the table presents two tests for normality. Both tests reject the hypothesis of normality.

10.4   Process capability indices

In assessing the process capability for variable data, two indices have recently gained popularity: C_p and C_{pk}. The first index is an indicator of the potential of a process to meet two-sided specifications with as few defects as possible. For symmetric specification limits, the full potential is actually achieved when the process is centered at the midpoint between the specification limits. In order to compute C_p, one simply divides the process tolerance by six standard deviations, i.e.:

$$C_p = (\text{Upper Limit} - \text{Lower Limit})/(6 * \text{Standard Deviation}) \qquad (10.4.1)$$

The numerator indicates how wide the specifications are, the denominator measures the width of the process. Under normal assumptions, the denominator is a range of values that accounts for 99.73% of the observations from a centered process, operating under stable conditions with variability only due to common causes. When $C_p = 1$, we expect that 0.27% of the observations to fall outside the specification limits. A target for many modern industries is to reach, on every process, a level of $C_p = 2$, which practically guarantees that under stable conditions, and for processes kept under control around the process nominal values, there will be no defective products ("zero defects"). With $C_p = 2$, the theoretical estimate under normal assumptions, allowing for a possible shift in the location of the process mean by as much as 1.5 standard deviations, is 3.4 cases per million observations outside specification limits.

Another measure of process capability is:

$$C_{pk} = \text{minimum}(C_{pu}, C_{pl}), \qquad (10.4.2)$$

where

$$C_{pu} = (\text{Upper limit} - \text{Process Mean})/(3 * \text{Standard Deviation})$$

and   (10.4.3)

$$C_{pl} = (\text{Process Mean} - \text{Lower limit})/(3 * \text{Standard Deviation}).$$

When the process mean is not centered midway between the specification limits, C_{pk} is different from C_p. Non-centered processes have their potential capability measured by C_p, and their actual capability measured by C_{pk}. As shown in Table 10.2, for the piston data, estimates of C_p and C_{pk} are $\hat{C}_p = 0.56$ and $\hat{C}_{pk} = 0.24$. This indicates that something could be gained by centering the piston cycle times around 0.3 seconds. Even if this is possible to achieve, there will still be observations outside the upper and lower limits, since the standard deviation is too large. The validity of the C_p and C_{pk} indices is questionable in cases where the measurements on X are not normally distributed, but have skewed distributions. The proper form of a capability index under non-normal conditions is yet to be developed (Kotz and Johnson 1994).

It is common practice to estimate C_p or C_{pk}, by substituting the sample mean, $\overline{X}$, and the sample standard deviation S, for the process mean, μ, and the process standard deviation σ, i.e.,

$$\hat{C}_{pu} = \frac{\xi_U - \overline{X}}{3S}, \quad \hat{C}_{pl} = \frac{\overline{X} - \xi_L}{3S} \quad (10.4.4)$$

and $\hat{C}_{pk} = \min(\hat{C}_{pu}, \hat{C}_{pl})$, where ξ_L and ξ_U are the lower and upper specification limits. The question is how close is $\hat{C}_{pk}$ to the true process capability value? We develop below confidence intervals for C_{pk}, which have confidence levels close to the nominal $(1 - \alpha)$ in large samples. The derivation of these intervals depends on the following results:

1. In a large size random sample from a normal distribution, the sampling distribution of S is approximately normal, with mean σ and variance $\sigma^2/2n$.
2. In a random sample from a normal distribution, the sample mean, $\overline{X}$, and the sample standard deviation S are independent.
3. If A and B are events such that $\Pr\{A\} = 1 - \alpha/2$ and $\Pr\{B\} = 1 - \alpha/2$ then $\Pr\{A \cap B\} \geq 1 - \alpha$. (This inequality is called the **Bonferroni inequality**.)

In order to simplify notation, let

$$\rho_1 = C_{pl}, \quad \rho_2 = C_{pu}, \quad \text{and} \quad \omega = C_{pk}.$$

Notice that since $\overline{X}$ is distributed like $N\left(\mu, \frac{\sigma^2}{n}\right)$, $\overline{X} - \xi_L$ is distributed like $N\left(\mu - \xi_L, \frac{\sigma^2}{n}\right)$. Furthermore, by the above results 1 and 2, the distribution of $\overline{X} - \xi_L - 3S\rho_1$ in large samples is like that of $N\left(0, \frac{\sigma^2}{n}\left(1 + \frac{9}{2}\rho_1^2\right)\right)$. It follows that, in large samples

$$\frac{(\overline{X} - \xi_L - 3S\rho_1)^2}{\frac{S^2}{n}\left(1 + \frac{9}{2}\rho_1^2\right)}$$

is distributed like $F[1, n - 1]$. Or,

$$\Pr\left\{\frac{(\overline{X} - \xi_L - 3S\rho_1)^2}{\frac{S^2}{n}\left(1 + \frac{9}{2}\rho_1^2\right)} \leq F_{1-\alpha/2}[1, n - 1]\right\} = 1 - \alpha/2. \quad (10.4.5)$$

Table 10.3 *Confidence limits for C_{pk}, at level $(1 - \alpha)$*

Lower limit	Upper limit	Condition
$\rho_{1,\alpha}^{(L)}$	$\rho_{1,\alpha}^{(U)}$	$\rho_{1,\alpha}^{(U)} < \rho_{2,\alpha}^{(L)}$
$\rho_{1,\alpha}^{(L)}$	$\rho_{1,\alpha}^{(U)}$	$\rho_{1,\alpha}^{(L)} < \rho_{2,\alpha}^{(L)} < \rho_{1,\alpha}^{(U)} < \rho_{2,\alpha}^{(U)}$
$\rho_{1,\alpha}^{(L)}$	$\rho_{2,\alpha}^{(U)}$	$\rho_{1,\alpha}^{(L)} < \rho_{2,\alpha}^{(L)} < \rho_{2,\alpha}^{(U)} < \rho_{1,\alpha}^{(U)}$
$\rho_{2,\alpha}^{(L)}$	$\rho_{1,\alpha}^{(U)}$	$\rho_{2,\alpha}^{(L)} < \rho_{1,\alpha}^{(L)} < \rho_{1,\alpha}^{(U)} < \rho_{2,\alpha}^{(U)}$
$\rho_{2,\alpha}^{(L)}$	$\rho_{2,\alpha}^{(U)}$	$\rho_{2,\alpha}^{(L)} < \rho_{1,\alpha}^{(L)} < \rho_{2,\alpha}^{(U)} < \rho_{1,\alpha}^{(U)}$
$\rho_{2,\alpha}^{(L)}$	$\rho_{2,\alpha}^{(U)}$	$\rho_{2,\alpha}^{(U)} < \rho_{1,\alpha}^{(L)}$

Thus, let $\rho_{1,\alpha}^{(L)}$ and $\rho_{1,\alpha}^{(U)}$ be the two real roots (if they exist) of the quadratic equation in ρ_1

$$(\overline{X} - \xi_L)^2 - 6S\rho_1(\overline{X} - \xi_L) + 9S^2\rho_1^2 = F_{1-\alpha/2}[1, n-1]\frac{S^2}{n}\left(1 + \frac{9}{2}\rho_1^2\right). \tag{10.4.6}$$

Equivalently, $\rho_{1,\alpha}^{(L)}$ and $\rho_{1,\alpha}^{(U)}$ are the two real roots $(\rho_{1,\alpha}^{(L)} \le \rho_{1,\alpha}^{(U)})$ of the quadratic equation

$$9S^2\left(1 - \frac{F_{1-\alpha/2}[1, n-1]}{2n}\right)\rho_1^2 - 6S(\overline{X} - \xi_L)\rho_1 + \left((\overline{X} - \xi_L)^2 - \frac{F_{1-\alpha/2}[1, n-1]}{S^2 n}\right) = 0. \tag{10.4.7}$$

Substituting in this equation $(\overline{X} - \xi_L) = 3S\hat{C}_{pl}$, we obtain the equation

$$\left(1 - \frac{F_{1-\alpha/2}[1, n-1]}{2n}\right)\rho_1^2 - 2\hat{C}_{pl}\rho_1 + \left(\hat{C}_{pl}^2 - \frac{F_{1-\alpha/2}[1, n-1]}{9n}\right) = 0. \tag{10.4.8}$$

We assume that n satisfies $n > \frac{F_{1-\alpha}[1, n-1]}{2}$. Under this condition $1 - \frac{F_{1-\alpha/2}[1, n-1]}{2n} > 0$ and the two real roots of the quadratic equation are

$$\rho_{1,\alpha}^{(U,L)} = \frac{\hat{C}_{pl} \pm \sqrt{\frac{F_{1-\alpha/2}[1, n-1]}{n}\left(\frac{\hat{C}_{pl}^2}{2} + \frac{1}{9}\left(1 - \frac{F_{1-\alpha/2}^2[1, n-1]}{2n}\right)\right)}}{\left(1 - \frac{F_{1-\alpha/2}[1, n-1]}{2n}\right)}. \tag{10.4.9}$$

From the above inequalities, it follows that $(\rho_{1,\alpha}^{(L)}, \rho_{1,\alpha}^{(U)})$ is a confidence interval for ρ_1 at confidence level $1 - \alpha/2$.

Similarly, $(\rho_{2,\alpha}^{(L)}, \rho_{2,\alpha}^{(U)})$ is a confidence interval for ρ_2, at confidence level $1 - \alpha/2$, where $\rho_{2,\alpha}^{(U,L)}$ are obtained by replacing $\hat{C}_{pl}$ by $\hat{C}_{pu}$ in the above formula of $\rho_{1,\alpha}^{(U,L)}$. Finally, from the Bonferroni inequality and the fact that $C_{pk} = \min\{C_{pl}, C_{pu}\}$ we obtain that confidence limits for C_{pk}, at level of confidence $(1 - \alpha)$ are given in Table 10.3.

Example 10.1. In the present example, we illustrate the computation of the confidence interval for C_{pk}. Suppose that the specification limits are $\xi_L = -1$ and $\xi_U = 1$. Suppose that $\mu = 0$ and $\sigma = 1/3$. In this case, $C_{pk} = 1$. We simulate now, using R, MINITAB, or JMP a sample of size $n = 20$, from a normal distribution with mean $\mu = 0$ and standard deviation

$\sigma = 1/3$. We obtain a random sample with $\overline{X} = 0.01366$ and standard deviation $S = 0.3757$. For this sample, $\hat{C}_{pl} = 0.8994$ and $\hat{C}_{pu} = 0.8752$. Thus, the estimate of C_{pk} is $\hat{C}_{pk} = 0.8752$. For $\alpha = 0.05$, $F_{0.975}[1, 19] = 5.9216$. Obviously, $n = 20 > \frac{F_{0.975}[1,19]}{2} = 2.9608$. According to the formula,

$$\rho_{1,0.05}^{(U,L)} = \frac{0.894 \pm \sqrt{\frac{5.9216}{20}\left(\frac{(0.8994)^2}{2} + \frac{1 - \frac{5.9216}{40}}{9}\right)^{1/2}}}{1 - \frac{5.9216}{40}}.$$

Thus, $\rho_{1,0.05}^{(L)} = 0.6845$ and $\rho_{1,0.05}^{(U)} = 1.5060$. Similarly, $\rho_{2,0.05}^{(L)} = 0.5859$ and $\rho_{2,0.05}^{(U)} = 1.4687$. Therefore, the confidence interval, at level 0.95, for C_{pk} is $(0.5859, 1.4687)$.

10.5   Seven tools for process control and process improvement

In this section, we review seven tools that have proven extremely effective in helping organizations control processes and implement process improvement projects. Some of these tools were presented in the preceding chapters. For completeness, all the tools are briefly reviewed with references given to earlier chapters.

The preface to the English edition of the famous text by Ishikawa (1986) on Quality Control states: "the book was written to introduce quality control practices in Japan which contributed tremendously to the country's economic and industrial development." The Japanese work force did indeed master an elementary set of tools that helped them improve processes. Seven of the tools were nicknamed the "magnificent seven" and they are: The flow chart, the check sheet, the run chart, the histogram, the Pareto chart, the scatterplot, and the cause and effect diagram.

10.5.1   Flow charts

Flow charts are used to describe a process being studied or to describe a desired sequence of a new, improved process. Often this is the first step taken by a team looking for ways to improve a process. The differences between how a process could work and how it actually does work exposes redundancies, misunderstandings, and general inefficiencies.

10.5.2   Check sheets

Check Sheets are basic manual data collection mechanisms. They consist of forms designed to tally the total number of occurrences of certain events by category. They are usually the starting point of data collection efforts. In setting up a check sheet, one needs to agree on the categories definitions, the data collection time frame, and the actual data collection method. An example of a check sheet is provided in Figure 10.15.

10.5.3   Run charts

Run charts are employed to visually represent data collected over time. They are also called connected time plots. Trends and consistent patterns are easily identified on run charts. Example of a run chart is given in Figure 10.2.

Defect	Period 7				Period 8				Total
	Wk 1	Wk 2	Wk 3	Wk 4	Wk 1	Wk 2	Wk 3	Wk 4	
Contamination	II		IIII	II	II	III	I	III	17
Mixed Product			I	II	III		I		7
Wrong Product	III			I	II		II	IIII	12
Electrical	I	IIII	I	THL	II	II	THL	II	22
Bent Leads	II		I			I	II		6
Chips		II							2
Permanency	I		II			I			4
Total	9	6	9	10	9	7	11	9	70

Figure 10.15   *A typical heck sheet.*

10.5.4   Histograms

The histogram was presented in Section 2.4 as a graphical display of the distribution of measurements collected as a sample. It shows the frequency or number of observations of a particular value or within a specified group. Histograms are used extensively in process capability studies to provide clues about the characteristics of the process generating the data. However, as we saw in Section 10.3, they ignore information on the order by which the data was collected.

10.5.5   Pareto charts

Pareto charts are used extensively in modern organizations. These charts help to focus on the important few causes for trouble. When observations are collected and classified into different categories using valid and clear criteria, one can construct a Pareto chart. The Pareto chart is a display, using bar graphs sorted in descending order, of the relative importance of events such as errors, by category. The importance can be determined by the frequency of occurrence or weighted, for example, by considering the product of occurrence and cost. Superimposed on the bars is a cumulative curve that helps point out the important few categories that contain most of cases. Pareto charts are used to choose the starting point for problem-solving, monitor changes, or identify the basic cause of a problem. Their usefulness stems from the Pareto principle which states that in any group of factors contributing to a common effect, a relative few (20%) account for most of the effect (80%). A Pareto chart of software errors found in testing a PBX electronic switch is presented in Figure 10.16. Errors are labeled according to the software unit where they occurred. For example the "EKT" (electronic key telephone) category makes up 6.5% of the errors. What can we learn from this about the software development process? The "GEN," "VHS," and "HI" categories account for over 80% of the errors. These are the causes of problems on which major improvements efforts should initially concentrate. Section 10.4.2 discusses a statistical test for comparing Pareto charts. Such tests are necessary if one wants to distinguish between differences that can be attributed to random noise and significant differences that should be investigated for identifying an assignable cause.

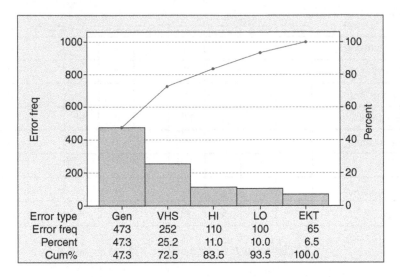

Figure 10.16   *Pareto chart of software errors (MINITAB).*

10.5.6   Scatterplots

Scatterplots are used to exhibit what happens to one variable, when another variable changes. Such information is needed in order to test a theory or make forecasts. For example one might want to verify the theory that the relative number of errors found in engineering drawings declines with increasing drawing sizes.

10.5.7   Cause and effect diagrams

Cause and effect diagrams (also called **fishbone charts** or **Ishikawa diagrams**) are used to identify, explore, and display all the possible causes of a problem or event. The diagram is usually completed in a meeting of individuals who have first-hand knowledge of the problem investigated. Figure 10.17 shows a cause and effect diagram listing causes for falls of hospitalized patients. A typical group to convene for completing such a diagram consists of nurses, physicians, administrative staff, housekeeping personnel, and physiotherapists. It is standard practice to weight the causes by impact on the problem investigated and then initiate projects to reduce the harmful effects of the main causes. Cause and effect diagrams can be derived after data was collected and presented, for example using a Pareto chart or be entirely based on collective experience without supporting data.

The successful efforts of previous improvement are data driven. In attempting to reduce levels of defects, or output variability, a team will typically begin by collecting data and charting the process. Flow charts and check sheets are used in these early stages. Run charts, histograms, and Pareto charts can then be prepared from the data collected on check sheets or otherwise. Diagnosing the current process is carried out, using in addition scatter plots and cause and effect diagrams. Once solutions for improvement are implemented, their impact can be assessed using run charts, histograms, and Pareto charts. A statistical test for comparing Pareto charts is presented next.

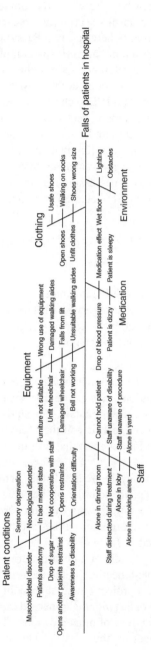

Figure 10.17 *A cause and effect diagram of patient falls during hospitalization (JMP).*

10.6   Statistical analysis of Pareto charts

Pareto charts are often compared over time or across processes. In such comparisons, one needs to know whether differences between two Pareto charts should be attributed to random variation or to special significant causes. In this section, we present a statistical test that is used to flag statistically significant differences between two Pareto charts (Kenett 1991). Once the classification of observations into different categories is completed, we have the actual number of observations, per category. The reference Pareto chart is a Pareto chart constructed in an earlier time period or on a different, but comparable, process. Other terms for the reference Pareto chart are the benchmark or standard Pareto chart. The proportion of observations in each category of the reference Pareto chart is the expected proportion. We expect to find these proportions in Pareto charts of data collected under the conditions of the reference Pareto chart. The expected number of observations in the different categories of the Pareto chart is computed by multiplying the total number of observations in a Pareto chart by the corresponding expected proportion. The standardized residuals are assessing the significance of the deviations between the new Pareto chart and the reference Pareto chart. The statistical test relies on computation of standardized residuals:

$$Z_i = (n_i - Np_i)/[Np_i(1 - p_i)]^{1/2}, \quad i = 1, \ldots, K, \tag{10.6.1}$$

where

$N =$ the total number of observations in Pareto chart
$p_i =$ the proportion of observations in category i, in reference Pareto chart
$Np_i =$ the expected number of observations in category i, given a total of N observations
$n_i =$ the actual number of observations in category i.

In performing the statistical test, one assumes that observations are independently classified into distinct categories. The actual classification into categories might depend on the data gathering protocol. Typically, the classification relies on the first error-cause encountered. A different test procedure could therefore produce different data. The statistical test presented here is more powerful than the standard Chi-squared test. It will therefore recognize differences between a reference Pareto chart and a current Pareto chart that will not be determined significant by the Chi-squared test.

In order to perform the statistical analysis, we first list the error categories in a fixed, unsorted order. A natural order to use is the alphabetic order of the categories' names. This organization of the data is necessary in order to permit meaningful comparisons. The test itself consists of seven steps. The last step being an interpretation of the results. To demonstrate these steps, we use data on timecard errors presented in Table 10.4.

The data comes from a monitoring system of timecard entries in a medium size company with 15 departments. During a management meeting, the human resources manager of the company was asked to initiate an improvement project aimed at reducing timecard errors. The manager asked to see a reference Pareto chart of last months' timecard errors by department. Departments # 6, 7, 8, and 12 were responsible for 46% of timecard errors. The manager appointed a special improvement team to learn the causes for these errors. The team recommended to change the format of the time card. The new format was implemented throughout the company. Three weeks later, a new Pareto chart was prepared from 346 newly reported timecard errors. A statistical analysis of the new Pareto chart was performed in order to determine what department had a significant change in its relative contribution of timecard errors.

Table 10.4   *Timecard errors data in 15 departments*

Department #	Reference Pareto	Current Pareto
1	23	14
2	42	7
3	37	85
4	36	19
5	17	23
6	50	13
7	60	48
8	74	59
9	30	2
10	25	0
11	10	12
12	54	14
13	23	30
14	24	20
15	11	0

The steps in applying the statistical test are as follows:

1. Compute for each department its proportion of observations in the reference Pareto chart:

$$p_1 = 23/516 = 0.04457$$

$$\vdots$$

$$p_{15} = 11/516 = 0.0213$$

2. Compute the total number of observations in the new Pareto chart:

$$N = 14 + 7 + 85 + \cdots + 20 = 346.$$

3. Compute the expected number of observations in department # i, $E_i = Np_i$, $i = 1, \ldots, 15$.

$$E_1 = 346 \times 0.04457 = 15.42$$

$$\vdots$$

$$E_{15} = 346 \times 0.0213 = 7.38$$

4. Compute the standardized residuals: $Z_i = (N_i - Np_i)/(Np_i(1 - p_i))^{1/2}$, $i = 1, \ldots, 15$.

$$Z_1 = (14 - 15.42)/[15.42(1 - 0.04457)]^{1/2} = -0.37$$

$$\vdots$$

$$Z_{15} = (0 - 7.38)/[7.38(1 - 0.0213)]^{1/2} = -2.75$$

5. Look up Table 10.5 for $K = 15$. Interpolate between $K = 10$ and $K = 20$. For $\alpha = 0.01$ significance level, the critical value is approximately $(3.10 + 3.30)/2 = 3.20$.

Table 10.5   *Critical values for standardized residuals*

	Significance level		
K	10%	5%	1%
4	1.95	2.24	2.81
5	2.05	2.32	2.88
6	2.12	2.39	2.93
7	2.18	2.44	2.99
8	2.23	2.49	3.04
9	2.28	2.53	3.07
10	2.32	2.57	3.10
20	2.67	2.81	3.30
30	2.71	2.94	3.46

Table 10.6   *Table of standardized residuals for the timecards error data*

Department #	Pareto	E_i	Z_i
1	14	5.42	−0.37
2	7	5.09	−4.16*
3	85	4.80	12.54*
4	19	4.74	−1.08
5	23	3.32	3.49*
6	13	5.50	−3.73*
7	48	5.96	1.30
8	59	6.52	1.44
9	2	4.35	−4.16*
10	0	3.99	−4.20*
11	12	2.56	2.06
12	14	5.69	−3.90*
13	30	3.84	3.80*
14	20	3.92	1.00
15	0	2.69	−2.75

6. Identify categories with standardized residuals larger, in absolute value, than 3.20. Table 10.6 indicates with a star the departments were the proportion of errors was significantly different from that in the reference Pareto.

7. Departments # 2, 3, 5, 6, 9, 10, 13, and 14 are flagged with a ∗ that indicates significant changes between the new Pareto data from the reference Pareto chart. In category 3, we expected 24.81 occurrences, a much smaller number than the actual 85.

The statistical test enables us to systematically compare two Pareto charts with the same categories. Focusing on the differences between Pareto charts complements the analysis of trends and changes in overall process error levels. Increases or decreases in such error levels may result from changes across all error categories. On the other hand, there may be no changes in error levels but significant changes in the mix of errors across categories. The statistical analysis reveals such changes. Another advantage of the statistical procedure is that it can apply to different time frames. For example, the reference Pareto can cover a period of one year and the current Pareto can span a period of three weeks.

The critical values are computed on the basis of the **Bonferroni inequality** approximation. This inequality states that, since we are examining simultaneously K standardized residuals, the overall significance level is not more than K times the significance level of an individual comparison. Dividing the overall significance level of choice by K, and using the normal approximation produces the critical values in Table 10.5. For more details on this procedure, see Kenett (1991).

10.7   The Shewhart control charts

The Shewhart control charts is a detection procedure in which every h units of time a sample of size n is drawn from the process. Let θ denote a parameter of the distribution of the observed random sample $x_1, \ldots, x_n$. Let $\hat{\theta}_n$ denote an appropriate estimate of θ. If θ_0 is a desired operation level for the process, we construct around θ_0 two limits UCL and LCL. As long as $\text{LCL} \leq \hat{\theta}_n \leq$ UCL, we say that the process is under statistical control.

More specifically, suppose that $x_1, x_2, \ldots$ are normally distributed and independent. Every h hours (time units) a sample of n observations is taken.

Suppose that when the process is under control $x_i \sim N(\theta_0, \sigma^2)$. Suppose that σ^2 is known. We set $\hat{\theta}_n \equiv \bar{x}_n = \frac{1}{n} \sum_{j=1}^{n} x_j$. The control limits are

$$\text{UCL} = \theta_0 + 3 \frac{\sigma}{\sqrt{n}}$$

$$\text{LCL} = \theta_0 - 3 \frac{\sigma}{\sqrt{n}} \tag{10.7.1}$$

The warning limits are set at

$$\text{UWL} = \theta_0 + 2 \frac{\sigma}{\sqrt{n}}$$

$$\text{LWL} = \theta_0 - 2 \frac{\sigma}{\sqrt{n}} \tag{10.7.2}$$

Notice that:

 (i) The samples are independent.
 (ii) All $\bar{x}_n$ are identically $N\left(\theta_0, \frac{\sigma^2}{n}\right)$ as long as the process is under control.
(iii) If α is the probability of observing $\bar{x}_n$ outside the control limits, when $\theta = \theta_0$; then $\alpha = .0027$. We expect one every $N = 370$ samples to yield a value of $\bar{x}_n$ outside the control limits.
(iv) We expect about 5% of the $\bar{x}_n$ points to lie outside the warning limits, when the process is under control. Thus, although for testing once the null hypothesis $H_0 : \theta = \theta_0$ against $H_1 : \theta \neq \theta_0$, we may choose a level of significance $\alpha = 0.05$ and use the limits UWL, LWL as rejection limits, in the control case the situation is equivalent to that of simultaneously (or repeatedly) testing many hypotheses. For this reason we have to consider a much smaller α, like $\alpha = 0.0027$, of the 3-sigma limits.
 (v) In most of the practical applications of the Shewhart 3-sigma control charts, the samples taken are of small size, $n = 4$ or $n = 5$, and the frequency of samples is high (h small).

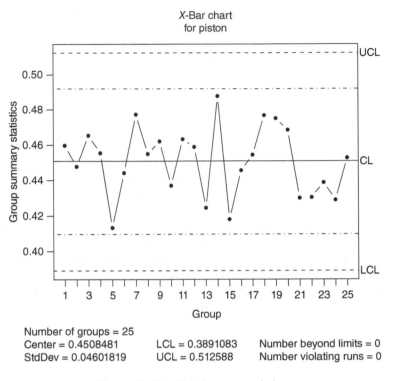

Figure 10.18   *Shewhart control chart.*

Shewhart recommended such small samples in order to reduce the possibility that a shift in θ will happen during sampling. On the other hand, if the samples are picked very frequently, there is a higher chance to detect a shift early. The question of how frequently to sample, and what should be the sample size is related to the idea of rational subgroups discussed earlier in Section 10.3. An economic approach to the determination of rational subgroups will be presented in Section 9.3.1. Figure 10.18 shows a control chart, with warning lines, tracking the piston cycle time at 25 successive time points. Each point represents an average of 4 consecutive measurements. The piston performance level, in the tracked period, is stable.

We provide now formulae for the control limits of certain Shewhart type control charts.

10.7.1   Control charts for attributes

We consider here control charts when the control statistic is the sample fraction defectives $\hat{p}_i = \frac{x_i}{n_i}$ $i = 1, \ldots, N$. Here n_i is the size of the ith sample and x_i is the number of defective items in the ith sample. It is desired that the sample size, n_i, will be the same over the samples.

Given N samples, we estimate the common parameter θ by $\hat{\theta} = \frac{\sum_{i=1}^{N} x_i}{\sum_{i=1}^{N} n_i}$. The upper and lower control limits are

$$LCL = \hat{\theta} - 3\sqrt{\frac{\hat{\theta}(1 - \hat{\theta})}{n}}$$

$$UCL = \hat{\theta} + 3\sqrt{\frac{\hat{\theta}(1 - \hat{\theta})}{n}}. \tag{10.7.3}$$

Table 10.7   *Number of defects in daily samples (sample size is n = 100)*

Sample/ day	# of defects	Sample/ day	# of defects
i	x_i	i	x_i
1	6	16	6
2	8	17	4
3	8	18	6
4	13	19	8
5	6	20	2
6	6	21	7
7	9	22	4
8	7	23	4
9	1	24	2
10	8	25	1
11	5	26	5
12	2	27	15
13	4	28	1
14	5	29	4
15	4	30	1
		31	5

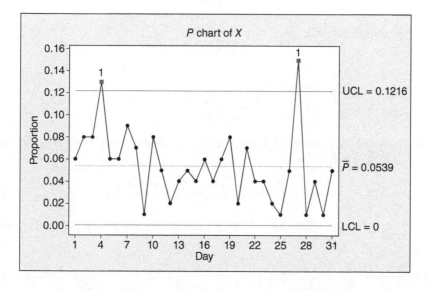

Figure 10.19   *p-chart for January data (MINITAB).*

In Table 10.7, we present, for example, the number of defective items found in random samples of size $n = 100$, drawn daily from a production line.

In Figure 10.19, we present the control chart for the data of Table 10.7. We see that there is indication that the fraction defectives in two days were significantly high, but the process on the whole remained under control during the month. Deleting these two days we can revise the control chart, by computing a modified estimate of θ. We obtain a new value of $\hat{\theta} = 139/2900 = 0.048$. This new estimator yields a revised upper control limit

$$UCL' = 0.112.$$

10.7.2   Control charts for variables

$\overline{X}$-charts

After the process has been observed for k sampling periods, we can compute estimates of the process mean and standard deviation. The estimate of the process mean is

$$\overline{\overline{X}} = \frac{1}{k}\sum_{i=1}^{k}\overline{X}_i.$$

This will be the centerline for the control chart. The process standard deviation can be estimated either using the average sample standard deviation

$$\hat{\sigma} = \overline{S}/c(n) \tag{10.7.4}$$

where

$$\overline{S} = \frac{1}{k}\sum_{i=1}^{k}S_i,$$

or the average sample range,

$$\hat{\sigma} = \overline{R}/d(n), \tag{10.7.5}$$

where

$$\overline{R} = \frac{1}{k}\sum_{i=1}^{k}R_i.$$

The factors $c(n)$ and $d(n)$ guarantee that we obtain unbiased estimates of σ. We can show, for example, that $E(\overline{S}) = \sigma c(n)$, where

$$c(n) = \left[\Gamma(n/2)/\Gamma\left(\frac{n-1}{2}\right)\right]\sqrt{2/(n-1)}. \tag{10.7.6}$$

Moreover $E\{R_n\} = \sigma d(n)$, where from the theory of order-statistics (see Section 3.7) we obtain that

$$d(n) = \frac{n(n-1)}{2\pi}\int_0^\infty y \int_{-\infty}^\infty \exp\left\{-\frac{x^2 + (y+x)^2}{2}\right\}[\Phi(x+y) - \Phi(x)]^{n-2}\,dx\,dy. \tag{10.7.7}$$

In Table 10.8, we present the factors $c(n)$ and $d(n)$ for $n = 2, 3, \ldots, 10$.

The control limits are now computed as

$$UCL = \overline{\overline{X}} + 3\hat{\sigma}/\sqrt{n}$$

and

$$\tag{10.7.8}$$

$$LCL = \overline{\overline{X}} - 3\hat{\sigma}/\sqrt{n}.$$

Table 10.8 *Factors c(n) and d(n) for estimating σ*

n	c(n)	d(n)
2	0.7979	1.2838
3	0.8862	1.6926
4	0.9213	2.0587
5	0.9400	2.3259
6	0.9515	2.5343
7	0.9594	2.7044
8	0.9650	2.8471
9	0.9693	2.9699
10	0.9727	3.0774

Despite the wide use of the sample ranges for estimating the process standard deviation, this method is neither very efficient nor robust. It is popular only because the sample range is easier to compute than the sample standard deviation. However, since many hand calculators now have built-in programs for computing the sample standard deviation, the computational advantage of the range should not be considered. In any case, the sample ranges should not be used when the sample size is greater than 10.

We illustrate the construction of an $\overline{X}$ chart for the data in Table 10.9.

These measurements represent the length (in cm) of the electrical contacts of relays in samples of size five, taken hourly from the running process. Both the sample standard deviation and the sample range are computed for each sample, for the purposes of illustration. The centerline for the control chart is $\overline{\overline{X}} = 2.005$. From Table 10.8, we find for $n = 5$, $c(5) = 0.9400$. Let

$$A_1 = 3/(c(5)\sqrt{n}) = 1.427.$$

The control limits are given by

$$\text{UCL} = \overline{\overline{X}} + A_1\overline{S} = 2.186$$

and

$$\text{LCL} = \overline{\overline{X}} - A_1\overline{S} = 1.824.$$

The resulting control chart is shown in Figure 10.20. If we use the **sample ranges** to determine the control limits, we first find that $d(5) = 2.326$ and

$$A_2 = 3/(d(5)\sqrt{n}) = 0.577.$$

This gives us control limits of

$$\text{UCL}' = \overline{\overline{X}} + A_2\overline{R} = 2.142$$

$$\text{LCL}' = \overline{\overline{X}} - A_2\overline{R} = 1.868.$$

Table 10.9   *20 Samples of five electric contact lengths*

Hour i	x_1	x_2	x_3	x_4	x_5	$\bar{X}$	S	R
1	1.9890	2.1080	2.0590	2.0110	2.0070	2.0348	0.04843	0.11900
2	1.8410	1.8900	2.0590	1.9160	1.9800	1.9372	0.08456	0.21800
3	2.0070	2.0970	2.0440	2.0810	2.0510	2.0560	0.03491	0.09000
4	2.0940	2.2690	2.0910	2.0970	1.9670	2.1036	0.10760	0.30200
5	1.9970	1.8140	1.9780	1.9960	1.9830	1.9536	0.07847	0.18300
6	2.0540	1.9700	2.1780	2.1010	1.9150	2.0436	0.10419	0.26300
7	2.0920	2.0300	1.8560	1.9060	1.9750	1.9718	0.09432	0.23600
8	2.0330	1.8500	2.1680	2.0850	2.0230	2.0318	0.11674	0.31800
9	2.0960	2.0960	1.8840	1.7800	2.0050	1.9722	0.13825	0.31600
10	2.0510	2.0380	1.7390	1.9530	1.9170	1.9396	0.12552	0.31200
11	1.9520	1.7930	1.8780	2.2310	1.9850	1.9678	0.16465	0.43800
12	2.0060	2.1410	1.9000	1.9430	1.8410	1.9662	0.11482	0.30000
13	2.1480	2.0130	2.0660	2.0050	2.0100	2.0484	0.06091	0.14300
14	1.8910	2.0890	2.0920	2.0230	1.9750	2.0140	0.08432	0.20100
15	2.0930	1.9230	1.9750	2.0140	2.0020	2.0014	0.06203	0.17000
16	2.2300	2.0580	2.0660	2.1990	2.1720	2.1450	0.07855	0.17200
17	1.8620	2.1710	1.9210	1.9800	1.7900	1.9448	0.14473	0.38100
18	2.0560	2.1250	1.9210	1.9200	1.9340	1.9912	0.09404	0.20500
19	1.8980	2.0000	2.0890	1.9020	2.0820	1.9942	0.09285	0.19100
20	2.0490	1.8790	2.0540	1.9260	2.0080	1.9832	0.07760	0.17500
					Average:	2.0050	0.09537	0.23665
						$\bar{\bar{X}}$	$\bar{S}$	$\bar{R}$

S-Charts and R-charts

As discussed earlier, control of the process variability can be as important as control of the process mean. Two types of control charts are commonly used for this purpose: an R-chart, based on sample ranges, and an S-chart, based on sample standard deviations. Since ranges are easier to compute than standard deviations, R-charts are probably more common in practice. The R-chart is not very efficient. In fact, its efficiency declines rapidly as the sample size increases and the sample range should not be used for a sample size greater than 5. However, we shall discuss both types of charts.

To construct control limits for the S-chart, we will use a normal approximation to the sampling distribution of the sample standard deviation, S. This means that we will use control limits

$$\text{LCL} = \bar{S} - 3\hat{\sigma}_s$$

and   (10.7.9)

$$\text{UCL} = \bar{S} + 3\hat{\sigma}_s,$$

where $\hat{\sigma}_s$ represents an estimate of the standard deviation of S. This standard deviation is

$$\sigma_s = \sigma/\sqrt{2(n-1)}.$$ (10.7.10)

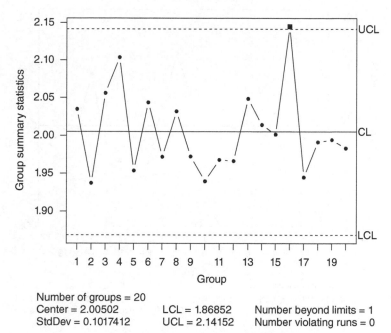

Number of groups = 20
Center = 2.00502   LCL = 1.86852   Number beyond limits = 1
StdDev = 0.1017412   UCL = 2.14152   Number violating runs = 0

Figure 10.20   *$\bar{X}$-control chart for contact data.*

Using the unbiased estimate $\hat{\sigma} = \bar{S}/c(n)$, we obtain

$$\hat{\sigma}_s = \bar{S}/(c(n)\sqrt{2(n-1)}) \qquad (10.7.11)$$

and hence the control limits

$$LCL = \bar{S} - 3\bar{S}/(c(n)\sqrt{2(n-1)}) = B_3\bar{S}$$

and $(10.7.12)$

$$UCL = \bar{S} + 3\bar{S}/(c(n)\sqrt{2(n-1)}) = B_4\bar{S}.$$

The factors B_3 and B_4 can be determined from Table 10.8.
  Using the electrical contact data in Table 10.9, we find

$$\text{centerline} = \bar{S} = 0.095,$$
$$LCL = B_3\bar{S} = 0,$$

and

$$UCL = B_4\bar{S} = 2.089(0.095) = 0.199.$$

The *S*-chart is given in Figure 10.21.

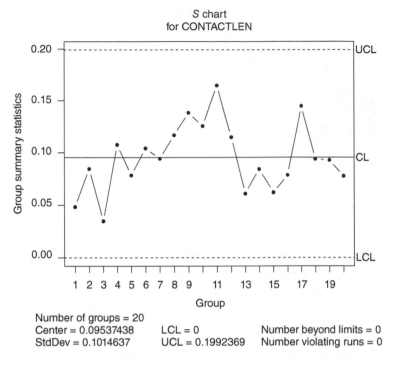

Figure 10.21   *S-chart for contact data.*

An R-chart is constructed using similar techniques, with a centerline $= \overline{R}$, and control limits:

$$LCL = D_3\overline{R},$$

and   (10.7.13)

$$UCL = D_4\overline{R},$$

where

$$D_3 = \left(1 - \frac{3}{d(n)\sqrt{2(n-1)}}\right)^+ \text{ and } D_4 = \left(1 + \frac{3}{d(n)\sqrt{2(n-1)}}\right).$$ (10.7.14)

Using the data of Table 10.9 we find

$$\text{centerline} = \overline{R} = 0.237,$$

$$LCL = D_3\overline{R} = 0,$$

and

$$UCL = D_4\overline{R} = (2.114)(0.237) = 0.501.$$

The R-chart is shown in Figure 10.22.

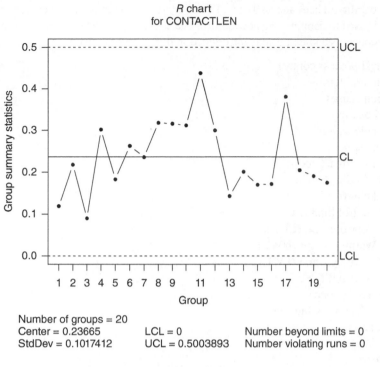

Figure 10.22 *R-Chart for contact data.*

The decision of whether to use an R-chart or S-chart to control variability ultimately depends on which method works best in a given situation. Both methods are based on several approximations. There is, however, one additional point that should be considered. The average value of the range of n variables depends to a great extent on the sample size n. As n increases, the range increases. The R-chart based on 5 observations per sample will look quite different from an R-chart based on 10 observations. For this reason, it is difficult to visualize the variability characteristics of the process directly from the data. On the other hand, the sample standard deviation, S, used in the S-chart, is a good estimate of the process standard deviation σ. As the sample size increases, S will tend to be even closer to the true value of σ. The process standard deviation is the key to understanding the variability of the process. An alternative approach, originally proposed by Fuchs and Kenett (1987), is to determine control limits on the basis of tolerance intervals derived either analytically (Section 4.5) or through nonparametric bootstrapping (Section 4.10). We discuss this approach, in the multivariate case, in Section 12.5.

10.8 Chapter highlights

Competitive pressures are forcing many management teams to focus on process control and process improvement, as an alternative to screening and inspection. This chapter discusses techniques used effectively in industrial organizations that have adopted such ideas as concepts. Classical control charts, quality control and quality planning tools are presented along with modern SPC procedures including new statistical techniques for constructing confidence intervals of process capability indices and analyzing Pareto charts. Throughout the chapter, a software piston simulator is used to demonstrate how control charts are set up and used in real-life applications. The main concepts and definitions introduced in this chapter include:

- Statistical process control
- Chronic problems
- Common causes
- Special causes
- Assignable causes
- Sporadic spikes
- External feedback loops
- Internal feedback loops
- Control charts
- Lower control limit (LCL)
- Upper Control Limit (UCL)
- Upper Warning Limit (UWL)
- Lower Warning Limit (LWL)
- Process Capability Study
- Rational Subgroups
- Process Capability Indexes
- Flow Charts
- Check Sheets
- Run Charts
- Histograms
- Pareto Charts
- Scatter Plots
- Cause and Effect Diagrams
- Control Charts for Attributes
- Control Charts for Variables
- Cumulative Sum Control Charts
- Average Run Length

10.9   Exercises

10.1 Use R, MINITAB, or JMP and file **OELECT.csv** to chart the individual electrical outputs of the 99 circuits. Do you observe any trend or nonrandom pattern in the data? (Use under SPC the option of Individual chart. For Mu and Sigma use "historical" values, which are $\bar{X}$ and S.)

10.2 Chart the individual variability of the length of steel rods, in **STEELROD.csv** file. Is there any perceived assignable cause of nonrandomness?

10.3 Examine the chart of the previous Exercise for possible patterns of nonrandomness.

10.4 Test the dat in file **OTURB2.csv** for lack of randomness. In this file we have three columns. In the first we have the sample size. In the second and third we have the sample means and standard deviation. If you use MINITAB, you can chart the individual means. For the historical mean use the mean of column $c2$. For historical standard deviation use $(\hat{\sigma}^2/5)^{1/2}$, where $\hat{\sigma}^2$ is the pooled sample **variance**.

10.5 A sudden change in a process lowers the process mean by one standard deviation. It has been determined that the quality characteristic being measured is approximately normally distributed and that the change had no effect on the process variance.

a. What percentage of points are expected to fall outside the control limits on the $\overline{X}$ chart if the subgroup size is 4?

b. Answer the same question for subgroups of size 6.

c. Answer the same question for subgroups of size 9.

10.6 Make capability analysis of the electric output (V) of 99 circuits in data file **OELECT.csv**, with target value of $\mu_0 = 220$ and LSL $= 210$, USL $= 230$.

10.7 Estimate the capability index C_{pk} for the output of the electronic circuits, based on data file **OELECT.csv** when LSL $= 210$ and USL $= 230$. Determine the point estimate as well as its confidence interval, with confidence level 0.95.

10.8 Estimate the capability index for the steel rods, given in data file **STEELROD.csv**, when the length specifications are $\xi_L = 19$ and $\xi_U = 21$ [cm] and the level of confidence is $1 - \alpha = 0.95$.

10.9 The specification limits of the piston cycle times are 0.3 ± 0.2 seconds. Generate 20 cycle times at the lower level of the 7 control parameters.

a. Compute C_p and C_{pk}.

b. Compute a 95% confidence interval for C_{pk}.
       Generate 20 cycle times at the upper level of the 7 control factors.

c. Recompute C_p and C_{pk}.

d. Recompute a 95% confidence interval for C_{pk}.

e. Is there a significant difference in process capability between lower and upper operating levels in the piston simulator?

10.10 A fiber manufacturer has a large contract which stipulates that its fiber, among other properties, have tensile strength greater than 1.800 [grams/fiber] in 95% of the fiber used. The manufacturer states the standard deviation of the process is 0.015 grams.

a. Assuming a process under statistical control, what is the smallest nominal value of the mean that will assure compliance with the contract?

b. Given the nominal value in part (a) what are the control limits of $\overline{X}$ and S charts for subgroups of size 6?

c. What is the process capability, if the process mean is $\mu = 1.82$?

10.11 The output voltage of a power supply is specified as 350 ± 5 V DC. Subgroups of four units are drawn from every batch and submitted to special quality control tests. The data from 30 subgroups on output voltage produced $\sum_{i=1}^{30} \overline{X} = 10,500.00$ and $\sum_{i=1}^{30} R_i = 86.5$

a. Compute the control limits for $\overline{X}$ and R.

b. Assuming statistical control and a normal distribution of output voltage, what properties of defective product is being made?

c. If the power supplies are set to a nominal value of 350 V, what is now the proportion of defective products?

d. Compute the new control limits for $\overline{X}$ and R.

e. If these new control limits are used but the adjustment to 350 V is not carried out, what is the probability that this fact will not be detected on the first subgroup?

f. What is the process capability before and after the adjustment of the nominal value to 350 V? Compute both C_p and C_{pk}.

10.12 The following data were collected in a circuit pack production plant during October

	Number of nonconformities
Missing component	293
Wrong component	431
Too much solder	120
Insufficient solder	132
Failed component	183

An improvement team recommended several changes that were implemented in the first week of November. The following data were collected in the second week of November.

	Number of nonconformities
Missing component	34
Wrong component	52
Too much solder	25
Insufficient solder	34
Failed component	18

    a. Construct Pareto charts of the nonconformities in October and the second week of November.

    b. Has the improvement team produced significant differences in the type of nonconformities?

10.13 Control charts for $\overline{X}$ and R are maintained on total soluble solids produced at 20°C in parts per million (ppm). Samples are drawn from production containers every hour and tested in a special test device. The test results are organized into subgroups of $n = 5$ measurements, corresponding to 5 hours of production. After 125 hours of production we find that $\sum_{i=1}^{25} \overline{X}_i = 390.8$ and $\sum_{i=1}^{25} R_i = 84$. The specification on the process states that containers with more than 18 ppm of total soluble solids should be reprocessed.

    a. Compute an appropriate capability index.

    b. Assuming a normal distribution and statistical control, what proportion of the sample measurements are expected to be out of spec?

    c. Compute the control limits for $\overline{X}$ and R.

10.14 Part I: Run the piston simulator at the lower levels of the 7 piston parameters and generate 100 cycle times. Add 2.0 to the last 50 cycle times.

    a. Compute control limits of $\overline{X}$ and R by constructing subgroups of size 5, and analyze the control charts.

    Part II: Assign a random number, R_i, from $U(0, 1)$, to each cycle time. Sort the 100 cycle times by R_i, $i = 1, \ldots, 100$.

   b. Recompute the control limits of $\overline{X}$ and R and reanalyze the control charts.

   c. Explain the differences between (a) and (b).

10.15 Part I: Run the piston simulator by specifying the 7 piston parameters within their acceptable range. Record the 7 operating levels you used and generate 20 subgroups of size 5.

   a. Compute the control limits for $\overline{X}$ and S.

        Part II: Rerun the piston simulator at the same operating conditions and generate 20 subgroups of size 10.

   b. Recompute the control limits for $\overline{X}$ and S.

   c. Explain the differences between (a) and (b).

11

Advanced Methods of Statistical Process Control

Following Chapter 10, we present in this chapter more advanced methods of statistical process control (SPC). We start with testing whether data collected over time is randomly distributed around a mean level, or whether there is a trend or a shift in the data. The tests which we consider are nonparametric **run tests**. This is followed with a section on modified Shewhart type control charts for the mean. Modifications of Shewhart charts were introduced as SPC tools, in order to increase the power of sequential procedures to detect change. Section 11.3 is devoted to the problem of determining the size and frequency of samples for proper statistical control of processes by Shewhart control charts. In Section 11.4, we introduce an alternative control tool, based on cumulative sums we develop and study the famous **CUSUM** procedures based on Page's control schemes. In Section 11.5, Bayesian detection procedures are presented. Section 11.6 is devoted to procedures of process control which track the process level. Section 11.7 introduces tools from engineering control theory, which are useful in automatically controlled processes.

11.1 Tests of randomness

In performing process capability analysis (see Chapter 10) or analyzing retroactively data for constructing a control chart, the first thing we would like to test is whether these data are randomly distributed around their mean. This means that the process is statistically stable and only common causes affect the variability. In this section, we discuss such tests of randomness.

Consider a sample $x_1, x_2, \ldots, x_n$, where the index of the values of x indicates some kind of ordering. For example, x_1 is the first observed value of X, x_2 is the second observed value, etc., while x_n is the value observed last. If the sample is indeed random, there should be no significant relationship between the values of X and their position in the sample. Thus, tests of randomness usually test the hypothesis that all possible configurations of the x's are equally probable, against the alternative hypothesis that some significant clustering of members takes place. For example, suppose that we have sequence of 5 0's and 5 1's. The ordering 0, 1, 1, 0, 0, 0, 1, 1, 0, 1 seems to be random, while the ordering 0, 0, 0, 0, 0, 1, 1, 1, 1, 1 seems, conspicuously, not to be random.

Modern Industrial Statistics: With Applications in R, MINITAB and JMP, Third Edition.
Ron S. Kenett and Shelemyahu Zacks.
© 2021 John Wiley & Sons, Ltd. Published 2021 by John Wiley & Sons, Ltd.

11.1.1   Testing the number of runs

In a sequence of m_1 0's and m_2 1's, we distinguish between **runs** of 0's, i.e., an uninterrupted string of 0's, and **runs** of 1's. Accordingly, in the sequence 0 1 1 1 0 0 1 0 1 1, there are 4 0's and 6 1's and there are 3 runs of 0's and 3 runs of 1's, i.e., a total of 6 runs. We denote the total number of runs by R. The probability distribution of the total number of runs, R, is determined under the model of randomness. It can be shown that, if there are m_1 0's and m_2 1's, then

$$\Pr\{R = 2k\} = \frac{2 \binom{m_1 - 1}{k - 1} \binom{m_2 - 1}{k - 1}}{\binom{n}{m_2}} \tag{11.1.1}$$

and

$$\Pr\{R = 2k + 1\} = \frac{\binom{m_1 - 1}{k - 1} \binom{m_2 - 1}{k} + \binom{m_1 - 1}{k} \binom{m_2 - 1}{k - 1}}{\binom{n}{m_2}}. \tag{11.1.2}$$

Here, n is the sample size, $m_1 + m_2 = n$.

One alternative to the hypothesis of randomness is that there is a tendency for **clustering** of the 0's (or 1's). In such a case, we expect to observe longer runs of 0's (or 1's) and, consequently, a smaller number of total runs. In this case, the hypothesis of randomness is rejected if the total number of runs, R, is too small. On the other hand, there could be an alternative to randomness which is the reverse of clustering. This alternative is called "**mixing**." For example, the following sequence of 10 0's and 10 1's is completely mixed, and is obviously not random:

$$0, \ 1, \ 0, \ 1, \ 0, \ 1, \ 0, \ 1, \ 0, \ 1, \ 0, \ 1, \ 0, \ 1, \ 0, \ 1, \ 0, \ 1, \ 0, \ 1.$$

The total number of runs here is $R = 20$. Thus, if there are too many runs, one should also reject the hypothesis of randomness. Consequently, if we consider the null hypothesis H_0 of randomness against the alternative H_1 of clustering, the lower (left) tail of the distribution should be used for the rejection region. If the alternative, H_1, is the hypothesis of mixing, then the upper (right) tail of the distribution should be used. If the alternative is either clustering or mixing, the test should be two-sided.

We test the hypothesis of randomness by using the test statistic R, which is the total number of runs. The critical region for the one-sided alternative that there is clustering, is of the form:

$$R \le R_\alpha,$$

where R_p is the pth-quantile of the null distribution of R. For the one-sided alternative of mixing, we reject H_0 if $R \ge R_{1-\alpha}$. In cases of large samples, we can use the normal approximations

$$R_\alpha = \mu_R - z_{1-\alpha} \sigma_R$$

and

$$R_{1-\alpha} = \mu_R + z_{1-\alpha} \sigma_R,$$

where

$$\mu_R = 1 + 2 m_1 m_2 / n \tag{11.1.3}$$

and

$$\sigma_R = [2m_1m_2(2m_1m_2 - n)/n^2(n - 1)]^{1/2} \qquad (11.1.4)$$

are the mean and standard deviation, respectively, of R under the hypothesis of randomness. We can also use the normal distribution to approximate the P-value of the test. For one-sided tests, we have

$$\alpha_L = \Pr\{R \leq r\} \cong \Phi((r - \mu_R)/\sigma_R) \qquad (11.1.5)$$

and

$$\alpha_U = \Pr\{R \geq r\} \cong 1 - \Phi((r - \mu_R)/\sigma_R), \qquad (11.1.6)$$

where r is the observed number of runs. For the two-sided alternative, the P-value of the test is approximated by

$$\alpha' = \begin{cases} 2\alpha_L, & \text{if } R < \mu_R \\ 2\alpha_U, & \text{if } R > \mu_R. \end{cases} \qquad (11.1.7)$$

11.1.2   Runs above and below a specified level

The runs test for the randomness of a sequence of 0's and 1's can be applied to test whether the values in a sequence, which are continuous in nature, are randomly distributed. We can consider whether the values are above or below the sample average or the sample median. In such a case, every value above the specified level will be assigned the value 1, while all the others will be assigned the value 0. Once this is done, the previous runs test can be applied.

For example, suppose that we are given a sequence of $n = 30$ observations and we wish to test for randomness using the number of runs, R, above and below the median, M_e. There are

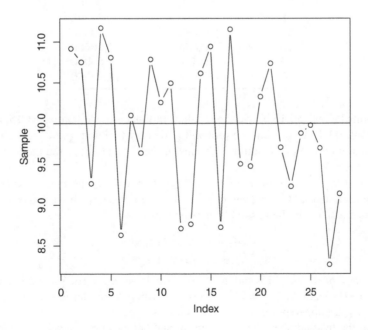

Figure 11.1   *Random normal sequence and the runs above and below the level 10.*

Table 11.1 *Distribution of R, in a random sample of size n = 30, m_1 = 15*

R	p.d.f.	c.d.f.
2	0.00000	0.00000
3	0.00000	0.00000
4	0.00000	0.00000
5	0.00002	0.00002
6	0.00011	0.00013
7	0.00043	0.00055
8	0.00171	0.00226
9	0.00470	0.00696
10	0.01292	0.01988
11	0.02584	0.04572
12	0.05168	0.09739
13	0.07752	0.17491
14	0.11627	0.29118
15	0.13288	0.42407
16	0.15187	0.57593
17	0.13288	0.70882
18	0.11627	0.82509
19	0.07752	0.90261
20	0.05168	0.95428
21	0.02584	0.98012
22	0.01292	0.99304
23	0.00470	0.99774
24	0.00171	0.99945
25	0.00043	0.99987
26	0.00011	0.99998
27	0.00002	1.00000
28	0.00000	1.00000
29	0.00000	1.00000
30	0.00000	1.00000

15 observations below and 15 above the median. In this case, we take $m_1 = 15$, $m_2 = 15$, and $n = 30$. In Table 11.1, we present the probability density function (p.d.f.) and cumulative distribution function (c.d.f.) of the number of runs, R, below and above the median, of a random sample of size $n = 30$.

For a level of significance of $\alpha = 0.05$ if $R \leq 10$ or $R \geq 21$, the two sided test rejects the hypothesis of randomness. Critical values for a two-sided runs test, above and below the median, can be obtained also by the large sample approximation

$$R_{\alpha/2} = \mu_R - z_{1-\alpha/2}\sigma_R$$
$$R_{1-\alpha/2} = \mu_R + z_{1-\alpha/2}\sigma_R \qquad (11.1.8)$$

Substituting $m = m_1 = m_2 = 15$ and $\alpha = 0.05$, we have $\mu_R = 16$, $\sigma_R = 2.69$, $z_{0.975} = 1.96$. Hence, $R_{\alpha/2} = 10.7$ and $R_{1-\alpha/2} = 21.3$. Thus, according to the large sample approximation, if $R \leq 10$ or $R \geq 22$, the hypothesis of randomness is rejected.

This test of the total number of runs, R, above and below a given level (e.g., the mean or the median of a sequence) can be performed by using R, MINITAB, or JMP.

Example 11.1.   In the present example, we have used MINITAB to perform a run test on a simulated random sample of size $n = 28$ from the normal distribution $N(10, 1)$. The test is of runs above and below the distribution mean 10. We obtain a total of $R = 14$ with $m_1 = 13$ values below and $m_2 = 15$ values above the mean. In Figure 11.1, we present this random sequence. The MINITAB analysis given below shows that one can accept the hypothesis of randomness.

```
MTB > print C1
C1
```

								Count
10.917	10.751	9.262	11.171	10.807	8.630	10.097	9.638	08
10.785	10.256	10.493	8.712	8.765	10.613	10.943	8.727	16
11.154	9.504	9.477	10.326	10.735	9.707	9.228	9.879	24
9.976	9.699	8.266	9.139					28

```
MTB > Runs 10.0 c1.

  C1
K = 10.0000
THE OBSERVED NO. OF RUNS = 14
THE EXPECTED NO. OF RUNS = 14.9286
13 OBSERVATIONS ABOVE K 15 BELOW
          THE TEST IS SIGNIFICANT AT 0.7193
          CANNOT REJECT AT ALPHA = 0.05
```

The MINITAB test is one sided. H_0 is rejected if R is large.
    In R, we use the following commands:

```
> data(RNORM10)
> X <- ifelse(RNORM10 <= 10,
            yes="l",
            no="u")
> X <- as.factor(X)
> library(tseries)
> runs.test(X,
         alternative="less")

        Runs Test

data:  X
Standard Normal = -0.35956, p-value = 0.3596
alternative hypothesis: less

> rm(X)
```

11.1.3   Runs up and down

Tests of the total number of runs above or below a specified level may not be sufficient in cases where the alternative hypothesis to randomness is cyclical fluctuations in the level of the process. For example, a sequence may show consistent fluctuations up and down, as in the following example:

$$-1, -0.75, -0.50, -0.25, \ 0, \ 0.5, \ 1, \ 0.5, \ 0.25, -0.75, \ \ldots.$$

Here we see a steady increase from -1 to 1 and then a steady decrease. This sequence is obviously not random, and even the previous test of the total number of runs above and below 0 will reject the hypothesis of randomness. If the development of the sequence is not as conspicuous as that above, as, for example, in the sequence

$$-1, -0.75, -0.50, \ 1, \ 0.5, -0.25, \ 0, \ 0.25, -0.25, \ 1, \ 0.5, \ 0.25, -0.75, \ \ldots,$$

the runs test above and below 0 may not reject the hypothesis of randomness. Indeed, in the present case, if we replace every negative number by 0 and every nonnegative number by $+1$, we find that $m_1 = 6$, $m_2 = 7$, and $R = 7$. In this case, the exact value of α_L is 0.5 and the hypothesis of randomness is not rejected.

To test for possible cyclical effects, we use a test of **runs up and down**. Let $x_1, x_2, \ldots, x_n$ be a given sequence, and let us define

$$y_i = \begin{cases} +1, & \text{if } x_i < x_{i+1} \\ -1, & \text{if } x_i \geq x_{i+1} \end{cases}$$

for $i = 1, \ldots, n-1$. We count, then, the total number of runs, R, up and down. A run up is a string of $+1$'s, while a run down is a string of -1's. In the previous sequence, we have the following values of x_i and y_i:

x_i	−1.00	−0.75	−0.50	1.00	.050	−0.25	0.00	0.25	−0.25	1.00	0.50	0.25	−0.75
y_i		1	1	1	−1	−1	1	1	−1	1	−1	−1	−1

We thus have a total of $R^* = 6$ runs, 3 up and 3 down, with $n = 13$.

To test the hypothesis of randomness based on the number of runs up and down, we need the null distribution of R^*.

When the sample size is sufficiently large, we can use the normal approximation

$$R_\alpha^* = \mu_{R^*} - z_{1-\alpha}\sigma_{R^*}$$

and

$$R_{1-\alpha}^* = \mu_{R^*} + z_{1-\alpha}\sigma_{R^*},$$

where

$$\mu_{R^*} = (2n-1)/3 \tag{11.1.9}$$

and

$$\sigma_{R^*} = [(16n - 29)/90]^{1/2}. \tag{11.1.10}$$

The attained significance levels are approximated by

$$\alpha_L^* = \Phi((r^* - \mu_{R^*})/\sigma_{R^*}) \tag{11.1.11}$$

and

$$\alpha_U^* = 1 - \Phi((r^* - \mu_{R^*})/\sigma_{R^*}). \tag{11.1.12}$$

Example 11.2.   The sample in data file **YARNSTRG.csv** contains 100 values of log-yarn strength. In this sample, there are $R^* = 64$ runs up or down, 32 runs up and 32 runs down. The expected value of R^* is

$$\mu_{R^*} = \frac{199}{3} = 66.33,$$

and its standard deviation is

$$\sigma_{R^*} = \left(\frac{1600 - 29}{90}\right)^{1/2} = 4.178.$$

The attained level of significance is

$$\alpha_L = \Phi\left(\frac{64 - 66.33}{4.178}\right) = 0.289.$$

The hypothesis of randomness is not rejected.

11.1.4   Testing the length of runs up and down

In the previous sections, we have considered tests of randomness based on the total number of runs. If the number of runs is small relative to the size of the sequence, n, we obviously expect some of the runs to be rather long. We shall now consider the question of just how long the runs can be under a state of randomness.

Consider runs up and down, and let R_k ($k = 1, 2, \ldots$) be the total number of runs, up or down, of length greater than or equal to k. Thus, $R_1 = R^*$, R_2 is the total number of runs, up or down, of length 2 or more, etc. The following are formulas for the expected values of each R_k, i.e., $E\{R_k\}$, under the assumption of randomness. Each expected value is expressed as a function of the size n of the sequence (Table 11.2).

Table 11.2   *Expected values of R_k as a function of n*

k	$E\{R_k\}$
1	$(2n - 1)/3$
2	$(3n - 5)/12$
3	$(4n - 11)/60$
4	$(5n - 19)/360$
5	$(6n - 29)/2520$
6	$(7n - 41)/20{,}160$
7	$(8n - 55)/181{,}440$

In general, we have

$$E\{R_k\} = \frac{2[n(k+1) - k^2 - k + 1]}{(k+2)!}, \quad 1 \le k \le n - 1. \tag{11.1.13}$$

If $k \ge 5$, we have $E\{R_k\} \doteq V\{R_k\}$ and the Poisson approximation to the probability distribution of R_k is considered good, provided $n > 20$. Thus, if $k \ge 5$, according to the Poisson approximation, we find

$$\Pr\{R_k \ge 1\} \doteq 1 - \exp(-E\{R_k\}). \tag{11.1.14}$$

For example, if $n = 50$, we present $E\{R_k\}$ and $\Pr\{R_k \ge 1\}$ in the following table.

k	$E\{R_k\}$	$\Pr\{R_k \ge 1\}$
5	0.1075	0.1020
6	0.0153	0.0152
7	0.0019	0.0019

We see in the above table that the probability to observe even 1 run, up or down, of length 6 or more is quite small. This is the reason for the rule of thumb, to **reject the hypothesis of randomness if a run is of length 6 or more**. This and other rules of thumb were presented in Chapter 10 for ongoing process control.

11.2   Modified Shewhart control charts for $\overline{X}$

The modified Shewhart control chart for $\overline{X}$, to detect possible shifts in the means of the parent distributions gives a signal to stop, whenever the sample means $\overline{X}$ fall outside the control limits $\theta_0 \pm a\dfrac{\sigma}{\sqrt{n}}$, or whenever a **run** of r sample means fall outside the warning limits (all on the same side) $\theta_0 \pm w\dfrac{\sigma}{\sqrt{n}}$.

We denote the modified scheme by (a, w, r). For example, 3-σ control charts, with warning lines at 2-σ and a run of $r = 4$ is denoted by $(3, 2, 4)$. If $r = \infty$, the scheme $(3, 0, \infty)$ is reduced to the common Shewhart 3-σ procedure. Similarly, the scheme $(a, 3, 1)$ for $a > 3$, is equivalent to the Shewhart 3-σ control charts. A control chart for a $(3, 1.5, 2)$ procedure is shown in Figure 11.2. The means are of samples of size 5. There is no run of length 2 or more between the warning and action limits.

The **run length** (RL), of a control chart, is the number of samples taken until an "out of control" alarm is given. The **average run length** (ARL) of an (a, w, r) plan is smaller than that of the simple Shewhart a-σ procedure. We denote the ARL of an (a, w, r) procedure by ARL (a, w, r). Obviously, if w and r are small, we will tend to stop too soon, even when the process is under control. For example, if $r = 1$, $w = 2$ then any procedure $(a, 2, 1)$ is equivalent to Shewhart 2-σ procedure, which stops on the average every 20 samples, when the process is under control. Weindling (1967) and Page (1962) derived the formula for the average run length ARL(a, w, r). Page used the theory of runs, while Weindling used another theory (Markov chains theory). An excellent expository paper discussing the results was published by Weindling et al. (1970).

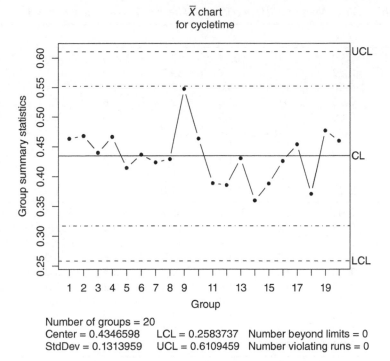

Number of groups = 20
Center = 0.4346598   LCL = 0.2583737   Number beyond limits = 0
StdDev = 0.1313959   UCL = 0.6109459   Number violating runs = 0

Figure 11.2 *A modified Shewhart X̄-chart.*

The basic formula for the determination of the average run length is

$$\text{ARL}_\theta(a, w, r) = \left[P_\theta(a) + H_\theta^r(a, w)\frac{1 - H_\theta(a, w)}{1 - H_\theta^r(a, w)} + L_\theta^r(a, w)\frac{1 - L_\theta(a, w)}{1 - L_\theta^r(a, w)} \right]^{-1}, \qquad (11.2.1)$$

where:

$$P_\theta(a) = P_\theta \left\{ \overline{X} \le \theta_0 - a\frac{\sigma}{\sqrt{n}} \right\} + P_\theta \left\{ \overline{X} \ge \theta_0 + a\frac{\sigma}{\sqrt{n}} \right\}$$

$$H_\theta(a, w) = P_\theta \left\{ \theta_0 + w\frac{\sigma}{\sqrt{n}} \le \overline{X} \le \theta_0 + a\frac{\sigma}{\sqrt{n}} \right\}$$

$$L_\theta(a, w) = P_\theta \left\{ \theta_0 - a\frac{\sigma}{\sqrt{n}} \le \overline{X} \le \theta_0 - w\frac{\sigma}{\sqrt{n}} \right\}. \qquad (11.2.2)$$

In Table 11.3, we present some values of ARL(a, w, r) for $a = 3$, $w = 1(0.5)2.5$, $r = 2(1)7$, when the samples are of size $n = 5$ from a normal distribution, and the shift in the mean is of size $\delta\sigma$. We see in the table that the procedures $(3, 1, 7)$, $(3, 1.5, 5)$, $(3, 2, 3)$, and $(3, 2.5, 2)$ yield similar ARL functions. However, these modified procedures are more efficient than the Shewhart 3-σ procedure. They all have close ARL values when $\delta = 0$, but when $\delta > 0$, their ARL values are considerably smaller than the Shewhart's procedure.

Table 11.3 Values of $ARL(a, w, r)a = 3.00$ against $\delta = (\mu_1 - \mu_0)/\sigma$, $n = 5$

w	$\delta \backslash r$	2	3	4	5	6	7
1.0	0.00	22.0	107.7	267.9	349.4	366.9	369.8
	0.25	5.1	15.7	39.7	76.0	107.2	123.5
	0.50	1.8	3.8	7.0	11.6	17.1	22.4
	0.75	0.9	1.6	2.5	3.6	4.7	5.9
	1.00	0.8	1.3	1.8	2.3	2.8	3.3
	1.25	1.0	1.4	1.8	2.0	2.2	2.3
	1.50	1.1	1.4	1.5	1.5	1.6	1.6
	1.75	1.1	1.2	1.2	1.2	1.2	1.2
	2.00	1.1	1.1	1.1	1.1	1.1	1.1
	2.25	1.0	1.0	1.0	1.0	1.0	1.0
1.5	0.00	93.1	310.1	365.7	370.1	370.4	370.4
	0.25	17.0	63.1	112.5	129.2	132.5	133.0
	0.50	4.5	11.4	20.3	28.0	31.5	32.3
	0.75	1.8	3.6	5.7	7.6	9.0	9.3
	1.00	1.2	2.0	2.7	3.4	3.8	4.1
	1.25	1.1	1.6	2.0	2.2	2.3	2.3
	1.50	1.2	1.4	1.5	1.6	1.6	1.6
	1.75	1.1	1.2	1.2	1.2	1.2	1.2
	2.00	1.1	1.1	1.1	1.1	1.1	1.1
	2.25	1.0	1.0	1.0	1.0	1.0	1.0
2.0	0.00	278.0	367.8	370.3	370.4	370.4	370.4
	0.25	61.3	123.4	132.5	133.1	133.2	133.2
	0.50	12.9	26.3	32.1	33.2	33.4	33.4
	0.75	4.2	7.6	9.6	10.4	10.7	10.7
	1.00	2.1	3.2	3.9	4.3	4.4	4.5
	1.25	1.5	2.0	2.2	2.3	2.4	2.4
	1.50	1.3	1.5	1.5	1.6	1.6	1.6
	1.75	1.2	1.2	1.2	1.2	1.2	1.2
	2.00	1.1	1.1	1.1	1.1	1.1	1.1
	2.25	1.0	1.0	1.0	1.0	1.0	1.0
2.5	0.00	364.0	370.3	370.4	370.4	370.4	370.4
	0.25	121.3	132.9	133.2	133.2	133.2	133.2
	0.50	23.1	33.1	33.4	33.4	33.4	33.4
	0.75	8.5	10.5	10.7	10.3	10.3	10.3
	1.00	3.6	4.3	4.5	4.5	4.5	4.5
	1.25	2.0	2.3	2.4	2.4	2.4	2.4
	1.50	1.4	1.5	1.6	1.6	1.6	1.6
	1.75	1.2	1.2	1.2	1.2	1.2	1.2
	2.00	1.1	1.1	1.1	1.1	1.1	1.1
	2.25	1.0	1.0	1.0	1.0	1.0	1.0

11.3   The size and frequency of sampling for Shewhart control charts

In the present section, we discuss the importance of designing the sampling procedure for Shewhart control charts. We start with the problem of the economic design of sampling for $\overline{X}$ charts.

11.3.1   The economic design for $\overline{X}$-charts

Duncan (1956, 1971, 1978) studied the question of optimally designing the $\overline{X}$ control charts. We show here, in a somewhat simpler fashion, how this problem can be approached. More specifically, assume that we sample from a normal population, and that σ^2 is known. A shift of size $\delta = (\theta_1 - \theta_0)/\sigma$ or larger should be detected with high probability.

Let c [\$/hour] be the hourly cost of a shift in the mean of size δ. Let d[\$] be the cost of sampling (and testing the items). Assuming that the time of shift from θ_0 to $\theta_1 = \theta_0 + \delta\sigma$ is exponentially distributed with mean $1/\lambda$ [hour], and that a penalty of 1[\$] is incurred for every unneeded inspection, the total expected cost is

$$K(h, n) \doteq \frac{ch + dn}{1 - \Phi(3 - \delta\sqrt{n})} + \frac{1 + dn}{\lambda h}. \tag{11.3.1}$$

This function can be minimized with respect to h and n, to determine the optimal sample size and frequency of sampling. Differentiating partially with respect to h and equating to zero, we obtain the formula of the optimal h, for a given n, namely

$$h^0 = \left(\frac{1 + d \cdot n}{c\lambda} \right)^{1/2} (1 - \Phi(3 - \delta\sqrt{n}))^{1/2}. \tag{11.3.2}$$

However, the function $K(h, n)$ is increasing with n, due to the contribution of the second term on the RHS. Thus, for this expected cost function we take every h^0 hours a sample of size $n = 4$. Some values of h^0 are:

δ	d	c	λ	h^0
2	0.5	3.0	0.0027	14.4
1	0.1	30.0	0.0027	1.5

For additional reading on this subject, see Gibra (1971).

11.3.2   Increasing the sensitivity of p-charts

The **operating characteristic function** for a Shewhart p-chart, is the probability, as a function of p, that the statistic $\hat{p}_n$ falls between the lower control limit (LCL) and upper control limit (UCL). Thus, the operating characteristic of a p-chart, with control limits $p_0 \pm 3\sqrt{\dfrac{p_0(1 - p_0)}{n}}$ is

$$\text{OC}(p) = \text{Pr}_\theta \left\{ p_0 - 3\sqrt{\frac{p_0(1 - p_0)}{n}} < \hat{p}_n < p_0 + 3\sqrt{\frac{p_0(1 - p_0)}{n}} \right\}, \tag{11.3.3}$$

where $\hat{p}_n$ is the proportion of defective items in the sample. $n \times \hat{p}_n$ has the binomial distribution, with c.d.f. $B(j; n, p)$. Accordingly,

$$OC(p) = B(np_0 + 3\sqrt{np_0(1-p_0)}; \ n, p) \\ - B(np_0 - 3\sqrt{np_0(1-p_0)}; \ n, p). \tag{11.3.4}$$

For large samples, we can use the normal approximation to $B(j; n, p)$ and obtain

$$OC(p) \cong \Phi\left(\frac{(UCL-p)\sqrt{n}}{\sqrt{p(1-p)}}\right) - \Phi\left(\frac{(LCL-p)\sqrt{n}}{\sqrt{p(1-p)}}\right). \tag{11.3.5}$$

The value of the $OC(p)$ at $p = p_0$ is $2\Phi(3) - 1 = 0.997$. The values of $OC(p)$ for $p \neq p_0$ are smaller. A typical $OC(p)$ function looks as in Figure 11.3.

When the process is in control with process fraction defective p_0, we have $OC(p_0) = 0.997$; otherwise $OC(p) < 0.997$. The probability that we will detect a change in quality to level p_1, with a **single** point outside the control limits, is $1 - OC(p_1)$. As an example, suppose we have estimated p_0 as $\bar{p} = 0.15$ from past data. With a sample of size $n = 100$, our control limits are

$$UCL = 0.15 + 3((0.15)(0.85)/100)^{1/2} = 0.257$$

and

$$LCL = 0.15 - 3((0.15)(0.85)/100)^{1/2} = 0.043.$$

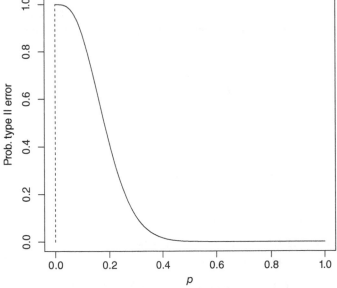

Figure 11.3 *Typical OC curve for a p-chart.*

Table 11.4 *Operating characteristic values for p-chart with $\overline{p} = 0.15$ and n = 100*

p	$OC(p)$	$1 - OC(p)$	$1 - [OC(p)]^5$
0.05	0.6255	0.3745	0.9043
0.10	0.9713	0.0287	0.1355
0.15	0.9974	0.0026	0.0130
0.20	0.9236	0.0764	0.3280
0.25	0.5636	0.4364	0.9432
0.30	0.1736	0.8264	0.9998
0.40	0.0018	0.9982	1.0000

In Table 11.4, we see that it is almost certain that a single point will fall outside the control limits when $p = 0.40$, but it is unlikely that it will fall there when $p = 0.20$. However, if the process fraction defective remains at the $p = 0.20$ level for several measurement periods, the probability of detecting the shift increases. The probability that at least one point falls outside the control limits when $p = 0.20$ for five consecutive periods is

$$1 - [OC(0.20)]^5 = 0.3279.$$

The probability of detecting shifts in the fraction defective is even greater than 0.33 if we apply run tests on the data.

The OC curve can also be useful for determining the required sample size for detecting, with high probability, a change in the process fraction defective in a single measurement period. To see this, suppose that the system is in control at level p_0, and we wish to detect a shift to level p_t with specified probability, $1 - \beta$. For example, to be 90% confident that the sample proportion will be outside the control limits immediately after the process fraction defective changes to p_t, we require that

$$1 - OC(p_t) = 0.90.$$

We can solve this equation to find that the required sample size is

$$n \doteq \frac{(3\sqrt{p_0(1 - p_0)} + z_{1-\beta}\sqrt{p_t(1 - p_t)})^2}{(p_t - p_0)^2}. \tag{11.3.6}$$

If we wish that with probability $(1 - \beta)$, the sample proportion will be outside the limits at least once within k sampling periods, when the precise fraction defective is p_t, the required sample size is

$$n \doteq \frac{(3\sqrt{p_0(1 - p_0)} + z_{1-b}\sqrt{p_t(1 - p_t)})^2}{(p_t - p_0)^2}, \tag{11.3.7}$$

where $b = \beta^{1/k}$.

These results are illustrated in Table 11.5 for a process with $p_0 = 0.15$.

It is practical to take at each period a small sample of $n = 5$. We see in Table 11.5 that in this case, a change from 0.15 to 0.40 would be detected within five periods with probability of 0.9. To detect smaller changes requires larger samples.

Table 11.5 *Sample size required for probability*
0.9 of detecting a shift to level p_t from level
$p_0 = 0.15$ (in one period and within five periods)

p_t	One period	Five periods
0.05	183	69
0.10	847	217
0.20	1003	156
0.25	265	35
0.30	122	14
0.40	46	5

11.4   Cumulative sum control charts

11.4.1   Upper Page's scheme

When the process level changes from a past or specified level, we expect that a control procedure will trigger an "alarm." Depending on the size of the change and the size of the sample, it may take several sampling periods before the alarm occurs. A method that has a smaller ARL than the standard Shewhart control charts, for detecting certain types of changes, is the **cumulative sum** (or CUSUM) control chart which was introduced by Barnard (1959) and Page (1954).

CUSUM charts differ from the common Shewhart control chart in several respects. The main difference is that instead of plotting the individual value of the statistic of interest, such as X, $\overline{X}$, S, R, p, or c, a statistic based on the cumulative sums is computed and tracked. By summing deviations of the individual statistic from a target value, T, we get a consistent increase, or decrease, of the cumulative sum when the process is above, or below, the target. In Figure 11.4, we show the behavior of the cumulative sums

$$S_t = \sum_{i=1}^{t}(X_i - 10) \tag{11.4.1}$$

of data simulated from a normal distribution with mean

$$\mu_t = \begin{cases} 10, & \text{if } t \le 20 \\ 13, & \text{if } t > 20 \end{cases}$$

and $\sigma_t = 1$ for all t.

We see that as soon as the shift in the mean of the data occurred, a pronounced drift in S_t started. Page (1954) suggested to detect an upward shift in the mean by considering the sequence

$$S_t^+ = \max\{S_{t-1}^+ + (X_k - K^+), 0\}, \quad t = 1, 2, \ldots, \tag{11.4.2}$$

where $S_0^+ \equiv 0$, and decide that a shift has occurred, as soon as $S_t^+ > h^+$. The statistics X_t, $t = 1, 2, \ldots$, upon which the (truncated) cumulative sums are constructed, could be means of samples of n observations, standard deviations, sample proportions, or individual observations. In the following section, we will see how the parameters K^+ and h^+ are determined. We will see that if X_t are means of samples of size n, with process variance σ^2, and if the desired process mean is θ_0 while the maximal tolerated process mean is θ_1, $\theta_1 - \theta_0 > 0$, then

$$K^+ = \frac{\theta_0 + \theta_1}{2} \quad \text{and} \quad h^+ = -\frac{\sigma^2 \log \alpha}{n(\theta_1 - \theta_0)}. \tag{11.4.3}$$

$0 < \alpha < 1.$

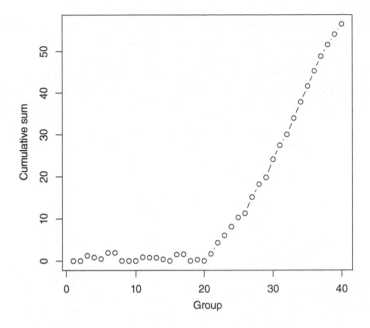

Figure 11.4 *A plot of cumulative sums with drift after t = 20.*

Example 11.3. The above procedure of Page, is now illustrated. The data in Table 11.6 represents the number of computer crashes per month, due to power failures experienced at a computer center, over a period of 28 months. After a crash, the computers are made operational with an "Initial Program Load." We refer to the data as the IPL data set.

Table 11.6 *Number of monthly computer crashes due to power failures*

t	X_t	t	X_t	t	X_t
1	0	11	0	21	0
2	2	12	0	22	1
3	0	13	0	23	3
4	0	14	0	24	2
5	3	15	0	25	1
6	3	16	2	26	1
7	0	17	2	27	3
8	0	18	1	28	5
9	2	19	0		
10	1	20	0		

Power failures are potentially very harmful. A computer center might be able to tolerate such failures when they are far enough apart. If they become too frequent, one might decide to invest in an uninterruptable power supply. It seems intuitively clear from Table 11.6 that computer crashes due to power failures become more frequent. Is the variability in failure rates due to chance alone (common causes) or can it be attributed to special causes that should be investigated? Suppose that the computer center can tolerate, at the most, an average of one power failure in three weeks (21 days) or $30/21 = 1.43$ crashes per month. It is desirable that there will be less than 1 failure per 6 weeks, or 0.71 per month. In Table 11.7, we show the computation of Page's statistics S_t^+, with $K^+ = \frac{1}{2}(0.71 + 1.43) = 1.07$.

Table 11.7   The S_t^+ statistics for the IPL data

t	X_t	$X_t - 1.07$	S_t^+
1	0	−1.07	0
2	2	0.93	0.93
3	0	−1.07	0
4	0	−1.07	0
5	3	1.93	1.93
6	3	1.93	3.86
7	0	−1.07	2.79
8	0	−1.07	1.72
9	2	0.93	2.65
10	1	−0.07	2.58
11	0	−1.07	1.51
12	0	−1.07	0.44
13	0	−1.07	0
14	0	−1.07	0
15	0	−1.07	0
16	2	0.93	0.93
17	2	0.93	1.86
18	1	−0.07	1.79
19	0	−1.07	0.72
20	0	−1.07	0
21	0	−1.07	0
22	1	−0.07	0
23	3	1.93	1.93
24	2	0.93	2.86
25	1	−0.07	2.79
26	1	−0.07	2.72
27	3	1.93	4.65
28	5	3.93	8.58

For $\alpha = 0.05$, $\sigma = 1$, and $n = 1$, we obtain the critical level $h^+ = 4.16$. Thus, we see that the first time an alarm is triggered is after the 27th month. In Figure 11.5, we present the graph of S_t^+ versus t. This graph is called a **CUSUM Chart**.

We see in Figure 11.5 that although S_6^+ is close to 4, the graph falls back toward zero, and there is no alarm triggered until the 27th month.

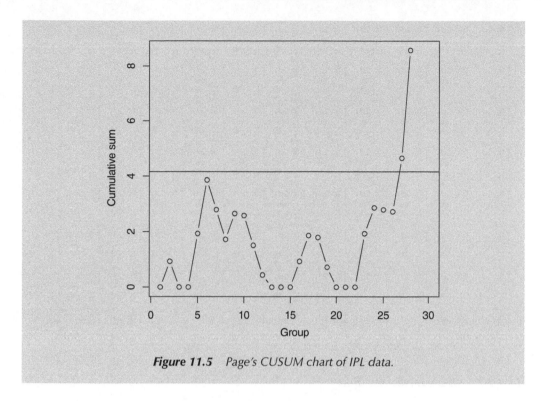

Figure 11.5   *Page's CUSUM chart of IPL data.*

11.4.2   Some theoretical background

Generally, if $X_1, X_2, \ldots$ is a sequence of independent and identically distributed (i.i.d.) random variables (continuous or discrete), having a p.d.f. $f(x; \theta)$, and we wish to test two simple hypotheses: $H_0 : \theta = \theta_0$ versus $H_1 : \theta = \theta_1$, with Type I and Type II error probabilities α and β, respectively, the Wald sequential probability ratio test (SPRT) is a sequential procedure which, after t observations, $t \geq 1$, considers the likelihood ratio

$$\Lambda(X_1, \ldots, X_t) = \prod_{i=1}^{t} \frac{f(X_i; \theta_1)}{f(X_i; \theta_0)}. \qquad (11.4.4)$$

If $\dfrac{\beta}{1-\alpha} < \Lambda(X_1, \ldots, X_t) < \dfrac{1-\beta}{\alpha}$, then another observation is taken; otherwise, sampling terminates. If $\Lambda(X_1, \ldots, X_t) < \dfrac{\beta}{1-\alpha}$, then H_0 is accepted; and if $\Lambda(X_1, \ldots, X_t) > \frac{1-\beta}{\alpha}$, H_0 is rejected.

In an upper control scheme, we can consider only the upper boundary, by setting $\beta = 0$. Thus, we can decide that the true hypothesis is H_1, as soon as

$$\sum_{i=1}^{t} \log \frac{f(X_i; \theta_1)}{f(X_i; \theta_0)} \geq -\log \alpha.$$

We will examine now the structure of this testing rule in a few special cases.

11.4.2.1   *Normal distribution*

We consider X_i to be normally distributed with **known** variance σ^2 and mean θ_0 or θ_1. In this case

$$\log \frac{f(X_i; \theta_1)}{f(X_i; \theta_0)} = -\frac{1}{2\sigma^2} \{(X_i - \theta_1)^2 - (X_i - \theta_0)^2\}$$

$$= \frac{\theta_1 - \theta_0}{\sigma^2} \left(X_i - \frac{\theta_0 + \theta_1}{2} \right). \tag{11.4.5}$$

Thus, the criterion

$$\sum_{i=1}^{t} \log \frac{f(X_i; \theta_1)}{f(X_i; \theta_0)} \geq -\log \alpha$$

is equivalent to

$$\sum_{i=1}^{t} \left(X_i - \frac{\theta_0 + \theta_1}{2} \right) \geq -\frac{\sigma^2 \log \alpha}{\theta_1 - \theta_0}.$$

For this reason, we use in the upper Page control scheme $K^+ = \dfrac{\theta_0 + \theta_1}{2}$, and $h^+ = -\dfrac{\sigma^2 \log \alpha}{\theta_1 - \theta_0}$.
If X_t is an average of n independent observations, then we replace σ^2 by σ^2/n.

11.4.2.2   *Binomial distributions*

Suppose that X_t has a binomial distribution $B(n, \theta)$. If $\theta \leq \theta_0$, the process level is under control. If $\theta \geq \theta_1$ the process level is out of control ($\theta_1 > \theta_0$). Since

$$f(x; \theta) = \binom{n}{x} \left(\frac{\theta}{1 - \theta} \right)^x (1 - \theta)^n,$$

$$\sum_{i=1}^{t} \log \frac{f(X_i; \theta_1)}{f(X_i; \theta_0)} \geq -\log \alpha \text{ if,}$$

$$\sum_{i=1}^{t} \left(X_i - \frac{n \log \left(\frac{1 - \theta_0}{1 - \theta_1} \right)}{\log \left(\frac{\theta_1}{1 - \theta_1} \cdot \frac{1 - \theta_0}{\theta_0} \right)} \right) \geq -\frac{\log \alpha}{\log \left(\frac{\theta_1}{1 - \theta_1} \cdot \frac{1 - \theta_0}{\theta_0} \right)}. \tag{11.4.6}$$

Accordingly, in an upper Page's control scheme, with binomial data, we use

$$K^+ = \frac{n \log \left(\frac{1 - \theta_0}{1 - \theta_1} \right)}{\log \left(\frac{\theta_1}{1 - \theta_1} \cdot \frac{1 - \theta_0}{\theta_0} \right)} \tag{11.4.7}$$

and

$$h^+ = -\frac{\log \alpha}{\log \left(\frac{\theta_1}{1 - \theta_1} \cdot \frac{1 - \theta_0}{\theta_0} \right)}. \tag{11.4.8}$$

11.4.2.3　*Poisson distributions*

When the statistics X_t have Poisson distribution with mean λ, then for specified levels λ_0 and λ_1, $0 < \lambda_0 < \lambda_1 < \infty$,

$$\sum_{i=1}^{t} \log \frac{f(X_i; \lambda_1)}{f(X_i; \lambda_0)} = \log\left(\frac{\lambda_1}{\lambda_0}\right) \sum_{i=1}^{t} X_i - t(\lambda_1 - \lambda_0). \tag{11.4.9}$$

It follows that the control parameters are

$$K^+ = \frac{\lambda_1 - \lambda_0}{\log(\lambda_1/\lambda_0)} \tag{11.4.10}$$

and

$$h^+ = -\frac{\log \alpha}{\log(\lambda_1/\lambda_0)}. \tag{11.4.11}$$

11.4.3　Lower and two-sided Page's scheme

In order to test whether a significant drop occurred in the process level (mean), we can use a lower page scheme. According to this scheme, we set $S_0^- \equiv 0$ and

$$S_t^- = \min\{S_{t-1}^- + (X_t - K^-), 0\}, \quad t = 1, 2, \ldots. \tag{11.4.12}$$

Here, the CUSUM values S_t^- are either zero or negative. We decide that a shift down in the process level, from θ_0 to θ_1, $\theta_1 < \theta_0$, occurred as soon as $S_t^- < h^-$. The control parameters K^- and h^- are determined by the formula of the previous section by setting $\theta_1 < \theta_0$.

Example 11.4. In file **COAL.csv**, one can find data on the number of coal mine disasters (explosions) in England, per year, for the period 1850–1961. These data are plotted in Figure 11.6. It seems that the average number of disasters per year dropped after 40 years

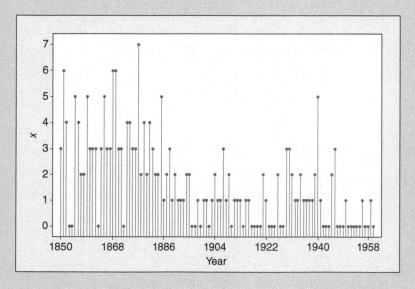

Figure 11.6　*Number of yearly coal mine disasters in England (MINITAB).*

from 3 to 2 and later settled around an average of one per year. We apply here the lower Page's scheme to see when do we detect this change for the first time. It is plausible to assume that the number of disasters per year, X_t, is a random variable having a Poisson distribution. We therefore set $\lambda_0 = 3$ and $\lambda_1 = 1$. The formula of the previous section, with K^+ and h^+ replaced by K^- and h^- yield, for $\alpha = 0.01$, $K^- = \dfrac{\lambda_1 - \lambda_0}{\log(\lambda_1/\lambda_0)} = 1.82$, and $h^- = -\dfrac{\log(0.01)}{\log(1/3)} = -4.19$.

In Table 11.8, we find the values of X_t, $X_t - K^-$ and S_t^- for $t = 1, \ldots, 50$. We see that $S_t^- < h^-$ for the first time at $t = 47$. The graph of S_t^- versus t is plotted in Figure 11.7.

Table 11.8 *Page's lower control scheme for the coal mine disasters data*

t	X_t	$X_t - K^-$	S_t^-
1	3	1.179	0
2	6	4.179	0
3	4	2.179	0
4	0	−1.820	−1.820
5	0	−1.820	−3.640
6	5	3.179	−0.461
7	4	2.179	0
8	2	0.179	0
9	2	0.179	0
10	5	3.179	0
11	3	1.179	0
12	3	1.179	0
13	3	1.179	0
14	0	−1.820	−1.820
15	3	1.179	−0.640
16	5	3.179	0
17	3	1.179	0
18	3	1.179	0
19	6	4.179	0
20	6	4.179	0
21	3	1.179	0
22	3	1.179	0
23	0	−1.820	−1.820
24	4	2.179	0
25	4	2.179	0
26	3	1.179	0
27	3	1.179	0
28	7	5.179	0
29	2	0.179	0
30	4	2.179	0
31	2	0.179	0
32	4	2.179	0
33	3	1.179	0
34	2	0.179	0
35	2	0.179	0

(continued)

Table 11.8   *(continued)*

t	X_t	$X_t - K^-$	S_t^-
36	5	3.179	0
37	1	−0.820	−0.820
38	2	0.179	−0.640
39	3	1.179	0
40	1	−0.820	−0.820
41	2	0.179	−0.640
42	1	−0.820	−1.461
43	1	−0.820	−2.281
44	1	−0.820	−3.102
45	2	0.179	−2.922
46	2	0.179	−2.743
47	0	−1.820	−4.563
48	0	−1.820	−6.384
49	1	−0.820	−7.204
50	0	−1.820	−9.025

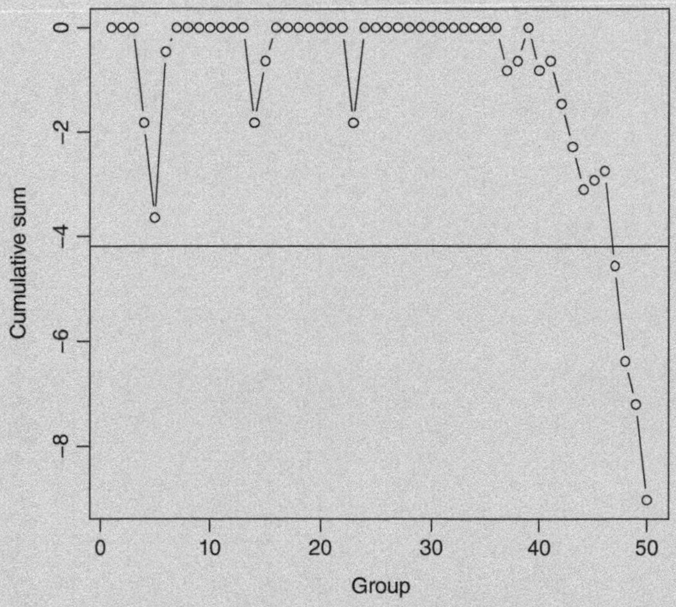

Figure 11.7   *Page's lower CUSUM control chart.*

If we wish to control simultaneously against changes in the process level in either upward or downward directions we use an upper and lower Page's schemes together, and trigger an alarm as soon as either $S_t^+ > h^+$ or $S_t^- < h^-$. Such a two sided scheme is denoted by the four control parameters (K^+, h^+, K^-, h^-).

Example 11.5. Yashchin (1991) illustrates the use of a two-sided Page's control scheme on data, which are the difference between the thickness of the grown silicon layer and its target value. He applied the control scheme ($K^+ = 3, h^+ = 9, K^- = -2, h^- = -5$). We present the values of X_t, S_t^+, and S_t^- in Table 11.9. We see in this table that $S_t^+ > h^+$ for the first time at $t = 40$. There is an indication that a significant drift upward in the level of thickness occurred.

Table 11.9   *Computation of (S_t^+, S_t^-) in a two-sided control scheme*

t	X_t	$X_t - K^+$	$X_t - K^-$	S_t^+	S_t^-
1	−4	−7	−2	0	−2
2	−1	−4	1	0	−1
3	3	0	5	0	0
4	−2	−5	0	0	0
5	−2.5	−5.5	−0.5	0	−0.5
6	−0.5	−3.5	1.5	0	0
7	1.5	−1.5	3.5	0	0
8	−3	−6	−1	0	−1
9	4	1	6	1	0
10	3.5	0.5	5.5	1.5	0
11	−2.5	−5.5	−0.5	0	−0.5
12	−3	−6	−1	0	−1.5
13	−3	−6	−1	0	−2.5
14	−0.5	−3.5	1.5	0	−1
15	−2.5	−5.5	−0.5	0	−1.5
16	1	−2	3	0	0
17	−1	−4	1	0	0
18	−3	−6	−1	0	−1
19	1	−2	3	0	0
20	4.5	−2	6.5	1.5	0
21	−3.5	−6.5	−1.5	0	−1.5
22	−3	−6	−1	0	−2.5
23	−1	−4	1	0	−1.5
24	4	1	6	1	0
25	−0.5	−3.5	1.5	0	0
26	−2.5	−5.5	−0.5	0	−0.5
27	4	1	6	1	0
28	−2	−5	0	0	0
29	−3	−6	−1	0	−1
30	−1.5	−4.5	0.5	0	−0.5
31	4	1	6	1	0
32	2.5	−0.5	4.5	0.5	0
33	−0.5	−3.5	1.5	0	0
34	7	4	9	4	0
35	5	2	7	6	0
36	4	1	6	7	0
37	4.5	1.5	6.5	8.5	0
38	2.5	−0.5	4.5	8	0
39	2.5	−0.5	4.5	7.5	0
40	5	2	3	9.5	0

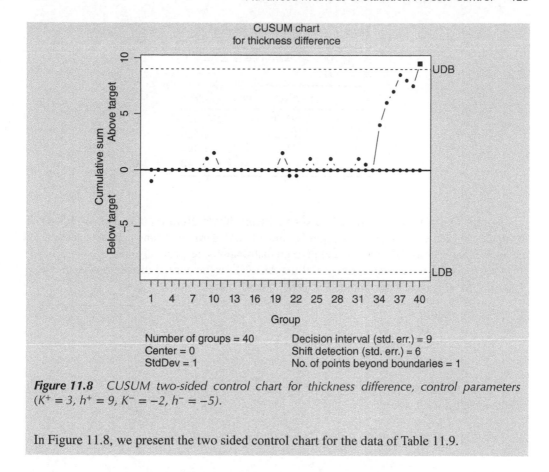

Figure 11.8 *CUSUM two-sided control chart for thickness difference, control parameters* ($K^+ = 3, h^+ = 9, K^- = -2, h^- = -5$).

In Figure 11.8, we present the two sided control chart for the data of Table 11.9.

The two-sided Page's control scheme can be boosted by changing the values of S_0^+ and S_0^- to nonzero. These are called **headstart values**. The introduction of nonzero headstarts was suggested by Lucas and Crosier (1982), in order to bring the history of the process into consideration, and accelerate the initial response of the scheme. Lucas (1982) suggested also to combine the CUSUM scheme with the Shewhart control chart. If any X_t value exceeds an upper limit UCL, or falls below a lower limit LCL, an alarm should be triggered.

11.4.4   Average run length, probability of false alarm, and conditional expected delay

The **run length** (RL) is defined as the number of time units until either $S_t^+ > h_t^+$ or $S_t^- < h^-$, for the first time. We have seen already that the **ARL** is an important characteristic of a control procedure, when there is either no change in the mean level (ARL(0)) or the mean level has shifted to $\mu_1 = \mu_0 + \delta\sigma$, before the control procedure started (ARL(δ)). When the shift from μ_0 to μ_1 occurs at some change point τ, $\tau > 0$, then we would like to know what is the probability of false alarm (PFA), i.e., that the RL is smaller than τ, and the conditional expected RL, given that RL $> \tau$. It is difficult to compute these characteristics of the Page control scheme analytically. The theory required for such an analysis is quite complicated (see Yashchin 1985). We provide computer programs which approximate these characteristics numerically, by simulation.

Table 11.10   *ARL(δ) estimates for the normal distribution, $\mu = \delta$, $\sigma = 1$ NR = 100, ($K^+ = 1$, $h^+ = 3$, $K^- = -1$, $h^- = -3$)*

δ	ARL	2$*$SE
0	1225.0	230.875
0.5	108.0	22.460
1.0	18.7	3.393
1.5	7.1	0.748

Programs **cusumArl**, **cusumPfaCedNorm**, **cusumPfaCedBinom**, and **cusumPfaCedPois** compute the average run length, ARL, probability of false alarm, PFA, and conditional expected delay, CED, for normal, binomial, and Poisson distributions, respectively.

In Table 11.10, we present estimates of the ARL(δ) for the normal distribution, with NR = 100 runs. SE = standard-deviation(RL)/$\sqrt{\text{NR}}$.

```
> cusumArl(mean= 0.0,
          N=100,
          limit=5000,
          seed=123)

      ARL Std. Error
 930.5500    125.8908

> cusumArl(mean= 0.5,
          N=100,
          limit=5000,
          seed=123)

      ARL Std. Error
 101.67000    14.01848

> cusumArl(mean= 1.0,
          N=100,
          limit=5000,
          seed=123)

      ARL Std. Error
 18.190000    2.313439

> cusumArl(mean= 1.5,
          N=100,
          limit=5000,
          seed=123)

      ARL Std. Error
 6.4500000    0.6968501
```

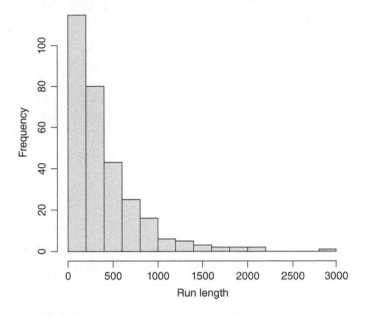

Figure 11.9   *Histogram of RL for $\mu = 10$, $\sigma = 5$, $K^+ = 12$, $h^+ = 29$, $K^- = 8$, $h^- = -29$.*

In Figure 11.9, we present the histogram of the RLs corresponding to the two-sided control scheme ($K^+ = 1$, $h^+ = 3$, $K^- = -1$, $h^- = -3$), in the normal case with $\mu = 0$ and $\sigma = 1$. The distribution of RL is very skewed.

Program **cusumArl** can be used also to determine the values of the control parameters h^+ and h^- so that a certain ARL(0) is attained. For example, if we use the Shewhart 3-sigma control charts for the sample means in the normal case, the probability that, under no shift in the process level, a point will fall outside the control limits is 0.0026, and ARL(0) = 385. Suppose we wish to devise a two-sided CUSUM control scheme, when $\mu_0 = 10$, $\sigma = 5$, $\mu_1^+ = 14$, $\mu_1^- = 6$. We obtain $K^+ = 12$, $K^- = 8$. If we take $\alpha = 0.01$, we obtain $h^+ = \dfrac{-25 \times \log(0.01)}{4} = 28.78$. Program **cusumArl** yields, for the parameters $\mu = 10$, $\sigma = 5$, $K^+ = 12$, $h^+ = 29$, $K^- = 8$, $h^- = -29$ the estimate ARL(0) = 464 $\pm$ 99.3. If we use $\alpha = 0.05$, we obtain $h^+ = 18.72$. Under the control parameters (12,18.7,8,−18.7), we obtain ARL(0) = 67.86 $\pm$ 13.373. We can now run the program for several $h^+ = -h^-$ values to obtain an ARL(0) estimate close to 385 (see Table 11.11). The value in Figure 11.9 ARL is 398.84 with SE of 32.599.

Thus, $h^+ = 29$ would yield a control scheme having an ARL(0) close to that of a Shewhart 3σ scheme.

Program **cusumArl** computes the estimates of the ARL(δ) for the binomial distribution. To illustrate, consider the case of the binomial distribution $B(n, \theta)$ with $n = 100$, $\theta = 0.05$. A two-sided Page's control scheme, protecting against a shift above $\theta_1^+ = 0.07$ or below $\theta_1^- = 0.03$, can use the control parameters $K^+ = 5.95$, $h^+ = 12.87$, $K^- = 3.92$, $h^- = -8.66$.

Table 11.11   *ARL(0) estimates for $\mu = 10$, $\sigma = 5$, $K^+ = 12$, $K^- = 8$, $h^+ = -h^-$*

h^+	18.72	25	27.5	30
ARL(0)	67.86 $\pm$ 13.37	186 $\pm$ 35.22	319 $\pm$ 65.16	412.96 $\pm$ 74.65

```
> cusumArl(size=100,
          prob=0.05,
          kp=5.95,
          km=3.92,
          hp=12.87,
          hm=-8.66,
          randFunc=rbinom,
          N=100,
          limit=2000,
          seed=123)


       ARL Std. Error
   347.6700    47.0467
```

The program **cusumArl** yields, for NR = 100 runs, the estimate ARL(0) = 371.2 ± 63.884. Furthermore, for $\delta = \theta_1 / \theta_0$, we obtain for the same control scheme

$$\text{ARL}\left(\frac{6}{5}\right) = 40.5 \pm 7.676, \quad \text{ARL}\left(\frac{7}{5}\right) = 11.4 \pm 1.303.$$

Similarly, program **cusumArl** can be used to estimate the ARL(δ) in the Poisson case. For example, suppose that X_t has a Poisson distribution with mean $\lambda_0 = 10$. We wish to control the process against shifts in λ greater than $\lambda_1^+ = 15$ or smaller than $\lambda_1^- = 7$. We use the control parameters $K^+ = 12.33$, $h^+ = 11.36$, $K^- = 8.41$, $h^- = -12.91$. The obtained estimate is ARL(0) = 284.2 ± 54.648.

```
> cusumArl(lambda=10,
          kp=12.33,
          km=8.41,
          hp=11.36,
          hm=-12.91,
          randFunc=rpois,
          N=100,
          limit=2000,
          seed=123)


       ARL Std. Error
   300.92000    40.54301
```

We can use now program **cusumPfaCedNorm** to estimate PFA and CED if a change in the mean of magnitude $\delta\sigma$ occurs at time τ. In Table 11.12, we present some estimates obtained from this program.

Table 11.12 *Estimates of PFA and CED, normal distribution $\mu_0 = 0$, $\sigma = 1$, control parameters ($K^+ = 1$, $h^+ = 3$, $K^- = -1$, $h^- = -3$), $\tau = 100$, NR = 500*

δ	PFA	CED
0.5	0.07	107.8 ± 20.75
1	0.07	16.23 ± 10.87
1.5	0.06	6.57 ± 9.86

```
> cusumPfaCedNorm(mean1=0.5,
                  tau=100,
                  N=100,
                  limit=1000,
                  seed=123)


      PFA          CED Std. Error
   0.0700    107.8495     20.7520

> cusumPfaCedNorm(mean1=1.0,
                  tau=100,
                  N=100,
                  limit=1000,
                  seed=123)


      PFA          CED Std. Error
   0.07000    15.40860     11.94727

> cusumPfaCedNorm(mean1=1.5,
                  tau=100,
                  N=100,
                  limit=1000,
                  seed=123)


      PFA          CED Std. Error
  0.070000    6.021505   10.982638
```

11.5   Bayesian detection

The Bayesian approach to the problem of detecting changes in distributions can be described in the following terms. Suppose that we decide to monitor the stability of a process with a statistic T, having a distribution with p.d.f. $f_T(t; \theta)$, where θ designates the parameters on which the distribution depends (process mean, variance, etc.). The statistic T could be the mean, $\overline{X}$, of a random sample of size n; the sample standard-deviation, S, or the proportion defectives in the sample. A sample of size n is drawn from the process at predetermined epochs.

Let T_i $(i = 1, 2, \dots)$ denote the monitoring statistic at the i-th epoch. Suppose that m such samples were drawn and that the statistics $T_1, T_2, \dots, T_m$ are independent. Let $\tau = 0, 1, 2, \dots$ denote the location of the point of change in the process parameter θ_0, to $\theta_1 = \theta_0 + \Delta$. τ is called the **change-point** of θ_0. The event $\{\tau = 0\}$ signifies that all the n samples have been drawn after the change-point. The event $\{\tau = i\}$, for $i = 1, \dots, m - 1$, signifies that the change-point occurred between the i-th and $(i + 1)$st sampling epoch. Finally, the event $\{\tau = m^+\}$ signifies that the change-point has not occurred before the first m sampling epochs.

Given $T_1, \dots, T_m$, the **likelihood function** of τ, for specified values of θ_0 and θ_1, is defined as

$$L_m(\tau; T_1, \dots, T_m) = \begin{cases} \displaystyle\prod_{i=1}^{m} f(T_i; \theta_1), & \tau = 0 \\[2ex] \displaystyle\prod_{i=1}^{\tau} f(T_i; \theta_0) \prod_{j=\tau+1}^{m} f(T_j; \theta_1), & 1 \le \tau \le m - 1 \\[2ex] \displaystyle\prod_{i=1}^{m} f(T_i; \theta_0), & \tau = m^+. \end{cases} \tag{11.5.1}$$

A **maximum likelihood** estimator of τ, given $T_1, \dots, T_m$, is the argument maximizing $L_m(\tau; T_1, \dots, T_m)$.

In the Bayesian framework, the statistician gives the various possible values of τ nonnegative weights, which reflect his belief where the change-point could occur. High weight expresses higher confidence. In order to standardize the approach, we will assume that the sum of all weights is one, and we call these weights, the **prior probabilities** of τ. Let $\pi(\tau)$, $\tau = 0, 1, 2, \dots$ denote the prior probabilities of τ. If the occurrence of the change-point is a realization of some random process, the following modified-geometric prior distribution could be used

$$\pi_m(\tau) = \begin{cases} \pi, & \text{if } \tau = 0 \\[1.5ex] (1 - \pi)p(1 - p)^{i-1}, & \text{if } \tau = i, (i = 1, \dots, m - 1) \\[1.5ex] (1 - \pi)(1 - p)^{m-1}, & \text{if } \tau = m^+, \end{cases} \tag{11.5.2}$$

where $0 < \pi < 1$, $0 < p < 1$ are prior parameters. Applying Bayes formula, we convert the prior probabilities $\pi(t)$ after observing $T_1, \dots, T_m$ to **posterior probabilities**. Let π_m denote the posterior probability of the event $\{\tau \le m\}$, given $T_1, \dots, T_m$. Using the above modified-geometric prior distribution, and employing Bayes theorem, we obtain the formula

$$\pi_m = \frac{\dfrac{\pi}{(1 - \pi)(1 - p)^{m-1}} \displaystyle\prod_{j=1}^{m} R_j + \dfrac{p}{(1 - p)^{m-1}} \displaystyle\sum_{i=1}^{m-1} (1 - p)^{i-1} \prod_{j=i+1}^{m} R_j}{\dfrac{\pi}{(1 - \pi)(1 - p)^{m-1}} \displaystyle\prod_{j=1}^{m} R_j + \dfrac{p}{(1 - p)^{m-1}} \displaystyle\sum_{i=1}^{m-1} (1 - p)^{i-1} \prod_{j=i+1}^{m} R_j + 1}, \tag{11.5.3}$$

where

$$R_j = \frac{f(T_j; \theta_1)}{f(T_j; \theta_0)}, \quad j = 1, 2, \dots. \tag{11.5.4}$$

A Bayesian detection of a change-point is a procedure which detects a change as soon as $\pi_m \geq \pi^*$, where π^* is a value in $(0, 1)$, close to 1.

The above procedure can be simplified, if we believe that the monitoring starts when $\theta = \theta_0$ (i.e., $\pi = 0$) and p is very small, we can represent π_m then, approximately, by

$$\tilde{\pi}_m = \frac{\sum_{i=1}^{m-1} \prod_{j=i+1}^{m} R_j}{\sum_{i=1}^{m-1} \prod_{j=i+1}^{m} R_j + 1}. \tag{11.5.5}$$

The statistic

$$W_m = \sum_{i=1}^{m-1} \prod_{j=i+1}^{m} R_j \tag{11.5.6}$$

is called the **Shiryaev–Roberts** (SR) statistic. Notice that $\tilde{\pi}_m \geq \pi^*$ if, $W_m \geq \frac{\pi^*}{1 - \pi^*}$. $\frac{\pi^*}{1 - \pi^*}$ is called the **stopping threshold**. Thus, for example, if the Bayes procedure is to "flag" a change as soon as $\tilde{\pi}_m \geq 0.95$, the procedure which "flags" as soon as $W_m \geq 19$, is equivalent.

We illustrate now the use of the SR statistic in the special case of monitoring the mean θ_0 of a process. The statistic T is the sample mean, $\overline{X}_n$, based on a sample of n observations. We will assume that $\overline{X}_n$ has a normal distribution $N\left(\theta_0, \frac{\sigma}{\sqrt{n}}\right)$, and at the change-point, θ_0 shifts to $\theta_1 = \theta_0 + \delta\sigma$. It is straightforward to verify that the likelihood ratio is

$$R_j = \exp\left\{ -\frac{n\delta^2}{2\sigma^2} + \frac{n\delta}{\sigma^2}(\overline{X}_j - \theta_0) \right\}, \quad j = 1, 2, \dots. \tag{11.5.7}$$

Accordingly, the SR statistic is

$$W_m = \sum_{i=1}^{m-1} \exp\left\{ \frac{n\delta}{\sigma^2} \sum_{j=i+1}^{m} (\overline{X}_j - \theta_0) - \frac{n\delta^2(m - i)}{2\sigma^2} \right\}. \tag{11.5.8}$$

Example 11.6.   We illustrate the procedure numerically. Suppose that $\theta_0 = 10$, $n = 5$, $\delta = 2$, $\pi^* = 0.95$, and $\sigma = 3$. The stopping threshold is 19. Suppose that $\tau = 10$. The values of $\overline{X}_j$ have the normal distribution $N\left(10, \frac{3}{\sqrt{5}}\right)$ for $j = 1, \dots, 10$ and $N\left(10 + \delta\sigma, \frac{3}{\sqrt{5}}\right)$ for $j = 11, 12, \dots$. In Table 11.13 we present the values of W_m.

Table 11.13   *Values of W_m for $\delta = 0.5(0.5)2.0$, $n = 5$, $\tau = 10$, $\sigma = 3$, $\pi^* = 0.95$*

m	$\delta = 0.5$	$\delta = 1.0$	$\delta = 1.5$	$\delta = 2.0$
2	0.3649	0.0773	0.0361	0.0112
3	3.1106	0.1311	0.0006	0.0002
4	3.2748	0.0144	0.0562	0.0000
5	1.1788	0.0069	0.0020	0.0000

(continued)

Table 11.13   *(continued)*

m	$\delta = 0.5$	$\delta = 1.0$	$\delta = 1.5$	$\delta = 2.0$
6	10.1346	0.2046	0.0000	0.0291
7	14.4176	0.0021	0.0527	0.0000
8	2.5980	0.0021	0.0015	0.0000
9	0.5953	0.6909	0.0167	0.0000
10	0.4752	0.0616	0.0007	0.0001
11	1.7219	5.6838	848.6259	1538.0943
12	2.2177	73.8345		
13	16.3432			
14	74.9618			

We see that the SR statistic detects the change-point quickly if δ is large.

The larger is the critical level $w^* = \pi^*/(1 - \pi^*)$, the smaller will be the frequency of detecting the change-point before it happens (false alarm). Two characteristics of the procedure are of interest:

 (i) the **probability of false alarm** (PFA); and
(ii) the **conditional expected delay**, (CED) given that the alarm is given after the change-point.

Programs **shroArlPfaCedNorm** and **shroArlPfaCedPois** estimate the ARL(0) of the procedure for the normal and Poisson cases. Programs **shroArlPfaCedNorm** and **shroArlPfaCed-Pois** estimate the PFA and CED of these procedures. In Table 11.14, we present simulation estimates of the PFA and CED for several values of δ. The estimates are based on 100 simulation runs.

```
> shroArlPfaCedNorm(mean0=10,
                    sd=3,
                    n=5,
                    delta=0.5,
                    tau=10,
                    w=99,
                    seed=123)


         ARL      Std. Error               PFA
   16.970000        0.267378          0.000000
         CED  CED-Std. Error
    6.970000        1.570188
```

Table 11.14   *Estimates of PFA and CED for $\mu_0 = 10$, $\sigma = 3$, $n = 5$, $\tau = 10$, stopping threshold = 99*

	$\delta = 0.5$	$\delta = 1.0$	$\delta = 1.5$	$\delta = 2.0$
PFA	0.00	0.04	0.05	0.04
CED	21.02	7.19	4.47	3.41

```
> shroArlPfaCedNorm(mean0=10,
                     sd=3,
                     n=5,
                     delta=1.0,
                     tau=10,
                     w=99,
                     seed=123)
```

ARL	Std. Error	PFA
12.6700000	0.1183681	0.0100000
CED	CED-Std. Error	
2.7070707	1.2529530	

```
> shroArlPfaCedNorm(mean0=10,
                     sd=3,
                     n=5,
                     delta=1.5,
                     tau=10,
                     w=99,
                     seed=123)
```

ARL	Std. Error	PFA
11.45000000	0.09420722	0.03000000
CED	CED-Std. Error	
1.57731959	1.16579980	

```
> shroArlPfaCedNorm(mean0=10,
                     sd=3,
                     n=5,
                     delta=2.0,
                     tau=10,
                     w=99,
                     seed=123)
```

ARL	Std. Error	PFA
11.0000000	0.1048809	0.0300000
CED	CED-Std. Error	
1.1649485	1.1280707	

We see that if the amount of shift δ is large ($\delta > 1$), then the CED is small. The estimates of PFA are small due to the large threshold value. Another question of interest is, what is the average run length (ARL) when there is no change in the mean. We estimated the ARL(0), for the same example of normally distributed sample means using function shroArlPfaCedNorm. 100 independent simulation runs were performed. In Table 11.15, we present the estimated values of ARL(0), as a function of the stopping threshold.

```
> shroArlPfaCedNorm(mean0=10,
                     sd=3,
                     n=5,
                     delta=2.0,
                     w=19,
                     seed=123)
```

```
> shroArlPfaCedNorm(mean0=10,
                    sd=3,
                    n=5,
                    delta=2.0,
                    w=50,
                    seed=123)
> shroArlPfaCedNorm(mean0=10,
                    sd=3,
                    n=5,
                    delta=2.0,
                    w=99,
                    seed=123)
```

Thus, the procedure based on the Shiryaev–Roberts detection is sensitive to changes, while in a stable situation (no changes), it is expected to run long till an alarm is given. Figure 11.10 shows a box plot of the RL with stopping threshold of 99, when there is no change. For more details on data analytic aspects of the Shiryaev–Roberts procedure, see Kenett and Pollak (1996).

11.6   Process tracking

Process **tracking** is a procedure which repeatedly estimates certain characteristics of the process which is being monitored. The CUSUM detection procedure, as well as that of Shiryaev–Roberts, are designed to provide warning quickly after changes occur. However, at

Table 11.15   *Average run length of Shiryaev–Roberts procedure, $\mu_0 = 10$, $\delta = 2$, $\sigma = 3$, $n = 5$*

Stopping Threshold	ARL(0)
19	49.37 ± 10.60
50	100.81 ± 19.06
99	224.92 ± 41.11

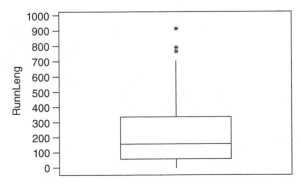

Figure 11.10   *Box and whisker plot of 100 run lengths of the Shiryaev–Roberts procedure normal distribution $\mu_0 = 10$, $\delta = 2$, $\sigma = 3$, $n = 5$, stopping threshold 99.*

times of stopping these procedures do not provide direct information on the current location of the process mean (or the process variance). In the Shewhart $\overline{X}$-bar control chart, each point provides an estimate of the process mean at that specific time. The precision of these estimates is generally low, since they are based on small samples. One may suggest that, as long as there is no evidence that a change in the process mean has occurred, an average of all previous sample means can serve as an estimator of the current value of the process mean. Indeed, if after observing m samples, each of size n, the grand average $\overline{\overline{X}}_m = \frac{1}{m}(\overline{X}_1 + \cdots + \overline{X}_m)$ has the standard error $\frac{\sigma}{\sqrt{nm}}$, while the standard error of the last mean, $\overline{X}_m$, is only $\sigma/\sqrt{n}$. It is well established by statistical estimation theory that, as long as the process mean μ_0 does not change, $\overline{\overline{X}}_m$ is the best (minimum variance) unbiased estimator of $\mu_m = \mu_0$. On the other hand, if μ_0 has changed to $\mu_1 = \mu_0 + \delta\sigma$, between the τ-th and the $(\tau+1)$-st sample, where $\tau < m$, the grand mean $\overline{\overline{X}}_m$ is a biased estimator of μ_1 (the current mean). The expected value of $\overline{\overline{X}}_m$ is $\frac{1}{m}(\tau\mu_0 + (m-\tau)\mu_1) = \mu_1 - \frac{\tau}{m}\delta\sigma$. Thus, if the change-point, τ, is close to m, the bias of $\overline{\overline{X}}$ can be considerable. The bias of the estimator of the current mean, when $1 < \tau < m$, can be reduced by considering different types of estimators. In this chapter, we focus attention on three procedures for tracking and monitoring the process mean. The exponentially weight moving average (EWMA) procedure, the Bayes estimation of the current mean (BECM), and the Kalman filter.

11.6.1 The EWMA procedure

The **exponentially weighted moving averages** chart is a control chart for the process mean which at time t ($t = 1, 2, \ldots$) plots the statistic

$$\hat{\mu}_t = (1 - \lambda)\hat{\mu}_{t-1} + \lambda\overline{X}_t, \tag{11.6.1}$$

where $0 < \lambda < 1$, and $\hat{\mu}_0 = \mu_0$ is the initial process mean. The Shewhart $\overline{X}$-chart is the limiting case of $\lambda = 1$. Small values of λ give high weight to the past data. It is customary to use the values of $\lambda = 0.2$ or $\lambda = 0.3$.

By repeated application of the recursive formula we obtain

$$\hat{\mu}_t = (1 - \lambda)^2\hat{\mu}_{t-2} + \lambda(1 - \lambda)\overline{X}_{t-1} + \lambda\overline{X}_t$$

$$= \cdots$$

$$= (1 - \lambda)^t\mu_0 + \lambda\sum_{i=1}^{t}(1 - \lambda)^{t-i}\overline{X}_i. \tag{11.6.2}$$

We see in this formula that $\hat{\mu}_t$ is a weighted average of the first t means $\overline{X}_1, \ldots, \overline{X}_t$ and μ_0, with weights which decrease geometrically, as $t - i$ grows.

Let τ denote the epoch of change from μ_0 to $\mu_1 = \mu_0 + \delta\sigma$. As in the previous section, $\{\tau = i\}$ implies that

$$E\{\overline{X}_j\} = \begin{cases} \mu_0, & \text{for } j = 1, \ldots, i \\ \mu_1, & \text{for } j = i+1, i+2, \ldots. \end{cases} \tag{11.6.3}$$

Accordingly, the expected value of the statistic $\hat{\mu}_t$ (an estimator of the current mean μ_t) is

$$E\{\hat{\mu}_t\} = \begin{cases} \mu_0, & \text{if } t \le \tau \\ \mu_1 - \delta\sigma(1-\lambda)^{t-\tau}, & \text{if } t > \tau. \end{cases} \tag{11.6.4}$$

We see that the bias of $\hat{\mu}_t$, $-\delta\sigma(1-\lambda)^{t-\tau}$, decreases to zero geometrically fast as t grows above τ. This is a faster decrease in bias than that of the grand mean, $\overline{\overline{X}}_t$, which was discussed earlier.

The variance of $\hat{\mu}_t$ can be easily determined, since $\overline{X}_1, \overline{X}_2, \ldots, \overline{X}_t$ are independent and $\text{Var}\{\overline{X}_j\} = \dfrac{\sigma^2}{n}, j = 1, 2, \ldots$. Hence,

$$\text{Var}\{\hat{\mu}_t\} = \frac{\sigma^2}{n}\lambda^2 \sum_{i=1}^{t}(1-\lambda)^{2(t-i)}$$

$$= \frac{\sigma^2}{n}\lambda^2 \frac{1-(1-\lambda)^{2t}}{1-(1-\lambda)^2}. \tag{11.6.5}$$

This variance converges to

$$\text{Avar}\{\hat{\mu}_t\} = \frac{\sigma^2}{n}\frac{\lambda}{2-\lambda}, \tag{11.6.6}$$

as $t \to \infty$. An **EWMA-Control** chart for monitoring shifts in the mean is constructed in the following manner. Starting at $\hat{\mu}_0 = \mu_0$, the points $(t, \hat{\mu}_t)$, $t = 1, 2, \ldots$ are plotted. As soon as these points cross either one of the control limits

$$\text{CL} = \mu_0 \pm L\frac{\sigma}{\sqrt{n}}\sqrt{\frac{\lambda}{2-\lambda}}, \tag{11.6.7}$$

an alarm is given that the process mean has shifted.

In Figure 11.11, we present an EWMA-chart with $\mu_0 = 10$, $\sigma = 3$, $n = 5$, $\lambda = 0.2$, and $L = 2$. The values of $\hat{\mu}_t$ indicate that a shift in the mean took place after the eleventh sampling epoch. An alarm for change is given after the fourteenth sample.

As in the previous sections, we have to characterize the efficacy of the EWMA-chart in terms of PFA and CED when a shift occurs, and the ARL when there is no shift. In Table 11.16, we present estimates of PFA and CED based on 1000 simulation runs. The simulations were from normal distributions, with $\mu_0 = 10$, $\sigma = 3$, and $n = 5$. The change-point was at $\tau = 10$. The shift was from μ_0 to $\mu_1 = \mu_0 + \delta\sigma$. The estimates $\hat{\mu}_t$ were determined with $\lambda = 0.2$. We see in this table that if we construct the control limits with the value of $L = 3$ then the PFA is very small, and the CED is not large.

The estimated ARL values for this example are

L	2	2.5	3.0
ARL	48.7	151.36	660.9

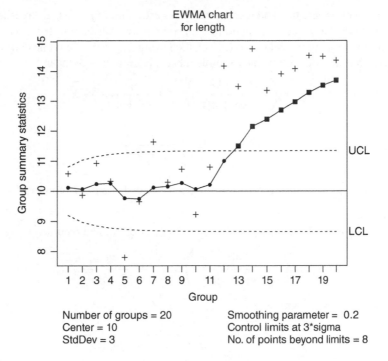

Figure 11.11   *EWMA-chart, $\mu_0 = 10$, $\sigma = 3$, $\delta = 0$, $n = 5$, $\lambda = 0.2$, and $L = 2$.*

Table 11.16   *Simulation estimates of PFA and CED of an EWMA-chart*

			CED		
L	PFA	$\delta = 0.5$	$\delta = 1.0$	$\delta = 1.5$	$\delta = 2.0$
2	0.168	3.93	2.21	1.20	1.00
2.5	0.043	4.35	2.67	1.41	1.03
3	0.002	4.13	3.36	1.63	1.06

11.6.2   The BECM procedure

In the present section, we present a Bayesian procedure for estimating the current mean μ_t ($t = 1, 2, \ldots$). Let $\overline{X}_1, \overline{X}_2, \ldots, \overline{X}_t$, $t = 1, 2, \ldots$ be means of samples of size n. The distribution of $\overline{X}_i$ is $N\left(\mu_i, \dfrac{\sigma}{\sqrt{n}}\right)$, where σ is the process standard deviation. We will assume here that σ is known and fixed throughout all sampling epochs. This assumption is made in order to simplify the exposition. In actual cases, one has to monitor also whether σ changes with time.

If the process mean stays stable throughout the sampling periods then

$$\mu_1 = \mu_2 = \cdots = \mu_t = \mu_0.$$

Let us consider this case first and present the Bayes estimator of μ_0. In the Bayesian approach, the model assumes that μ_0 itself is random, with some prior distribution. If we assume that the prior distribution of μ_0 is normal, say $N(\mu^*, \tau)$, then using Bayes theorem one can show that the posterior distribution of μ_0, given the t sample means, is normal with mean

$$\hat{\mu}_{B,t} = \left(1 - \frac{nt\tau^2}{\sigma^2 + nt\tau^2}\right)\mu^* + \frac{nt\tau^2}{\sigma^2 + nt\tau^2} \to \overline{\overline{X}}_t \tag{11.6.8}$$

and variance

$$w_t^2 = \tau^2\left(1 - \frac{nt\tau^2}{\sigma^2 + nt\tau^2}\right), \tag{11.6.9}$$

where $\to \overline{\overline{X}}_t = \frac{1}{t}\sum_{i=1}^{t}\overline{X}_i$. The mean $\hat{\mu}_{B,t}$ of the posterior distribution is commonly taken as the Bayes estimator of μ_0 (see Chapter 4).

It is interesting to notice that $\hat{\mu}_{B,t}$, $t = 1, 2, \ldots$ can be determined recursively by the formula

$$\hat{\mu}_{B,t} = \left(1 - \frac{nw_{t-1}^2}{\sigma^2 + nw_{t-1}^2}\right)\hat{\mu}_{B,t-1} + \frac{nw_{t-1}^2}{\sigma^2 + nw_{t-1}^2}\overline{X}_t, \tag{11.6.10}$$

where $\hat{\mu}_{B,0} = \mu^*$, $w_0^2 = \tau^2$ and

$$w_t^2 = \frac{\sigma^2 w_{t-1}^2}{\sigma^2 + nw_{t-1}^2}. \tag{11.6.11}$$

This recursive formula resembles that of the EWMA estimator. The difference here is that the weight λ is a function of time, i.e.,

$$\lambda_t = \frac{nw_{t-1}^2}{\sigma^2 + nw_{t-1}^2}. \tag{11.6.12}$$

From the above recursive formula for w_t^2, we obtain that $w_t^2 = \lambda_t \frac{\sigma^2}{n}$, or $\lambda_t = \lambda_{t-1}/(1 + \lambda_{t-1})$, $t = 2, 3, \ldots$ where $\lambda_1 = nt\tau^2/(\sigma^2 + nt\tau^2)$. The procedures become more complicated if change points are introduced. We discuss in the following section a dynamic model of change.

11.6.3   The Kalman filter

In this section, we present a model of dynamic changes in the observed sequence of random variables, and a Bayesian estimator of the current mean, called the **Kalman filter**.

At time t, let Y_t denote an observable random variable, having mean μ_t. We assume that μ_t may change at random from one time epoch to another, according to the model

$$\mu_t = \mu_{t-1} + \Delta_t, \quad t = 1, 2, \ldots$$

where Δ_t, $t = 1, 2, \ldots$ is a sequence of i.i.d. random variables having a normal distribution $N(\delta, \sigma_2)$. Furthermore, we assume that $\mu_0 \sim N(\mu_0^*, w_0)$, and the observation equation is

$$Y_t = \mu_t + \epsilon_t, \quad t = 1, 2, \ldots$$

where ϵ_t are i.i.d. $N(0, \sigma_\epsilon)$. According to this dynamic model, the mean at time t (the current mean) is normally distributed with mean

$$\hat{\mu}_t = B_t(\hat{\mu}_{t-1} + \delta) + (1 - B_t)Y_t, \tag{11.6.13}$$

where

$$B_t = \frac{\sigma_e^2}{\sigma_\epsilon^2 + \sigma_2^2 + w_{t-1}^2}, \tag{11.6.14}$$

$$w_t^2 = B_t(\sigma_2^2 + w_{t-1}^2). \tag{11.6.15}$$

The posterior variance of μ_t is w_t^2. $\hat{\mu}_t$ is the Kalman filter.

If the prior parameters σ_ϵ^2, σ_2^2, and δ are unknown, we could use a small portion of the data to estimate these parameters.

According to the dynamic model, we can write

$$y_t = \mu_0 + \delta t + \epsilon_t^*, \quad t = 1, 2, \dots$$

where $\epsilon_t^* = \sum_{i=1}^{t} [(\Delta_i - \delta) + \epsilon_i]$. Notice that $E\{\epsilon_t^*\} = 0$ for all t and $V\{\epsilon_t^*\} = t(\sigma_2^2 + \sigma_\epsilon^2)$. Let $U_t = y_t / \sqrt{t}, t = 1, 2, \dots$, then we can write the regression model

$$U_t = \mu_0 x_{1t} + \delta x_{2t} + \eta_t, \quad t = 1, 2, \dots$$

where $x_{1t} = 1/\sqrt{t}$, and $x_{2t} = \sqrt{t}$ and η_t, $t = 1, 2 \dots$ are independent random variables, with $E\{\eta_t\} = 0$ and $V\{\eta_t\} = (\sigma_2^2 + \sigma_\epsilon^2)$.

Using the first m points of (t, y_t) and fitting, by the method of least-squares (see Chapter 4), the regression equation of U_t against (x_{1t}, x_{2t}), we obtain estimates of μ_0, δ and of $(\sigma_2^2 + \sigma_\epsilon^2)$. Estimate of σ_ϵ^2 can be obtained, if y_t are group means, by estimating within groups variance, otherwise we assume a value for σ_ϵ^2, smaller than the least squares estimate of $\sigma_2^2 + \sigma_\epsilon^2$. We illustrate this now by example.

Example 11.7.   In Figure 11.12, we present the Dow-Jones financial index for the 300 business days of 1935 (file **DOJO1935.csv**). The Kalman filter estimates of the current means are plotted in this figure too. These estimates were determined by the formula

$$\hat{\mu}_t = B_t(\hat{\mu}_{t-1} + \delta) + (1 - B_t)y_t, \tag{11.6.16}$$

where the prior parameters were computed as suggested above, on the basis of the first $m = 20$ data points. The least squares estimates of μ_0, δ, and $\sigma_2^2 + \sigma_\epsilon^2$ are, respectively, $\hat{\mu}_0 = 127.484$, $\hat{\delta} = 0.656$, and $\hat{\sigma}_2^2 + \hat{\sigma}_\epsilon^2 = 0.0731$. For $\hat{\sigma}_\epsilon^2$ we have chosen the value 0.0597 and for w_0^2 the value 0.0015. The first 50 values of the data, y_t, and the estimate $\hat{\mu}_t$, are given in Table 11.17.

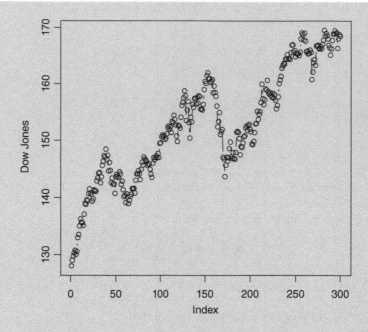

Figure 11.12   *The daily Dow-Jones financial index for 1935.*

Table 11.17   *The Dow-Jones index for the first 50 days of 1935, and the KF estimates*

t	y_t	$\hat{\mu}_t$	t	y_t	$\hat{\mu}_t$
1	128.06	128.0875	26	141.31	141.5413
2	129.05	128.8869	27	141.2	141.8236
3	129.76	129.6317	28	141.07	141.9515
4	130.35	130.3119	29	142.9	142.7171
5	130.77	130.8927	30	143.4	143.3831
6	130.06	130.9880	31	144.25	144.1181
7	130.59	131.2483	32	144.36	144.6189
8	132.99	132.3113	33	142.56	144.2578
9	133.56	133.1894	34	143.59	144.4178
10	135.03	134.2893	35	145.59	145.2672
11	136.26	135.4378	36	146.32	146.0718
12	135.68	135.9388	37	147.31	146.9459
13	135.57	136.2108	38	147.06	147.3989
14	135.13	136.2161	39	148.44	148.1991
15	137.09	136.9537	40	146.65	148.0290
16	138.96	138.1156	41	147.37	148.1923
17	138.77	138.7710	42	144.61	147.2604
18	139.58	139.4843	43	146.12	147.2434
19	139.42	139.8704	44	144.72	146.7082

(continued)

Table 11.17   (continued)

t	y_t	$\hat{\mu}_t$	t	y_t	$\hat{\mu}_t$
20	140.68	140.5839	45	142.59	145.5755
21	141.47	141.3261	46	143.38	145.1632
22	140.78	141.5317	47	142.34	144.5157
23	140.49	141.5516	48	142.35	144.1145
24	139.35	141.1370	49	140.72	143.2530
25	139.74	141.0238	50	143.58	143.7857

The Kalman filter is a state space dynamic linear model. Such models are available in JMP in **Analyze > Specialized Modeling > Time Series Forecast**. JMP automatically recommends a state space model that fits the data.

11.6.4   Hoadley's QMP

B. Hoadley (1981) introduced at Bell Laboratories a **quality measurement plan** (QMP), which employs Bayesian methods of estimating the current mean of a process. This QMP provides reporting capabilities of large data sets and, in a certain sense, is an improvement over the Shewhart 3-sigma control. These plans were implemented throughout Western Electric Co. in the late 1980s. The main idea is that the process mean does not remain at a constant level, but changes at random every time period according to some distribution. This framework is similar to that of the Kalman filter, but was developed for observations X_t having Poisson distributions with means λ_t ($t = 1, 2, \ldots$), and where $\lambda_1, \lambda_2, \ldots$ are independent random variables having a common gamma distribution $G(v, \Lambda)$. The parameters v and Λ are **unknown**, and are estimated from the data. At the end of each period, a box plot is put on a chart. The centerline of the box plot represents the posterior mean of λ_t, given past observations. The lower and upper sides of the box represent the 0.05th and 0.95th quantiles of the posterior distribution of λ_t. The lower whisker starts at the 0.01th quantile of the posterior distribution and the upper whisker ends at the 0.99th quantile of that distribution. These box plots are compared to a desired quality level.

We have seen in Section 4.8.3 that if X_t has a Poisson distribution $P(\lambda_t)$, and λ_t has a gamma distribution $G(v, \Lambda)$ then the posterior distribution of λ_t, given X_t, is the gamma distribution $G\left(v + X_t, \dfrac{\Lambda}{1 + \Lambda}\right)$. Thus, the Bayes estimate of λ_t, for a squared error loss, is the posterior expectation

$$\hat{\lambda}_t = (v + X_t)\frac{\Lambda}{1 + \Lambda}. \tag{11.6.17}$$

Similarly, the p-th quantile of the posterior distribution is

$$\lambda_{t,p} = \frac{\Lambda}{1 + \Lambda} G_p(v + X_t, 1), \tag{11.6.18}$$

where $G_p(v + X_t, 1)$ is the p-th quantile of the standard gamma distribution $G(v + X_t, 1)$. We remark that if v is an integer then

$$G_p(v + X_t, 1) = \frac{1}{2}\chi_p^2[2(v + X_t)]. \tag{11.6.19}$$

We assumed that $\lambda_1, \lambda_2, \ldots$ are independent and identically distributed. This implies that $X_1, X_2, \ldots$ are independent, having the same **negative-binomial** predictive distribution, with predictive expectation

$$E\{X_t\} = v\Lambda \tag{11.6.20}$$

and predictive variance

$$V\{X_t\} = v\Lambda(1 + \Lambda). \tag{11.6.21}$$

We therefore can estimate the prior parameters v and Λ by the consistent estimators

$$\hat{\Lambda}_T = \left(\frac{S_T^2}{\overline{X}_T} - 1\right)^{+} \tag{11.6.22}$$

and

$$\hat{v}_T = \frac{\overline{X}_T}{\hat{\Lambda}_T}, \tag{11.6.23}$$

where $\overline{X}_T$ and S_T^2 are the sample mean and sample variance of $X_1, X_2, \ldots, X_T$. For determining $\hat{\lambda}_t$ and $\lambda_{t,p}$, we can substitute $\hat{\Lambda}_T$ and $\hat{v}_T$ in the above equations, with $T = t - 1$. We illustrate this estimation method, called **parametric empirical Bayes** method, in the following example.

Example 11.8. In file **SOLDEF.csv**, we present results of testing batches of circuit boards for defects in solder points, after wave soldering. The batches include boards of similar design. There were close to 1000 solder points on each board. The results X_t are number of defects per 10^6 points (**PPM**). The quality standard is $\lambda^0 = 100$ (PPM). λ_t values below λ^0 represent high quality soldering. In this data file, there are $N = 380$ test results. Only 78 batches had an X_t value greater than $\lambda^0 = 100$. If we take UCL $= \lambda^0 + 3\sqrt{\lambda^0} = 130$, we see that only 56 batches had X_t values greater than the UCL. All runs of consecutive X_t values greater than 130 are of length not greater than 3. We conclude therefore that the occurrence of low quality batches is sporadic, caused by common causes. These batches are excluded from the analysis. In Table 11.18, we present the X_t values and the associated values of $\overline{X}_{t-1}$, S_{t-1}^2, $\hat{\Lambda}_{t-1}$, and $\hat{v}_{t-1}$, associated with $t = 10, \ldots, 20$. The statistics $\overline{X}_{t-1}$ etc. are functions of $X_1, \ldots, X_{t-1}$.

Table 11.18   *Number of defects (PPM) and associated statistics for the SOLDEF data*

t	X_t	$\overline{X}_{t-1}$	S_{t-1}^2	$\hat{\Lambda}_{t-1}$	$\hat{v}_{t-1}$
10	29	23.66666	75.55555	2.192488	10.79443
11	16	24.20000	70.56000	1.915702	12.63244
12	31	23.45454	69.70247	1.971811	11.89492
13	19	24.08333	68.24305	1.833621	13.13429
14	18	23.69230	64.82840	1.736263	13.64556
15	20	23.28571	62.34693	1.677475	13.88140
16	103	23.06666	58.86222	1.551830	14.86416
17	31	28.06250	429.5585	14.30721	1.961423
18	33	28.23529	404.7681	13.33553	2.117296
19	12	28.50000	383.4722	12.45516	2.288207
20	46	27.63157	376.8642	12.63889	2.186233

Table 11.19   *Empirical Bayes estimates of λ_t and $\lambda_{t,p}$, $p = 0.01, 0.05, 0.95, 0.99$*

		Quantiles of posterior distributions			
t	$\hat{\lambda}_t$	0.01	0.05	0.95	0.99
10	27.32941	18.40781	21.01985	34.45605	37.40635
11	18.81235	12.34374	14.23760	24.59571	26.98991
12	28.46099	19.59912	22.19367	35.60945	38.56879
13	20.79393	13.92388	15.93527	26.82811	29.32615
14	20.08032	13.52340	15.44311	25.95224	28.38310
15	21.22716	14.44224	16.42871	27.22614	29.70961
16	71.67607	57.21276	61.44729	82.53656	87.03260
17	30.80809	16.17445	20.45885	39.63539	43.28973
18	32.66762	17.93896	22.25118	41.73586	45.48995
19	13.22629	2.578602	5.696005	18.98222	21.36506
20	44.65323	28.14758	32.98006	55.23497	59.61562

In Table 11.19, we present the values of $\hat{\lambda}_t$ and the quantiles $\lambda_{t,0.01}$ for $p = 0.01, 0.05, 0.95$ and 0.99.

11.7   Automatic process control

Certain production lines are fully automated, such as in chemical industries, paper industries, automobile industry. In such production lines, it is often possible to build in feedback and control mechanism, so that if there is indication that the process mean or standard deviation change significantly, then a correction is made automatically via the control mechanism. If μ_t denotes the level of the process mean at time t, and u_t denotes the control level at time t, the **dynamic liner model** (DLM) of the process mean is

$$\mu_t = \mu_{t-1} + \Delta_t + bu_{t-1}, \quad t = 1, 2, \ldots \tag{11.7.1}$$

and the observations equation, is as before

$$Y_t = \mu_t + \epsilon_t, \quad t = 1, 2, \ldots. \tag{11.7.2}$$

Δ_t is a random disturbance in the process evolution. The recursive equation of the DLM is linear, in the sense that the effect on μ_t of u_{t-1} is proportional to u_{t-1}. The control could be on a vector of several variables, whose level at time t is given by a vector $\mathbf{u}_t$. The question is, how to determine the levels of the control variables? This question of optimal control of systems, when the true level μ_t of the process mean is not known exactly, but only estimated from the observed values of Y_t, is a subject of studies in the field of **stochastic control**. We refer the reader to the book of Aoki (1989). The reader is referred also to the paper by Box and Kramer (1992).

It is common practice, in many industries, to use the **proportional rule** for control. That is, if the process level (mean) is targeted at μ_0, and the estimated level at time t is $\hat{\mu}_t$, then

$$u_t = -p(\hat{\mu}_t - \mu_0), \tag{11.7.3}$$

where p is some factor, which is determined by the DLM, by cost factors, etc. This rule is not necessarily optimal. It depends on the objectives of the optimization. For example, suppose that the DLM with control is

$$\mu_t = \mu_{t-1} + bu_{t-1} + \Delta_t, \quad t = 1, 2, \ldots, \tag{11.7.4}$$

where the process mean is set at μ_0 at time $t = 0$. Δ_T is a random disturbance, having a normal distribution $N(\delta, \sigma)$. The process level μ_t, is estimated by the Kalman filter, which was described in the previous section. We have the option to adjust the mean, at each time period, at a cost of $c_A u^2$ [\$]. On the other hand, at the end of T periods, we pay a penalty of \$ $c_d(\mu_T - \mu_0)^2$, for the deviation of μ_T from the target level. In this example, the optimal levels of u_t, for $t = 0, \ldots, T-1$, are given by

$$u_t^0 = -\frac{bq_{t+1}}{c_A + q_{t+1}b^2}(\hat{\mu}_t - \mu_0), \tag{11.7.5}$$

where

$$q_T = c_d$$

and, for $t = 0, \ldots, T-1$

$$q_t = \frac{c_A q_{t+1}}{c_A + q_{t+1}b^2}. \tag{11.7.6}$$

These formulae are obtained as special cases from general result given in Aoki (1989, p. 128). Thus, we see that the values that u_t obtains, under the optimal scheme, are proportional to $-(\hat{\mu}_t - \mu_0)$, but with varying factor of proportionality,

$$p_t = bq_{t+1}/(c_A + q_{t+1}b^2). \tag{11.7.7}$$

In Table 11.20, we present the optimal values of p_t for the case of $c_A = 100$, $c_d = 1000$, $b = 1$, and $T = 15$.

Table 11.20   *Factors of proportionality in optimal control*

t	q_t	p_t
15	–	–
14	90.909	0.909
13	47.619	0.476
12	32.258	0.323
11	24.390	0.244
10	19.608	0.196
9	16.393	0.164
8	14.085	0.141
7	12.346	0.124
6	10.989	0.110
5	9.901	0.099
4	9.009	0.090
3	8.264	0.083
2	7.634	0.076
1	7.092	0.071

If the penalty for deviation from the target is cumulative, we wish to minimize the total expected penalty function, namely

$$J_t = c_d \sum_{t=1}^{T} E\{(\mu_t - \mu_0)^2\} + c_A \sum_{t=0}^{T-1} u_t^2. \tag{11.7.8}$$

The optimal solution in this case is somewhat more complicated than the above rule, and it is also not one with fixed factor of proportionality p. The method of obtaining this solution is called **dynamic programming**. We do not present here this optimization procedure. The interested reader is referred to Aoki (1989). We just mention that the optimal solution using this method yields for example that the last control is at the level (when $b = 1$) of

$$u_{T-1}^0 = -\frac{cd}{c_A + c_d}(\hat{\mu}_{T-1} - \mu_0). \tag{11.7.9}$$

The optimal control at $t = T - 2$ is

$$u_{T-2}^0 = -\frac{cd}{c_A + 2c_d}(\hat{\mu}_{T-2} - \mu_0), \tag{11.7.10}$$

and so on.

We conclude this section mentioning that a simple but reasonable method of automatic process control is to use the EWMA-chart, and whenever the trend estimates, $\hat{\mu}_t$, are above or below

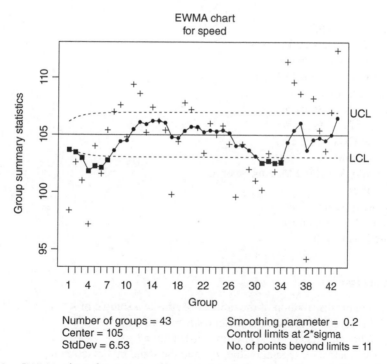

**EWMA chart
for speed**

Number of groups = 43
Center = 105
StdDev = 6.53

Smoothing parameter = 0.2
Control limits at 2*sigma
No. of points beyond limits = 11

Figure 11.13   *EWMA chart for average film speed in subgroups of n = 5 film rolls, $\mu_0 = 105$, $\sigma = 6.53$, $\lambda = 0.2$.*

the upper or lower control limits, then a control is applied of size

$$u = -(\hat{\mu}_t - \mu_0). \tag{11.7.11}$$

In Figure 11.13, we present the results of such a control procedure on the film speed (file **FILMSP.csv**) in a production process of coating film rolls. This EWMA-chart was constructed with $\mu_0 = 105$, $\sigma = 6.53$, $\lambda = 0.2$, $L = 2$, and $n = 5$. Notice that at the beginning the process was out of control. After a remedial action, the process returned to a state of control. At time 30 it drifted downward, but was corrected again.

11.8   Chapter highlights

Tests of randomness, called the run tests, are discussed. These tests are required as a first step in checking the statistical stability of a process. Modifications of the 3-sigma control limits that include warning limits are presented as well as control limits based on economic considerations. Particular attention is given to the determination of frequency and size of subgroups used in the process control system. The theory of cumulative sum control charts, CUSUM, is introduced and the main results are given. Special computer programs are given for the estimation of PFA, CED, and expected RL. Special sections on modern topics including Bayesian detection, process tracking, multivariate control charts, and automatic process control introduce the readers to nonstandard techniques and applications.

The main concepts and definitions introduced in this chapter include:

- Run Tests
- Average Run Length
- Operating Characteristic Functions
- Multivariate Control Charts
- Cumulative Sum Control Charts
- Bayesian Detection
- Shiryaev–Roberts Statistic
- Probability of False Alarm
- Conditional Expected Delay
- Process Tracking
- Exponentially Weighted Moving Average
- Kalman Filter
- Quality Measurement Plan
- Automatic Process Control
- Dynamic Programming

11.9   Exercises

11.1 Generate the distribution of the number of runs in a sample of size $n = 25$, if the number of elements above the sample mean is $m_2 = 10$.
   (i) What are Q_1, M_e, and Q_3 of this distribution?
   (ii) Compute the expected value, μ_R, and the standard deviation σ_R.
   (iii) What is $\Pr\{10 \le R \le 16\}$?
   (iv) Determine the normal approximation to $\Pr\{10 \le R \le 16\}$.

11.2 Use MINITAB to perform a run test on the simulated cycle times from the pistons, which are in data file **CYCLT.csv**. Is the number of runs above the mean cycle time significantly different than its expected value?

11.3 (i) What is the expected number of runs up or down, in a sample of size 50?

(ii) Compute the number of runs up or down in the cycle time data (**CYCLT.csv**).

(iii) Is this number significantly different than expected?

(iv) What is the probability that a random sample of size 50 will have at least one run of size greater or equal to 5?

11.4 Analyze the observations in **YARNSTRG.csv** for runs.

11.5 Run the piston simulator at the upper level of the seven control parameters and generate 50 samples of size 5. Analyze the output for runs in both $\bar{X}$ and S charts.

11.6 (i) Run the piston simulator at the upper level of the seven control parameters and generate 50 samples of size 5 (both $\bar{X}$ and S charts).

(ii) Repeat the exercise allowing T to change over time (mark radio-button as YES).

(iii) Compare the results in (i) and (ii) with those of Exercise 11.4.

11.7 Construct a p-chart for the fraction of defective substrates received at a particular point in the production line. One thousand ($n = 1000$) substrates are sampled each week. Remove data for any week for which the process is not in control. Be sure to check for runs as well as points outside the control limits. Construct the revised p-chart and be sure to check for runs again.

Week	No. Def.	Week	No. Def.
1	18	16	38
2	14	17	29
3	9	18	35
4	25	19	24
5	27	20	20
6	18	21	23
7	21	22	17
8	16	23	20
9	18	24	19
10	24	25	17
11	20	26	16
12	19	27	10
13	22	28	8
14	22	29	10
15	20	30	9

11.8 Substrates were inspected for defects on a weekly basis, on two different production lines. The weekly sample sizes and the number of defectives are indicated below in the data set. Plot the data below and indicate which of the lines is not in a state of statistical control. On what basis do you make your decision?

Use R, MINITAB, or JMP to construct control charts for the two production lines.

Note: When the sample size is not the same for each sampling period, we use **variable** control limits. If $X(i)$ and $n(i)$ represent the number of defects and sample size, respectively, for sampling period i, then the upper and lower control limits for the i-th period are

$$UCL_i = \bar{p} + 3(\bar{p}(1 - \bar{p})/n_i)^{1/2}$$

and

$$LCL_i = \bar{p} - 3(\bar{p}(1 - \bar{p})/n_i)^{1/2}$$

where

$$\bar{p} = \sum X(i) / \sum n(i)$$

is the centerline for the control chart.

	Line 1		Line 2	
Week	X_i	n_i	X_i	n_i
1	45	7920	135	2640
2	72	6660	142	2160
3	25	6480	16	240
4	25	4500	5	120
5	33	5840	150	2760
6	35	7020	156	2640
7	42	6840	140	2760
8	35	8460	160	2980
9	50	7020	195	2880
10	55	9900	132	2160
11	26	9180	76	1560
12	22	7200	85	1680

11.9 In designing a control chart for the fraction defectives p, a random sample of size n is drawn from the productions of each day (very large lot). How large should n be so that the probability of detecting a shift from $p_0 = 0.01$ to $p_t = 0.05$, within a 5 day period, will not be smaller than 0.8?

11.10 The following data represent dock-to-stock cycle times for a certain type of shipment (class D). Incoming shipments are classified according to their "type," which is determined by the size of the item and the shipment, the type of handling required, and the destination of the shipment. Samples of five shipments per day are tracked from their initial arrival to their final destination, and the time it takes for this cycle to be complete is noted. The samples are selected as follows: at five preselected times during the day, the next class D shipment to arrive is tagged and the arrival time and identity of the shipment are recorded. When the shipment reaches its final destination, the time is again recorded. The difference between these times is the cycle time. The cycle time is always recorded for the day of arrival.

Dock to stock cycle times

Day	Times				
1	27	43	49	32	36
2	34	29	34	31	41
3	36	32	48	35	33
4	31	41	51	51	34
5	43	35	30	32	31
6	28	42	35	40	37
7	38	37	41	34	44
8	28	44	44	34	50
9	44	36	38	44	35
10	30	43	37	29	32
11	36	40	50	37	43
12	35	36	44	34	32
13	48	49	44	27	32
14	45	46	40	35	33
15	38	36	43	38	34
16	42	37	40	42	42
17	44	31	36	42	39
18	32	28	42	39	27
19	41	41	35	41	44
20	44	34	39	30	37
21	51	43	36	50	54
22	52	50	50	44	49
23	52	34	38	41	37
24	40	41	40	23	30
25	34	38	39	35	33

(i) Construct $\overline{X}$ and S-charts from the data. Are any points out of control? Are there any trends in the data? If there are points beyond the control limits, assume that we can determine special causes for the points, and recalculate the control limits, excluding those points which are outside the control limits.

(ii) Use a t-test to decide whether the mean cycle time for days 21 and 22 was significantly greater than 45.

(iii) Make some conjectures about possible causes of unusually long cycle times. Can you think of other appropriate data that might have been collected, such as the times at which the shipments reached intermediate points in the cycle? Why would such data be useful?

11.11 Consider the modified Shewhart control chart for sample means, with $a = 3$, $w = 2$, and $r = 4$. What is the ARL of this procedure when $\delta = 0, 1, 2$ and the sample size is $n = 10$?

11.12 Repeat the previous exercise for $a = 3$, $w = 1$, $r = 15$; when $n = 5$ and $\delta = 0.5$.

11.13 Suppose that a shift in the mean is occurring at random, according to an exponential distribution with mean of 1 hour. The hourly cost is US\$ 100 per shift of size $\delta = \dfrac{\mu_1 - \mu_0}{\sigma}$.

The cost of sampling and testing is $d =$ US\$ 10 per item. How often should samples of size $n = 5$ be taken, when lifts of size $\delta \geq 1.5$ should be detected?

11.14 Compute the $OC(p)$ function, for a Shewhart 3-sigma control chart for p, based on samples of size $n = 20$, when $p_0 = 0.10$. (Use the formula for exact computations.)

11.15 How large should the sample size n be, for a 3-sigma control chart for p, if we wish that the probability of detecting a shift from $p_0 = 0.01$ to $p_t = 0.05$ be $1 - \beta = 0.90$?

11.16 Suppose that a measurement X, of hardness of brackets after heat treatment, has a normal distribution. Every hour a sample of n units is drawn and a $\bar{X}$ chart with control limits $\mu_0 \pm 3\sigma/\sqrt{n}$ is used. Here μ_0 and σ are the assumed process mean and standard deviation. The OC function is

$$OC(\delta) = \Phi(3 - \delta\sqrt{n}) + \Phi(3 + \delta\sqrt{n}) - 1$$

where $\delta = (\mu - \mu_0)/\sigma$ is the standardized deviation of the true process mean from the assumed one.

  (i) How many hours, on the average, would it take to detect a shift in the process mean of size $\delta = 1$, when $n = 5$?

 (ii) What should be the smallest sample size, n, so that a shift in the mean of size $\delta = 1$ would be on the average detected in less than 3 hours?

(iii) One has two options: to sample $n_1 = 5$ elements every hour or to sample $n_2 = 10$ elements every 2 hours. Which one would you choose? State your criterion for choosing between the two options and make the necessary computations.

11.17 Electric circuits are designed to have an output of 220 (V, DC). If the mean output is above 222 (V, DC) you wish to detect such a shift as soon as possible. Examine the sample of data file **OELECT.csv** for such a shift. For this purpose construct a CUSUM upward scheme with K^+ and h^+ properly designed (consider for h^+ the value $\alpha = 0.001$). Each observation is of sample of size $n = 1$. Is there an indication of a shift in the mean?

11.18 Estimate the probability of false alarm and the conditional expected delay in the Poisson case, with a CUSUM scheme. The parameters are $\lambda_0 = 15$ and $\lambda_1 = 25$. Use $\alpha = 0.001$, $\tau = 30$.

11.19 A CUSUM control scheme is based on sample means.

  (i) Determine the control parameters K^+, h^+, K^-, h^-, when $\mu_0 = 100$, $\mu_1 = 110$, $\sigma = 20$, $n = 5$, $\alpha = 0.001$.

 (ii) Estimate the PFA and CED, when the change point is at $\tau = 10, 20, 30$.

(iii) How would the properties of the CUSUM change if each sample size is increased from 5 to 20.

11.20 Show that the Shiryaev–Roberts statistic W_n, for detecting a shift in a Poisson distribution from a mean λ_0 to a mean $\lambda_1 = \lambda_0 + \delta$ is

$$W_m = (1 + W_{m-1})R_m$$

where $W_0 \equiv 0$, $R_m = \exp\{-\delta + x_m \log(\rho)\}$, and $\rho = \lambda_1/\lambda_0$.

11.21 Analyze the data in file **OELECT1.csv**, with an EWMA control chart with $\lambda = 0.2$.

11.22 Analyze the variable diameters in the data file **ALMPIN.csv** with an EWMA control chart with $\lambda = 0.2$. Explain how you would apply the automatic process control technique described at the end of Section 9.7.

11.23 Construct the Kalman filter for the Dow-Jones daily index, which is given in the data file **DOW1941.csv**.

12

Multivariate Statistical Process Control

12.1 Introduction

Univariate control charts track observations on one dimension. Multivariate data is much more informative than a collection of one dimensional variables. Simultaneously accounting for variation in several variables requires both an overall measure of departure of the observation from the targets as well as an assessment of the data covariance structure. Multivariate control charts were developed for that purpose. We present here the construction of multivariate control charts with the multivariate data on aluminum pins, which were introduced in Chapter 5, file **ALMPIN.csv**.

The following is the methodology for constructing a multivariate control chart. We first use the first 30 cases of the data file as a base sample. The other 40 observations will be used as observations on a production process which we wish to control.

The observations in the base sample provide estimates of the means, variance, and covariances of the six variables being measured. Let $\bar{X}_i$ denote the mean of variable X_i ($i = 1, \ldots, p$) in the base sample.

Let S_{ij} denote the covariance between X_i and X_j ($i, j = 1, \ldots, p$), namely,

$$S_{ij} = \frac{1}{n-1} \sum_{l=1}^{n} (X_{il} - \bar{X}_{i\cdot})(X_{jl} - \bar{X}_{j\cdot}). \qquad (12.1.1)$$

Notice that S_{ii} is the sample variance of X_i ($i = 1, \ldots, p$). Let $\mathbf{S}$ denote the $p \times p$ covariance matrix, i.e.

$$\mathbf{S} = \begin{bmatrix} S_{11} & S_{12} & \cdots & S_{1p} \\ S_{21} & S_{22} & \cdots & S_{2p} \\ \vdots & & & \\ S_{p1} & S_{p2} & \cdots & S_{pp} \end{bmatrix} \qquad (12.1.2)$$

Notice that $S_{ij} = S_{ji}$ for every i, j. Thus, $\mathbf{S}$ is a **symmetric** and positive definite matrix.

Let $\mathbf{M}$ denote the $(p \times 1)$ vector of sample means, whose transpose is

$$\mathbf{M}' = (\bar{X}_{1\cdot}, \ldots, \bar{X}_{p\cdot}).$$

Modern Industrial Statistics: With Applications in R, MINITAB and JMP, Third Edition.
Ron S. Kenett and Shelemyahu Zacks.
© 2021 John Wiley & Sons, Ltd. Published 2021 by John Wiley & Sons, Ltd.

Finally, we compute the inverse of $\mathbf{S}$, namely $\mathbf{S}^{-1}$. This inverse exists, unless one (or some) of the variable(s) is (are) linear combinations of the others. Such variables should be excluded.

Suppose now that every time unit we draw a sample of size m ($m \geq 1$) from the production process, and observe on each element the p variables of interest. In order to distinguish between the sample means from the production process to those of the base sample, we will denote by $\bar{Y}_{i.}(t), t = 1, 2, \ldots$ the sample mean of variable X_i from the sample at time t. Let $\mathbf{Y}(t)$ be the vector of these p means, i.e. $\mathbf{Y}'(t) = (\bar{Y}_{1.}(t), \ldots, \bar{Y}_{p.}(t))$. We construct now a control chart, called the T^2-**Chart**. The objective is to monitor the means $\mathbf{Y}(t)$, of the samples from the production process, to detect when a significant change from $\mathbf{M}$ occurs. We assume that the covariances do not change in the production process. Thus, for every time period t, $t = 1, 2, \ldots$ we compute the T^2 statistics

$$\mathbf{T}_t^2 = (\mathbf{Y}(t) - \mathbf{M})'\mathbf{S}^{-1}(\mathbf{Y}(t) - \mathbf{M}). \tag{12.1.3}$$

It can be shown that as long as the process mean and covariance matrix are the same as those of the base sample,

$$T^2 \sim \frac{(n-1)p}{n-p}F[p, n-p]. \tag{12.1.4}$$

Accordingly, we set up the (upper) control limit for T^2 at

$$\text{UCL} = \frac{(n-1)p}{n-p}F_{0.997}[p, n-p]. \tag{12.1.5}$$

If a point $T(t)$ falls above this control limit, there is an indication of a significant change in the mean vector in the baseline data and, after investigations, we might decide to remove such points.

After establishing the baseline control limits, Upper Prediction Limit, UPL, is used in follow up monitoring is computed so as to account for the number of observations in the baseline phase. The UPL for monitoring is:

$$\text{UPL} = \frac{(n-1)(n+1)p}{n(n-p)}F_{0.997}[p, n-p]. \tag{12.1.6}$$

Example 12.1.   The base sample consists of the first 30 rows of data file **ALMPIN.csv**. The mean vector of the base sample is

$$\mathbf{M}' = (9.99, 9.98, 9.97, 14.98, 49.91, 60.05).$$

The covariance matrix of the base sample is $\mathbf{S}$, where $10^3\mathbf{S}$ is

$$\begin{bmatrix} 0.1826 & 0.1708 & 0.1820 & 0.1826 & -0.0756 & -0.0054 \\ & 0.1844 & 0.1853 & 0.1846 & -0.1002 & -0.0377 \\ & & 0.2116 & 0.1957 & -0.0846 & 0.0001 \\ & & & 0.2309 & -0.0687 & -0.0054 \\ & & & & 1.3179 & 1.0039 \\ & & & & & 1.4047 \end{bmatrix}$$

(Since **S** is symmetric we show only the upper matrix). The inverse of **S** is

$$
S^{-1} = \begin{bmatrix}
53191.3 & -22791.0 & -17079.7 & -9343.4 & 145.0 & -545.3 \\
 & 66324.2 & -28342.7 & -10877.9 & 182.0 & 1522.8 \\
 & & 50553.9 & -6467.9 & 853.1 & -1465.1 \\
 & & & 25745.6 & -527.5 & 148.6 \\
 & & & & 1622.3 & -1156.1 \\
 & & & & & 1577.6
\end{bmatrix}
$$

We compute now for the last 40 rows of this data file the T_t^2 values. We consider as though each one of these rows is a vector of a sample of size one taken every 10 minutes. In Table 12.1, we present these 40 vectors and their corresponding T^2 values.

Table 12.1 *Dimensions of aluminum pins in a production process and their T^2 value*

X_1	X_2	X_3	X_4	X_5	X_6	T^2
10.00	9.99	9.99	14.99	49.92	60.03	3.523
10.00	9.99	9.99	15.00	49.93	60.03	6.983
10.00	10.00	9.99	14.99	49.91	60.02	6.411
10.00	9.99	9.99	14.99	49.92	60.02	4.754
10.00	9.99	9.99	14.99	49.92	60.00	8.161
10.00	10.00	9.99	15.00	49.94	60.05	7.605
10.00	9.99	9.99	15.00	49.89	59.98	10.299
10.00	10.00	9.99	14.99	49.93	60.01	10.465
10.00	10.00	9.99	14.99	49.94	60.02	10.771
10.00	10.00	9.99	15.00	49.86	59.96	10.119
10.00	9.99	9.99	14.99	49.90	59.97	11.465
10.00	10.00	10.00	14.99	49.92	60.00	14.317
10.00	10.00	9.99	14.98	49.91	60.00	13.675
10.00	10.00	10.00	15.00	49.93	59.98	20.168
10.00	9.99	9.98	14.98	49.90	59.98	8.985
9.99	9.99	9.99	14.99	49.88	59.98	9.901
10.01	10.01	10.01	15.01	49.87	59.97	14.420
10.00	10.00	9.99	14.99	49.81	59.91	15.998
10.01	10.00	10.00	15.01	50.07	60.13	30.204
10.01	10.00	10.00	15.00	49.93	60.00	12.648
10.00	10.00	10.00	14.99	49.90	59.96	19.822
10.01	10.01	10.01	15.00	49.85	59.93	21.884
10.00	9.99	9.99	15.00	49.83	59.98	9.535
10.01	10.01	10.00	14.99	49.90	59.98	18.901
10.01	10.01	10.00	15.00	49.87	59.96	13.342
10.00	9.99	9.99	15.00	49.87	60.02	5.413
9.99	9.99	9.99	14.98	49.92	60.03	8.047
9.99	9.98	9.98	14.99	49.93	60.03	5.969

(Continued)

Table 12.1   (continued)

X_1	X_2	X_3	X_4	X_5	X_6	T^2
9.99	9.99	9.98	14.99	49.89	60.01	4.645
10.00	10.00	9.99	14.99	49.89	60.01	5.674
9.99	9.99	9.99	15.00	50.04	60.15	23.639
10.00	10.00	10.00	14.99	49.84	60.03	10.253
10.00	10.00	9.99	14.99	49.89	60.01	5.674
10.00	9.99	9.99	15.00	49.88	60.01	5.694
10.00	10.00	9.99	14.99	49.90	60.04	4.995
9.90	9.89	9.91	14.88	49.99	60.14	82.628
10.00	9.99	9.99	15.00	49.91	60.04	4.493
9.99	9.99	9.99	14.98	49.92	60.04	7.211
10.01	10.01	10.00	15.00	49.88	60.00	8.737
10.00	9.99	9.99	14.99	49.95	60.10	3.421

For example T_1^2 of the table is computed according to the formula

$$\mathbf{T}_1^2 = (\mathbf{Y}(1) - \mathbf{M})'\mathbf{S}^{-1}(\mathbf{Y}(1) - \mathbf{M}) = 3.523.$$

The 40 values of T_t^2 of Table 12.1 are plotted in Figure 12.1. The UCL in this chart is UPL = 34.56.

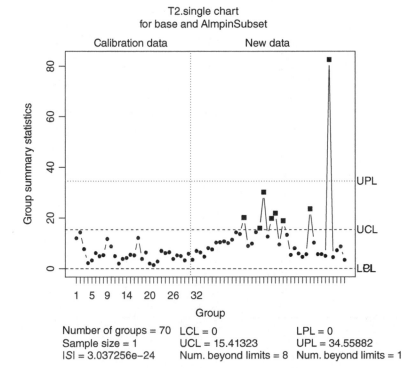

Figure 12.1 T^2-chart for aluminum pins.

We remark here that the computations can be performed by the following R-program. Note that R labels the control limit for the base (the first 30 observations), UCL, and for the ongoing monitoring (the following 40 observations), UPL.

```
> data(ALMPIN)
> Base <- ALMPIN[1:30,]
> MeansBase <- colMeans(Base)
> CovBase <- cov(Base)
> AlmpinSubset <- ALMPIN[-(1:30),]
> library(qcc)
> Mqcc <- mqcc(data=Base,
               type="T2.single",
               center=MeansBase,
               cov=CovBase,
               pred.limits = TRUE,
               newdata = AlmpinSubset,
               confidence.level = 0.997,
               add.stats = TRUE,
               plot=TRUE)
> summary(Mqcc)

Call:
mqcc(data = Base, type = "T2.single", center = MeansBase, cov =
    CovBase, pred.limits = TRUE, newdata = AlmpinSubset,
    confidence.level = 0.997, plot = TRUE, add.stats = TRUE)


T2.single chart for Base

Summary of group statistics:
    Min.   1st Qu.    Median       Mean    3rd Qu.
 1.421054  3.834135  5.273873  5.800000  6.472311
    Max.
14.348835


Number of variables:   6
Number of groups:   30
Group sample size:   1


Center:
    diam1      diam2      diam3    capDiam    lenNocp
 9.986333   9.978667   9.974333  14.976333  49.907333
    lenWcp
60.047667


Covariance matrix:
                  diam1                diam2
diam1     0.000182643678    0.00017080460
diam2     0.000170804598    0.00018436782
diam3     0.000181954023    0.00018528736
```

```
capDiam   0.000182643678   0.00018459770
lenNocp  -0.000075632184  -0.00010022989
lenWcp   -0.000005402299  -0.00003770115
                     diam3           capDiam
diam1     0.0001819540230   0.000182643678
diam2     0.0001852873563   0.000184597701
diam3     0.0002116091954   0.000195747126
capDiam   0.0001957471264   0.000230919540
lenNocp  -0.0000845977011  -0.000068735632
lenWcp    0.0000001149425  -0.000005402299
                   lenNocp            lenWcp
diam1     -0.00007563218  -0.0000054022989
diam2     -0.00010022989  -0.0000377011494
diam3     -0.00008459770   0.0000001149425
capDiam   -0.00006873563  -0.0000054022989
lenNocp    0.00137195402   0.0010039080460
lenWcp     0.00100390805   0.0014047126437
|S|:   3.037256e-24

Summary of group statistics in AlmpinSubset:
    Min.  1st Qu.   Median    Mean  3rd Qu.     Max.
 3.42051  5.90029  9.71798 12.52126 13.83571 82.62795

Number of groups:   40
Group sample size:   1

Control limits:
 LCL       UCL
   0  15.41323

Prediction limits:
 LPL       UPL
   0  34.55882
```

For a comprehensive treatment of multivariate quality control with MINITAB applications and detailed listing of MINITAB macros, the reader is referred to the book of Fuchs and Kenett (1998).

12.2   A review multivariate data analysis

Chapters 10 and 11 present applications of statistical process control (SPC) to measurements in one dimension. To extend the approach to measurements consisting of several dimensions lead us to multivariate statistical process control (MSPC), as introduced in Section 12.1. MSPC requires applications of methods and tools of multivariate data analysis presented in Chapter 5. In this section, we expand on the material presented in Chapter 5 and use the components placement data of Examples 5.1 and 5.2. The case study consists of displacement coordinates of 16 components placed by a robot on a printed circuit board. Overall, there are 26 printed circuit boards and therefore a total of 416 placed components (see **PLACE.csv**). The components' coordinates are measured in three dimensions, representing deviations with respect to the target in the horizontal, vertical, and angular dimensions. The measured variables are labeled x-dev, y-dev, and theta-dev. The placement of components on the 26 boards was part of a validation test designed to fine tune the placement software in order to minimize the placement deviations.

Figure 12.2 presents a scatterplot matrix of x-dev, y-dev, and theta-dev with nonparametric densities providing a visual display of the two dimensional distribution densities. On the y-dev x-dev scatterplot, there are clearly three groups of boards. In Section 5.1.2, we identified them with box plots and confirmed the classification with coding and redrawing the scatterplot (Figure 5.6).

Figure 12.2 presents the histograms of x-dev, y-dev, and theta-dev with the low values of x-dev highlighted. Through dynamic linking, the figure highlights the corresponding values for y-dev and theta-dev. We can see that components position on the left of the target tend to be also placed below the target, with some components in this group being positioned on target in the vertical direction (the group of y-dev between -0.001 and 0.002).

To further characterize the components placed on the left of the target (negative x-dev), we draw a plot of x-dev and y-dev versus the board number (Figure 12.3). On each board, we get 16 measurements of x-dev (circle) and y-dev (cross). The highlighted points correspond to the

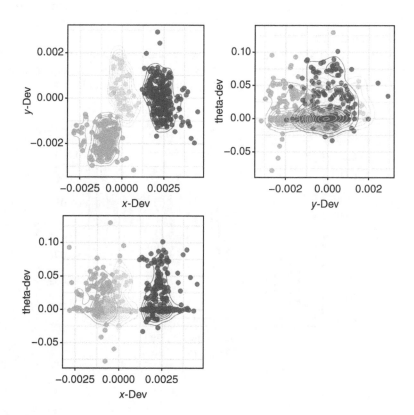

Figure 12.2 Scatterplot matrix of placement data with nonparametric densities.

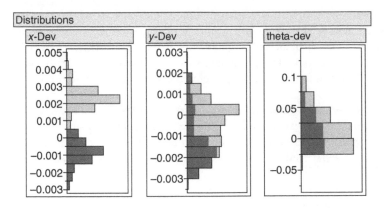

Figure 12.3   *Histograms of x-dev, y-dev, and theta-dev, with dynamic linking (JMP).*

highlighted values in Figure 12.3. One can see from Figure 12.3 that up to board number 9 we have components placed to the right and below the target. In boards 10, 11, and 12 there has been a correction in component placement which resulted in components being placed above the target in the vertical direction and on target in the horizontal direction. These are the components in the histogram of y-dev in Figure 12.2 with the high trailing values between −0.001 and 0.002 mentioned above (see also Figure 12.4). For these first 12 circuit boards, we do not notice any specific pattern in theta-dev.

Software such as R, MINITAB, and JMP provide visualization and exploration technologies that complement the application of MSPC. We will use the placement data example again in Section 12.4 to demonstrate the application of a multivariate control chart.

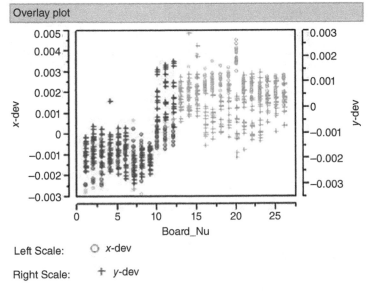

Figure 12.4   *Plot of x-dev and y-dev versus circuit board number (JMP).*

12.3   Multivariate process capability indices

Chapter 10 introduced process capability studies that are a prerequisite to the setup of control limits in control charts. Section 8.3 presented several univariate process capability indices such as C_p and C_{pk} that are used to characterize the performance of a process by comparing the quality attributes specifications to the process variability. These indices map processes in terms of their ability to deliver high critical quality parameters. In this section, we focus on multivariate data and develop several multivariate process capability indices.

As in the univariate case, the multivariate capability indices are based on the multivariate normal distribution. In Chapter 3, we introduced the bivariate normal distribution, whose p.d.f. is given in (3.6.6). We introduce first the m-variate normal distribution as a joint distribution of a vector $\mathbf{X}' = (X_1, \ldots, X_m)$ of m random variable. The expected value of such a vector is the vector $\boldsymbol{\mu}' = (E\{X_1\}, \ldots, E\{X_m\})$. The covariance matrix of this vector is an $m \times m$, symmetric, positive definite matrix $\boldsymbol{\Sigma} = (\sigma_{ij}; i, j = 1, \ldots, m)$, where $\sigma_{ij} = \mathrm{cov}(X_i, X_j)$. The multivariate normal vector is denoted by $N(\boldsymbol{\mu}, \boldsymbol{\Sigma})$. The joint p.d.f. of $\mathbf{X}$ is

$$f(\mathbf{x}; \boldsymbol{\mu}, \boldsymbol{\Sigma}) = \frac{1}{(2\pi)^{m/2}|\boldsymbol{\Sigma}|^{1/2}} \exp\left\{-\frac{1}{2}(\mathbf{x} - \boldsymbol{\mu})'\boldsymbol{\Sigma}^{-1}(\mathbf{x} - \boldsymbol{\mu})\right\}. \tag{12.3.1}$$

In the multivariate normal distribution, the marginal distribution of X_i is normal $N(\mu_i, \sigma_i^2)$, $i = 1, \ldots, m$.

We describe now some of the multivariate capability indices. The reader is referred to papers of Chen (1994), Haridy et al. (2011), and Jalili et al. (2012).

A **tolerance region** (TR) is a region around a **target** point, **T**. We will consider here two possible regions: (i) a hyper-rectangular, RTR, and (ii) a sphere, CTR. In case (i), the region is specified by parameters $(\delta_1, \ldots, \delta_m)$, and is the set

$$\text{RTR} = \{\mathbf{x} : |x_i - T_i| \le \delta_i, i = 1, \ldots, m\}. \tag{12.3.2}$$

In case (ii), TR is

$$\text{CTR} = \{\mathbf{x} : |\mathbf{X} - \mathbf{T}| \le r\}, \tag{12.3.3}$$

where r is the radius of a sphere centered at **T**.

In the paper of Jalili et al. (2012), a TR, which is the largest ellipsoidal region in the RTR (12.3.2), is considered. In the present section, we will confine attention to (12.3.2) and (12.3.3). Chen (1994) suggested the following multivariate capability index (MC$_p$). Suppose that the TR is CTR, with a specified radius r. Let r_α be the value of r, for which $P\{|\mathbf{X} - T| \le r_\alpha\} = 1 - \alpha$, with $\alpha = 0.0027$. Then, the MC$_p$ index is

$$\text{MC}_p = \frac{r}{r_\alpha}. \tag{12.3.4}$$

In the case of a RTR, Chen suggested the index

$$\text{MC}_p = \frac{\delta_s}{\delta_\alpha}, \tag{12.3.5}$$

where δ_α is the value for which

$$P\left\{\max\left\{\frac{|X_i - T_i|}{\delta_i}, i = 1, \ldots, m\right\} \le \delta_\alpha\right\} = 1 - \alpha.$$

$\delta_s = \max\{\delta_i, i = 1, \ldots, m\}$. We will show later how to compute these indices in the case of $m = 2$. In the special case where $\mathbf{X} \sim N(\mathbf{T}, \sigma^2 I)$, i.e. all the m components of $\mathbf{X}$ are independent with $\mu_i = T_i$ and $\sigma_i^2 = \sigma^2$ we have a simple solution. If $\delta_i = \delta$ for all i, then

$$\max\left\{\frac{|X_i - T_i|}{\delta}, i = 1, \ldots, m\right\} \sim \left(\frac{\sigma}{\delta}\right)|Z|_{(m)}, \quad \text{where } |Z|_{(m)} = \max\{|Z_i|, i = 1, \ldots, m\} \text{ and}$$

$Z \sim N(0,1)$. Thus,

$$P\left\{\frac{\sigma}{\delta}|Z|_{(m)} \le y\right\} = P\{|Z| \le y\delta/\sigma\}^m$$

$$= \left(2\Phi\left(\frac{y\delta}{\sigma}\right) - 1\right)^m.$$

Hence, δ_α is a solution of $\left(2\Phi\left(\dfrac{\delta_\alpha\delta}{\sigma}\right) - 1\right)^m = 1 - \alpha$, or

$$\delta_\alpha = \frac{\sigma}{\delta}\Phi^{-1}\left(\frac{1}{2} + \frac{1}{2}(1-\alpha)^{1/m}\right). \tag{12.3.6}$$

Haridy et al. (2011) suggested a different type of index, based on the principal components of Σ.

Let H' be an orthogonal matrix, whose column vectors are the orthogonal eigenvectors of Σ. Let $\lambda_1 \ge \lambda_2 \ge \cdots \ge \lambda_m > 0$ be the corresponding eigenvalues. Recall that

$$H\Sigma H' = \begin{pmatrix} \lambda_1 & & 0 \\ & \ddots & \\ 0 & & \lambda_m \end{pmatrix}. \tag{12.3.7}$$

The transformed vector

$$\mathbf{Y} = H(\mathbf{X} - \boldsymbol{\mu}), \tag{12.3.8}$$

is called the **principal components vector**.

The distribution of $\mathbf{Y}$ is that of $N(\mathbf{0}, \Lambda)$, where $\Lambda = \text{diag}\{\lambda_i, i = 1, \ldots, m\}$. Here, λ_i is the variance of Y_i. Also, $Y_1, \ldots, Y_m$ are independent.

The vector of upper specification limits in the RTR is $\mathbf{U} = \mathbf{T} + \boldsymbol{\delta}$. The corresponding vector of lower specification limits is $\mathbf{L} = \mathbf{T} - \boldsymbol{\delta}$. These vectors are transformed into $\mathbf{U}^* = H\boldsymbol{\delta}$ and $\mathbf{L}^* = -H\boldsymbol{\delta}$.

Suppose that $\mathbf{X}_1, \ldots, \mathbf{X}_n$ is a random sample from the process. These n vectors are independent and identically distributed. The maximum likelihood estimator of $\boldsymbol{\mu}$ is $\mathbf{M} = \frac{1}{n}\sum_{i=1}^n \mathbf{X}_i$. An estimator of Σ is the sample covariance matrix S (see (12.1.1)). Let $\hat{H}'$ and $\hat{\Lambda}$ be the corresponding matrices of eigenvectors and eigenvalues of S. The estimated vectors of the principal components are

$$\hat{\mathbf{Y}}_i = \hat{H}(\mathbf{X}_i - \mathbf{M}), \quad i = 1, \ldots, n. \tag{12.3.9}$$

Let $\{\hat{Y}_{1j}, \ldots, \hat{Y}_{n,j}\}$ be the sample of the j-th $(j = 1, \ldots, m)$ principal components. Let

$$C_{p,pc_j} = \frac{U^* - L^*}{6\hat{\sigma}_{y_j}}, \quad j = 1, \ldots, m, \tag{12.3.10}$$

where $\hat{\sigma}_{y_j}^2 = \frac{1}{n-1}\sum_{i=1}^n (\hat{Y}_{ij} - \bar{Y}_j)^2$. The MCP index, based on the principal components is

$$\text{MCP} = \left(\prod_{j=1}^m C_{p,pc_j}\right)^{1/m}. \tag{12.3.11}$$

We derive now explicit formula for an RTR, when $m = 2$ (bivariate normal distribution). The distribution of $\mathbf{X}$ is $N\left(\begin{bmatrix} \xi \\ \eta \end{bmatrix}, \begin{bmatrix} \sigma_1^2 & \rho\sigma_1\sigma_2 \\ \bullet & \sigma_2^2 \end{bmatrix}\right)$. The conditional distribution of X_2, given X_1, is $N\left(\eta + \rho\frac{\sigma_2}{\sigma_1}(X_1 - \xi), \sigma_2^2(1 - \rho^2)\right)$. Accordingly, the probability that $\mathbf{X}$ belongs to the rectangular tolerance region (RTR) with specified $\boldsymbol{\delta} = (\delta_1, \delta_2)'$, is

$$P\{\mathbf{X} \in \text{RTR}(\boldsymbol{\delta})\} = \int_{T_1 - \delta_1 - \xi_1/\sigma_1}^{(T_1 + \delta_1 - \xi_1)/\sigma_1} \phi(z) \left[\Phi\left(\frac{T_2 + \delta_2 - (\eta + \rho\frac{\sigma_2}{\sigma_1}(z - (\xi - T_1)))}{\sigma_2(1 - \rho^2)^{1/2}}\right) \right.$$
$$\left. -\Phi\left(\frac{T_2 - \delta_2 - (\eta + \rho\frac{\sigma_2}{\sigma_1}(z - (\xi - T_1)))}{\sigma_2(1 - \rho^2)^{1/2}}\right) \right] dz. \qquad (12.3.12)$$

In particular, if $T_1 = \xi$ and $T_2 = \eta$ then

$$P\{\mathbf{X} \in \text{RTR}(\boldsymbol{\delta})\} = \int_{-\frac{\delta_1}{\sigma_1}}^{\frac{\delta_1}{\sigma_1}} \phi(z) \left[\Phi\left(\frac{\sigma_1\delta_2 - \rho\sigma_2 z}{\sigma_1\sigma_2(1 - \rho^2)^{1/2}}\right) \right.$$
$$\left. -\Phi\left(\frac{-\sigma_1\delta_2 - \rho\sigma_2 z}{\sigma_1\sigma_2(1 - \rho^2)^{1/2}}\right) \right] dz. \qquad (12.3.13)$$

$\phi(z) = \frac{1}{\sqrt{2\pi}} \exp(-\frac{1}{2}z^2)$ is the standard normal density.

If TR is circular, CRC(r), with radius r, $(\xi, \eta) = (T_1, T_2) = \mathbf{0}$, then

$$P\{\mathbf{X} \in \text{CRC}(r)\} = P\{\lambda_1^{-1}Y_1^2 + \lambda_2^{-1}Y_2^2 \leq r\}, \qquad (12.3.14)$$

where λ_1 and λ_2 are the eigenvalues of $\mathbf{\Sigma} = \begin{pmatrix} \sigma_1^2 & \rho\sigma_1\sigma_2 \\ \bullet & \sigma_2^2 \end{pmatrix}$, and Y_1^2, Y_2^2 are independent, having a $\chi^2[1]$ distribution. Thus,

$$P\{\mathbf{X} \in \text{CRC}(r)\} = \frac{1}{\sqrt{2\pi}} \int_0^{\sqrt{\lambda r}} x^{-1/2}e^{-x/2} \cdot [2\Phi((\lambda_1, \lambda_2 r^2 - \lambda_2 x)^{1/2}) - 1]dx. \qquad (12.3.15)$$

We compute now the capability index MC_p according to (12.3.5), for the **ALMPIN.csv** data. We restrict attention to the last two variables in the data set, i.e. Lengthncp and Lengthwcp. The sample consists of the first 30 data vectors, described in Section 12.1. We use equation (12.3.13) with the $r = 0.7377$, $\sigma_1^2 = 1.3179$, and $\sigma_2^2 = 1.4047$. We assume that $T_1 = \xi = 49.91$ and $T_2 = \eta = 60.05$. From equation (12.3.13), we find that for $\boldsymbol{\delta} = (3.5, 3.5)'$ $P\mathbf{X} \in \text{RTR}(\boldsymbol{\delta})\} = 0.9964$. If the tolerance region is specified by $\boldsymbol{\delta}_s = (1.5, 1.5)'$ then we get index $\text{MC}_p = \frac{1.5}{3.5} = 0.4286$.

If TR is circular, we get according to (12.3.15) $r_\alpha = 4.5$. Thus, $\text{MC}_p = \frac{1.5}{4.5} = 0.333$. We compute now the MCP index (12.3.11), for the 6-dimensional vector of the **ALMPIN.csv** data. We base the estimation on the last 37 vectors of the data set. Notice that the expected value of the principal components $\mathbf{Y}$ is zero. Hence, we consider the upper tolerance limit for Y to be USL $= 0.015$ and the lower tolerance limit to be LSL $= -0.015$. For these specifications, we get MCP $= 0.4078$. If we increase the tolerance limits to ± 0.03 we obtain MCP $= 0.8137$.

12.4   Advanced applications of multivariate control charts

12.4.1   Multivariate control charts scenarios

The Hotelling T^2 chart introduced in Section 12.2 plots the T^2 statistic, which is the squared standardized distance of a vector from a target point (see equation (12.1.3)). Values of T^2 represent equidistant vectors along a multidimensional ellipse centered at the target vector point. The chart has an upper control limit (UCL) determined by the F distribution (see equation (12.1.4)). Points exceeding UCL are regarded as an out-of-control signal. The charted T^2 statistic is a function that reduces multivariate observations into a single value while accounting for the covariance matrix. Out-of-control signals on the T^2 chart trigger an investigation to uncover the causes for the signal.

The setup of an MSPC chart is performed by a process capability study. The process capability study period is sometimes referred to as **phase I**. The ongoing control using control limits determined in phase I is then called **phase II**. The distinction between these two phases is important.

In setting MSPC charts, one meets several alternative scenarios derived from the characteristics of the reference sample and the appropriate control procedure. These include:

1. internally derived targets
2. using an external reference sample
3. externally assigned targets
4. measurements units considered as batches.

We proceed to discuss these four scenarios.

12.4.2   Internally derived targets

Internally derived targets are a typical scenario for process capability studies. The parameters to be estimated include the vector of process means, the process covariance matrix, and the control limit for the control chart. Consider a process capability study with a base sample of size n of p-dimensional observations, $X_1, X_2, \ldots, X_n$. When the data are grouped and k subgroups of observations of size m are being monitored, $n = km$, the covariance matrix estimator, S_p can be calculated as the pooled covariances of the subgroups. In that case, for the jth subgroup, the Hotelling T^2 statistic is then given by:

$$T^2 = \mathbf{m}(\bar{X}_j - \bar{\bar{X}})' S_p^{-1} (\bar{X}_j - \bar{\bar{X}}), \tag{12.4.1}$$

where $\bar{X}_j$ is the mean of the jth subgroup, $\bar{\bar{X}}$ is the overall mean, and S_p^{-1} is the inverse of the pooled estimated covariance matrix. The UCL for this case is

$$\text{UCL} = \frac{p(k-1)(m-1)}{k(m-1) - p + 1} F_\alpha[p, k(m-1) - p + 1]. \tag{12.4.2}$$

When the data is ungrouped, and individual observations are analyzed, the estimation of the proper covariance matrix and control limits requires further consideration. Typically in this case, the covariance matrix is estimated from the pooled individual observations as $S = \frac{1}{n-1} \sum_{i=1}^{n} (X_i - \bar{X})(X_i - \bar{X})'$, where $\bar{X}$ is the mean of the n observations. The corresponding T^2 statistic for the ith observation, $i = 1, \ldots, n$ is given by $T^2 = (X_i - \bar{X})' S^{-1} (X_i - \bar{X})$. In this case, $(X_i - \bar{X})$ and S are not independently distributed and the appropriate upper control for T^2 is based on the beta

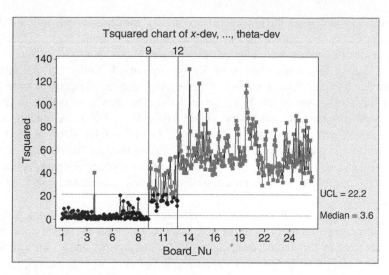

Figure 12.5 *T^2 control chart of x-dev, y-dev, and theta-dev with control limits and correlation structure set up with data from first 9 printed circuit boards (MINITAB).*

distribution with

$$UCL = \frac{(n-1)^2}{n} B_{1-\alpha/2}\left(\frac{p}{2}, \frac{n-p-1}{2}\right) \tag{12.4.3}$$

and

$$LCL = \frac{(n-1)^2}{n} B_{\alpha/2}\left(\frac{p}{2}, \frac{n-p-1}{2}\right), \tag{12.4.4}$$

where $B_\alpha(v_1, v_2)$ is the $(1-\alpha)$th quantile of the beta distribution with v_1 and v_2 as parameters. While theoretically the lower control limit (LCL) can be calculated as above, in most circumstances LCL is set to zero. Figure 12.6 presents the T^2 Hotelling Control Chart for the placement data used in Figure 12.2.

In Figure 12.6, Phase I was conducted over the first 9 printed circuit boards. The implication is that the first 144 observations are used to derive estimates of the means and covariances of x-dev, y-dev, and theta-dev and, with these estimates, the UCL is determined.

The chart in Figure 12.5 indicates an out of control point at observations 55 from board 4 due to very low horizontal and vertical deviations (x-dev = −0.0005200, y-dev = 0.0002500) and an extreme deviations in theta (theta-dev = 0.129810). In the components inserted on the first 9 boards, the average and standard deviations (in bracket) of x-dev, y-dev, and theta-dev are, respectively: −0.001062 (0.000602), −0.001816 (0.000573), +0.01392 (0.02665).

Referring again to Figure 12.5, we see a deviation in performance after board 9 with a significant jump past after board 12. We already studied what happened on boards 10–12 and know that the shift is due to a correction in the vertical direction, increasing the values of y-dev to be around zero (see Section 12.2).

12.4.3   Using an external reference sample

Consider again a process yielding independent observations $X_1, X_2, \ldots$ of a p-dimensional random variable $\mathbf{X}$, such as the quality characteristics of a manufactured item or process measurements. Initially, when the process is "in control," the observations follow a distribution F, with density f. We now assume that we have a "reference" sample $X_1, \ldots, X_n$ of F from an in-control period. To control the quality of the produced items, multivariate data is monitored for potential change in the distribution of $\mathbf{X}$, by sequentially collecting and analyzing the observations X_i. At some time $t = n + k$, k time units after n, the process may run out of control and the distribution of the X_i's changes to G. Our aim is to detect, in phase II, the change in the distribution of subsequent observations X_{n+k}, $k \geq 1$, as quickly as possible, subject to a bound $\alpha \in (0, 1)$ on the probability of raising a false alarm at each time point $t = n + k$ (that is, the probability of erroneously deciding that the distribution of X_{n+k} is not F). The reference sample $X_1, \ldots, X_n$ does not incorporate the observations X_{n+k} taken after the "reference" stage, even if no alarm is raised, so that the rule is conditional only on the reference sample.

When the data in phase II is grouped, and the reference sample from historical data includes k subgroups of observations of size m, $n = km$, with the covariance matrix estimator S_p calculated as the pooled covariances of the subgroups, the T^2 for a new subgroup of size m with mean $\bar{Y}$ is given by $T^2 = \mathbf{m}(\bar{Y} - \bar{\bar{X}})' S_p^{-1} (\bar{Y} - \bar{\bar{X}})$, and the UCL is given by $\text{UCL} = \dfrac{p(k+1)(m-1)}{k(m-1) - p + 1} F_\alpha[p, k(m-1) - p + 1]$. Furthermore, if in phase I, l subgroups were outside the control limits and assignable causes were determined, those subgroups are omitted from the computation of $\bar{\bar{X}}$ and S_p^{-1}, and the control limits for this case are $\text{UCL} = \dfrac{p(k-l+1)(m-1)}{(k-l)m-1) - p + 1} F_\alpha[p, (k-l/(m-1) - p + 1]$. The T^2 control charts constructed in phase I, and used both in phase I and in phase II, are the multivariate equivalent of the Shewhart control chart. Those charts, as well as some more advanced ones, simplify the calculations down to single-number criteria and produce a desired Type I error or in-control run length.

While we focused on the reference sample provided by phase I of the multivariate process control, other possibilities can occur as well. In principle, the reference in-control sample can also originate from historical data. In this case, the statistical analysis will be the same but this situation has to be treated with precaution since both the control limits and the possible correlations between observations may shift.

12.4.4   Externally assigned targets

If all parameters of the underlying multivariate distribution are known and externally assigned, the T^2 value for a single multivariate observation of dimension p is computed as

$$T^2 = (\mathbf{Y} - \boldsymbol{\mu})' \Sigma^{-1} (\mathbf{Y} - \boldsymbol{\mu}), \tag{12.4.5}$$

where $\boldsymbol{\mu}$ and Σ are the expected value and covariance matrix, respectively.

The probability distribution of the T^2 statistic is a χ^2 distribution with p degrees of freedom. Accordingly, the 0.95 UCL for T^2 is $\text{UCL} = \chi^2_{v, 0.95}$. When the data are grouped in subgroups of size m, and both $\boldsymbol{\mu}$ and Σ are known, the T^2 value of the mean vector $\bar{Y}$ is $T^2 = m(\bar{Y} - \boldsymbol{\mu})' \Sigma^{-1} (\bar{Y} - \boldsymbol{\mu})$ with the same UCL as above.

If only the expected value of the underlying multivariate distribution, $\boldsymbol{\mu}$, is known and externally assigned, the covariance matrix has to be estimated from the tested sample. The T^2 value for a single multivariate observation of dimension p is computed as $T^2 = (\mathbf{Y} - \boldsymbol{\mu})'S^{-1}(\mathbf{Y} - \boldsymbol{\mu})$, where $\boldsymbol{\mu}$ is the expected value and S is the estimate of the covariance matrix $\boldsymbol{\Sigma}$, estimated either as the pooled contribution of the individual observations, i.e. $S = \frac{1}{n-1}\sum_{i=1}^{n}(X_i - \bar{X})(X_i - \bar{X})'$, or by a method which accounts for possible lack of independence between observations.

In this case, the 0.95 UCL for T^2 is $\text{UCL} = \dfrac{p(n-1)}{n(n-p)}F_{0.95}[p, n-p]$.

When the tested observations are grouped, the mean vector of a subgroup with m observations (a rational sample) will have the same expected value as the individual observations, $\boldsymbol{\mu}$, and a covariance matrix $\boldsymbol{\Sigma}/m$. The covariance matrix $\boldsymbol{\Sigma}$ can be estimated by S or as S_p obtained by pooling the covariances of the k subgroups.

When $\boldsymbol{\Sigma}$ is estimated by S, the T^2 value of the mean vector $\bar{Y}$ of m tested observations is $T^2 = m(\bar{Y} - \boldsymbol{\mu})'S^{-1}(\bar{Y} - \boldsymbol{\mu})$ and the 0.95 UCL is

$$\text{UCL} = \frac{p(m-1)}{m-p}F_{0.95}[p, m-p]. \tag{12.4.6}$$

When $\boldsymbol{\Sigma}$ is estimated by S_p, the 0.95 UCL of $T^2 = m(\bar{Y} - \boldsymbol{\mu})'S_p^{-1}(\bar{Y} - \boldsymbol{\mu})$ is

$$\text{UCL} = \frac{pk(m-1)}{k(m-1)-p+1}F_{0.95}[p, k(m-1)-p+1]. \tag{12.4.7}$$

12.4.5   Measurement units considered as batches

In the semiconductor industry, production is typically organized in batches or production lots. In such cases, the quality-control process can be performed either at the completion of the batch or sequentially, in a curtailed inspection, aiming at reaching a decision as soon as possible. When the quality-control method used is reaching a decision at the completion of the process, the possible outcomes are (1) determine the production process to be in statistical control and accept the batch or (2) stop the production flow because of a signal that the process is out of control. On the other hand, in a curtailed inspection, based on a statistical stopping rule, the results from the first few items tested may suffice to stop the process prior to the batch completion.

Consider a batch of size n, with the tested items $\mathbf{Y}_1, \dots, \mathbf{Y}_n$. The curtailed inspection tests the items sequentially. Assume that the targets are specified, either externally assigned or from a reference sample or batch. With respect to those targets, let $V_i = 1$ if the T^2 of the ordered ith observation exceeds the critical value κ and $V_i = 0$, otherwise. For the ith observation, the process is considered to be in control if for a prespecified P, say $P = 0.95$, $\Pr(V_i = 0) \geq P$. Obviously, the inspection will be curtailed only at an observation i for which $V_i = 1$ (not necessarily the first).

Let $N(g) = \sum_{i=1}^{g} V_i$ be the number of rejection up to the gth tested item. For each number of individual rejections U (out of n), $R(U)$ denotes the minimal number of observations allowed up to the Uth rejection, without rejecting the overall null hypothesis. Thus, for each U, $R(U)$ is the minimal integer value such that under the null hypothesis, $\Pr\left(\sum_{i=1}^{R(U)} V_i \leq U\right) \geq \alpha$. For fixed U, the random variable $\sum_{i=1}^{U} V_i$ has a negative binomial distribution, and we can compute $R(N(g))$ from the inverse of the negative binomial distribution. For example when $n = 13$, $P = 0.95$, and $\alpha = 0.01$, the null hypothesis is rejected if the second rejection occurred at or before the third observation, or if the third rejection occurred at or before the ninth observation, and so on.

12.4.6   Variable decomposition and monitoring indices

Data in batches is naturally grouped, but even if quality control is performed on individual items, grouping the data into rational consequent subgroups may yield relevant information on within subgroups variability, in addition to deviations from targets. In the jth subgroup (or batch) of size n_j, the individual T_{ij}^2 values, $i = 1, \ldots, n_j$ are given by $T_{ij}^2 = (\mathbf{Y}_{ij} - \boldsymbol{\theta})' G^{-1} (\mathbf{Y}_{ij} - \boldsymbol{\theta})$. When the targets are externally assigned, then $\boldsymbol{\theta} = \boldsymbol{\mu}$. If the covariance matrix is also externally assigned then $G = \Sigma$, otherwise G is the covariance matrix estimated either from the tested or from the reference sample. In the case of targets derived from an external reference sample $\boldsymbol{\theta} = \mathbf{m}$ and $G = S$, where $\mathbf{m}$ and S are the mean and the covariance matrix of a reference sample of size n. Within the jth subgroup, let us denote the mean of the subgroup observations by $\bar{Y}_j$ and the mean of the target values in the jth subgroup by $\boldsymbol{\theta}_j$.

The sum of the individual T_{ij}^2 values, $T_{0j}^2 = \sum_{i=1}^{\bar{n}} T_{ij}^2$ can be decomposed into two measurements of variability, one representing the deviation of the subgroup mean from the multivariate target denoted by T_{Mj}^2, and the other measuring the internal variability within the subgroup, denoted by T_{Dj}^2. The deviation of the subgroup mean from the multivariate target is estimated by $T_{Mj}^2 = (\bar{Y}_j - \boldsymbol{\theta}_j)' G^{-1} (\bar{Y}_j - \boldsymbol{\theta}_j)$, while the internal variability within the subgroup is estimated by

$$T_{Dj}^2 = (\mathbf{Y}_{ij} - \bar{Y}_j)' G^{-1} (\mathbf{Y}_{ij} - \bar{Y}_j), \quad \text{with} \quad T_{0j}^2 = (n-1)T_{Mj}^2 + T_{Dj}^2. \tag{12.4.8}$$

Since asymptotically, T_{Mj}^2 and T_{Dj}^2 have a χ^2 distribution with p and $(n-1)p$ degrees of freedom, respectively, one can further compute two indices, I_1 and I_2, to determine whether the overall variability is mainly due to the distances between the means of the tested subgroup from targets or to the within subgroup variability. The indices are relative ratios of the normalized versions of the two components of T_{0j}^2, i.e. $I_1 = I_1^* / (I_1^* + I_2^*)$, and $I_2 = I_2^* / (I_1^* + I_2^*)$, where $I_1^* = T_{Mj/p}^2$ and $I_2^* = T_{Dj}^2 / [(n-1)p]$. We can express the indices in terms of the original T^2 statistics as, $I_1 = (n-1)T_{Mj}^2 / [(n-1)T_{Mj}^2 + T_{Dj}^2]$ and $I_2 = T_{Dj}^2 / [(n-1)T_{Mj}^2 + T_{Dj}^2]$. Tracking these indices provides powerful monitoring capabilities.

12.5   Multivariate tolerance specifications

Multivariate TRs are based on estimates of quantiles from a multivariate distribution with parameters either known or estimated from the data (John 1963). Setting up a process control scheme, on the basis of tolerance regions, involves estimating the level set $\{f \geq c\}$ of the density f which generates the data, with a prespecified probability content $1 - \alpha$. With this approach, originally proposed in Fuchs and Kenett (1987), the rejecting region is $X_{n+1} \in \{f \geq c\}$. This method provides an exact false alarm probability of α. Since f is usually unknown, the population TR $\{f \geq c\}$ needs to be estimated by an estimator of f. A similar approach was adopted by the Food and Drug Administration to determine equivalence of a drug product tablet before and after a change in manufacturing processes such as introduction of new equipment, a transfer of operations to another site or the scaling up of production to larger vessels. The equivalence is evaluated by comparing tablet dissolution profiles of a batch under test with dissolution profiles of tablets from a reference batch and allowing for at most a 15% difference. We expand on this example using the procedure proposed by Tsong et al. (1996).

When comparing the dissolution data of a new product and a reference approved product, the goal is to assess the similarity between the mean dissolution values at several observed sample time points. The decision of accepting or rejecting the hypothesis that the two batches have similar dissolution profiles, i.e. are bioequivalent, is based on determining if the difference in mean dissolution values between the test and reference products is no larger than the maximum expected difference between any two batches of the approval product. When dissolution value is measured at a single time point, the confidence interval of the true difference between the two batches is compared with prespecified similarity limits. When dissolution values are measured at several time points, the Mahalanobis D_M^2 defined above can be used to compare the overall dissolution profiles.

The important property of the Mahalanobis D^2 is that differences at points with low variability are given a higher weight than differences at points with higher variability. This ensures that the experimental noise is properly addressed.

Let $X_1 = (x_{11}x_{12}, \ldots, x_{1p})$ and $X_2 = (x_{21}, x_{22}, \ldots, x_{2p})$ represent the mean dissolution values at p time instances of the reference and the batch under test, respectively. These means can correspond to a different number of replicates, say n and m.

The Mahalanobis distance between any two vectors $\mathbf{X}_1$ and $\mathbf{X}_2$, having the same dispersion matrix $\mathbf{\Sigma}$ is

$$D_M(\mathbf{X}_1, \mathbf{X}_2) = ((\mathbf{X}_1 - \mathbf{X}_2)'\mathbf{\Sigma}^{-1}(\mathbf{X}_1 - \mathbf{X}_2))^{1/2}. \tag{12.5.1}$$

If we estimate $\mathbf{\Sigma}$ by covariance matrices S_1 and S_2, we substitute for $\mathbf{\Sigma}$ in (12.5.1) the pooled estimator, S_{pooled}. A confidence region for the difference $\mathbf{\Delta} = \mu_1 - \mu_2$, between the expected value of the batch and the reference populations, at confidence level $1 - \alpha$, is

$$CR = \{\mathbf{Y} : (\mathbf{Y} - (\mathbf{X}_1 - \mathbf{X}_2))'S_{\text{pooled}}^{-1}(\mathbf{Y} - (\mathbf{X}_1 - \mathbf{X}_2)') \leq KF_{1-\alpha}[p, 2n - p - 1]\}, \tag{12.5.2}$$

where p is the dimension of the vectors, and

$$K = \frac{4(n-1)p}{n(2n-p-1)}. \tag{12.5.3}$$

To demonstrate the procedure, we use an example where Y is the percent dissolution of a tablet, measured at two time instances, 15 minutes and 90 minutes (see Table 12.2). Calculations with MINITAB can be performed using *Calc>Matrices>Arithmetic*.

Table 12.2 *Dissolution data of reference and batch under test*

	Batch	Tablet	15	90
1	REF	1	65.58	93.14
2	REF	2	67.17	88.01
3	REF	3	65.56	86.83
4	REF	4	66.51	88.00
5	REF	5	69.06	89.70
6	REF	6	69.77	88.88
7	TEST	1	47.77	92.39
8	TEST	2	49.46	89.93
9	TEST	3	47.76	90.19
10	TEST	4	49.72	94.12
11	TEST	5	52.68	93.80
12	TEST	6	51.01	94.45

```
> data(DISS)
> mahalanobisT2(DISS[, c("batch", "min15", "min90")],
                factor.name="batch",
                compare.to=c(15, 15))

$coord
          min15       min90
LCR     14.55828  -2.810681
Center  17.54167  -3.386667
UCR     20.52506  -3.962653

$mahalanobis
      LCR     Center        UCR
 8.664794 10.440449 12.216104

$mahalanobis.compare
          [,1]
[1,] 9.630777
```

A scatter plot of the data shows the difference between test and reference. At 15 minutes, dissolution is lower in the tested batch than the reference; at 90 minutes, this is reversed (see Figure 12.6). Our tested material therefore starts dissolving than the reference but then things change and it reaches high dissolution levels faster than the reference.

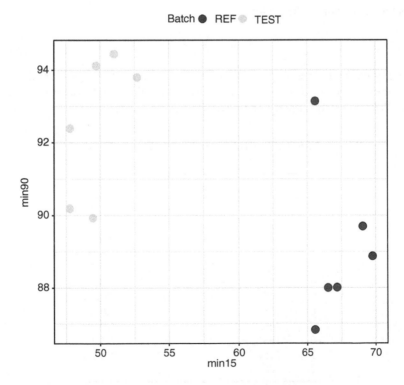

Figure 12.6   *Scatterplot of reference and batch under test.*

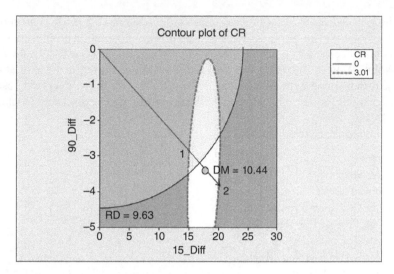

Figure 12.7 *Difference between dissolution of batch under test and reference at 15 and 90 minutes (MINITAB).*

For this data, $n = 6, p = 2, K = 1.35, F_{2,19,0.90} = 3.01, (X_2 - X_1) = (17.54, -3.39)$ and $D_M = 10.44$. A contour plot with the limits of CR set at 3.01 is presented in Figure 12.7.

The center of the ellipsoid is set at $(17.54, -3.39)$ and, as mentioned above, at that point, $D_M = 10.44$. The line from the origin connecting to this point is $Y = -0.193X$. It crosses the ellipse first at $(15.03, -2.9)$ labeled as "1" on Figure 12.7 and then at $(20.05, -3.87)$ labeled as "2" with D_M values of $D_M^l = 8.95$ and $D_M^u = 11.93$, respectively.

To determine equivalence, with a 15% buffer, we consider the contour corresponding to results within this buffer. The D_M value for these point RD $= Sqrt[(15, 15)'S_{pooled}^{-1}(15, 15)] = 9.63$.

Since $D_M^u > $ RD, we have a confidence region for the true difference in mean dissolution that exceeds the 15% buffer. We therefore declare the batch under test not to be equivalent to the reference.

An index that can be used to assess process capability in terms of equivalence between reference and batch under test is Ceq $=$ RD$/D_M^u$. To determine the batch under test equivalent to the reference, we need to show that Ceq > 1.

12.6 Chapter highlights

As was discussed in Chapter 5, multivariate observations require special techniques for visualization and analysis. Chapter 10 expands Chapter 5 and presents techniques for MSPC based on the Mahalanobis T^2 chart. Like in previous chapters, examples of MSPC using R, MINITAB, and JMP are provided. Section 12.3 introduces the reader to multivariate extensions of process capability indices. These are expansions of the capability indices presented in Chapter 10 that are not available in MINITAB or JMP. A special role is played in this context by the concept of multivariate TRs which extends the tolerance intervals introduced in Section 4.5. Section 12.4 considers four scenarios for setting up and running MSPC: (1) internally derived targets, (2) using an external reference sample, (3) externally assigned targets, and (4) measurements units considered as batches. These four cases cover most practical applications of MSPC. Two subsections cover the special cases of measurement units considered as batches and a variable decomposition of indices used for process monitoring. Section 12.5 is a special application of MSPC to the monitoring of bioequivalence of drug product dissolution profiles. In this application, tablets manufactured by a generic drug company are compared to the original product at several dissolution times. The Food and Drug Administration allows for a gap of at most 15%, a requirement that define multivariate specification limits. We show how TR is used in such cases. More on multivariate applications in pharmaceuticals will be discussed in Chapter 13 on Quality by Design.

The main concepts and tools introduced in this chapter include:

- Mean Vector
- Covariance Matrix
- Mahalanobis T2
- Multivariate Statistical Process Control
- Multivariate Process Capability Indices
- Multivariate Tolerance Region
- Hyper-rectangular Tolerance Regions
- Circular Tolerance Regions
- Principal Components
- Internal Targets
- Reference Sample
- External Targets
- Batches
- Monitoring Indices

12.7   Exercises

12.1 In data file **TSQ.csv**, we find 368 T^2 values corresponding to the vectors (x, y, θ) in the **PLACE.csv** file. The first $n = 48$ vectors in **PLACE.csv** file were used as a base sample, to compute the vector of means **m** and the covariance matrix S. The T^2 values are for the other individual vectors ($m = 1$). Plot the T^2 values in the file **TSQ.csv**. Compute the UCL and describe from the plot what might have happened in the placement process generating the (x, y, θ) values.

12.2 Prove that if **X** has a multivariate normal distribution, $(N_v(\boldsymbol{\mu}, \boldsymbol{\sigma}))$, then $(\mathbf{X} - \boldsymbol{\mu})' \boldsymbol{\Sigma}^{-1} (\mathbf{X} - \boldsymbol{\mu})$ has a χ^2 distribution with v degrees of freedom where $R = \chi^2_{1-p}[v]$ is the corresponding $(1 - p)$ quantile of the χ^2 distribution with v degrees of freedom.

12.3 Sort the dataset CAR.csv by variable **cyl**, indicating the number of cylinders in a car, and run a T^2 chart with internally derived targets for the variables **turn, hp, mpg**, with separate computations for cars with 4, 6, and 8 cylinders. If you use MINITAB use the option Stages. How is the number of cylinders affecting the overall performance of the cars.

12.4 Sort the dataset CAR.csv by variable **origin**, indicating the country of origin, and run a T^2 chart with internally derived targets for the variables **turn, hp, mpg**, with separate computations for cars from 1 = US; 2 = Europe; 3 = Asia. If you use MINITAB use the option Stages. How is the country of origin affecting the overall performance of the cars.

12.5 Load the dataset GASOL.csv and compute a T^2 chart for $x1$, $x2$, **astm, endPt, yield**. Design the chart with an external assigned target based on observations $12-24$. Compare the charts. Explain the differences.

12.6 Repeat exercise 12.5, but this time design he chart with an externally assigned target based on observations 25–32. Explain the computational difficulty.

12.7 Calculate control limits for grouped data with 20 subgroups of size 5 and 6 dimensions, with internally derived targets (equation (12.4.2)). How will the control limits change if you start monitoring a process with similar data.

12.8 Let $X_1 = (x_{11}, x_{12}, \ldots, x_{1p})$ and $X_2 = (x_{21}, x_{22}, \ldots, x_{2p})$ represent the mean dissolution values of tablets at p time instances of a reference product and a batch under test, respectively. The Mahalanobis distance T^2, between X_1 and X_2, is defined here as $D_M = $ Sqrt$[(X_2 - X_1)'S_{\text{pooled}}^{-1}(X_2 - X_1)]$, where $S_{\text{pooled}} = (S_{\text{reference}} + S_{\text{test}})/2$, is the pooled covariance matrix of the reference and test samples. The confidence region, CR, of the difference between batch and reference consists of all vectors Y satisfying: $K[(Y - (X_2 - X_1)'S_{\text{pooled}}^{-1}(Y - (X_2 - X_1)] \leq F_{0.90}[p, 2n - p - 1]$ where $F_{0.90}[p, 2n - p - 1]$ is the 90th quantile of the F-distribution with degrees of freedom p and $(2n - p - 1)$. Prove that for measurements conducted at one time instance ($p = 1$) these formulae correspond to the confidence intervals presented in Chapter 4.

13

Classical Design and Analysis of Experiments

Experiments are used in industry to improve productivity, reduce variability, and obtain robust products and manufacturing processes. In this chapter, we study how to design and analyze experiments which are aimed at testing scientific or technological hypotheses. These hypotheses are concerned with the effects of procedures or treatments on the yield; the relationship between variables; the conditions under which a production process yields maximum output or other optimum results, etc. The chapter presents the classical methods of design of experiments. We start with an introductory section, which provides some examples and discusses the guiding principles in designing experiments.

13.1   Basic steps and guiding principles

The following are guiding principles which we follow in designing experiments. They are designed to ensure high information quality (InfoQ) of the study, as introduced in Chapter 1.

1. The **objectives** of the study should be well stated, and criteria established to test whether these objectives have been met.
2. The **response variable(s)** should be clearly defined so that the study objectives are properly translated. At this stage, measurement uncertainty should be established (see Chapter 2).
3. All factors which might effect the response variable(s) should be listed, and specified. We call these the **controllable factors**. This requires interactive brainstorming with content experts.
4. The type of **measurements** or observations on all variables should be specified.
5. The **levels** of the controllable factors to be tested should be determined.
6. A statistical **model** should be formulated concerning the relationship between the pertinent variables, and their error distributions. This can rely on prior knowledge or literature search.
7. An **experimental layout** or **experimental array** should be designed so that the inference from the gathered data will be:

Modern Industrial Statistics: With Applications in R, MINITAB and JMP, Third Edition.
Ron S. Kenett and Shelemyahu Zacks.
© 2021 John Wiley & Sons, Ltd. Published 2021 by John Wiley & Sons, Ltd.

(a) **valid**;
(b) **precise**;
(c) **generalizable**;
(d) **easy to obtain**.

8. The trials should be performed if possible in a **random order**, to avoid bias by factors which are not taken into consideration.

9. A **protocol** of execution should be prepared, as well as the method of analysis. The method of analysis and data collection depends on the design.

10. The execution of the experiment should carefully follow the protocol with proper documentation.

11. The results of the experiments should be carefully analyzed and reported ensuring proper documentation and traceability. Modern technology (like sweave) ensures that data, analysis, and conclusions are fully integrated and reproducible.

12. Confirmatory experiments should be conducted, to validate the inference (conclusions) of the main experiments.

We illustrate the above principles with two examples.

Example 13.1.   The first example deals with a problem of measuring weights of objects and is brought to illustrate what is an experimental layout (design) and why an optimal one should be chosen.

Step 1: Formulation of objectives

The **objective** is to devise a measurement plan that will yield weight estimates of chemicals with maximal precision, under a fixed number of four weighing operations.

Step 2: Description of response

The weight measurement device is a chemical balance, which has right and left pans. One or more objects can be put on either pan. The **response** variable Y is the measurement read on the scale of the chemical balance. This is equal to the total weight of objects on the right pan $(+)$ minus the total weight of objects on the left pan $(-)$, plus a measurement error.

Step 3: Controllable variables

Suppose we have four objects O_1, O_2, O_3, O_4, having unknown weights w_1, w_2, w_3, w_4. The controllable (influencing) variables are

$$X_{ij} = \begin{cases} 1, & \text{if } j\text{th object is put on } + \text{ pan} \\ & \text{in the } i\text{th measurement} \\ \\ -1, & \text{if } j\text{th object is put on } - \text{ pan} \\ & \text{in the } i\text{th measurement.} \end{cases}$$

$i, j = 1, 2, 3, 4.$

Step 4: Type of measurements

The response Y is measured on a continuous scale in an interval (y^*, y^{**}). The observations are a realization of continuous random variables.

Step 5: Levels of controllable variables

$X_{ij} = \pm 1$, as above.

Step 6: A statistical model

The measurement model is linear, i.e.,

$$Y_i = w_1 X_{i1} + w_2 X_{i2} + w_3 X_{i3} + w_4 X_{i4} + e_i$$

$i = 1, \ldots, 4$, where e_1, e_2, e_3, e_4 are independent random variables, with $E\{e_i\} = 0$ and $V\{e_i\} = \sigma^2$, $i = 1, 2, \ldots, 4$.

Step 7: Experimental layout

An experimental layout is represented by a 4×4 matrix

$$(X) = (X_{ij}; i, j = 1, \ldots, 4).$$

Such a matrix is called a **design matrix**.

Given a design matrix (X), and a vector of measurements $\mathbf{Y} = (Y_1, \ldots, Y_4)'$, we estimate $\mathbf{w} = (w_1, \ldots, w_4)'$ by

$$\hat{\mathbf{W}} = (L)\mathbf{Y},$$

where (L) is a 4×4 matrix. We say that the design is **valid**, if there exists a matrix L such that $E\{\hat{\mathbf{W}}\} = \mathbf{w}$. Any **nonsingular** design matrix (X) represents a valid design with $(L) = (X)^{-1}$. Indeed, $E\{\mathbf{Y}\} = (X)\mathbf{w}$. Hence

$$E\{\hat{\mathbf{W}}\} = (X)^{-1}E\{\mathbf{Y}\} = \mathbf{w}.$$

The **precision** of the design matrix (X) is measured by $\left(\sum_{i=1}^{4} V\{\hat{W}_i\} \right)^{-1}$. The problem is to find a design matrix $(X)^0$ which **maximizes the precision**.

It can be shown that an optimal design is given by the **orthogonal array**

$$(X)^0 = \begin{bmatrix} 1 & -1 & -1 & 1 \\ 1 & 1 & -1 & -1 \\ 1 & -1 & 1 & -1 \\ 1 & 1 & 1 & 1 \end{bmatrix}$$

or any row (or column) permutation of this matrix. Notice that in this design, in each one of the first three weighing operation (row) two objects are put on the left pan ($-$) and two on the right. Also, each object, excluding the first, is put twice on ($-$) and twice on ($+$). The weight estimates under this design as

$$\hat{\mathbf{W}} = \frac{1}{4} \begin{bmatrix} 1 & 1 & 1 & 1 \\ -1 & 1 & -1 & 1 \\ -1 & -1 & 1 & 1 \\ 1 & -1 & -1 & 1 \end{bmatrix} \begin{bmatrix} Y_1 \\ Y_2 \\ Y_3 \\ Y_4 \end{bmatrix}.$$

Moreover,

$$\sum_{i=1}^{4} V\{\hat{W}_i\} = \sigma^2.$$

The order of measurements is **random**.

Example 13.2. The second example illustrates a complex process, with a large number of factors which may effect the yield variables.

Wave soldering of circuit pack assemblies (CPA) is an automated process of soldering which, if done in an optimal fashion, can raise quality and productivity. The process, however, involves three phases and many variables. We analyze here the various steps required for designing an experiment to learn the effects of the various factors on the process. We follow the process description of Lin and Kacker (1989).

If the soldering process yields good results, the CPAs can proceed directly to automatic testing. This is a big savings in direct labor cost and increase in productivity. The wave soldering process (WSP) is in three phases. In Phase I, called **fluxing**, the solder joint surfaces are cleaned by the soldering flux, which also protects it against reoxidation. The fluxing lowers the surface tension for better solder wetting and solder joint formation.

Phase II of the WSP is the **soldering assembly**. This is performed in a cascade of wave soldering machine. After preheating the solution, the noncomponent side of the assembly is immersed in a solder wave for 1–2 seconds. All solder points are completed when the CPA exits the wave. Preheating must be gradual. The correct heating is essential to effective soldering. Also important is the conveyor speed and the conveyor's angle. The last phase, Phase III, of the process is that of **detergent cleaning**. The assembly is first washed in detergent solution, then rinsed in water and finally dried with hot air. The temperature of the detergent solution is raised to achieve effective cleaning and prevent excessive foaming. The rinse water is heated to obtain effective rinsing.

We list now the design steps:

1. *Objectives*: To find the effects of the various factors on the quality of wave soldering, and optimize the process.
2. *Response variables*: There are four yield variables
   (a) Insulation resistance
   (b) Cleaning characterization
   (c) Soldering efficiency
   (d) Solder mask cracking.
3. *Controllable variables*: There are 17 variables (factors) associated with the three phases of the process.

I. **Flux formulation**	II. **Wave Soldering**	III. **Detergent cleaning**
A. Type of activator	H. Amount of Flux	N. Detergent concentration
B. Amount of activator	I. Preheat time	O. Detergent temperature
C. Type of surfactant	J. Solder temperature	P. Cleaning conveyor speed
D. Amount of surfactant	K. Conveyor speed	Q. Rinse water temperature
E. Amount of antioxidant	L. Conveyor angle	
F. Type of solvent	M. Wave height setting	
G. Amount of solvent		

4. *Measurements*:
   (a) Insulation resistance test at 30 minutes, 1 and 4 days after soldering at
      • −35C, 90% RN, no bias voltage
      • −65C, 90% RH, no bias voltage
      • (continuous variable).

(b) Cleaning characterization: The amounts of residues on the board (continuous variable).

(c) Soldering efficiency: Visual inspection of no solder, insufficient solder, good solder, excess solder, and other defects (discrete variables).

(d) Solder mask cracking: Visual inspection of cracked spots on the solder mask (discrete variables).

5. *Levels of controllable factors*:

Factor	# Levels	Factor	# Levels
A	2	J	3
B	3	K	3
C	2	L	3
D	3	M	2
E	3	N	2
F	2	O	2
G	3	P	3
H	3	Q	2
I	3		

6. *The statistical model*: The response variables are related to the controllable variables by linear models having "main effects" and "interaction" parameters, as will be explained in Section 13.3.

7. *The experiment layout*: A fractional factorial experiment, as explained in Section 13.8, is designed. Such a design is needed, because a full factorial design contains $3^{10}2^7 = 7,558,272$ possible combinations of factor levels. A fractional replication design chooses a manageable fraction of the full factorial in a manner that allows valid inference, and precise estimates of the parameters of interest.

8. **Protocol of execution**: Suppose that it is decided to perform a fraction of $3^3 2^2 = 108$ trials at certain levels of the 17 factors. However, the setup of the factors takes time and one cannot perform more than four trials a day. The experiment will last 27 days. It is important to construct the design so that the important effects, to be estimated, will not be confounded with possible differences between days (blocks). The order of the trials within each day is randomized as well as, the trials which are assigned to different days. **Randomization** is an important component of the design, which comes to enhance its validity and generalizability. The execution protocol should clearly specify the order of execution of the trials.

13.2   Blocking and randomization

Blocking and randomization are devices in planning of experiments, which are aimed at increasing the precision of the outcome and ensuring the validity of the inference. Blocking is used to reduce errors. A block is a portion of the experimental material that is expected to be more homogeneous than the whole aggregate. For example, if the experiment is designed to test the effect of polyester coating of electronic circuits on their current output, the variability between circuits could be considerably bigger than the effect of the coating on the current output. In order to reduce this component of variance, one can block by circuit. Each circuit will be tested under two treatments: no-coating and coating. We first test the current output of a circuit without coating. Later we coat the circuit, and test again. Such a comparison of before and after a treatment, of the same units, is called **paired-comparison**.

Another example of blocking is the famous boy's shoes examples of Box et al. (1978, pp. 97). Two kinds of shoe soles' materials are to be tested by fixing the soles on n pairs of boys' shoes, and measuring the amount of wear of the soles after a period of actively wearing the shoes. Since there is high variability between activity of boys, if m pairs will be with soles of one type and the rest of the other, it will not be clear whether any difference that might be observed in the degree of wearout is due to differences between the characteristics of the sole material or to the differences between the boys. By blocking by pair of shoes, we can reduce much of the variability. Each pair of shoes is assigned the two types of soles. The comparison within each block is free of the variability between boys. Furthermore, since boys use their right or left foot differently, one should assign the type of soles to the left or right shoes at random. Thus, the treatments (two types of soles) are assigned within each block at random.

Other examples of blocks could be machines, shifts of production, days of the week, operators, etc.

Generally, if there are t treatments to compare, and b blocks, and if all t treatments can be performed within a single block, we assign all the t treatments to each block. The order of applying the treatments within each block should be **randomized**. Such a design is called a **randomized complete block design**. We will see later how a proper analysis of the yield can validly test for the effects of the treatments.

If not all treatments can be applied within each block it is desirable to assign treatments to blocks in some balanced fashion. Such designs, to be discussed later, are called **balanced incomplete block designs** (BIBD).

Randomization within each block is important also to validate the assumption that the error components in the statistical model are independent. This assumption may not be valid if treatments are not assigned at random to the experimental units within each block. In experiments compaing a treatment to a placebo, randomization of experimenatl units between both options is key to the support of causality claims derived from the analysis of the collected data. Randomization will be discussed in Section 13.4.

13.3   Additive and nonadditive linear models

Seventeen factors which might influence the outcome in WSP are listed in Example 13.2. Some of these factors, such as type of activator (A), or type of surfactant (C) are categorical variables. The number of levels listed for these factors was 2. That is, the study compares the effects of two types of activators and two types of surfactants.

If the variables are continuous, like amount of activator (B), we can use a regression linear model to represent the effects of the factors on the yield variables. Such models will be discussed later (Section 13.7). In the present section, linear models which are valid for both categorical or continuous variables are presented.

For the sake of explanation, let us start first with a simple case, in which the response depends on one factor only. Thus, let A designate some factor, which is applied at different levels, $A_1, \ldots, A_a$. These could be a categories. The levels of A are also called "**treatments**."

Suppose that at each level of A we make n independent repetitions (replicas) of the experiment. Let Y_{ij}, $i = 1, \ldots, a$ and $j = 1, \ldots, n$ denote the observed yield at the jth replication of level A_i.

We model the random variables Y_{ij} as

$$Y_{ij} = \mu + \tau_i^A + e_{ij}, \quad i = 1, \ldots, a, \, j = 1, \ldots, n, \qquad (13.3.1)$$

where μ and $\tau_1^A, \ldots, \tau_a^A$ are unknown parameters, satisfying

$$\sum_{i=1}^{a} \tau_i^A = 0. \qquad (13.3.2)$$

$e_{ij}, \, i = 1, \ldots, a, j = 1, \ldots, n$, are independent random variables such that,

$$E\{e_{ij}\} = 0 \quad \text{and} \quad V\{e_{ij}\} = \sigma^2, \qquad (13.3.3)$$

for all $i = 1, \ldots, a; j = 1, \ldots, n$.
   Let

$$\bar{Y}_i = \frac{1}{n} \sum_{j=1}^{n} Y_{ij}, \quad i = 1, \ldots, a.$$

The expected values of these means are

$$E\{\bar{Y}_i\} = \mu + \tau_i^A, \quad i = 1, \ldots, k. \qquad (13.3.4)$$

   Let

$$\bar{\bar{Y}} = \frac{1}{k} \sum_{i=1}^{k}, \; \bar{Y}_i. \qquad (13.3.5)$$

This is the mean of all $N = k \times n$ observations (the grand mean), since $\sum_{i=1}^{a} \tau_i^A = 0$, we obtain that

$$E\{\bar{\bar{Y}}\} = \mu. \qquad (13.3.6)$$

The parameter τ_i^A is called the **main effect** of A at level i.
   If there are two factors, A and B, at a and b levels, respectively, there are $a \times b$ **treatment combinations** (A_i, B_j), $i = 1, \ldots, a, j = 1, \ldots, b$. Suppose also that n independent replicas are made at each one of the treatment combinations. The yield at the kth replication of treatment combination (A_i, B_j) is given by

$$Y_{ijk} = \mu + \tau_i^A + \tau_j^B + \tau_{ij}^{AB} + e_{ijk}. \qquad (13.3.7)$$

The error terms e_{ijk} are independent random variables satisfying

$$E\{e_{ijl}\} = 0, \quad V\{e_{ijl}\} = \sigma^2, \qquad (13.3.8)$$

for all $i = 1, \ldots, a, j = 1, \ldots, b, k = 1, \ldots, n$.
   We further assume that

$$\sum_{j=1}^{b} \tau_{ij}^{AB} = 0, \quad i = 1, \ldots, a$$

$$\sum_{i=1}^{a} \tau_{ij}^{AB} = 0, \quad j = 1, \ldots, b.$$

$$\sum_{i=1}^{a} \tau_i^A = 0, \tag{13.3.9}$$

$$\sum_{j=1}^{b} \tau_j^B = 0.$$

τ_i^A is the **main effect** of A at level i, τ_j^B is the **main effect** of B at level j, and τ_{ij}^{AB} is the **interaction effect** at (A_i, B_j).

If all the interaction effects are zero then the model reduces to

$$Y_{ijk} = \mu + \tau_i^A + \tau_j^B + e_{ijk}. \tag{13.3.10}$$

Such a model is called **additive**. If not all the interaction components are zero then the model is called **nonadditive**.

This model is generalized in a straight forward manner to include a larger number of factors. Thus, for three factors, there are three types of main effect terms, τ_i^A, τ_j^B, and τ_k^C; three types of interaction terms τ_{ij}^{AB}, τ_{ik}^{AC} and τ_{jk}^{BC}; and one type of interaction τ_{ijk}^{ABC}.

Generally, if there are p factors, there are 2^p types of parameters,

$$\mu, \tau_i^A, \tau_j^B, \ldots, \tau_{ij}^{AB}, \tau_{ik}^{AC}, \ldots, \tau_{ijk}^{ABC}, \ldots$$

etc. Interaction parameters between two factors are called first-order interactions. Interaction parameters between three factors are called second-order interactions, and so on. In particular, modeling it is often assumed that all interaction parameters of higher than first order are zero.

13.4   The analysis of randomized complete block designs

13.4.1   Several blocks, two treatments per block: paired comparison

As in the shoe soles example, or the example of the effect of polyester coating on circuits output, there are two treatments applied in each one of n blocks. The linear model can be written as

$$Y_{ij} = \mu + \tau_i + \beta_j + e_{ij}, \quad i = 1, 2; \ j = 1, \ldots, n \tag{13.4.1}$$

where τ_i is the effect of the ith treatment and β_j is the effect of the jth block. e_{ij} is an independent random variable, representing the experimental random error or deviation. It is assumed that $E\{e_{ij}\} = 0$ and $V\{e_{ij}\} = \sigma_e^2$. Since we are interested in testing whether the two treatments have different effects, the analysis is based on the within block differences

$$D_j = Y_{2j} - Y_{1j} = \tau_2 - \tau_1 + e_j^*, \quad j = 1, \ldots, n \tag{13.4.2}$$

The error terms e_j^* are independent random variables with $E\{e_j^*\} = 0$ and $V\{e_j^*\} = \sigma_d^2$, $j = 1, \ldots, n$ where $\sigma_d^2 = 2\sigma_e^2$.

An unbiased estimator of σ_d^2 is

$$S_d^2 = \frac{1}{n-1} \sum_{j=1}^{n} (D_j - \bar{D}_n)^2, \tag{13.4.3}$$

where $\bar{D}_n = \frac{1}{n} \sum_{j=1}^{n} D_j$. The hypotheses to be tested are:

$$H_0 : \delta = \tau_2 - \tau_1 = 0$$

against

$$H_1 : \delta \neq 0.$$

13.4.1.1   The t-test

Most commonly used is the t-test, in which H_0 is tested by computing the test statistic

$$t = \frac{\sqrt{n}\,\bar{D}_n}{S_d}. \tag{13.4.4}$$

If $e_1^*, \ldots, e_n^*$ are i.i.d., normally distributed then, under the null hypothesis, t has a t-distribution with $(n - 1)$ degrees of freedom. In this case, H_0 is rejected if

$$|t| > t_{1-\alpha/2}[n - 1],$$

where α is the selected level of significance.

13.4.1.2   Randomization tests

A randomization test for paired comparison, constructs a **reference distribution** of all possible averages of the differences that can be obtained by randomly assigning the sign + or − to the value of D_i. It computes then an average difference $\bar{D}$ for each one of the 2^n sign assignments.

The P-value of the test, for the two-sided alternative, is determined according to this reference distribution, by

$$P = \Pr\{\bar{Y} \geq \text{Observed } \bar{D}\}.$$

For example, suppose we have four differences, with values $1.1, 0.3, -0.7,$ and -0.1. The mean is $\bar{D}_4 = 0.15$. There are $2^4 = 16$ possible ways of assigning a sign to $|D_i|$ (Table 13.1). Let $X_i = \pm 1$ and $\bar{Y} = \frac{1}{4} \sum_{i=1}^{4} X|D_i|$

Under the reference distribution, all these possible means are equally probable. The P-value associated with the observed $\bar{D} = 0.15$ is $P = \frac{7}{15} = 0.47$. If the number of pairs (blocks) n is large the procedure becomes cumbersome, since we have to determine all the 2^n sign assignments. If $n = 20$ there are $2^{20} = 1,048,576$ such assignments. We can however estimate the P-value by taking a RSWR from this reference distribution. This can be easily done by using the MINITAB macro listed below (type it and save it as **RPCOMP.MTB**):

```
Random k1 C2;
Integer 1 2.
Let C3 = 2 * (C2 − 1.5)
Let k2 = mean(C1 * C3)
stack C4 k2 C4
end
```

Table 13.1   Sign assignments and values of $\bar{Y}$

Signs				D
−1	−1	−1	−1	−0.55
1	−1	−1	−1	0
−1	1	−1	−1	−0.4
1	1	−1	−1	0.15
−1	−1	1	−1	−0.20
1	−1	1	−1	0.35
−1	1	1	−1	−0.05
1	1	1	−1	0.50
−1	−1	−1	1	−0.50
1	−1	−1	1	0.05
−1	1	−1	1	−0.35
1	1	−1	1	0.2
−1	−1	1	1	−0.15
1	−1	1	1	0.40
−1	1	1	1	0
1	1	1	1	0.55

In order to execute it, we first set $k1 = n$, by the command

MTB> Let $k1 = n$

where n is the sample size. In column $C1$, we set the n values of the observed differences, $D_1, \ldots, D_n$. Initiate column $C4$ by

MTB> Let $C4(1) = \text{mean}(C1)$.

After executing this macro M times, we estimate the P-value by the proportion of cases in $C4$, whose value is greater or equal to that of mean $(C1)$. In R, this is performed with the following commands:

```
> X <- c(1.1, 0.3, -0.7, -0.1)
> M <- 200
> set.seed(123)
> Di <- matrix(sample(x=c(-1,1),
                       size=length(X)*M,
                       replace=TRUE),
               nrow=M)
> Xi <- matrix(X,
               nrow=M,
               ncol=length(X),
               byrow=TRUE)
> sum(rowMeans(Di*Xi) >= mean(X))/M

[1] 0.34

> rm(X, M, Di, Xi)
```

Example 13.3. We analyze here the results of the shoe soles experiment, as reported in Box et al. (1978, p. 100). The observed differences in the wear of the soles, between type B and type A, for $n = 10$ children, are:

$$0.8, \quad 0.6, \quad 0.3, \quad -0.1, \quad 1.1, \quad -0.2, \quad 0.3, \quad 0.5, \quad 0.5, \quad 0.3.$$

The average difference is $\bar{D}_{10} = 0.41$.
A t-test of H_0, after setting the differences in column $C3$, is obtained by the MINITAB command,

MTB> T Test 0.0 $C3$;
SUBC> Alternative 0.

The result of this t-test is

	N	MEAN	STDEV	SE MEAN	T	P VALUE
TEST OF MU = 0.000 VS MU N.E. 0.000						
$C3$	10	0.410	0.387	0.122	3.35	0.0086

Executing macro **RPCOMP.MTB** $M = 200$ times on the data in column $C1$, gave 200 values of $\bar{D}$, whose stem and leaf plot is (Figure 13.1)

```
Character Stem-and-Leaf Display
Stem-and-leaf of C4 N = 200
Leaf Unit = 0.010
     3    -3    955
     6    -3    111
    17    -2    99755555555
    22    -2    33311
    44    -1    9999999997777775555555
    63    -1    3333333333111111111
    88    -0    99999999977777777775555555
   (17)   -0    33333333333311110
    95     0    11111111111333333333
    75     0    55555555557777777799999999
    50     1    111133333
    41     1    5555555577779999999999
    19     2    113
    16     2    555777
    10     3    11133
     5     3    779
     2     4    13
```

Figure 13.1 *Stem-and-leaf plot of 200 random difference averages.*

According to this, the *P*-value is estimated by

$$\hat{P} = \frac{2}{200} = 0.01.$$

This estimate is almost the same as the *P*-value of the *t*-test. An equivalent analysis in R is performed using:

```
> X <- c(0.8, 0.6, 0.3, -0.1, 1.1, -0.2, 0.3, 0.5, 0.5, 0.3)
> M <- 200
> set.seed(123)
> Di <- matrix(sample(x=c(-1,1),
                      size=length(X)*M,
                      replace=TRUE),
             nrow=M)
> Xi <- matrix(X,
              nrow=M,
              ncol=length(X),
              byrow=TRUE)
> Means <- rowMeans(Di*Xi)
> sum(rowMeans(Di*Xi) >= mean(X))/M

[1] 0.005

> stem(Means)

  The decimal point is 1 digit(s) to the left of the |

  -4 | 1
  -3 | 7
  -3 | 3111
  -2 | 9999997775555555
  -2 | 333111111
  -1 | 99999999977777777755555
  -1 | 333311111111111
  -0 | 9999999999999977777755555
  -0 | 333333331111
   0 | 1111113333333333333333
   0 | 555557777777799999999
   1 | 1113333
   1 | 55555555557777799999
   2 | 11111133
   2 | 555557799
   3 | 133
   3 | 5579
   4 | 3

> rm(Di, Xi, M, X, Means)
```

13.4.2   Several blocks, t treatments per block

As said earlier, the randomized complete block designs (RCBD) are those in which each block contains all the t treatments. The treatments are assigned to the experimental units in each block at random. Let b denote the number of blocks. The linear model for these designs is

$$Y_{ij} = \mu + \tau_i + \beta_j + e_{ij}, \quad i = 1, \ldots, t \ \ j = 1, \ldots, b \tag{13.4.5}$$

where Y_{ij} is the yield of the ith treatment in the jth block. The main effect of the ith treatment is τ_i, and the main effect of the jth block is β_j. It is assumed that the effects are additive (no interaction). Under this assumption, each treatment is tried only once in each block. The different blocks serve the role of replicas. However, since the blocks may have additive effects, β_j, we have to adjust for the effects of blocks in estimating σ^2. This is done as shown in the ANOVA table below. A nonparametirc approach to ANOVA based on bootstrapping is presented in Section 4.11.5.2.

Further assume that, e_{ij} are the error random variables with $E\{e_{ij}\} = 0$ and $V\{e_{ij}\} = \sigma^2$ for all (i,j). The ANOVA for this model is presented in Table 13.2.

Here,

$$SST = \sum_{i=1}^{t} \sum_{j=1}^{b} (Y_{ij} - \bar{\bar{Y}})^2, \tag{13.4.6}$$

$$SSTR = b \sum_{i=1}^{t} (\bar{Y}_{i.} - \bar{\bar{Y}})^2, \tag{13.4.7}$$

$$SSBL = t \sum_{j=1}^{b} (\bar{Y}_{.j} - \bar{\bar{Y}})^2, \tag{13.4.8}$$

and

$$SSE = SST - SSTR - SSBL. \tag{13.4.1}$$

$$\bar{Y}_{i.} = \frac{1}{b} \sum_{j=1}^{b} Y_{ij}, \quad \bar{Y}_{.j} = \frac{1}{t} \sum_{i=1}^{t} Y_{ij} \tag{13.4.9}$$

and $\bar{\bar{Y}}$ is the grand mean.

The significance of the treatment effects is tested by the F-statistic

$$F_t = \frac{MSTR}{MSE}. \tag{13.4.10}$$

Table 13.2   *ANOVA table for RCBD*

Source of variation	DF	SS	MS	E{MS}
Treatments	$t - 1$	SSTR	MSTR	$\sigma^2 + \dfrac{b}{t-1} \displaystyle\sum_{i=1}^{t} \tau_i^2$
Blocks	$b - 1$	SSBL	MSBL	$\sigma^2 + \dfrac{t}{b-1} \displaystyle\sum_{j=1}^{b} \beta_j^2$
Error	$(t-1)(b-1)$	SSE	MSE	σ^2
Total	$tb - 1$	SST	–	

The significance of the block effects is tested by

$$F_b = \frac{MSBL}{MSE}. \tag{13.4.11}$$

These statistics are compared with the corresponding $(1 - \alpha)$th quantiles of the F-distribution. Under the assumption that $\sum_{i=1}^{t} \tau_i = 0$, the main effects of the treatments are estimated by

$$\hat{\tau}_i = \bar{Y}_{i.} - \bar{\bar{Y}}, \quad i = 1, \ldots, t. \tag{13.4.12}$$

These are least-squares estimates. Each such estimation is a linear contrast

$$\hat{\tau}_i = \sum_{i'=1}^{t} c_{ii'} \bar{Y}_{i'.}, \tag{13.4.13}$$

where

$$c_{ii'} = \begin{cases} 1 - \dfrac{1}{t}, & \text{if } i = i' \\[2ex] -\dfrac{1}{t}, & \text{if } i \neq i'. \end{cases} \tag{13.4.14}$$

Hence,

$$\begin{aligned} V\{\hat{\tau}_i\} &= \frac{\sigma^2}{b} \sum_{i'=1}^{t} c_{ii'}^2 \\ &= \frac{\sigma^2}{b} \left(1 - \frac{1}{t}\right), \quad i = 1, \ldots, t. \end{aligned} \tag{13.4.15}$$

An unbiased estimator of σ^2 is given by MSE. Thus, simultaneous confidence intervals for τ_i $(i = 1, \ldots, t)$, according to the Scheffé method, are

$$\hat{\tau}_i \pm S_\alpha \left(\frac{MSE}{b} \left(1 - \frac{1}{t}\right)\right)^{1/2}, \quad i = 1, \ldots, t \tag{13.4.16}$$

where

$$S_\alpha = ((t-1)F_{1-\alpha}[t-1, (t-1)(b-1)])^{1/2}.$$

Example 13.4. In Example 5.12 we estimated the effects of hybrids on the resistance in cards. We have $t = 6$ hybrids (treatments) on a card, and 32 cards. We can test now whether there are significant differences between the cards, by considering the cards as blocks, and using the ANOVA for RCBD. In this case, $b = 32$. Using Two Way ANOVA in MINITAB and data file **HADPAS.csv**. The ANOVA table obtained is (Table 13.3)

Table 13.3   *ANOVA for hybrid data*

Source	DF	SS	MS	F
Hybrids	5	1,780,741	356148	105.7
Cards	31	2,804,823	90478	26.9
Error	155	522,055	3368	–
Total	191	5,107,619	–	–

Since $F_{.99}[5,155] = 2.2725$ and $F_{.99}[31,155] = 1.5255$, both the treatment effects and the card effects are significant.

The estimator of σ, $\hat{\sigma}_p = (MSE)^{1/2}$, according to the above ANOVA, is $\hat{\sigma}_p = 58.03$. Notice that this estimator is considerably smaller than the one of Example 5.16. This is due to the variance reduction effect of the blocking.

The simultaneous confidence intervals, at level of confidence 0.95, for the treatment effects (the average hybrid measurements minus the grand average of 1965.2) are:

Hybrid 1:	178.21 ± 31.05;
Hybrid 6:	48.71 ± 31.05;
Hybrid 5:	15.36 ± 31.05;
Hybrid 2:	-62.39 ± 31.05;
Hybrid 4:	-64.79 ± 31.05;
Hybrid 3:	-114.86 ± 31.05.

Accordingly, the effects of Hybrid 2 and Hybrid 4 are not significantly different, and that of Hybrid 5 is not significantly different from zero. In R we obtain:

```
> data(HADPAS)
> HADPAS$diska <- as.factor(HADPAS$diska)
> HADPAS$hyb <- as.factor(HADPAS$hyb)
> AovH <- aov(res3 ~ diska + hyb,
          data=HADPAS)
> summary(AovH)

            Df  Sum Sq  Mean Sq  F value  Pr(> F)
diska       31  2804823   90478    26.86  < 2e-16  ***
hyb          5  1780741  356148   105.74  < 2e-16  ***
Residuals  155   522055    3368
---
Signif. codes:
0 '***' 0.001 '**' 0.01 '*' 0.05 '.' 0.1 ' ' 1

> tail(confint(AovH), 5)

          2.5 %      97.5 %
hyb2 -269.2543  -211.9332
hyb3 -321.7231  -264.4019
hyb4 -271.6606  -214.3394
hyb5 -191.5043  -134.1832
hyb6 -158.1606  -100.8394

> rm(AovH)
```

13.5   Balanced incomplete block designs

As mentioned before, it is often the case that the blocks are not sufficiently large to accommodate all the t treatments. For example, in testing the wearout of fabric, one uses a special machine (Martindale wear tester) which can accommodate only four pieces of clothes simultaneously. Here the block size is fixed at $k = 4$, while the number of treatments t, is the number of types of cloths to be compared. **Balanced Incomplete Block Designs** (BIBD) are designs which assign t treatment to b blocks of size k ($k < t$) in the following manner.

1. Each treatment is assigned only once to any one block.
2. Each treatment appears in r blocks. r is the number of replicas.
3. Every pair of two different treatments appears in λ blocks.
4. The order of treatments within each block is randomized.
5. The order of blocks is randomized.

According to these requirements there are, altogether, $N = tr = bk$ trials. Moreover, the following equality should hold

$$\lambda(t - 1) = r(k - 1). \tag{13.5.1}$$

The question is how to design a BIBD, for a given t and k. One can obtain a BIBD by the complete combinatorial listing of the $\binom{t}{k}$ selections without replacements of k out of t letters. In this case, the number of blocks is

$$b = \binom{t}{k}. \tag{13.5.2}$$

The number of replicas is $r = \binom{t-1}{k-1}$, and $\lambda = \binom{t-2}{k-2}$. The total number trials is

$$N = tr = t\binom{t-1}{k-1} = \frac{t!}{(k-1)!(t-k)!} = k\binom{t}{k} \tag{13.5.3}$$
$$= kb.$$

Such designs of BIBD are called **combinatoric designs**. They might be however too big. For example, if $t = 8$ and $k = 4$ we are required to have $\binom{8}{4} = 70$ blocks. Thus, the total number of trials is $N = 70 \times 4 = 280$ and $r = \binom{7}{3} = 35$. Here, $\lambda = \binom{6}{2} = 15$.

There are advanced algebraic methods which can yield smaller designs for $t = 8$ and $k = 4$. Box et al. (1978, pp. 272) list a BIBD of $t = 8$, $k = 4$ in $b = 14$ blocks. Here $N = 14 \times 4 = 56$, $r = 7$ and $\lambda = 3$.

It is not always possible to have a BIBD smaller in size than a complete combinatoric design. Such a case is $t = 8$ and $k = 5$. Here the smallest number of blocks possible is $\binom{8}{5} = 56$, and $N = 56 \times 5 = 280$.

The reader is referred to Box et al. (1978, pp. 270–274) for a list of some useful BIBD's for $k = 2, \ldots, 6$, $t = k, \ldots, 10$. Let B_i denote the set of treatments in the ith block. For example, if block 1 contains the treatments 1, 2, 3, 4 then $B_1 = \{1, 2, 3, 4\}$. Let Y_{ij} be the yield of treatment $j \in B_i$. The effects model is

$$Y_{ij} = \mu + \beta_i + \tau_j + e_{ij}, \quad i = 1, \ldots, b \; j \in B_i \tag{13.5.4}$$

$\{e_{ij}\}$ are random experimental errors, with $E\{e_{ij}\} = 0$ and $V\{e_{ij}\} = \sigma^2$ all (i, j). The block and treatment effects, $\beta_1, \ldots, \beta_b$ and $\tau_1, \ldots, \tau_t$ satisfy the constraints $\sum_{j=1}^{t} \tau_j = 0$ and $\sum_{i=1}^{b} \beta_i = 0$.

Let T_j be the set of all indices of blocks containing the jth treatment. The least squares estimates of the treatment effects are obtained in the following manner.

Let $W_j = \sum_{i \in T_j} Y_{ij}$ be the sum of all Y values under the jth treatment. Let W_j^* be the sum of the values in all the r blocks which contain the jth treatment, i.e., $W_j^* = \sum_{i \in T_j} \sum_{l \in B_i} Y_{il}$. Compute

$$Q_j = kW_j - W_j^*, \quad j = 1, \ldots, t. \tag{13.5.5}$$

The LSE of τ_j is

$$\hat{\tau}_j = \frac{Q_j}{t\lambda}, \quad j = 1, \ldots, t. \tag{13.5.6}$$

Notice that $\sum_{j=1}^{t} Q_j = 0$. Thus, $\sum_{j=1}^{t} \hat{\tau}_j = 0$. Let $\bar{\bar{Y}} = \frac{1}{N} \sum_{i=1}^{b} \sum_{l \in B_i} Y_{il}$. The **adjusted** treatment average is defined as $\bar{Y}_j^* = \bar{\bar{Y}} + \hat{\tau}_j, j = 1, \ldots, t$. The ANOVA for a BIBD is given in Table 13.4. Here,

$$SST = \sum_{i=1}^{b} \sum_{l \in B_i} Y_{il}^2 - \left(\sum_{i=1}^{b} \sum_{l \in B_i} Y_{il} \right)^2 / N; \tag{13.5.7}$$

$$SSBL = \frac{1}{k} \sum_{i=1}^{b} \left(\sum_{l \in B_i} Y_{il} \right)^2 - N\bar{\bar{Y}}^2; \tag{13.5.8}$$

$$SSTR = \frac{1}{\lambda kt} \sum_{j=1}^{t} Q_j^2 \tag{13.5.9}$$

Table 13.4 *ANOVA for a BIBD*

Source of variation	DF	SS	MS	E{MS}
blocks	$b-1$	SSBL	MSBL	$\sigma^2 + \dfrac{t}{b-1} \sum_{i=1}^{b} \beta_i^2$
Treatments adjusted	$t-1$	SSTR	MSTR	$\sigma^2 + \dfrac{b}{t-1} \sum_{j=1}^{t} \tau_j^2$
Error	$N-t-b+1$	SSE	MSE	σ^2
Total	$N-1$	SST	–	–

and

$$SSE = SST - SSBL - SSTR. \qquad (13.5.10)$$

The significance of the treatments effects is tested by the statistic

$$F = \frac{MSTR}{MSE}. \qquad (13.5.11)$$

Example 13.5.   Six different adhesives ($t = 6$) are tested for the bond strength in a lamination process, under curing pressure of 200 [psi]. Lamination can be done in blocks of size $k = 4$.
A combinatoric design will have $\binom{6}{4} = 15$ blocks, with $r = \binom{5}{3} = 10$, $\lambda = \binom{4}{2} = 6$ and $N = 60$. The treatment indices of the 15 blocks are (Table 13.5)

Table 13.5   *Block sets*

i	B_i	i	B_i
1	1, 2, 3, 4	9	1, 3, 5, 6
2	1, 2, 3, 5	10	1, 4, 5, 6
3	1, 2, 3, 6	11	2, 3, 4, 5
4	1, 2, 4, 5	12	2, 3, 4, 6
5	1, 2, 4, 6	13	2, 3, 5, 6
6	1, 2, 5, 6	14	2, 4, 5, 6
7	1, 3, 4, 5	15	3, 4, 5, 6
8	1, 3, 4, 6		

The observed bond strength in these trials are (Table 13.6):

Table 13.6   *Values of Y_{il}, $l \in B_i$*

i	Y_{il}	i	Y_i
1	24.7, 20.8, 29.4, 24.9	8	23.1, 29.3, 27.1, 34.4
2	24.1, 20.4, 29.8, 30.3	9	22.0, 29.8, 31.9, 36.1
3	23.4, 20.6, 29.2, 34.4	10	22.8, 22.6, 33.2, 34.8
4	23.2, 20.7, 26.0, 30.8	11	21.4, 29.6, 24.8, 31.2
5	21.5, 22.1, 25.3, 35.4	12	21.3, 28.9, 25.3, 35.1
6	21.4, 20.1, 30.1, 34.1	13	21.6, 29.5, 30.4, 33.6
7	23.2, 28.7, 24.9, 31.0	14	20.1, 25.1, 32.9, 33.9
		15	30.1, 24.0, 30.8, 36.5

The grand mean of the bond strength is $\bar{\bar{Y}} = 27.389$. The sets T_j and the sums W_j, W_j^* are (Table 13.7)

Table 13.7   *The set T_j and the statistics W_j, W_j^*, Q_j*

j	T_j	W_j	W_j^*	Q_j
1	1, 2, 3, 4, 5, 6, 7, 8, 9, 10	229.536	1077.7	−159.56
2	1, 2, 3, 4, 5, 6, 11, 12, 13, 14	209.023	1067.4	−231.31
3	1, 2, 3, 7, 8, 9, 11, 12, 13, 15	294.125	1107.60	68.90
4	1, 4, 5, 7, 8, 10, 11, 12, 14, 15	249.999	1090.90	−90.90
5	2, 4, 6, 7, 9, 10, 11, 13, 14, 15	312.492	1107.50	142.47
6	3, 5, 6, 8, 9, 10, 12, 13, 14, 15	348.176	1123.80	268.90

The ANOVA table is (Table 13.8)

Table 13.8   *ANOVA for BIBD*

Source	DF	SS	MS	F
Blocks	14	161.78	11.556	23.99
Treat. Adj.	5	1282.76	256.552	532.54
Error	40	19.27	0.48175	–
Total	59	1463.81		

The adjusted mean effects of the adhesives are (Table 13.9)

Table 13.9   *Mean effects and their S.E.*

Treatment	$\bar{Y}_i^*$	S.E.$\{\bar{Y}_i^*\}$
1	22.96	1.7445
2	20.96	1.7445
3	29.33	1.7445
4	24.86	1.7445
5	31.35	1.7445
6	34.86	1.7445

The variance of each adjusted mean effect is

$$V\{\bar{Y}_j^*\} = \frac{k\sigma^2}{t\lambda}, \quad j = 1, \dots, t. \tag{13.5.12}$$

Thus, the S.E. of $\bar{Y}_i^*$ is

$$\text{S.E.}\{\bar{Y}_j^*\} = \left(\frac{k\,MSE}{t\lambda}\right)^{1/2}, \quad j = 1, \dots, t. \tag{13.5.13}$$

It seems that there are two homogeneous groups of treatments $\{1, 2, 4\}$ and $\{3, 5, 6\}$.

13.6   Latin square design

Latin square designs are such that we can block for two error inducing factors in a balanced fashion, and yet save considerable amount of trials.

Suppose that we have t treatments to test, and we wish to block for two factors. We assign the blocking factors t levels (the number of treatments) in order to obtain squared designs. For example, suppose that we wish to study the effects of four new designs of keyboards for desktop computers. The design of the keyboard might have effect on the speed of typing or on the number of typing errors. Noisy factors are typist or type of job. Thus, we can block by typist and by job. We should pick at random four typists and four different jobs. We construct a square with four rows and four columns for the blocking factors.

Let A, B, C, and D denote the 4 keyboard designs. We assign the letters to the cells of the above square so that

1. each letter appears exactly once in a row;
2. each letter appears exactly once in a column.

Finally, the order of performing these trials is random. Notice that a design which contains all the combinations of typist, job and keyboard spans over $4 \times 4 \times 4 = 64$ combinations. Thus, the latin square design saves many trials. However, it is based on the assumption of no interactions between the treatments and the blocking factors. That is, in order to obtain valid analysis, the model relating the response to the factor effects should be **additive**, i.e.,

$$Y_{ijk} = \mu + \beta_i + \gamma_j + \tau_k + e_{ijk}, \quad i, j, k = 1, \ldots, t \tag{13.6.1}$$

where μ is the grand mean, β_i are the row effects, γ_j are the column effects, and τ_k are the treatment effects. The experimental error variables are $\{e_{ijk}\}$, with $E\{e_{ijk}\} = 0$ and $V\{e_{ijk}\} = \sigma^2$ for all (i, j). Furthermore,

$$\sum_{i=1}^{t} \beta_i = \sum_{j=1}^{t} \gamma_j = \sum_{k=1}^{t} \tau_k = 0. \tag{13.6.2}$$

The Latin square presented in Table 13.10 is not unique. There are other 4×4 Latin squares. For example,

$$
\begin{array}{cccc}
A & B & C & D \\
D & C & B & A \\
B & A & D & C \\
C & D & A & B
\end{array}
\qquad
\begin{array}{cccc}
A & B & C & D \\
C & D & A & B \\
D & C & B & A \\
B & A & D & C
\end{array}
.
$$

A few Latin square designs, for $t = 3, \ldots, 9$ are given in Box et al. (1978, pp. 261–262).

If we perform only one replication of the Latin square, the ANOVA for testing the main effects is (Table 13.11)

Table 13.10   *A 4×4 Latin square*

	Job 1	Job 2	Job 3	Job 4
Typist 1	A	B	C	D
Typist 2	B	A	D	C
Typist 3	C	D	A	B
Typist 4	D	C	B	A

Table 13.11 *ANOVA for a Latin square, one replication*

Source	DF	SS	MS	F
Treatments	$t-1$	SSTR	MSTR	MSTR/MSE
Rows	$t-1$	SSR	MSR	MSR/MSE
Columns	$t-1$	SSC	MSL	MSC/MSE
Error	$(t-1)(t-2)$	SSE	MSE	–
Total	t^2-1	SST	–	–

Table 13.12 *ANOVA for replicated Latin square*

Source	D.F.	SS	MS	F
Treatments	$t-1$	SSTR	MSTR	MSTR/MSE
Rows	$t-1$	SSR	MSR	–
Columns	$t-1$	SSC	MSC	–
Replicas	$r-1$	SSREP	MSREP	–
Error	$(t-1)[r(t+1)-3])$	SSE	MSE	–
Total	$r(t^2-1)$	SST	–	–

Formulae for the various SS terms will be given below. At this time, we wish to emphasize that if t is small, say $t = 3$, then the number of DF for *SSE* is only 2. This is too small. The number of DF for the error SS can be increased by performing replicas. One possibility is to perform the same Latin square r times independently, and as similarly as possible. However, significant differences between replicas may emerge. The ANOVA, for r identical replicas is as in Table 13.12.

Notice that now we have rt^2 observations. Let $T_{....}$ and $Q_{....}$ be the sum and sum of squares of all observations. Then,

$$SST = Q_{....} - T_{....}^2/rt^2. \tag{13.6.3}$$

Let $T_{i...}$ denote the sum of rt observations in the ith row of all r replications. Then

$$SSR = \frac{1}{tr}\sum_{i=1}^{t}T_{i...}^2 - \frac{T_{....}^2}{rt^2}. \tag{13.6.4}$$

Similarly, let $T_{.j.}$ and $T_{..k.}$ be the sums of all rt observations in column j of all replicas, and treatment k of all r replicas, then

$$SSC = \frac{1}{rt}\sum_{j=1}^{t}T_{.j.}^2 - \frac{T_{....}^2}{rt^2} \tag{13.6.5}$$

and

$$SSTR = \frac{1}{rt}\sum_{k=1}^{t}T_{..k.}^2 - \frac{T_{....}^2}{rt^2}. \tag{13.6.6}$$

Finally, let $T_{...l}$ ($l = 1, \ldots, r$) denote the sum of all t^2 observations in the lth replication. Then,

$$SSREP = \frac{1}{t^2} \sum_{l=1}^{r} t^2_{...l} - \frac{T^2_{.....}}{rt^2}. \tag{13.6.7}$$

The pooled sum of squares for error is obtained by

$$SSE = SSD - SSR - SSC - SSTR - SSREP. \tag{13.6.8}$$

Notice that if $t = 3$ and $r = 3$, the number of DF for SSE increases from 2 (when $r = 1$) to 18 (when $r = 3$).

The most important hypothesis is that connected with the main effects of the treatments. This we test with the statistic

$$F = MSTR/MSE. \tag{13.6.9}$$

Example 13.6. Five models of keyboards (treatments) were tested in a latin square design in which the blocking factors are typist and job. Five typists were randomly selected from a pool of typists of similar capabilities. Five typing jobs were selected. Each typing job had 4000 characters. The yield, Y_{ijk}, is the number of typing errors found at the ith typist, jth job under the kth keyboard. The Latin square design used is presented in Table 13.13.

Table 13.13   *Latin square design, t = 5*

Typist	Job				
	1	2	3	4	5
1	A	B	C	D	E
2	B	C	D	E	A
3	C	D	E	A	B
4	D	E	A	B	C
5	E	A	B	C	D

The five keyboards are denoted by the letters A, B, C, D, and E. The experiment spanned over five days. In each day, a typist was assigned a job at random (from those not yet tried). The keyboard used is the one associated with the job. Only one job was tried in a given day. The observed number of typing errors (per 4000 characters) are (Table 13.14):
Figures 13.2–13.4 present boxplots of the error rates for the three different factors. The only influencing factor is the typist effect, as shown in Figure 13.4.
The total sum of squares is

$$Q = 30636.$$

Thus,

$$SST = 30636 - \frac{798^2}{25} = 5163.84.$$

Table 13.14   *Number of typing errors*

Typist	Job 1	Job 2	Job 3	Job 4	Job 5	Row Sums
1	*A* 20	*B* 18	*C* 25	*D* 17	*E* 20	100
2	*B* 65	*C* 40	*D* 55	*E* 58	*A* 59	277
3	*C* 30	*D* 27	*E* 35	*A* 21	*B* 27	140
4	*D* 21	*E* 15	*A* 24	*B* 16	*C* 18	94
5	*E* 42	*A* 38	*B* 40	*C* 35	*D* 32	187
Column Sum	178	138	179	147	156	798

Sums	*A*	*B*	*C*	*D*	*E*
Keyboard	162	166	148	152	170

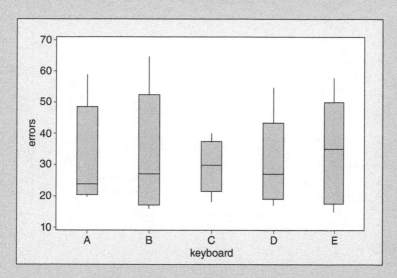

Figure 13.2   *Effect of keyboard on error rate (MINITAB).*

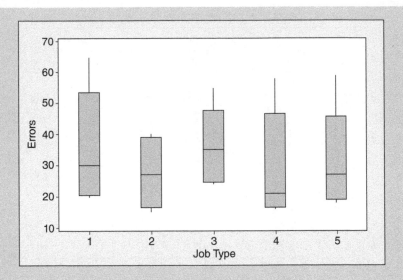

Figure 13.3   *Effect of job type on error rate (MINITAB).*

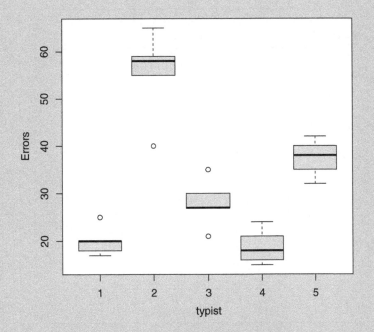

Figure 13.4   *Effect of typist on error rate.*

Similarly,

$$SSR = \frac{1}{5}(100^2 + 277^2 + \cdots + 187^2) - \frac{798^2}{25}$$

$$= 4554.64$$

$$SSC = \frac{1}{5}(178^2 + 138^2 + \cdots + 156^2) - \frac{798^2}{25}$$

$$= 270.641$$

and

$$SSTR = \frac{1}{5}(162^2 + 166^2 + \cdots + 170^2) - \frac{798^2}{25}$$

$$= 69.4395.$$

The analysis of variance, Table 13.15, is

Table 13.15 *ANOVA for keyboard Latin Square experiment*

Source	DF	SS	MS	F
Typist	4	4554.640	1138.66	50.772
Job	4	270.641	67.66	3.017
Keyboard	4	69.439	17.3598	0.774
Error	12	269.120	22.4267	–
Total	24	5163.840	–	–

The null hypothesis that the main effects of the keyboards are zero cannot be rejected. The largest source of variability in this experiment were the typists. The different jobs contributed also to the variability. The P-value for the F test of Jobs is 0.062.

13.7   Full factorial experiments

13.7.1   The structure of factorial experiments

Full factorial experiments are those in which complete trials are performed of all the combinations of the various factors at all their levels. For example, if there are five factors, each one tested at three levels, there are altogether $3^5 = 243$ treatment combinations. All these 243 treatment combinations are tested. The full factorial experiment may also be replicated several times. The order of performing the trials is random.

In full factorial experiments, the number of levels of different factors do not have to be the same. Some factors might be tested at two levels and others at three or four levels. Full factorial, or certain fractional factorials which will be discussed later, are necessary, if the statistical model is not additive. In order to estimate or test the **effects** of **interactions**, one needs to perform factorial experiments, full or fractional. In a full factorial experiment, all the main effects and interactions can be tested or estimated. Recall that if there are p factors $A, B, C, \ldots$ there are p types of main effects, $\binom{p}{2}$ types of pairwise interactions $AB, AC, BC, \ldots, \binom{p}{3}$ interactions between three factors, ABC, ABD, and so on. On the whole there are, together with the grand mean μ, 2^p types of parameters.

In the following section, we discuss the structure of the ANOVA for testing the significance of main effects and interaction. This is followed by a section on the estimation problem. In Sections 13.7.4 and 13.7.5, we discuss the structure of full factorial experiments with 2 and 3 levels per factor, respectively.

13.7.2   The ANOVA for full factorial designs

The analysis of variance for full factorial designs is done for testing the hypotheses that main-effects or interaction parameters are equal to zero. We present the ANOVA for a two-factor situation, factor A at a levels and factor B at b levels. The method can be generalized to any number of factors.

The structure of the experiment is such that all $a \times b$ treatment combinations are tested. Each treatment combination is repeated n times. The model is

$$Y_{ijk} = \mu + \tau_i^A + \tau_j^B + \tau_{ij}^{AB} + e_{ijk}, \tag{13.7.1}$$

$i = 1, \ldots, a; j = 1, \ldots, b; k = 1, \ldots, n$. e_{ijk} are independent random variables $E\{e_{ijh}\} = 0$ and $V\{e_{ijk}\} = \sigma^2$ for all i, j, k. Let

$$\bar{Y}_{ij} = \frac{1}{n} \sum_{k=1}^{n} Y_{ijk} \tag{13.7.2}$$

$$\bar{Y}_{i.} = \frac{1}{b} \sum_{j=1}^{b} \bar{Y}_{ij}, \quad i = 1, \ldots, a \tag{13.7.3}$$

$$\bar{Y}_{.j} = \frac{1}{a} \sum_{i=1}^{a} Y_{ij}, \quad j = 1, \ldots, b \tag{13.7.4}$$

and

$$\bar{\bar{Y}} = \frac{1}{ab} \sum_{i=1}^{a} \sum_{j=1}^{b} \bar{Y}_{ij}. \tag{13.7.5}$$

The ANOVA partitions first the total sum of squares of deviations from $\bar{\bar{Y}}$, i.e.,

$$SST = \sum_{i=1}^{a} \sum_{j=1}^{b} \sum_{k=1}^{n} (Y_{ijk} - \bar{\bar{Y}})^2 \tag{13.7.6}$$

to two components

$$SSW = \sum_{i=1}^{a} \sum_{j=1}^{b} \sum_{k=1}^{n} (Y_{ijl} - \bar{Y}_{ij})^2 \tag{13.7.7}$$

and

$$SSB = n \sum_{i=1}^{a} \sum_{j=1}^{b} (\bar{Y}_{ij} - \bar{\bar{Y}})^2. \tag{13.7.8}$$

It is straightforward to show that

$$SST = SSW + SSB. \tag{13.7.9}$$

In the second stage, the sum of squares of deviations SSB is partitioned to three components SSI, $SSMA$, $SSMB$, where

$$SSI = n \sum_{i=1}^{a} \sum_{j=1}^{b} (\bar{Y}_{ij} - \bar{Y}_{i.} - \bar{Y}_{.j} + \bar{\bar{Y}})^2, \tag{13.7.10}$$

$$SSMA = nb \sum_{i=1}^{a} (\bar{Y}_{i.} - \bar{\bar{Y}})^2 \tag{13.7.11}$$

Table 13.16   *Table of ANOVA for a 2-factor factorial experiment*

Source of variation	DF	SS	MS	F
A	$a - 1$	SSMA	MSA	F_A
B	$b - 1$	SSMB	MSB	F_B
AB	$(a - 1)(b - 1)$	SSI	MSAB	F_{AB}
Between	$ab - 1$	SSB	–	–
Within	$ab(n - 1)$	SSW	MSW	–
Total	$N - 1$	SST	–	–

and

$$SSMB = na \sum_{j=1}^{b} (\bar{Y}_j - \bar{\bar{Y}})^2, \tag{13.7.12}$$

i.e.,

$$SSB = SSI + SSMA + SSMB. \tag{13.7.13}$$

All these terms are collected in a table of ANOVA (Table 13.16).
Thus,

$$MSA = \frac{SSMA}{k - 1}, \tag{13.7.14}$$

$$MSB = \frac{SSMB}{m - 1}, \tag{13.7.15}$$

and

$$MSAB = \frac{SSI}{(k - 1)(m - 1)}, \tag{13.7.16}$$

$$MSW = \frac{SSW}{km(n - 1)}. \tag{13.7.17}$$

Finally, we compute the F-statistics

$$F_A = \frac{MSA}{MSW}, \tag{13.7.18}$$

$$F_B = \frac{MSB}{MSW} \tag{13.7.19}$$

and

$$F_{AB} = \frac{MSAB}{MSW}. \tag{13.7.20}$$

F_A, F_B, and F_{AB} are test statistics to test, respectively, the significance of the main effects of A, the main effects of B and the interactions AB.

If $F_A < F_{1-\alpha}[k - 1, km(n - 1)]$, the null hypothesis

$$H_0^A : \tau_1^A = \cdots = \tau_k^A = 0.$$

cannot be rejected.

If $F_B < F_{1-\alpha}[m-1, km(n-1)]$, the null hypothesis

$$H_0^B : \tau_1^B = \cdots = \tau_m^B = 0.$$

cannot be rejected.
    Also, if

$$F_{AB} < F_{1-\alpha}[(k-1)(m-1), km(n-1)],$$

we cannot reject the null hypothesis

$$H_0^{AB} : \tau_{11}^{AB} = \cdots = \tau_{km}^{AB} = 0.$$

The ANOVA for two factors can be performed by MINITAB. We illustrate this estimation and testing in the following example.

Example 13.7.   In Chapter 10, we have introduced the piston example. Seven prediction factors for the piston cycle time were listed. These are

A: Piston weight, 30–60 [kg]
B: Piston surface area, 0.005–0.020 [m^2]
C: Initial gas volume, 0.002 - 0.010 [m^3]
D: Spring coefficient, 1,000–5,000 [N/m]
E: Atmospheric pressure, 90,000–100,000 [N/m^2]
F: Ambient temperature, 290–296 [°K]
G: Filling gas temperature, 340–360[°K].

We are interested to test the effects of the piston weight (A) and the spring coefficient (D) on the cycle times (seconds). For this purpose we designed a factorial experiment at three levels of A, and three levels of D. The levels are

$$A_1 = 30 \text{ [kg]}, \quad A_2 = 45 \text{ [kg] and } A_3 = 60 \text{ [kg]}.$$

The levels of factor D (spring coefficient) are $D_1 = 1,500$ [N/m], $D_2 = 3,000$ [N/m] and $D_3 = 4,500$ [N/m]. Five replicas were performed at each treatment combination ($n = 5$). The data can be obtained by using the JMP or R piston simulator. The five factors which were not under study were kept at the levels $B = 0.01$ [m^2], $C = 0.005$ [m^3], $E = 95,000$ [N/m^2], $F = 293$ [°K] and $G = 350$ [°K].

```
> library(DoE.base)
> Factors <- list(
    m=c(30, 45, 60),
    k=c(1500, 3000, 4500))
> FacDesign <- fac.design(
    factor.names=Factors,
    randomize=TRUE,
    replications=5,
    repeat.only=TRUE)
> Levels <- data.frame(
    lapply(
```

```
      lapply(FacDesign,
             as.character),
      as.numeric),
   s=0.01,
   v0=0.005,
   p0=95000,
   t=293,
   t0=350)
> Ps <- pistonSimulation(m=Levels$m,
                         s=Levels$s,
                         v0=Levels$v0,
                         k=Levels$k,
                         p0=Levels$p0,
                         t=Levels$t,
                         t0=Levels$t0,
                         each=1,
                         seed=123)
> FacDesign <- add.response(
    design=FacDesign,
    response=Ps$seconds)
> summary(
    aov(Ps.seconds ~ m*k,
        data=FacDesign))

            Df Sum Sq Mean Sq F value    Pr(> F)
m            2 0.1769  0.0884   2.362      0.109
k            2 1.1402  0.5701  15.227 0.0000161 ***
m:k          4 0.0901  0.0225   0.602      0.664
Residuals   36 1.3478  0.0374
---
Signif. codes:
0 '***' 0.001 '**' 0.01 '*' 0.05 '.' 0.1 ' ' 1

> rm(Levels, Factors)
```

We start with the two-way ANOVA presented in Table 13.17.

Table 13.17   *Two-way ANOVA for cycle time*

Source	DF	SS	MS	F	p
Spr_Cof	2	1.01506	0.50753	8.66	0.001
Pist_Wg	2	0.09440	0.04720	0.81	0.455
Spr_Cof*Pist_Wg	4	0.06646	0.01662	0.28	0.887
Error	36	2.11027	0.05862		
Total	44	3.28619			

Figure 13.5 shows the effect of the factor spring coefficient on cycle time. Spring coefficient at 4500 $[N/m]$ reduces the mean cycle time and its variability. The boxplot at the right margin is for the combined samples.

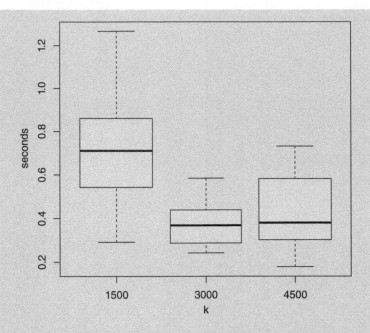

Figure 13.5   *Effect of spring coefficient on cycle time.*

Figure 13.6 is an **interaction-plot** showing the effect of piston weight on the mean cycle time, at each level of the spring coefficient.

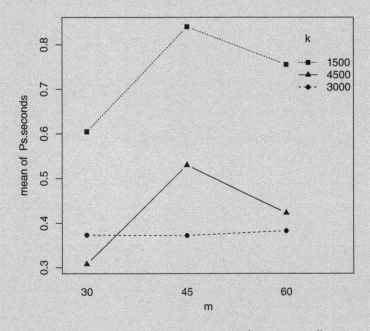

Figure 13.6   *Interaction plot of piston weight spring coefficient.*

The P-values are computed with the appropriate F-distributions. We see in the ANOVA table that only the main effects of the spring coefficient (D) are significant. Since the effects of the piston weight (A) and that of the interaction are not significant, we can estimate σ^2 by a pooled estimator, which is

$$\hat{\sigma}^2 = \frac{SSW + SSI + SSMA}{36 + 4 + 2} = \frac{2.2711}{42}$$

$$= 0.0541.$$

To estimate the main effects of D we pool all data from samples having the same level of D together. We obtain pooled samples of size $n_p = 15$. The means of the cycle time for these samples are

	D_1	D_2	D_3	Grand
$\bar{Y}$	0.743	0.509	0.380	0.544
Main Effects	0.199	−0.035	−0.164	−

The standard error of these main effects is $\text{S.E.}\{\hat{\tau}_j^D\} = \dfrac{0.23259}{\sqrt{15}} \sqrt{\dfrac{1}{2}} = 0.0425.$

Since we estimate on the basis of the pooled samples, and the main effects $\hat{\tau}_j^D$ ($j = 1, 2, 3$) are contrasts of 3 means, the coefficient S_α for the simultaneous confidence intervals has the formula

$$S_\alpha = (2F_{.95}[2, 42])^{1/2}$$
$$= \sqrt{2 \times 3.22} = 2.538.$$

The simultaneous confidence intervals for τ_j^D, at $\alpha = 0.05$, are

	Lower Limit	Upper Limit
τ_1^D:	0.0911	0.3069
τ_2^D:	−0.1429	0.0729
τ_3^D:	−0.2619	−0.0561

We see that the confidence interval for τ_2^D covers zero. Thus, $\hat{\tau}_2^D$ is **not significant**. The significant main effects are $\hat{\tau}_1^D$ and $\hat{\tau}_3^D$.

13.7.3   Estimating main effects and interactions

In the present section, we discuss the estimation of the main effects and interaction parameters. Our presentation is confined to the case of two factors A and B, which are at a and b levels, respectively. The number of replicas of each treatment combinations is n. We further assume that the errors $\{e_{ijk}\}$ are i.i.d., having a normal distribution $N(0, \sigma^2)$.

Let

$$\bar{Y}_{ij} = \frac{1}{n}\sum_{l=1}^{n} Y_{ijl} \tag{13.7.21}$$

and

$$Q_{ij} = \sum_{l=1}^{n}(Y_{ijl} - \bar{Y}_{ij})^2, \tag{13.7.22}$$

$i = 1, \ldots, a; j = 1, \ldots, b$. It can be shown that the **least squares** estimators of τ_i^A, τ_j^B and τ_{ij}^{AB} are, respectively,

$$\begin{aligned}\hat{\tau}_{i.}^A &= \bar{Y}_{i.} - \bar{\bar{Y}}, \quad i = 1, \ldots, a \\ \hat{\tau}_{j}^B &= \bar{Y}_{.j} - \bar{\bar{Y}}, \quad j = 1, \ldots, b\end{aligned} \tag{13.7.23}$$

and

$$\hat{\tau}_{ij}^{AB} = \bar{Y}_{ij} - \bar{Y}_{i.} - \bar{Y}_{.j} + \bar{\bar{Y}}, \tag{13.7.24}$$

where

$$\bar{Y}_{i.} = \frac{1}{m}\sum_{j=1}^{m} \bar{Y}_{ij}, \tag{13.7.25}$$

and

$$\bar{Y}_{.j} = \frac{1}{k}\sum_{i=1}^{k} \bar{Y}_{ij}. \tag{13.7.26}$$

Furthermore, an unbiased estimator of σ^2 is

$$\hat{\sigma}^2 = \frac{\sum_{i=1}^{a}\sum_{j=1}^{b} Q_{ij}}{ab(n-1)}. \tag{13.7.27}$$

The standard errors of the estimators of the interactions are

$$\text{S.E.}\{\hat{\tau}_{ij}^{AB}\} = \frac{\hat{\sigma}}{\sqrt{n}}\left(\left(1 - \frac{1}{a}\right)\left(1 - \frac{1}{b}\right)\right)^{1/2}, \tag{13.7.28}$$

for $i = 1, \ldots, a; j = 1, \ldots, b$. The standard errors of the estimators of the main effects are

$$\text{S.E.}\{\hat{\tau}_i^A\} = \frac{\hat{\sigma}}{\sqrt{nb}}\left(1 - \frac{1}{a}\right)^{1/2}, \quad i = 1, \ldots, a \tag{13.7.29}$$

and

$$\text{S.E.}\{\hat{\tau}_j^B\} = \frac{\hat{\sigma}}{\sqrt{na}}\left(1 - \frac{1}{b}\right)^{1/2}, \quad j = 1, \ldots, b. \tag{13.7.30}$$

Confidence limits at level $(1 - \alpha)$ for such a parameter are obtained by

$$\begin{aligned}\hat{\tau}_i^A &\pm S_\alpha \cdot \text{S.E.}\{\hat{\tau}_i^A\} \\ \hat{\tau}_j^B &\pm S_\alpha \cdot \text{S.E.}\{\hat{\tau}_j^B\}\end{aligned} \tag{13.7.31}$$

and

$$\hat{\tau}_{ij}^{AB} \pm S_\alpha \text{S.E.}\{\hat{\tau}_{ij}^{AB}\} \tag{13.7.32}$$

where

$$S_\alpha = ((ab - 1)F_{1-\alpha}[ab - 1, ab(n - 1)])^{1/2}.$$

Multiplying the S_α guarantees that all the confidence intervals are simultaneously covering the true parameters with probability $(1 - \alpha)$. Any confidence interval which covers the value zero implies that the corresponding parameter is not significantly different than zero.

13.7.4 2^m factorial designs

2^m factorial designs are full factorials of m factors, each one at two levels. The levels of the factors are labelled as "Low" and "High" or 1 and 2. If the factors are categorical then the labelling of the levels is arbitrary and the values of the main effects and interaction parameters depend on this arbitrary labeling. We will discuss here experiments in which the levels of the factors are measured on a continuous scale, like in the case of the factors effecting the piston cycle time. The levels of the ith factor ($i = 1, \ldots, m$) are fixed at x_{i1} and x_{i2}, where $x_{i1} < x_{i2}$.

By simple transformation, all factor levels can be reduced to

$$c_i = \begin{cases} +1, & \text{if } x = x_{i2} \\ -1, & \text{if } x = x_{i1} \end{cases} \quad , i = 1, \ldots, m.$$

In such a factorial experiment, there are 2^m possible treatment combinations. Let $(i_1, \ldots, i_m)$ denote a treatment combination, where $i_1, \ldots, i_m$ are indices, such that

$$i_j = \begin{cases} 0, & \text{if } c_i = -1 \\ 1, & \text{if } c_i = 1. \end{cases}$$

Thus, if there are $m = 3$ factors, the number of possible treatment combinations is $2^3 = 8$. These are given in Table 13.18.

The index v of the standard order, is given by the formula

$$v = \sum_{j=1}^{m} i_j 2^{j-1}. \tag{13.7.33}$$

Table 13.18 *Treatment combinations of a 2^3 experiment*

v	i_1	i_2	i_3
0	0	0	0
1	1	0	0
2	0	1	0
3	1	1	0
4	0	0	1
5	1	0	1
6	0	1	1
7	1	1	1

Notice that v ranges from 0 to $2^m - 1$. This produces tables of the treatment combinations for a 2^m factorial design, arranged in a standard order (see Table 13.19). A full factorial experiment is a combination of fractional factorial designs. In R, we obtain a fraction of a full factorial design with:

```
> library(FrF2)
> FrF2(nfactors=5, resolution=5)

     A  B  C  D  E
1  -1  1  1 -1  1
2   1  1  1 -1 -1
3  -1  1 -1  1  1
4   1 -1 -1  1  1
5  -1 -1  1 -1 -1
6  -1 -1 -1 -1  1
7   1  1  1  1  1
8   1 -1  1 -1  1
9  -1  1  1  1 -1
10  1  1 -1 -1  1
11 -1 -1 -1  1 -1
12 -1 -1  1  1  1
13  1  1 -1  1 -1
14  1 -1  1  1 -1
15 -1  1 -1 -1 -1
16  1 -1 -1 -1 -1
class=design, type= FrF2
```

which is a half fractional replications of a 2^5 designs as will be explained in Section 13.8. In Table 13.19, we present the design of a 2^5 full factorial experiment derived from the R application:

```
> Design <- fac.design(nlevels=2,
                         nfactors=5)
> head(Design, 3)

  A B C D E
1 1 2 2 2 1
2 2 2 1 1 2
3 1 1 1 1 1

> tail(Design, 3)

   A B C D E
30 1 1 2 1 2
31 2 1 2 2 2
32 1 1 1 2 1

> rm(Design)
```

Table 13.19 *The labels in standard order for a 2^5 factorial design.*

v	l_1	l_2	l_3	l_4	l_5	v	l_1	l_2	l_3	l_4	l_5
0	1	1	1	1	1	16	1	1	1	1	2
1	2	1	1	1	1	17	2	1	1	1	2
2	1	2	1	1	1	18	1	2	1	1	2
3	2	2	1	1	1	19	2	2	1	1	2
4	1	1	2	1	1	20	1	1	2	1	2
5	2	1	2	1	1	21	2	1	2	1	2
6	1	2	2	1	1	22	1	2	2	1	2
7	2	2	2	1	1	23	2	2	2	1	2
8	1	1	1	2	1	24	1	1	1	2	2
9	2	1	1	2	1	25	2	1	1	2	2
10	1	2	1	2	1	26	1	2	1	2	2
11	2	2	1	2	1	27	2	2	1	2	2
12	1	1	2	2	1	28	1	1	2	2	2
13	2	1	2	2	1	29	2	1	2	2	2
14	1	2	2	2	1	30	1	2	2	2	2
15	2	2	2	2	1	31	2	2	2	2	2

Table 13.20 *Treatment means in a 2^2 design.*

	Factor A		Row
Factor B	1	2	Means
1	$\bar{Y}_0$	$\bar{Y}_1$	$\bar{Y}_{1.}$
2	$\bar{Y}_2$	$\bar{Y}_3$	$\bar{Y}_{2.}$
Column			
Means	$\bar{Y}_{.1}$	$\bar{Y}_{.2}$	$\bar{\bar{Y}}$

Let Y_v, $v = 0, 1, \ldots, 2^m - 1$, denote the yield of the vth treatment combination. We discuss now the estimation of the main effects and interaction parameters. Starting with the simple case of 2 factors, the variables are presented schematically, in Table 13.20.

According to our previous definition, there are four main effects τ_1^A, τ_2^A, τ_1^B, τ_2^B and four interaction effects τ_{11}^{AB}, τ_{12}^{AB}, τ_{21}^{AB}, τ_{22}^{AB}. But since $\tau_1^A + \tau_2^A = \tau_1^B + \tau_2^B = 0$, it is sufficient to represent the main effects of A and B by τ_2^A and τ_2^B. Similarly, since $\tau_{11}^{AB} + \tau_{12}^{AB} = 0 = \tau_{11}^{AB} + \tau_{21}^{AB}$ and $\tau_{12}^{AB} + \tau_{22}^{AB} = 0 = \tau_{21}^{AB} + \tau_{22}^{AB}$, it is sufficient to represent the interaction effects by τ_{22}^{AB}.

The main effect τ_2^A is estimated by

$$\hat{\tau}_2^A = \bar{Y}_{.2} - \bar{\bar{Y}} =$$

$$= \frac{1}{2}(\bar{Y}_1 + \bar{Y}_3) - \frac{1}{4}(\bar{Y}_0 + \bar{Y}_1 + \bar{Y}_2 + \bar{Y}_3)$$

$$= \frac{1}{4}(-\bar{Y}_0 + \bar{Y}_1 - \bar{Y}_2 + \bar{Y}_3).$$

The estimator of τ_2^B is

$$\hat{\tau}_2^B = \bar{Y}_{2.} - \bar{\bar{Y}}$$

$$= \frac{1}{2}(\bar{Y}_2 + \bar{Y}_3) - \frac{1}{4}(\bar{Y}_0 + \bar{Y}_1 + \bar{Y}_2 + \bar{Y}_3)$$

$$= \frac{1}{4}(-\bar{Y}_0 - \bar{Y}_1 + \bar{Y}_2 + \bar{Y}_3).$$

Finally, the estimator of τ_{22}^{AB} is

$$\hat{\tau}_{22}^{AB} = \bar{Y}_3 - \bar{Y}_{2.} - \bar{Y}_{.2} + \bar{\bar{Y}}$$

$$= \bar{Y}_3 - \frac{1}{2}(\bar{Y}_2 + \bar{Y}_3) - \frac{1}{2}(\bar{Y}_1 + \bar{Y}_3)$$

$$+ \frac{1}{4}(\bar{Y}_0 + \bar{Y}_1 + \bar{Y}_2 + \bar{Y}_3)$$

$$= \frac{1}{4}(\bar{Y}_0 - \bar{Y}_1 - \bar{Y}_2 + \bar{Y}_3).$$

The parameter μ is estimated by the grand mean $\bar{\bar{Y}} = \frac{1}{4}(\bar{Y}_0 + \bar{Y}_1 + \bar{Y}_2 + \bar{Y}_3)$. All these estimators can be presented in a matrix form as

$$\begin{bmatrix} \hat{\mu} \\ \hat{\tau}_2^A \\ \hat{\tau}_2^B \\ \hat{\tau}_{22}^{AB} \end{bmatrix} = \frac{1}{4} \begin{bmatrix} 1 & 1 & 1 & 1 \\ -1 & 1 & -1 & 1 \\ -1 & -1 & 1 & 1 \\ 1 & -1 & -1 & 1 \end{bmatrix} \cdot \begin{bmatrix} \bar{Y}_0 \\ \bar{Y}_1 \\ \bar{Y}_2 \\ \bar{Y}_3 \end{bmatrix}.$$

The indices in a 2^2 design are given in the following 4×2 matrix

$$D_{2^2} = \begin{bmatrix} 1 & 1 \\ 2 & 1 \\ 1 & 2 \\ 2 & 2 \end{bmatrix}.$$

The corresponding C coefficients are the 2nd and 3rd columns in the matrix

$$C_{2^2} = \begin{bmatrix} 1 & -1 & -1 & 1 \\ 1 & 1 & -1 & -1 \\ 1 & -1 & 1 & -1 \\ 1 & 1 & 1 & 1 \end{bmatrix}.$$

The 4th column of this matrix is the product of the elements in the 2nd and 3rd columns. Notice also that the linear model for the yield vector is

$$\begin{bmatrix} Y_0 \\ Y_1 \\ Y_2 \\ Y_3 \end{bmatrix} = \begin{bmatrix} 1 & -1 & -1 & 1 \\ 1 & 1 & -1 & -1 \\ 1 & -1 & 1 & -1 \\ 1 & 1 & 1 & 1 \end{bmatrix} \begin{bmatrix} \mu \\ \tau_2^A \\ \tau_2^B \\ \tau_{22}^{AB} \end{bmatrix} + \begin{bmatrix} e_1 \\ e_2 \\ e_2 \\ e_4 \end{bmatrix},$$

where e_1, e_2, e_3, and e_4 are independent random variables, with $E\{e_i\} = 0$ and $V\{e_i\} = \sigma^2$, $i = 1, 2, \ldots, 4$.

Let $\mathbf{Y}^{(4)} = (Y_0, Y_1, Y_2, Y_3)'$, $\theta^{(4)} = (\mu, \tau_2^A, \tau_2^B, \tau_{22}^{AB})'$ and $\mathbf{e}^{(4)} = (e_1, e_2, e_3, e_4)'$ then the model is

$$\mathbf{Y}^{(4)} = C_{2^2}\theta^{(4)} + \mathbf{e}^{(4)}.$$

This is the usual linear model for multiple regression. The least squares estimator of $\theta^{(4)}$ is

$$\hat{\theta}^{(4)} = [C'_{2^2}C_{2^2}]^{-1}C'_{2^2}\mathbf{Y}^{(4)}.$$

The matrix C_{2^2} has orthogonal column (row) vectors and

$$C'_{2^2}C_{2^2} = 4I_4,$$

where I_4 is the identity matrix of rank 4. Therefore,

$$\hat{\theta}^{(4)} = \frac{1}{4}C'_{2^2}\mathbf{Y}^{(4)}$$

$$= \frac{1}{4}\begin{bmatrix} 1 & 1 & 1 & 1 \\ -1 & 1 & -1 & 1 \\ -1 & -1 & 1 & 1 \\ 1 & -1 & -1 & 1 \end{bmatrix}\begin{bmatrix} \bar{Y}_0 \\ \bar{Y}_1 \\ \bar{Y}_2 \\ \bar{Y}_3 \end{bmatrix}.$$

This is identical with the solution obtained earlier.

The estimators of the main effects and interactions are the least squares estimators, as has been mentioned before.

This can now be generalized to the case of m factors. For a model with m factors, there are 2^m parameters. The mean μ, m main effects $\tau^1, \ldots, \tau^m$, $\binom{m}{2}$ first-order interactions τ^{ij}, $i \neq j = 1, \ldots, m$, $\binom{m}{3}$ second-order interactions τ^{ijk}, $i \neq j \neq k$, etc. We can now order the parameters in a standard manner in the following manner. Each one of the 2^m parameters can be represented by a binary vector $(j_1, \ldots, j_m)$, where $j_i = 0, 1$ $(i = 1, \ldots, m)$. The vector $(0, 0, \ldots, 0)$ represents the grand mean μ. A vector $(0, 0, \ldots, 1, 0, \ldots, 0)$ where the 1 is the ith component, represents the main effect of the ith factor $(i = 1, \ldots, m)$. A vector with two ones, at the ith and jth component $(i = 1, \ldots, m-1; j = i+1, \ldots, m)$ represent the first-order interaction between factor i and factor j. A vector with three ones, at i, j, k components, represents the second-order interaction between factors i, j, k, etc.

Let $\omega = \sum_{i=1}^m j_i 2^{i-1}$ and β_ω be the parameter represented by the vector with index ω. For example β_3 corresponds to $(1, 1, 0, \ldots, 0)$, which represents the first- order interaction between factors 1 and 2.

Let $\mathbf{Y}^{(2^m)}$ be the yield vector, whose components are arranged in the standard order, with index $v = 0, 1, 2, \ldots, 2^m - 1$. Let C_{2^m} be the matrix of coefficients, that is obtained recursively by the equations

$$C_2 = \begin{bmatrix} 1 & -1 \\ 1 & 1 \end{bmatrix}, \tag{13.7.34}$$

and

$$C_{2^l} = \begin{bmatrix} C_{2^{l-1}} & -C_{2^{l-1}} \\ C_{2^{l-1}} & C_{2^{l-1}} \end{bmatrix}, \tag{13.7.35}$$

$l = 2, 3, \ldots, m$. Then, the linear model relating $\mathbf{Y}^{(2m)}$ to $\beta^{(2m)}$ is

$$\mathbf{Y}^{(2^m)} = C_{2^m} \cdot \beta^{(2^m)} + \mathbf{e}^{(2^m)}, \tag{13.7.36}$$

where

$$\beta^{(2^m)} = (\beta_0, \beta_1, \ldots, \beta_{2^m-1})'.$$

Since the column vectors of C_{2^m} are orthogonal, $(C_{2^m})'C_{2^m} = 2^m I_{2^m}$, the least squares estimator (LSE) of $\beta^{(2^m)}$ is

$$\hat{\beta}^{(2^m)} = \frac{1}{2^m}(C_{2^m})'Y^{(2^m)}. \tag{13.7.37}$$

Accordingly, the LSE of β_ω is

$$\hat{\beta}_\omega = \frac{1}{2^m}\sum_{v=0}^{2^m-1} c_{(v+1),(\omega+1)}^{(2^m)} Y_v, \tag{13.7.38}$$

where $c_{ij}^{(2^m)}$ is the ith row and jth column element of C_{2^m}, i.e., multiply the components of $Y^{(2^m)}$ by those of the column of C_{2^m}, corresponding to the parameter β_ω, and divide the sum of products by 2^m.

We do not have to estimate all the 2^m parameters, but can restrict attention only to parameters of interest, as will be shown in the following example.

Since $c_{ij}^{(2^m)} = \pm 1$, the variance of $\hat{\beta}_\omega$ is

$$V\{\hat{\beta}_\omega\} = \frac{\sigma^2}{2^m}, \quad \text{for all } \omega = 0, \ldots, 2^m - 1. \tag{13.7.39}$$

Finally, if every treatment combination is repeated n times, the estimation of the parameters is based on the means $\bar{Y}_v$ of the n replications. The variance of $\hat{\beta}_\omega$ becomes

$$V\{\hat{\beta}_\omega\} = \frac{\sigma^2}{n2^m}. \tag{13.7.40}$$

The variance σ^2 can be estimated by the pooled variance estimator, obtained from the between replication variance within each treatment combinations. That is, if $Y_{vj}, j = 1, \ldots, n$, are the observed values at the vth treatment combination then

$$\hat{\sigma}^2 = \frac{1}{(n-1)2^m}\sum_{v=1}^{2^m}\sum_{j=1}^{n}(Y_{vj} - \bar{Y}_v)^2. \tag{13.7.41}$$

Example 13.8.   In Example 13.7 we studied the effects of two factors on the cycle time of a piston in a gas turbine, keeping all the other five factors fixed. In the present example we perform a 2^5 experiment with the piston varying factors A, B, C, D and F at two levels, keeping the atmospheric pressure (factor E) fixed at 90,000 [N/m^2] and the filling gas temperature (factor G) at 340 [°K]. The two levels of each factor are those specified, in Example 13.7, as the limits of the experimental range. Thus, for example, the low level of piston weight (factor A) is 30 [kg] and its high level is 60 [kg]. The treatment combinations are listed in Table 13.21.

The number of replications is $n = 5$. Denote the means $\bar{Y}_v$ and the standard deviations, S_v, of the five observations in each treatment combination. We obtain the value $\hat{\sigma}^2 = 0.02898$ and the estimated variance of all L.S.E. of the parameters is

$$\hat{V}\{\hat{\beta}_\omega\} = \frac{\hat{\sigma}^2}{5 \times 32} = 0.0001811,$$

Table 13.21 *Labels of treatment combinations and average response.*

A	B	C	D	F	$\bar{Y}$
1	1	1	1	1	0.929
2	1	1	1	1	1.111
1	2	1	1	1	0.191
2	2	1	1	1	0.305
1	1	2	1	1	1.072
2	1	2	1	1	1.466
1	2	2	1	1	0.862
2	2	2	1	1	1.318
1	1	1	2	1	0.209
2	1	1	2	1	0.340
1	2	1	2	1	0.123
2	2	1	2	1	0.167
1	1	2	2	1	0.484
2	1	2	2	1	0.690
1	2	2	2	1	0.464
2	2	2	2	1	0.667
1	1	1	1	2	0.446
2	1	1	1	2	0.324
1	2	1	1	2	0.224
2	2	1	1	2	0.294
1	1	2	1	2	1.067
2	1	2	1	2	1.390
1	2	2	1	2	0.917
2	2	2	1	2	1.341
1	1	1	2	2	0.426
2	1	1	2	2	0.494
1	2	1	2	2	0.271
2	2	1	2	2	0.202
1	1	2	2	2	0.482
2	1	2	2	2	0.681
1	2	2	2	2	0.462
2	2	2	2	2	0.649

or standard error of S.E.$\{\hat{\beta}_\omega\} = 0.01346$. As an estimate of the main effect of A we obtain the value $\hat{\beta}_1 = 0.0871$. In the following tables derived using R, we present the LSE's of all the 5 main effects and 10 first-order interactions. The S.E. values in Table 13.22 are the standard errors of the estimates and the t values are $t = \frac{\text{LSE}}{\text{SE}}$.

```
> Factors <- list(
    m=c(30, 60),
40
    s=c(0.005, 0.02),
    v0=c(0.002, 0.01),
    k=c(1000, 5000),
    t=c(290, 296))
```

```
> FacDesign <- fac.design(
    factor.names=Factors,
    randomize=TRUE,
    replications=5,
    repeat.only=TRUE)
> Levels <- data.frame(
    lapply(
        lapply(FacDesign, as.character),
        as.numeric),
    p0=90000,
    t0=340, stringsAsFactors=F)
> Ps <- pistonSimulation(m=Levels$m,
                        s=Levels$s,
                        v0=Levels$v0,
                        k=Levels$k,
                        p0=Levels$p0,
                        t=Levels$t,
                        t0=Levels$t0,
                        each=1,
                        seed=123)
> FacDesign <- add.response(
    design=FacDesign,
    response=Ps$seconds)
> summary(
    aov(Ps.seconds ~ (m+s+v0+k+t)^2,
            data=FacDesign))
```

	Df	Sum Sq	Mean Sq	F value	Pr(>F)	
m	1	1.524	1.524	27.049	6.69e-07	***
s	1	4.507	4.507	80.012	1.67e-15	***
v0	1	1.998	1.998	35.472	1.89e-08	***
k	1	3.495	3.495	62.048	7.45e-13	***
t	1	0.101	0.101	1.798	0.1820	
m:s	1	0.205	0.205	3.633	0.0587	.
m:v0	1	0.006	0.006	0.113	0.7378	
m:k	1	0.097	0.097	1.722	0.1915	
m:t	1	0.009	0.009	0.160	0.6898	
s:v0	1	0.151	0.151	2.685	0.1035	
s:k	1	1.042	1.042	18.489	3.13e-05	***
s:t	1	0.066	0.066	1.180	0.2791	
v0:k	1	0.097	0.097	1.719	0.1919	
v0:t	1	0.000	0.000	0.001	0.9779	
k:t	1	0.050	0.050	0.887	0.3478	
Residuals	144	8.112	0.056			

```
---
Signif. codes:
0 '***' 0.001 '**' 0.01 '*' 0.05 '.' 0.1 ' ' 1

> rm(Levels, Ps, Factors)
```

Table 13.22   *LSE of main effects and interactions*

Effect	LSE	S.E.	t
A	0.08781	0.01346	6.52*
B	−0.09856	0.01346	−7.32**
C	0.24862	0.01346	18.47**
D	−0.20144	0.01346	−14.97**
F	−0.02275	0.01346	−1.69
AB	0.00150	0.01346	0.11
AC	0.06169	0.01346	4.58*
AD	−0.02725	0.01346	−2.02
AF	−0.02031	0.01346	−1.51
BC	0.05781	0.01346	4.29*
BD	0.04850	0.01346	3.60*
BF	0.03919	0.01346	2.91*
CD	−0.10194	0.01346	−7.57**
CF	0.02063	0.01346	1.53
DF	0.05544	0.01346	4.12*

Values of t which are greater in magnitude than 2.6 are significant at $\alpha = 0.02$. If we wish, however, that all 15 tests have simultaneously a level of significance of $\alpha = 0.05$ we should use as critical value the Scheffé coefficient $\sqrt{32 \times F_{.95}[32,128]} = 7.01$, since all the LSE are contrasts of 32 means. In Table 13.22 we marked with one * the t values greater in magnitude

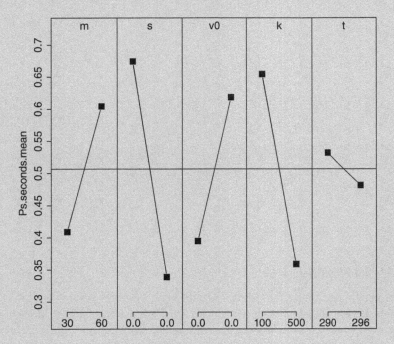

Figure 13.7   *Main effects plot.*

than 2.6, and with ** those greater than 7. The above estimates of the main effects and first-order interactions can be also obtained by running a multiple regression of $C11$ on 15 predictors, in $C6-C10$ and $C13-C22$.

When we execute this regression, we obtain $R^2 = 0.934$. The variance around the regression surface is $s^2_{y|x} = 0.02053$. This is significantly greater than $\hat{\sigma}^2/5 = 0.005795$. This means that there might be significant high-order interactions, which have not been estimated.

Figure 13.7 is a graphical display of the main effects of factors A, B, C, D and F. The left limit of a line shows the average response at a low level and the right limit that at a high level. Factors C and D seem to have the highest effect, as is shown by the t-values in Table 13.22. Figure 13.8 shows the two-way interactions of the various factors. Interaction $C * D$ is the most pronounced.

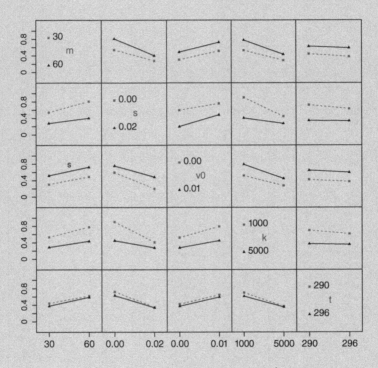

Figure 13.8   *Two-way interaction plots.*

13.7.5 3^m factorial designs

We discuss in the present section estimation and testing of model parameters, when the design is full factorial, of m factors each one at $p = 3$ levels. We assume that the levels are measured on a continuous scale, and are labelled Low, Medium, and High. We introduce the indices i_j $(j = 1, \ldots, m)$, which assume the values 0, 1, 2 for the Low, Medium, and High levels, correspondingly, of each factor. Thus, we have 3^m treatment combinations, represented by vectors of indices $(i_1, i_2, \ldots, i_m)$. The index v of the **standard order** of treatment combination is

$$v = \sum_{j=1}^{m} i_j 3^{j-1}. \tag{13.7.42}$$

This index ranges from 0 to $3^m - 1$. Let $\bar{Y}_v$ denote the yield of n replicas of the vth treatment combination, $n \geq 1$.

Since we obtain the yield at three levels of each factor we can, in addition to the linear effects estimate also the quadratic effects of each factor. For example, if we have $m = 2$ factors, we can use a multiple regression method to fit the model

$$Y = \beta_0 + \beta_1 x_1 + \beta_2 x_1^2 + \beta_3 x_2 + \beta_4 x_1 x_2$$
$$+ \beta_5 x_1^2 x_2 + \beta_6 x_2^2 + \beta_7 x_1 x_2^2 + \beta_8 x_1^2 x_2^2 + e. \tag{13.7.43}$$

This is a quadratic model in two variables. β_1 and β_3 represent the linear effects of x_1 and x_2. β_2 and β_6 represent the quadratic effects of x_1 and x_2. The other coefficients represent interaction effects. β_4 represents the linear $\times$ linear interaction, β_5 represents the quadratic $\times$ linear interaction, etc. We have two main effects for each factor (linear and quadratic) and 4 interaction effects.

Generally, if there are m factors we have, in addition to β_0, $2m$ parameters for main effects (linear and quadratic) $2^2 \binom{m}{2}$ parameters for interactions between 2 factors, $2^3 \binom{m}{3}$ interactions between 3 factors, etc. Generally, we have 3^m parameters, where

$$3^m = \sum_{j=0}^{m} 2^j \binom{m}{j}.$$

As in the case of 2^m models, each parameter in a 3^m model is represented by a vector of m indices $(\lambda_1, \lambda_2, \ldots, \lambda_m)$ where $\lambda_j = 0, 1, 2$. Thus, for example the vector $(0, 0, \ldots, 0)$ represent the grand mean $\mu = \gamma_0$. A vector $(0, \ldots, 0, 1, 0, \ldots, 0)$ with 1 at the ith component represents the linear effect of the ith factor. Similarly, $(0, 0, \ldots, 0, 2, 0, \ldots, 0)$ represents the quadratic effect of the ith factor. Two indices equal to 1 and all the rest zero, represent the linear $\times$ linear interaction of the ith and jth factor, etc. The standard order of the parameters is

$$\omega = \sum_{j=1}^{m} \lambda_j 3^{j-1}, \quad \omega = 0, \ldots, 3^m - 1.$$

If m is not too large, it is also customary to label the factors by the letters $A, B, C, \ldots$ and the parameters by $A^{\lambda_1} B^{\lambda_2} C^{\lambda_3} \cdots$. In this notation, a letter to the zero power is omitted. In Table 13.23, we list the parameters of a 3^3 system.

It is simple to transform the x-values of each factor to

$$X_j = \begin{cases} -1, & \text{if } i_j = 0 \\ 0, & \text{if } i_j = 1 \\ 1, & \text{if } i_j = 2. \end{cases}$$

Table 13.23   *The main effects and interactions of a 3^3 factorial*

ω	Parameter	Indices	ω	Parameter	Indices
0	Mean	(0,0,0)	15	B^2C	(0,2,1)
1	A	(1,0,0)	16	AB^2C	(1,2,1)
2	A^2	(2,0,0)	17	A^2B^2C	(2,2,1)
3	B	(0,1,0)	18	C^2	(0,0,2)
4	AB	(1,1,0)	19	AC^2	(1,0,2)
5	A^2B	(2,1,0)	20	A^2C^2	(2,0,2)
6	B^2	(0,2,0)	21	BC^2	(0,1,2)
7	AB^2	(1,2,0)	22	ABC^2	(1,1,2)
8	A^2B^2	(2,2,0)	23	A^2BC^2	(2,1,2)
9	C	(0,0,1)	24	B^2C^2	(0,2,2)
10	AC	(1,0,1)	25	AB^2C^2	(1,2,2)
11	A^2C	(2,0,1)	26	$A^2B^2C^2$	(2,2,2)
12	BC	(0,1,1)			
13	ABC	(1,1,1)			
14	A^2BC	(2,1,1)			

However, the matrix of coefficients X that is obtained, when we have quadratic and interaction parameters, is not orthogonal. This requires then the use of the computer to obtain the least squares estimators, with the usual multiple regression program. Another approach is to redefine the effects so that the statistical model will be linear with a matrix having coefficients obtained by the method of orthogonal polynomials (see Draper and Smith (1981), pp. 166). Thus, consider the model

$$\mathbf{Y}^{(3^m)} = \Psi_{(3^m)}\boldsymbol{\gamma}^{(3^m)} + \mathbf{e}^{(3^m)}, \tag{13.7.44}$$

where

$$\mathbf{Y}^{(3^m)} = (\bar{Y}_0, \ldots, \bar{Y}_{3^m-1})',$$

and $\mathbf{e}^{(3^m)} = (e_0, \ldots, e_{3^m-1})'$ is a vector of random variables with

$$E\{e_v\} = 0, \quad V\{e_v\} = \sigma^2 \quad \text{all} \quad v = 0, \ldots, 3^m - 1.$$

Moreover

$$\Psi_{(3)} = \begin{bmatrix} 1 & -1 & 1 \\ 1 & 0 & -2 \\ 1 & 1 & 1, \end{bmatrix} \tag{13.7.45}$$

and for $m \geq 2$,

$$\Psi_{(3^m)} = \begin{bmatrix} \Psi_{(3^{m-1})} & -\Psi_{(3^{m-1})} & \Psi_{(3^{m-1})} \\ \Psi_{(3^{m-1})} & 0 & -2\Psi_{(3^{m-1})} \\ \Psi_{(3^{m-1})} & \Psi_{(3^{m-1})} & \Psi_{(3^{m-1})} \end{bmatrix}. \tag{13.7.46}$$

The matrices $\Psi_{(3^m)}$ have orthogonal column vectors and

$$(\Psi_{(3^m)})'(\Psi_{(3^m)}) = \Delta_{(3^m)}, \tag{13.7.47}$$

where $\Delta_{(3^m)}$ is a diagonal matrix whose diagonal elements are equal to the sum of squares of the elements in the corresponding column of $\Psi_{(3^m)}$. For example, for $m = 1$,

$$\Delta_{(3)} = \begin{pmatrix} 3 & 0 & 0 \\ 0 & 2 & 0 \\ 0 & 0 & 6 \end{pmatrix}.$$

For $m = 2$ we obtain

$$\Psi_{(9)} = \begin{bmatrix} 1 & -1 & 1 & -1 & 1 & -1 & 1 & -1 & 1 \\ 1 & 0 & -2 & -1 & 0 & 2 & 1 & 0 & -2 \\ 1 & 1 & 1 & -1 & -1 & -1 & 1 & 1 & 1 \\ 1 & -1 & 1 & 0 & 0 & 0 & -2 & 2 & -2 \\ 1 & 0 & -2 & 0 & 0 & 0 & -2 & 0 & 4 \\ 1 & 1 & 1 & 0 & 0 & 0 & -2 & -2 & -2 \\ 1 & -1 & 1 & 1 & -1 & 1 & 1 & -1 & 1 \\ 1 & 0 & -2 & 1 & 0 & -2 & 1 & 0 & -2 \\ 1 & 1 & 1 & 1 & 1 & 1 & 1 & 1 & 1 \end{bmatrix}$$

and

$$\Delta_{(9)} = \begin{bmatrix} 9 & & & & & & & & \\ & 6 & & & & & & & \\ & & 18 & & & 0 & & & \\ & & & 6 & & & & & \\ & & & & 4 & & & & \\ & & & & & 12 & & & \\ & & 0 & & & & 18 & & \\ & & & & & & & 12 & \\ & & & & & & & & 36 \end{bmatrix}.$$

Thus, the LSE of $\gamma^{(3^m)}$ is

$$\hat{\gamma}^{(3^m)} = \Delta_{(3^m)}^{-1}(\Psi_{(3^m)})' \mathbf{Y}^{(3^m)}. \tag{13.7.48}$$

These LSE are best linear unbiased estimators and

$$V\{\hat{\gamma}_\omega\} = \frac{\sigma^2}{n \sum_{i=1}^{3^m} (\Psi_{i,\omega+1}^{(3^m)})^2}. \tag{13.7.49}$$

If the number of replicas, n, is greater than 1 then σ^2 can be estimated by

$$\hat{\sigma}^2 = \frac{1}{3^m(n-1)} \sum_{v=0}^{3^m-1} \sum_{l=1}^{n} (Y_{vl} - \bar{Y}_v)^2. \tag{13.7.50}$$

If $n = 1$, we can estimate σ^2 if it is known a priori that some parameters γ_ω are zero. Let Λ_0 be the set of all parameters which can be assumed to be negligible. Let K_0 be the number of elements of Λ_0. If $\omega \in \Lambda_0$ then $\hat{\gamma}_\omega^2 \left(\sum_{j=1}^{3^m} (\Psi_{i,\omega+1}^{(3^m)})^2 \right)$ is distributed like $\sigma^2 \chi^2[1]$. Therefore, an unbiased estimator of σ^2 is

$$\hat{\hat{\sigma}}^2 = \frac{1}{k_0} \sum_{\omega \in \Lambda_0} \hat{\gamma}_\omega^2 \left(\sum_{j=1}^{3^m} (\Psi_{j,\omega+1}^{(3^m)})^2 \right). \tag{13.7.51}$$

Example 13.9.   Oikawa and Oka (1987) reported the results of a 3^3 experiment to investigate the effects of three factors A, B, C on the stress levels of a membrane Y. The data is given in file **STRESS.csv**. The first three columns of the data file provide the levels of the three factors, and column 4 presents the stress values. To analyze this data with R we apply:

```
> data(STRESS)
> summary(
    lm(stress ~ (A+B+C+I(A^2)+I(B^2)+I(C^2))^3,
      data=STRESS))

Call:
lm.default(formula = stress ~ (A + B + C + I(A^2) + I(B^2) +
    I(C^2))^3, data = STRESS)

Residuals:
ALL 27 residuals are 0: no residual degrees of freedom!

Coefficients: (15 not defined because of singularities)
```

	Estimate	Std. Error	t value
(Intercept)	191.8000	NA	NA
A	38.5000	NA	NA
B	-46.5000	NA	NA
C	63.0000	NA	NA
I(A^2)	0.2000	NA	NA
I(B^2)	14.0000	NA	NA
I(C^2)	-27.3000	NA	NA
A:B	-32.7500	NA	NA
A:C	26.4500	NA	NA
A:I(A^2)	NA	NA	NA
A:I(B^2)	13.1500	NA	NA
A:I(C^2)	-0.5000	NA	NA
B:C	-44.0000	NA	NA
B:I(A^2)	0.7000	NA	NA
B:I(B^2)	NA	NA	NA
B:I(C^2)	21.6000	NA	NA
C:I(A^2)	8.2000	NA	NA
C:I(B^2)	6.0000	NA	NA
C:I(C^2)	NA	NA	NA
I(A^2):I(B^2)	-3.6000	NA	NA
I(A^2):I(C^2)	-5.2500	NA	NA
I(B^2):I(C^2)	-4.0000	NA	NA
A:B:C	37.7375	NA	NA
A:B:I(A^2)	NA	NA	NA
A:B:I(B^2)	NA	NA	NA
A:B:I(C^2)	-16.7875	NA	NA
A:C:I(A^2)	NA	NA	NA

A:C:I(B^2)	-3.2625	NA	NA
A:C:I(C^2)	NA	NA	NA
A:I(A^2):I(B^2)	NA	NA	NA
A:I(A^2):I(C^2)	NA	NA	NA
A:I(B^2):I(C^2)	-0.5875	NA	NA
B:C:I(A^2)	-10.7875	NA	NA
B:C:I(B^2)	NA	NA	NA
B:C:I(C^2)	NA	NA	NA
B:I(A^2):I(B^2)	NA	NA	NA
B:I(A^2):I(C^2)	4.4875	NA	NA
B:I(B^2):I(C^2)	NA	NA	NA
C:I(A^2):I(B^2)	-0.5875	NA	NA
C:I(A^2):I(C^2)	NA	NA	NA
C:I(B^2):I(C^2)	NA	NA	NA
I(A^2):I(B^2):I(C^2)	1.5875	NA	NA

	Pr(>|t|)
(Intercept)	NA
A	NA
B	NA
C	NA
I(A^2)	NA
I(B^2)	NA
I(C^2)	NA
A:B	NA
A:C	NA
A:I(A^2)	NA
A:I(B^2)	NA
A:I(C^2)	NA
B:C	NA
B:I(A^2)	NA
B:I(B^2)	NA
B:I(C^2)	NA
C:I(A^2)	NA
C:I(B^2)	NA
C:I(C^2)	NA
I(A^2):I(B^2)	NA
I(A^2):I(C^2)	NA
I(B^2):I(C^2)	NA
A:B:C	NA
A:B:I(A^2)	NA
A:B:I(B^2)	NA
A:B:I(C^2)	NA
A:C:I(A^2)	NA
A:C:I(B^2)	NA
A:C:I(C^2)	NA

```
A:I(A^2):I(B^2)              NA
A:I(A^2):I(C^2)              NA
A:I(B^2):I(C^2)              NA
B:C:I(A^2)                   NA
B:C:I(B^2)                   NA
B:C:I(C^2)                   NA
B:I(A^2):I(B^2)              NA
B:I(A^2):I(C^2)              NA
B:I(B^2):I(C^2)              NA
C:I(A^2):I(B^2)              NA
C:I(A^2):I(C^2)              NA
C:I(B^2):I(C^2)              NA
I(A^2):I(B^2):I(C^2)         NA


Residual standard error: NaN on 0 degrees of freedom
Multiple R-squared:         1,          Adjusted R-squared:    NaN
F-statistic:   NaN on 26 and 0 DF,  p-value: NA

> summary(
    aov(stress ~ (A+B+C)^3 +I(A^2)+I(B^2)+I(C^2),
        data=STRESS))

            Df Sum Sq Mean Sq F value    Pr(>F)
A            1  36315   36315 378.470 1.47e-12 ***
B            1  32504   32504 338.751 3.43e-12 ***
C            1  12944   12944 134.904 3.30e-09 ***
I(A^2)       1    183     183   1.911 0.185877
I(B^2)       1   2322    2322  24.199 0.000154 ***
I(C^2)       1   4536    4536  47.270 3.73e-06 ***
A:B          1   3290    3290  34.289 2.44e-05 ***
A:C          1   6138    6138  63.971 5.56e-07 ***
B:C          1    183     183   1.910 0.185919
A:B:C        1     32      32   0.338 0.569268
Residuals   16   1535      96
---
Signif. codes:
0 '***' 0.001 '**' 0.01 '*' 0.05 '.' 0.1 ' ' 1
```

The values of the LSE and their standard errors are given in Table 13.24. The formula for the variance of an LSE is

$$V\{\hat{\gamma}_\omega\} = \sigma^2/n \left(\sum_{i=1}^{3^3} (\Psi^{(3^3)}_{i,\omega+1})^2 \right),\tag{13.7.52}$$

where $\psi^{(3^3)}_{i,j}$ is the coefficient of $\Psi_{(3^3)}$ in the ith row and jth column.

Table 13.24 *The LSE of the parameters of the 3^3 system*

Parameter	LSE	S.E.	Significance
A	44.917	2.309	
A^2	−1.843	1.333	n.s.
B	−42.494	2.309	
AB	−16.558	2.828	
A^2B	−1.897	1.633	n.s.
B^2	6.557	1.333	
AB^2	1.942	1.633	n.s.
A^2B^2	−0.171	0.943	n.s.
C	26.817	2.309	
AC	22.617	2.828	
A^2C	0.067	1.633	n.s.
BC	−3.908	2.828	n.s.
ABC	2.013	3.463	n.s.
A^2BC	1.121	1.999	n.s.
B^2C	−0.708	1.633	n.s.
AB^2C	0.246	1.099	n.s.
A^2B^2C	0.287	1.154	n.s.
C^2	−9.165	1.333	
AC^2	−4.833	1.633	
A^2C^2	0.209	0.943	n.s.
BC^2	2.803	1.633	n.s.
ABC^2	−0.879	1.999	n.s.
A^2BC^2	0.851	1.154	n.s.
B^2C^2	−0.216	0.943	n.s.
AB^2C^2	0.287	1.154	n.s.
$A^2B^2C^2$	0.059	0.666	n.s.

Suppose that from technological considerations we decide that all interaction parameters involving quadratic components are negligible (zero). In this case we can estimate σ^2 by $\hat{\sigma}^2$. In the present example, the set Λ_0 contains 16 parameters, i.e.,

$$\Lambda_0 = \{A^2B, AB^2, A^2B^2, A^2C, A^2BC, B^2C, AB^2C, A^2B^2C, AC^2,$$
$$A^2C^2, BC^2, ABC^2, A^2BC, B^2C^2, AB^2C^2, A^2B^2C^2\}.$$

Thus, $K_0 = 16$ and the estimator $\hat{\sigma}^2$ has 16 degrees of freedom. The estimate of σ^2 is $\hat{\sigma}^2 = 95.95$. The estimates of the standard errors (S.E.) in Table 13.24 use this estimate. All the nonsignificant parameters are denoted by n.s. In Figures 13.9 and 13.10, we present the main effects and interaction plots.

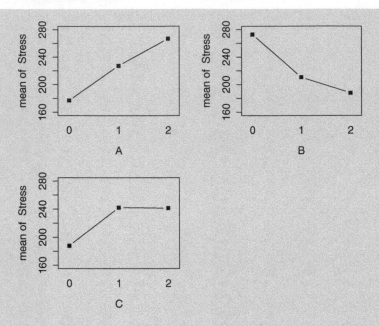

Figure 13.9   *Main effects plot for 3^3 design.*

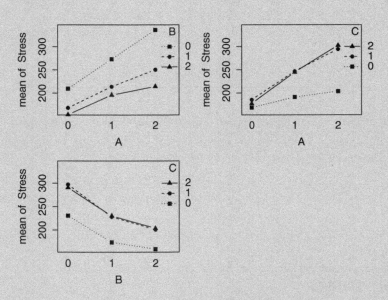

Figure 13.10   *Interaction plots for 3^3 design.*

13.8   Blocking and fractional replications of 2^m factorial designs

Full factorial experiments with large number of factors might be impractical. For example, if there are $m = 12$ factors, even at $p = 2$ levels, the total number of treatment combinations is $2^{12} = 4096$. This size of an experiment is generally not necessary, because most of the high-order interactions might be negligible and there is no need to estimate 4096 parameters. If only main effects and first-order interactions are considered, a priori of importance, while all the rest are believed to be negligible, we have to estimate and test only $1 + 12 + \binom{12}{2} = 79$ parameters. A fraction of the experiment, of size $2^7 = 128$ would be sufficient. Such a fraction can be even replicated several times. The question is, how do we choose the fraction of the full factorial in such a way that desirable properties of orthogonality, equal variances of estimators, etc. will be kept, and the parameters of interest will be estimable unbiasedly.

The problem of fractioning the full factorial experiment arises also when the full factorial cannot be performed in one block, but several blocks are required to accommodate all the treatment conditions. For example, a 2^5 experiment is designed, but only $8 = 2^3$ treatment combinations can be performed in any given block (day, machine, etc.) We have to design the fractions that will be assigned to each block in such a way that, if there are significant differences between the blocks, the block effects will not confound or obscure parameters of interest. We start with a simple illustration of the fractionization procedure, and the properties of the ensuing estimators.

Consider 3 factors A, B, and C at 2 levels. We wish to partition the $2^3 = 8$ treatment combinations to two fractions of size $2^2 = 4$. Let $\lambda_i = 0, 1$ ($i = 1, 2, 3$) and let $A^{\lambda_1} B^{\lambda_2} C^{\lambda_3}$ represent the eight parameters. One way of representing the treatment combinations, when the number of factors is not large, is by using low case letters $a, b, c, \ldots$. The letter a indicates that factor A is at the High level ($i_1 = 1$), similarly about other factors. The absence of a letter indicates that the corresponding factor is at Low level. The symbol (1) indicates that all levels are Low. Thus, the treatment combinations and the associated coefficients $c_{ij}^{(2^3)}$ are shown in Table 13.25.

Suppose now that the treatment combinations should be partitioned to two fractional replications (blocks) of size 4. We have to choose a parameter, called a **defining parameter**, according to which the partition will be done. This defining parameter is in a sense sacrificed. Since its effects will be either confounded with the block effects, or unestimable if only one block of trials is performed. Thus, let us choose the parameter ABC, as a defining parameter. Partition the treatment combinations to two blocks, according to the signs of the coefficients corresponding

Table 13.25   *A 2^2 factorial*

| Treatments | Main effects | | | Defining parameter |
	A	B	C	ABC
(1)	−1	−1	−1	−1
a	1	−1	−1	1
b	−1	1	−1	1
ab	1	1	−1	−1
c	−1	−1	1	1
ac	1	−1	1	−1
bc	−1	1	1	−1
abc	1	1	1	1

to ABC. These are the products of the coefficients in the A, B, and C columns. Thus, two blocks are obtained

$$B_- = \{(1), ab, ac, bc\},$$
$$B_+ = \{a, b, c, abc\}.$$

If 2^m treatment combinations are partitioned to $2^k = 2$ blocks, we say that the **degree of fractionation** is $k = 1$, the fractional replication is of size 2^{m-k}, and the design is $1/2^k$ fraction of a full factorial. If, for example, $m = 5$ factors and we wish to partition to 4 blocks of 8, the degree of fractionization is $k = 2$. Select $k = 2$ parameters to serve as defining parameters, e.g., ACE and BDE, and partition the treatment combinations according to the signs ± 1 of the coefficients in the ACE and BDE columns. This becomes very cumbersome if m and k are large. Function fac.design performs this partitioning and prints into a file the block which is requested. We will return to this later. It is interesting to check now what are the properties of estimators in the 2^{3-1} fractional replication, if only the block B_- was performed. The defining parameter was ABC.

Let $Y(1)$ be the response of treatment combination (1), this is Y_0 in the standard order notation, let $Y(a)$ be the response of "a," etc. The results of performing B_-, with the associated coefficients of parameters of interest can be presented in the following manner (Table 13.26).

We see that the six columns of coefficients are orthogonal to each other, and each column has $2 -1$'s and $2 +1$'s. The L.S.E. of the above parameters are orthogonal contrasts, given by

$$\bar{A} = \frac{1}{4}(-Y(1) + Y(ab) + Y(ac) - Y(bc)),$$

$$\hat{B} = \frac{1}{4}(-Y(1) + Y(ab) - Y(ac) + Y(bc)),$$

etc. The variances of all these estimators, when $n = 1$, are equal to $\frac{\sigma^2}{4}$. However, the estimators might be **biased**. The expected value of the first estimator is

$$E\{\bar{A}\} = \frac{1}{4}(-E\{Y(1)\} + E\{Y(ab)\} + E\{Y(ac)\} - E\{Y(bc)\}).$$

Now,

$$E\{Y(1)\} = \mu - A - B - C + AB + AC + BC - ABC,$$

$$E\{Y(ab)\} = \mu + A + B - C + AB - AC - BC - ABC,$$

$$E\{Y(ac)\} = \mu + A - B + C - AB + AC - BC - ABC,$$

and

$$E\{Y(bc)\} = \mu - A + B + C - AB - AC + BC - ABC.$$

Table 13.26 *Coefficients and response for several treatment combinations (t.c.)*

t.c.	A	B	C	AB	AC	BC	Y
(1)	−1	−1	−1	1	1	1	Y(1)
ab	1	1	−1	1	−1	−1	Y(ab)
ac	1	−1	1	−1	1	−1	Y(ac)
bc	−1	1	1	−1	−1	1	Y(bc)

Collecting all these terms, the result is

$$E\{\overline{A}\} = A - BC.$$

Similarly, one can show that

$$E\{\hat{B}\} = B - AC,$$

$$E\{\hat{C}\} = C - AB,$$

$$E\{\widehat{AB}\} = AB - C,$$

etc. The LSE of all the parameters are biased, unless $AB = AC = BC = 0$. The bias terms are called **aliases**. The aliases are obtained by multiplying the parameter of interest by the defining parameter, when any letter raised to the power 2 is eliminated, e.g.,

$$A \otimes ABC = A^2 BC = BC.$$

The sign of the alias is the sign of the block. Since we have used the block B_-, all the aliases appear above with a negative sign. The general rules for finding the aliases in 2^{m-k} designs is as follows.

To obtain a 2^{m-k} fractional replication, one needs k defining parameters. The multiplication operation of parameters was illustrated above. The k defining parameters should be **independent**, in the sense that none can be obtained as a product of the other ones. Such independent defining parameters are called **generators**. For example, to choose 4 defining parameters, when the factors are A, B, C, D, E, F, G, H, choose first two parameters, like $ABCH$ and $ABEFG$. The product of these two is $CEFGH$. In the next step choose, for the third defining parameter, any one which is different than $\{ABCH, ABEFG, CEFGH\}$. Suppose one chooses $BDEFH$. The three independent parameters $ABCH$, $ABEFG$, and $BDEFH$ generate a subgroup of eight parameters, including the mean μ. These are:

μ	$BDEFH$
$ABCH$	$ACDEF$
$ABEFG$	$ADGH$
$CEFGH$	$BCDG$

Finally, to choose a 4th independent defining parameter, one can choose any parameter which is not among the eight listed above. Suppose that the parameter $BCEFH$ is chosen. Now we obtain a subgroup of $2^4 = 16$ defining parameter, by adding to the eight listed above their products with $BCEFH$. Thus, this subgroup is

μ	$BCEFH$
$ABCH$	AEF
$ABEFG$	$ACGH$
$CEFGH$	BG
$BDEFH$	CD
$ACDEF$	$ABDH$
$ADGH$	$ABCDEFG$
$BCDG$	$DEFGH$

Notice that this subgroup includes, excluding the mean, two first-order interactions CD and BG. This shows that the choice of defining parameters was not a good one. Since the aliases

Table 13.27   *The aliases to the main effects in a 2^{8-4} design, the generators are ABCH, ABEFG, BDEFH, and BCEFH*

Main Effects	Aliases
A	BCH, BEFG, ACEFGH, ABDEFH, CDEF, DGH, ABCDG, ABCEFH, **EF**, CGH, ABG, ACD, BDH, BCDEFG, ADEFGH
B	ACH, AEFG, BCEFGH, DEFH, ABCDEF, ABDGH, CDG, CEFH, ABEF, ABCGH, **G**, BCD, ADH, ACDEFG, BDEFGH
C	ABH, ABCEFG, EFGH, BCDEFH, ADEF, ACDGH, BDG, BEFH, ACEF, AGH, BCG, **D**, ABCDH, ABDEFG, CDEFGH
D	ABCDH, ABDEFG, CDEFGH, BEFH, ACEF, AGH, BCG, BCDEFH, ADEF, ACDGH, BDG, **C**, ABH, ABCEFG, EFGH
E	ABCEH, ABFG, CFGH, BDFH, ACDF, ADEGH, BCDEG, BCFH, **AF**, ACEGH, BEG, CDE, ABDEH, ABCDFG, DFGH
F	ABCFH, ABEG, CEGH, BDEH, ACDE, ADFGH, BCDFG, BCEH, **AE**, ACFGH, BFG, CDF, ABDFH, ABCDEG, DEGH
G	ABCGH, ABEF, CEFH, BDEFGH, ACDEFG, ADH, BCD, BCEFGH, AEFG, ACH, **B**, CDG, ABDGH, ABCDEF, DEFH
H	ABC, ABEFGH, CEFG, BDEF, ACDEFH, ADG, BCDGH, BCEF, AEFH, ACG, BGH, CDH, ABD, ABCDEFGH, DEFG

which will be created by these defining parameters will include main effects and other low-order interactions.

Given a subgroup of defining parameters, the aliases of a given parameter are obtained by **multiplying** the parameter by the defining parameters. In Table 13.27, we list the aliases of the eight main effects, with respect to the above subgroup of 2^4 defining parameters.

We see in this table that most of the aliases to the main effects are high-order interactions (that are generally negligible). However, among the aliases to A there is *EF*. Among the aliases to B, there is the main effect G. Among the aliases to C there is D, etc. This design is not good since it may yield strongly biased estimators. The **resolution of a 2^{m-k} design is the length of the smallest word (excluding μ) in the subgroup of defining parameters**. For example, if in a 2^{8-4} design we use the following four generators $BCDE$, $ACDF$, $ABCG$, and $ABDH$, we obtain the 16 defining parameters $\{\mu, BCDE, ACDF, ABEF, ABCG, ADEG, BDFG, CEFG, ABDH, ACEH, BDFH,$ $DEFH, CDGH, BEGH, AFGH, ABCDEFGH\}$. The length of the smallest word, excluding μ, among these defining parameters is four. Thus, the present 2^{8-4} design is a **resolution IV** design. In this design, all aliases of main effects are second-order interactions or higher (words of length greater or equal to three). Aliases to first-order interactions are interactions of first order or higher. The present design is obviously better, in terms of resolution, than the previous one (which is of resolution II). We should always try to get resolution IV or higher. If the degree of fractionation is too high there may not exist resolution IV designs. For example, in 2^{6-3} and 2^{7-4}, 2^{9-5}, 2^{10-6} and 2^{11-7} we have only resolution III designs. One way to reduce the bias is to choose several fractions at random. For example in a 2^{11-7} we have $2^7 = 128$ blocks of size $2^4 = 16$. If we execute only one block, the best we can have is a resolution III. In this case, some main effects are biased (confounded) with some first-order interactions. If one chooses n blocks at **random**

(RSWOR) out of the 128 possible ones, and compute the average estimate of the effects, the bias is reduced to zero, but the variance of the estimators is increased.

To illustrate this, suppose that we have a 2^{6-2} design with generators $ABCE$ and $BCDF$. This will yield a resolution IV design. There are four blocks and the corresponding bias terms of the LSE of A are

block	
0	$-BCE - ABCDF + DEF$
1	$BCE - ABCDF - DEF$
2	$-BCE + ABCDF - DEF$
3	$BCE + ABCDF + DEF$

If we choose one block at random, the expected bias is the average of the four terms above, which is zero. The total variance of $\overline{A}$ is $\frac{\sigma^2}{16}$ + Variance of conditional bias = $\frac{\sigma^2}{16}$ + [$(BCE)^2$ + $(ABCDF)^2 + (DEF)^2]/4$.

Example 13.10. In the present example we illustrate the construction of fractional replications. The case that is illustrated is a 2^{8-4} design. Here we can construct 16 fractions, each one of size 16. As discussed before, four generating parameters should be specified. Let these be $BCDE, ACDF, ABCG, ABDH$. These parameters generate resolution 4 design where the degree of fractionation, $k=4$. The blocks can be indexed $0, 1, \ldots, 15$. Each index is determined by the signs of the four generators, which determine the block. Thus, the signs $(-1, -1, 1, 1)$ correspond to $(0, 0, 1, 1)$ which yields the index $\sum_{j=1}^{4} i_j 2^{j-1} = 12$. The index of generator 1 ($BCDE = A^0 B^1 C^1 D^1 E^1 F^0 G^0 H^0$) is $0, 1, 1, 1, 1, 0, 0, 0$, for generator 2: $1, 0, 1, 1, 0, 1, 0, 0$; for generator 3: $1, 1, 1, 0, 0, 0, 1, 0$ and for generator 4: $1, 1, 0, 1, 0, 0, 0, 1$.
In Table 13.28, two blocks derived with R are printed.

```
> Gen <- matrix(c(
    0,1,1,1,1,0,0,0,
    1,0,1,1,0,1,0,0),
                nrow=2,
                byrow=TRUE)
> head(
    fac.design(nlevels=2,
                nfactors=8,
                blocks=4,
                block.gen=Gen))

  Blocks A B C D E F G H
1        1 2 2 1 2 1 1 2 1
2        1 1 1 2 2 1 1 2 2
3        1 2 2 1 1 2 2 1 1
4        1 1 1 1 1 1 1 1 2
5        1 2 1 2 1 2 1 1 2
6        1 1 2 1 1 2 1 1 2

> rm(Gen)
```

Table 13.28   Blocks of 2^{8-4} designs

Block				0				Block				1			
1	1	1	1	1	1	1	1	1	2	2	2	1	1	1	1
1	2	2	2	2	1	1	1	1	1	1	1	2	1	1	1
2	1	2	2	1	2	1	1	2	2	1	1	1	2	1	1
2	2	1	1	2	2	1	1	2	1	2	2	2	2	1	1
2	2	2	1	1	1	2	1	2	1	1	2	1	1	2	1
2	1	1	2	2	1	2	1	2	2	2	1	2	1	2	1
1	2	1	2	1	2	2	1	1	1	2	1	1	2	2	1
1	1	2	1	2	2	2	1	1	2	1	2	2	2	2	1
2	2	1	2	1	1	1	2	2	1	2	1	1	1	1	2
2	1	2	1	2	1	1	2	2	2	1	2	2	1	1	2
1	2	2	1	1	2	1	2	1	1	1	2	1	2	1	2
1	1	1	2	2	2	1	2	1	2	2	1	2	2	1	2
1	1	2	2	1	1	2	2	1	2	1	1	1	1	2	2
1	2	1	1	2	1	2	2	1	1	2	2	2	1	2	2
2	1	1	1	1	2	2	2	2	2	2	2	1	2	2	2
2	2	2	2	2	2	2	2	2	1	1	1	2	2	2	2

In Box et al. (1978, pp. 410) there are recommended generators for 2^{m-k} designs. Some of these generators are given in Table 13.29.

The LSE of the parameters is performed by writing first the columns of coefficients $c_{i,j} = \pm 1$ corresponding to the design, multiplying the coefficients by the Y values, and dividing by 2^{m-k}.

Table 13.29   Some generators for 2^{m-k} designs

		m		
k	5	6	7	8
1	ABCDE	ABCDEF	ABCDEFGH	ABCDEFGH
2	ABD	ABCE	ABCDF	ABCDG
	ACE	BCDF	ABDEG	ABEFH
3		ABD	ABCE	ABCF
		ACD	BCDF	ABDG
		BCF	ACDG	BCDEH
4			ABD	BCDE
			ACE	ACDF
			BCF	ABCG
			ABCG	ABDH

13.9   Exploration of response surfaces

The functional relationship between the yield variable Y and the experimental variables $(x_1, \ldots, x_k)$ is modeled as

$$Y = f(x_1, \ldots, x_k) + e,$$

where e is a random variable with zero mean and a finite variance, σ^2. The set of points $\{f(x_1, \ldots, x_k), x_i \in D_i, i = 1, \ldots, k\}$, where $(D_1, \ldots, D_k)$ is the experimental domain of the x-variables, is called a **response surface**. Two types of response surfaces were discussed before, the **linear**

$$f(x_1, \ldots, x_k) = \beta_0 + \sum_{i=1}^{k} \beta_i x_i \qquad (13.9.1)$$

and the **quadratic**

$$f(x_1, \ldots, x_k) = \beta_0 + \sum_{i=1}^{k} \beta_i x_i + \sum_{i=1}^{k} \beta_{ii} x_i^2 + \sum \sum_{i \neq j} \beta_{ij} x_i x_j. \qquad (13.9.2)$$

Response surfaces may be of complicated functional form. We assume here that in local domains of interest, they can be approximated by linear or quadratic models.

Researchers are interested in studying, or exploring, the nature of response surfaces, in certain domains of interest, for the purpose of predicting future yield, and in particular for optimizing a process, by choosing the x-values to maximize (or minimize) the expected yield (or the expected loss). In the present section, we present special designs for the exploration of quadratic surfaces, and for the determination of optimal domains (conditions). Designs for quadratic models are called **second-order designs**. We start with the theory of second-order designs, and conclude with the optimization process.

13.9.1   Second-order designs

Second-order designs are constructed in order to estimate the parameters of the quadratic response function

$$E\{Y\} = \beta_0 + \sum_{i=1}^{k} \beta_i x_i + \sum_{i=1}^{k} \beta_{ii} x_i^2 + \sum_{i=1}^{k-1} \sum_{j=i+1}^{k} \beta_{ij} x_i x_j. \qquad (13.9.3)$$

In this case, the number of regression coefficients is $p = 1 + 2k + \binom{k}{2}$. We will arrange the vector β in the form

$$\beta' = (\beta_0, \beta_{11}, \ldots, \beta_{kk}, \beta_1, \ldots, \beta_k, \beta_{12}, \ldots, \beta_{1k}, \beta_{23}, \ldots, \beta_{2k}, \ldots, \beta_{n-1,k}).$$

Let N be the number of x-points. The design matrix takes the form

$$(X) = \begin{bmatrix} 1 & x_{11}^2 & \cdots & x_{1k}^2 & x_{11} & \cdots & x_{1k} & x_{11}x_{12} & \cdots & x_{1,k-1}x_{1,k} \\ 1 & x_{21}^2 & & x_{2k}^2 & x_{21} & & x_{2k} & x_{21}x_{22} & & \\ 1 & x_{31}^2 & & x_{3k}^2 & x_{31} & & x_{3k} & & & \\ \vdots & \vdots & & \vdots & \vdots & & \vdots & \vdots & & \vdots \\ 1 & x_{N1}^2 & & x_{Nk}^2 & x_{N1} & \cdots & x_{Nk} & x_{N1}x_{N2} & \cdots & x_{N,k-1}x_{N,k} \end{bmatrix}$$

Impose on the x-values the conditions:

(i) $\sum_{j=1}^{N} x_{ji} = 0, i = 1, \ldots, k$

(ii) $\sum_{j=1}^{N} x_{ji}^3 = 0, i = 1, \ldots, k$

(iii) $\sum_{j=1}^{N} x_{ji}^2 x_{jl} = 0, i \neq l$

(iv)

$$\sum_{j=1}^{N} x_{ji}^2 = b, i = 1, \ldots, k$$

(13.9.4)

(v) $\sum_{j=1}^{N} x_{ji}^2 x_{jl}^2 = c, i \neq l$

(vi) $\sum_{j=1}^{N} x_{ji}^4 = c + d.$

The matrix $(S) = (X)'(X)$ can be written in the form

$$(S) = \begin{bmatrix} (U) & 0 \\ 0 & (B) \end{bmatrix},$$

(13.9.5)

where (U) is the $(k + 1) \times (k + 1)$ matrix

$$(U) = \begin{bmatrix} N & b\mathbf{1}'_k \\ b\mathbf{1}_k & d\mathbf{I}_k + c\mathbf{J}_k \end{bmatrix}$$

(13.9.6)

and (B) is a diagonal matrix of order $\frac{k(k+1)}{2}$

$$(B) = \begin{bmatrix} b & & & & & & \\ & \ddots & & & & & \\ & & b & & 0 & & \\ & & & c & & & \\ & & & & c & & \\ & 0 & & & & \ddots & \\ & & & & & & c \end{bmatrix}.$$

(13.9.7)

One can verify that

$$(U)^{-1} = \begin{bmatrix} p & q\mathbf{1}'_k \\ q\mathbf{1}_k & t\mathbf{I}_k + s\mathbf{J}_k \end{bmatrix},$$

(13.9.8)

where

$$p = \frac{d + kc}{N(d + kc) - b^2 k},$$

$$q = -\frac{b}{N(d + kc) - b^2 k},$$

$$t = \frac{1}{d} \tag{13.9.9}$$

$$s = \frac{b^2 - Nc}{d[N(d + kc) - b^2 k]}.$$

Notice that U is singular if $N(d + kc) = b^2 k$. We therefore say that **the design is nonsingular** if

$$N \neq \frac{b^2 k}{d + kc}.$$

Furthermore, if $N = b^2/c$ then $s = 0$. In this case, the design is called **orthogonal**.
Let $\mathbf{x}^{0'} = (x_1^0, \ldots, x_k^0)$ be a point in the experimental domain, and

$$\xi^{0'} = (1, (x_1^0)^2, \ldots, (x_k^0)^2, x_1^0, \ldots, x_k^0, x_1^0 x_2^0, x_1^0 x_3^0, \ldots, x_{k-1}^0 x_k^0).$$

The variance of the predicted response at $\mathbf{x}^0$ is

$$V\{\overline{\mathbf{Y}}(\mathbf{x}^0)\} = \sigma^2 \xi^{0'}
\begin{bmatrix}
(U)^{-1} & & & & \mathbf{0} & & & \\
& & b^{-1} & & & 0 & & \\
& & & \ddots & & & & \\
\mathbf{0} & & & & b^{-1} & & & \\
& & & & & c^{-1} & & \\
& 0 & & & & & \ddots & \\
& & & & & & & c^{-1}
\end{bmatrix}
\xi^0$$

$$= \sigma^2 \left[p + \frac{1}{b} \sum_{i=1}^{k} (x_i^0)^2 + (t + s) \sum_{i=1}^{k} (x_i^0)^4 \right.$$

$$+ \frac{1}{c} \xrightarrow[h<j]{} \sum \sum (x_h^0)^2 (x_j^0)^2 + 2b \sum_{i=1}^{k} (x_i^0)^2$$

$$\left. + 2s \xrightarrow[h<j]{} \sum \sum (x_h^0)^2 (x_j^0)^2 \right]$$

$$= \sigma^2 \left[p + \rho^2 \left(2b + \frac{1}{b} \right) + (t + s) \sum_{i=1}^{k} (x_i^0)^4 \right.$$

$$\left. + 2 \left(s + \frac{1}{2c} \right) \xrightarrow[h<j]{} \sum \sum (x_h^0)^2 (x_j^0)^2 \right], \tag{13.9.10}$$

where $\rho^2 = \sum_{i=1}^{k} (x_i^0)^2$. Notice that

$$\rho^4 = \left(\sum_{i=1}^{k} (x_i^0)^2 \right)^2 = \sum_{i=1}^{k} (x_i^0)^4 + 2 \xrightarrow[h<j]{} \sum \sum (x_h^0)^2 (x_j^0)^2.$$

Thus, if $d = 2c$ then $t + s = s + \frac{1}{2c}$ and

$$V\{\overline{Y}(x^0)\} = \sigma^2 \left[p + \rho^2 \left(2b + \frac{1}{b} \right) + (t + s)\rho^4 \right]. \tag{13.9.11}$$

Such a design $(d = 2c)$ is called **rotatable**, since $V\{\overline{Y}(x^0)\}$ is constant for all points x^0 on the circumference of a circle of radius ρ, centered at the origin.

13.9.2   Some specific second-order designs

3^k-Designs

Consider a factorial design of k factors, each one at three levels $-1, 0, 1$. In this case, the number of points is $N = 3^k$. Obviously, $\sum_{j=1}^{3^k} x_{ji} = 0$ for all $i = 1, \ldots, k$. Also $\sum_{j=1}^{3} x_{ji}^3 = 0$ and $\sum_{j=1}^{3^k} x_{ji}^2 x_{jk} = 0$, $i \neq k$.

$$b = \sum_{j=1}^{3^k} x_{ji}^2 = \frac{2}{3} 3^k = 2 \cdot 3^{k-1}.$$

$$c = \sum_{j=1}^{3^k} x_{ji}^2 x_{jl}^2 = \frac{2}{3} b = 4 \cdot 3^{k-2}. \tag{13.9.12}$$

$$c + d = b.$$

Hence,

$$d = 2 \cdot 3^{k-1} - 4 \cdot 3^{k-2} = 2 \cdot 3^{k-2}. \tag{13.9.13}$$

$b^2 = 4 \cdot 3^{2k-2}$ and $N \cdot c = 3^k \cdot 4 \cdot 3^{k-2} = 4 \cdot 3^{2k-2}$. Thus, $Nc = b^2$. The design is orthogonal. However, $d \neq 2 \cdot c$. Thus, the design is not rotatable.

Central composite designs

A central composite design is one in which we start with $n_c = 2^k$ points of a factorial design, in which each factor is at levels -1 and $+1$. To these points, we add $n_a = 2k$ axial points which are at a fixed distance α from the origin. These are the points

$$(\pm\alpha, 0, \ldots, 0), (0, \pm\alpha, 0, \ldots, 0), \ldots, (0, 0, \ldots, 0, \pm\alpha).$$

Finally, put n_0 points at the origin. These n_0 observations yield an estimate of the variance σ^2. Thus, the total number of points is $N = 2^k + 2k + n_0$. In such a design,

$$b = 2^k + 2\alpha^2,$$

$$c = 2^k, \tag{13.9.14}$$

$$c + d = 2^k + 2\alpha^4, \quad \text{or}$$

$$d = 2\alpha^4.$$

The rotatability condition is $d = 2c$. Thus, the design is rotatable if

$$\alpha^4 = 2^k, \quad \text{or}$$
$$\alpha = 2^{k/4}. \tag{13.9.15}$$

For this reason, in central composite designs, with $k = 2$ factors we use $\alpha = \sqrt{2} = 1.414$. For $k = 3$ factors we use $\alpha = 2^{3/4} = 1.6818$. For rotatability and orthogonality, the following should be satisfied

$$n_0 + 2^k = \frac{4\alpha^2(2^k + \alpha^2)}{2^k}, \tag{13.9.16}$$

and since $\alpha^2 = 2^{k/2}$ (for rotatability)

$$n_0 = 4(2^{k/2} + 1) - 2^k. \tag{13.9.17}$$

Thus, if $k = 2$, the number of points at the origin is $n_0 = 8$. For $k = 4$ we need $n_0 = 4$ points at the origin. For $k = 3$ there is no rotatability, since $4(2^{3/2} + 1) - 8 = 7.313$.

Example 13.11. In Example 13.7, the piston simulation experiment was considered. We tested there, via a 3^2 experiment, the effects of piston weight (factor A) and the spring coefficient (factor D). It was found that the effects of the spring coefficient were significant, while the piston weight had no significant effect on the cycle time. We will conduct now a central composite design of four factors, in order to explore the response surface. The factors chosen are piston surface area (factor B); initial gas volume (factor C); spring coefficient (factor D), and filling gas temperature (factor G) (Table 13.30).

Table 13.30 *Factors and level in piston simulator experiment.*

Factor	Levels				
Piston surface area	.0075	.01	.0125	.015	.0175
Initial gas volume	.0050	.00625	.0075	.00875	.0100
Spring coefficient	1000	2000	3000	4000	5000
Filling gas temperature	340	345	350	355	360
Code	−2	−1	0	1	2

The experiment is performed with the piston simulator designed to attain both orthogonality and rotatability. Since $k = 4$, we have $\alpha = 2$ and $n_0 = 4$. The number of replications is $n_r = 30$. The experimental design and response outcomes are presented in Table 13.31. Figure 13.11 presents the main effects plot for the above four factors. The spring coefficient and the filling gas temperature have similar main effects. The cycle time average is monotonically decreasing with increasing levels of these two factors.

In Table 13.32, we present the results of regression analysis of the mean cycle time $\overline{Y}$ on 14 predictors $x_1^2, x_2^2, x_3^2, x_4^2, x_1, x_2, x_3, x_4, x_1 x_2, x_1 x_3, x_1 x_4, x_2 x_3, x_2 x_4, x_3 x_4$, where x_1 corresponds to factor B, x_2 to factor C, x_3 to factor D, and x_4 to factor G.

Table 13.31   *The central composite design and the mean and standard deviations of cycle time*

CODE		LEVELS			
B	C	D	G	$\bar{Y}$	STD.
−1.00	−1.00	−1.00	−1.00	0.671	0.2328
1.00	−1.00	−1.00	−1.00	0.445	0.1771
−1.00	1.00	−1.00	−1.00	0.650	0.2298
1.00	1.00	−1.00	−1.00	0.546	0.2228
−1.00	−1.00	1.00	−1.00	0.534	0.1650
1.00	−1.00	1.00	−1.00	0.410	0.1688
−1.00	1.00	1.00	−1.00	0.534	0.1257
1.00	1.00	1.00	−1.00	0.495	0.1388
−1.00	−1.00	−1.00	1.00	0.593	0.2453
1.00	−1.00	−1.00	1.00	0.542	0.2266
−1.00	1.00	−1.00	1.00	0.602	0.2185
1.00	1.00	−1.00	1.00	0.509	0.1977
−1.00	−1.00	1.00	1.00	0.480	0.1713
1.00	−1.00	1.00	1.00	0.411	0.1658
−1.00	1.00	1.00	1.00	0.435	0.1389
1.00	1.00	1.00	1.00	0.438	0.1482
2.00	0.00	0.00	0.00	0.458	0.1732
−2.00	0.00	0.00	0.00	0.635	0.1677
0.00	2.00	0.00	0.00	0.570	0.1569
0.00	−2.00	0.00	0.00	0.481	0.1757
0.00	0.00	2.00	0.00	0.428	0.1064
0.00	0.00	−2.00	0.00	0.742	0.3270
0.00	0.00	0.00	2.00	0.496	0.2029
0.00	0.00	0.00	−2.00	0.549	0.1765
0.00	0.00	0.00	0.00	0.490	0.1802
0.00	0.00	0.00	0.00	0.468	0.1480
0.00	0.00	0.00	0.00	0.481	0.1636
0.00	0.00	0.00	0.00	0.557	0.1869

We see in Table 13.32 that only factor D (spring coefficient) has a significant quadratic effect. Factors B, D, and G have significant linear effects. The interaction effects of B with C, D, and G will also be added. Thus, the response surface can be approximated by the equation,

$$Y = 0.499 - 0.0440x_1 - 0.0604x_3 - 0.0159x_4 + 0.0171x_3^2$$
$$+ 0.0148x_1x_2 + 0.0153x_1x_3 + 0.0177x_1x_4.$$

The contour lines for the mean cycle time, corresponding to $x_2 = x_4 = 0$ are shown in Figure 13.12.

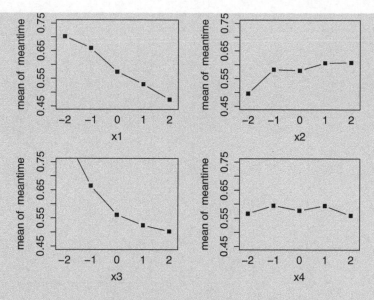

Figure 13.11   *Main effects plot.*

Table 13.32   *MINITAB regression analysis*

	Predictor	Coef	Stdev	t-ratio	p
μ	Constant	0.499000	0.018530	26.92	0.000
β_{11}	x_1^2	0.007469	0.007567	0.99	0.342
β_{22}	x_2^2	0.002219	0.007567	0.29	0.774
β_{33}	x_3^2	0.017094	0.007567	2.26	0.042
β_{44}	x_4^2	0.001469	0.007567	0.19	0.849
β_1	x_1	−0.044042	0.007567	−5.82	0.000
β_2	x_2	0.012542	0.007567	1.66	0.121
β_3	x_3	−0.060375	0.007567	−7.98	0.000
β_4	x_4	−0.015875	0.007567	−2.10	0.056
β_{12}	x_1x_2	0.014813	0.009267	1.60	0.134
β_{13}	x_1x_3	0.015313	0.009267	1.65	0.122
β_{14}	x_1x_4	0.017687	0.009267	1.91	0.079
β_{23}	x_2x_3	0.000688	0.009267	0.07	0.942
β_{24}	x_2x_4	−0.012937	0.009267	−1.40	0.186
β_{34}	x_3x_4	−0.008937	0.009267	−0.96	0.352
	$s = 0.03707$	$R - sq = 90.4\%$		$R - sq(adj) = 80.0\%$	

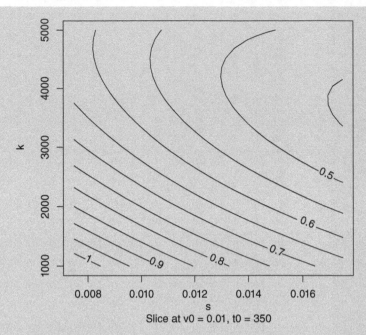

Slice at v0 = 0.01, t0 = 350

Figure 13.12   *Contour lines of the response surface* $Y = 0.583 - 0.0448x_1 - 0.0576x_3 - 0.0056x_1^2 + 0.0079x_3^2 + 0.0060x_1x_3$.

```
> library(rsm)
> s  <- c(0.0075, 0.01, 0.0125, 0.015, 0.0175)
> v0 <- c(0.0050, 0.00625, 0.0075, 0.00875, 0.0100)
> k  <- c(1000, 2000, 3000, 4000, 5000)
> t0 <- c(340, 345, 350, 355, 360)
> Ccd <- ccd(basis=4,
             n0=4,
             alpha=2,
             coding=list(x1 ~ -5 + s*400,
                         x2 ~ -6 + v0*800,
                         x3 ~ -3 + k*0.001,
                         x4 ~ -70 + t0*0.2),
             randomize=FALSE)
> head(Ccd)

  run.order std.order    s      v0      k   t0 Block
1         1         1    1 0.010 0.00625 2000 345     1
2         2         2    2 0.015 0.00625 2000 345     1
3         3         3    3 0.010 0.00875 2000 345     1
4         4         4    4 0.015 0.00875 2000 345     1
5         5         5    5 0.010 0.00625 4000 345     1
6         6         6    6 0.015 0.00625 4000 345     1
```

Data are stored in coded form using these coding formulas ...

```
x1 ~ -5 + s * 400
x2 ~ -6 + v0 * 800
x3 ~ -3 + k * 0.001
x4 ~ -70 + t0 * 0.2

> Levels <- as.data.frame(
    decode.data(Ccd))[, c("s", "v0", "k", "t0")]
> Ps <- pistonSimulation(m=rep(60, nrow(Levels)),
                         s=Levels$s,
                         v0=Levels$v0,
                         k=Levels$k,
                         p0=rep(110000, nrow(Levels)),
                         t=rep(296, nrow(Levels)),
                         t0=Levels$t0,
                         each=30,
                         seed=123)
> Ps <- simulationGroup(Ps, 30)
> Ccd$meantime <- aggregate(Ps["seconds"],
                            by=Ps["group"],
                            FUN=mean)$seconds
> Rsm <- rsm(meantime ~ SO(x1, x2, x3, x4),
             data= Ccd)
> summary(Rsm)

Call:
rsm(formula = meantime ~ SO(x1, x2, x3, x4), data = Ccd)

              Estimate    Std. Error    t value    Pr(>|t|)
(Intercept)   0.55651498  0.00895672    62.1338    < 2.2e-16
x1           -0.06273744  0.00517116   -12.1322    8.508e-10
x2            0.01732095  0.00517116     3.3495    0.003801
x3           -0.07605865  0.00517116   -14.7082    4.227e-11
x4           -0.00086087  0.00517116    -0.1665    0.869748
x1:x2         0.01063617  0.00633336     1.6794    0.111360
x1:x3         0.01540720  0.00633336     2.4327    0.026322
x1:x4        -0.00398071  0.00633336    -0.6285    0.538010
x2:x3         0.00097643  0.00633336     0.1542    0.879289
x2:x4        -0.00634481  0.00633336    -1.0018    0.330483
x3:x4        -0.00099975  0.00633336    -0.1579    0.876431
x1^2          0.00780867  0.00466122     1.6752    0.112182
x2^2         -0.00124712  0.00466122    -0.2676    0.792262
x3^2          0.02990712  0.00466122     6.4161    6.389e-06
x4^2          0.00155039  0.00466122     0.3326    0.743492

(Intercept) ***
x1          ***
x2          **
x3          ***
```

```
x4
x1:x2
x1:x3          *
x1:x4
x2:x3
x2:x4
x3:x4
x1^2
x2^2
x3^2          ***
x4^2
---
Signif. codes:
0 '***' 0.001 '**' 0.01 '*' 0.05 '.' 0.1 ' ' 1


Multiple R-squared: 0.9618,          Adjusted R-squared: 0.9304
F-statistic: 30.59 on 14 and 17 DF, p-value: 0.000000002791


Analysis of Variance Table


Response: meantime
                    Df    Sum Sq   Mean Sq  F value
FO(x1, x2, x3, x4)   4  0.240520  0.060130  93.6921
TWI(x1, x2, x3, x4)  6  0.006537  0.001090   1.6976
PQ(x1, x2, x3, x4)   4  0.027809  0.006952  10.8327
Residuals           17  0.010910  0.000642
Lack of fit         10  0.006863  0.000686   1.1871
Pure error           7  0.004047  0.000578
                          Pr(>F)
FO(x1, x2, x3, x4)  2.391e-11
TWI(x1, x2, x3, x4) 0.1821633
PQ(x1, x2, x3, x4)  0.0001499
Residuals
Lack of fit         0.4223317
Pure error


Stationary point of response surface:
        x1            x2            x3            x4
 5.65531833    1.91720834   -0.02501298   11.45274099


Stationary point in original units:
          s              v0              k
   0.02663830     0.00989651   2974.98701820
          t0
 407.26370496
```

```
Eigenanalysis:
eigen() decomposition
$values
[1]   0.0325114175   0.0091700709   0.0009851376
[4]  -0.0046475718

$vectors
             [,1]          [,2]          [,3]          [,4]
x1  -0.31297494    0.7394155    0.4733013   -0.36234958
x2  -0.06696307    0.4770755   -0.1256402    0.86725401
x3  -0.94645473   -0.2948946   -0.1089955    0.07335251
x4   0.04226194   -0.3724256    0.8650552    0.33345567

> rm(k, s, t0, v0)
```

13.9.3   Approaching the region of the optimal yield

Very often, the purpose for fitting a response surface is to locate the levels of the factors, which yield optimal results.

Initially, one might be far from the optimal regions. A series of small experiments may be performed, in order to move toward the optimal region. Thus, we start with simple first-order experiments, like 2^k factorial, and fit to the results a linear model of the form

$$\overline{Y} = b_0 + b_1 x_1 + \cdots + b_k x_k.$$

We wish to determine now a new point, ξ^* say, whose distance from the center of the 1st stage experiment (say 0) is R, and with maximal (or minimal) predicted yield. The predicted yield at $\xi^*($ is

$$\hat{y}^* = b_0 + \sum_{i=1}^{k} b_i \xi_i^*.$$

To find ξ^*, we differentiate the Lagrangian

$$L = b_0 + \sum_{i=1}^{k} b_i \xi_i^* + \lambda \left(R^2 - \sum_{i=1}^{k} (\xi_i^*)^2 \right)$$

with respect to ξ_i^* $(i = 1, \ldots, k)$ and λ. The solution is

$$\xi_i^* = R \frac{b_i}{\sqrt{\sum_{i=1}^{k} b_i^2}}, \quad i = 1, \ldots, k.$$

The direction of the **steepest accent** (descent) is in the direction of the normal (perpendicular) to the contours of equal response.

At the second stage, we perform experiments at a few points along the direction of the steepest ascent (at $R_1, R_2, \ldots$) until there is no further increase in the mean yield. We then enter the third stage, at which we perform a second-order design, centered at a new region, where the optimal conditions seem to prevail.

```
> steepest(Rsm,
           dist=seq(0, 2.5, by=0.5),
           descent=TRUE)

Path of steepest descent from ridge analysis:
    dist     x1      x2     x3      x4 |        s         v0
1   0.0  0.000   0.000  0.000   0.000 | 0.012500  0.00750000
2   0.5  0.348  -0.140  0.331   0.011 | 0.013370  0.00732500
3   1.0  0.742  -0.415  0.525   0.025 | 0.014355  0.00698125
4   1.5  1.120  -0.795  0.598   0.020 | 0.015300  0.00650625
5   2.0  1.460  -1.228  0.611  -0.015 | 0.016150  0.00596500
6   2.5  1.762  -1.673  0.597  -0.077 | 0.016905  0.00540875
       k      t0 |  yhat
1   3000 350.000 | 0.557
2   3331 350.055 | 0.512
3   3525 350.125 | 0.478
4   3598 350.100 | 0.447
5   3611 349.925 | 0.417
6   3597 349.615 | 0.387
```

13.9.4   Canonical representation

The quadratic response function

$$\overline{Y} = b_0 + \sum_{i=1}^{k} b_i x_i + \sum_{i=1}^{k} b_{ii} x_i^2 + 2 \xrightarrow{i<j} \sum \sum b_{ij} x_i x_j, \tag{13.9.18}$$

can be written in the matrix form

$$\overline{Y} = b_0 + \mathbf{b}'\mathbf{x} + \mathbf{x}'\mathbf{B}\mathbf{x}, \tag{13.9.19}$$

where $\mathbf{x}' = (x_1, \ldots, x_k)$, $\mathbf{b}' = (b_1, \ldots, b_k)$ and

$$\mathbf{B} = \begin{bmatrix} b_{11} & b_{12} & \cdots & b_{1k} \\ b_{12} & \ddots & & \vdots \\ \vdots & & \ddots & \vdots \\ b_{1k} & \cdots & \cdots & b_{kk} \end{bmatrix}.$$

Let $\boldsymbol{\nabla}\overline{Y}$ be the gradient of $\overline{Y}$, i.e.,

$$\boldsymbol{\nabla}\overline{Y} = \frac{\partial}{\partial \mathbf{x}}\overline{Y} = \mathbf{b} + 2B\mathbf{x}.$$

Let $\mathbf{x}^0$ be a point at which the gradient is zero, namely

$$\mathbf{x}^0 = -\frac{1}{2}\mathbf{B}^{-1}\mathbf{b}, \tag{13.9.20}$$

assuming that the matrix **B** is nonsingular. Making the transformation (change of origin to $\mathbf{x}^0$) $\mathbf{z} = \mathbf{x} - \mathbf{x}^0$, we obtain

$$
\begin{aligned}
\overline{Y} &= b_0 + (\mathbf{x}^0 + \mathbf{z})'\mathbf{b} + (\mathbf{x}^0 + \mathbf{z})'\mathbf{B}(\mathbf{x} + \mathbf{z}) \\
&= \overline{Y}_0 + \mathbf{z}'\mathbf{Bz},
\end{aligned}
\tag{13.9.21}
$$

where $\overline{Y}_0 = b_0 + \mathbf{b}'\mathbf{x}^0$.

The matrix B is real symmetric. Thus, there exists an orthogonal matrix **H** (see Appendix I), which consists of the normalized eigenvectors of B, such that

$$
\mathbf{HBH}' = \begin{pmatrix} \lambda_{} & & \\ 0 & \ddots & 0 \\ & & \lambda_k \end{pmatrix},
$$

where λ_i ($i = 1, \ldots, k$) are the eigenvalues of B. We make now a new transformation (rotation), namely,

$$
\mathbf{w} = \mathbf{HZ}.
$$

Since H is orthogonal, $\mathbf{z} = \mathbf{H}'\mathbf{w}$ and

$$
\mathbf{z}'\mathbf{Bz} = \mathbf{w}'\mathbf{HBH}'\mathbf{w}
$$

$$
= \sum_{i=1}^{k} \lambda_i w_i^2.
$$

In these new coordinates,

$$
\overline{Y} = \overline{Y}_0 + \sum_{i=1}^{k} \lambda_i w_i^2.
\tag{13.9.22}
$$

This representation of the quadratic surface is called the **canonical form**. We see immediately that if $\lambda_i > 0$ for all $i = 1, \ldots, k$, then $\overline{Y}_0$ is a point of **minimum**. If $\lambda_i < 0$ for all $i = 1, \ldots, k$ then $\overline{Y}_0$ is a **maximum**. If some eigenvalues are positive and some are negative, then $\overline{Y}_0$ is a **saddle point**.

The following examples of second-order equations are taken from Box et al. (1978), pp. 527–530).

1) Simple maximum:

$$
\begin{aligned}
\overline{Y} &= 83.57 + 9.39x_1 + 7.12x_2 - 7.44x_1^2 - 3.71x_2^2 - 5.80x_1x_2 \\
&= 87.69 - 902w_1^2 - 2.13w_2^2.
\end{aligned}
$$

2) Minimax:

$$
\begin{aligned}
\overline{Y} &= 84.29 + 11.06x_1 + 4.05x_2 - 6.46x_1^2 - 0.43x_2^2 - 9.38x_1x_2 \\
&= 87.69 - 9.02w_1^2 + 2.13w_2^2.
\end{aligned}
$$

3) Stationary ridge:

$$
\begin{aligned}
\overline{Y} &= 83.93 + 10.23x_1 + 5.59x_2 - 6.95x_1^2 - 2.07x_2^2 - 7.59x_1x_2 \\
&= 87.69 - 9.02w_1^2 + 0.00w_2^2.
\end{aligned}
$$

4) Rising ridge:

$$\overline{Y} = 82.71 + 8.80x_1 + 8.19x_2 - 6.95x_1^2 - 2.07x_2^2 - 7.59x_1x_2$$
$$= 87.69 - 9.02w_1^2 + 2.97w_2.$$

```
> canonical.path(Rsm,
                 dist=seq(0, 2.5, by=0.5),
                 descent=TRUE)

  dist    x1    x2     x3      x4 |          s
1  0.0 5.655 1.917 -0.025 11.453 | 0.0266375
2  0.5 5.474 2.351  0.012 11.619 | 0.0261850
3  1.0 5.293 2.784  0.048 11.786 | 0.0257325
4  1.5 5.112 3.218  0.085 11.953 | 0.0252800
5  2.0 4.931 3.652  0.122 12.120 | 0.0248275
6  2.5 4.749 4.085  0.158 12.286 | 0.0243725
            v0    k      t0 |  yhat
1 0.00989625 2975 407.265 | 0.392
2 0.01043875 3012 408.095 | 0.391
3 0.01098000 3048 408.930 | 0.387
4 0.01152250 3085 409.765 | 0.381
5 0.01206500 3122 410.600 | 0.373
6 0.01260625 3158 411.430 | 0.363
```

13.10   Chapter highlights

The chapter covers the range of classical experimental designs including complete block designs, Latin squares, full and fractional factorial designs with factors at two and three levels. The basic approach to the analysis is through modeling the response variable and computing ANOVA tables. Particular attention is also given to the generation of designs using R.

The main concepts and definitions introduced in this chapter include:

- Response Variable
- Controllable Factor
- Factor Level
- Statistical Model
- Experimental Array
- Blocking
- Randomization
- Block Designs
- Main Effects
- Interactions
- Analysis of Variance
- Latin Squares
- Factorial Designs
- Fractional Factorial Designs

13.11   Exercises

13.1 Describe a production process familiar to you, like baking of cakes, or manufacturing concrete. List the pertinent variables. What is (are) the response variable(s)? Classify the variables which effect the response to noise variables and control variables. How many levels would you consider for each variable?

13.2 Different types of adhesive are used in a lamination process, in manufacturing a computer card. The card is tested for bond strength. In addition to the type of adhesive, a factor which might influence the bond strength is the curing pressure (currently at 200 psi). Follow the basic steps of experimental design to set a possible experiment for testing the effects of adhesives and curing pressure on the bond strength.

13.3 Provide an example where blocking can reduce the variability of a product.

13.4 Three factors A, B, C are tested in a given experiment, designed to assess their effects on the response variable. Each factor is tested at 3 levels. List all the main effects and interactions.

13.5 Let x_1, x_2 be two quantitative factors and Y a response variable. A regression model $Y = \beta_0 + \beta_1 x_1 + \beta_2 x_2 + \beta_{12} x_1 x_2 + e$ is fitted to the data. Explain why β_{12} can be used as an interaction parameter.

13.6 Consider the ISC values for times t_1, t_2 and t_3 in data file **SOCELL.csv**. Make a paired comparison for testing whether the mean ISC in time t_2 is different from that in time t_1, by using a t-test.

13.7 Use macro **RPCOMP.MTB** to perform a randomization test for the differences in the ISC values of the solar cells in times t_2 and t_3 (data file **SOCELL.csv**).

13.8 Box et al. (1978, p. 209) give the following results of four treatments A, B, C, D in penicillin manufacturing in five different blends (blocks).

	Treatments			
blends	A	B	C	D
1	89	88	97	94
2	84	77	92	79
3	81	87	87	85
4	87	92	89	84
5	79	81	80	88

Perform an ANOVA to test whether there are significant differences between the treatments or between the blends.

13.9 Eight treatments $A, B, C, \ldots, H$ were tested in a BIBD of 28 blocks, $k = 2$ treatments per block, $r = 7$ and $\lambda = 1$. The results of the experiments are

block	Treatments				block	Treatments			
1	A	38	B	30	15	D	11	G	24
2	C	50	D	27	16	F	37	H	39
3	E	33	F	28	17	A	23	F	40

block	Treatments			block	Treatments		
4	G	62	H	30	18	B	20	D	14
5	A	37	C	25	19	C	18	H	10
6	B	38	H	52	20	E	22	G	52
7	D	89	E	89	21	A	66	G	67
8	F	27	G	75	22	B	23	F	46
9	A	17	D	25	23	C	22	E	28
10	B	47	G	63	24	D	20	H	40
11	C	32	F	39	25	A	27	H	32
12	E	20	H	18	26	B	10	E	40
13	A	5	E	15	27	C	32	G	33
14	B	45	C	38	28	D	18	F	23

Make an ANOVA to test the significance of the block effects, treatment effects, If the treatment effects are significant, make multiple comparisons of the treatments.

13.10 Four different methods of preparing concrete mixtures A, B, C, D were tested, these methods consisted of two different mixture ratios of cement to water and two blending duration). The four methods (treatments) were blocks in four batches and four days, according to a Latin square design. The concrete was poured to cubes and tested for compressive strength [Kg/cm^2] after 7 days of storage in special rooms with 20°C temperature and 50% relative humidity. The results are:

Days	batches			
	1	2	3	4
1	A	B	C	D
	312	299	315	290
2	C	A	D	B
	295	317	313	300
3	B	D	A	C
	295	298	312	315
4	D	C	B	A
	313	314	299	300

Are the differences between the strength values of different treatments significant? [Perform the ANOVA.]

13.11 Repeat the experiments described in Example 13.7 at the low levels of factors B, C, E, F, and G. Perform the ANOVA for the main effects and interaction of spring coefficient and piston weight on the cycle time. Are your results different from those obtained in the example?

13.12 For the data in Example 13.7, compute the least squares estimates of the main effects on the means and on the standard deviations.

13.13 A 2^4 factorial experiment gave the following response values, arranged in standard order: 72, 60, 90, 80, 65, 60, 85, 80, 60, 50, 88, 82, 58, 50, 84, 75.
  (i) Estimate all possible main effects.
  (ii) Estimate σ^2 under the assumption that all the interaction parameters are zero.
  (iii) Determine a confidence interval for σ^2 at level of confidence 0.99.

13.14 A 3^2 factorial experiment, with $n = 3$ replications, gave the following observations:

	A_1	A_2	A_3
	18.3	17.9	19.1
B_1	17.9	17.6	19.0
	18.5	16.2	18.9
	20.5	18.2	22.1
B_2	21.1	19.5	23.5
	20.7	18.9	22.9
	21.5	20.1	22.3
B_3	21.7	19.5	23.5
	21.9	18.9	23.3

Perform an ANOVA to test the main effects and interactions. Break the between treatments sum of squares to one degree of freedom components. Use the Scheffé S_α coefficient to determine which effects are significant.

13.15 Construct a 2^{8-2} fractional replication, using the generators $ABCDG$ and $ABEFH$. What is the resolution of this design? Write the aliases to the main effects, and to the first-order interactions with the factor A.

13.16 Consider a full factorial experiment of $2^6 = 64$ runs. It is required to partition the runs to 8 blocks of 8. The parameters in the group of defining parameters are confounded with the effects of blocks, and are not estimable. Show which parameters are not estimable if the blocks are generated by ACE, $ABEF$ and $ABCD$.

13.17 A 2^2 factorial design is expanded by using 4 observations at 0. The design matrix and the response are:

X_1	X_2	Y
-1	-1	55.8
-1	-1	54.4
1	-1	60.3
1	-1	60.9
-1	1	63.9
-1	1	64.4
1	1	67.9
1	1	68.5

X_1	X_2	Y
0	0	61.5
0	0	62.0
0	0	61.9
0	0	62.4

(i) Fit a response function of the form: $Y = \beta_0 + \beta_1 X_1 + \beta_2 X_2 + \beta_{12} X_1 X_2 + e$, and plot its contour lines.

(ii) Estimate the variance σ^2 and test the goodness of fit of this model.

13.18 The following represents a design matrix and the response for a control composite design

X_1	X_2	Y
1.0	0.000	95.6
0.5	0.866	77.9
−0.5	0.866	76.2
−1.0	0	54.5
−0.5	−0.866	63.9
0.5	−0.866	79.1
0	0	96.8
0	0	94.8
0	0	94.4

(i) Estimate the response function and its stationary point.

(ii) Plot contours of equal response, in two dimensions.

(iii) Conduct an ANOVA.

14

Quality by Design

Factorial designs discussed in the previous chapter, were developed in the 1930s by R.A. Fisher and F. Yates in Rothamsted agricultural station in Britain. Fractional replications were developed in the 1940s by D. Finney, also in Rothamsted. After World War II, these experimental design methods were applied to industrial problems in the Imperial Chemical Laboratories (ICL) in Britain. The objective of agronomists is to find treatment combinations which lead to maximal yield in agricultural product growth. The chemical engineer wishes to find the right combinations of pressure, temperature, and other factors, which lead to a maximal amount of the product coming out of a reactor. The objective in manufacturing engineering is, on the other hand, to design the process so that the products will be as close as possible to some specified target, without much fluctuations over time.

Flaws in engineering design can cause severe problems over time, here is an example. A small British electronics company called Encrypta designed an ingenious electronic seal for lorries and storerooms. Industrial versions of D-size batteries were used to drive the circuit and numeric display. Encrypta started to receive defective seals returned by customers. A failure mode analysis revealed that, when dropped on a hard surface, the batteries would heat up and cause a short circuit. Encrypta won 30,000 lb in compensations from the batteries manufacturer and switched to Vidor batteries made by Fuji who passed the test. Encrypta found that the D-batteries failed on dropping because a cloth-like separation inside ruptures. This produced an active chemicals mix that discharged the battery. Fuji uses a tough, rolled separator which eliminates the problem. (New Scientist, 119, September 15, 1988, p. 39.)

In this chapter, we discuss methods and tools compatible with the top of the Quality Ladder mentioned in Chapter 9. At that level, the organization has adopted a Quality by Design (QbD) approach which requires high organizational maturity. An example of QbD is the comprehensive quality engineering approach developed in the 1950s by the Japanese engineer Genichi Taguchi. Taguchi labeled his methodology **off-line quality control**. The basic ideas of off-line quality control originated while Taguchi was working at the Electrical Communications Laboratory (ECL) of the Nippon Telephone and Telegraph Company (NTT). Taguchi's task was to help Japanese engineers develop high-quality products with raw materials of poor quality, outdated

Modern Industrial Statistics: With Applications in R, MINITAB and JMP, Third Edition.
Ron S. Kenett and Shelemyahu Zacks.
© 2021 John Wiley & Sons, Ltd. Published 2021 by John Wiley & Sons, Ltd.

manufacturing equipment, and an acute shortage of skilled engineers. Central to his approach is the application of statistically designed experiments. Taguchi's impact on Japan has expanded to a wide range of industries. He won the 1960 Deming prize for application of quality as well as three Deming prizes for literature on quality in 1951, 1953, and 1984. In 1959, the Japanese company NEC followed Taguchi's methods and ran 402 such experiments. In 1976, Nippon Denso, which is a 20,000-employee company producing electrical parts for automobiles, reported to have run 2700 experiments using the Taguchi off-line quality control method. Off-line quality control was first applied in the West to Integrated Circuit manufacturing (see Phadke *et al.* 1983). Applications of off-line quality control range now from the design of automobiles, copiers and electronic systems to cash-flow optimization in banking, improvements in computer response times and runway utilization in an airport. Another industry that has adopted QbD, to ensure high quality and reduced inefficiencies, is the pharmaceutical industry. The chapter will cover both the Taguchi methods and the application of QbD in Pharmaceutical companies.

14.1   Off-line quality control, parameter design, and the Taguchi method

Kacker (1985), Dehand (1989), Phadke (1989), John (1990), Box *et al.* (1988) and others explain the Taguchi methodology for off-line experimentation. We provide here a concise summary of this approach.

The performance of products or processes is typically quantified by performance measures. Examples include measures such as piston cycle time, yield of a production process, output voltage of an electronic circuit, noise level of a compressor or response times of a computer system. These performance measures might be affected by several factors that have to be set at specific levels to get desired results. For example, the piston simulator introduced in previous chapters has seven factors that can be used to control the piston cycle time. The aim of off-line quality control is to determine the factor-level combination that gives the least variability to the appropriate performance measure, while keeping the mean value of the measure on target. The goal is to control **both** accuracy and variability. In the next section, we discuss an optimization strategy that solves this problem by minimizing various loss functions.

14.1.1   Product and process optimization using loss functions

Optimization problems of products or processes can take many forms that depend on the objectives to be reached. These objectives are typically derived from customer requirements. Performance parameters such as dimensions, pressure, or velocity usually have a target or nominal value. The objective is to reach the target within a range bounded by upper specification limits (USLs) and lower specification limits (LSL). We call such cases "**nominal is best**." Noise levels, shrinkage factors, amount of wear, and deterioration are usually required to be as low as possible. We call such cases "**the smaller the better**." When we measure strength, efficiency, yields, or time to failure, our goal is, in most cases, to reach the maximum possible levels. Such cases are called "**the larger the better**." These three types of cases require different objective (target) functions to optimize. Taguchi introduced the concept of **loss function** to help determine the appropriate optimization procedure.

When "nominal is best" specification limits are typically two sided with an USL and a LSL. These limits are used to differentiate between conforming and nonconforming products. Nonconforming products are usually fixed, retested and sometimes downgraded or simply scrapped. In all cases, defective products carry a loss to the manufacturer. Taguchi argues that only products on target should carry no loss. Any deviation carries a loss which is not always immediately perceived by the customer or production personnel. Taguchi proposes a quadratic function as a simple approximation to a graduated loss function that measures loss on a continuous scale. A quadratic loss function has the form

$$L(y, M) = K(y - M)^2, \tag{14.1.1}$$

where y is the value of the performance characteristic of a product, M is the target value of this characteristic, and K is a positive constant, which yields monetary or other utility value to the loss. For example, suppose that $(M - \Delta, M + \Delta)$ is the **customer's tolerance interval** around the target. When y falls out of this interval, the product has to be repaired or replaced at a cost of $\$ A$. Then, for this product,

$$A = K\Delta^2 \tag{14.1.2}$$

or

$$K = A/\Delta^2. \tag{14.1.3}$$

The **manufacturer's tolerance interval** is generally tighter than that of the customer, namely $(M - \delta, M + \delta)$, where $\delta < \Delta$. One can obtain the value of δ in the following manner. Suppose the cost to the manufacturer to repair a product that exceeds the customer's tolerance, before shipping the product, is $\$B, B < A$. Then

$$B = \left(\frac{A}{\Delta^2}\right)(Y - M)^2,$$

or

$$Y = M \pm \Delta\left(\frac{B}{A}\right)^{1/2}. \tag{14.1.4}$$

Thus,

$$\delta = \Delta\left(\frac{B}{A}\right)^{1/2}. \tag{14.1.5}$$

The manufacturer should reduce the variability in the product performance characteristic so that process capability C_{pk} for the tolerance interval $(M - \delta, M + \delta)$ should be high. See Figure 14.1 for a schematic presentation of these relationships.

Notice that the expected loss is

$$E\{L(Y, M)\} = K(\text{Bias}^2 + \text{Variance}), \tag{14.1.6}$$

where $\text{Bias} = \mu - M$, $\mu = E\{Y\}$ and $\text{Variance} = E\{(Y - \mu)^2\}$. Thus, the objective is to have a manufacturing process with μ as close as possible to the target M, and variance, σ^2, as small as possible ($\sigma < \frac{\delta}{3}$ so that $C_{pk} > 1$). Recall that Variance + Bias^2 is the mean squared error, MSE. Thus, when "normal is best," the objective should be to minimize the MSE.

Objective functions for cases of "the bigger the better" or "the smaller the better" depend on the case under consideration. In cases where the performance measure is the life length of a product, the objective might be to design the product to maximize the expected life length. In the literature, we may find the objective of minimizing $\frac{1}{n}\sum\frac{1}{y^j}$, which is an estimator of $E\left\{\frac{1}{Y}\right\}$. This parameter, however, may not always exist (e.g., when Y has an exponential distribution), and this objective function might be senseless.

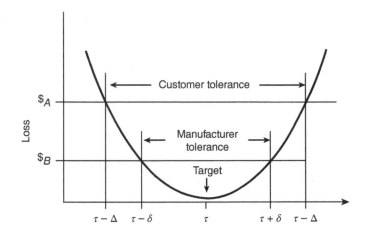

Figure 14.1   *Quadratic loss and tolerance intervals.*

14.1.2   Major stages in product and process design

A major challenge to industry is to reduce variability in products and processes. The previous section dealt with measuring the impact of such variability. In this section, we discuss methods for actually reducing variability. Design of products or processes involves two main steps: designing the system and setting tolerances. System design is the stage where engineering skills, innovation, and technology are pooled together to create a basic design. Once the design is ready to go into production, one has to specify tolerances of parts and subassemblies so that the product or process meets its requirements. Loose tolerances are typically less expensive than tight tolerances. Taguchi proposed to change the classical approach to the design of products and processes and add an intermediate stage of **parameter design**. Thus, the three major stages in designing a product or a process are:

   I. *System design*: This is when the product architecture and technology are determined.
  II. *Parameter design*: At this stage, a planned optimization program is carried out in order to minimize variability and costs.
 III. *Tolerance design*: Once the optimum performance is Determined, tolerances should be specified, so that the product or process stays within specifications. The setting of optimum values of the tolerance factors is called tolerance design.

Table 14.1 (adapted from Phadke (1989)) shows the relationships between the type of problems experienced in industrial products and processes and the various design phases.

14.1.3   Design parameters and noise factors

Taguchi classifies the variables which affect the performance characteristics into two categories: **design parameters** and **source of noise**. All factors which cause variability are included in the source of noise. Sources of noise are classified into two categories: **external sources** and **internal sources**. External sources are those external to the product, like environmental conditions (temperature, humidity, dust, etc.); human variations in operating the product and other similar

Table 14.1 Noise factors and design phases

Activity	Design phase	Leverage on noise factors				Comments
		External	Manufacturing imperfection	Natural deterioration		
Product design	(a) System design	High	High	High		Involves innovation to reduce sensitivity to all noise factors.
	(b) Parameter design	High	High	High		Most important step for reducing sensitivity to all noise factors.
	(c) Tolerance design	High	High	High		Method for selecting most economical grades of materials, components and manufacturing equipment, and operating environment for the product.
Manufacturing process design	(a) Concept design	Low	High	Low		Involves innovation to reduce the effect of manufacturing imperfections.
	(b) Parameter design	Low	High	Low		Important for reducing sensitivity of unit-to-unit variation to manufacturing variations.
	(c) Tolerance design	Low	High	Low		Method for determining tolerance on manufacturing process parameters.
Manufacturing	(a) Concept design	Low	High	Low		Method of detecting problems when they occur and correcting them.
	(b) Parameter design	Low	High	Low		Method of compensating for known problems.
	(c) Tolerance design	Low	High	Low		Last alternative, useful when process capability is poor.
Customer usage	Warranty and repair	Low	Low	Low		

factors. Internal sources of variability are those connected with manufacturing imperfections and product degradation or natural deterioration.

The design parameters, on the other hand, are controllable factors which can be set at predetermined values (level). The product designer has to specify the values of the design parameters to achieve the objectives. This is done by running an experiment which is called **parameter design**. In manufacturing conditions, the values of these parameters may slightly vary from values determined in the parameter design stage (the nominal ones). In **tolerance designs**, we test the effects of such variability and determine tolerances which yield the desired results at lower cost.

Example 14.1.   An R, L circuit is an electrical circuit of alternating current which obtains an input of voltage 100 [V] AC and frequency 55 [Hz]. The output current of the circuit is aimed at 10 [A], with tolerances of $\Delta = \pm 4$ [A]. There are four factors which influence the output y.

1. V: Input voltage [V];
2. f: Input frequency [Hz];
3. R: Resistance [Ω];
4. L: Self-inductance [H].

R and L are controllable factors, while V and f are noise factors. Assuming that V has a distribution between 90 and 110 [V] and f has a distribution between 55 and 65 [Hz]. R and L are design parameters. What should be values of R [Ω] and L [H] to obtain an output y distributed around the target of $M = 10$ [A], with minimal MSE and lowest cost? In the next section, we study how one can take advantage of the nonlinear relationship between the above factors to attain lower variability and high accuracy.

14.1.4   Parameter design experiments

In a parameter design experiment, we test the effects of the controllable factors and the noise factors on the performance characteristics of the product, in order to:

(a) make the product robust (insensitive) to environmental conditions;
(b) make the product insensitive to components variation;
(c) minimize the MSE about a target value.

We distinguish between two types of experiments, **physical experiments** and **computer-based** simulation experiments. In Chapter 13, we discuss this type of computer-based simulation experiments. Let $\theta = (\theta_1, \ldots, \theta_k)$ be the vector of design parameters. The vector of noise variables is denoted by $\mathbf{x} = (x_1, \ldots, x_m)$.

The response function $y \in f(\theta, \mathbf{x})$ involves in many situations the factors θ and $\mathbf{x}$ in a nonlinear fashion. The RL circuit described in Example 14.1 involves the four factors V, f, R, and L and the output y according to the nonlinear response function

$$y = \frac{V}{(R^2 + (2\pi f L)^2)^{1/2}}.$$

If the noise factors V and f had no variability, one could determine the values of R and L to always obtain a target value $y_0 = M$. The variability of V and f around their nominal values turns y to be a random variable, Y, with expected value μ and variance σ^2, which depend on the setting of the design parameter R and L and on the variances of V and f. The effects of the nonlinearity on the distribution of Y will be studied in Section 14.2. The objective of parameter design experiments is to take advantage of the effects of the nonlinear relationship. The strategy is to perform a factorial experiment to investigate the effects of the design parameters (controllable factors). If we learn from the experiments that certain design parameters effect the mean of Y but not its variance and, on the other hand, other design factors effect the variance but not the mean, we can use the latter group to reduce the variance of Y as much as possible, and then adjust the levels of the parameters in the first group to set μ close to the target M. We illustrate this approach in the following example.

Example 14.2. The data for the present example is taken from John (1990, p. 335). Three factors $A, B, and\ C$ (controllable) effect the output, Y, of a system. In order to estimate the main effects of $A, B, and\ C$, a 2^3 factorial experiment was conducted. Each treatment combination was repeated four times, at the "low" and "high" levels of two noise factors. The results are given in Table 14.2.

Table 14.2 *Response at a 2^3 factorial experiment*

Factor levels			Response			
A	B	C	y_1	y_2	y_3	y_4
−1	−1	−1	60.5	61.7	60.5	60.8
1	−1	−1	47.0	46.3	46.7	47.2
−1	1	−1	92.1	91.0	92.0	91.6
1	1	−1	71.0	71.7	71.1	70.0
−1	−1	1	65.2	66.8	64.3	65.2
1	−1	1	49.5	50.6	49.5	50.5
−1	1	1	91.2	90.5	91.5	88.7
1	1	1	76.0	76.0	78.3	76.4

The mean, $\bar{Y}$, and standard deviation, S, of Y at the eight treatment combinations are:

v	$\bar{Y}$	S
0	60.875	0.4918
1	46.800	0.3391
2	91.675	0.4323
3	70.950	0.6103
4	65.375	0.9010
5	50.025	0.5262
6	90.475	1.0871
7	76.675	0.9523

Regressing the column $\bar{Y}$ on the three orthogonal columns under $A, B, and \, C$ in Table 14.2, we obtain

$$\text{Mean} = 69.1 - 7.99A + 13.3B + 1.53C$$

with $R^2 = 0.991$. Moreover, the coefficient 1.53 of C is not significant (P value of 0.103). Thus, the significant main effects on the mean yield are of factors A and B only. Regressing the column of S on $A, B, and \, C$, we obtain the equation

$$\text{STD} = 0.655 - 0.073A + 0.095B + 0.187C$$

with $R^2 = 0.805$. Only the main effect of C is significant. Factors A and B have no effects on the standard deviation. The strategy is therefore to set the value of C at -1 (as small as possible) and the values of A and B to adjust the mean response to be equal to the target value M. If $M = 85$, we find A and B to solve the equation

$$69.1 - 7.99A + 13.3B = 85.$$

Letting $B = 0.75$ then $A = -0.742$. The optimal setting of the design parameters $A, B, and \, C$ is at $A = -0.742$, $B = 0.75$, and $C = -1$.

14.1.5   Performance statistics

As we have seen in the previous example, the performance characteristic y at various combinations of the design parameters is represented by the mean, $\bar{Y}$, and standard deviation, S, of the y values observed at various combinations of the noise factors. We performed the analysis first on $\bar{Y}$ and then on S to detect which design parameters influence $\bar{Y}$ but not S, and which influence S but not $\bar{Y}$. Let $\eta(\theta)$ denote the expected value of Y, as a function of the design parameters $\theta_1, \ldots, \theta_k$. Let $\sigma^2(\theta)$ denote the variance of Y as a function of θ. The situation described above corresponds to the case that $\eta(\theta)$ and $\sigma^2(\theta)$ are independent. The **objective** in setting the values of $\theta_1, \ldots, \theta_k$, is to minimize the mean squared error

$$\text{MSE}(\theta) = B^2(\theta) + \sigma^2(\theta), \tag{14.1.7}$$

where $B(\theta) = \eta(\theta) - M$. The **performance statistic** is an estimator of MSE (θ), namely,

$$\widehat{\text{MSE}}(\theta) = (\bar{Y}(\theta) - M)^2 + S^2(\theta). \tag{14.1.8}$$

If $\bar{Y}(\theta)$ and $S^2(\theta)$ depend on different design parameters, we perform the minimization in two steps. First, we minimize $S^2(\theta)$ and then $\hat{B}^2(\theta) = (\bar{Y}(\theta) - M)^2$. If $\eta(\theta)$ and $\sigma^2(\theta)$ are **not** independent, the problem is more complicated.

Taguchi recommends to devise a function of $\eta(\theta)$ and $\sigma(\theta)$, which is called a **signal to noise** (SN) **ratio** and maximize an estimator of this SN function. Taguchi devised a large number of such performance statistics. In particular, Taguchi recommended to maximize the performance statistic

$$\eta = 10 \log \left(\frac{\bar{Y}^2}{S^2} - \frac{1}{n} \right), \tag{14.1.9}$$

which is being used in many studies. As can be shown by the method of the next section, the variance of $\log(\bar{Y})$ in large samples is approximately

$$V\{\log(\bar{Y})\} \cong \frac{\sigma^2}{n\mu^2}. \tag{14.1.10}$$

Thus,

$$-\log V\{\log(\bar{Y})\} = \log\left(\frac{\mu^2}{\sigma^2}\right) + \log n.$$

In the case of a normal distribution of Y, $\dfrac{\bar{Y}^2}{S^2} - \dfrac{1}{n}$ is an **unbiased** estimator of $\dfrac{\mu^2}{\sigma^2}$. Thus,

$$10\log\left(\frac{\bar{Y}^2}{S^2} - \frac{1}{n}\right) \tag{14.1.11}$$

is an estimator of $-10\log(nV\{\log(\bar{Y})\})$, although not an unbiased one. It is difficult to give any other justification to the performance statistic η(SN ratio). One has to be careful, since maximizing this SN might achieve bad results if $\eta(\theta)$ is far from the target M. Thus, if the objective is to set the design parameters to obtain means close to M and small standard deviations, one should minimize the MSE and not necessarily maximize the above SN ratio.

14.2   The effects of non-linearity

As mentioned in the previous section, the response function $f(\theta, \mathbf{x})$ might be nonlinear in θ and $\mathbf{x}$, an example was given for the case of output current of an *RL* circuit, namely

$$Y = \frac{V}{(R^2 + (2\pi f L)^2)^{1/2}}.$$

This is a nonlinear function of the design parameters R and L and the noise factor f. We have assumed that V and f are random variables, R and L are constant parameters. The output current is the random variable Y. What is the expected value and variance of Y? Generally, one can estimate the expected value of Y and its variance by simulation, using the function $f(\theta, \mathbf{X})$, and the assumed joint distribution of $\mathbf{X}$. An approximation to the expected value and the variance of Y can be obtained by the following method.

Let the random variables $X_1, \ldots, X_k$ have expected values $\xi_1, \ldots, \xi_k$ and variance covariance matrix

$$V = \begin{bmatrix} \sigma_1^2 & \sigma_{12} & \cdots & \sigma_{1k} \\ \sigma_{21} & & & \vdots \\ \vdots & \ddots & & \vdots \\ \sigma_{k1} & \sigma_{k2} & \cdots & \sigma_k^2 \end{bmatrix}.$$

Assuming that $f(\theta, \mathbf{X})$ can be expanded into a Taylor series around the means $\boldsymbol{\xi}_1 = (\xi_1, \ldots, \xi_k)$, we obtain the approximation

$$f(\theta, \mathbf{X}) \cong f(\theta, \boldsymbol{\xi}) + \sum_{i=1}^{k}(x_i - \xi_i)\frac{\partial}{\partial x_i}f(\theta, \boldsymbol{\xi}) + \frac{1}{2}(\mathbf{X} - \boldsymbol{\xi})'H(\theta, \boldsymbol{\xi})(\mathbf{X} - \boldsymbol{\xi}), \tag{14.2.1}$$

where $H(\theta, \xi)$ is a $k \times k$ matrix of second order partial derivatives, evaluated at ξ_i with (i,j)th element equal to

$$H_{ij}(\theta, \xi) = \frac{\partial^2}{\partial x_i \partial x_j} f(\theta, \xi), \quad i, j = 1, \ldots, k. \tag{14.2.2}$$

Thus, the expected value of $f(\theta, \mathbf{X})$ is approximated by

$$E\{f(\theta, \mathbf{X})\} \cong f(\theta, \xi) + \frac{1}{2} \sum_{i=1}^{k} \sum_{j=1}^{i} \sigma_{ij} H_{ij}(\theta, \xi), \tag{14.2.3}$$

and the variance of $f(\theta, \mathbf{X})$ is approximated by

$$V\{f(\theta, \mathbf{X})\} \cong \sum_{i=1}^{k} \sum_{j=1}^{k} \sigma_{ij} \frac{\partial}{\partial x_i} f(\theta, \xi) \frac{\partial}{\partial x_j} f(\theta, \xi). \tag{14.2.4}$$

As seen in these approximations, if the response variable Y is a nonlinear function of the random variables $X_1, \ldots, X_m$ its expected value depends also on the variances and covariances of the X's. This is not the case if Y is a linear function of the x's. Moreover, in the linear case the formula for $V\{Y\}$ is exact.

Example 14.3. Consider the function

$$Y = \frac{V}{(R^2 + (2\pi f L)^2)^{1/2}},$$

where $R = 5.0$ [Ω] and $L = 0.02$ [H]. V and f are independent random variables having normal distributions:

$$V \sim N(100, 9),$$

$$f \sim N(55, 25/9).$$

Notice that

$$\frac{\partial y}{\partial v} = \frac{1}{(R^2 + (2\pi fL)^2)^{1/2}},$$

$$\frac{\partial y}{\partial f} = -4V(R^2 + (2\pi fL)^2)^{-3/2} \pi^2 L^2 f,$$

$$\frac{\partial^2 y}{\partial v^2} = 0,$$

$$\frac{\partial^2 y}{\partial v \partial f} = -4(R^2 + (2\pi fL)^2)^{-3/2} \pi^2 L^2 f.$$

Also,

$$\frac{\partial^2 y}{\partial f^2} = -4V(R^2 + (2\pi fL)^2)^{-3/2} \pi^2 L^2$$

$$+ 48V(R^2 + (2\pi fL)^2)^{-5/2} \pi^4 L^4 f^2.$$

Substituting in these derivatives the value, of R and L, and the expected values of V and f, we obtain

$$\frac{\partial y}{\partial v} = \frac{1}{8.5304681} = 0.11723,$$

$$\frac{\partial y}{\partial f} = -0.13991,$$

$$\frac{\partial^2 y}{\partial v^2} = 0,$$

$$\frac{\partial^2 y}{\partial v \partial f} = -0.0013991,$$

$$\frac{\partial^2 y}{\partial f^2} = -0.0025438 + 0.0050098$$

$$= 0.002466.$$

Accordingly, an approximation for $E\{Y\}$ is given by

$$E\{Y\} \cong 11.722686 + \frac{1}{2}(9 \times 0 + 2.7778 \times 0.002466) = 11.7261.$$

The variance of Y is approximated by

$$V\{Y\} \cong 9 \times (0.11723)^2 + \frac{25}{9} \times (-0.13991)^2 = 0.17806.$$

To check the goodness of these approximations, we do the following simulation using MINITAB.

We simulate $N = 500$ normal random variables having mean 100 and standard deviation 3 into $C1$. Similarly, 500 normal random variables having mean 55 and standard deviation 1.67 are simulated into $C2$. In $C3$, we put the values of Y. This is done by the program

```
MTB> Random 500 C1;
SUBC> Normal 100 3.
MTB> Random 500 C2;
SUBC> Normal 55 1.67.
MTB> let k1 = 8 * ATAN(1)
MTB> let C3 = C1/sqrt(25 + (k1 * 0.02 * C2) ** 2)
MTB> mean C3
MTB> stan C3
```

```
> set.seed(123)
> X1 <- rnorm(500,
              mean=100,
              sd=3)
> X2 <- rnorm(500,
              mean=55,
              sd=1.67)
```

```
> K <- 8 * atan(1)
> Y <- X1/sqrt(25 + (K * 0.02 * X2)^2)
> mean(Y)
```

[1] 11.73927

```
> var(Y)
```

[1] 0.1822364

```
> rm(X1, X2, K, Y)
```

The results obtained are $\bar{Y}_{500} = 11.687$ and $S^2_{500} = 0.17123$. The analytical approximations are very close to the simulation estimates. Actually, a 0.95 confidence interval for $E\{Y\}$ is given by $\bar{Y}_{500} \pm 2\dfrac{S_{500}}{\sqrt{500}}$, which is $(11.650, 11.724)$. The result of the analytical approximation, 11.7261, is only slightly above the upper confidence limit. The approximation is quite good.

It is interesting to estimate the effects of the design parameters R and L on $E\{Y\}$ and $V\{Y\}$. We conduct a small experiment on the computer for estimating $E\{Y\}$ and $V\{Y\}$ by a 3^2 factorial experiment. The levels of R and L as recommended by G. Taguchi, in his review paper (see Ghosh (1990), pp. 1–34) are

	0	1	2
R	0.05	5.00	9.50
L	0.01	0.02	0.03

In each treatment combination, we simulate 500 y-values. The results are given in Table 14.3.

Table 14.3   *The means, variances, and MSE of Y in a 3^2 experiment*

R	L	$\bar{Y}_{500}$	S^2_{500}	MSE
0	0	28.943	1.436	360.27
0	1	14.556	0.387	21.14
0	2	9.628	0.166	0.30
1	0	16.441	0.286	41.77
1	1	11.744	0.171	3.21
1	2	8.607	0.109	1.88
2	0	9.891	0.087	0.10
2	1	8.529	0.078	2.24
2	2	7.119	0.064	8.36

The objective is to find the combinations of R and L which yield minimum MSE $= (E\{Y\} - M)^2 + V\{Y\}$, where M is the target of 10 [Ω]. It seems that the best setting of the design parameters is $R = 9.5$ [Ω] and $L = 0.01$ [H]. The combination $R = 0.05$ [Ω] and $L = 0.03$ [H] also yields very small MSE. One should choose the least expensive setting.

14.3 Taguchi's designs

In order to simulate the effect of noise factors, Taguchi advocates the combination of two experimental arrays, an **inner array** and an **outer array**. The inner array is used to determine factor-level combinations of factors that can be controlled by the designer of the product or process. The outer array is used to generate the variability due to noise factors that is experienced by the product or process under optimization, in its day to day operation.

The experimental arrays used by Taguchi are **orthogonal array** designs. An example of a design with 15 factors at two levels each, using 16 experiments is given in Table 14.4. The levels are indicated by 1 and 2, and the first row consists of all factors at level 1. This experimental array was introduced in Chapter 13 as a 2^{15-11} fractional factorial design which is a fully saturated design with 15 factors and 16 experimental runs. The corresponding full factorial design consists of $2^{15} = 32,768$ experiments. Taguchi labeled several experimental arrays using a convenient notation and reproduced them in tables that were widely distributed among engineers (see Taguchi and Taguchi 1987). The availability of these tables made it convenient for practitioners to design and run such experiments.

One can note from Table 14.4 that if we run the experiment using the order of the experiment array we will find it convenient to assign to column 1 a factor which is difficult to change from level 1 to level 2. For example if changing the temperature of a solder bath requires 5 hours the assignment of temperature to column 1 would require one change only. Column 15 on the other hand has nine changes between levels. Taguchi recommends that in some cases randomization be abandoned for the benefit of simplicity and cost. If we choose to run the experiment in the order of the experimental array, we can reduce the practical difficulties in running the experiment by proper assignment of factors to columns. An easily changed factor can get assigned to column 15 with low penalty. However assigning a factor that is difficult to change to column 15 might make the whole experiment impractical.

Table 14.4 *Factor-level combinations of 15 factors at two levels each in an $L_{16}(2^{15})$ orthogonal array*

Trial	Columns														
	1	2	3	4	5	6	7	8	9	10	11	12	13	14	15
1	1	1	1	1	1	1	1	1	1	1	1	1	1	1	1
2	1	1	1	1	1	1	1	2	2	2	2	2	2	2	2
3	1	1	1	2	2	2	2	1	1	1	1	2	2	2	2
4	1	1	1	2	2	2	2	2	2	2	2	1	1	1	1
5	1	2	2	1	1	2	2	1	1	2	2	1	1	2	2
6	1	2	2	1	1	2	2	2	2	1	1	2	2	1	1
7	1	2	2	2	2	1	1	1	1	2	2	2	2	1	1
8	1	2	2	2	2	1	1	2	2	1	1	1	1	2	2
9	2	1	2	1	2	1	2	1	2	1	2	1	2	1	2
10	2	1	2	1	2	1	2	2	1	2	1	2	1	2	1
11	2	1	2	2	1	2	1	1	2	1	2	2	1	2	1
12	2	1	2	2	1	2	1	2	1	2	1	1	2	1	2
13	2	2	1	1	2	2	1	1	2	2	1	1	2	2	1
14	2	2	1	1	2	2	1	2	1	1	2	2	1	1	2
15	2	2	1	2	1	1	2	1	2	2	1	2	1	1	2
16	2	2	1	2	1	1	2	2	1	1	2	1	2	2	1

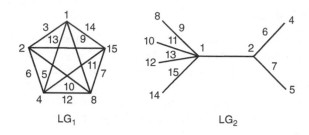

Figure 14.2   *Two linear graphs for $L_{16}(2^{15})$.*

In order to simulate the noise factors, we can design a second experiment using an external array. In some cases, noise cannot be directly simulated and the external array consists of replicating the internal array experiments over a specified length of time or amount of material. In Section 14.6, we describe two such experiments. The first experiment deals with a speedometer cable where the percentage of shrinkage is measured on several pieces of cable taken from various parts of a spool and running a heat test. The external array simply consists of the sampled parts of the spool. The second experiment deals with optimizing the response time of a computer system. Here the inner array experiment was carried out on an operational system so that the variability induced by various users were not specifically simulated. A retrospective study verified that there was no bias in user methods.

The design given in Table 14.4 can be used for up to 15 factors. It allows us to compute estimates of main effects, provided there are **no interactions** of any order between them. If, on the other hand, **all** first-order interactions can be potentially significant, this design cannot be used with more than five factors. (The resolution of the above design is III.) In order to assist the engineer with the correct choice of columns from the table of orthogonal arrays, Taguchi devised a graphical method of presenting the columns of an orthogonal array table which are confounded with first-order interactions of some factors. These graphs are called **linear graphs**.

In Figure 14.2, we present two linear graphs associated with Table 14.4. Linear graph LG_1 corresponds to the case where **all** interactions might be significant. The graph has 5 vertices and 10 lines connecting the vertices. The factors A, B, C, D, and E are assigned to the columns with numbers at the vertices. Thus, the assignment of the factors to columns are, according to this linear graph.

Factor	A	B	C	D	E
Column	1	2	4	8	15

We also see that column 3 can be used to estimate the interaction AB. Column 6 for the interaction BC, column 5 for the interaction AC, etc.

The second linear graph, LG_2, represents the case where only some interactions are significant. These are AB, AC, AD, AE FG, and FH. In this case, we can perform the 16 trials experiment with 8 factors and assign them to columns

Factor	A	B	C	D	E	F	G	H
Column	1	8	10	12	14	2	4	5

The columns that can be used to estimate the interactions are 3, 6, 7, 9, 11, 13, and 15.

Although the emphasis in Taguchi's methodology is on estimating main effects, one should not forget that interactions might exist. It is better to be cautious and not to over saturate a small design with too many factors. Recall that when fractional replications are used, the estimates of main effects might be confounded. We wish to choose a design with sufficient resolution (see Chapter 13), and this may require sufficiently large fractions. The table that we presented (Table 14.4) is a fraction of a 2^{15} factorial experiment. Taguchi also prepared tables of orthogonal arrays for 3^n factorial experiments, and for mixtures of factors with 2 and 3 levels ($2^m \times 3^k$ factorials). The reader is referred to the tables of Taguchi and Taguchi (1987) and also Appendix C of Phadke (1989).

14.4 Quality by design in the pharmaceutical industry

14.4.1 Introduction to quality by design

A product or process in the pharmaceutical industry is well understood when all critical sources of variability are identified and explained, variability is proactively managed, and product quality attributes can be accurately and reliably predicted. Drug manufacturing processes must meet current good manufacturing practices (cGMP) to ensure that drug products meet safety and efficacy requirements. Traditionally, the pharmaceutical industry has performed process validation studies on three batches. This approach, however, does not represent routine manufacturing and therefore is unlikely to cover all potential sources of variability (e.g., raw materials, operators, shifts, reactor vessels). The Office of New Drug Quality Assessment at the Food and Drug Administration (FDA) has identified this issue as a challenge to the regulatory process and launched a QbD initiative with a focus on product and process understanding (Nasr 2007).

QbD in the pharmaceutical industry is a systematic approach to development of drug products and drug manufacturing processes that begins with predefined objectives, emphasizes product and process understanding, and sets up process control based on sound science and quality risk management. In the traditional approach, product quality and performance are achieved predominantly by restricting flexibility in the manufacturing process and by end product testing. Under the QbD paradigm, pharmaceutical quality is assured by understanding and controlling manufacturing and formulation variables. End product testing is used to confirm the quality of the product and is not part of the ongoing consistency assurance and/or process control. The Food and Drug Administration (FDA) and the International Conference on Harmonization of Technical Requirements for Registration of Pharmaceuticals for Human Use (ICH) are promoting QbD in an attempt to curb rising development costs and regulatory barriers to innovation and creativity (Kenett and Kenett 2008). ICH guidelines published several guidelines that define QbD for both the pharmaceutical and biopharmaceutical industry (ICH Q8-Q11). The implementation of QbD involves the application of statistical Design of Experiments (DoE) described in Chapter 13 in the development of products, processes, analytical methods, and pharmaceutical formulations (Rathore and Mhatre 2009; Faltin *et al.* 2012). This section introduces QbD with a focus on the application of statistically designed experiments in this context.

Under QbD, statistically designed experiments are used for efficiently and effectively investigating potential main effects and interactions among process and product factors. The mathematical model derived from such designed experiments are then used together with the acceptable boundaries of critical quality attributes (CQA) to define a design space for a given process step. The normal operating range which is embedded within the design space yields quality attribute measurements that fall within the lower control limits (LCL) and upper control limits (UCL) representing the process performance (see Chapters 10–12). When the LCL and UCL fall well within the LSL and USL, the process step is predicted to be highly capable of delivering product that meets the requirements of subsequent steps in the process. Excursions outside the normal operating range are expected to deliver product with quality attributes that are acceptable for further processing, as long as the operating parameters are held to limits defined by the design space. When operations are affected by several quality attributes, the design space for the unit operation is obtained from overlays of the design spaces derived from analyses of multiple attributes, or from a multivariate analysis of the system.

The ICH Q10 guideline for Quality Systems indicates that controls for a product consist not only of process controls and final specifications for drug substance and drug product but also controls associated with the raw materials, excipients, container and closure, manufacturing equipment, and facility. It is a state of control in which all of the "planned controls" work together to ensure that the product delivered to the patient meets the patient's needs. Design space boundaries, as described above, are an integral part of a comprehensive control strategy. The control strategy for a product is expected to evolve through the product lifecycle. The purpose of a control strategy for a product is to ensure that sufficient controls are in place to maintain the risks associated with the product at a tolerable level. Risk management and control strategy principles are described in ICH Q9 (For a comprehensive treatment of operational risks see Kenett and Raanan (2010)). A well-designed control strategy that results from appropriate leveraging of QbD principles, then, leads to reliable product quality and patient safety profiles. The steps to develop a QbD drug application consist of: (1) Determine the quality target product profile, (2) define the CQA, (3) conduct a risk assessment to identify potential critical process parameters, (4) conduct statistically designed of experiments (DoE) to identify actual critical process parameters, (5) determine an appropriate control strategy, and (6) revise the risk assessment. In the next section, we present a QbD case study focusing on the setting up of a design space.

14.4.2   A quality by design case study – the full factorial design

The case study is a steroid lotion formulation of a generic product designed to match the properties of an existing brand using in vitro tests. In vitro release is one of several standard methods which can be used to characterize performance characteristics of a finished topical dosage form. Important changes in the characteristics of a drug product formula or the chemical and physical properties of the drug it contains should show up as a difference in drug release. Release is theoretically proportional to the square root of time when the formulation in question is in control of the release process because the release is from a receding boundary. In vitro release method for topical dosage forms described in SUPAC (1997) is based on an open chamber diffusion cell system such as a Franz cell system, fitted usually with a synthetic membrane. The test product is placed on the upper side of the membrane in the open donor chamber of the diffusion cell and a sampling fluid is placed on the other side of the membrane in a receptor cell. Diffusion of drug from the topical product to and across the membrane is monitored by assay of sequentially collected samples of the receptor fluid. A plot of the amount of drug released per unit area (mcg/cm^2) against the square root of time yields a straight line, the slope of which represents the release rate. This release rate measure is formulation-specific and can be used to monitor

Table 14.5   *Risk assessment of manufacturing process variables*

CQA	Manufacturing Process Variables				
	Temperature of reaction	Blending time	Cooling temperature	Cooling time	Order of ingredient addition
Appearance	High	High	Low	Low	High
Viscosity	High	High	Low	High	High
Assay	Low	Low	Low	Low	Low
In vitro permeability	High	High	Low	High	High

product quality. The typical in vitro release testing apparatus has six cells where the tested and brand products are being compared. A 90% confidence interval for the ratio of the median in vitro release rate in the tested and brand products are computed, and expressed in percentage terms. If the interval falls within the limits of 75–133.33%, the tested and brand products are considered equivalent.

An initial risk assessment mapped risks in meeting specifications of CQA. Table 14.5 presents expert opinions on the impact of manufacturing process variables on various CQAs. Cooling temperature was considered to have low impact and the order of ingredient addition was determined using risk contamination considerations. Later, both these factors were not studied in setting up the process design space.

The responses that will be considered in setting up the process design space include eight quality attributes: (1) assay of active ingredient, (2) in vitro permeability lower confidence interval, (3) in vitro permeability upper confidence interval, (4) 90*th* percentile of particle size, (5) assay of material A, (6) assay of material B, (7) viscosity, and (8) pH values. Three process factors are considered: (1) temperature of reaction, (2) blending time, and (3) cooling time.

In order to elicit the effect of the three factors on the eight responses, we will use a full factorial experiment with two center points (see Table 14.6 which presents the experimental array in standard order). In order to study the performance of this design, we use the JMP Prediction Variance Profile that shows the ratio of the prediction variance to the error variance, also called the relative variance of prediction, at various factor level combinations. Relative variance is minimized at the center of the design. As expected, if we choose to use a half fraction replication with four experimental runs on the edge of the cube, instead of the eight points full factorial, our variance will double (see Figure 14.3).

The data with the experimental arrays are available in the files "QbD Experiment.jmp" and "QbD Experiment.mtb." The two center points in the experimental array allow us to test for nonlinearity in the response surface by comparing the average responses at the center points, e.g., for viscosity (4135.5) with the average of the responses on the corners of the cube (4851.6). For a visual display of viscosity responses at the corners of the cube see Figure 14.4. From this figure, we can see that increasing temperature clearly increases viscosity. The difference between the averages of 716.1, in an analysis considering third-order interactions as noise, is not found significant at the 1% level of significance (Table 14.7). None of the other effects were found significant at that level. The bottom part of Table 14.7 repeats the analysis without the test for nonlinearity. Table 14.8 is an identical analysis of viscosity with JMP.

Table 14.6   *Full factorial design with two center points*

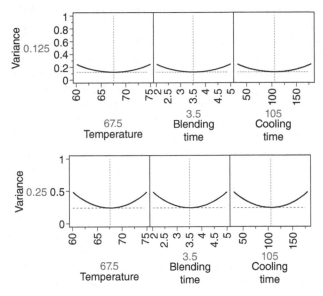

	Pattern	Temp	Blending Time	Cooling Time
1	----	60	2	30
2	+--	75	2	30
3	-+-	60	5	30
4	++-	75	5	30
5	000	67.5	3.5	105
6	000	67.5	3.5	105
7	--+	60	2	180
8	+-+	75	2	180
9	-++	60	5	180
10	+++	75	5	180

Source: JMP.

Figure 14.3   *Prediction variance profile for full factorial design of Table 14.6 (top) and half fraction design (bottom) (JMP).*

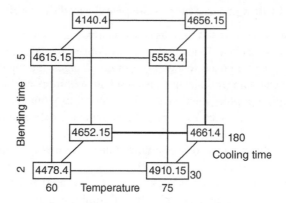

Figure 14.4 *Cube display of viscosity responses (JMP).*

Table 14.7 *Analysis of viscosity using second-order interaction model with and without test for nonlinearity (MINITAB)*

Term	Effect	Coefficient	SE coefficient	T	P
Constant		4851.6	72.13	67.27	0.000
Temperature	473.7	236.9	72.13	3.28	0.082
Blending time	65.8	32.9	72.13	0.46	0.693
Cooling time	−361.8	−180.9	72.13	−2.51	0.129
Temperature ↓ Blending time	253.3	126.6	72.13	1.76	0.221
Temperature ↓ Cooling time	−211.2	−105.6	72.13	−1.46	0.281
Blending time ↓ Cooling time	−324.2	−162.1	72.13	−2.25	0.154
Ct Pt		−716.1	161.28	−4.44	0.047
Term	Effect	Coefficient	SE coefficient	T	P
Constant		4708.4	173.6	27.13	0.000
Temperature	473.7	236.9	194.1	1.22	0.309
Blending time	65.8	32.9	194.1	0.17	0.876
Cooling time	−361.8	−180.9	194.1	−0.93	0.420
Temperature ↓ Blending time	253.3	126.6	194.1	0.65	0.561
Temperature ↓ Cooling time	−211.2	−105.6	194.1	−0.54	0.624
Blending time ↓ Cooling time	−324.2	−162.1	194.1	−0.84	0.465

Table 14.8 *Analysis of viscosity using second order interaction model (JMP)*

Term	Estimate	Std error	t Ratio	Prob>|t|
Temperature(60.75)	236.875	194.0543	1.22	0.3094
Cooling time(30.180)	−180.875	194.0543	−0.93	0.4200
Blending time*Cooling time	−162.125	194.0543	−0.84	0.4648
Temperature*Blending time	126.625	194.0543	0.65	0.5606
Temperature *Cooling time	−105.625	194.0543	−0.54	0.6241
Blending time(2.5)	32.875	194.0543	0.17	0.8763

14.4.3   A quality by design case study – the profiler and desirability function

The design space we are seeking is simultaneously addressing requirements on eight responses named: (1) active Assay, (2) in-vitro lower, (3) in-vitro upper, (4) D90, (5) A assay, (6) B assay, (7) viscosity, and (8) pH. Our goal is to identify operating ranges of temperature, blending time, and cooling time that guarantee that all eight responses are within specification limits. To achieve this objective, we apply a popular solution called the desirability function (Derringer and Suich 1980). Other techniques exist such as principal components analysis and nonlinear principal components (see Figini *et al.* 2010).

In order to combine the eight responses simultaneously, we first compute a desirability function using the characteristics of each response $Y_i(x), i = 1, \ldots, 8$. For each response, $Y_i(x)$, the univariate desirability function $d_i(Y_i)$ assigns numbers between 0 and 1 to the possible values of Y_i, with $d_i(Y_i) = 0$ representing a completely undesirable value of Y_i and $d_i(Y_i) = 1$ representing a completely desirable or ideal response value. The desirability functions for the eight responses are presented graphically in Figure 14.5. For Active assay, we want to be above 95% and up to 105%. Assay values below 95% yield desirability of zero, assay above 105% yield desirability of 1. For In vitro upper, we do not want to be above 133%. Our target for D90 is 1.5 with results above 2 and below 1 having zero desirability. The desirability functions scale the various responses to a value between 0 and 1. The design space can be assessed by an overall desirability index using the geometric mean of the individual desirabilities: *DesirabilityIndex* = $[d_1(Y_1) * d_2(Y_2) * \cdots d_k(Y_k)]^{\frac{1}{k}}$ with k denoting the number of measures (In our case $k = 8$). Notice that if any response Yi is completely undesirable ($d_i(Y_i) = 0$), then the overall desirability is zero. From Figure 14.5, we can see that setting Temperature = 65, Blending time = 2.5, and Cooling time = 150 gives us an overall Desirability index = 0.31. The JMP software allows you to introduce variability in the factor levels, as implemented in the Piston Simulator. In Figure 14.5, we apply three normal distributions for inducing variability in the settings of Temperature, Blending time, and Cooling time. This variability is then transferred to the eight responses and to the overall desirability index. The JMP output allows to simulate responses and visualize the impact of variability in factor level combinations to variability in response. Viscosity and In-Vitro Upper show the smallest variability relative to the experimental range.

14.4.4   A quality by design case study – the design space

To conclude the analysis we apply the JMP Contour Profiler to the experimental data fitting a model with main effects and two way interactions. The overlay surface is limited by the area with In vitro upper being above 133, which is not acceptable. As a design space we identify operating regions with blending time below 2.5 minutes, cooling time above 150 minutes and temperature ranging from 60 to 75 °C. Once approved by the regulator, these areas of operations are defined as the normal range of operation. Under QbD any changes within these regions do not require preapproval, only post change notification. This change in regulatory strategy is considered a breakthrough in traditional inspection doctrines and provide a significant regulatory relief.

An essential component of QbD submissions to the FDA is the design of a control strategy. Control is established by determining expected results and tracking actual results in the context of expected results. The expected results are used to set up upper and lower control limits. The use of simulations, as presented in Figure 14.6 can be used for this purpose. A final step in a QbD submission is to revise the risk assessment analysis. At this stage, the experts agreed that

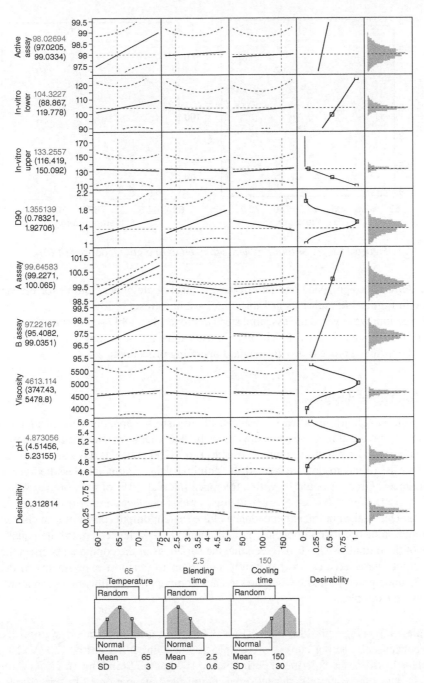

Figure 14.5 *Prediction profiler with individual and overall desirability function and variability in factor levels: Temperature = 65, Blending time = 2.5 and Cooling time = 150.*

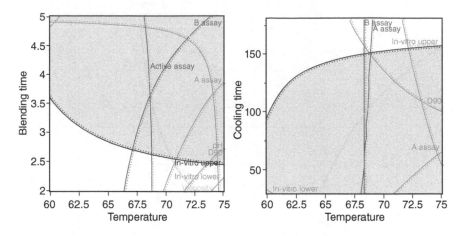

Figure 14.6   *Contour profiler with overlap of eight responses (JMP).*

with the defined design space and an effective control strategy accounting for the variability presented in Figure 14.5, all risks in Table 14.5 have been reset as low.

In this section, we covered the essential steps in preparing a QbD submission. We focused on the application of statistically designed experiments and show how they can be used to achieve robust and optimized process design standard operating procedures.

14.5   Tolerance designs

Usually parts which are installed in systems, like resistors, capacitors, transistors, and other parts of mechanical nature, have some deviations in their characteristics from the nominal ones. For example, a resistor with a nominal resistance of 8200 [Ω] will have an actual resistance value which is a random deviate around the nominal value. Parts are classified according to their tolerances. Grade A could be with a tolerance interval $\pm 1\%$ of the nominal value. Grade B of $\pm 5\%$, grade C of $\pm 10\%$, etc. Parts with high-grade tolerances are more expensive than low grade ones. Due to the nonlinear dependence of the system output (performance characteristic) on the input values of its components, not all component variances contribute equally to the variance of the output. We have also seen that the variances of the components effect the means of the output characteristics. It is therefore important to perform experiments to determine which tolerance grade should be assigned to each component. We illustrate such a problem in the following example.

Example 14.4.   Taguchi (1987, Vol. 1, pp. 379) describes a tolerance design for a circuit which converts alternating current of 100 [V] AC to a direct current of 220 [V] DC. This example is based on an experiment performed in 1974 at the Shin Nippon Denki Company.

The output of the system, Y, depends in a complicated manner on 17 factors. The R simulator `powerCircuitSimulation` was designed to experiment with this system.

In this example, we use `powerCircuitSimulation` to execute a fractional replication of 2^{13-8} to investigate the effects of two tolerance grades of 13 components, 10 resistors, and 3 transistors, on the output of the system. The two design levels for each factor are the two tolerance grades. For example, if we specify for a given factor a tolerance of 10%, then the experiment at level 1 will use at level 1 a tolerance of 5% and at level 2 a tolerance of 10%. The value of a given factor is simulated according to a normal distribution with mean at the nominal value of that factor. The standard deviation is 1/6 of the length of the tolerance interval. For example, if the nominal value for factor A is 8200 [Ω], and the tolerance level is 10%, the standard deviation for level 1 is 136.67 [Ω] and for level 2 is 273.33 [Ω].

As mentioned earlier, the control factors are 10 resistors, labeled $A-J$ and 3 transistors labeled $K-M$. The nominal levels of these factors are:

$$A = 8200, \quad B = 220,000, \quad C = 1000, \quad D = 33,000, \quad E = 56,000, \quad F = 5600,$$
$$G = 3300, \quad H = 58.5, \quad I = 1000, \quad J = 120, \quad K = 130, \quad L = 100, \quad M = 130$$

The levels of the 13 factors in the 2^{13-8} fractional replicate are given in Table 14.9.

Table 14.9 *Factor levels for the 2^{13-8} design*

Run	A	B	C	D	E	F	G	H	I	J	K	L	M
1	1	1	1	1	1	1	1	1	1	1	1	1	1
2	2	2	2	2	2	2	2	2	1	1	1	1	1
3	2	1	2	2	1	1	2	1	2	2	1	1	1
4	1	2	1	1	2	2	1	2	2	2	1	1	1
5	1	1	2	1	2	2	2	1	2	1	2	1	1
6	2	2	1	2	1	1	1	2	2	1	2	1	1
7	2	1	1	2	2	2	1	1	1	2	2	1	1
8	1	2	2	1	1	1	2	2	1	2	2	1	1
9	2	2	2	1	2	1	1	1	2	1	1	2	1
10	1	1	1	2	1	2	2	2	2	1	1	2	1
11	1	2	1	2	2	1	2	1	1	2	1	2	1
12	2	1	2	1	1	2	1	2	1	2	1	2	1
13	2	2	1	1	1	2	2	1	1	1	2	2	1
14	1	1	2	2	2	1	1	2	1	1	2	2	1
15	1	2	2	2	1	2	1	1	2	2	2	2	1
16	2	1	1	1	2	1	2	2	2	2	2	2	1
17	2	2	2	1	2	1	1	1	2	1	1	1	2
18	1	1	1	2	1	2	2	2	2	1	1	1	2
19	1	2	1	2	2	1	2	1	1	2	1	1	2
20	2	1	2	1	1	2	1	2	1	2	1	1	2
21	2	2	1	1	1	2	2	1	1	1	2	1	2
22	1	1	2	2	2	1	1	2	1	1	2	1	2
23	1	2	2	2	1	2	1	1	2	2	2	1	2
24	2	1	1	1	2	1	2	2	2	2	2	1	2
25	1	1	1	1	1	1	1	1	1	1	1	2	2
26	2	2	2	2	2	2	2	1	1	1	1	2	2
27	2	1	2	2	1	1	2	1	2	2	1	2	2
28	1	2	1	1	2	2	1	2	2	2	1	2	2
29	1	1	2	1	2	2	2	1	2	1	2	2	2
30	2	2	1	2	1	1	1	2	2	1	2	2	2
31	2	1	1	2	2	2	1	1	1	2	2	2	2
32	1	2	2	1	1	1	2	2	1	2	2	2	2

We perform this experiment on the computer, using program `powerCircuitSimulation`. We wish to find a treatment combination (run) which yields a small MSE at low cost per circuit. We will assume that grade B parts (5% tolerance) cost $1 and grade C parts (10% tolerance) cost $0.5. In order to obtain sufficiently precise estimates of the MSE, we perform at each run a simulated sample of size $n = 100$. The results of this experiment are given in Table 14.10.

Table 14.10   *Performance characteristics of tolerance design experiment*

Run	$\bar{Y}$	STD	MSE	TC
1	219.91	3.6420	13.2723	13
2	219.60	7.5026	56.4490	9
3	220.21	5.9314	35.2256	10
4	220.48	7.3349	54.0312	10
5	219.48	4.8595	23.8851	10
6	219.82	6.3183	39.9533	10
7	219.61	6.0647	36.9327	10
8	219.40	5.2205	27.6136	10
9	220.29	5.6093	31.5483	10
10	218.52	6.5635	45.2699	10
11	219.71	4.0752	16.6914	10
12	220.27	5.6723	32.2479	10
13	220.74	5.8068	34.2665	10
14	219.93	5.4065	29.2351	10
15	219.92	5.6605	32.0477	9
16	219.71	6.9693	48.6552	9
17	219.93	5.1390	26.4142	10
18	221.49	6.6135	45.9585	10
19	219.98	4.1369	17.1143	10
20	220.10	6.5837	43.3551	10
21	220.65	6.0391	36.8932	10
22	219.38	5.7089	32.9759	10
23	220.26	6.2068	38.5920	9
24	219.97	6.3469	40.2840	9
25	220.53	4.0378	16.5847	12
26	220.20	6.6526	44.2971	8
27	220.22	5.4881	30.1676	9
28	219.48	6.1564	38.1717	9
29	219.60	5.1583	26.7681	9
30	220.50	6.3103	40.0699	9
31	221.43	5.8592	36.3751	9
32	220.22	5.2319	27.4212	9

```
> library(FrF2)
> Factors <- list(
    tlA = c(5, 10), tlB = c(5, 10), tlC = c(5, 10),
    tlD = c(5, 10), tlE = c(5, 10), tlF = c(5, 10),
    tlG = c(5, 10), tlH = c(5, 10), tlI = c(5, 10),
```

```
    tlJ = c(5, 10), tlK = c(5, 10), tlL = c(5, 10),
    tlM = c(5, 10))
> FrDesign <- FrF2(nruns=32,
                  factor.names=Factors,
                  randomize=TRUE,
                  replications=100,
                  repeat.only=TRUE)
> FrDesign[c(1, 101, 201), ]

    tlA tlB tlC tlD tlE tlF tlG tlH tlI tlJ tlK tlL
1     5   5   5   5   5   5   5   5   5   5   5   5
101   5  10  10  10  10   5   5   5  10   5   5  10
201   5   5   5   5  10   5   5   5   5  10  10  10
    tlM
1     5
101   5
201  10


> Levels <- data.frame(
    rsA = 8200, rsB = 2,20,000, rsC = 1000,
    rsD = 33000, rsE = 56,000, rsF = 5600,
    rsG = 3300, rsH = 58.5, rsI = 1000,
    rsJ = 120, trK = 130, trL = 100, trM = 130,
    lapply(
      lapply(FrDesign,
             as.character),
      as.numeric))
> Ps <- powerCircuitSimulation(
    rsA = Levels$rsA, rsB = Levels$rsB, rsC = Levels$rsC,
    rsD = Levels$rsD, rsE = Levels$rsE, rsF = Levels$rsF,
    rsG = Levels$rsG, rsH = Levels$rsH, rsI = Levels$rsI,
    rsJ = Levels$rsJ, trK = Levels$trK, trL = Levels$trL,
    trM = Levels$trM, tlA = Levels$tlA, tlB = Levels$tlB,
    tlC = Levels$tlC, tlD = Levels$tlD, tlE = Levels$tlE,
    tlF = Levels$tlF, tlG = Levels$tlG, tlH = Levels$tlH,
    tlI = Levels$tlI, tlJ = Levels$tlJ, tlK = Levels$tlK,
    tlL = Levels$tlL, tlM = Levels$tlM,
    each=1,
    seed=123)
> FrDesign <- add.response(
    design=FrDesign,
    response=Ps$volts)
> Ps <- simulationGroup(Ps, 100)
> X <- aggregate(x=Ps["volts"],
                 by=Ps["group"],
                 FUN=mean)
> names(X) <- c("run", "mean")
> X2 <- aggregate(x=Ps["volts"],
                 by=Ps["group"],
```

```
                    FUN=sd)
> names(X2) <- c("run", "sd")
> X <- merge(X, X2, by="run")
> X2 <- aggregate(
    rowSums(ifelse(Ps[,14:26] == 10,
                   yes=0.5,
                   no=1)),
    by=Ps["group"],
    FUN=mean)
> names(X2) <- c("run", "tc")
> X <- merge(X, X2)
> rownames(X) <- X$run
> head(X[order(X$sd),], 6)

   run     mean        sd    tc
1    1 230.0750  3.049538  13.0
18  18 229.9883  4.271285   9.5
3    3 230.9698  4.608221  10.5
26  26 230.0474  4.782348  10.0
32  32 229.2853  4.915447  10.0
13  13 230.4899  4.943400  10.0

> rm(Levels, Factors, FrDesign, Ps, X, X2)
```

We see that the runs having small MSE are 1, 11, 19, and 25. Among these, the runs with the smallest total cost (TC) are 11 and 19. The MSE of run 11 is somewhat smaller than that of run 19. The difference, however, is not significant. We can choose either combination of tolerance levels for the manufacturing of the circuits.

14.6   Case studies

14.6.1   The Quinlan experiment

This experiment was carried out at Flex Products in Midvale Ohio (Quinlan 1985). Flex Products is a subcontractor of General Motors, manufacturing mechanical speedometer cables. The basic cable design has not changed for 15 years and General Motors had experienced many disappointing attempts at reducing the speedometer noise level. Flex products decided to apply the off-line quality control and involve in the project customers, production personnel, and engineers with experience in the product and manufacturing process. A large experiment involving 15 factors was designed and completed. The data showed that much improvement could be gained by few simple changes. The results were dramatic and the loss per unit was reduced from $2.12 to $0.13 by changing the braid type, the linear material, and the braiding tension.

We proceed to describe the experiment using an eight points template:

1. *Problem definition*: The product under investigation is an extruded thermoplastic speedometer casing used to cover the mechanical speedometer cable on automobiles. Excessive shrinkage of the casing is causing noise in the mechanical speedometer cable assembly.

2. *Response variable*: The performance characteristic in this problem is the post extrusion shrinkage of the casing. The percent shrinkage is obtained by measuring approximately 600 mm of casing that has been properly conditioned (*A*), placing that casing in a 2-hour heat soak in an air circulating oven, reconditioning the sample and measuring the length (*B*). Shrinkage is computed as: Shrinkage = $100 \times (A - B)/A$.

3. *Control factors*:

    Liner process:

        A: Liner O.D.
        B: Liner die
        C: Liner material
        D: Liner line speed

    Wire braiding:

        E: Wire braid type
        F: Braiding tension
        G: Wire diameter
        H: Liner tension
        I: Liner temperature

    Coating process:

        J: Coating material
        K: Coating dye type
        L: Melt temperature
        M: Screen pack
        N: Cooling method
        O: Line speed

4. *Factor levels*: Existing (1) – Changed (2)
5. *Experimental array*: $L_{16}(2^{15})$ Orthogonal array.
6. *Number of replications*: Four random samples of 600 mm from the 3000 ft manufactured at each experimental run.
7. *Data analysis:* Signal to noise ratios (SN) are computed for each experimental run and analyzed using main effect plots and an ANOVA. Savings are derived from Loss function computations.

The signal to noise formula used by Quinlan is $\eta = -10\log_{10}\left(\dfrac{1}{n}\sum_{i=1}^{n} y_i^2\right)$.

For example, experimental run number 1 produced shrinkage factors of: 0.49, 0.54, 0.46, 0.45. The SN is 6.26. The objective is to maximize the SN by proper setup of the 15 controllable factors.

Table 14.11 shows the factor levels and the SN values, for all 16 experimental runs.

Notice that Quinlan, by using the orthogonal array $L_{16}(2^{15})$ for all the 15 factors, assumes that there are no significant interactions. If this assumption is correct then, the main effects of the 15 factors are:

Factor	*A*	*B*	*C*	*D*	*E*	*F*	*G*	*H*
Main effect	−1.145	0.29	1.14	−0.86	3.60	1.11	2.37	−0.82

Table 14.11   *Factor levels and SN values*

Run	A	B	C	D	E	F	G	H	I	J	K	L	M	N	O	SN
1	1	1	1	1	1	1	1	1	1	1	1	1	1	1	1	6.26
2	1	1	1	1	1	1	1	2	2	2	2	2	2	2	2	4.80
3	1	1	1	2	2	2	2	1	1	1	1	2	2	2	2	21.04
4	1	1	1	2	2	2	2	2	2	2	2	1	1	1	1	15.11
5	1	2	2	1	1	2	2	1	1	2	2	1	1	2	2	14.03
6	1	2	2	1	1	2	2	2	2	1	1	2	2	1	1	16.69
7	1	2	2	2	2	1	1	1	1	2	2	2	2	1	1	12.91
8	1	2	2	2	2	1	1	2	2	1	1	1	1	2	2	15.05
9	2	1	2	1	2	1	2	1	2	1	2	1	2	1	2	17.67
10	2	1	2	1	2	1	2	2	1	2	1	2	1	2	1	17.27
11	2	1	2	2	1	2	1	1	2	1	2	2	1	2	1	6.82
12	2	1	2	2	1	2	1	2	1	2	1	1	2	1	2	5.43
13	2	2	1	1	2	2	1	1	2	2	1	1	2	2	1	15.27
14	2	2	1	1	2	2	1	2	1	1	2	2	1	1	2	11.2
15	2	2	1	2	1	1	2	1	2	2	1	2	1	1	2	9.24
16	2	2	1	2	1	1	2	2	1	1	2	1	2	2	1	4.68

Factor	I	J	K	L	M	N	O
Main effect	0.49	−0.34	−1.19	0.41	0.22	0.28	0.22

Figure 14.7 presents the main effects plot for this experiment. Factors E and G seem to be most influential. These main effects, as defined in Chapter 11, are the regression coefficients of SN on the design coefficients ± 1. As mentioned in Chapter 13, these are sometimes called "half effects." Only the effects of factors E and G are significant. If the assumption of no-interaction is wrong, and all the first-order interactions are significant then, as shown in the linear graph LG_1 in Figure 14.2, only the effects of factors A, B, D, H and O are not confounded. The effects of the other factors are confounded with first- order interactions. The main effect of factor E is confounded with the interaction AD, that of G is confounded with HO. In order to confirm the first hypothesis, that all interactions are negligible, an additional experiment should be performed, in which factors E and G will be assigned columns which do not represent possible interactions (like columns 1 and 2 of Table 14.4). The results of the additional experiment should reconfirm the conclusions of the original experiment.

8. *Results*: As a result of Quinlan's analysis, factors E and G were properly changed. This reduced the average shrinkage index from 26% to 5%. The shrinkage standard deviation was also reduced, from 0.05 to 0.025. This was considered a substantial success in quality improvement.

14.6.2   Computer response time optimization

The experiment described here was part of an extensive effort to optimize a UNIX operating system running on a VAX 11-780 machine (Pao *et al.* 1985). The machine had 48 user terminal

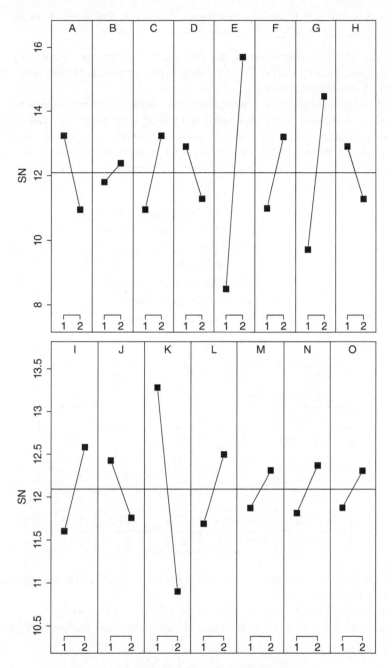

Figure 14.7 *Main effects plot for Quinlan experiment.*

ports, two remote job entry links, four megabytes of memory, and five disk drives. The typical number of users logged on at a given time was between 20 and 30.

1. *Problem definition*: Users complained that the system performance was very poor, especially in the afternoon. The objective of the improvement effort was to both minimize response time and reduce variability in response.
2. *Response variable*: In order to get an objective measurement of the response time two specific representative commands called **standard** and **trivial** were used. The **standard** command consisted of creating, editing and removing a file. The **trivial** command was the UNIX system "date" command. Response times were measured by submitting these commands every 10 minutes and clocking the time taken for the system to complete their execution.
3. **Control factors**:

A: Disk drives
B: File distribution
C: Memory size
D: System buffers
E: Sticky bits
F: KMCs used
G: INODE table entries
H: Other system tables

4. **Factor levels**:

Factor	Levels		
A: RM05 & RP06	4 & 1		4 & 2
B: File distribution	a	b	c
C: Memory size (MB)	4	3	3.5
D: System buffers	1/5	1/4	1/3
E: Sticky bits	0	3	8
F: KMCs used	2		0
G: INODE table entries	400	500	600
H: Other system tables	a	b	c

5. *Experimental array*: The design was an orthogonal array $L_{18}(3^8)$. This and the mean response are given in the following table.

Each mean response in Table 14.12 is over $n = 96$ measurements.
6. *Data analysis*: The measure of performance characteristic used was the S/N ratio

$$\eta = -10\log_{10}\left(\frac{1}{n}\sum_{i=1}^{n} y_i^2\right),$$

where y_i is the ith response time.

Table 14.12 Factor levels and mean response*

	F	B	C	D	E	A	G	H	Mean (seconds)	SN
1	1	1	1	1	1	1	1	1	4.65	−14.66
2	1	1	2	2	2	2	2	2	5.28	−16.37
3	1	1	3	3	3	3	3	3	3.06	−10.49
4	1	2	1	1	2	2	3	3	4.53	−14.85
5	1	2	2	2	3	3	1	1	3.26	−10.94
6	1	2	3	3	1	1	2	2	4.55	−14.96
7	1	3	1	2	1	3	2	3	3.37	−11.77
8	1	3	2	3	2	1	3	1	5.62	−16.72
9	1	3	3	1	3	2	1	2	4.87	−14.67
10	2	1	1	3	3	2	2	1	4.13	−13.52
11	2	1	2	1	1	3	3	2	4.08	−13.79
12	2	1	3	2	2	1	1	3	4.45	−14.19
13	2	2	1	2	3	1	3	2	3.81	−12.89
14	2	2	2	3	1	2	1	3	5.87	−16.75
15	2	2	3	1	2	3	2	1	3.42	−11.65
16	2	3	1	3	2	3	1	2	3.66	−12.23
17	2	3	2	1	3	1	2	3	3.92	−12.81
18	2	3	3	2	1	2	3	1	4.42	−13.71

*) Factor *A* had only 2 levels. All levels 3 in the table were changed to level 2.

Figure 14.8 is the main effects plot of these eight factors. The linear and quadratic effects of the factors were found to be

Factor	Linear	Quadratic
A	0.97	–
B	0.19	−0.15
C	−1.24	−1.32
D	−0.37	−1.23
E	1.72	1.86
F	0.44	–
G	0.17	−0.63
H	0.05	1.29

We see that factors having substantial effects are *A*, *C*, *D*, *E*, and *H*. As a result, the number of disk drives were changed to 4 and 2. The system buffers were changed from 1/3 to 1/4. Number of sticky bits were changed from 0 to 8. After introducing these changes, the average response time dropped from 6.15 (seconds) to 2.37 (seconds) with a substantial reduction in response times variability.

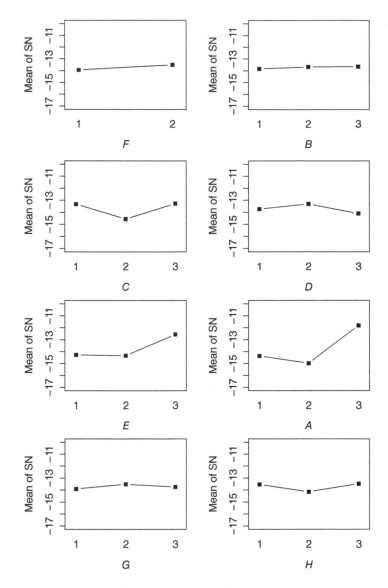

Figure 14.8   *Main effects plot.*

14.7   Chapter highlights

Quality is largely determined by decisions made in the early planning phases of products and processes. A particularly powerful technique for making optimal design decisions is the statistically designed experiment. The chapter covers the basics of experimental designs in the context of engineering and economic optimization problems. Taguchi's loss function, signal to noise ratios, factorial models, and orthogonal arrays are discussed using case studies and simple examples. A special section is dedicated to the application of QbD in the pharmaceutical industry. QbD is supported internationally by the International Conference on

Harmonization of Technical Requirements for Registration of Pharmaceuticals for Human Use (ICH) and by the Food and Drug administration (FDA).

The main concepts and definitions introduced in this chapter include:

- Design of Experiments
- Robust Design
- Quality Planning
- Quality Engineering
- Off-Line Quality Control
- Loss Functions
- Parameter Design
- Tolerance Design
- Response Surfaces
- Mixture Designs
- Inner Array
- Outer Array
- Linear Graph
- Signal to Noise
- Performance Measures
- Quality by Design (QbD)
- Design Space
- Control Strategy
- Risk Management
- Critical Quality Attributes (CQA)
- ICH Guidelines Q8-Q11
- Desirability Function
- Current Good Manufacturing Practices (cGMP)
- Prediction Profiler
- Desirability Function

14.8 Exercises

14.1 The objective is to find the levels of the factors of the turbo piston, which yield average cycle time of 0.45 [seconds]. Execute program `pistonSimulation` or the JMP addin, with sample size $n = 100$.

  (i) Determine which treatment combination yields the smallest $\text{MSE} = (\bar{Y} - 0.45)^2 + S^2$.

  (ii) Determine which treatment combination yields the largest SN ratio, $\eta = 10 \log_{10} \left(\dfrac{Y^2}{S^2} - \dfrac{1}{100} \right)$? What is the MSE at this treatment combination?

  The five factors which are varied are: piston weight, piston surface area, initial gas volume, spring coefficient, and ambient temperature. The factors atmospheric pressure and filling gas temperature are kept constant at the midrange level.

14.2 Run program `pistonSimulation` or the JMP addin with sample size of $n = 100$ and generate the sample means and standard deviation of the $2^7 = 128$ treatment combinations of a full factorial experiment, for the effects on the piston cycle time. Perform regression analysis to find which factors have significant effects on the signal to noise ratio $SN = \log((\bar{X}/S)^2)$.

14.3 Let (X_1, X_2) have joint distribution with means (ξ_1, ξ_2) and covariance matrix

$$V = \begin{pmatrix} \sigma_1^2 & \sigma_{12} \\ \sigma_{12} & \sigma_2^2 \end{pmatrix}.$$

Find approximations to the expected values and variances of:
  (i) $Y = X_1/X_2$;
  (ii) $Y = \log(X_1^2/X_2^2)$;
  (iii) $Y = (X_1^2 + X_2^2)^{1/2}$.

14.4 The relationship between the absorption ratio Y of a solid image in a copied paper and the light intensity X is given by the function

$$Y = 0.0782 + \frac{0.90258}{1 + 0.6969X^{-1.4258}}.$$

Assuming that X has the gamma distribution $G(1, 1.5)$, approximate the expected value and variance of Y.

14.5 Let $\bar{X}_n$ and S_n^2 be the mean and variance of a random sample of size n from a normal distribution $N(\mu, \sigma)$. We know that $\bar{X}_n$ and S_n^2 are independent, $\bar{X}_n \sim N\left(\mu, \dfrac{\sigma}{\sqrt{n}}\right)$ and $S_n^2 \sim \dfrac{\sigma^2}{n-1}\chi^2[n-1]$. Find an approximation to the expected value and variance of $Y = \log\left(\dfrac{\bar{X}_n^2}{S_n^2}\right)$.

14.6 An experiment based on an L_{18} orthogonal array involving eight factors, gave the following results (see Phadke *et al.* 1983). Each run had $n = 5$ replications.

Run	Factors								$\bar{X}$	S
	1	2	3	4	5	6	7	8		
1	1	1	1	1	1	1	1	1	2.500	0.0827
2	1	1	2	2	2	2	2	2	2.684	0.1196
3	1	1	3	3	3	3	3	3	2.660	0.1722
4	1	2	1	1	2	2	3	3	1.962	0.1696
5	1	2	2	2	3	3	1	1	1.870	0.1168
6	1	2	3	3	1	1	2	2	2.584	0.1106
7	1	3	1	2	1	3	2	3	2.032	0.0718
8	1	3	2	3	2	1	3	1	3.267	0.2101
9	1	3	3	1	3	2	1	2	2.829	0.1516
10	2	1	1	3	3	2	2	1	2.660	0.1912
11	2	1	2	1	1	3	3	2	3.166	0.0674
12	2	1	3	2	2	1	1	3	3.323	0.1274

Run	Factors								$\bar{X}$	S
	1	2	3	4	5	6	7	8		
13	2	2	1	2	3	1	3	2	2.576	0.0850
14	2	2	2	3	1	2	1	3	2.308	0.0964
15	2	2	3	1	2	3	2	1	2.464	0.0385
16	2	3	1	3	2	3	1	2	2.667	0.0706
17	2	3	2	1	3	1	2	3	3.156	0.1569
18	2	3	3	2	1	2	3	1	3.494	0.0473

Analyze the effects of the factors of the SN ratio $\eta = \log(\bar{X}/S)$.

14.7 Using `pistonSimulation` or the JMP addin perform a full factorial (2^7), a 1/8 (2^{7-3}), 1/4 (2^{7-2}) and 1/2 (2^{7-1}) fractional replications of the cycle time experiment. Estimate the main-effects of the seven factors with respect to SN $= \log(\bar{X}/S)$ and compare the results obtained from these experiments.

14.8 To see the effect of the variances of the random variables on the expected response, in nonlinear cases, execute `pistonSimulation` or the JMP addin, with $n = 20$, and compare the output means to the values in Table 14.5.

14.9 Run program `powerCircuitSimulation` with 1% and 2% tolerances, and compare the results to those of Table 12.8.

15

Computer Experiments

15.1 Introduction to computer experiments

Experimentation via computer modeling has become very common in many areas of science and technology. In computer experiments, physical processes are simulated by running a computer code that generates output data for given input values. In physical experiments, data is generated directly from a physical process. In both physical and computer experiments, a study is designed to answer specific research questions and appropriate statistical methods are needed to design the experiment and to analyze the resulting data. Chapters 13 and 14 present such methods and many examples. In this chapter, we focus on computer experiments and specific design and analysis methods relevant to such experiments.

Because of experimental error, a physical experiment will produce a different output for different runs at the same input settings. Computer experiments are deterministic and the same inputs will always result in the same output. Thus, none of the traditional principles of blocking, randomization, and replication can be used in the design and analysis of computer experiments data. On the other hand, computer experiments use extensively random number generators, these are described in Section 15.6.

Computer experiments consist of a number of runs of a simulation code, and factor-level combinations correspond to a subset of code inputs. By considering computer runs as a realization of a stochastic process, a statistical framework is available both to design the experimental points and to analyze the responses. A major difference between computer numerical experiments and physical experiments is the logical difficulty in specifying a source of randomness for computer experiments.

The complexity of the mathematical models implemented in the computer programs can, by themselves, build equivalent sources of random noise. In complex code, a number of parameters and model choices gives the user many degrees of freedom that provide potential variability to the outputs of the simulation. Examples include different solution algorithms (i.e., implicit or explicit methods for solving differential systems), approach to discretization intervals, and convergence thresholds for iterative techniques. In this very sense, an experimental error can be considered in the statistical analysis of computer experiments. The nature of the experimental

Modern Industrial Statistics: With Applications in R, MINITAB and JMP, Third Edition.
Ron S. Kenett and Shelemyahu Zacks.

error in both physical and simulated experiments is our ignorance about the phenomena and the intrinsic error of the measurements. Real-world phenomena are too complex for the experimenter to keep under control by specifying all the factors affecting the response of the experiment. Even if it were possible, the physical measuring instruments, being not ideal, introduce problems of accuracy and precision (see Section 2.2). Perfect knowledge would be achieved in physical experiments only if all experimental factors can be controlled and measured without any error. Similar phenomena occur in computer experiments. A complex code has several degrees of freedom in its implementation that are not controllable.

A specific case where randomness is introduced to computer experiments consists of the popular finite element method (FEM) programs. These models are applied in a variety of technical sectors such as electromagnetics, fluid-dynamics, mechanical design, and civil design. The FEM mathematical models are based on a system of partial differential equations defined on a time–space domain for handling linear or nonlinear, steady-state or dynamic problems. FEM software can deal with very complex shapes as well as with a variety of material properties, boundary conditions, and loads. Applications of FEM simulations require subdivision of the space-domain into a finite number of subdomains, named finite elements, and solving the partial differential system within each subdomain, letting the field-function to be continuous on its border.

Experienced FEM practitioners are aware that results of complex simulations (complex shapes, nonlinear constitutive equations, dynamic problems, contacts among different bodies, etc.) can be sensitive to the choice of manifold model parameters. Reliability of FEM results is a critical issue for the single simulation and even more for a series of computer experiment. The model parameters used in the discretization of the geometry are likely to be the most critical. Discretization of the model geometry consists in a set of points (nodes of the mesh) and a set of elements (two-dimensional patches or three-dimensional volumes) defined through a connectivity matrix whose rows list the nodes enclosing the elements. Many degrees of freedom are available to the analyst when defining a mesh on a given model. Changing the location and the number of nodes, the shape and the number of elements an infinity of meshes are obtained. Any of them will produce different results. How can we model the effects of different meshes on the experimental response? In principle, the finer the discretization, the better the approximation of numerical solution, even if numerical instabilities may occur using very refined meshes. Within a reasonable approximation, a systematical effect can be assigned to mesh density; it would be a fixed-effect factor if it is included in the experiment. A number of topological features (node locations, element shape), which the analyst has no meaningful effect to assign to, are generators of random variability. One can assume that they are randomized along the experiment or random-effect factors with nuisance variance components if they are included as experimental factors.

Mesh selection has also a direct economical impact as computational complexity grows with the power of the number of the elements. In the case of computer experiments, the problem of balancing reliability and cost of the experiment needs to be carefully addressed. In principle, for any output of a numerical code, the following deterministic model holds:

$$y = f(\mathbf{x}) + g(\mathbf{x}; \mathbf{u}), \qquad (15.1.1)$$

where the function f represents the dependence of the output y on the vector x of experimental factors, and g describes the contribution of parameters, u, which are necessary for the setup of the computer model. Since the function g may have interactions with engineering parameters,

x is also an argument of function g. Generally, an engineer is interested in the estimation of function f while he considers g as a disturbance. In general, two options are available for analyzing computer experiments: (1) considering the model parameters as additional experimental factors or (2) fixing them along the whole experiment. The first option allows the estimation of the deterministic model written in (15.3.1). This is a good choice since the influence of both engineering and model parameters on the experimental response can be evaluated. This requires however an experimental effort that cannot be often affordable. Keeping every model parameter at a fixed value in the experiment, only the first term f of model (15.1.1) can be estimated. This results in a less expensive experiment but has two dangerous drawbacks: (1) the presence of effects of model parameters on the function g in (15.1.1) can cause a bias in the response and (2) the estimates of the effects of engineering parameters are distorted by the interactions between model and engineering parameters according to the function g. A different approach is to randomize along the experiment those model parameters whose effects can reasonably be assumed to be normal random variables with zero average. In this case, the underlying model becomes a stochastic one:

$$y = f(\mathbf{x}) + g^*(\mathbf{x}; \mathbf{u}) + \epsilon, \tag{15.1.2}$$

where g^* in (15.1.2) is a function that represents the mixed contribution between engineering and fixed-effects model parameters, after random-effects model parameters have been accounted for in building the experimental error. Any model parameter that is suspected to have a substantial interaction with some engineering parameters should be included as experimental factor so that the systematic deviation of effects of such engineering parameters is prevented. Randomization of model parameters yields two simultaneous benefits. On the one hand, the model has acquired a random component equivalent to the experimental error of physical experiments; and in this way, the rationale of replications is again justified so that a natural measurement scale for effects is introduced and usual statistical significance tests can be adopted. On the other hand, without any increase of experimental effort, possible interactions between randomized model parameters and engineering parameters do not give rise to distortion of effects of engineering parameters

Table 15.1   *Different models for computer experiments*

Option	Model	Model nature	Advantages	Disadvantages
Fixed model factors	3:mprescripts$y = f(\mathbf{x})$	Deterministic	Inexpensive	Possible bias and distortion of effects of engineering parameters.
Model of factor effects is included in the experiment	$y = f(\mathbf{x}) + g(\mathbf{x}; \mathbf{u})$	Deterministic	More accurate. Systematic effect of $\mathbf{u}$ can be discovered.	More programming is required.
Randomizing model parameters. with random effects	$y = f(\mathbf{x}) + g^*(\mathbf{x}; \mathbf{u}) + \epsilon$	Stochastic	Possibility of calibrating experimental error.	Even more programming is required.

or experimental factors. Moreover, it becomes possible to tune the experimental error of the computer experiment to that of the experimental error of a related physical experiment. In the case where several u parameters are present, it is likely that the normality assumption for random errors is reasonable. Table 15.1, adapted from Romano and Vicario (2002), summarizes a variety of approaches to computer experiments that are presented below in some detail.

One of the modeling methods applied to computer experiments data is Kriging also called Gaussian process models. Section 15.3 is dedicated to Kriging methods for data analysis. Throughout this chapter, we refer to the piston simulator of Example 2.1 and later as Examples 4.13, 4.23, 4.24, and 4.25. A JMP script implementing this simulator is described in the Book Appendix. An R version of this simulator, **pistonSimulation**, is included in the `mistat` package. We describe next the mathematical foundation of the piston simulator.

Example 15.1.   The piston cycle time data introduced in Example 2.1 is generated by software simulating a piston moving within a cylinder. The piston's linear motion is transformed into circular motion by connecting a linear rod to a disk. The faster the piston moves inside the cylinder, the quicker the disk rotation and therefore the faster the engine will run. The piston's performance is measured by the time it takes to complete one cycle, in seconds. The purpose of the simulator is to study the causes of variability in piston cycle time. The following factors (listed below with their units and ranges) affect the piston's performance:

- M = Piston weight (kg), 30–60
- S = Piston surface area (m^2), 0.005–0.020
- V_0 = Initial gas volume (m^3), 0.002–0.010
- k = Spring coefficient (N/m), 1000–5000
- P_0 = Atmospheric pressure (N/m^2), 9×10^4–11×10^4
- T = Ambient temperature (K), 290–296
- T_0 = Filling gas temperature (K), 340–360

These factors affect the cycle time via a chain of nonlinear equations:

$$\text{Cycle Time} = 2\pi \sqrt{\frac{M}{k + S^2 \frac{P_0 V_0}{T_0} \frac{T}{V^2}}}, \tag{15.1.3}$$

where

$$V = \frac{S}{2k}\left(\sqrt{A^2 + 4k\frac{P_0 V_0}{T_0}T} - A\right) \text{ and } A = P_0 S + 19.62 M - \frac{kV_0}{S}. \tag{15.1.4}$$

Randomness in Cycle Time is induced by generating observations for factors set up around design points with noise added to the nominal values. Figure 15.1 shows the operating panel of the piston simulator add-in within the JMP application. To change the factor-level combinations simply move the sliders left or right. To install it, after installing JMP, download the file **com.jmp.cox.ian.piston.jmpaddin** from the book website and double click on. This will open up a "Piston Simulator" Add-In on the JMP top ruler.

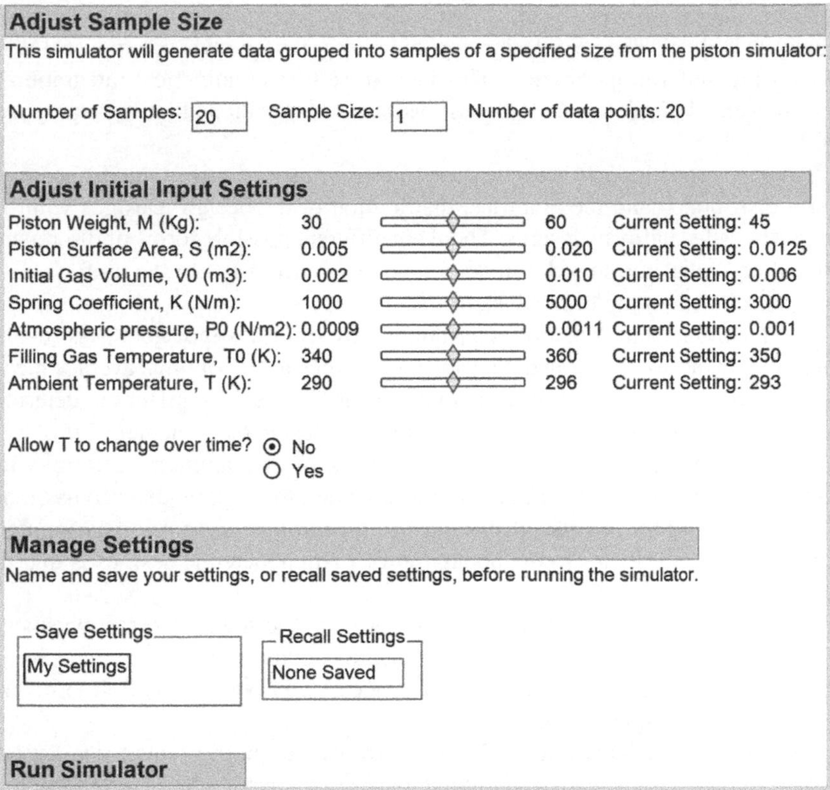

Adjust Sample Size

This simulator will generate data grouped into samples of a specified size from the piston simulator:

Number of Samples: `20`   Sample Size: `1`   Number of data points: 20

Adjust Initial Input Settings

Piston Weight, M (Kg):	30	60	Current Setting: 45
Piston Surface Area, S (m2):	0.005	0.020	Current Setting: 0.0125
Initial Gas Volume, V0 (m3):	0.002	0.010	Current Setting: 0.006
Spring Coefficient, K (N/m):	1000	5000	Current Setting: 3000
Atmospheric pressure, P0 (N/m2):	0.0009	0.0011	Current Setting: 0.001
Filling Gas Temperature, T0 (K):	340	360	Current Setting: 350
Ambient Temperature, T (K):	290	296	Current Setting: 293

Allow T to change over time? ⊙ No
⃝ Yes

Manage Settings

Name and save your settings, or recall saved settings, before running the simulator.

Save Settings
`My Settings`

Recall Settings
`None Saved`

Run Simulator

Figure 15.1 *Operating panel of the piston simulator add-in within the JMP application (JMP).*

We can run the simulator by manually setting up the factor-level combinations or by using statistically designed experimental arrays. In this chapter, the arrays we will refer to are called, in general, space filling experiments. The simulator was used in the context of Statistical Process Control (Chapter 10). We use it here in the context of statistically designed computer experiments. The next section deals with designing computer experiments. We will discuss the space filling designs that are specific to computer experiments where the factor-level combinations can be set freely, without physical constraints at specific levels. The section after that, Section 15.3, deals with models used in the analysis of computer experiments. These models are called Kriging, Dace, or Gaussian process models. They will be introduced at a general level designed to provide basic understanding of their properties, without getting into their theoretical development. The JMP piston simulator is fully integrated with the JMP DOE experimental design features.

15.2 Designing computer experiments

Experimentation via computer modeling has become very common. We introduce here two popular designs for such experiments: the uniform design and the Latin hypercube design.

Suppose that the experimenter wants to estimate μ, the overall mean of the response y on the experimental domain X. The best design for this purpose is one whose empirical distribution approximates the uniform distribution. This idea arose first in numerical integration methods for high-dimensional problems, called quasi-Monte Carlo methods that were proposed in the early 1960s.

The discrepancy function, $D(\cdot)$, or measure of uniformity, quantifies the difference between the uniform distribution and the empirical distribution of the design. Designs with minimum discrepancy are called uniform designs. There are different possible forms of discrepancy functions, depending on the norm used to measure the difference between the uniform distribution and the empirical distribution of the design.

In general, the discrepancy function is a Kolmogorov–Smirnov type goodness-of-fit statistic. For estimating μ in the overall mean model, the uniform design has optimal average mean-square error assuming random h and optimal maximum mean-square error assuming deterministic h. This implies that the uniform design is a kind of robust design (see Chapter 12).

Latin hypercube designs are easy to generate. They achieve maximum uniformity in each of the univariate margins of the design region, thus allowing the experimenter to use models that are capable of capturing the complex dependence of the response variable on the input variables. Another reason that contributes to the popularity of Latin hypercube designs is that they have no repeated runs.

In computer experiments, repeated runs do not provide additional information since running a deterministic computer code twice yields the identical output. Latin hypercube designs are a very large class of designs that, however, do not necessarily perform well in terms of criteria such as orthogonality or space filling.

An $n \times m$ matrix $D = (d_{ij})$ is called a Latin hypercube design of n runs for m factors if each column of D is a permutation of $1, \ldots, n$. What makes Latin hypercube designs distinctly different from other designs is that every factor in a Latin hypercube design that has the same number of levels as the run size.

Let $y = f(x_1, \ldots, x_m)$ be a real-valued function with m variables defined on the region given by $0 \le x_j \le 1$ for $j = 1, \ldots, m$. The function represents the deterministic computer model in the case of computer experiments or the integrand in the case of numerical integration. There are two natural ways of generating design points based on a given Latin hypercube. The first is through

$$x_{ij} = (d_{ij} - 0.5)/n,$$

with the n points given by $(x_{i1}, \ldots, x_{im})$ with $i = 1, \ldots, n$. The other is through

$$x_{ij} = (d_{ij} - u_{ij})/n,$$

with the n points given by $(x_{i1}, \ldots, x_{im})$ with $i = 1, \ldots, n$, where u_{ij} are independent random variables with a common uniform distribution on $(0, 1]$. The difference between the two methods can be seen as follows. When projected onto each of the m variables, both methods have the property that one and only one of the n design points fall within each of the n small intervals defined by $[0, 1/n), [1/n, 2/n), \ldots, [(n - 1)/n, 1]$. The first method gives the mid-points of these intervals, whereas the second method gives the points that are uniformly distributed in their corresponding intervals. Figure 15.2 presents two Latin hypercube designs of $n = 5$ runs for $m = 2$ factors.

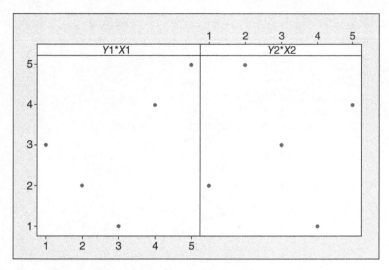

Figure 15.2   *Two Latin hypercube designs (D1 left and D2 right) with 5 runs and 2 factors (MINITAB).*

Although they are both Latin hypercube designs, design D2 provides a higher coverage of the design region than design D1. This raises the need of developing specific methods for selecting better Latin hypercube designs. Basic Latin hypercube designs are very easy to generate. By simply combining several permutations of $1, \ldots, n$, one obtains a Latin hypercube design. There is no restriction whatsoever on the run size n and the number m of factors. Since a Latin hypercube design has n distinct levels in each of its factors, it achieves the maximum uniformity in each univariate margin. Two useful properties follow from this simple fact: (1) the maximum number of levels, a Latin hypercube design presents the experimenter with the opportunity of modeling the complex dependence of the response variable on each of the input variables and (2) there is no repeated levels in each factor. Since running computer code twice at the same setting of input variables produces the same output, using repeated runs in computer experiments is necessarily a waste of resources.

By definition, a Latin hypercube does not guarantee any property in two or higher dimensional margins. It is therefore up to the user to find the "right permutations" so that the resulting design has certain desirable properties in two or higher dimensions. One simple strategy is to use a random Latin hypercube design in which the permutations are selected randomly. This helps eliminate the possible systematic patterns in the resulting design but there is no guarantee that the design will perform well in terms of other useful design criteria. A Latin hypercube design will provide a good coverage of the design region if all the points are farther apart, i.e., no two points are too close to each other. This idea can be formally developed using the maximin distance criterion, according to which designs should be selected by maximizing $\min_{i=j} d(p_i, p_j)$, where $d(p_i, p_j)$ denotes the distance between design points p_i and p_j. Euclidean distance is commonly used but other distance measures are also useful.

Example 15.2.   To design a space filling experiment with the piston simulator on needs to click the "Make Table of Inputs" in the Piston Simulator JMP add in and load factors in the DOE "Space Filling Design" window. This leads us to Figure 15.3. In that window, the response cycle time (Y) has been set to a target of 0.5 seconds and specification limits of 0.4 and 0.6 seconds.

```
> library(lhs)
> set.seed(123)
> Des <- maximinLHS(n=14, k=7)
> Des[, 1] <- Des[, 1] * (60-30) + 30
> Des[, 2] <- Des[, 2] * (0.02-0.005) + 0.005
> Des[, 3] <- Des[, 3] * (0.01-0.002) + 0.002
> Des[, 4] <- Des[, 4] * (5000-1000) + 1000
> Des[, 5] <- Des[, 5] * (110000-90000) + 90000
> Des[, 6] <- Des[, 6] * (296-290) + 290
> Des[, 7] <- Des[, 7] * (360-340) + 340
> Ps <- pistonSimulation(m=Des[,1],
                  s=Des[,2],
                  v0=Des[,3],
                  k=Des[,4],
                  p0=Des[,5],
                  t=Des[,6],
                  t0=Des[,7],
                  each=50, seed = 123)
> Ps <- simulationGroup(Ps, 50)
> aggregate(Ps[, !names(Ps) %in% "group"], by=Ps["group"], mean)
```

	group	m	s	v0	k
1	1	36.62212	0.007681687	0.007192159	3720.667
2	2	48.48374	0.011804155	0.008447430	3334.485
3	3	35.88351	0.010191667	0.003954534	1708.363
4	4	49.46867	0.013364038	0.006907016	2501.092
5	5	30.64397	0.014768569	0.006460968	3011.366
6	6	57.03120	0.018620736	0.002115426	2895.657
7	7	42.32771	0.019180645	0.007976814	2011.303
8	8	44.82227	0.008294597	0.009684653	3897.172
9	9	54.59089	0.014573907	0.003562142	2223.918
10	10	46.21526	0.015882680	0.009265574	4880.866
11	11	53.00704	0.005693276	0.005903555	1332.865
12	12	33.98024	0.006257340	0.004792323	4170.372
13	13	59.80531	0.010378651	0.003115797	1122.461
14	14	38.69225	0.017344265	0.005171781	4530.218

	p0	t	t0	seconds
1	95492.76	291.0658	352.5617	0.4491319
2	97514.47	294.8791	345.7469	0.5255135
3	105747.35	292.5744	349.1670	0.5078945
4	92660.57	291.7159	348.3927	0.5240666
5	100477.98	293.4264	345.2105	0.3599614
6	103759.70	290.4748	355.6064	0.2943296
7	99352.04	294.0366	355.9603	0.4303809
8	104823.49	292.4505	359.0226	0.5339606

```
9   101694.10  295.4611  341.1489  0.4576193
10   90612.77  293.8023  344.2699  0.4220562
11  108043.96  294.5462  353.3177  0.9140116
12   96458.35  295.8955  357.1446  0.4106764
13   93799.46  291.6488  342.8460  0.8205386
14  109613.37  290.2626  350.2119  0.2964288
```

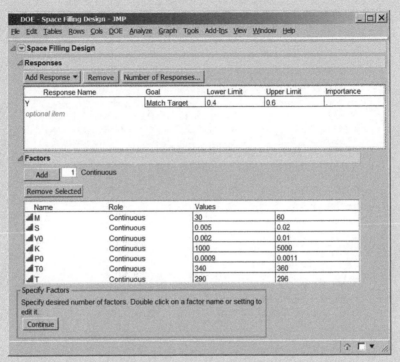

Figure 15.3 *Setting of responses and factors in space filling design panel for piston simulator (JMP) clicking on "Continue" opens a new window with several experimental designs (Figure 15.4).*

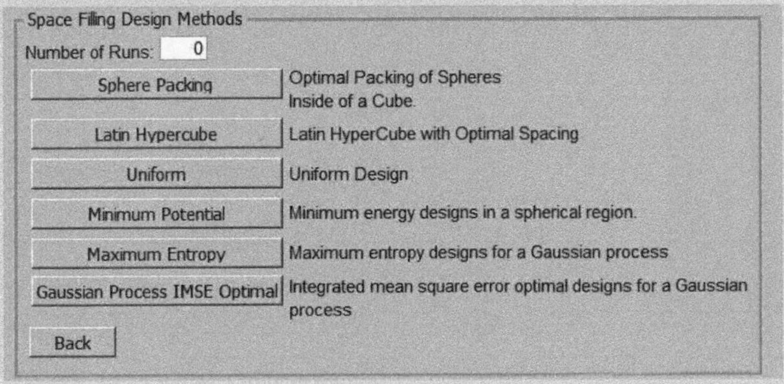

Figure 15.4 *Specifying design in space filling design panel for piston simulator (JMP).*

Clicking on Latin hypercube produces the design shown in Figure 15.5. A graphical display of this space filing design with seven factors is presented in Figure 15.6. As seen in Figure 15.4, other designs are also available. In Section 15.3, we will run the piston simulator at each one of the 14 experimental runs.

Space Filling Latin Hypercube

◢ **Factor Settings**

Run	M	S	V0	K	P0	T0	T
1	53.07692	0.01769	0.00938	2538.462	0.00107	343.0769	291.8462
2	41.53846	0.00962	0.00262	1615.385	0.00099	341.5385	293.6923
3	39.23077	0.00615	0.00385	4384.615	0.00105	344.6154	290.0000
4	30.00000	0.01885	0.00631	3461.538	0.00098	340.0000	294.1538
5	50.76923	0.01654	0.00200	5000.000	0.00095	346.1538	292.3077
6	36.92308	0.01308	0.00815	4076.923	0.00090	350.7692	290.9231
7	34.61538	0.01077	0.00323	1923.077	0.00101	360.0000	294.6154
8	57.69231	0.00500	0.00446	3153.846	0.00092	355.3846	292.7692
9	46.15385	0.00731	0.01000	1000.000	0.00096	347.6923	293.2308
10	55.38462	0.01538	0.00569	3769.231	0.00104	358.4615	290.4615
11	43.84615	0.00846	0.00754	4692.308	0.00102	349.2308	295.0769
12	32.30769	0.01423	0.00877	2230.769	0.00110	356.9231	291.3846
13	48.46154	0.02000	0.00508	2846.154	0.00093	353.8462	296.0000
14	60.00000	0.01192	0.00692	1307.692	0.00108	352.3077	295.5385

▷ **Design Diagnostics**

Design Table

[Make Table]   [Back]

Figure 15.5   *Latin hypercube design for piston simulator (JMP).*

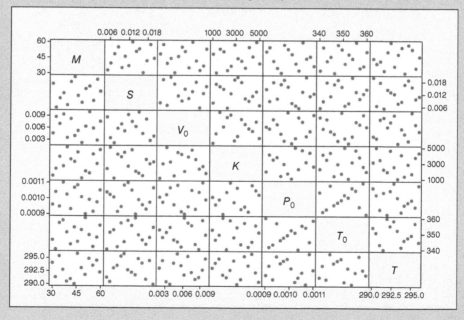

Figure 15.6   *Latin hypercube design for piston simulator (MINITAB).*

The next section is focused on models used for analyzing computer experiments.

15.3 Analyzing computer experiments

As already mentioned in Section 15.1, Kriging was developed for modeling spatial data in Geostatistics. Matheron (1963) named this method after D. G. Krige, a South African mining engineer who in the 1950s developed empirical methods for estimating true ore grade distributions based on sample ore grades. At the same time, the same ideas were developed in meteorology under L. S. Gandin (1963) in the Soviet Union. Gandin named the method Optimal Interpolation. The central feature of Kriging models is that spatial trends can be modeled using spatial correlation structures, similar to time series models, in which observations are assumed to be dependent. Spatial models, however, need to be more flexible than time series models, as there is dependence in a multitude of directions. In general, the approach is a method of optimal spatial linear prediction based on minimum-mean-squared-error. The use of Kriging for modeling data from computer experiments was originally labeled Design and Analysis of Computer Experiments (DACE) by Sacks et al. (1989). Kriging models are also known as Gaussian Process models (e.g., in JMP). Computer experiments may have many input variables, whereas spatial models have just 2 or 3. The DACE algorithm uses a model that treats the deterministic output of a computer code as the realization of a stochastic process. This nonparametric model simultaneously identifies important variables and builds a predictor that adapts to nonlinear and interaction effects in the data.

Assume there is a single scalar output $y(x)$, which is a function of a d-dimensional vector of inputs, x. The deterministic response $y(x)$ is treated as a realization of a random function

$$Y(x) = \beta + Z(x). \tag{15.3.1}$$

The random process $Z(x)$ is assumed to have mean 0 and covariance function

$$\text{Cov}(Z(x_i), Z(x_j)) = \sigma^2 R(x_i, x_j) \tag{15.3.2}$$

between $Z(x_i)$ and $Z(x_j)$ at two vector-valued inputs x_i and x_j, where σ^2 is the process variance and $R(x_i, x_j)$ is the correlation.

DACE is using the correlation function:

$$R(x_i, x_j) = \prod_{k=1}^{d} \exp(-\theta_k \mid x_{ik} - x_{jk} \mid^{p_k}), \tag{15.3.3}$$

where $\theta_k \geq 0$ and $0 \leq p_k \leq 2$.

The basic idea behind this covariance is that values of Y for points "near" each other in the design space should be more highly correlated than for points "far" from each other. Thus, we should be able to estimate the value of $Y(x)$ at a new site by taking advantage of observed values at sites that have a high correlation with the new site. The parameters in the correlation function determine which of the input variables are important in measuring the distance between two points. For example, a large value of θ_k means that only a small neighborhood of values on this variable is considered to be "close" to a given input site and will typically correspond to an input with a strong effect. In this model, the covariance structure is specified via R rather than by the variogram, as is traditionally done in Geostatistics.

All the unknown parameters are estimated using maximum likelihood estimation (MLEs). Since the global maximization is very problematic from a computational perspective, a pseudo maximization algorithm is applied using a "stepwise" approach, where at each step the parameters for one input factor are "free" and all the rest are equal.

Given the correlation parameters θ and p, the MLE of β is

$$\hat{\beta} = (\mathbf{J}'\mathbf{R}_D^{-1}\mathbf{J})^{-1}(\mathbf{J}'\mathbf{R}_D^{-1}y). \tag{15.3.4}$$

where J is a vector of ones and R_D is the $n \times n$ matrix of correlations $R(x_i, x_j)$.

The generalized least squares estimator, and the MLE, of σ^2 is

$$\hat{\sigma}^2 = (y - \mathbf{J}\hat{\beta})'\mathbf{R}_D^{-1}(y - \mathbf{J}\hat{\beta})/n. \tag{15.3.5}$$

The best linear unbiased predictor (BLUP) at an untried x is

$$\hat{y}(x) = \hat{\beta} + \mathbf{r}'(x)\hat{\mathbf{R}}_D^{-1}(y - \underline{J}\hat{\beta}), \tag{15.3.6}$$

where $r(x) = [R(x_1, x), \ldots, R(x_n, x)]'$ is the vector of correlations between Z's at the design points and at the new point x. The BLUP interpolates the observed output at sites x that are in the training data.

Example 15.3. We invoke again the piston simulator by applying the JMP add-in application. On the Latin hypercube design of Figure 15.5, we run the simulator by clicking the "Run Simulator" add in. This adds a column Y of cycle times to the JMP table for values of the factors determined by the experimental array. The box on the top right in JMP now includes two models, a screening model and a Gaussian (Kriging) model. Clicking on the red arrows in the JMP report we can edit these models or simply run them.

```
> library(DiceEval)
> data(LATHYPPISTON)
> Dice <- modelFit(LATHYPPISTON[, !names(LATHYPPISTON) %in%
                   "seconds"],
                   LATHYPPISTON[,"seconds"],
                   type = "Kriging",
                   formula=~ .,
                   control=list(trace=FALSE))
> Dice$model

Call:
km(formula = ..1, design = data[, 1:f], response = data[, f +
    1], control = ..2)

Trend  coeff.:
              Estimate
  (Intercept)    -5.5479
            m     0.0047
            s   -18.7901
           v0   120.7077
            k     0.0001
           p0   675.7679
            t     0.0303
           t0    -0.0129

Covar. type  : matern5_2
```

```
Covar. coeff.:
                Estimate
    theta(m)     18.6575
    theta(s)      0.0200
    theta(v0)     0.0024
    theta(k)   6516.2661
    theta(p0)     0.0000
    theta(t)      3.6451
    theta(t0)     5.0568


Variance estimate: 0.01418025

> Dice <- modelFit(scale(x=LATHYPPISTON[, !names(LATHYPPISTON) %in%
                   "seconds"]),
                   LATHYPPISTON[,"seconds"],
                   type = "Kriging",
                   formula= ~ .,
                   control=list(trace=FALSE))
> Dice$model

Call:
km(formula = ..1, design = data[, 1:f], response = data[, f +
    1], control = ..2)

Trend  coeff.:
                Estimate
  (Intercept)     0.4910
            m     0.0363
            s    -0.0795
           v0     0.3178
            k     0.1405
           p0     0.0256
            t     0.0712
           t0    -0.0738


Covar. type  : matern5_2
Covar. coeff.:
                Estimate
    theta(m)      6.2152
    theta(s)      0.8028
    theta(v0)     0.7725
    theta(k)      6.2152
    theta(p0)     6.2152
    theta(t)      6.2152
    theta(t0)     6.2152


Variance estimate: 0.01488352
```

Running the screening model produces Figure 15.7. We can see that the initial gas volume, V_0, and the spring coefficient, K, are the only two significant factors. This was determined by fitting a screening design polynomial model of main effects and two-way interactions. The half normal plot of the effects and their interactions confirm this finding. Clicking "Run mode" will produce a marginal regression analysis of the factors highlighted by the color bar. In the setup of Figure 15.7, this produces an analysis of the quadratic effect of S, the piston surface area.

Running the Gaussian model produces Figure 15.8 which presents the estimated model parameters and an assessment of the goodness of fit using the jackknife, a technique very similar to the bootstrapping presented in Chapter 4. The jackknife is considering parameters estimates for the full data set without one observation, for all observations. With this approach

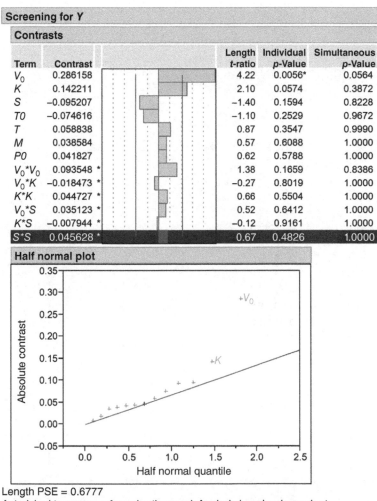

Figure 15.7 *Half normal plot of main effects and interactions of factors affecting cycle time (JMP) piston simulator Latin hypercube experiment (JMP).*

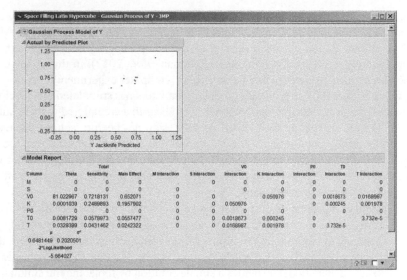

Figure 15.8 *Estimates of DACE parameters in cycle time piston simulator Latin hypercube experiment (JMP).*

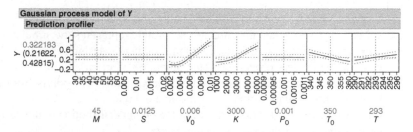

Figure 15.9 *JMP profiler showing marginal effect of factors in cycle time piston simulator Latin hypercube experiment (JMP).*

one can compare the observed value with the value predicted from the model as shown in the top of Figure 15.8. Points lying on the line of equality ($Y = X$) indicate a good fit of the model since the observed data points are well predicted by the model.

In Figure 15.9, we show the JMP profiler which, again, confirms that only V_0 and K are having an effect on the average cycle time.

15.4 Stochastic emulators

Traditional engineering practice augments deterministic design system predictions with factors of safety or design margins to provide some assurance of meeting requirements in the presence of uncertainty and variability in modeling assumptions, boundary conditions, manufacturing, materials, and customer usage. Modern engineering practice is implementing Quality by Design (QbD) methods to account for probability distributions of component or system performance characteristics. Chapter 12 provided several such examples, including the robust design approach developed by Genichi Taguchi in Japan. At Pratt and Whitney, in the United States, Grant Reinman and his team developed a methodology labeled design for variation (DFV) that incorporates the same principles (Reinman *et al.*, 2012). In this chapter, we focus on an essential element of modern QbD engineering, computer experiments.

The new experimental framework of computer simulators has stimulated the development of new types of experimental designs and methods of analysis that are tailored to these studies. The guiding idea in computer simulation experimental design has been to achieve nearly uniform coverage of the experimental region. The most commonly used design has been the so-called Latin hypercube presented in Section 15.2. In Latin hypercube designs, each factor is given a large number of levels, an option that is virtually impossible in physical experiments but very easy when experimenting on a simulator.

In using computer experiments for robust design problems, outcome variation is induced via uncertainty in the inputs. The most direct way to assess such variation is to generate simulator output for a moderate to large sample of input settings (see Section 15.1). However, if the simulator is slow and/or expensive, such a scheme may not be practical. The stochastic emulator paradigm, also called metamodel, provides a simple solution by replacing the simulator with an emulator for the bulk of the computations. The key steps of the stochastic emulator approach are as follows:

1. Begin with a Latin hypercube (or other space-filling) design of moderate size.
2. Use the simulator to generate data at points in the design.
3. Model the simulator data to create an emulator, called the stochastic emulator.
4. Use cross-validation to verify that the emulator accurately represents the simulator.
5. Generate a new space-filling design. Each configuration in this design is a potential nominal setting at which we will assess properties of the output distribution.
6. At each configuration in the new design, sample a large number of points from the noise factors and compute output data from the stochastic emulator.
7. Construct statistical models that relate features of the output distribution to the design factor settings. These models might themselves be emulators.

This approach can dramatically reduce the overall computational burden by using the stochastic emulator, rather than the simulator, to compute the results in Step 6. Stochastic emulators are a primary QbD tools in organizations that have successfully incorporated simulation experiments in the design of drug products, analytical methods, and scale-up processes.

15.5   Integrating physical and computer experiments

Information from expert opinion, computer experiments, and physical experiments can be combined in a simple regression model of the form:

$$\mathbf{Y} = f(\mathbf{X}, \beta) + \epsilon. \tag{15.5.1}$$

In this model, $\mathbf{X}$ represents the design space corresponding, and the vector β represents the values of the model coefficients, and $\mathbf{Y}$ represents the k observations, for example of method resolution. This is achieved by modeling physical experimental data as:

$$\mathbf{Y}_p \sim N(\mathbf{X}_p \beta, \sigma^2 \mathbf{I}), \tag{15.5.2}$$

where σ^2 is the experimental variance representing the uncertainty of responses due to experimental conditions and measurement system.

Instead of relying solely on the physical experiments to establish the distribution of the response in the design space, we start by first eliciting estimates from expert opinion and, later, add results from computer experiments. Results from physical experiments are then superimposed on these two sources of information. Suppose there are e expert opinions. Expert opinions on the values of β can be described as quantiles of:

$$\mathbf{Y}_0 \sim N(\mathbf{X}_0 \beta + \delta_0, \sigma^2 \Sigma_0), \tag{15.5.3}$$

where δ_0 is the expert specific location bias.

Assuming the following prior distributions for the unknown parameters β and σ^2:

$$\beta \mid \sigma^2 \sim N(\mu_0, \sigma^2 \mathbf{C}_0), \tag{15.5.4}$$

$$\sigma^2 \sim IG(\alpha_0, \gamma_0), \tag{15.5.5}$$

where $N(\mu, \sigma^2)$ stands for a normal distribution and $IG(\alpha, \gamma)$ is the inverse gamma distribution that we will meet again in Section 17.1. Using Bayes's theorem, the resulting rior distribution of β becomes:

$$\pi(\beta \mid \sigma^2, \eta, \mathbf{y}_0) \sim N((\mathbf{X}_0' \Sigma_0^{-1} \mathbf{X}_0 + \mathbf{C}_0^{-1})^{-1}\mathbf{z}, \sigma^2(\mathbf{X}_0' \Sigma_0^{-1} \mathbf{X}_0 + \mathbf{C}_0^{-1})^{-1}) \tag{15.5.6}$$

with

$$\mathbf{z} = \mathbf{X}_0' \Sigma_0'(\mathbf{y}_0 - \delta_0) + \mathbf{C}_0^{-1}\mu. \tag{15.5.7}$$

The computer experimental data can be described as:

$$\mathbf{Y}_c \sim N(\mathbf{X}_c \beta + \delta_c, \sigma^2 \sigma_c). \tag{15.5.8}$$

Combining these results with the expert opinion posteriors we derive a second posterior distribution and then adding estimates from physical experiments trough Markov Chain Monte Carlo we calculate the final distribution for β.

$$\text{Stage 1 } (\mathbf{Y}_0) \rightarrow \text{Stage 2 } (\mathbf{Y}_0 + \mathbf{Y}_c) \rightarrow \text{Stage 3 } (\mathbf{Y}_0 + \mathbf{Y}_c + \mathbf{Y}_p). \tag{15.5.9}$$

A related approach called "variable fidelity experiments" has been proposed in [8] to combine results from experiments conducted at various levels of sophistication.

Consider for example combining simple calculations in Excel, to results from a mixing simulation software and actual physical mixing experiments.

The combined model is:

$$Y(\mathbf{x}, l) = \mathbf{f}_1(\mathbf{x})'\beta_1 + \mathbf{f}_1(\mathbf{x})'\beta_2 + Z_{\text{sys}}(\mathbf{x}, l) + \epsilon_{\text{means}}(l), \tag{15.5.10}$$

where $l = 1, \ldots, m$ is fidelity level of the experimental system, $Z_{\text{sys}}(\mathbf{x}, l)$, is the systematic error and $\epsilon_{\text{means}}(l)$ is the random error ($l = 1$ corresponds to the real system). There are also primary terms and potential terms, only the primary terms, $\mathbf{f}_1(\mathbf{x})$, are included in the regression model.

Assuming that the covariance matrix $\mathbf{V}$ is known and $\mathbf{Y}$ is a vector that contains data from n experiments, the GLS estimator of β_1 is:

$$\hat{\beta}_1 = (\mathbf{X}_1' \mathbf{V}^{-1} \mathbf{X}_1)^{-1}\mathbf{X}_1' \mathbf{V}^{-1}\mathbf{Y}. \tag{15.5.11}$$

Both the integrated model, combining expert opinion with simulation and physical experiments and the variable fidelity level experiments have proven useful in practical applications where experiments are conducted in different conditions and prior experience has been accumulated.

15.6   Simulation of random variables

15.6.1   Basic procedures

Simulation is an artificial technique of generating on the computer a sequence of random numbers, from a given distribution, in order to explore the properties of a random phenomenon. Observation of random phenomena often takes long time of data collection, until we have sufficient information for analysis. For example, consider patients arriving at random times to a hospital, to obtain a certain treatment in clinical trials. It may take several months until we have a large enough sample for analysis. Suppose that we assume that the epochs of arrival of the patients follow a Poisson process. If we can generate on the computer a sequence of random times, which follow a Poisson process, we might be able to predict how long the trial will continue. Physically, we can create random numbers by flipping a "balanced" coin many times, throwing dice or shuffling cards. One can use a Geiger counter to count how many particles are emitted from a decaying radio-active process. All these methods are slow and cannot be used universally. The question is, how can we generate on the computer random numbers, following a given distribution.

The key for random numbers generation is the well-known result that, if a random variable X has a continuous distribution F then $F(X)$ is uniformly distributed on the interval $(0, 1)$. Accordingly, if one can generate at random a variable $U \sim R(0, 1)$ then the random variable $X = F^{-1}(U)$ is distributed according to F. In order to generate a uniformly distributed random variables, computer programs like R, JMP, MINITAB and others generally apply an algorithm that yields, after a while (asymptotically), uniformly distributed results. The R function for the uniform random variable $R(\alpha, \beta)$ is runif($1, \alpha, \beta$). The standard uniform on $(0,1)$ can be obtained with runif(1).

We provide below a few examples of random numbers generation:

The exponential distribution with mean $1/\lambda$ is $F(x) = 1 - \exp\{-\lambda x\}$, for $0 \leq x \leq \infty$. Thus, if U has a uniform distribution on $(0,1)$ then $X = -(1/\lambda) \ln(1 - U)$ has an exponential distribution with mean $1/\lambda$. In R, we can generate an exponential (λ) by the function rgamma($1,1$)$/\lambda$._

If X is distributed according to exponential with $\lambda = 1$, then $X^{1/\alpha}$ is distributed like Weibull with shape parameter α, and scale parameter 1.

In R, we can generate such a random variable with the function rweibull($1,\alpha$). If there is also a scale parameter β, we write β *rweibull($1,\alpha$).

A standard normal random variable can be generated by the Uth quantiles $\Phi^{-1}(U)$. In R, these random variables are generated by the

$$rnorm(1, \mu, \sigma) \sim \mu + \sigma * rnorm(1).$$

Random numbers following a discrete distribution can be similarly generated. For example, generating a random number from the binomial $(20,.6)$ can be done by the R function $rbinom(1, 20, .6)$, or $qbinom(runif(1), 20, .6)$.

If we wish to generate a random sample of n realizations, we write the number n at the beginning instead of 1. For example rbinom(20,10,.5).

15.6.2   Generating random vectors

Generating random vectors following a given multivariate distribution is done in stages. Suppose we are given a k-dimensional vector having a multivariate distribution, with density $p(x_1, x_2, \ldots, x_k)$. If the variables are mutually independent, we can generate independently k variables according to their marginal densities. This is the simplest case. On the other hand, if the variables are dependent but exchangeables (all permutations of the components yield the same joint density function), we generate the first variable according to its marginal density. Given a realization of X_1, say x_1, we generate a value of X_2 according to the conditional density $p(x_2|x_1)$, then the value of X_3 according to the conditional density $p(x_3|x_1, x_2)$, and so on. We illustrate this process on the bivariate normal distribution. This distribution has five parameters $\mu = E\{X\}, \sigma_1^2 = V\{X\}, \eta = E\{Y\}, \sigma_2^2 = V\{Y\}$,, and $\rho = \text{Corr}(X, Y)$. In Step 1, we generate $X \sim N(\mu, \sigma_1^2)$. In Step 2, we generate $Y \sim N(\eta + (\sigma_1\rho/\sigma_2)(X - \mu), \sigma_2^2(1 - \rho^2))$. In Figure 15.10, we present the generation of 1000 standard bivariate normal vectors $(\mu = \eta = 0, \sigma_1 = \sigma_2 = 1, \rho_1 = 0.8)$, following the R function stanbinorm.

```
> stanbinorm <- function(rho, Ns){
    X <- rnorm(Ns)
    Temp <- rnorm(Ns)
    Y <- rho * X + sqrt(1-rho^2) * Temp
    return(data.frame(x=X, y=Y))
  }
> set.seed(123)
> plot(stanbinorm(0.5, 1000))
>
```

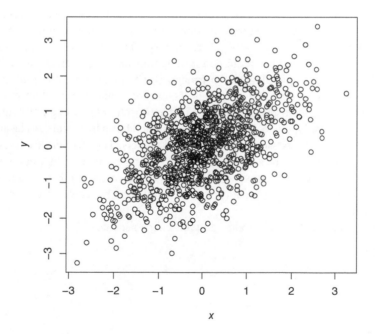

Figure 15.10   *Scatter plot of 1000 standard bivariate normal vectors.*

Package `mvtnorm` computes multivariate normal and t probabilities, quantiles, random deviates, and densities.

15.6.3   Approximating integrals

Let $g(X, Y)$ be a function of a random vector (X, Y) having a bivariate distribution, with a joint p.d.f. $p(x, y)$. The expected value of $g(X, Y)$, if $E\{|g(X, Y)|\} < \infty$, is

$E\{g(X,Y)\} = \int_{-\infty}^{\infty} \int_{-\infty}^{\infty} g(x, y)p(x, y)dx \, dy$.

According to the strong Law of Large Numbers, under the above condition of absolute integrability,

$E\{g(X, Y)\} = \lim_{n \to \infty} \frac{1}{n} \sum_{j=1}^{n} g(X_j, Y_j)$, a.s.

Accordingly, the mean of $g(X, Y)$ in very large samples approximates its expected value. The following is an example.

Let $g(X, Y) = \exp(X + Y)$, where $\{(X_j, Y_j), j = 1, \ldots, n\}$ is a sample from the bivariate standard normal distribution, discussed in Part II. Using moment generating functions of the normal distribution, we obtain that

$E\{e^{\theta(X+Y)}\} = E\{e^{\theta X} E\{e^{\theta Y} | X\}\} = \exp\{(1 + \rho)\theta^2\}$.

Thus, for $\theta = 1$ and $\rho = 0.8$ we obtain the exact result that $E\{e^{X+Y}\} = 6.049647$.

On the other hand, for a sample of size 100 we obtain mean($\exp\{X + Y\}, n = 100$) = 7.590243. This sample size is not sufficiently large. For $n = 1000$, we obtain mean($\exp\{X + Y\}, n = 1000$) = 5.961779. This result is already quite close to the exact value.

15.7   Chapter highlights

Computer experiments are integrated in modern product and service development activities. Technology is providing advanced digital platforms for studying various properties of suggested designs, without the need to physically concretize them. This chapter is about computer experiments and the special techniques required for designing such experiments and analyzing their outcomes. A specific example of such experiments is the piston simulator used throughout the book to demonstrate statistical concepts and tools. In this simulator, random noise is induced on the control variables themselves, a nonstandard approach in modeling physical phenomena. The experiments covered include space filling designs and Latin hypercubes. The analysis of the experimental outputs is based on Kriging or DACE models. The chapter discusses the concept of a stochastic emulator where a model derived from the simulation outputs is used to optimize the design in a robust way. A special section is discussing several approaches to integrate the analysis of computer and physical experiments followed by a section on the generation of random numbers.

The main concepts and tools introduced in this chapter include:

- simulation
- space Filling Designs
- latin Hypercubes
- kriging
- Metamodel
- Emulator
- Stochastic Emulator

- Physical Experiments
- Bayesian Hierarchical Model
- Fidelity Level

15.8 Exercises

15.1 Exercise 3.21 is called the birthday problem. We will revisit it using a computer simulation. The JAVA applet is available at https://www.geogebra.org/m/shGNfyfF simulates the birthday problem. Show that if there are more than 22 people in the party, the probability is greater than 1/2 that at least 2 will have birthdays on the same day.

15.2 The Deming funnel experiment was designed to show that an inappropriate reaction to common cause variation will make matters worse. Common cause and special causes affecting processes over time have been discussed in Part III. In the actual demonstration, a funnel is placed above a circular target. The objective is to drop a marble through the funnel as close to the target as possible. A pen or pencil is used to mark the spot where the marble actually hits. Usually, 20 or more drops are performed in order to establish the pattern and extent of variation about the target. The funnel represents common causes affecting a system. Despite the operator's best efforts, the marble will not land exactly on the target each time. The operator can react to this variability in one of four ways: (1) do not move the funnel, (2) measure the distance the hit is from the target and move the funnel an equal distance, but in the opposite direction (error relative to the previous position), (3) measure the distance the hit is from the target and move the funnel this distance in the opposite direction, starting at the target (error relative to the target), and (4) move the funnel to be exactly over the location of the last hit. Use R, MINITAB or JMP to compare these four strategies using simulation data. A MINITAB macro simulating the funnel is available from http://www.minitab.com/enAU/support/macros/default.aspx?action=code&id=25.

15.3 Design a 50 runs experimental array for running the piston simulator using the six options available in JMP (Sphere Packing, Latin Hypercube, Uniform Design, Maximum Potential, Maximum Entropy, and Gaussian Process IMSE Optimal). Compare the designs.

15.4 Fit a Gaussian Process model to data generated by the six designs listed in Exercise 15.1 and compare the MSE of the model fits.

15.5 Using a Uniform Design, generate a Stochastic Emulator for the piston simulator in order to get 0.2 seconds cycle time with minimal variability.

15.6 Using a Latin Hypercube Design, generate a Stochastic Emulator for the piston simulator in order to achieve 0.2 seconds cycle time with minimal variability. Compare your results to what you got in Exercise 15.5.

16

Reliability Analysis

Industrial products are considered to be of high quality, if they conform to their design specifications and appeal to the customer. However, products can fail after a while, due to degradation over time or to some instantaneous shock. A system or a component of a system is said to be **reliable** if it continues to function, according to specifications, for a long time. Reliability of a product is a dynamic notion, over time. We say that a product is highly reliable if the probability that it will function properly for a specified long period, is close to 1. As will be defined later, the reliability function, $R(t)$, is the probability that a product will continue functioning at least t units of time.

We distinguish between the reliability of systems which are unrepairable and that of repairable systems. A repairable system, after failure, goes through a period of repair and then returns to function normally. Highly reliable systems need less repair. Repairable systems which need less repair are more available to operate, and are therefore more desirable. **Availability** of a system at time t is the probability that the system will be up and running at time t. To increase the availability of repairable systems, maintenance procedures are devised. Maintenance is designed to prevent failures of a system by periodic replacement of parts, tuning, cleaning, etc. It is very important to develop maintenance procedures, based on the reliability properties of the components of systems, which are cost-effective and helpful to the availability of the systems.

One of the intriguing features of failure of components and systems is their random nature. We consider therefore the length of time that a part functions till failure as a random variable, called the **life length** of the component or the system. The distribution functions of life length variables are called **life distributions**. The role of statistical reliability theory is to develop methods of estimating the characteristics of life distributions from failure data and to design experiments called life tests. An interesting subject connected to life testing is **accelerated life testing**. Highly reliable systems may take a long time till failure (TTF). In accelerated life tests, early failures are induced by subjecting the systems to higher than normal stress. In analyzing the results of such experiments, one has to know how to relate failure distributions under stressful conditions to those under normal operating conditions. The present chapter provides

Modern Industrial Statistics: With Applications in R, MINITAB and JMP, Third Edition.
Ron S. Kenett and Shelemyahu Zacks.
© 2021 John Wiley & Sons, Ltd. Published 2021 by John Wiley & Sons, Ltd.

the foundations to the theoretical and practical treatment of the subjects mentioned above. For additional readings, see Zacks (1992).

The following examples illustrate the importance of reliability analysis and modifications (improvements) for industry.

1. *Florida power and light*: A reduction of power plant outage rate from 14% to less than 4% has generated $300 million savings to the consumer, on an investment of $5 million for training and consulting. Customer service interruptions dropped from 100 minutes/year to 42 minutes/year.

2. *Tennessee Valley Authority (TVA)*: The Athens Utilities Board is one of 160 power distributors supplied by TVA with a service region of 100 miles^2, 10,000 customers, and a peak load of 80 MW. One year's worth of trouble service data was examined in three South Athens feeders. The primary circuit failure rate was 15.3 failures/year/mile, restoring service using automatic equipment took, on the average, 3 minutes/switch, while manual switching requires approximately 20 minutes. Line repair generally takes 45 minutes. The average outage cost for an industrial customer in the United States is $11.87/kWh. Without automation, the yearly outage cost for a 6000 kW load/year is, on the average, $540K. The automation required to restore service in 3 minutes in South Athens costs about $35K. Automation has reduced outage costs to $340K. These improvements in reliability of the power supply have therefore produced an average return on investment of $9.7 for every dollar invested in automation.

3. *AT&T*: An original plan for a transatlantic telephone cable called for three spares to back up each transmitter in the 200 repeaters that would relay calls across the seabed. A detailed reliability analysis with SUPER (System Used for Prediction and Evaluation of Reliability) indicated that one spare is enough. This reduced the cost of the project by 10% – and AT&T won the job with a bid just 5% less than that of its nearest competitor.

4. *AVX*: The levels of reliability now achieved by tantalum capacitors, along with their small size and high stability is promoting their use in many applications that are electrically and environmentally more aggressive than in the past. The failure rates are 0.67 FIT (failures in 10^9 component hours) with shorts contributing approximately 67% of the total.

5. *Siemens*: Broadband transmission systems use a significant number of microwave components and these are expected to work without failure from first switch-on. The 565 Mbit coaxial repeater uses 30 diodes and transistor functions in each repeater which adds up to 7000 SP87-11 transistors along the 250 km link. The link must not fail within 15 years, and redundant circuits is not possible because of the complex circuitry. Accelerated life testing has demonstrated that the expected failure rate of the SP87-11 transistor is less than 1 FIT, thus meeting the 15 years requirement.

6. *National semiconductor*: A single-bit error in microelectronic device can cause an entire system crash. In developing the BiCmos III component one-third of the design team were assigned the job of improving the component's reliability. Accelerated life tests under high temperature and high humidity (145°C, 85% relative humidity and under bias) proved the improved device to have a failure rate below 100 FIT. In a system using 256-kbit BiCmos III static random-access memories, this translates to less than one failure in 18 years.

7. *Lockheed*: Some 60% of the cost of military aircraft now goes for its electronic systems, and many military contracts require the manufacturer to provide service at a fixed price for product defects that occur during the warranty period. Lockheed Corporation produces switching logic units used in the US Navy S-3A antisubmarine aircraft to distribute communications within and outside the aircraft. These units were high on the Pareto of component

failures. They were therefore often removed for maintenance, thereby damaging the chassis. The mean time between failures for the switching logic units was approximately 100 hours. Changes in the design and improved screening procedures increased the mean time between failures to 500 hours. The average number of units removed each week from nine aircraft dropped from 1.8 to 0.14.

16.1 Basic notions

16.1.1 Time categories

The following time categories play an important role in the theory of reliability, availability, and maintainability of systems.

I. *Usage related time categories*:
   1. **Operating Time** is the time interval during which the system is in actual operation.
   2. **Scheduled operating time** is the time interval during which the system is required to properly operate.
   3. **Free time** is the time interval during which the system is scheduled to be off duty.
   4. **Storage time** is the time interval during which a system is stored as a spare part.
II. *Equipment condition time categories*:
   1. **Up time** is the time interval during which the system is operating or ready for operation.
   2. **Down time** is the time interval out of the scheduled operating time during which the system is in state of failure (inoperable).
      Down time is the sum of
      (i) administrative time
      (ii) active repair time
      (iii) logistic time (repair suspension due to lack of parts).
III. *Indices*:

$$\textbf{Scheduled Operating Time} = \text{operating time} + \text{down time}$$

$$\textbf{Intrinsic Availability} = \frac{\text{operating time}}{\text{operating time} + \text{active repair time}}$$

$$\textbf{Availability} = \frac{\text{operating time}}{\text{operating time} + \text{down time}}$$

$$\textbf{Operational Readiness} = \frac{\text{Up time}}{\text{total calendar time}}$$

Example 16.1. A machine is scheduled to operate for two shifts a day (8 hours each shift), 5 days a week. During the last 48 weeks, the machine was "down" five times. The average down time is partitioned into

1. Average administrative time = 9 [hours]
2. Average repair time = 30 [hours]
3. Average logistic time = 7.6 [hours].

Thus, the total down time in the 48 weeks is

$$\text{down time} = 5 \times (9 + 30 + 7.6) = 233 \text{ [hours]}.$$

The total scheduled operating time is $48 \times 16 \times 5 = 3840$ [hours]. Thus, total operating time $= 3607$ [hours]. The indices of availability and intrinsic availability are

$$\text{Availability} = \frac{3607}{3840} = 0.9393.$$

$$\text{Intrinsic Availability} = \frac{3607}{3607 + 150} = 0.9601.$$

Finally the operational readiness of the machine is

$$\text{Operational Readiness} = \frac{8064 - 233}{8064} = 0.9711.$$

16.1.2   Reliability and related functions

The length of life (lifetime) of a (product) system is the length of the time interval, T, from the initial activation of it till its failure. If a system is switched on and off, we consider the total active time of the system till its failure. T is a nonnegative random variable. The distribution of T is called a **life distribution**. We generally assume that T is a continuous random variable, having a p.d.f. $f_T(t)$ and c.d.f. $F_T(t)$. The **reliability function** of a (product system) is defined as

$$R(t) = \Pr\{T \geq t\}$$

$$= 1 - F_T(t), \quad t \geq 0. \tag{16.1.1}$$

The expected life length of a product is called the **mean time till failure** (MTTF). This quantity is given by

$$\mu = \int_0^\infty t f_T(t) dt$$

$$= \int_0^\infty R(t) dt. \tag{16.1.2}$$

The instantaneous **hazard function** of a product, called also the **failure rate** function is defined as

$$h(t) = \frac{f(t)}{R(t)}, \quad t \geq 0. \tag{16.1.3}$$

Notice that $h(t)$ and $f(t)$ have the dimension of $1/T$. That is, if T is measured in hours, the dimension of $h(t)$ is [1/hour].

Notice that $h(t) = \dfrac{d}{dt} \log(R(t))$. Accordingly,

$$R(t) = \exp\left\{ -\int_0^t h(u) du \right\}. \tag{16.1.4}$$

The function

$$H(t) = \int_0^t h(u)du \qquad (16.1.5)$$

is called the **cumulative hazard rate** function.

Example 16.2. In many applications of reliability theory the exponential distribution with mean μ is used for T. In this case

$$f_T(t) = \frac{1}{\mu}\exp\{-t/\mu\}, \quad t \geq 0$$

and

$$R(t) = \exp\{-t/\mu\}, \quad t \geq 0.$$

In this model, the reliability function diminishes from 1 to 0 exponentially fast, relative to μ. The hazard rate function is

$$h(t) = \frac{\frac{1}{\mu}\cdot\exp\{-t/\mu\}}{\exp\{-t/\mu\}} = \frac{1}{\mu}, \quad t \geq 0.$$

That is, the exponential model is valid for cases where the hazard rate function is a constant independent of time. If the MTTF is $\mu = 100$ [hours], we expect 1 failure/100 [hours], i.e., $h(t) = \frac{1}{100}\left[\frac{1}{hr}\right]$.

16.2 System reliability

In the present section, we will learn how to compute the reliability function of a system, as a function of the reliability of its components (modules). Thus, if we have a system comprised of k subsystems (components or modules) having reliability functions $R_1(t), \ldots, R_k(t)$, the reliability of the system is given by

$$R_{sys}(t) = \psi(R_1(t), \ldots, R_k(t)); \quad t \geq 0. \qquad (16.2.1)$$

The function $\psi(\cdot)$ is called a **structure function**. It reflects the functional relationship between the subsystems and the system. In the present section, we discuss some structure functions of simple systems. We will also assume that the random variables $T_1, \ldots, T_k$, representing the life length of the subsystems, are **independent**.

Consider a system having two subsystems (modules) C_1 and C_2. We say that the subsystems are connected in series, if a failure of either one of the subsystems causes immediate failure of the system. We represent this series connection by a **block diagram**, as in Figure 16.1. Let I_i ($i = 1, \ldots, k$) be indicator variables, assuming the value 1 if the component C_i does not fail during a specified time interval $(0, t_0)$. If C_i fails during $(0, t_0)$ then $I_i = 0$. A **series structure function** of k components is

$$\psi_s(I_1, \ldots, I_k) = \prod_{i=1}^k I_i. \qquad (16.2.2)$$

The expected value of I_i is

$$E\{I_i\} = \Pr\{I_i = 1\} = R_i(t_0). \qquad (16.2.3)$$

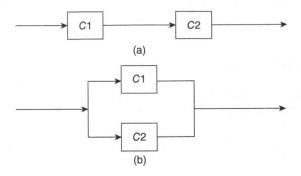

Figure 16.1   *Block diagrams for systems (a) in series and (b) in parallel.*

Thus, if the system is connected in series then, since $T_1, \ldots, T_k$ are independent,

$$R_{\text{sys}}^{(s)}(t_0) = E\{\psi_s(I_1, \ldots, I_k)\} = \prod_{i=1}^{k} R_i(t_0)$$

$$= \psi_s(R_1(t_0), \ldots, R_k(t_0)). \tag{16.2.4}$$

Thus, the system reliability function for subsystems connected in series, is given by $\psi_s(R_1, \ldots, R_k)$, where $R_1, \ldots, R_k$ are the reliability values of the components.

A system comprised of k subsystems is said to be **connected in parallel**, if the system fails the instant **all** subsystems fail. In a parallel connection, it is sufficient that one of the subsystems will function for the whole system to function.

The structure function for parallel connection is

$$\psi_p(I_1, \ldots, I_k) = 1 - \prod_{i=1}^{k}(1 - I_i). \tag{16.2.5}$$

The reliability function for a system in parallel is, in the case of independence,

$$R_{\text{sys}}^{(p)}(t_0) = E\{\psi_p(I_1, \ldots, I_k)\}$$

$$= 1 - \prod_{i=1}^{k}(1 - R_i(t_0)). \tag{16.2.6}$$

Example 16.3.   A computer card has 200 components, which should function correctly. The reliability of each component, for a period of 200 hours of operation, is $R = 0.9999$. The components are independent of each other. What is the reliability of the card, for this time period? Since all the components should function we consider a series structure function. Thus, the system reliability for $t_0 = 200$ [hours] is

$$R_{\text{sys}}^{(s)}(t_0) = (0.9999)^{200} = 0.9802.$$

Thus, despite the fact that each component is unlikely to fail, there is a probability of 0.02 that the card will fail within 200 hours. If each of the components has only a reliability of

0.99, the card reliability is

$$R_{\text{sys}}^{(s)}(t_0) = (0.99)^{200} = 0.134.$$

This shows why it is so essential in the electronic industry to demand from the vendors of the components most reliable products.

Suppose that there is on the card room for some redundancy. It is therefore decided to use the parts having reliability of $R = 0.99$ and duplicate each component in a parallel structure. The parallel structure of duplicated components is considered a module. The reliability of each module is $R_M = 1 - (1 - 0.99)^2 = 0.9999$. The reliability of the whole system is again

$$R_{\text{sys}}^{(s)} = (R_M)^{200} = 0.9802.$$

Thus, by changing the structure of the card, we can achieve the 0.98 reliability with 200 pairs of components, each with reliability value of 0.99.

Systems may have more complicated structures. In Figure 16.2, we see the block diagram of a system consisting of five components. Let $R_1, R_2, \ldots, R_5$ denote the reliability values of the five components $C_1, \ldots, C_5$, respectively. Let M_1 be the module consisting of components C_1 and C_2, and let M_2 be the module consisting of the other components. The reliability of M_1 for some specified time interval is

$$R_{M_1} = R_1 R_2.$$

The reliability of M_2 is

$$
\begin{aligned}
R_{M_2} &= R_3(1 - (1 - R_4)(1 - R_5)) \\
&= R_3(R_4 + R_5 - R_4 R_5) \\
&= R_3 R_4 + R_3 R_5 - R_3 R_4 R_5.
\end{aligned}
$$

Finally, the system reliability for that block diagram is

$$
\begin{aligned}
R_{\text{sys}} &= 1 - (1 - R_{M_1})(1 - R_{M_2}) \\
&= R_{M_1} + R_{M_2} - R_{M_1} R_{M_2} \\
&= R_1 R_2 + R_3 R_4 + R_3 R_5 - R_3 R_4 R_5 \\
&\quad - R_1 R_2 R_3 R_4 - R_1 R_2 R_3 R_5 + R_1 R_2 R_3 R_4 R_5.
\end{aligned}
$$

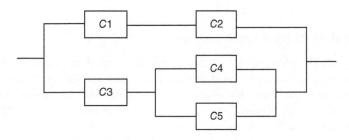

Figure 16.2 *A parallel–series structure.*

Another important structure function is that of k **out of** n subsystems. In other words, if a system consists of n subsystems, it is required that **at least** k, $1 \leq k < n$ subsystems will function, throughout the specified time period, in order that the system will function. Assuming independence of the lifetimes of the subsystems, we can construct the reliability function of the system, by simple probabilistic considerations. For example, if we have three subsystems having reliability values, for the given time period, of R_1, R_2, R_3 and at least 2 out of the 3 should function, then the system reliability is

$$R_{\text{sys}}^{2(3)} = 1 - (1 - R_1)(1 - R_2)(1 - R_3) - R_1(1 - R_2)(1 - R_3)$$
$$- R_2(1 - R_1)(1 - R_3) - R_3(1 - R_1)(1 - R_2)$$
$$= R_1 R_2 + R_1 R_3 + R_2 R_3 - 2R_1 R_2 R_3.$$

If all the subsystems have the same reliability value R, for a specified time period, then the reliability function of the system, in a k out of n structure, can be computed by using the binomial c.d.f. $B(j; n, R)$, i.e.

$$R_{\text{sys}}^{k(n)} = 1 - B(k - 1; n, R). \tag{16.2.7}$$

Example 16.4.   A cooling system for a reactor has three identical cooling loops. Each cooling loop has two identical pumps connected in parallel. The cooling system requires that 2 out of the 3 cooling loops operate successfully. The reliability of a pump over the life span of the plant is $R = 0.6$. We compute the reliability of the cooling system.

First, the reliability of a cooling loop is

$$R_{\text{cl}} = 1 - (1 - R)^2 = 2R - R^2$$
$$= 1.2 - 0.36 = 0.84.$$

Finally, the system reliability is

$$R_{\text{sys}}^{2(3)} = 1 - B(1; 3, 0.84) = 0.9314.$$

This reliability can be increased by choosing pumps with higher reliability. If the pump reliability is 0.9, the loop's reliability is 0.99 and the system's reliability is 0.9997.

The reader is referred to Zacks (1992, Ch. 3) for additional methods of computing systems reliability.

16.3   Availability of repairable systems

Repairable systems alternate during their functional life through cycles of up phase and down phase. During the up phase the system functions as required, till it fails. At the moment of failure, the system enters the down phase. The system remains in this down phase until it is repaired and activated again. The length of time the system is in the up phase is called the **time till failure** (TTF). The length of time the system is in the down phase is called the **time till repair** (TTR). Both TTF and TTR are modeled as random variables, T and S, respectively. We assume here that T and S are independent. The **cycle time** is the random variable $C = T + S$.

The process in which the system goes through these cycles is called a **renewal process**. Let $C_1, C_2, C_3, \ldots$ be a sequence of cycles of a repairable system. We assume that $C_1, C_2, \ldots$ are i.i.d. random variables.

Let $F(t)$ be the c.d.f. of the TTF, and $G(t)$ the c.d.f. of the TTR. Let $f(t)$ and $g(t)$ be the corresponding p.d.f. Let $K(t)$ denote the c.d.f. of C. Since T and S are independent random variables,

$$K(t) = \Pr\{C \le t\}$$

$$= \int_0^t f(x)P\{S \le t - x\}dx \qquad (16.3.1)$$

$$= \int_0^t f(x)G(t - x)dx.$$

Assuming that $G(0) = 0$, differentiation of $K(t)$ yields the p.d.f. of the cycle time, $k(t)$, namely

$$k(t) = \int_0^t f(x)g(t - x)dx. \qquad (16.3.2)$$

The operation of getting $k(t)$ from $f(t)$ and $g(t)$ is called a **convolution**.

The **Laplace transform** of an integrable function $f(t)$, on $0 < t < \infty$, is defined as

$$f^*(s) = \int_0^\infty e^{-ts}f(t)dt, \quad s \ge 0. \qquad (16.3.3)$$

Notice that if $f(t)$ is a p.d.f. of a nonnegative continuous random variable, then $f^*(s)$ is its moment generating function (m.g.f.) at $-s$. Since $C = T + S$, and T, S are independent, the m.g.f. of C is $M_C(u) = M_T(u)M_S(u)$, for all $u \le u^*$ at which these m.g.f. exist. In particular, if $k^*(s)$ is the Laplace transform of $k(t)$,

$$k^*(s) = f^*(s)g^*(s), \quad s \ge 0. \qquad (16.3.4)$$

Example 16.5. Suppose that T is exponentially distributed like $E(\beta)$, and S is exponentially distributed like $E(\gamma)$; $0 < \beta, \gamma < \infty$, i.e.,

$$f(t) = \frac{1}{\beta} \exp\{-t/\beta\},$$

$$g(t) = \frac{1}{\gamma} \exp\{-t/\gamma\}.$$

The p.d.f. of C is

$$k(t) = \int_0^t f(x)g(t - x)dx$$

$$= \begin{cases} \dfrac{1}{\beta - \gamma}(e^{-t/\beta} - e^{-t/\gamma}), & \text{if } \beta \ne \gamma \\ \dfrac{t}{\beta^2}e^{-t/\beta}, & \text{if } \beta = \gamma. \end{cases}$$

The corresponding Laplace transforms are

$$f^*(s) = (1 + s\beta)^{-1},$$

$$g^*(s) = (1 + s\gamma)^{-1},$$

$$k^*(s) = (1 + s\beta)^{-1}(1 + s\gamma)^{-1}.$$

Let $N_F(t)$ denote the number of failures of a system during the time interval $(0, t]$. Let $W(t) = E\{N_F(t)\}$. Similarly, let $N_R(t)$ be the number of repairs during $(0, t]$ and $V(t) = E\{N_R(t)\}$. Obviously, $N_R(t) \le N_F(t)$ for all $0 < t < \infty$.

Let $A(t)$ denote the probability that the system is up at time t. $A(t)$ is the **availability function** of the system. In unrepairable systems, $A(t) = R(t)$.

Let us assume that $W(t)$ and $V(t)$ are differentiable, and let $w(t) = W'(t)$, $v(t) = V'(t)$.

The **failure intensity function** of repairable systems is defined as

$$\lambda(t) = \frac{w(t)}{A(t)}, \quad t \ge 0. \tag{16.3.5}$$

Notice that if the system is unrepairable then $W(t) = F(t)$, $w(t) = f(t)$, $A(t) = R(t)$ and $\lambda(t)$ is the hazard function $h(t)$. Let $Q(t) = 1 - A(t)$ and $v(t) = V'(t)$. The **repair intensity function** is

$$\mu(t) = \frac{v(t)}{Q(t)}, \quad t \ge 0. \tag{16.3.6}$$

The function $V(t) = E\{N_R(t)\}$ is called the **renewal function**. Notice that

$$\Pr\{N_R(t) \ge n\} = \Pr\{C_1 + \cdots + C_n \le t\}$$

$$= K_n(t), \quad t \ge 0 \tag{16.3.7}$$

where $K_n(t)$ is the c.d.f. of $C_1 + \cdots + C_n$.

The renewal function is, since $N_R(t)$ is a nonnegative random variable,

$$V(t) = \sum_{n=1}^{\infty} \Pr\{N_r(t) \ge n\}$$

$$= \sum_{n=1}^{\infty} K_n(t). \tag{16.3.8}$$

Example 16.6. Suppose that $TTF \sim E(\beta)$ and that the repair is instantaneous. Then, C is distributed like $E(\beta)$ and $K_n(t)$ is the c.d.f. of $G(n, \beta)$, i.e.,

$$K_n(t) = 1 - P\left(n - 1; \frac{t}{\beta}\right), \quad n = 1, 2, \ldots$$

where $P(j; \lambda)$ is the c.d.f. of a Poisson random variable with mean λ. Thus, in the present case,

$$V(t) = \sum_{n=1}^{\infty} \left(1 - P\left(n - 1; \frac{t}{\beta} \right) \right)$$

$$= E\left\{ \text{Pois}\left(\frac{t}{\beta} \right) \right\} = \frac{t}{\beta}, \quad t \geq 0.$$

Here $\text{Pois}\left(\frac{t}{\beta} \right)$ designates a random variable having a Poisson distribution with mean t/β. At time t, $0 < t < \infty$, there are two possible events:

E_1: The first cycle is not yet terminated;
E_2: The first cycle has terminated at some time before t.

Accordingly, $V(t)$ can be written as

$$V(t) = K(t) + \int_0^t k(x)V(t - x)dx. \tag{16.3.9}$$

The derivative of $V(t)$ is called the **renewal density**. Let $v(t) = V'(t)$. Since $V(0) = 0$, we obtain by differentiating this equation, that

$$v(t) = k(t) + \int_0^t k(x)v(t - x)dx. \tag{16.3.10}$$

Let $v^*(s)$ and $k^*(s)$ denote the Laplace transforms of $v(t)$ and $k(t)$, respectively. Then, from the above equation

$$v^*(s) = k^*(s) + k^*(s)v^*(s), \tag{16.3.11}$$

or, since $k^*(s) = f^*(s)g^*(s)$,

$$v^*(s) = \frac{f^*(s)g^*(s)}{1 - f^*(s)g^*(s)}. \tag{16.3.12}$$

The renewal density $v(t)$ can be obtained by inverting $v^*(s)$.

Example 16.7. As before, suppose that the TTF is $E(\beta)$ and that the TTR is $E(\gamma)$. Let $\lambda = \frac{1}{\beta}$ and $\mu = \frac{1}{\gamma}$

$$f^*(s) = \frac{\lambda}{\lambda + s},$$

and

$$g^*(s) = \frac{\mu}{\mu + s}.$$

Then

$$v^*(s) = \frac{\lambda\mu}{s^2 + (\lambda + \mu)s}$$

$$= \frac{\lambda\mu}{\lambda + \mu}\left(\frac{1}{s} - \frac{1}{s + \lambda + \mu}\right).$$

$\frac{1}{s}$ is the Laplace transform of 1, and $\frac{\lambda+\mu}{s+\lambda+\mu}$ is the Laplace transform of $E\left(\frac{1}{\lambda+\mu}\right)$. Hence

$$v(t) = \frac{\lambda\mu}{\lambda + \mu} - \frac{\lambda\mu}{\lambda + \mu}e^{-t(\lambda+\mu)}, \quad t \geq 0.$$

Integrating $v(t)$, we obtain the renewal function

$$V(t) = \frac{\lambda\mu}{\lambda + \mu}t - \frac{\lambda\mu}{(\lambda + \mu)^2}(1 - e^{-t(\lambda+\mu)}),$$

$0 \leq t < \infty$.
In a similar fashion, we can show that

$$W(t) = \frac{\lambda\mu}{\lambda + \mu}t + \frac{\lambda^2}{(\lambda + \mu)^2}(1 - e^{-t(\lambda+\mu)}),$$

$0 \leq t < \infty$.
Since $W(t) > V(t)$ if, and only if, the last cycle is still incomplete, and the system is down, the probability, $Q(t)$, that the system is down at time t is

$$Q(t) = W(t) - V(t)$$

$$= \frac{\lambda}{\lambda + \mu} - \frac{\lambda}{\lambda + \mu}e^{-t(\lambda+\mu)}, \quad t \geq 0.$$

Thus, the availability function is

$$A(t) = 1 - Q(t)$$

$$= \frac{\mu}{\lambda + \mu} + \frac{\lambda}{\lambda + \mu}e^{-t(\lambda+\mu)}, \quad t \geq 0.$$

Notice that the availability at large values of t is approximately

$$\lim_{t \to \infty} A(t) = \frac{\mu}{\lambda + \mu} = \frac{\beta}{\beta + \gamma}.$$

The availability function $A(t)$ can be determined from $R(t)$ and $v(t)$ by solving the equation

$$A(t) = R(t) + \int_0^t v(x)R(t - x)dx. \tag{16.3.13}$$

The Laplace transform of this equation is

$$A^*(s) = \frac{R^*(s)}{1 - f^*(s)g^*(s)}, \quad 0 < s < \infty. \tag{16.3.14}$$

This theory can be useful in assessing different system structures, with respect to their availability. The following asymptotic (large t approximations) results are very useful. Let μ and σ^2 be the mean and variance of the cycle time.

1.
$$\lim_{t \to \infty} \frac{V(t)}{t} = \frac{1}{\mu}. \tag{16.3.15}$$

2.
$$\lim_{t \to \infty} (V(t+a) - V(t)) = \frac{a}{\mu}, \quad a > 0. \tag{16.3.16}$$

3.
$$\lim_{t \to \infty} \left(V(t) - \frac{t}{\mu} \right) = \frac{\sigma^2}{2\mu^2} - \frac{1}{2}. \tag{16.3.17}$$

If the p.d.f. of C, $k(t)$, is continuous, then

4.
$$\lim_{t \to \infty} v(t) = \frac{1}{\mu}. \tag{16.3.18}$$

5.
$$\lim_{t \to \infty} \Pr \left\{ \frac{N_R(t) - t/\mu}{(\sigma^2 t/\mu^3)^{1/2}} \leq z \right\} = \Phi(z). \tag{16.3.19}$$

6.
$$A_\infty = \lim_{T \to \infty} \frac{1}{T} \int_0^T A(t) dt = \frac{E\{TTF\}}{E\{TTF\} + E\{TTR\}}. \tag{16.3.20}$$

According to (1), the expected number of renewals, $V(t)$, is approximately t/μ, for large t. According to (2), we expect approximately a/μ renewals in a time interval of length $(t, t+a)$, when t is large. The third result (3) says that t/μ is an under (over) estimate, for large t, if the squared coefficient of variation σ^2/μ^2, of the cycle time is large (smaller) than 1. The last three properties can be interpreted in a similar fashion. We illustrate these asymptotic properties with examples.

Example 16.8. Consider a repairable system. The TTF [hour] has a gamma distribution like $G(2,100)$. The TTR [hour] has a Weibull distribution $W(2, 2.5)$. Thus, the expected TTF is $\mu_T = 200$ [hours], and the expected TTR is $\mu_s = 2.5 \times \Gamma \left(\frac{3}{2} \right) = 1.25\sqrt{\pi} = 2.2$ [hours]. The asymptotic availability is

$$A_\infty = \frac{200}{202.2} = 0.989.$$

That is, in the long run, the proportion of total availability time is 98.9%.

The expected cycle time is $\mu_c = 222.2$ and the variance of the cycle time is

$$\sigma_c^2 = 2 \times 100^2 + 6.25 \left[\Gamma(2) - \Gamma^2 \left(\frac{3}{2} \right) \right]$$

$$= 20,000 + 1.34126 = 20,001.34126.$$

Thus, during 2000 [hours] of scheduled operation, we expect close to $\frac{2000}{202.2} \cong 10$ renewal cycles. The probability that $N_R(2000)$ will be less than 11 is

$$\Pr \{ N_R(2000) \leq 11 \} \cong \Phi \left(\frac{2}{1.91} \right) = \Phi(1.047) = 0.8525.$$

An important question is, what is the probability, for large values of t, that we will find the system operating, and will continue to operate without a failure for at least u additional time units. This function is called the **asymptotic operational reliability**, and is given by

$$R_\infty(u) = A_\infty \cdot \frac{\int_u^\infty R(u)du}{\mu_T}, \quad 0 \le u, \qquad (16.3.21)$$

where $R(u) = 1 - F_T(u)$.

Example 16.9.   We continue discussing the case of Example 16.8. In this case,

$$F_T(u) = \Pr\{G(2, 100) \le u\}$$

$$= \Pr\left\{G(2, 1) \le \frac{u}{100}\right\}$$

$$= 1 - P\left(1; \frac{u}{100}\right) = 1 - e^{-u/100} - \frac{u}{100}e^{-u/100},$$

and

$$R(u) = e^{-u/100} + \frac{u}{100}e^{-u/100}.$$

Furthermore, $\mu_T = 200$ and $A_\infty = 0.989$. Hence

$$R_\infty(u) = 0.989 \cdot \frac{\int_u^\infty \left(1 + \frac{x}{100}\right)e^{-x/100}\,dx}{200}$$

$$= \frac{98.9}{200}\left(2 + \frac{u}{100}\right)e^{-u/100}.$$

Thus, $R_\infty(0) = 0.989$, $R_\infty(100) = 0.546$, and $R_\infty(200) = 0.268$.

Before concluding the present section, we introduce two R applications, `availDis` and `renewDis` which provide the empirical bootstrap distribution (EBD) of the number of renewals in a specified time interval and the EBD of the asymptotic availability index A_∞, based on observed samples of failure times and repair times. These programs provide computer aided estimates of the renewal distribution and of the precision of A_∞. We illustrate this in the following example.

Example 16.10.   Consider again the renewal process described in Example 16.8. Consider $n = 50$ observed values of i.i.d. TTF from $G(2,100)$ and $n = 50$ observed repair times. We run `renewDis` 1000 times to obtain an EBD of the number of renewals in 1000 [hours]. The program yields that the mean number of renewals for 1000 hours of operation is 6.128. This is the bootstrap estimate of $V(1000)$. The asymptotic approximation is $1000/202.2 = 4.946$. The bootstrap confidence interval for $V(1000)$ at 0.95 level of confidence is (4,9). This confidence interval covers the asymptotic approximation. Accordingly, the bootstrap estimate of 6.13 is not significantly different from the asymptotic approximation.

```
> set.seed(123)
> Ttf <- rgamma(50,
                shape=2,
                scale=100)
> Ttr <- rgamma(50,
                shape=2,
                scale=1)
> AvailEbd <- availDis(ttf=Ttf,
                       ttr=Ttr,
                       n=1000,
                       seed=123)
```

```
The estimated MTTF from ttf is 153.01
The estimated MTTR from ttr is 1.91
The estimated asymptotic availability is 0.9877
availability EBD
Min.    :0.9813
1st Qu.:0.9864
Median :0.9875
Mean    :0.9875
3rd Qu.:0.9887
Max.    :0.9914
```

```
> RenewEbd <- renewDis(ttf=Ttf,
                       ttr=Ttr,
                       time=1000,
                       n=1000)
```

```
The estimated MEAN NUMBER Of RENEWALS is 8.21
number of renewals EBD
   Min. 1st Qu.  Median    Mean 3rd Qu.     Max.
  4.000   7.000   8.000   8.213   9.000  16.000
```

```
> rm(AvailEbd, RenewEbd, Ttf, Ttr)
```

Additional topics of interest are maintenance, repairability, and availability. The objective is to increase the availability by instituting maintenance procedures and by adding stand-by systems and repairment. The question is what is the optimal maintenance period, how many stand-by systems and repairment to add. The interested reader is referred to Zacks (1992, Ch. 4) and Gertsbakh (1989).

In the following sections, we discuss statistical problems associated with reliability assessment, when one does not know definitely the model and the values of its parameters.

16.4   Types of observations on *TTF*

The proper analysis of data depends on the type of observations available. Dealing with TTF and TTR random variables, we wish to have observations which give us the exact length of time interval from activation (failure) of a system (component) till its failure (repair). However, one can find that proper records have not been kept, and instead one can find only the number of failures (repairs) in a given period of time. These are discrete random variables rather than the continuous ones under investigation. Another type of problem typical to reliability studies is that some observations are **censored**. For example, if it is decided to put n identical systems on test for a specified length of time t^*, we may observe only a random number, K_n, of failures in the time interval $(0, t^*]$. On the other $n - K_n$ systems which did not fail we have only partial information, i.e., their TTF is greater than t^*. The observations on these systems are called **right censored**. In the above example, n units are put on test at the same time. The censoring time t^* is a fixed time. Sometimes we have observations with **random censoring**. This is the case when we carry a study for a fixed length of time t^* [years], but the units (systems) enter the study at random times between 0 and t^*, according to some distribution.

Suppose that a unit enters the study at the random time τ, $0 < \tau < t^*$, and its TTF is T. We can observe only $W = \min(T, t^* - \tau)$. Here the censoring time is the random variable $t^* - \tau$. An example of such a situation is when we sell a product under warranty. The units of this product are sold to different customers at random times during the study period $(0, t^*)$. Products which fail are brought back for repair. If this happens during the study period, we have an uncensored observation on the TTF of that unit; otherwise the observation is censored, i.e., $W = t^* - \tau$.

The censored observations described above are time censored. Another type of censoring is **frequency censoring**. This is done when n units are put on test at the same time, but the test is terminated the instant the rth failure occurs. In this case, the length of the test is the rth order statistic of failure times $T_{n,r}$ ($r = 1, \ldots, n$). Notice that $T_{n,r} = T_{(r)}$, where $T_{(1)} < T_{(2)} < \cdots < T_{(n)}$ are the order statistics of n i.i.d. TTF's. If T is distributed exponentially, $E(\beta)$, for example, the expected length of the experiment is

$$E\{T_{n,r}\} = \beta \left(\frac{1}{n} + \frac{1}{n-1} + \cdots + \frac{1}{n-r+1} \right).$$

There may be substantial time saving if we terminate the study at the rth failure, when $r < n$. For example, in the exponential case, with $E\{T\} = \beta = 1000$ [hours] and $n = 20$,

$$E\{T_{20,20}\} = 1000 \times \left(1 + \frac{1}{2} + \frac{1}{3} + \cdots + \frac{1}{20} \right) = 3597.7 \text{ [hours]}.$$

On the other hand, for $r = 10$ we have $E\{T_{20,10}\} = 668.8$ [hours]. Thus, a frequency censored experiment, with $r = 10$ and $n = 20$, $\beta = 1000$ lasts on the average only 19% of the time length of an uncensored experiment. We will see later how one can determine the optimal n and r for estimating the mean TTF (MTTF) in the exponential case.

16.5   Graphical analysis of life data

In this section, we discuss some graphical procedures to fit a life distribution to failure data, and obtain estimates of the parameters from the graphs.

Let $t_1, t_2, \ldots, t_n$ be n uncensored observation on i.i.d. random variables $T_1, \ldots, T_n$, having some life distribution $F(t)$.

The empirical c.d.f., given $t_1, \ldots, t_n$, is defined as

$$F_n(t) = \frac{1}{n} \sum_{i=1}^{n} I\{t_i \le t\}, \tag{16.5.1}$$

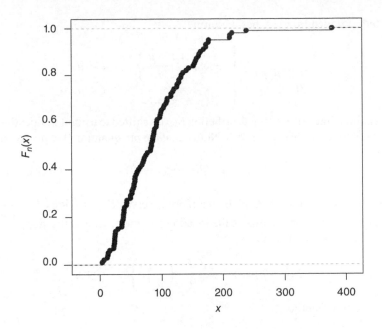

Figure 16.3 *The empirical c.d.f. of a random sample of 100 variables from W(1.5, 100).*

where $I\{t_i \leq t\}$ is the indicator variable, assuming the value 1 if $t_i \leq t$, and the value 0 otherwise. A theorem in probability theory states that the empirical c.d.f. $F_n(t)$ converges to $F(t)$, as $n \to \infty$.

In Figure 16.3, we present the empirical c.d.f. of a random sample of 100 variables having the Weibull distribution $W(1.5, 100)$. Since $F_n(t_{(i)}) = \dfrac{i}{n}$ for $i = 1, 2, \ldots, n$, the $\dfrac{i}{n}$-th quantile of $F_n(t)$ is the ordered statistic $t_{(i)}$. Accordingly, if $F(t)$ has some specific distribution, the scattergram of $\left(F^{-1}\left(\dfrac{i}{n}\right), t_{(i)} \right)$ $(i = 1, \ldots, n)$ should be around a straight line with slope 1. The plot of $t_{(i)}$ versus $F^{-1}\left(\dfrac{i}{n}\right)$ is called a *Q–Q* **plot** (quantile versus quantile). The *Q–Q* plot is the basic graphical procedure to test whether a given sample of failure times is generated by a specific life distribution. Since $F^{-1}(1) = \infty$ for the interesting life distribution, the quantile of F is taken at $\dfrac{i}{n+1}$ or at some other $\dfrac{i+\alpha}{n+\beta}$, which gives better plotting position for a specific distribution. For the normal distribution, $\dfrac{i - 3/8}{n + 1/4}$ is used.

If the distribution depends on location and scale parameters, we plot $t_{(i)}$ against the quantiles of the standard distribution. The intercept and the slope of the line fitted through the points yield estimates of these location and scale parameters. For example, suppose that $t_1, \ldots, t_n$ are values of a sample from a $N(\mu, \sigma^2)$ distribution. Thus, $t_{(i)} \approx \mu + \sigma \Phi^{-1}\left(\dfrac{i}{n}\right)$. Thus, if we plot $t_{(i)}$ against $\Phi^{-1}\left(\dfrac{i}{n}\right)$ we should have points around a straight line whose slope is an estimate of σ and intercept an estimate of μ.

We focus attention here on three families of life distributions.

1. The exponential or shifted exponential;
2. The Weibull; and
3. The log normal.

The **shifted-exponential** c.d.f. has the form

$$F(t; \mu, \beta) = \begin{cases} 1 - \exp\left\{\dfrac{t - \mu}{\beta}\right\}, & t \geq \mu \\ 0, & t < \mu. \end{cases} \tag{16.5.2}$$

The starting point of the exponential distribution $E(\beta)$ is shifted to a point μ. Location parameters of interest in reliability studies are $\mu \geq 0$. Notice that the pth quantile, $0 < p < \infty$, of the shifted exponential is

$$t_p = \mu + \beta(-\log(1 - p)). \tag{16.5.3}$$

Accordingly, for exponential Q–Q plots, we plot $t_{(i)}$ versus $E_{i,n} = -\log\left(1 - \dfrac{i}{n+1}\right)$. Notice that in this plot, the intercept estimates the location parameter μ, and the slope estimates β. In the Weibull case, $W(v, \beta)$, the c.d.f. is

$$F(t; v, \beta) = 1 - \exp\left\{-\left(\frac{t}{\beta}\right)^v\right\}, \quad t \geq 0. \tag{16.5.4}$$

Thus, if t_p is the pth quantile,

$$\log t_p = \log \beta + \frac{1}{v}\log(-\log(1 - p)). \tag{16.5.5}$$

For this reason, we plot $\log t_{(i)}$ versus

$$W_{i,n} = \log\left(-\log\left(1 - \frac{i}{n+1}\right)\right), \quad i = 1, \ldots, n. \tag{16.5.6}$$

The slope of the straight line estimates $1/v$ and the intercept estimates $\log \beta$. In the log normal case, we plot $\log t_{(i)}$ against $\Phi^{-1}\left(\dfrac{i - 3/8}{n + 1/4}\right)$.

Example 16.11.   In Figure 16.4, we present the Q–Q plot of 100 values generated at random from an exponential distribution $E(5)$. We fit a straight line through the origin to the points by the method of least-squares. A linear regression routine provides the line

$$\hat{x} = 5.94 * E.$$

Accordingly, the slope of a straight line fitted to the points provides an estimate of the true mean and standard deviation, $\beta = 5$. An estimate of the median is

$$\hat{x}(0.693) = 0.693 \times 5.9413 = 4.117.$$

The true median is $Me = 3.465$.
In Figure 16.5, we provide a probability plot of $n = 100$ values generated from a Weibull distribution with parameters $v = 2$ and $\beta = 2.5$.

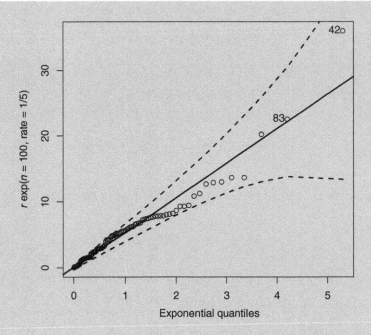

Figure 16.4 *Q–Q Plot of a sample of 100 values from E(5).*

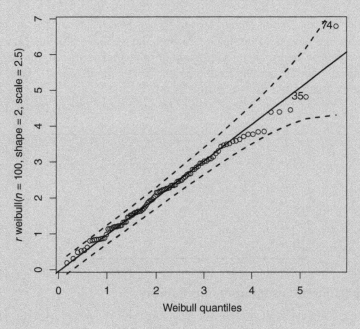

Figure 16.5 *Q–Q plot of a sample of 100 from W(2, 2.5).*

Least-squares fitting of a straight line to these points yields the line

$$\hat{y} = 0.856 + 0.479W.$$

Accordingly, we obtain the following estimates:

$$\hat{v} = 1/0.479 = 2.087,$$

$$\hat{\beta} = \exp(0.856) = 2.354,$$

$$\text{Median} = \exp(0.856 - (0.3665)(0.479))$$

$$= 1.975.$$

The true median is equal to $\beta(\ln 2)^{1/2} = 2.081$. The estimate of the mean is

$$\hat{\mu} = \hat{\beta}\Gamma(1 + 0.479)$$

$$= \hat{\beta} \times 0.479 \times \Gamma(0.479) = 2.080.$$

The true mean is $\mu = \beta\Gamma(1.5) = 2.216$. Finally, an estimate of the standard deviation is

$$\hat{\sigma} = \hat{\beta}(\Gamma(1.958) - \Gamma^2(1.479))^{1/2}$$

$$= \hat{\beta}[0.958 \times \Gamma(0.958) - (0.479 \times \Gamma(0.479))^2]^{1/2} = 1.054.$$

The true value is $\sigma = \beta(\Gamma(2) - \Gamma^2(1.5))^{1/2} = 1.158$.

If observations are censored from the left or from the right, we plot the quantiles only from the uncensored part of the sample. The plotting positions take into consideration the number of censored values from the left and from the right. For example, if $n = 20$ and the 2 smallest observations are censored, the plotting positions are

i	$t_{(i)}$	$\dfrac{i}{n+1}$
1	–	–
2	–	–
3	$t_{(3)}$	$\dfrac{3}{21}$
$\vdots$		
20	t_{20}	$\dfrac{20}{21}$

16.6   Nonparametric estimation of reliability

A nonparametric method called the Kaplan–Meier method yields an estimate, called the **product limit** (**PL**) estimate of the reliability function, without an explicit reference to the life distribution. The estimator of the reliability function at time t will be denoted by $\hat{R}_n(t)$ when n is the number of units put on test at time $t = 0$. If all the failure times $0 < t_1 < t_2 < \cdots < t_n < \infty$

are known then the PL estimator is equivalent to

$$\hat{R}_n(t) = 1 - F_n(t), \tag{16.6.1}$$

where $F_n(t)$ is the empirical c.d.f. defined earlier.

In some cases, either random or nonrandom censoring or withdrawals occur, and we do not have complete information on the exact failure times. Suppose that $0 < t_1 < t_2 < \cdots < t_k < \infty$, $k \leq n$, are the failure times and $w = n - k$ is the total number of withdrawals.

Let $I_j = (t_{j-1}, t_j)$, $j = 1, \ldots, k+1$, with $t_0 = 0$, $t_{k+1} = \infty$, be the time intervals between recorded failures. Let W_j be the number of withdrawals during the time interval I_j. The PL estimator of the reliability function is then

$$\hat{R}_n(t) = I\{t < t_1\} + \sum_{i=2}^{k+1} I\{t_{i-1} \leq t \leq t_i\} \prod_{j=1}^{i-1} \left(1 - \frac{1}{n_{j-1} - w_j/2}\right), \tag{16.6.2}$$

where $n_0 = n$, and n_l is the number of operating units just prior to the failure time t_l.

Usually, when units are tested in the laboratory under controlled conditions, there may be no withdrawals. This is not the case, however, if tests are conducted in field conditions, and units on test may be lost, withdrawn, or destroyed for reasons different than the failure phenomenon under study.

Suppose now that systems are installed in the field as they are purchased (random times). We decide to make a follow-up study of the systems for a period of two years. TTF of systems participating in the study is recorded. We assume that each system operates continuously from the time of installment until its failure. If a system has not failed by the end of the study period, the only information available is the length of time it has been operating. This is a case of multiple censoring. At the end of the study period, we have the following observations $\{(T_i, \delta_i),$ $i = 1, \ldots, n\}$, where n is the number of systems participating in the study; T_i is the length of operation of the ith system (TTF or time till censoring); $\delta_i = 1$ if ith observation is not censored and $\delta_i = 0$ otherwise.

Let $T_{(1)} \leq T_{(2)} \leq \cdots \leq T_{(n)}$ be the order statistic of the operation times and let $\delta_{j_1}, \delta_{j_2}, \ldots, \delta_{j_n}$ be the δ-values corresponding to the ordered T values where j_i is the index of the ith order statistic $T_{(i)}$, i.e., $T_{(i)} = T_j$ $(i = 1, \ldots, n)$.

The PL estimator of $R(t)$ is given by

$$\hat{R}_n(t) = I\{t < T_{(1)}\}$$

$$+ \sum_{i=1}^{n} I\{T_{(i)} \leq T_{(i+1)}\} \prod_{j=1}^{i} \left(1 - \frac{\delta_j}{n - j + 1}\right). \tag{16.6.3}$$

Another situation prevails in the laboratory or in field studies when the exact failure times cannot be recorded. Let $0 < t_1 < t_2 < \cdots < t_k < \infty$ be fixed inspection times. Let w_i be the number of withdrawals and f_i the number of failures in the time interval I_i $(i = 1, \ldots, k+1)$. In this case, the formula is modified to be

$$\hat{R}_n(t) = I\{t < t_1\}$$

$$+ \sum_{i=2}^{k+1} I\{t_i \leq t < t_{i+1}\} \prod_{j=1}^{i-1} \left(1 - \frac{f_j}{n_{j-1} - \frac{w_j}{2}}\right). \tag{16.6.4}$$

This version of the estimator of $R(t)$, when the inspection times are fixed (not random failure times), is called the **actuarial estimator**.

In the following examples, we illustrate these estimators of the reliability function.

Example 16.12.   A machine is tested before shipping it to the customer for a one-week period (120 [hours]) or till its first failure, whichever comes first. Twenty such machines were tested consecutively. In Table 16.1, we present the ordered time till failure or time till censor (TTF/TTC) of the 20 machines, the factors $(1 - \delta_i/(n - i + 1))$ and the PL estimator $\hat{R}(t_i)$, $i = 1, \ldots, 20$.

Table 16.1   *Failure times [hour] and PL estimates*

i	$T_{(i)}$	$\left(1 - \dfrac{\delta_i}{n-i+1}\right)$	$\hat{R}(T_i)$
1	4.787715	0.95	0.95
2	8.378821	0.9473684	0.9
3	8.763973	0.9444444	0.85
4	13.77360	0.9411765	0.8
5	29.20548	0.9375	0.75
6	30.53487	0.9333333	0.7
7	47.96504	0.9285714	0.65
8	59.22675	0.9230769	0.6
9	60.66661	0.9166667	0.55
10	62.12246	0.9090909	0.5
11	67.06873	0.9	0.45
12	92.15673	0.8888889	0.4
13	98.09076	0.875	0.35
14	107.6014	0.8571429	0.3
15	120	1	0.3
16	120	1	0.3
17	120	1	0.3
18	120	1	0.3
19	120	1	0.3
20	120	1	0.3

16.7   Estimation of life characteristics

In Chapter 4, we studied the estimation of parameters of distributions, and of functions of these parameters. We discussed point estimators and confidence intervals. In particular, we discussed unbiased estimators, least-squares estimators, maximum likelihood estimators, and Bayes estimators. All these methods of estimation can be applied in reliability studies. We will discuss in the present section the maximum likelihood estimation of the parameters of some common life distributions, like the exponential and the Weibull, and some nonparametric techniques, for censored and uncensored data.

16.7.1 Maximum likelihood estimators for exponential TTF distribution

We start with the case of **uncensored observations**. Thus, let $T_1, T_2, \ldots, T_n$ be i.i.d. random variables distributed like $E(\beta)$. Let $t_1, \ldots, t_n$ be their sample realization (random sample). The likelihood function of β, $0 < \beta < \infty$, is

$$L(\beta; \mathbf{t}) = \frac{1}{\beta^n} \exp \left\{ -\frac{1}{\beta} \sum_{i=1}^{n} t_i \right\}. \tag{16.7.1}$$

It is easy to check that the maximum likelihood estimator (MLE) of β is the sample mean

$$\hat{\beta}_n = \overline{T}_n = \frac{1}{n} \sum_{i=1}^{n} T_i. \tag{16.7.2}$$

$\overline{T}_n$ is distributed like $G\left(n, \frac{\beta}{n}\right)$. Thus, $E\{\hat{\beta}_n\} = \beta$ and $V\{\hat{\beta}_n\} = \frac{\beta^2}{n}$. From the relationship between the Gamma and the χ^2 distributions, we have that $\hat{\beta}_n \sim \frac{\beta}{2^n} \chi^2[2n]$. Thus, a $(1 - \alpha)$ level **confidence interval** for β, based on the MLE $\hat{\beta}_n$ is

$$\left(\frac{2n\hat{\beta}_n}{\chi^2_{1-\alpha/2}[2n]}, \frac{2n\hat{\beta}_n}{\chi^2_{\alpha/2}[2n]} \right). \tag{16.7.3}$$

For large samples, we can use the normal approximation $\hat{\beta}_n \pm z_{1-\alpha/2} \frac{\hat{\beta}_n}{\sqrt{n}}$.

Example 16.13. The failure times of 20 electric generators (in [hour]) are:

121.5	1425.5	2951.2	5637.9
1657.2	592.1	10609.7	9068.5
848.2	5296.6	7.5	2311.1
279.8	7201.9	6853.7	6054.3
1883.6	6303.9	1051.7	711.5.

Exponential probability plotting of these data yields a scatter around the line with slope of 3866.17 and $R^2 = 0.95$. The exponential model fits the failure times quite well. The MLE estimator of the MTTF, β, yields $\hat{\beta}_{20} = 3543.4$ [hours]. Notice that the MLE is different, but not significantly from the above graphical estimate of β. Indeed, the standard error of $\hat{\beta}_{20}$ is S.E. $= \hat{\beta}_{20}/\sqrt{20} = 732.6$.

```
> data(FAILTIME)
> library(survival)
> SuRe <- survreg(
    Surv(time=FAILTIME) ~ 1 ,
    dist = "exponential")
> summary(SuRe)

Call:
survreg(formula = Surv(time = FAILTIME) ~ 1, dist =
                 "exponential")
```

```
               Value Std. Error    z      p
(Intercept) 8.173         0.224 36.5 <2e-16


Scale fixed at 1


Exponential distribution
Loglik(model)= -183.5    Loglik(intercept only)= -183.5
Number of Newton-Raphson Iterations: 5
n= 20


> confint(SuRe)


              2.5 %    97.5 %
(Intercept) 7.734572 8.611095
```

Confidence interval, at level of 0.95, for β is given by (2355.2, 5684.9). The normal approximation to the confidence interval is (2107.6, 4979.1). The sample size is not sufficiently large for the normal approximation to be effective.

When the observations are **time censored** by a fixed constant t^*, let K_n denote the number of uncensored observations. K_n is a random variable having the binomial distribution $B(n, 1 - \exp\{-t^*/\beta\})$. Let $\hat{p}_n = \dfrac{K_n}{n}$. $\hat{p}_n$ is a consistent estimator of $1 - \exp\{-t^*/\beta\}$. Hence, a consistent estimator of β is

$$\tilde{\beta}_n = -t^* / \log(1 - \hat{p}_n). \qquad (16.7.4)$$

This estimator is not an efficient one, since it is not based on observed failures. Moreover, if $K_n = 0$, $\tilde{\beta}_n = 0$. Using the expansion method shown in Section 13.2, we obtain that the asymptotic variance of $\tilde{\beta}_n$ is

$$AV\{\tilde{\beta}_n\} \cong \frac{\beta^4}{nt^{*2}} \cdot \frac{1 - e^{-t^*/\beta}}{e^{-t^*/\beta}}. \qquad (16.7.5)$$

The likelihood function of β in this time censoring case is

$$L(\beta; K_n, \mathbf{T}_n) = \frac{1}{\beta^{K_n}} \exp\left\{-\frac{1}{\beta}\left(\sum_{i=1}^{K_n} T_i + t^*(n - K_n)\right)\right\}. \qquad (16.7.6)$$

Also here, if $K_n = 0$, the MLE of β does not exist. If $K_n \geq 1$, the MLE is

$$\hat{\beta}_n = \frac{S_{n,K_n}}{K_n}, \qquad (16.7.7)$$

where $S_{n,K_n} = \sum_{i=1}^{K_n} T_i + (n - K_n)t^*$, is the **total time on test** of the n units.

The theoretical evaluation of the properties of the MLE $\hat{\beta}_n$ is complicated. We can, however, get information on its behavior by simulation.

Example 16.14.   For a sample of size $n = 50$, with $\beta = 1000$ and $t^* = 2000$, $\Pr\{K_{50} = 0\} = \exp\{-100\} \doteq 0$. Thus, we expect that $\hat{\beta}_n$ will exist in all the simulation runs. For 100 MLEs of β, obtained by this simulation we have a mean $= 1025.5$, a median $= 996.7$ and a standard deviation $= 156.9$. The standard deviation of $\tilde{\beta}_n$, according to the previous formula, with the above values of n, β, and t^*, is 178.7.

```
> library(boot)
> FAILTIME[FAILTIME >= 7000] <- 7000 # Censor data at 7000
> X <- data.frame(
    time= FAILTIME,
    event=ifelse(FAILTIME < 7000,
                  yes=1,
                  no=0))
> head(X, 8)


    time event
1   121.5     1
2 1425.5     1
3 2951.2     1
4 5637.9     1
5 1657.2     1
6   592.1     1
7 7000.0     0
8 7000.0     0


> B <- boot(data=X,
            statistic=function(x, i){
              coefficients(
                survreg(
                  Surv(
                    time=x[i,1],
                    event=x[i,2]) ~ 1 ,
                  dist = "exponential"))
              },
            R = 100)
> boot.ci(B,
          conf=0.95,
          type="perc")


BOOTSTRAP CONFIDENCE INTERVAL CALCULATIONS
Based on 100 bootstrap replicates

CALL :
boot.ci(boot.out = B, conf = 0.95, type = "perc")

Intervals :
Level      Percentile
95%    ( 7.718,   8.911 )
Calculations and Intervals on Original Scale
Some percentile intervals may be unstable


> rm(B)
```

For further reading on the properties of MLE under time censoring, see Zacks (1992, pp. 125).

Under **frequency censoring** the situation is simpler. Suppose that the censoring is at the rth failure. The total time on test is $S_{n,r} = \sum_{i=1}^{r} T_{(i)} + (n-r)T_{(r)}$. In this case,

$$S_{n,r} \sim \frac{\beta}{2}\chi^2[2r] \qquad (16.7.8)$$

and the MLE $\hat{\beta}_{n,r} = \dfrac{S_{n,r}}{r}$ is an unbiased estimator of β, with variance

$$V\{\hat{\beta}_{n,r}\} = \frac{\beta^2}{r}.$$

Or,

$$\text{S.E.}\{\hat{\beta}_{n,r}\} = \frac{\beta}{\sqrt{r}}. \qquad (16.7.9)$$

If we wish to have a certain precision, so that $\text{S.E.}\{\hat{\beta}_{n,r}\} = \gamma\beta$ then $r = \frac{1}{\gamma^2}$. Obviously $n \geq r$.

Suppose that we pay for the test c_2 \$ per unit and c_1 \$ per time unit, for the duration of the test. Then, the total cost of the test is

$$TK_{n,r} = c_1 T_{n,r} + c_2 n. \qquad (16.7.10)$$

For a given r, we choose n to minimize the expected total cost. The resulting formula is

$$n^0 \doteq \frac{r}{2}\left(1 + \left(1 + \frac{4c_1}{rc_2}\beta\right)^{1/2}\right). \qquad (16.7.11)$$

The problem is that the optimal sample size n^0 depends on the unknown β. If one has some prior estimate of β, it could be used to determine a good starting value for n.

Example 16.15.   Consider a design of a life testing experiment with frequency censoring and exponential distribution of the TTF. We require that $\text{S.E.}\{\hat{\beta}_{n,r}\} = 0.2\beta$. Accordingly, $r = \left(\frac{1}{0.2}\right)^2 = 25$. Suppose that we wish to minimize the total expected cost, at $\beta = 100$ [hours], where $c_1 = c_2 = 2$ \$. Then,

$$n^0 \doteq \frac{25}{2}\left(1 + \left(1 + \frac{4}{25}100\right)^{1/2}\right) = 64.$$

The expected duration of this test is

$$E\{T_{64,25}\} = 100\sum_{i=1}^{25}\frac{1}{65-i} = 49.0 \text{ [hours]}.$$

16.7.2 Maximum likelihood estimation of the Weibull parameters

Let $t_1, \ldots, t_n$ be uncensored failure times of n random variables having a Weibull distribution $W(v, \beta)$. The likelihood function of (v, β) is

$$L(v, \beta; \mathbf{t}) = \frac{v^n}{\beta^{nv}} \left(\prod_{i=1}^{n} t_i \right)^{v-1} \exp \left\{ -\sum_{i=1}^{n} \left(\frac{t_i}{\beta} \right)^v \right\}, \tag{16.7.12}$$

$0 < \beta, v < \infty$. The MLE of v and β are the solutions $\hat{\beta}_n$, $\hat{v}_n$ of the equations

$$\hat{\beta}_n = \left(\frac{1}{n} \sum_{i=1}^{n} t_i^{\hat{v}_n} \right)^{1/\hat{v}_n}, \tag{16.7.13}$$

and

$$\hat{v}_n = \left[\frac{\sum_{i=1}^{n} t_i^{\hat{v}_n} \log t_i}{\sum_{i=1}^{n} t_i^{\hat{v}_n}} - \frac{1}{n} \sum_{i=1}^{n} \log(t_i) \right]^{-1}. \tag{16.7.14}$$

All logarithms are on base e (ln).

The equation for $\hat{v}_n$ is solved iteratively by the recursive equation

$$\hat{v}^{(j+1)} = \left[\frac{\sum_{i=1}^{n} t_i^{\hat{v}^{(j)}} \log(t_i)}{\sum_{i=1}^{n} t_i^{\hat{v}^{(j)}}} - \frac{1}{n} \sum_{i=1}^{n} \log(t_i) \right]^{-1}, \quad j = 0, 1, \ldots, \tag{16.7.15}$$

where $\hat{v}^{(0)} = 1$.

To illustrate, we simulated a sample of $n = 50$ failure times from $W(2.5, 10)$. In order to obtain the MLE, we have to continue the iterative process until the results converge. We show here the obtained values, as functions of the number of iterations.

# Iterations	$\hat{\beta}$	$\hat{v}$
10	11.437	2.314
20	9.959	2.367
30	9.926	2.368
40	9.925	2.368

It seems that 40 iterations yield sufficiently accurate solutions.

Confidence intervals for $\hat{v}_n$, $\hat{\beta}_n$ can be determined, for large samples, by using large sample approximation formulae for the standard errors of the MLE, which are (see Zacks (1992, p. 147))

$$SE\{\hat{\beta}_n\} \cong \frac{\hat{\beta}_n}{\sqrt{n}\, \hat{v}_n} \cdot 1.053 \tag{16.7.16}$$

and

$$SE\{\hat{v}_n\} \cong 0.780 \frac{\hat{v}_n}{\sqrt{n}}. \tag{16.7.17}$$

The large sample confidence limits are

$$\hat{\beta}_n \pm z_{1-\alpha/2}\text{S.E.}\{\hat{\beta}_n\}, \tag{16.7.18}$$

and

$$\hat{v}_n \pm z_{1-\alpha/2}\text{S.E.}\{\hat{v}_n\}. \tag{16.7.19}$$

In the above numerical example, we obtained the MLE $\hat{\beta}_{50} = 9.925$ and $\hat{v}_{50} = 2.368$. Using these values, we obtain the large sample approximate confidence intervals, with level of confidence $1 - \alpha = 0.95$, to be $(8.898, 11.148)$ for β and $(1.880, 2.856)$ for v.

We can obtain bootstrapping confidence intervals. The bootstrap confidence limits are the $\alpha/2$th and $(1 - \alpha/2)$th quantile of the simulated values which produced confidence intervals $(8.378, 11.201)$ for β and $(1.914, 3.046)$ for v. The difference between these confidence intervals and the large sample approximation ones is not significant.

Maximum likelihood estimation in censored cases is more complicated and will not be discussed here. Estimates of v and β in the censored case can be obtained from the intercept and slope of the regression line in the Q–Q plot.

```
> B <- boot (
    data=X,
    statistic=function(x, i){
      coefficients(
        survreg(
          Surv(
            time=x[i,1],
            event=x[i,2]) ~ 1 ,
          dist="weibull"))
    },
    R = 100)
> boot.ci(B,
          conf=0.95,
          type="perc")


BOOTSTRAP CONFIDENCE INTERVAL CALCULATIONS
Based on 100 bootstrap replicates

CALL :
boot.ci(boot.out = B, conf = 0.95, type = "perc")

Intervals :
Level      Percentile
95%    ( 7.691,   8.682 )
Calculations and Intervals on Original Scale
Some percentile intervals may be unstable

> rm(B)
```

16.8   Reliability demonstration

Reliability demonstration is a procedure for testing whether the reliability of a given device (system) at a certain age is sufficiently high. More precisely, a time point t_0 and a desired reliability R_0 are specified, and we wish to test whether the reliability of the device at age t_0, $R(t_0)$, satisfies the requirement that $R(t_0) \geq R_0$. If the life distribution of the device is completely known, including all parameters, there is no problem of reliability demonstration – one computes $R(t_0)$ exactly and determines whether $R(t_0) \geq R_0$. If, as is generally the case, either the life distribution or its parameters are unknown, then the problem of reliability demonstration is that of obtaining suitable data and using them to test the statistical hypothesis that $R(t_0) \geq R_0$ versus the alternative that $R(t_0) < R_0$. Thus, the theory of testing statistical hypotheses provides the tools for reliability demonstration. In the present section, we review some of the basic notions of hypothesis testing as they pertain to reliability demonstration.

In the following subsections, we develop several tests of interest in reliability demonstration. We remark here that procedures for obtaining confidence intervals for $R(t_0)$, which were discussed in the previous sections, can be used to test hypotheses. Specifically, the procedure involves computing the upper confidence limit of a $(1 - 2\alpha)$-level confidence interval for $R(t_0)$ and comparing it with the value R_0. If the upper confidence limit exceeds R_0 then the null hypothesis $H_0 : R(t_0) > R_0$ is accepted, otherwise it is rejected. This test will have a significance level of α.

For example, if the specification of the reliability at age $t = t_0$ is $R = 0.75$ and the confidence interval for $R(t_0)$, at level of confidence $\gamma = 0.90$, is $(0.80, 0.85)$, the hypothesis H_0 can be immediately accepted at a level of significance of $\alpha = (1 - \gamma)/2 = 0.05$. There is a duality between procedures for testing hypotheses and for confidence intervals.

16.8.1   Binomial testing

A random sample of n devices is put on life test simultaneously. Let J_n be the number of failures in the time interval $[0, t_0)$, and $K_n = n - J_n$. We have seen that $K_n \sim B(n, R(t_0))$. Thus, if H_0 is true, i.e., $R(t_0) \geq R_0$, the values of K_n will tend to be larger, in a probabilistic sense. Thus, one tests H_0 by specifying a critical value C_α and rejecting H_0 whenever $K_n \leq C_\alpha$. The critical value C_α is chosen as the largest value satisfying

$$F_B(C_\alpha; n, R_0) \leq \alpha.$$

The OC function of this test, as a function of the true reliability R, is

$$OC(R) = \Pr\{K_n > C_\alpha \mid R(t_0) = R\}$$
$$= 1 - F_B(C_\alpha; n, R). \tag{16.8.1}$$

If n is large, then one can apply the normal approximation to the Binomial c.d.f.. In these cases, we can determine C_α to be the integer most closely satisfying

$$\Phi\left(\frac{C_\alpha + 1/2 - nR_0}{(nR_0(1 - R_0))^{1/2}}\right) = \alpha. \tag{16.8.2}$$

Generally, this will be given by

$$C_\alpha = \text{integer closest to } \{nR_0 - 1/2 - z_{1-\alpha}(nR_0(1 - R_0))^{1/2}\}, \tag{16.8.3}$$

where $z_{1-\alpha} = \Phi^{-1}(1 - \alpha)$. The OC function of this test in the large sample case is approximated by

$$OC(R) \cong \Phi \left(\frac{nR - C_\alpha - 1/2}{(nR(1 - R))^{1/2}} \right). \tag{16.8.4}$$

The normal approximation is quite accurate whenever $n > 9/(R(1 - R))$.

If in addition to specifying α, we specify that the test have Type II error probability β, when $R(t_0) = R_1$, then the normal approximation provides us with a formula for the necessary sample size:

$$n \doteq \frac{(z_{1-\alpha}\sigma_0 + z_{1-\beta}\sigma_1)^2}{(R_1 - R_0)^2}, \tag{16.8.5}$$

where $\sigma_i^2 = R_i(1 - R_i)$, $i = 0, 1$.

Example 16.16. Suppose that we wish to test at significance level $\alpha = 0.05$ the null hypothesis that the reliability at age 1000 [hours] of a particular system is at least 85%. If the reliability is 80% or less, we want to limit the probability of accepting the null hypothesis to $\beta = 0.10$. Our test is to be based on K_n, the number of systems, out of a random sample of n, surviving at least 1000 hours of operation. Setting $R_0 = 0.85$ and $R_1 = 0.80$, we have $\sigma_0 = 0.357$, $\sigma_1 = 0.4$, $z_{0.95} = 1.645$, $z_{0.90} = 1.282$. Substituting above we obtain that the necessary sample size is $n = 483$. The critical value is $C_{0.05} = 397$.

We see that in binomial testing one may need very large samples to satisfy the specifications of the test. If in the above problem we reduce the sample size to $n = 100$, then $C_{0.05} = 79$. However, now the probability of accepting the null hypothesis when $R = 0.80$ is $OC(0.8) = \Phi(0.125) = 0.55$, which is considerably higher than the corresponding probability of 0.10 under $n = 483$.

16.8.2   Exponential distributions

Suppose that we know that the life distribution is exponential $E(\beta)$, but β is unknown. The hypotheses

$$H_0 : R(t_0) \geq R_0$$

versus

$$H_1 : R(t_0) < R_0$$

can be rephrased in terms of the unknown parameter, β, as

$$H_0 : \beta \geq \beta_0$$

versus

$$H_1 : \beta < \beta_0$$

where $\beta_0 = -t_0/\ln R_0$. Let $t_1, \ldots, t_n$ be the values of a (complete) random sample of size n. Let $\bar{t}_n = \frac{1}{n} \sum_{i=1}^n t_i$. The hypothesis H_0 is rejected if $\bar{t}_n < C_\alpha$, where

$$C_\alpha = \frac{\beta_0}{2n} \chi_\alpha^2[2n]. \tag{16.8.6}$$

The OC function of this test, as a function of β, is

$$OC(\beta) = \Pr\{\bar{t}_n > C_\alpha \mid \beta\}$$

$$= \Pr\left\{\chi^2[2n] > \frac{\beta_0}{\beta}\chi^2_\alpha[2n]\right\}. \tag{16.8.7}$$

If we require that at $\beta = \beta_1$ the OC function of the test will assume the value γ, then the sample size n should satisfy

$$\frac{\beta_0}{\beta_1}\chi^2_\alpha[2n] \geq \chi^2_{1-\gamma}[2n].$$

The quantiles of $\chi^2[2n]$, for $n \geq 15$, can be approximated by the formula

$$\chi^2_p[2n] \cong \frac{1}{2}(\sqrt{4n} + z_p)^2. \tag{16.8.8}$$

Substituting this approximation and solving for n, we obtain the approximation

$$n \cong \frac{1}{4}\frac{(z_{1-\gamma} + z_{1-\alpha}\sqrt{\zeta})^2}{(\sqrt{\zeta} - 1)^2}, \tag{16.8.9}$$

where $\zeta = \beta_0/\beta_1$.

Example 16.17. Suppose that in Example 16.16, we know that the system lifetimes are exponentially distributed. It is interesting to examine how many systems would have to be tested in order to achieve the same error probabilities as before, if our decision were now based on $\bar{t}_n$.

Since $\beta = -t/\ln R(t)$, the value of the parameter β under $R(t_0) = R(1000) = 0.85$ is $\beta_0 = -1000/\ln(0.85) = 6153$ [hours], while its value under $R(t_0) = 0.80$ is $\beta_1 = -1000/\ln(0.80) = 4481$ [hours]. Substituting these values into (9.3.5), along with $\alpha = 0.05$ and $\gamma = 0.10$ (γ was denoted by β in Example 16.16), we obtain the necessary sample size $n \cong 87$.

Thus we see that the additional knowledge that the lifetime distribution is exponential, along with the use of complete lifetime data on the sample, allows us to achieve a greater than fivefold increase in efficiency in terms of the sample size necessary to achieve the desired error probabilities.

We remark that if the sample is censored at the rth failure then all the formulae developed above apply after replacing n by r, and $\bar{t}_n$ by $\hat{\beta}_{n,r} = T_{n,r}/r$.

Example 16.18. Suppose that the reliability at age $t = 250$ [hours] should be at least $R_0 = 0.85$. Let $R_1 = 0.75$. The corresponding values of β_0 and β_1 are 1538 and 869 [hours], respectively. Suppose that the sample is censored at the $r = 25$th failure. Let $\hat{\beta}_{n,r} = T_{n,r}/25$ be the MLE of β. H_0 is rejected, with level of significance $\alpha = 0.05$, if

$$\hat{\beta}_{n,r} \leq \frac{1538}{50}\chi^2_{0.05}[50] = 1069 \text{ [hours]}.$$

The Type II error probability of this test, at $\beta = 869$, is

$$
\begin{aligned}
\mathrm{OC}(869) &= \Pr\left\{\chi^2[50] > \frac{1538}{869}\chi^2_{0.05}[50]\right\} \\
&= \Pr\{\chi^2[50] > 61.5\} \\
&\doteq 1 - \Phi\left(\frac{61.5 - 50}{\sqrt{100}}\right) \\
&= 0.125.
\end{aligned}
$$

Sometimes in reliability demonstration an overriding concern is keeping the number of items tested to a minimum, subject to whatever accuracy requirements are imposed. This could be the case, for example, when testing very complex and expensive systems. In such cases, it may be worthwhile applying a sequential testing procedure, where items are tested one at a time in sequence until the procedure indicates that testing can stop and a decision be made. Such an approach would also be appropriate when testing prototypes of some new design, which are being produced one at a time at a relatively slow rate.

In Chapter 11, we have introduced the Wald SPRT for testing hypotheses with binomial data. Here we reformulate this test for reliability testing.

16.8.2.1   The SPRT for binomial data

Without any assumptions about the lifetime distribution of a device, we can test hypotheses concerning $R(t_0)$ by simply observing whether or not a device survives to age t_0. Letting K_n represent the number of devices among n randomly selected ones surviving to age t_0, we have $K_n \sim B(n, R(t_0))$. The likelihood ratio is given by

$$
\lambda_n = \left(\frac{1 - R_1}{1 - R_0}\right)^n \left(\frac{R_1(1 - R_0)}{R_0(1 - R_1)}\right)^{K_n}. \tag{16.8.10}
$$

Thus,

$$
\ln \lambda_n = n \ln\left(\frac{1 - R_1}{1 - R_0}\right) - K_n \ln\left(\frac{R_0(1 - R_1)}{R_1(1 - R_0)}\right).
$$

It follows that the SPRT can be expressed in terms of K_n as follows:

Continue sampling if $-h_1 + sn < K_n < h_2 + sn$,

Accept H_0 if $K_n \geq h_2 + sn$,

Reject H_0 if $K_n \leq -h_1 + sn$,

where

$$
\begin{cases}
s = \ln\left(\dfrac{1 - R_1}{1 - R_0}\right) / \ln\left(\dfrac{R_0(1 - R_1)}{R_1(1 - R_0)}\right), \\[2ex]
h_1 = \ln\left(\dfrac{1 - \gamma}{\alpha}\right) / \ln\left(\dfrac{R_0(1 - R_1)}{R_1(1 - R_0)}\right), \\[2ex]
h_2 = \ln\left(\dfrac{1 - \alpha}{\gamma}\right) / \ln\left(\dfrac{R_0(1 - R_1)}{R_1(1 - R_0)}\right).
\end{cases}
\tag{16.8.11}
$$

α and γ are the prescribed probabilities of Type I and Type II errors. Note that if we plot K_n versus n, the accept and reject boundaries are parallel straight lines with common slope s and intercepts h_2 and $-h_1$, respectively.

The OC function of this test is expressible (approximately) in terms of an implicit parameter ψ. Letting

$$R^{(\psi)} = \begin{cases} \dfrac{1 - \left(\dfrac{1 - R_1}{1 - R_0}\right)^{\psi}}{\left(\dfrac{R_1}{R_0}\right)^{\psi} - \left(\dfrac{1 - R_1}{1 - R_0}\right)^{\psi}}, & \psi \neq 0 \\[2em] s, & \psi = 0, \end{cases} \qquad (16.8.12)$$

we have that the OC function at $R(t_0) = R^{(\psi)}$ is given by

$$\mathrm{OC}(R^{(\psi)}) \approx \begin{cases} \dfrac{\left(\dfrac{1 - \gamma}{\alpha}\right)^{\psi} - 1}{\left(\dfrac{1 - \gamma}{\alpha}\right)^{\psi} - \left(\dfrac{\gamma}{1 - \alpha}\right)^{\psi}}, & \psi \neq 0 \\[3em] \dfrac{\ln\left(\dfrac{1 - \gamma}{\alpha}\right)}{\ln\left(\dfrac{(1 - \alpha)(1 - \gamma)}{\alpha\gamma}\right)}, & \psi = 0. \end{cases} \qquad (16.8.13)$$

It is easily verified that for $\psi = 1$, $R^{(\psi)}$ equals R_0 and $\mathrm{OC}(R^{(\psi)})$ equals $1 - \alpha$, while for $\psi = -1$, $R^{(\psi)}$ equals R_1 and $\mathrm{OC}(R^{(\psi)})$ equals γ.

The expected sample size, or **average sample number** (ASN), as a function of $R^{(\psi)}$, is given by

$$\mathrm{ASN}(R^{(\psi)}) \approx \begin{cases} \dfrac{\ln\dfrac{1 - \gamma}{\alpha} - \mathrm{OC}(R^{(\psi)})\ln\left(\dfrac{(1 - \alpha)(1 - \gamma)}{\alpha\gamma}\right)}{\ln\dfrac{1 - R_1}{1 - R_0} - R^{(\psi)}\ln\left(\dfrac{R_0(1 - R_1)}{R_1(1 - R_0)}\right)}, & \psi \neq 0 \\[3em] \dfrac{h_1 h_2}{s(1 - s)}, & \psi = 0. \end{cases} \qquad (16.8.14)$$

The ASN function will typically have a maximum at some value of R between R_0 and R_1, and decrease as R moves away from the point of maximum in either direction.

Example 16.19. Consider Example 16.17, where we had $t = 1000$ [hours], $R_0 = 0.85$, $R_1 = 0.80$, $\alpha = 0.05$, $\gamma = 0.10$. Suppose now that systems are tested sequentially, and we apply the SPRT based on the number of systems still functioning at 1000 [hours]. The parameters of the boundary lines are $s = 0.826$, $h_1 = 8.30$, and $h_2 = 6.46$.

The OC and ASN functions of the test are given in Table 16.2, for selected values of ψ.

Compare the values in the ASN column to the sample size required for the corresponding fixed-sample test, $n = 483$. It is clear that the SPRT effects a considerable saving in sample size, particularly when $R(t_0)$ is less than R_1 or greater than R_0. Note also that the maximum ASN value occurs when $R(t_0)$ is near s.

Table 16.2   *OC and ASN values for the SPRT for Example 16.19*

ψ	$R^{(\psi)}$	$OC(R^{(\psi)})$	$ASN(R^{(\psi)})$
−2.0	0.7724	0.0110	152.0
−1.8	0.7780	0.0173	167.9
−1.6	0.7836	0.0270	186.7
−1.4	0.7891	0.0421	208.6
−1.2	0.7946	0.0651	234.1
−1.0	0.8000	0.1000	263.0
−0.8	0.8053	0.1512	294.2
−0.6	0.8106	0.2235	325.5
−0.4	0.8158	0.3193	352.7
−0.2	0.8209	0.4357	370.2
0.0	0.8259	0.5621	373.1
0.2	0.8309	0.6834	360.2
0.4	0.8358	0.7858	334.8
0.6	0.8406	0.8629	302.6
0.8	0.8453	0.9159	269.1
1.0	0.8500	0.9500	238.0
1.2	0.8546	0.9709	210.8
1.4	0.8590	0.9833	187.8
1.6	0.8634	0.9905	168.6
1.8	0.8678	0.9946	152.7
2.0	0.8720	0.9969	139.4

16.8.2.2   The SPRT for exponential lifetimes

When the lifetime distribution is known to be exponential, we have seen the increase in efficiency gained by measuring the actual failure times of the parts being tested. By using a sequential procedure based on these failure times, further gains in efficiency can be achieved.

Expressing the hypotheses in terms of the parameter β of the lifetime distribution $E(\beta)$, we wish to test $H_0 : \beta \geq \beta_0$ versus $H_1 : \beta < \beta_0$, with significance level α and Type II error probability γ, when $\beta = \beta_1$, where $\beta_1 < \beta_0$. Letting $\mathbf{t}_n = (t_1, \ldots, t_n)$ be the times till failure of the first n parts tested, the likelihood ratio statistic is given by

$$\lambda_n(\mathbf{t}_n) = \left(\frac{\beta_0}{\beta_1}\right)^n \exp\left(-\left(\frac{1}{\beta_1} - \frac{1}{\beta_0}\right) \sum_{i=1}^{n} t_i\right). \tag{16.8.15}$$

Thus,

$$\ln \lambda_n(\mathbf{t}_n) = n \ln(\beta_0/\beta_1) - \left(\frac{1}{\beta_1} - \frac{1}{\beta_0}\right) \sum_{i=1}^{n} t_i.$$

The SPRT rules are accordingly,

$$\text{Continue sampling if} - h_1 + sn < \sum_{i=1}^{n} t_i < h_2 + sn,$$

$$\text{Accept } H_0 \text{ if } \sum_{i=1}^{n} t_i \geq h_2 + sn,$$

$$\text{Reject } H_0 \text{ if } \sum_{i=1}^{n} t_i \leq -h_1 + sn,$$

where

$$s = \ln(\beta_0/\beta_1)/\left(\frac{1}{\beta_1} - \frac{1}{\beta_0}\right),$$

$$h_1 = \ln((1 - \gamma)/\alpha)/\left(\frac{1}{\beta_1} - \frac{1}{\beta_0}\right),$$

$$h_2 = \ln((1 - \alpha)/\gamma)/\left(\frac{1}{\beta_1} - \frac{1}{\beta_0}\right). \tag{16.8.16}$$

Thus, if we plot $\sum_{i=1}^{n} t_i$ versus n, the accept and reject boundaries are again parallel straight lines.

As before, let ψ be an implicit parameter, and define

$$\beta^{(\psi)} = \begin{cases} \dfrac{(\beta_0/\beta_1)^{\psi} - 1}{\psi \left(\dfrac{1}{\beta_1} - \dfrac{1}{\beta_0} \right)}, & \psi \neq 0 \\[4mm] \dfrac{\ln(\beta_0/\beta_1)}{\dfrac{1}{\beta_1} - \dfrac{1}{\beta_0}}, & \psi = 0. \end{cases} \tag{16.8.17}$$

Then, the OC and ASN functions are approximately given by

$$\text{OC}(\beta^{(\psi)}) \approx \begin{cases} \dfrac{((1 - \gamma)/\alpha)^{\psi} - 1}{((1 - \gamma)/\alpha)^{\psi} - (\gamma/(1 - \alpha))^{\psi}}, & \psi \neq 0 \\[4mm] \dfrac{\ln((1 - \gamma)/\alpha)}{\ln((1 - \alpha)(1 - \gamma)/\alpha\gamma)}, & \psi = 0 \end{cases} \tag{16.8.18}$$

and

$$\text{ASN}(\beta^{(\psi)}) \approx \begin{cases} \dfrac{\ln((1 - \gamma)/\alpha) - \text{OC}(\beta^{(\psi)})\ln((1 - \alpha)(1 - \gamma)/\alpha\gamma)}{\ln(\beta_0/\beta_1) - \beta^{(\psi)}\left(\dfrac{1}{\beta_1} - \dfrac{1}{\beta_0}\right)}, & \psi \neq 0, \\[4mm] \dfrac{h_1 h_2}{s^2}, & \psi = 0 \end{cases} \tag{16.8.19}$$

Note that when $\psi = 1$, $\beta^{(\psi)}$ equals β_0, while when $\psi = -1$, $\beta^{(\psi)}$ equals β_1.

Example 16.20.   Continuing Example 16.17, recall we had $\alpha = 0.05$, $\gamma = 0.10$, $\beta_0 = 6153$, $\beta_1 = 4481$. The parameters of the boundaries of the SPRT are $s = 5229$, $h_1 = 47{,}662$, and $h_2 = 37{,}124$. The OC and ASN functions, for selected values of ψ, are given in Table 16.3.

Table 16.3   *OC and ASN values for the SPRT for Example 16.20*

ψ	$\beta^{(\psi)}$	$OC(\beta^{(\psi)})$	$ASN(\beta^{(\psi)})$
−2.0	3872	0.0110	34.3
−1.8	3984	0.0173	37.1
−1.6	4101	0.0270	40.2
−1.4	4223	0.0421	43.8
−1.2	4349	0.0651	47.9
−1.0	4481	0.1000	52.4
−0.8	4618	0.1512	57.1
−0.6	4762	0.2235	61.4
−0.4	4911	0.3193	64.7
−0.2	5067	0.4357	66.1
0.0	5229	0.5621	64.7
0.2	5398	0.6834	60.7
0.4	5575	0.7858	54.8
0.6	5759	0.8629	48.1
0.8	5952	0.9159	41.5
1.0	6153	0.9500	35.6
1.2	6363	0.9709	30.6
1.4	6582	0.9833	26.4
1.6	6811	0.9905	22.9
1.8	7051	0.9946	20.1
2.0	7301	0.9969	17.8

In Example 16.17, we saw that the fixed-sample test with the same α and γ requires a sample size of $n = 87$. Thus, in the exponential case as well, we see that the SPRT can result in substantial savings in sample size.

It is obviously impractical to perform a sequential test, similar to the one described in the above example by running one system, waiting till it fails, renewing it and running it again and again until a decision can be made. In the above example, if the MTTF of the system is close to the value of $\beta_0 = 6153$ [hours], it takes on the average about 256 days between failures, and on the average about 36 failures till decision is reached. This trial may take over 25 years. There are three ways to overcome this problem. The first is to put on test several systems simultaneously. Thus, if in the trial described in the above example 25 systems are tested simultaneously, the expected duration of the test will be reduced to one year. Another way is to consider a test based on a continuous time process, not on discrete samples of failure times. The third possibility of reducing the expected test duration is to perform **accelerated life testing**. In the following sections, we discuss these alternatives.

16.8.2.3   The SPRT for Poisson processes

Suppose that we put n systems on test starting at $t = 0$. Suppose also that any system which fails is instantaneously renewed, and at the renewal time it is as good as new. In addition we assume that the life characteristics of the systems are identical, the TTF of each system is **exponential** (with the same β) and failures of different systems are **independent** of each other.

Under these assumptions, the number of failures in each system, in the time interval $(0, t]$ is a Poisson random variable with mean λt, where $\lambda = 1/\beta$.

Let $X_n(t) =$ total number of failures among all the n systems during the time interval $(0, t]$. $X_n(t) \sim \text{Pos}(n\lambda t)$, and the collection $\{X_n(t); \ 0 < t < \infty\}$ is called a **Poisson process**. We add the initial condition that $X_n(0) = 0$.

The random function $X_n(t)$, $0 < t < \infty$, is a nondecreasing step function which jumps one unit at each random failure time of the system. The random functions $X_n(t)$ satisfy:

(i) $X_n(t), \sim \text{Pos}(n\lambda t)$, all $0 < t < \infty$;
(ii)   For any $t_1 < t_2$, $X_n(t_2) - X_n(t_1)$ is independent of $X_n(t_1)$;
(iii)   For any $t_1, t_2, 0 < t_1 < t_2 < \infty$, $X_n(t_2) - X_n(t_1) \sim \text{Pos}(n\lambda(t_2 - t_1))$.

We develop now the SPRT based on the random functions $X_n(t)$.

The hypotheses $H_0 : \beta \geq \beta_0$ versus $H_1 : \beta \leq \beta_1$, for $0 < \beta_1 < \beta_0 < \infty$, are translated to the hypotheses $H_0 : \lambda \leq \lambda_0$ versus $H_1 : \lambda \geq \lambda_1$ where $\lambda = 1/\beta$. The likelihood ratio at time t is

$$\Lambda(t; X_n(t)) = \left(\frac{\lambda_1}{\lambda_0}\right)^{X_n(t)} \exp\{-nt(\lambda_1 - \lambda_0)\}. \tag{16.8.20}$$

The test continues as long as the random graph of $(T_n(t), X_n(t))$ is between the two linear boundaries

$$b_U(t) = h_2 + sT_n(t), \quad 0 \leq t < \infty$$

and

$$b_L(t) = -h_1 + sT_n(t), \quad 0 \leq t < \infty,$$

where $T_n(t) = nt$ is the total time on test at t,

$$h_1 = \frac{\ln\left(\dfrac{1-\alpha}{\gamma}\right)}{\ln\left(\dfrac{\lambda_1}{\lambda_0}\right)},$$

$$h_2 = \frac{\ln\left(\dfrac{1-\gamma}{\alpha}\right)}{\ln\left(\dfrac{\lambda_1}{\lambda_0}\right)}, \tag{16.8.21}$$

and

$$s = \frac{\lambda_1 - \lambda_0}{\ln\left(\dfrac{\lambda_1}{\lambda_0}\right)}.$$

The instant $X_n(t)$ jumps above $b_U(t)$ the test terminates and H_0 is rejected; on the other hand, the instant $X_n(t) = b_L(t)$ the test terminates and H_0 is accepted. Acceptance of H_0 entails that the reliability meets the specified requirement. Rejection of H_0 may lead to additional engineering modification to improve the reliability of the system.

The OC function of this sequential test is the same as that in the exponential case. Let τ denote the random time of termination. It can be shown that $\Pr_\lambda\{\tau < \infty\} = 1$ for all $0 < \lambda < \infty$. The expected deviation of the test is given approximately by

$$E_\lambda\{\tau\} = \frac{1}{\lambda n}E_\lambda\{X_n(\tau)\}, \tag{16.8.22}$$

where

$$E_\lambda\{X_n(\tau)\} \cong \begin{cases} \dfrac{h_2 - \mathrm{OC}(\lambda)(h_1 + h_2)}{1 - s/\lambda}, & \text{if}\lambda \neq s \\[2ex] h_1 h_2, & \text{if}\lambda = s. \end{cases} \tag{16.8.23}$$

It should be noticed that the last formula yields the same values as the formula in the exponential case for $\lambda = 1/\beta^{(\psi)}$. The SPRT of the previous section can terminate only after a failure, while the SPRT based on $X_n(t)$ may terminate while crossing the lower boundary $b_L(t)$, before a failure occurs.

The minimal time required to accept H_0 is $\tau_0 = h_1/ns$. In the case of Example 16.19, with $n = 20$, $\tau_0 = 9.11536/(20 \times 0.0001912) = 2383.2$ [hours]. That is, over 99 days of testing without any failure. The SPRT may be, in addition, frequency censored by fixing a value x^* so that, as soon as $X_n(t) \geq x^*$ the test terminates and H_0 is rejected. In Example 16.19 we see that the expected number of failures at termination may be as large as 66. We can censor the test at $x^* = 50$. This will reduce the expected duration of the test, but will increase the probability of a Type I error, α. Special programs are available for computing the operating characteristics of such censored tests, but these are beyond the scope of the present text.

16.9   Accelerated life testing

It is often impractical to test highly reliable systems, or components, under normal operating conditions, because no failures may be observed during long periods of time. In accelerated life testing, the systems are subjected to higher than normal stress conditions in order to generate failures. The question is how to relate failure distributions, under higher than normal stress conditions, to those under normal conditions?

Accelerated life testing is used by engineers in testing materials such as food and drugs, lubricants, concrete and cement, building materials, and nuclear reactor materials. The stress conditions are generally, mechanical load, vibrations, high temperatures, high humidity, high contamination, etc. Accelerated testing is used for semiconductors including transistors, electronic devices such as diodes, random access memories, plastic encapsulants. The reader is referred to Nelson (1992) for a survey of methods and applications. The statistical methodology of accelerated life testing is similar to the methods described earlier in this chapter, including graphical analysis and maximum likelihood estimation. The reader is referred to Nelson (1992), Mann et al. (1974), for details. We describe below some of the models used to relate failures under various stress conditions.

16.9.1   The Arrhenius temperature model

This model is widely used when the product failure time is sensitive to high temperature. Applications include electrical insulations and dielectric (see Goba 1969); solid state and semiconductors (Peck and Trapp 1978); battery cells; lubricants and greases; plastics; and incandescent light filaments.

The **Arrhenius law** states that the rate of simple chemical reaction depends on temperature as follows:

$$\lambda = A \exp\{-E/(kT)\}, \tag{16.9.1}$$

where E is the activation energy (in electron volts; k is the Boltzmann constant, 8.6171×10^{-5} eV °C; T is the absolute Kelvin temperature ($273.16 +$ °C); A is a product parameter, which depends on the test conditions and failure characteristics. In applying the Arrhenius model to failure times distribution, we find the Weibull–Arrhenius life distribution, in which the scale parameter β of the Weibull distribution is related to temperature, T, according to the function

$$\beta(T) = \lambda \exp\left\{A + \frac{B}{T}\right\}, \quad B > 0, \tag{16.9.2}$$

where λ, A, and B are fitted empirically to the data.

16.9.2 Other models

Another model is called the **log normal Arrhenius** model, in which the log failure time is normally distributed with mean $A + B/T$ and variance σ^2. According to this model, the expected failure time is $\exp\{A + B/T + \sigma^2/2\}$. Another model prevalent in the literature, relates the expected failure time to a stress level V according to the **inverse power model**

$$\text{MTTF}(V) = \frac{C}{V^p}, \quad C > 0. \tag{16.9.3}$$

The statistical data analysis methodology is to fit an appropriate model to the data, usually by maximum likelihood estimation, and then predict the MTTF of the system under normal conditions, or some reliability or availability function. Tolerance intervals, for the predicted value should be determined.

16.10 Burn-in procedures

Many products show relatively high frequency of early failures. For example, if a product has an exponential distribution of the TTF with MTTF of $\beta = 10,000$ [hours], we do not expect more than 2% of the product to fail within the first 200 [hours]. Nevertheless, many products designed for high value of MTTF show a higher than expected number of early failures. This phenomenon led to the theory that the hazard rate function of products is typically a U-shaped function. In its early life the product is within a phase with monotone decreasing hazard rate. This phase is called the "infant mortality" phase. After this phase the product enters a phase of "maturity" in which the hazard rate function is almost constant. Burn-in procedures are designed to screen (burn) the weak products within the plant, by setting the product to operate for several days, in order to give the product a chance to fail in the plant and not in the field, where the loss due to failure is high.

How long should a burn-in procedure last? Jensen and Petersen (1982) discuss this and other issues, in designing burn-in procedures. We refer the reader to this book for more details. We present here only some basic ideas.

Burn-in procedures discussed by Jensen and Petersen are based on a model of a mixed life distribution. For example, suppose that experience shows that the life distribution of a product is

Weibull, $W(v, \beta_1)$. A small proportion of units manufactured may have generally short life, due to various reasons, which is given by another Weibull distribution, say $W(v, \beta_0)$, with $\beta_0 < \beta_1$. Thus, the life distribution of a randomly chosen product has a distribution which is a mixture of $W(v, \beta_0)$ and $W(v, \beta_1)$ i.e.,

$$F(t) = 1 - \left[p \exp \left\{ -\left(\frac{t}{\beta_0}\right)^v \right\} + (1-p) \exp \left\{ -\left(\frac{t}{\beta_1}\right)^v \right\} \right], \tag{16.10.1}$$

for $t > 0$. The objective of the burn-in is to let units having the $W(v, \beta_0)$ distribution an opportunity to fail in the plant. The units which do not fail during the burn-in have, for their remaining life, a life distribution closer to the desired $W(v, \beta_1)$.

Suppose that a burn-in continues for t^* time units. The conditional distribution of the TTF T, given that $\{T > t^*\}$ is

$$F^*(t) = \frac{\int_{t^*}^{t} f(u)du}{1 - F(t^*)}, \quad t \geq t^*. \tag{16.10.2}$$

The c.d.f. $F^*(t)$, of units surviving the burn-in, start at t^*, i.e., $F^*(t^*) = 0$ and has MTTF

$$\beta^* = t^* + \int_{t^*}^{\infty} (1 - F^*(t))dt. \tag{16.10.3}$$

We illustrate this in the following example on mixtures of exponential life times.

Example 16.21. Suppose that a product is designed to have an exponential life distribution, with mean of $\beta = 10,000$ [hours]. A proportion $p = 0.05$ of the products come out of the production process with a short MTTF of $\gamma = 100$ [hours]. Suppose that all products go through a burn-in for t^* [hour].

The c.d.f. of the TTF of units which did not fail during the burn-in is

$$F^*(t) = 1 - \frac{0.05 \exp \left\{ -\frac{t}{100} \right\} + 0.95 \exp \left\{ -\frac{t}{10,000} \right\}}{0.05 \exp \left\{ -\frac{200}{100} \right\} + 0.95 \exp \left\{ -\frac{200}{10,000} \right\}}$$

$$= 1 - \frac{1}{0.93796} \left[0.05 \exp \left\{ -\frac{t}{100} \right\} + 0.95 \exp \left\{ -\frac{t}{10,000} \right\} \right]$$

for $t \geq 200$. The MTTF, for units surviving the burn-in is thus

$$\beta^* = 200 + \frac{1}{0.93796} \int_{200}^{\infty} \left(0.05 \exp \left\{ -\frac{t}{100} \right\} + 0.95 \exp \left\{ -\frac{t}{10,000} \right\} \right) dt$$

$$= 200 + \frac{5}{0.93796} \exp \left\{ -\frac{200}{100} \right\} + \frac{9500}{0.93796} \exp \left\{ -\frac{200}{10,000} \right\}$$

$$= 10,128.53 \text{ [hours]}.$$

A unit surviving 200 hours of burn-in is expected to operate an additional 9928.53 hours in the field. The expected life of these units without the burn-in is $0.05 \times 200 + 0.95 \times 10,000 = 9510$ [hours]. The burn-in of 200 hours in the plant, is expected to increase the mean life of the product in the field by 418 hours. Whether this increase in the MTTF justifies the burn-in depends on the relative cost of burn-in in the plant to the cost of failures in the field. The proportion p of "short life" units plays also an important role. If this proportion is $p = 0.1$ rather than 0.05, the burn-in increases the MTTF in the field from 9020 hours to 9848.95 hours. One can easily verify that for $p = 0.2$, if the income for an hour of operation of one unit in the field is $C_p = 5\$$ and the cost of the burn-in per unit is 0.15$/hour, then the length of burn-in which maximizes the expected profit is about 700 hours.

16.11   Chapter highlights

The previous chapter dwelled on design decisions of product and process developers that are aimed at optimizing the quality and robustness of products and processes. This chapter is looking and performance over time and discusses basic notions of repairable and nonrepairable systems. Graphical and nonparametric techniques are presented together with classical parametric techniques for estimating life distributions. Special sections cover reliability demonstrational procedures, sequential reliability testing, burn-in procedures and accelerated life testing. Design and testing of reliability is a crucial activity for organizations at the top of the Quality Ladder.

The main concepts and definitions introduced in this chapter include:

- Life Distributions
- Accelerated Life Testing
- Availability
- Time Categories
- Up Time
- Down Time
- Intrinsic Availability
- Operational Readiness
- Mean Time To Failure (MTTF)
- Reliability Function
- Failure Rate
- Structure Function
- Time Till Failure (TTF)
- Time Till Repair (TTR)
- Cycle Time
- Renewal Function
- Censored Data
- Product Limit (PL) estimator
- Average Sample Number (ASN)
- Sequential Probability Ratio Test (SPRT)
- Burn-In Procedure

16.12   Exercises

16.1 During 600 hours of manufacturing time a machine was up 510 hours. It had 100 failures which required a total of 11 hours of repair time. What is the MTTF of this machine? What is its mean time till repair, MTTR? What is the intrinsic availability?

16.2 The frequency distribution of the lifetime in a random sample of $n = 2000$ solar cells, under accelerated life testing is the following:

$t/10^3$ [hours]	0–1	1–2	2–3	3–4	4–5	5–
Prof. frequency	0.15	0.25	0.25	0.10	0.10	0.15

The relationship of the scale parameters of the life distributions, between normal and accelerated conditions is 10:1.
  (i) Estimate the reliability of the solar cells at age $t = 4.0$ [year].
  (ii) What proportion of solar cells are expected to survive 40,000 [hours] among those which survived 20,000 [hours]?

16.3 The c.d.f. of the lifetime [months] of an equipment is

$$F(t) = \begin{cases} t^4/20{,}736, & 0 \le t < 12, \\ 1, & 12 \le t. \end{cases}$$

  (i) What is the failure rate function of this equipment?
  (ii) What is the MTTF?
  (iii) What is the reliability of the equipment at 4 months?

16.4 The reliability of a system is

$$R(t) = \exp\{-2t - 3t^2\}, \quad 0 \le t < \infty.$$

  (i) What is the failure rate of this system at age $t = 3$?
  (ii) Given that the system reached the age of $t = 3$, what is its reliability for two additional time units?

16.5 An aircraft has four engines but can land using only two engines.
  (i) Assuming that the reliability of each engine, for the duration of a mission, is $R = 0.95$, and that engine failures are independent, compute the mission reliability of the aircraft.
  (ii) What is the mission reliability of the aircraft if at least one functioning engine must be on each wing?

16.6 (i) Draw a block diagram of a system having the structure function

$$R_{sys} = \psi_s(\psi_p(\psi_{M_1}, \psi_{M_2}), R_6), \quad \psi_{M_1} = \psi_p(R_1, R_2R_3), \quad \psi_{M_2} = \psi_2(R_4, R_5).$$

  (ii) Determine R_{sys} if all the components act independently, and have the same reliability $R = 0.8$.

16.7 Consider a system of n components in a series structure. Let $R_1, \ldots, R_n$ be the reliabilities of the components. Show that

$$R_{sys} \ge 1 - \sum_{i=1}^{n}(1 - R_i).$$

16.8 A 4 out of 8 system has identical components whose life lengths T [weeks] are independent and identically distributed like a Weibull $W\left(\frac{1}{2}, 100\right)$. What is the reliability of the system at $t_0 = 5$ weeks?

16.9 A system consists of a main unit and two standby units. The lifetimes of these units are exponential with mean $\beta = 100$ [hours]. The standby units undergo no failure while idle. Switching will take place when required. What is the MTTF of the system? What is the reliability function of this system?

16.10 Suppose that the TTF in a renewal cycle has a $W(\alpha, \beta)$ distribution and that the TTR has a lognormal distribution $LN(\mu, \sigma)$. Assume further that TTF and TTR are independent. What are the mean and standard deviation of a renewal cycle.

16.11 Suppose that a renewal cycle has the normal distribution $N(100, 10)$. Determine the p.d.f. of $N_R(200)$.

16.12 Let the renewal cycle C be distributed like $N(100, 10)$. Approximate $V(1000)$.

16.13 Derive the renewal density $v(t)$ for a renewal process with $C \sim N(100, 10)$.

16.14 Two identical components are connected in parallel. The system is not repaired until both components fail. Assuming that the TTF of each component is exponentially distributed, $E(\beta)$, and the total repair time is $G(2, \gamma)$, derive the Laplace transform of the availability function $A(t)$ of the system.

16.15 Simulate a sample of 100 TTF of a system comprised of two components connected in parallel, where the life distribution of each component (in hours) is $E(100)$. Similarly, simulate a sample of 100 repair times (in hours), having a $G(2, 1)$ distribution. Estimate the expected value and variance of the number of renewals in 2000 [hours].

16.16 In a given life test, $n = 15$ units are placed to operate independently. The time till failure of each unit has an exponential distribution with mean 2000 [hours]. The life test terminates immediately after the 10th failure. How long is the test expected to last?

16.17 If n units are put on test and their TTF are exponentially distributed with mean β, the time elapsed between the rth and $(r + 1)$st failure, i.e., $\Delta_{n,r} = T_{n,r+1} - T_{n,r}$, is exponentially distributed with mean $\beta/(n - r)$, $r = 0, 1, \ldots, n - 1$. Also, $\Delta_{n,0}, \Delta_{n,2}, \ldots, \Delta_{n,n-1}$ are independent. What is the variance of $T_{n,r}$? Use this result to compute the variance of the test length in the previous exercise.

16.18 Consider again the previous exercise. How would you estimate unbiasedly the scale parameter β, if the r failure times, $T_{n,1}, T_{n,2}, \ldots, T_{n,r}$ are given? What is the variance of this unbiased estimator?

16.19 Simulate a random sample of 100 failure times, following the Weibull distribution $W(2.5, 10)$. Draw a Weibull Probability plot of the data. Estimate the parameters of the distribution from the parameters of the linear regression fitted to the $Q-Q$ plot.

16.20 The following is a random sample of the compressive strength of 20 concrete cubes [kg/cm²].

94.9, 106.9, 229.7, 275.7, 144.5, 112.8, 159.3, 153.1, 270.6, 322.0,

216.4, 544.6, 266.2, 263.6, 138.5, 79.0, 114.6, 66.1, 131.2, 91.1

Make a lognormal $Q-Q$ plot of these data and estimate the mean and standard deviation of this distribution.

16.21 The following data represent the time till first failure [days] of electrical equipment. The data were censored after 400 days.

13, 157, 172, 176, 249, 303, 350, 400⁺, 400⁺.

(Censored values appear as x^+.) Make a Weibull Q-Q plot of these data and estimate the median of the distribution.

16.22 Make a PL (Kaplan–Meier) estimate of the reliability function of an electronic device, based on 50 failure times in file **ELECFAIL.csv**.

16.23 Assuming that the failure times in file **ELECFAIL.csv** come from an exponential distribution $E(\beta)$, compute the MLE of β and of $R(50; \beta) = \exp\{-50/\beta\}$. [The MLE of a function of a parameter is obtained by substituting the MLE of the parameter in the function.] Determine confidence intervals for β and for $R(50; \beta)$ at level of confidence 0.95.

16.24 The following are values of 20 random variables having an exponential distribution $E(\beta)$. The values are censored at $t^* = 200$.

$$96.88, \quad 154.24, \quad 67.44, \quad 191.72, \quad 173.36, \quad 200, \quad 140.81, \quad 200, \quad 154.71, \quad 120.73,$$

$$24.29, \quad 10.95, \quad 2.36, \quad 186.93, \quad 57.61, \quad 99.13, \quad 32.74, \quad 200, \quad 39.77, \quad 39.52.$$

Determine the MLE of β. Use β equal to the MLE, to estimate the standard deviation of the MLE, and to obtain confidence interval for β, at level $1 - \alpha = 0.95$. [This simulation is called an empirical Bootstrap.]

16.25 Determine n^0 and r for a frequency censoring test for the exponential distribution, where the cost of a unit is 10 times bigger than the cost per time unit of testing. We wish that S.E.$\{\hat{\beta}_n\} = 0.1\beta$, and the expected cost should be minimized at $\beta = 100$ [hours]. What is the expected cost of this test, at $\beta = 100$, when $c_1 = \$1$ [hour].

16.26 File **WEIBUL.csv** contains the values of a random sample of size $n = 50$ from a Weibull distribution.
  (i) Obtain MLE of the scale and shape parameters β and v.
  (ii) Use the MLE estimates of β and v, to obtain parametric bootstrap EBD of the distribution of $\hat{\beta}$, $\hat{v}$, with $M = 500$ runs. Estimate from this distribution the standard deviations of $\hat{\beta}$ and $\hat{v}$. Compare these estimates to the large sample approximations.

16.27 In binomial life testing by a fixed size sample, how large should the sample be in order to discriminate between $R_0 = 0.99$ and $R_1 = 0.90$, with $\alpha = \beta = 0.01$? [α and β denote the probabilities of error of Types I and II.]

16.28 Design the Wald SPRT for binomial life testing, in order to discriminate between $R_0 = 0.99$ and $R_1 = 0.90$, with $\alpha = \beta = 0.01$. What is the expected sample size, ASN, if $R = 0.9$?

16.29 Design a Wald SPRT for exponential life distribution, to discriminate between $R_0 = 0.99$ and $R_1 = 0.90$, with $\alpha = \beta = 0.01$. What is the expected sample size, ASN, when $R = 0.90$?

16.30 $n = 20$ computer monitors are put on accelerated life testing. The test is an SPRT for Poisson processes, based on the assumption that the TTF of a monitor, in those conditions, is exponentially distributed. The monitors are considered to be satisfactory if their MTBF is $\beta \geq 2000$ [hours] and considered to be unsatisfactory if $\beta \leq 1500$ [hours]. What is the expected length of the test if $\beta = 2000$ [hours].

16.31 A product has an exponential life time with MTTF $\beta = 1000$ [hours]. One percentage of the products come out of production with MTTF of $\gamma = 500$ [hours]. A burn in of $t^* = 300$ [hours] takes place. What is the expected life of units surviving the burn-in? Is such a long burn-in justified?

17

Bayesian Reliability Estimation and Prediction

It is often the case that some information is available on the parameters of the life distributions from prior experiments or prior analysis of failure data. The Bayesian approach provides the methodology for formal incorporation of prior information with the current data. We introduced Bayesian decision procedures in Section 8.4.

17.1   Prior and posterior distributions

Let $X_1, \ldots, X_n$ be a random sample from a distribution with a p.d.f. $f(x; \theta)$, where $\theta = (\theta_1, \ldots, \theta_k)$ is a vector of k parameters, belonging to a parameter space Θ. So far we have assumed that the point θ is an unknown constant. In the Bayesian approach, θ is considered a random vector having some specified distribution. The distribution of θ is called a **prior distribution** (see Section 4.8.1). The problem of which prior distribution to adopt for the Bayesian model is not easily resolvable, since the values of θ are not directly observable. The discussion of this problem is beyond the scope of the book.

Let $h(\theta_1, \ldots, \theta_k)$ denote the joint p.d.f. of $(\theta_1, \ldots, \theta_k)$, corresponding to the prior distribution. This p.d.f. is called the **prior p.d.f.** of θ. The joint p.d.f. of X and θ is

$$g(x, \theta) = f(x; \theta)h(\theta). \tag{17.1.1}$$

The marginal p.d.f. of X, which is called the **predictive p.d.f.**, is

$$f^*(x) = \int \cdots \int_{\Theta} f(x; \theta)h(\theta)d\theta_1 \cdots d\theta_k. \tag{17.1.2}$$

Furthermore, the conditional p.d.f. of θ given $X = x$ is

$$h(\theta \mid x) = g(x, \theta)/f^*(x). \tag{17.1.3}$$

Modern Industrial Statistics: With Applications in R, MINITAB and JMP, Third Edition.
Ron S. Kenett and Shelemyahu Zacks.
© 2021 John Wiley & Sons, Ltd. Published 2021 by John Wiley & Sons, Ltd.

This conditional p.d.f. is called the **posterior p.d.f.** of θ, given x. Thus, starting with a prior p.d.f., $h(\theta)$, we convert it, after observing the value of x, to the posterior p.d.f. of θ given x.

If $x_1, \ldots, x_n$ is a random sample from a distribution with a p.d.f. $f(x; \theta)$ then the posterior p.d.f. of θ, corresponding to the prior p.d.f. $h(\theta)$, is

$$h(\theta \mid \mathbf{x}) = \frac{\prod_{i=1}^{n} f(x_i; \theta) h(\theta)}{\int \cdots \int_{\Theta} \prod_{i=1}^{n} f(x_i; \theta) h(\theta) d\theta_1 \cdots d\theta_k}. \tag{17.1.4}$$

For a given sample, $\mathbf{x}$, the posterior p.d.f. $h(\theta \mid \mathbf{x})$ is the basis for most types of Bayesian inference.

Example 17.1.

I. **Binomial Distributions**

   $X \sim B(n; \theta), 0 < \theta < 1$.
   The p.d.f. of X is

$$f(x; \theta) = \binom{n}{x} \theta^x (1 - \theta)^{n-x}, \quad x = 0, \ldots, n.$$

Suppose that θ has a prior beta distribution, with p.d.f.

$$h(\theta; \nu_1, \nu_2) = \frac{1}{\mathrm{Be}(\nu_1, \nu_2)} \theta^{\nu_1 - 1} (1 - \theta)^{\nu_2 - 1}, \tag{17.1.5}$$

$0 < \theta < 1, 0 < \nu_1, \nu_2 < \infty$, where $\mathrm{Be}(a, b)$ is the **complete beta function**

$$\mathrm{Be}(a, b) = \int_0^1 x^{a-1} (1 - x)^{b-1} dx$$

$$= \frac{\Gamma(a)\Gamma(b)}{\Gamma(a + b)}.$$

The posterior p.d.f. of θ, given $X = x$, is

$$h(\theta \mid x) = \frac{1}{\mathrm{Be}(\nu_1 + x, \nu_2 + n - x)} \theta^{\nu_1 + x - 1} (1 - \theta)^{\nu_2 + n - x - 1}, \quad 0 < \theta < 1. \tag{17.1.6}$$

Notice that the posterior p.d.f. is also that of a beta distribution, with parameters $\nu_1 + x$ and $\nu_2 + n - x$. The expected value of the posterior distribution of θ, given $X = x$, is

$$\begin{aligned} E\{\theta \mid x\} &= \frac{1}{\mathrm{Be}(\nu_1 + x, \nu_2 + n - x)} \int_0^1 \theta^{\nu_1 + x} (1 - \theta)^{\nu_2 + n - x - 1} d\theta \\ &= \frac{\mathrm{Be}(\nu_1 + x + 1, \nu_2 + n - x)}{\mathrm{Be}(\nu_1 + x, \nu_2 + n - x)} = \frac{\nu_1 + x}{\nu_1 + \nu_2 + n}. \end{aligned} \tag{17.1.7}$$

The R function *binomial.beta.mix* in the LearnBayes package, computes the posterior distribution when the proportion p has a mixture of betas prior distribution. The inputs to this function are **probs**, the vector of mixing probabilities; **betapar**, a matrix of beta shape parameters where each row corresponds to a component of the prior; and **data**, the vector of the number of successes and number of failures in the sample. The example below consists of 12 Binomial trials. The beta prior first parameter can be interpreted as the number of successes in 12 trials. The output of the function is a list with two components

- **probs** is a vector of posterior mixing probabilities and **betapar** is a matrix containing the shape parameters of the updated beta posterior densities.

In the example below, the prior distribution of the binomial parameter, θ, is split evenly between two beta distributions, with expected values, 0.5 and 0.88, respectively. The Binomial experiment produced 10 successes in 12 trials, an estimated probability of success of 0.83, much closer to the second beta. A posteriori, the mix of beta distributions is 0.28 and 0.72, respectively, strongly favoring the second beta distribution. The expected values of the posterior beta distributions being 0.79 and 0.86, clearly more in line with the observed data.

```
> library(LearnBayes)
> Probs <- c(0.5, 0.5)
> BetaPar1<- c(1, 1)
> BetaPar2 <- c(15, 2)
> Betapar <- rbind(BetaPar1, BetaPar2)
> Data<- c(10, 2)
> binomial.beta.mix(probs=Probs,
                    betapar=Betapar,
                    data=Data)
```

```
$probs
 BetaPar1  BetaPar2
0.2845528 0.7154472


$betapar
          [,1] [,2]
BetaPar1   11    3
BetaPar2   25    4
```

II. Poisson Distributions

$X \sim P(\lambda), 0 < \lambda < \infty.$

The p.d.f. of X is

$$f(x; \lambda) = e^{-\lambda}\frac{\lambda^x}{x!}, \quad x = 0, 1, \ldots.$$

Suppose that the prior distribution of λ is the gamma distribution, $G(v, \tau)$. The prior p.d.f. is thus

$$h(\lambda; v, \tau) = \frac{1}{\tau^v\Gamma(v)}\lambda^{v-1}e^{-\lambda/\tau}. \tag{17.1.8}$$

The posterior p.d.f. of λ, given $X = x$, is

$$h(\lambda \mid x) = \frac{\lambda^{v+x-1}}{(\tau 1 + \tau)^{v+x}\Gamma(v + x)}e^{-\lambda(1+\tau)/\tau}. \tag{17.1.9}$$

That is, the posterior distribution of λ, given $X = x$, is $G(v + x, \frac{\tau}{1+\tau})$. The posterior expectation of λ, given $X = x$, is $(v + x)\tau/(1 + \tau)$.

The R function *poisson.gamma.mix* in the LearnBayes package, computes the posterior distribution of λ. The inputs to this function are similar to the inputs to the function *binomial.beta.mix* described above.

```
> LSinput <- c(0.5, 0.5)
> GammaPar1 <- c(1, 1)   # Gamma parameters are expressed as
>                        # shape and rate
>                        # scale is 1/rate
> GammaPar2 <- c(15, 2)
> Gammapar <- rbind(GammaPar1, GammaPar2)
> Data<- list(
    y=c(5),
    t=c(1))
> poisson.gamma.mix(probs=Probs,
                    gammapar=Gammapar,
                    data=Data)

$probs
GammaPar1 GammaPar2
0.1250978 0.8749022


$gammapar
          [,1] [,2]
GammaPar1    6    2
GammaPar2   20    3
```

III. **Exponential Distributions**

$X \sim E(\beta)$.

The p.d.f. of X is

$$f(x; \beta) = \frac{1}{\beta} e^{-x/\beta}.$$

Let β have an inverse-gamma prior distribution, $\mathrm{IG}(\nu, \tau)$. That is, $\frac{1}{\beta} \sim G(\nu, \tau)$. The prior p.d.f. is

$$h(\beta; \nu, \tau) = \frac{1}{\tau^{\nu} \Gamma(\nu) \beta^{\nu+1}} e^{-1/\beta\tau}. \tag{17.1.10}$$

Then, the posterior p.d.f. of β, given $X = x$, is

$$h(\beta \mid x) = \frac{(1 + x\tau)^{\nu+1}}{\tau^{\nu+1} \Gamma(\nu+1) \beta^{\nu+2}} e^{-(x+1/\tau)/\beta}. \tag{17.1.11}$$

That is, the posterior distribution of β, given $X = x$, is $\mathrm{IG}(\nu + 1, \frac{\tau}{1+x\tau})$. The posterior expectation of β, given $X = x$, is $(x + 1/\tau)/\nu$.

The likelihood function $L(\theta; \mathbf{x})$, is a function over a parameter space Θ. In the definition of the posterior p.d.f. of θ, given $\mathbf{x}$, we see that any factor of $L(\theta; \mathbf{x})$ which does not depend on θ is irrelevant. For example, the binomial p.d.f., under θ is

$$f(x; \theta) = \binom{n}{x} \theta^x (1 - \theta)^{n-x}, \quad x = 0, 1, \ldots, n,$$

$0 < \theta < 1$. The factor $\binom{n}{x}$ can be omitted from the likelihood function in Bayesian calculations. The factor of the likelihood which depends on θ is called the **kernel** of the likelihood. In the above binomial example, $\theta^x(1 - \theta)^{n-x}$ is the kernel of the binomial likelihood. If the prior p.d.f. of θ, $h(\theta)$, is of the same functional form (up to a proportionality factor which does not depend on θ) as that of the likelihood kernel, we call that prior p.d.f. a **conjugate** one. As shown in Example 17.1, the beta prior distributions are conjugate to the binomial model, the gamma prior distributions are conjugate to the Poisson model and the inverse-gamma priors are conjugate to the exponential model.

If a conjugate prior distribution is applied, the posterior distribution belongs to the conjugate family.

One of the fundamental problems in Bayesian analysis is that of the choice of a prior distribution of θ. From a Bayesian point of view, the prior distribution should reflect the prior knowledge of the analyst on the parameter of interest. It is often difficult to express the prior belief about the value of θ in a p.d.f. form. We find that analysts apply, whenever possible, conjugate priors whose means and standard deviations may reflect the prior beliefs. Another common approach is to use a "diffused," "vague," or Jeffrey's prior, which is proportional to $|I(\theta)|^{1/2}$, where $I(\theta)$ is the Fisher information function (matrix). For further reading on this subject, the reader is referred to Box and Tiao (1973), Good (1965), and Press (1989).

17.2   Loss functions and bayes estimators

In order to define Bayes estimators, we must first specify a **loss function**, $L(\hat{\theta}, \theta)$, which represents the cost involved in using the estimate $\hat{\theta}$ when the true value is θ. Often this loss is taken to be a function of the distance between the estimate and the true value, i.e. $|\hat{\theta} - \theta|$. In such cases, the loss function is written as

$$L(\hat{\theta}, \theta) = W(|\hat{\theta} - \theta|).$$

Examples of such loss functions are

$$\text{Squared-error loss}: \quad W(|\hat{\theta} - \theta|) = (\hat{\theta} - \theta)^2,$$
$$\text{Absolute-error loss}: \quad W(|\hat{\theta} - \theta|) = |\hat{\theta} - \theta|.$$

The loss function does not have to be symmetric. For example, we may consider the function

$$L(\hat{\theta}, \theta) = \begin{cases} \alpha(\theta - \hat{\theta}), & \text{if } \hat{\theta} \leq \theta \\ \beta(\hat{\theta} - \theta), & \text{if } \hat{\theta} > \theta \end{cases},$$

where α and β are some positive constants.

The **Bayes estimator** of θ, with respect to a loss function $L(\hat{\theta}, \theta)$, is defined as the value of $\hat{\theta}$ which minimizes the **posterior risk**, given x, where the posterior risk is the expected loss with respect to the posterior distribution. For example, suppose that the p.d.f. of X depends on several parameters $\theta_1, \ldots, \theta_k$, but we wish to derive a Bayes estimator of θ_1 with respect to the squared-error loss function. We consider the marginal posterior p.d.f. of θ_1, given $\mathbf{x}$, $h(\theta_1 \mid x)$. The posterior risk is

$$R(\hat{\theta}_1, \mathbf{x}) = \int (\hat{\theta}_1 - \theta_1)^2 h(\theta_1 \mid \mathbf{x}) d\theta_1.$$

It is easily shown that the value of $\hat{\theta}_1$ which minimizes the posterior risk $R(\hat{\theta}_1, \mathbf{x})$ is the **posterior expectation** of θ_1:

$$E\{\theta_1 \mid \mathbf{x}\} = \int \theta_1 h(\theta_1 \mid \mathbf{x}) d\theta_1.$$

If the loss function is $L(\hat{\theta}_1, \theta) = |\hat{\theta}_1 - \theta_1|$, the Bayes estimator of θ_1 is the **median** of the posterior distribution of θ_1 given **x**.

17.2.1   Distribution-free bayes estimator of reliability

Let J_n denote the number of failures in a random sample of size n, during the period $[0, t)$. The reliability of the device on test at age t is $R(t) = 1 - F(t)$, where $F(t)$ is the CDF of the life distribution. Let $K_n = n - J_n$. The distribution of K_n is the binomial $B(n, R(t))$. Suppose that the prior distribution of $R(t)$ is uniform on $(0, 1)$. This prior distribution reflects our initial state of ignorance concerning the actual value of $R(t)$.

The uniform distribution is a special case of the beta distribution with $v_1 = 1$ and $v_2 = 1$. Hence, according to Part I of Example 17.1, the posterior distribution of $R(t)$, given K_n, is a beta distribution with parameters $v_1 = K_n + 1$ and $v_2 = 1 + n - K_n$. Hence, the Bayes estimator of $R(t)$, with respect to the squared-error loss function, is

$$\hat{R}(t; K_n) = E\{R(t) \mid K_n\}$$
$$= \frac{K_n + 1}{n + 2}. \tag{17.2.1}$$

If the sample size is $n = 50$, and $K_{50} = 27$, the Bayes estimator of $R(t)$ is $\hat{R}(t; 27) = 28/52 = 0.538$. Notice that the MLE of $R(t)$ is $\hat{R}_{50} = 27/50 = 0.540$. The sample size is sufficiently large for the MLE and the Bayes estimator to be numerically close. If the loss function is $|\hat{R} - R|$, the Bayes estimator of R is the median of the posterior distribution of $R(t)$ given K_n, i.e. the median of the beta distribution with parameters $v_1 = K_n + 1$ and $v_2 = n - K_n + 1$.

Generally, if v_1 and v_2 are integers then the median of the beta distribution is

$$\text{Me} = \frac{v_1 F_{0.5}\left[2v_1, 2v_2\right]}{v_2 + v_1 F_{0.5}\left[2v_1, 2v_2\right]}, \tag{17.2.2}$$

where $F_{0.5}[j_1, j_2]$ is the median of the $F[j_1, j_2]$ distribution. Substituting $v_1 = K_n + 1$ and $v_2 = n - K_n + 1$ in (17.2.2), we obtain that the Bayes estimator of $R(t)$ with respect to the absolute error loss is

$$\hat{R}(t) = \frac{(K_n + 1)F_{0.5}\left[2K_n + 2, 2n + 2 - 2K_n\right]}{n + 1 - K_n + (K_n + 1)F_{0.5}\left[2K_n + 2, 2n + 2 - 2K_n\right]}. \tag{17.2.3}$$

Numerically, for $n = 50$, $K_n = 27$, $F_{0.5}[56, 48] = 1.002$, and $\hat{R}(t) = 0.539$. The two Bayes estimates are very close.

17.2.2   Bayes estimator of reliability for exponential life distributions

Consider a Type II censored sample of size n from an exponential distribution, $E(\beta)$, with censoring at the rth failure. Let $t_{(1)} \leq t_{(2)} \leq \cdots \leq t_{(r)}$ be the ordered failure times. For squared-error loss, the Bayes estimator of $R(t) = e^{-t/\beta}$ is given by

$$\hat{R}(t) = E\{R(t) \mid t_{(1)}, \ldots, t_{(r)}\}$$
$$= E\{e^{-t/\beta} \mid t_{(1)}, \ldots, t_{(r)}\}. \tag{17.2.4}$$

This conditional expectation can be computed by integrating $e^{-t/\beta}$ with respect to the posterior distribution of β, given $t_{(1)}, \ldots, t_{(r)}$.

Suppose that the prior distribution of β is IG(v, τ). One can easily verify that the posterior distribution of β given $t_{(1)}, \ldots, t_{(r)}$ is the inverted-gamma IG$\left(v + r, \frac{\tau}{1 + T_{n,r}\tau}\right)$, where $T_{n,r} = \sum_{i=1}^{r} t_{(i)} + (n - r)t_{(r)}$. Hence, the Bayes estimator of $R(t) = \exp(-t/\beta)$ is, for squared-error loss,

$$\hat{R}(t) = \frac{(1+T_{n,r}\tau)^{r+v}}{\tau^{r+v}\Gamma(r+v)} \int_0^\infty \frac{1}{\beta^{r+v+1}}$$

$$\cdot \exp\left(-\frac{1}{\beta}\left(T_{n,r} + \frac{1}{\tau} + t\right)\right) d\beta \tag{17.2.5}$$

$$= \left(\frac{1+T_{n,r}\tau}{1+(T_{n,r}+t)\tau}\right)^{r+v}.$$

Note that the estimator only depends on n through $T_{n,r}$.

In the following table, we provide a few values of the Bayes estimator $\hat{R}(t)$ for selected values of t, when $v = 3, r = 23, T_{n,r} = 2242$, and $\tau = 10^{-2}$, along with the corresponding MLE, which is

$$\text{MLE} = e^{-t/\hat{\beta}_{n,r}} = e^{-rt/T_{n,r}}.$$

t	50	100	150	200
$\hat{R}(t)$	0.577	0.337	0.199	0.119
MLE	0.599	0.359	0.215	0.129

If we have a series structure of k modules, and the TTF of each module is exponentially distributed, then formula (17.2.5) is extended to

$$\hat{R}_{\text{sys}}(t) = \prod_{i=1}^{k}\left(1 - \frac{t\tau_i}{1 + T_{n,r_i}^{(i)}\tau_i + t\tau_i}\right)^{r_i+v_i}, \tag{17.2.6}$$

where $T_{n,r_i}^{(i)}$ is the total time on test statistic for the ith module, r_i is the censoring frequency of the observations on the ith module, τ_i and v_i are the prior parameters for the ith module. As in (17.2.5) and (17.2.6) is the Bayes estimator for the squared-error loss, under the assumption that the MTTFs of the various modules are priorly independent. In a similar manner, one can write a formula for the Bayes estimator of the reliability of a system having a parallel structure.

17.3 Bayesian credibility and prediction intervals

Bayesian credibility intervals at level γ are intervals $C_\gamma(\mathbf{x})$ in the parameter space Θ, for which the posterior probability that $\theta \in C_\gamma(\mathbf{x})$ is at least γ, i.e.

$$\Pr\left\{\theta \in C_\gamma(\mathbf{x}) \mid \mathbf{x}\right\} \geq \gamma. \tag{17.3.1}$$

$\Pr\{E \mid \mathbf{x}\}$ denotes the posterior probability of the event E, given $\mathbf{x}$. The Bayesian credibility interval for θ, given $\mathbf{x}$, has an entirely different interpretation than that of the confidence intervals discussed in previous sections. While the confidence level of the classical confidence interval is based on the sample-to-sample variability of the interval, for fixed θ, the credibility level of the Bayesian credibility interval is based on the presumed variability of θ, for a fixed sample.

17.3.1   Distribution-free reliability estimation

In Section 17.2.1, we developed the Bayes estimator, with respect to squared-error loss, of the reliability at age t, $R(t)$, when the data available are the number of sample units which survive at age t, namely K_n. We have seen that the posterior distribution of $R(t)$, given K_n, for a uniform prior is the beta distribution with $v_1 = K_n + 1$ and $v_2 = n - K_n + 1$. The Bayesian credibility interval at level γ is the interval whose limits are the ϵ_1- and ϵ_2-quantiles of the posterior distribution, where $\epsilon_1 = (1 - \gamma)/2$, $\epsilon_2 = (1 + \gamma)/2$. These limits can be determined with aid of R, getting the quantile of the F-distribution, according to the formulae

$$\text{Lower limit} = \frac{(K_n + 1)}{(K_n + 1) + (n - K_n + 1)F_{\epsilon_2}\left[2n + 2 - 2K_n, 2K_n + 2\right]} \tag{17.3.2}$$

and

$$\text{Upper limit} = \frac{(K_n + 1)F_{\epsilon_2}\left[2K_n + 2, 2n + 2 - 2K_n\right]}{(n - K_n + 1) + (K_n + 1)F_{\epsilon_2}\left[2K_n + 2, 2n + 2 - 2K_n\right]}. \tag{17.3.3}$$

In Section 17.2.1, we considered the case of $n = 50$ and $K_n = 27$. For $\gamma = 0.95$ we need

$$F_{0.975}[48, 56] = 1.735$$

and

$$F_{0.975}[56, 48] = 1.746.$$

Thus, the Bayesian credibility limits obtained for $R(t)$ are 0.402 and 0.671. Recall the Bayes estimator was 0.538.

17.3.2   Exponential reliability estimation

In Section 17.2.2, we developed a formula for the Bayes estimator of the reliability function $R(t) = \exp(-t/\beta)$ for Type II censored data. We saw that if the prior on β is $\text{IG}(v, \tau)$ then the posterior distribution of β, given the data, is $\text{IG}(v + r, \tau/(1 + \tau T_{n,r}))$. Thus, γ level Bayes credibility limits for β are given by $\beta_{L,\gamma}$ (lower limit) and $B_{U,\gamma}$ (upper limit), where

$$\beta_{L,\gamma} = \frac{T_{n,r} + 1/\tau}{G_{\epsilon_2}(v + r, 1)} \tag{17.3.4}$$

and

$$\beta_{U,\gamma} = \frac{T_{n,r} + 1/\tau}{G_{\epsilon_1}(v + r, 1)}. \tag{17.3.5}$$

Moreover, if v is an integer then we can replace $G_p(v + r, 1)$ by $\frac{1}{2}\chi_p^2[2v + 2r]$. Finally, since $R(t) = \exp(-t/\beta)$ is an increasing function of β, the γ-level Bayes credibility limits for $R(t)$ are

$$R_{L,\gamma}(t) = \exp(-t/\beta_{L,\gamma}) \tag{17.3.6}$$

and

$$R_{U,\gamma}(t) = \exp(-t/\beta_{U,\gamma}). \tag{17.3.7}$$

If we consider the values $v = 3$, $r = 23$, $T_{n,r} = 2242$, and $\tau = 10^{-2}$, we need for $\gamma = 0.95$, $\chi^2_{0.025}[52] = 33.53$ and $\chi^2_{0.975}[52] = 73.31$. Thus,

$$\beta_{L,0.95} = 63.91 \quad \text{and} \quad \beta_{U,0.95} = 139.73.$$

The corresponding Bayesian credibility limits for $R(t)$, at $t = 50$, are $R_{L,0.95}(50) = 0.457$ and $R_{U,0.95}(50) = 0.699$.

17.3.3 Prediction intervals

In this Section 2.5, we introduced the notion of prediction intervals of level γ. This notion can be adapted to the Bayesian framework in the following manner.

Let $\mathbf{X}$ be a sample from a distribution governed by a parameter θ; we assume that θ has a prior distribution. Let $h(\theta \mid \mathbf{x})$ denote the posterior p.d.f. of θ, given $\mathbf{X} = \mathbf{x}$. $\mathbf{x}$ represents the values of a random sample already observed. We are interested in predicting the value of some statistic $T(\mathbf{Y})$ based on a future sample $\mathbf{Y}$ from the same distribution. Let $g(t; \theta)$ denote the p.d.f. of $T(\mathbf{Y})$ under θ. Then the **predictive distribution** of $T(\mathbf{Y})$, given $\mathbf{x}$, is

$$g^*(t \mid \mathbf{x}) = \int_\Theta g(t; \theta) h(\theta \mid \mathbf{x}) d\theta. \tag{17.3.8}$$

A **Bayesian prediction interval** of level γ for $T(Y)$, given $\mathbf{x}$, is an interval $(T_L(\mathbf{x}), T_U(\mathbf{x}))$ which contains a proportion γ of the predictive distribution, i.e. satisfying

$$\int_{T_L(\mathbf{x})}^{T_U(\mathbf{x})} g^*(t \mid \mathbf{x}) dt = \gamma. \tag{17.3.9}$$

Generally, the limits are chosen so that the tail areas are each $(1 - \gamma/2)$. We illustrate the derivation of a Bayesian prediction interval in the following example.

Example 17.2. Consider a device with an exponential lifetime distribution $E(\beta)$. We test a random sample of n of these, stopping at the rth failure. Suppose the prior distribution of β is IG(v, τ). Then, as seen in Section 17.2.2, the posterior distribution of β given the ordered failure times $t_{(1)}, \ldots, t_{(r)}$ is IG$\left(v + r, \frac{\tau}{1+T_{n,r}\tau}\right)$, where $T_{n,r} = \sum_{i=1}^r t_{(i)} + (n - r)t_{(r)}$.

Suppose we have an additional s such devices, to be used one at a time in some system, replacing each one immediately upon failure by another. We are interested in a prediction interval of level γ for T, the time until all s devices have been used up. Letting $\mathbf{Y} = (Y_1, \ldots, Y_s)$ be the lifetimes of the devices, we have $T(\mathbf{Y}) = \sum_{i=1}^s Y_i$. Thus, $T(\mathbf{y})$ has a $G(s, \beta)$ distribution. Substituting in (17.3.8), it is easily shown that the predictive p.d.f. of $T(\mathbf{Y})$, given $t_{(1)}, \ldots, t_{(r)}$, is

$$g^*(t \mid t_{(1)}, \ldots, t_{(r)}) = (B(s, v + r)(T_{n,r} + 1/\tau))^{-1}$$

$$\cdot \left(\frac{\tau}{t+T_{n,r}+1/\tau}\right)^{s-1} \left(\frac{T_{n,r}+1/\tau}{t+T_{n,r}+1/\tau}\right)^{r+v+1}. \tag{17.3.10}$$

Making the transformation

$$U = (T_{n,r} + 1/\tau)/(T(\mathbf{Y}) + T_{n,r} + 1/\tau)$$

one can show that the predictive distribution of U given $t_{(1)}, \ldots, t_{(r)}$ is the $\text{Be}(r + v, s)$ distribution. If we let $\text{Be}_{\epsilon_1}(r + v, s)$ and $\text{Be}_{\epsilon_2}(r + v, s)$ be the ϵ_1- and ϵ_2-quantiles of $\text{Be}(r + v, s)$, where $\epsilon_1 = (1 - \gamma)/2$ and $\epsilon_2 = (1 + \gamma)/2$, then the lower and upper Bayesian prediction limits for $T(\mathbf{Y})$ are

$$T_L = \left(T_{n,r} + \frac{1}{\tau}\right)\left(\frac{1}{\text{Be}_{\epsilon_2}(v + r, s)} - 1\right) \tag{17.3.11}$$

and

$$T_U = \left(T_{n,r} + \frac{1}{\tau}\right)\left(\frac{1}{\text{Be}_{\epsilon_1}(v + r, s)} - 1\right). \tag{17.3.12}$$

If v is an integer, the prediction limits can be expressed as

$$T_L = \left(T_{n,r} + \frac{1}{\tau}\right)\frac{s}{v + r}F_{\epsilon_1}[2s, 2v + 2r] \tag{17.3.13}$$

and

$$T_U = \left(T_{n,r} + \frac{1}{\tau}\right)\frac{s}{v + r}F_{\epsilon_2}[2s, 2v + 2r].$$

Formulae (17.3.12) and (17.3.13) have been applied in the following context.

Twenty computer monitors have been put on test at time $t_0 = 0$. The test was terminated at the sixth failure ($r = 6$). The total time on test was $T_{20,6} = 75,805.6$ [hours]. We wish to predict the time till failure [hour] of monitors which are shipped to customers. Assuming that TTF $\sim E(\beta)$ and ascribing β a prior $\text{IG}(5, 10^{-3})$ distribution, we compute the prediction limits T_L and T_U for $s = 1$, at level $\gamma = 0.95$.

In the present case, $2v + 2r = 22$ and $F_{0.025}[2, 22] = 1/F_{0.975}[22, 2] = 1/39.45 = 0.0253$. Moreover, $F_{0.975}[2, 22] = 4.38$. Thus,

$$T_L = 76,805.6\,\frac{1}{11} \times 0.0253 = 176.7 \text{ [hours]}$$

and

$$T_U = 76,805.6\,\frac{1}{11} \times 4.38 = 30,582.6 \text{ [hours]}.$$

We have high prediction confidence that a monitor in the field will not fail before 175 hours of operation.

17.3.4   Applications with JMP

In Section 16.7.2 we discussed the maximum likelihood estimation of the scale parameter β and the shape parameter v of the Weibull life distribution (see 16.7.12–16.7.15). We have seen that the estimation of the scale parameter, when the shape parameter is known is straight forward and simple. On the other hand, when the shape parameter is unknown, the estimation requires iterative solution which converges to the correct estimates. The analysis there was done for uncensored data. In reliability life testing we often encounter right censored data, when at termination of the study some systems have not yet failed. In this case the likelihood function has to be modified, and the maximum likelihood estimation might be more complicated.

In this section we show how to utilize JMP, in order to obtain interesting analysis of right censored data. We apply methods described in Chapter 16 and in this chapter to the data set **System failure.csv.** The data set **System failure.csv** consists of 208 observations on systems operating at 90 geographically dispersed sites. Twelve systems are new installed and are labeled as "Young." All the other systems are labeled "Mature." Out of the 208 observations, 68 (33%) report time stamps of a failure (uncensored). The other observations are censored, as indicated by the value 1 in the censor variable column.

A measure of time, the time stamp, is recorded for each observation in the data. This variable presented in operational units (activity time), at time of observation. The bigger the time, the longer the system performed. The observations with a value 0 of the censor variable, represent length of operation till failure of the systems.

As a first analysis we fit a Weibull model. The estimated Weibull scale and shape parameters are 4,350,747.2 and 0.72261367, respectively. Figure 17.1 presents the likelihood function with a hair-cross at these maximum likelihood estimates. Note that the notation in JMP is α for scale and β for shape which is reversed from the one used in Section 3.4.4 were we introduce the Weibull distribution which is α for shape and β for scale.

In Figure 17.1 we see contours of confidence regions. Asymptotically, for very large samples, the two maximum likelihood estimators are distributed like bivariate-normal variables. Since the present sample size is not very large, the contour lines are not exactly ellipsoidal, but close enough. One can obtain such regions by randomization methods, like bootstrapping. The colored key on the right-hand side of the figure indicates the proportion of estimates falling in each contoured region. The contour lines for 50–99% are drawn in the figure.

The distribution profiler in Figure 17.2 shows an estimate of the probability that a system fails before the designated 3,000,000 time units. At this designated time, the estimated probability is $P = 53.4\%$ This estimator has a 95% level confidence interval of (44.3–63.1%).

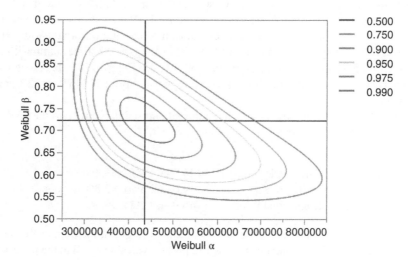

Figure 17.1 *Likelihood function contour plots of the Weibull parameters fit to the data.*

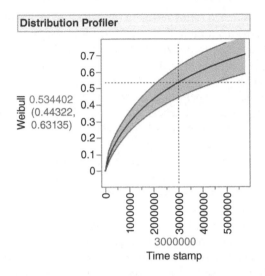

Figure 17.2   *Distribution profiler of Weibull failure rate with a hair-ross at 3,000,000 time units.*

We now apply Bayesian inference to this data by invoking different priors on the Weibull parameters. We start with the uniform joint prior distribution of the scale and shape parameters. The points in Figure 17.3a are the alpha and beta values according to the random uniform choice. The points in Figure 17.3b are the posterior estimates of alpha and beta according to the Weibull model. The HPD contours are shown on top of the scattered points. Figure 17.4a, b is similar but with a lognormal prior.

The calculations in this JMP applications are based on Markov chain Monte Carlo (MCMC) simulations. Specifically, the platform attempts a basic rejection sampler and if the rejection sampler cannot produce valid results, the platform uses a random walk Metropolis–Hastings algorithm (Singpurwalla 2006; Robert and Casella 2010; Insua et al. 2012; Gelman et al. 2013).

After choosing a prior, the initial values provided by JMP are estimates consistent with the maximum likelihood estimates. The parameter values listed as outputs are the joint HPD where the joint posterior density is maximized. The 95% HPD at 3,000,000 time stamp with a uniform prior is (44.0–62.5%) and the Bayes estimate is 52.8%, see Figure 17.4a, b, respectively.

The 95% HPD at 3,000,000 time stamp with a lognormal prior is (46.9–60.2%) and Bayes estimate is 53.4%.

Comparing the 95% parametric confidence interval and point estimate to the 95% Bayesian HPD and Bayes estimate at 3,000,000 time stamp we get:

Parametric estimate	(44.3–63.1%), 53.4%
Uniform prior	(44.0–62.5%), 52.8%
Lognormal prior	(46.9–60.2%), 53.4%

The uniform prior is reflecting lack of prior information. It reduced the point estimate. The lognormal is a more pointed prior and this produced a predictive value equal to the parametric point estimate with a shorter HPD than the corresponding confidence interval. Note that the interpretation of the HPD is probabilistic, making assertions on the posterior failure rate probabilities,

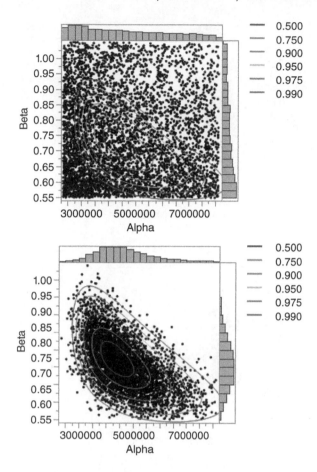

Figure 17.3   *Uniform prior and posterior distributions of Weibull parameters.*

while the confidence interval refers to a statement on the long-term coverage of the intervals constructed by this method.

To conclude Chapter 16 and in this chapter on reliability we refer again to **System failure.csv** and assess the impact on reliability of the maturity level of the system. As mentioned, we have 12 systems labeled as "Young," are they behaving differently?

In Figure 17.5 we show the application of **Fit Life by** X in JMP, where X is the variable indicating maturity level. We can clearly see a difference between the mature and young systems. As expected, mature systems have lower failure rates. This is often due to new product introduction early deployment problems affecting the young systems. The distribution profilers for mature and young systems are shown on Figure 17.6. At 3,000,000 time units of operation, the failure rate of mature systems is 53.1% with 95% confidence interval (43.8–63.1%), for young systems the estimates are 79.4% with 95% confidence interval (33.0–99.8%) a higher failure with a very wide uncertainty due to the small number of data points. Another feature available in JMP is to consider different failure modes by applying competing risk models where the occurrence of a failure is modeled as censoring on all other failure causes. For more information on reliability models implemented in JMP see Meeker and Escobar (1989) and the JMP documentation.

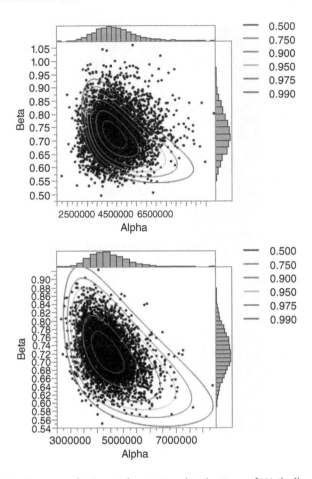

Figure 17.4   *Lognormal prior and posterior distributions of Weibull parameters.*

17.4   Credibility intervals for the asymptotic availability of repairable systems: the exponential case

Consider a repairable system. We take observations on n consecutive renewal cycles. It is assumed that in each renewal cycle, TTF $\sim E(\beta)$ and TTR $\sim E(\gamma)$. Let $t_1, \ldots, t_n$ be the values of TTF in the n cycles and $s_1, \ldots, s_n$ be the values of TTR. One can readily verify that the likelihood function of β depends on the statistic $U = \sum_{i=1}^{n} t_i$ and that of γ depends on $V = \sum_{i=1}^{n} s_i$. U and V are called the **likelihood** (or **minimal sufficient**) **statistics**. Let $\lambda = 1/\beta$ and $\mu = 1/\gamma$. The asymptotic availability is $A_\infty = \mu/(\mu + \lambda)$.

In the Bayesian framework, we assume that λ and μ are priorly independent, having prior gamma distributions $G(\nu, \tau)$ and $G(\omega, \zeta)$, respectively. One can verify that the posterior distributions of λ and μ, given U and V, are $G(n + \nu, U + \tau)$ and $G(n + \omega, V + \zeta)$, respectively. Moreover, λ and μ are posteriorly independent. Routine calculations yield that

$$\frac{\frac{1-A_\infty}{A_\infty}(U + \tau)}{\frac{1-A_\infty}{A_\infty}(U + \tau) + (V + \zeta)} \sim \mathrm{Be}(n + \nu, n + \omega),$$

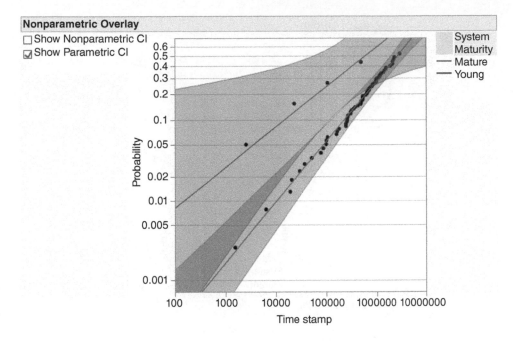

Figure 17.5   *Weibull analysis of young and mature systems.*

where $\text{Be}(p, q)$ denotes a random variable having a beta distribution, with parameters p and q, $0 < p, q < \infty$. Let $\epsilon_1 = (1 - \gamma)/2$ and $\epsilon_2 = (1 + \gamma)/2$. We obtain that the lower and upper limits of the γ-level credibility interval for A_∞ are A_{∞, ϵ_1} and A_{∞, ϵ_2} where

$$A_{\infty, \epsilon_1} = \left[1 + \frac{V + \zeta}{U + \tau} \cdot \frac{\text{Be}_{\epsilon_2}(n + v, n + \omega)}{\text{Be}_{\epsilon_1}(n + \omega, n + v)}\right]^{-1} \tag{17.4.1}$$

and

$$A_{\infty, \epsilon_2} = \left[1 + \frac{V + \zeta}{U + \tau} \cdot \frac{\text{Be}_{\epsilon_1}(n + v, n + \omega)}{\text{Be}_{\epsilon_2}(n + \omega, n + v)}\right]^{-1}. \tag{17.4.2}$$

where $\text{Be}_\epsilon(p, q)$ is the ϵth quantile of $\text{Be}(p, q)$. Moreover, the quantiles of the beta distribution are related to those of the F-distribution according to the following formulae:

$$\text{Be}_{\epsilon_2}(a_1, a_2) = \frac{\frac{a_1}{a_2} F_{\epsilon_2}[a_1, a_2]}{1 + \frac{a_1}{a_2} F_{\epsilon_2}[a_1, a_2]} \tag{17.4.3}$$

and

$$\text{Be}_{\epsilon_1}(a_1, a_2) = \frac{1}{1 + \frac{a_2}{a_1} F_{\epsilon_2}[a_2, a_1]}. \tag{17.4.4}$$

We illustrate these results in the following example.

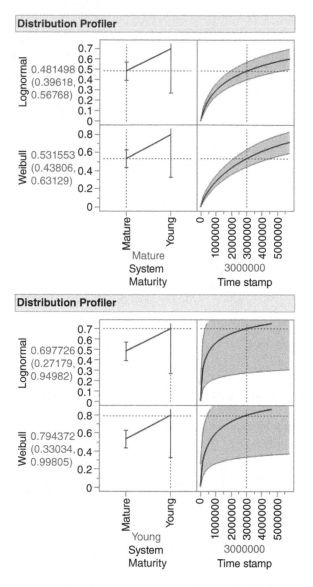

Figure 17.6 Distribution profiler of young and mature systems.

Example 17.3. Observations were taken on $n = 72$ renewal cycles of an insertion machine. It is assumed that TTF $\sim E(\beta)$ and TTR $\sim E(\gamma)$ in each cycle. The observations gave the values $U = 496.9$ [minutes] and $V = 126.3$ [minutes]. According to these values, the MLE of A_∞ is $\hat{A}_\infty = 496.9/(496.9 + 126.3) = 0.797$. Assume the gamma prior distributions for λ and μ, with $\nu = 2$, $\tau = 0.001$, $\omega = 2$ and $\zeta = 0.005$. We obtain from (17.4.3) and (17.4.4) for $\gamma = 0.95$,

$$\text{Be}_{0.025}(74, 74) = 0.3802, \quad \text{Be}_{0.975}(74, 74) = 0.6198.$$

Finally, the credibility limits obtained from (17.4.1) and (17.4.2) are $A_{\infty,0.025} = 0.707$, and $A_{\infty,0.975} = 0.865$. To conclude this example we remark that the Bayes estimator of A_∞, for the absolute deviation loss function, is the median of the posterior distribution of A_∞, given (U, V), namely $A_{\infty,0.5}$.

In the present example, $n + \nu = n + \omega = 74$. The $Be(74, 74)$ distribution is symmetric. Hence, $Be_{0.5}(74, 74) = 0.5$. To obtain the $A_{\infty,0.5}$, we solve the equation

$$\frac{\frac{1-A_{\infty,0.5}}{A_{\infty,0.5}}(U+\tau)}{1 - \frac{A_{\infty,0.5}}{A_{\infty,0.5}}(U+\tau) + (V+\zeta)} = Be_{0.5}(n+\nu, n+\omega).$$

In the present case, we get

$$A_{\infty,0.5} = \left(1 + \frac{V+\zeta}{U+\tau}\right)^{-1} = \frac{1}{1 + \frac{126.305}{496.901}} = 0.797.$$

This is equal to the value of the MLE.

17.5   Empirical bayes method

Empirical Bayes estimation is designed to utilize the information in large samples to estimate the Bayes estimator, without specifying the prior distribution. We introduce the idea in relation to estimating the parameter, λ, of a Poisson distribution.

Suppose that we have a sequence of independent trials, in each trial a value of λ (failure rate) is chosen from some prior distribution $H(\lambda)$, and then a value of X is chosen from the Poisson distribution $P(\lambda)$. If this is repeated n times we have n pairs $(\lambda_1, x_1), \ldots, (\lambda_n, x_n)$. The statistician, however, can observe only the values $x_1, x_2, \ldots, x_n$. Let $f_n(i)$, $i = 0, 1, 2, \ldots$, be the empirical p.d.f. of the observed variable X, i.e. $f_n(i) = \frac{1}{n}\sum_{j=1}^{n} I\{x_j = i\}$.

A new trial is to be performed. Let Y be the observed variable in the new trial. It is assumed that Y has a Poisson distribution with mean λ which will be randomly chosen from the prior distribution $H(\lambda)$. The statistician has to estimate the new value of λ from the observed value y of Y. Suppose that the loss function for erroneous estimation is the squared-error loss, $(\hat{\lambda} - \lambda)^2$. The Bayes estimator, if $H(\lambda)$ is known, is

$$E_H\{\lambda \mid y\} = \frac{\int_0^\infty \lambda^{y+1} e^{-\lambda} h(\lambda) d\lambda}{\int_0^\infty \lambda^y e^{-\lambda} h(\lambda) d\lambda}, \tag{17.5.1}$$

where $h(\lambda)$ is the prior p.d.f. of λ.

The predictive p.d.f. of Y, under H, is

$$f_H(y) = \frac{1}{y!} \int_0^\infty \lambda^y e^{-\lambda} h(\lambda) d\lambda. \tag{17.5.2}$$

The Bayes estimator of λ (17.5.1) can be written in the form

$$E_H\{\lambda \mid y\} = (y+1)\frac{f_H(y+1)}{f_H(y)}, \quad y = 0, 1, \ldots \tag{17.5.3}$$

The empirical p.d.f. $f_n(y)$ converges (by the Strong Law of Large Numbers) in a probabilistic sense, as $n \to \infty$, to $f_H(y)$. Accordingly, replacing $f_H(y)$ in (17.5.3) with $f_n(y)$ we obtain an estimator of $E_H\{\lambda \mid y\}$ based on the past n trials. This estimator is called an **empirical Bayes estimator** (EBE) of λ:

$$\hat{\lambda}_n(y) = (y+1)\frac{f_n(y+1)}{f_n(y)}, \quad y = 0, 1, \ldots \quad (17.5.4)$$

In the following example, we illustrate this estimation method.

Example 17.4. $n = 188$ batches of circuit boards were inspected for soldering defects. Each board has typically several hundred soldering points, and each batch contained several hundred boards. It is assumed that the number of soldering defects, X (per 10^5 points), has a Poisson distribution. In Table 17.1, we present the frequency distribution of X among the 188 observed batches.

Table 17.1 *Empirical distribution of number of soldering defects (per 100,000 points)*

x	0	1	2	3	4	5	6	7	8
$f(x)$	4	21	29	32	19	14	13	5	8

x	9	10	11	12	13	14	15	16	17	18
$f(x)$	5	9	1	2	4	4	1	4	2	1

x	19	20	21	22	23	24	25	26	Total
$f(x)$	1	1	1	1	2	1	2	1	188

Accordingly, if in a new batch the number of defects (per 10^5 points) is $y = 8$, the EBE of λ is $\hat{\lambda}_{188}(8) = 9 \times \frac{5}{8} = 5.625$ (per 10^5), or 56.25 (per 10^6 points), i.e. 56.25 PPM. After observing $y_{189} = 8$, we can increase $f_{188}(8)$ by 1, i.e. $f_{189}(8) = f_{188}(8) + 1$, and observe the next batch.

The above method of deriving an EBE can be employed for any p.d.f. $f(x; \theta)$ of a discrete distribution, such that

$$\frac{f(x+1; \theta)}{f(x; \theta)} = a(x) + b(x)\theta.$$

In such a case, the EBE of θ is

$$\hat{\theta}_n(x) = \frac{f_n(x+1)}{f_n(x)b(x)} - \frac{a(x)}{b(x)}. \quad (17.5.5)$$

Generally, however, it is difficult to obtain an estimator which converges, as n increases, to the value of the Bayes estimator. A **parametric** EB procedure is one in which, as part of the model, we assume that the prior distribution belongs to a parametric family, but the parameter of the prior distribution is consistently estimated from the past data. For example, if the model assumes that the observed TTF is $E(\beta)$ and that $\beta \sim IG(\nu, \tau)$, instead of specifying the values of ν and τ, we use the past data to estimate. We may obtain an estimator of $E\{\theta \mid T, \nu, \tau)$ which converges in a probabilistic sense, as n increases, to the Bayes estimator. An example of such a parametric EBE is given below.

Example 17.5. Suppose that $T \sim E(\beta)$ and β has a prior $IG(\nu, \tau)$. The Bayes estimator of the reliability function is given by (17.5.5).
Let $t_1, t_2, \ldots, t_n$ be past independent observations on T.
The expected value of T under the predictive p.d.f. is

$$E_{\tau,\nu}\{T\} = \frac{1}{\tau(\nu - 1)},\qquad (17.5.6)$$

provided $\nu > 1$. The second moment of T is

$$E_{\tau,\nu}\{T^2\} = \frac{2}{\tau^2(\nu - 1)(\nu - 2)},\qquad (17.5.7)$$

provided $\nu > 2$.
Let $M_{1,n} = \frac{1}{n}\sum_{i=1}^{n} t_i$ and $M_{2,n} = \frac{1}{n}\sum_{i=1}^{n} t_i^2$. $M_{1,n}$ and $M_{2,n}$ converge in a probabilistic sense to $E_{\tau,\nu}\{T\}$ and $E_{\tau,\nu}\{T^2\}$, respectively. We estimate τ and ν by the **method of moment equations**, by solving

$$M_{1,n} = \frac{1}{\hat{\tau}(\hat{\nu} - 1)}\qquad (17.5.8)$$

and

$$M_{2,n} = \frac{2}{\hat{\tau}^2(\hat{\nu} - 1)(\hat{\nu} - 2)}.\qquad (17.5.9)$$

Let $D_n^2 = M_{2,n} - M_{1,n}^2$ be the sample variance. Simple algebraic manipulations yield the estimators

$$\hat{\tau}_n = \frac{(D_n^2 - M_{1,n}^2)}{[M_{1,n}(D_n^2 + M_{1,n}^2)]},\qquad (17.5.10)$$

$$\hat{\nu}_n = \frac{2D_n^2}{D_n^2 - M_{1,n}^2},\qquad (17.5.11)$$

provided $D_n^2 > M_{1,n}^2$. It can be shown that for large values of n, $D_n^2 > M_{1,n}^2$ with high probability.
Substituting the empirical estimates $\hat{\tau}_n$ and $\hat{\nu}_n$ in (17.5.5) we obtain a parametric EBE of the reliability function.

For additional results on the EBE of reliability functions, see Martz and Waller (1982) and Tsokos and Shimi (1977).

17.6   Chapter highlights

This advance chapter presents reliability estimation and prediction from a Bayesian perspective. It introduces the reader to prior and posterior distributions used in Bayesian reliability inference, discusses loss functions and Bayesian estimators, and distribution free Bayes estimators of reliability. A special section is dedicated to Bayesian credibility and prediction intervals. A final section covers empirical Bayes methods. The main concepts and definitions introduced in this chapter include:

- Prior Distribution
- Predictive Distribution
- Posterior Distribution
- Beta Function
- Conjugate Distributions
- Bayes Estimator
- Posterior Risk
- Posterior Expectation
- Distribution Free Estimators
- Credibility Intervals
- Minimal Sufficient Statistics
- Empirical Bayes Method

17.7   Exercises

17.1 Suppose that the TTF of a system is a random variable having exponential distribution, $E(\beta)$. Suppose also that the prior distribution of $\lambda = 1/\beta$ is $G(2.25, 0.01)$.
   (i) What is the posterior distribution of λ, given $T = 150$ [hours]?
   (ii) What is the Bayes estimator of β, for the squared-error loss?
   (iii) What is the posterior SD of β?

17.2 Let $J(t)$ denote the number of failures of a device in the time interval $(0, t]$. After each failure the device is instantaneously renewed. Let $J(t)$ have a Poisson distribution with mean λt. Suppose that λ has a gamma prior distribution, with parameters $v = 2$ and $\tau = 0.05$.
   (i) What is the predictive distribution of $J(t)$?
   (ii) Given that $J(t)/t = 10$, how many failures are expected in the next time unit?
   (iii) What is the Bayes estimator of λ, for the squared-error loss?
   (iv) What is the posterior SD of λ?

17.3 The proportion of defectives, θ, in a production process has a uniform prior distribution on $(0, 1)$. A random sample of $n = 10$ items from this process yields $K_{10} = 3$ defectives.
   (i) What is the posterior distribution of θ?
   (ii) What is the Bayes estimator of θ for the absolute error loss?

17.4 Let $X \sim \mathcal{P}(\lambda)$ and suppose that λ has the Jeffrey improper prior $h(\lambda) = \frac{1}{\sqrt{\lambda}}$. Find the Bayes estimator for squared-error loss and its posterior SD.

17.5 Apply formula (17.2.3) to determine the Bayes estimator of the reliability when $n = 50$ and $K_{50} = 49$.

17.6 A system has three modules, M_1, M_2, M_3. M_1 and M_2 are connected in series and these two are connected in parallel to M_3, i.e.

$$R_{\text{sys}} = \psi_p(R_3, \psi_s(R_1, R_2)) = R_3 + R_1 R_2 - R_1 R_2 R_3,$$

where R_i is the reliability of module M_i. The TTFs of the three modules are independent random variables having exponential distributions with prior $IG(v_i, \tau_i)$ distributions of their MTTF. Moreover, $v_1 = 2.5$, $v_2 = 2.75$, $v_3 = 3$, $\tau_1 = \tau_2 = \tau_3 = 1/1000$. In separate independent trials of the TTF of each module we obtained the statistics $T_n^{(1)} = 4565$ [hours], $T_n^{(2)} = 5720$ [hours] and $T_n^{(3)} = 7505$ [hours], where in all three experiments $n = r = 10$. Determine the Bayes estimator of R_{sys}, for the squared-error loss.

17.7 $n = 30$ computer monitors were put on test at a temperature of 100 °F and relative humidity of 90% for 240 [hours]. The number of monitors which survived this test is $K_{30} = 28$. Determine the Bayes credibility interval for $R(240)$, at level $\gamma = 0.95$, with respect to a uniform prior on $(0, 1)$.

17.8 Determine a $\gamma = 0.95$ level credibility interval for $R(t)$ at $t = 25$ [hours] when TTF $\sim E(\beta)$, $\beta \sim IG(3, 0.01)$, $r = 27$, $T_{n,r} = 3500$ [hours].

17.9 Under the conditions of Exercise [17.3](ii) determine a Bayes prediction interval for the total life of $s = 2$ devices.

17.10 A repairable system has exponential TTF and exponential TTR, which are independent of each other. $n = 100$ renewal cycles were observed. The total times till failure were 10,050 [hours] and the total repair times were 500 [minutes]. Assuming gamma prior distributions for λ and μ with $v = \omega = 4$ and $\tau = 0.0004$ [hour], $\zeta = 0.01$ [minute], find a $\gamma = 0.95$ level credibility interval for A_∞.

17.11 In reference to Example 17.4, suppose that the data of Table 17.1 were obtained for a Poisson random variable where $\lambda_1, \ldots, \lambda_{188}$ have a gamma (v, τ) prior distribution.
   (i) What is the predictive distribution of the number of defects per batch?
   (ii) Find the formulae for the first two moments of the predictive distribution.
   (iii) Find, from the empirical frequency distribution of Table 17.1, the first two sample moments.
   (iv) Use the method of moment equations to estimate the prior parameters v and τ.
   (v) What is the Bayes estimator of λ if $X_{189} = 8$?

18

Sampling Plans for Batch and Sequential Inspection

18.1 General discussion

Sampling plans for product inspection are quality assurance schemes, designed to test whether the quality level of a product conforms with the required standards. These methods of quality inspection are especially important when products are received from suppliers or vendors on whom we have no other assessment of the quality level of their production processes. Generally, if a supplier has established procedures of statistical process control which assure the required quality standards are met (see Chapters 10–12), then sampling inspection of his shipments may not be necessary. However, periodic auditing of the quality level of certified suppliers might be prudent to ensure that these do not drop below the acceptable standards. Quality auditing or inspection by sampling techniques can also be applied within the plant, at various stages of the production process, e.g., when lots are transferred from one department to another.

In the present chapter, we discuss various sampling and testing procedures, designed to maintain quality standards. In particular, single, double, and sequential sampling plans for attributes and single sampling plans for continuous measurements are studied. We discuss also testing via tolerance limits. The chapter is concluded with a section describing some of the established standards, and in particular the Skip Lot procedure, which appears in modern standards.

We present here a range of concepts and tools associated with sampling inspection schemes. The methods presented below can be implemented with R applications or the MINITAB **Acceptance Sampling By Attributes** features available in the Stat Quality Tools window. Modern nomenclature is different from the one which was established for almost 50 years. A product unit which did not meet the quality specifications or requirements was called **defective**. This term has been changed recently to **nonconforming**. Thus, in early standards, like MIL-STD 105E and others, we find the term "defective items" and "number of defects." In modern standards like ANSI/ASQC Z1.4 and the international standard ISO 2859, the term used is nonconforming. We will use the two terms interchangeably. Similarly, the terms LTPD and LQL, which will be explained later, will be used interchangeably.

Modern Industrial Statistics: With Applications in R, MINITAB and JMP, Third Edition.
Ron S. Kenett and Shelemyahu Zacks.
© 2021 John Wiley & Sons, Ltd. Published 2021 by John Wiley & Sons, Ltd.

A **lot** is a collection of N elements which are subject to quality inspection. Accordingly, a lot is a finite real population of products. **Acceptance** of a lot is a quality approval, providing the "green light" for subsequent use of the elements of the lot. Generally, we refer to lots of raw material, of semi-finished or finished products, etc., which are purchased from vendors or produced by subcontractors. Before acceptance, a lot is typically subjected to quality inspection unless the vendor has been certified and its products are delivered directly, without inspection, to the production line. The purchase contracts typically specify the acceptable quality level and the method of inspection.

In general, it is expected that a lot contains no more than a certain percentage of nonconforming (defective) items, where the test conditions which classify an item as defective are usually well specified. One should decide if a lot has to be subjected to a complete inspection, item by item, or whether it is sufficient to determine acceptance using a sample from the lot. If we decide to inspect a sample, we must determine how large it is and what is the criterion for accepting or rejecting the lot. Furthermore, the performance characteristics of the procedures in use should be understood.

The **proportion of nonconforming items** in a lot is the ratio $p = M/N$, where M is the number of defective items in the whole lot and N is the size of the lot. If we choose to accept only lots with zero defectives, we have to inspect each lot completely, item by item. This approach is called 100% inspection. This is the case, for example, when the items of the lots are used in a critical or very expensive system. A communication satellite is an example of such a system. In such cases, the cost of inspection is negligible compared to the cost of failure. On the other hand, there are many situations in which complete inspection is impossible (e.g., destructive testing) or impractical (because of the large expense involved). In this situation, the two parties involved, the customer and its supplier, specify an **acceptable quality level** (AQL) and a **limiting quality level** (LQL). When the proportion defectives, p, in the lot is not larger than the AQL, the lot is considered good and should be accepted with high probability. If, on the other hand, the proportion defectives in the lot is greater than the LQL, the lot should be rejected with high probability. If p is between the AQL and the LQL, then either acceptance or rejection of the lot can happen with various probability levels.

How should the parties specify the AQL and LQL levels? Usually, the AQL is determined by the quality requirements of the customer, who is going to use the product. The producer of the product, which is the supplier, tries generally to demonstrate to the customer that his production processes maintain a capability level in accordance with the customer's or consumer's requirements. Both the AQL and LQL are specified in terms of proportions p_0 and p_t of nonconforming in the process.

The risk of rejecting a good lot, i.e., a lot with $p \leq$ AQL, is called the **producer's risk**, while the risk of accepting a bad lot, i.e., a lot for which $p \geq$ LQL, is called the **consumer's risk**. Thus, the problem of designing an acceptance sampling plan is that of choosing:

(1) the method of sampling,
(2) the sample size,
    and
(3) the acceptance criteria for testing the hypothesis

$$H_0 : p \leq \text{AQL},$$

against the alternative

$$H_1 : p \geq \text{LQL},$$

so that the probability of rejecting a good lot will not exceed a value α (the level of significance) and the probability of accepting a bad lot will not exceed β. In this context, α and β are called the **producer's risk** and the **consumer's risk**, respectively.

18.2   Single-stage sampling plans for attributes

A single-stage sampling plan for an attribute is an acceptance/rejection procedure for a lot of size N, according to which a random sample of size n is drawn from the lot, without replacement. Let M be the number of defective items (elements) in the lot, and let X be the number of defective items in the sample. Obviously, X is a random variable whose range is $\{0, 1, 2, \ldots, n^*\}$, where $n^* = \min(n, M)$. The distribution function of X is the hypergeometric distribution $H(N, M, n)$, (see Section 3.3.2) with the probability distribution function (p.d.f.)

$$h(x; N, M, n) = \frac{\binom{M}{x}\binom{N-M}{n-x}}{\binom{N}{n}}, \quad x = 0, \ldots, n^* \tag{18.2.1}$$

and the cumulative distribution function (c.d.f.)

$$H(x; N, M, n) = \sum_{j=0}^{x} h(j; N, M, n). \tag{18.2.2}$$

Suppose we consider a lot of $N = 100$ items to be acceptable if it has no more than $M = 5$ nonconforming items, and nonacceptable if it has more than $M = 10$ nonconforming items. For a sample of size $n = 10$, we derive the hypergeometric distribution $H(100, 5, 10)$ and $H(100, 10, 10)$:

From Tables 18.1 and 18.2 we see that, if such a lot is accepted whenever $X = 0$, the consumer's risk of accepting a lot which should be rejected is

$$\beta = H(0; 100, 10, 10) = 0.3305.$$

The producer's risk of rejecting an acceptable lot is

$$\alpha = 1 - H(0; 100, 5, 10) = 0.4162.$$

As before, let p_0 denote the AQL and p_t the LQL. Obviously, $0 < p_0 < p_t < 1$. Suppose that the decision is to accept a lot whenever the number of nonconforming X is not greater than c,

Table 18.1   *The p.d.f. and c.d.f. of H(100, 5, 10)*

j	$h(j; 100, 5, 10)$	$H(j; 100, 5, 10)$
0	0.5838	0.5838
1	0.3394	0.9231
2	0.0702	0.9934
3	0.0064	0.9997
4	0.0003	1.0000

Table 18.2   *The p.d.f. and c.d.f. of H(100, 10, 10)*

j	$h(j; 100, 10, 10)$	$H(j; 100, 10, 10)$
0	0.3305	0.3305
1	0.4080	0.7385
2	0.2015	0.9400
3	0.0518	0.9918
4	0.0076	0.9993
5	0.0006	1.0000

Table 18.3   *Sample size, n, and critical level, c, for single-stage acceptance sampling with N = 100, and α = β = 0.05*

p_0	p_t	n	c	p_0	p_t	n	c
0.01	0.05	65	1	0.03	0.05	92	3
0.01	0.08	46	1	0.03	0.08	71	3
0.01	0.11	36	1	0.03	0.11	56	3
0.01	0.14	29	1	0.03	0.14	37	2
0.01	0.17	24	1	0.03	0.17	31	2
0.01	0.20	20	1	0.03	0.20	27	2
0.01	0.23	18	1	0.03	0.23	24	2
0.01	0.26	16	1	0.03	0.26	21	2
0.01	0.29	14	1	0.03	0.29	19	2
0.01	0.32	13	1	0.03	0.32	13	1

i.e., $X \leq c$. c is called the **acceptance number**. For specified values of p_0, p_t, α, and β, we can determine n and c so that

$$\Pr\{X \leq c \mid p_0\} \geq 1 - \alpha \tag{18.2.3}$$

and

$$\Pr\{X \leq c \mid p_t\} \leq \beta. \tag{18.2.4}$$

Notice that n and c should satisfy the inequalities

$$H(c; N, M_0, n) \geq 1 - \alpha \tag{18.2.5}$$

$$H(c; N, M_t, n) \leq \beta \tag{18.2.6}$$

where $M_0 = [Np_0]$ and $M_t = [Np_t]$ and $[a]$ is the integer part of a. In Table 18.3, a few numerical results show how n and c depend on p_0 and p_t, when the lot is of size $n = 100$ and $\alpha = \beta = 0.05$. To achieve this in R, we apply the following commands:

```
> library(AcceptanceSampling)
> as.data.frame(
    find.plan(PRP=c(0.01,  0.95),
          CRP=c(0.08,  0.05),
          type="hypergeom",  N=100))

    n  c  r
1  46  1  2
```

We see that, even if the requirements are not very stringent, for example, when $p_0 = 0.01$ and $p_t = 0.05$, the required sample size is $n = 65$. If in such a sample there is more than 1 defective item, then the entire lot is rejected. Similarly, if $p_0 = 0.03$ and $p_t = 0.05$, then the required sample size is $n = 92$, which is almost the entire lot. On the other hand, if $p_0 = 0.01$ and p_t is greater than 0.20, we need no more than 20 items in the sample. If we relax the requirement concerning α and β and allow higher producer's and consumer's risks, the required sample size will be smaller, as shown in Table 18.4.

An important characterization of an acceptance sampling plan is given by its **operating-characteristic** (OC) function. This function, denoted by $OC(p)$, yields the probability of accepting a lot having proportion p of defective items. If we let $M_p = [Np]$, then we can calculate the OC function by

$$OC(p) = H(c; N, M_p, n). \tag{18.2.7}$$

In Table 18.5, we present a few values of the OC function for single-stage acceptance sampling, when the lot is of size $N = 100$; based on sample size $n = 50$ and acceptance number $c = 1$.

In Figure 18.1, we present the graph of the OC function, corresponding to Table 18.5.

Table 18.4 *Sample size, n, and critical level, c, for single-stage acceptance sampling, N = 100, α = 0.10, β = 0.20*

p_0	p_t	n	c	p_0	p_t	n	c
0.01	0.05	49	1	0.03	0.05	83	3
0.01	0.08	33	1	0.03	0.08	58	3
0.01	0.11	25	1	0.03	0.11	35	2
0.01	0.14	20	1	0.03	0.14	20	1
0.01	0.17	9	0	0.03	0.17	16	1
0.01	0.20	7	0	0.03	0.20	14	1
0.01	0.23	6	0	0.03	0.23	12	1
0.01	0.26	6	0	0.03	0.26	11	1
0.01	0.29	5	0	0.03	0.29	10	1
0.01	0.32	5	0	0.03	0.32	9	1

Table 18.5 *The OC function of a single-stage acceptance sampling plan, N = 100, n = 50, c = 1*

p	$OC(p)$	p	$OC(p)$
0.000	1.0000	0.079	0.0297
0.008	1.0000	0.086	0.0154
0.015	0.7525	0.094	0.0154
0.023	0.7525	0.102	0.0078
0.030	0.5000	0.109	0.0039
0.039	0.3087	0.118	0.0019
0.047	0.1811	0.126	0.0009
0.055	0.1811	0.133	0.0009
0.062	0.1022	0.141	0.0004
0.070	0.0559	0.150	0.0002

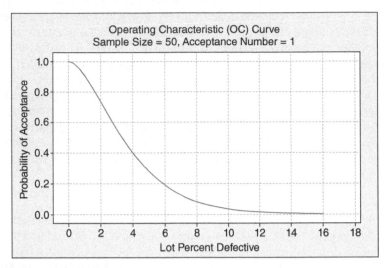

Figure 18.1   *Operating characteristics curve for a single-stage (MINITAB) acceptance sampling plan, N = 100, n = 50, c = 1*

18.3   Approximate determination of the sampling plan

If the sample size, n, is not too small, the c.d.f. of the hypergeometric distribution can be approximated by the normal distribution. More specifically, for large values of n, we have the following approximation

$$H(a; N, M, n) \doteq \Phi\left(\frac{a + 0.5 - nP}{\left(nPQ \left(1 - \frac{n}{N}\right)\right)^{1/2}} \right), \tag{18.3.1}$$

where $P = M/N$ and $Q = 1 - P$.

The first question to ask is: how large should n be? The answer to this question depends on how close we wish the approximation to be. Generally, if $0.2 < P < 0.8$, $n = 20$ is large enough to yield a good approximation, as illustrated in Table 18.6.

If $P < 0.2$ or $P > 0.8$, we usually need larger sample sizes to attain good approximation. We show now how the constants (n, c) can be determined. The two requirements to satisfy are $\text{OC}(p_0) = 1 - \alpha$ and $\text{OC}(p_t) = \beta$. These requirements are expressed approximately by the following two equations:

$$c + \frac{1}{2} - np_0 = z_{1-\alpha}\left(np_0 q_0 \left(1 - \frac{n}{N}\right)\right)^{1/2}$$

$$c + \frac{1}{2} - np_t = -z_{1-\beta}\left(np_t q_t \left(1 - \frac{n}{N}\right)\right)^{1/2}. \tag{18.3.2}$$

Approximate solutions to n and c, n^* and c^* respectively, are:

$$n^* \cong \frac{n_0}{1 + n_0/N}, \tag{18.3.3}$$

where

$$n_0 = \frac{(z_{1-\alpha}\sqrt{p_0 q_0} + z_{1-\beta}\sqrt{p_t q_t})^2}{(p_t - p_0)^2}, \tag{18.3.4}$$

Table 18.6 Hypergeometric c.d.f.'s and their normal approximations

a	$N = 100, M = 30, n = 20$ Hypergeometric	Normal	$N = 100, M = 50, n = 20$ Hypergeometric	Normal	$N = 100, M = 80, n = 20$ Hypergeometric	Normal
0	0.00030	0.00140	0.00000	0.00000	0.00000	0.00000
1	0.00390	0.00730	0.00000	0.00000	0.00000	0.00000
2	0.02270	0.02870	0.00000	0.00000	0.00000	0.00000
3	0.08240	0.08740	0.00040	0.00060	0.00000	0.00000
4	0.20920	0.20780	0.00250	0.00310	0.00000	0.00000
5	0.40100	0.39300	0.01140	0.01260	0.00000	0.00000
6	0.61510	0.60700	0.03920	0.04080	0.00000	0.00000
7	0.79540	0.79220	0.10540	0.10680	0.00000	0.00000
8	0.91150	0.91260	0.22700	0.22780	0.00000	0.00000
9	0.96930	0.97130	0.40160	0.40180	0.00000	0.00000
10	0.99150	0.99270	0.59840	0.59820	0.00060	0.00030
11	0.99820	0.99860	0.77300	0.77220	0.00390	0.00260
12	0.99970	0.99980	0.89460	0.89320	0.01810	0.01480
13	1.00000	1.00000	0.96080	0.95920	0.06370	0.06000
14			0.98860	0.98740	0.17270	0.17550
15			0.99750	0.99690	0.36470	0.37790
16			0.99960	0.99940	0.60840	0.62210
17			1.00000	0.99990	0.82420	0.82450
18					0.95020	0.94000
19					0.99340	0.98520

Table 18.7 Exact and approximate single-stage sampling plans for $\alpha = \beta = 0.05$, $N = 500, 1000, 2000$, $p_0 = 0.01$, $p_t = 0.03, 0.05$

N	Method	$p_0 = 0.01, p_t = 0.03$ n	c	$\hat{\alpha}$	$\hat{\beta}$	$p_0 = 0.01, p_t = 0.05$ n	c	$\hat{\alpha}$	$\hat{\beta}$
500	Exact	254	4	0.033	0.050	139	3	0.023	0.050
	Approx.	248	4	0.029	0.060	127	2	0.107	0.026
1000	Exact	355	6	0.028	0.050	146	3	0.045	0.049
	Approx.	330	5	0.072	0.036	146	3	0.045	0.049
2000	Exact	453	8	0.022	0.050	176	4	0.026	0.050
	Approx.	396	6	0.082	0.032	157	3	0.066	0.037

and

$$c^* \cong n^* p_0 - \frac{1}{2} + z_{1-\alpha} \sqrt{n^* p_0 q_0 (1 - n^*/N)}. \qquad (18.3.5)$$

In Table 18.7, we present several single-stage sampling plans, (n, c), and their approximations (n^*, c^*). We provide also the corresponding attained risk levels $\hat{\alpha}$ and $\hat{\beta}$. We see that the approximation provided for n and c yields risk levels which are generally close to the nominal ones.

18.4   Double-sampling plans for attributes

A double sampling plan for attributes is a two-stage procedure. In the first stage, a random sample of size n_1 is drawn, without replacement, from the lot. Let X_1 denote the number of defective items in this first stage sample. Then the rules for the second stage are the following: if $X_1 \leq c_1$, sampling terminates and the lot is accepted; if $X_1 \geq c_2$, sampling terminates and the lot is rejected; if X_1 is between c_1 and c_2, a second stage random sample, of size n_2, is drawn, without replacement, from the remaining items in the lot. Let X_2 be the number of defective items in this second-stage sample. Then, if $X_1 + X_2 \leq c_3$, the lot is accepted and if $X_1 + X_2 > c_3$ the lot is rejected.

Generally, if there are very few (or very many) defective items in the lot, the decision to accept or reject the lot can be reached after the first stage of sampling. Since the first stage samples are smaller than those needed in a single stage sampling a considerable saving in inspection cost may be attained.

In this type of sampling plan, there are five parameters to select, namely, n_1, n_2, c_1, c_2, and c_3. Variations in the values of these parameters affect the operating characteristics of the procedure, as well as the expected number of observations required (i.e., the total sample size). Theoretically, we could determine the optimal values of these five parameters by imposing five independent requirements on the OC function and the function of expected total sample size, called the **Average Sample Number** or **ASN-function**, at various values of p. However, to simplify this procedure, it is common practice to set $n_2 = 2n_1$ and $c_2 = c_3 = 3c_1$. This reduces the problem to that of selecting just n_1 and c_1. Every such selection will specify a particular double-sampling plan. For example, if the lot consists of $N = 150$ items, and we choose a plan with $n_1 = 20$, $n_2 = 40$, $c_1 = 2$, $c_2 = c_3 = 6$, we will achieve certain properties. On the other hand, if we set $n_1 = 20$, $n_2 = 40$, $c_1 = 1$, $c_2 = c_3 = 3$, the plan will have different properties.

The formula of the OC function associated with a double-sampling plan $(n_1, n_2, c_1, c_2, c_3)$ is

$$OC(p) = H(c_1; N, M_p, n_1)$$

$$+ \sum_{j=c_1+1}^{c_2-1} h(j; N, M_p, n_1) H(c_3 - j; N - n_1, M_p - j, n_2) \qquad (18.4.1)$$

where $M_p = [Np]$. Obviously, we must have $c_2 \geq c_1 + 2$, for otherwise the plan is a single-stage plan. The probability $\Pi(p)$ of stopping after the first stage of sampling is

$$\Pi(p) = H(c_1; N, M_p, n_1) + 1 - H(c_2 - 1; N, M_p, n_1)$$

$$= 1 - [H(c_2 - 1; N, M_p, n_1) - H(c_1; N, M_p, n_1)]. \qquad (18.4.2)$$

The expected total sample size, ASN, is given by the formula

$$ASN(p) = n_1 \Pi(p) + (n_1 + n_2)(1 - \Pi(p))$$

$$= n_1 + n_2 [H(c_2 - 1; N, M_p, n_1) - H(c_1; N, M_p, n_1)]. \qquad (18.4.3)$$

In Table 18.8, we present the OC function and the ASN function for the double-sampling plan $(20, 40, 2, 6, 6)$, for a lot of size $N = 150$.

We see from Table 18.8 that the double sampling plan illustrated here is not stringent. The probability of accepting a lot with 10% defectives is 0.89 and the probability of accepting a lot with 15% defectives is 0.39. If we consider the plan $(20, 40, 1, 3, 3)$ a more stringent procedure

Table 18.8 *The OC and ASN of a double sampling plan (20,40,2,6,6), N = 150*

p	OC(p)	ASN(p)	P	OC(p)	ASN(p)
0.000	1.0000	20.0	0.250	0.0714	41.4
0.025	1.0000	20.3	0.275	0.0477	38.9
0.050	0.9946	22.9	0.300	0.0268	35.2
0.075	0.9472	26.6	0.325	0.0145	31.6
0.100	0.7849	32.6	0.350	0.0075	28.4
0.125	0.5759	38.3	0.375	0.0044	26.4
0.150	0.3950	42.7	0.400	0.0021	24.3
0.175	0.2908	44.6	0.425	0.0009	22.7
0.200	0.1885	45.3	0.450	0.0004	21.6
0.225	0.1183	44.1	0.475	0.0002	21.1
			0.500	0.0001	20.6

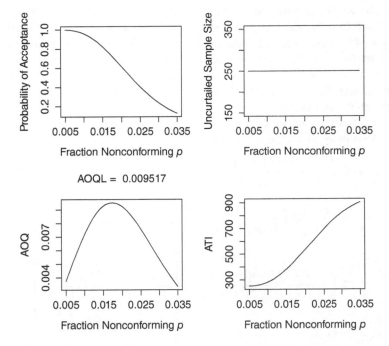

Figure 18.2 *AOQ curve for single sampling plan with N = 1000, n = 250 and c = 5*

is obtained, as shown in Table 18.9 and Figure 18.2. The probability of accepting a lot having 10% defectives has dropped to 0.39, and that of accepting a lot with 15% defectives has dropped to 0.15. Table 18.9 shows that the ASN is 23.1 when $p = 0.025$ (most of the time the sampling is terminated after the first stage), and the ASN is 29.1 when $p = 0.15$. The maximum ASN occurs around $p = 0.10$.

To determine an acceptable double sampling plan for attributes suppose, for example, that the population size is $N = 1000$. Define AQL = 0.01 and $LQL = 0.03$. If $n_1 = 200$, $n_2 = 400$,

Table 18.9 *The OC and ASN for the double sampling plan*
(20,40,1,3,3), N = 150

p	OC(p)	ASN(p)
0.000	1.0000	20.0
0.025	0.9851	23.1
0.050	0.7969	28.6
0.075	0.6018	31.2
0.100	0.3881	32.2
0.125	0.2422	31.2
0.150	0.1468	29.1
0.175	0.0987	27.3
0.200	0.0563	25.2
0.225	0.0310	23.5

$c_1 = 3$, $c_2 = 9$, $c_3 = 9$ then OC(0.01) = 0.9892, and OC(0.03) = 0.1191. Thus, $\alpha = 0.011$ and $\beta = 0.119$. The double-sampling plan with $n_1 = 120$, $n_2 = 240$, $c_1 = 0$, $c_2 = c_3 = 7$ yields $\alpha = 0.044$ and $\beta = 0.084$. For the last plan, the expected sample sizes are ASN(0.01) = 294 and ASN(0.03) = 341. These expected sample sizes are smaller than the required sample size of $n = 355$ in a single-stage plan. Moreover, with high probability, if $p \leq p_0$ or $p \geq p_t$, the sampling will terminate after the first stage with only $n_1 = 120$ observations. This is a factor of threefold decrease in the sample size, over the single sampling plan. There are other double-sampling plans which can do even better.

If the lot is very large, and we use large samples in stage one and stage two, the formulae for the OC and ASN function can be approximated by

$$OC(p) \cong \Phi\left(\frac{c_1 + 1/2 - n_1 p}{(n_1 pq(n1 - n_1/N))^{1/2}}\right)$$

$$+ \sum_{j=c_1+1}^{c_2-1}\left[\Phi\left(\frac{j + 1/2 - n_1 p}{(n_1 pq(1 - n_1/N))^{1/2}}\right) - \right.$$

$$\left.\Phi\left(\frac{j - 1/2 - n_1 p}{(n_1 pq(1 - n_1/N))^{1/2}}\right)\right] \cdot \Phi\left(\frac{c_3 - j + 1/2 - n_2 p}{\left(n_2 pq\left(1 - \dfrac{n_2}{Nn_1}\right)\right)^{1/2}}\right); \quad (18.4.4)$$

and

$$ASN(p) = n_1 + n_2\left[\Phi\left(\frac{c_2 - 1/2 - n_1 p}{\left[n_1 pq(1 - n_1/N)\right]^{1/2}}\right)\right.$$

$$\left. - \Phi\left(\frac{c_1 + 1/2 - n_1 p}{(n_1 pq(1 - n_1/N))^{1/2}}\right)\right]. \quad (18.4.5)$$

In Table 18.10, we present the OC and the ASN functions for double sampling from a population of size $N = 1000$, when the parameters of the plan are $(100, 200, 3, 6, 6)$. The exact values thus obtained are compared to the values obtained from the large sample approximation formulae. In the next section, the idea of double sampling is generalized in an attempt to reach acceptance decisions quicker and therefore at reduced costs.

Table 18.10   *The exact and approximate OC and ASN functions for the double sampling plan (100,200,3,6,6), N = 1000.*

	OC		ASN	
p	Exact	Approx.	Exact	Approx.
0.01	0.998	0.996	102.4	100.8
0.02	0.896	0.871	112.1	125.1
0.03	0.657	0.621	156.4	163.5
0.04	0.421	0.394	175.6	179.2
0.05	0.243	0.234	174.7	172.2
0.06	0.130	0.134	160.6	155.7
0.07	0.064	0.074	142.7	138.7
0.08	0.030	0.040	127.1	125.1
0.09	0.014	0.021	115.8	115.5

18.5   Sequential A/B testing

The A/B testing method is common practice for testing which treatment or action, A or B, is preferred by a customer. In randomized controlled trials, customers are seeing either A or B through random allocation. Eventually a decision is made and either A or B are chosen. In sequential A/B testing two alternative actions are presented before randomly picked customers. The goal is to maximize the expected reward. The reward is not certain in any case, and the probability of reward is unknown. This problem is similar to the classical "Two-Armed Bandit" problem. A gambler is standing in front of a slot machine and has the opportunity to try his luck in N trials. If he pulls the left hand the probability of reward is p_1, the probability of reward on the right hand is p_2. If the probabilities of success are known, the gambler will always pull the hand having the largest probability. What should be his strategy when the probabilities are unknown.

Much research was done on this problem in the seventies and the eighties. The reader is referred to the books of Gittins (1989) and Berry and Friestedt (1985). In the next section, we start with the One-Armed Bandit (OAB) for Bernoulli trials, and then discuss the Two-Armed Bandit (TAB) problem.

18.5.1   The one-armed Bernoulli bandits

The one armed Bernoulli bandit is a simpler case, where the probability of success in arm A is known, λ say. The probability p of success in arm B is unknown. It is clear that in this case we have to start with a sequence of trials on arm B (the learning phase), and move to arm A as soon as we are convinced that $p < \lambda$. The trials are Bernoulli trials namely, all trials on arm A or arm B are independent. The results of each trial are binary ($J = 1$ for success and $J = 0$ for failure). The probabilities of success in all trials at the same arm are equal.

Suppose that n trial have been performed on arm B. Let $X_n = \sum_{j=1}^{n} J_j$. The distribution of X_n is binomial, $B(n, p)$.

I. The Bayesian Strategy.
In a Bayesian framework, we start with a uniform prior distribution for p. The posterior distribution of p, given (n, X_n) is $Be(X_n + 1, n + 1 - X_n)$, .i.e.,

$$P\{p \leq \xi | n, X_n\} = [1/Be(X_n + 1, n + 1 - X_n)] \int_0^{\xi} u^{X_n}(1 - u)^{n - X_n} du, 0 < \xi < 1. \quad (18.5.1)$$

The function $Be(a, b) = \Gamma(a)\Gamma(b)/\Gamma(a + b)$ is called the complete beta function. The right-hand side of (18.5.1) is called the incomplete beta function ratio, and is denoted as $I_x(a, b)$. The predictive distribution of J_{n+1}, given X_n is $P\{J_{n+1} = 1|X_n\} = (X_n + 1)/(n + 2)$. The posterior probability that $\{p < \lambda\}$ is the λ-quantile of the above beta distribution, which is the incomplete beta function ratio $I_\lambda(X_n + 1, n + 1 - X_n)$.

A relatively simple Bayesian stopping rule for arm B is the first n greater than or equal to an initial sample size k, at which the posterior probability is greater than some specified value γ, i.e.,

$$M_\gamma = \min \{n \geq k : I_\lambda(X_n + 1, n + 1 - X_n) \geq \gamma\}. \tag{18.5.2}$$

Example 18.1.   In the present example, we simulate a Bayesian OAB for a Bernoulli process, in which $\lambda = 0.5$, and $N = 50$. If we choose to play all the N trials on arm A, our expected reward is $N\lambda = 25$ (reward units). On the other hand, in the OAB we start with an initial sample of size $k = 10$. We also wish to switch from arm B to arm A with confidence probability $\gamma = 0.95$. We illustrate the random process with two cases:

Case (i) $p = 0.6$. This probability is unknown. In this example, we use the R functions pbeta, qbeta, and rbeta for the beta distribution $I_x(a, b)$, for the p-th quantile $I^{-1}(a, b)$ and for a random choice from a distribution. The 0.1-quantile of the binomial B(10,0.6) is $qbinom(0.1, 10, 0.6) = 4$, and $pbeta(0.5, 5, 7) = 0.7256$. Thus, we expect that in 90% of the possible results we stay in the $k + 1$ trial at arm B. Notice that even if $X_{11} = 4$, we have $pbeta(0.5, 5, 8) = 0.8062$, and we stay with arm B. Thus with high probability we will stay with arm B for all the 50 trials, with an expected reward of $50 * 0.6 = 30$.

Case (ii) $p = 0.3$. We might get at random $X_{10} = rbinom(1, 10, 0.3) = 2$, In this case, $pbeta(0.5, 3, 9) = 0.9673$. We move immediately to arm A, with the expected reward of $2 + 40 * 0.5 = 22$.

In Table 18.11, we present the results of 1000 simulation runs. In each run, we recorded the mean value of the stopping time M_γ, its standard deviation, and the mean value of the expected reward and its standard deviation. The computations were done with program "simOAB."

We see in the table that, according to the present strategy, the cost of ignorance about p is pronounced only if $p < 0.5$. For example, if we knew that $p = 0.4$ we would have started with arm A, with expected reward of 25 rather than 21.263. The loss in expected reward when $p = 0.45$ is

Table 18.11   *Simulation estimates of the expected stopping time and the associated reward, for N = 50, λ = 0.5, k=10, γ = 0.95, Number of runs N_s = 1000*

p	$E\{M_\gamma\}$	std$\{M_\gamma\}$	$E\{Reward\}$	std$\{Reward\}$
0.4	18.619	11.545	21.263	0.845
0.45	38.908	15.912	23.055	2.701
0.5	44.889	12.109	25.167	3.323
0.55	47.530	8.991	27.357	3.568
0.6	49.013	6.013	29.685	3.518
0.7	49.887	2.026	35.199	3.363

not significant. The expected reward when $p \geq 0.5$ is not significantly different than what we could achieve if we knew p [ ].

II. The Bayesian Optimal Strategy for Bernoulli trials

As in the previous strategy, we assume that the prior distribution of p is uniform on $(0,1)$. The optimal strategy is determined by the **Dynamic Programming.** The principle of Dynamic Programming is to optimize the future possible trials, irrespective of what has been done in the past. We consider here as before a truncated game, in which only N trials are allowed. Thus, we start with the last trial, and proceed inductively backwards.

(i) Suppose we have already done $N - 1$ trials. The expected reward for the N-th trial is

$$R_N(X_{N-1}) = I\{\text{if arm } A \text{ is chosen} = \lambda\} + I\{\text{if arm } B \text{ is chosen}\}(X_{N-1} + 1)/(N + 1).$$
$$(18.5.3)$$

Thus, the maximal predictive expected reward for the last trial is

$$\rho^{(0)}(X_{N-1}) = \max \{\lambda, (X_{N-1} + 1)/(N + 1)\}$$
$$= \lambda I\{X_{N-1} < \lambda(N + 1) - 1\} + [(X_{N-1} + 1)/(N + 1)]I\{X_{N-1} \geq \lambda(N + 1) - 1\}.$$
$$(18.5.4)$$

(ii) After $N - 2$ trials, is we are at arm A we stay there, but if we are at arm B our predictive reward is

$$R_{N-1}(X_{N-2}) = 2\lambda I\{\text{if we are at arm } A\} + (X_{N-2} + 1)/N + E\{\rho^{(0)}(X_{N-2} + J_{N-1})|X_{N-2}\}$$
$$I\{\text{ if we chose arm } B\}. \qquad (18.5.5)$$

The maximal predictive expected reward is

$$\rho^{(1)}(X_{N-2}) = \max \{2\lambda, [(X_{N-2} + 1)/N]\rho^{(0)}(X_{N-2} + 1) + [(N - 1 - X_{N-1})/N]\rho^{(0)}(X_{N-2})$$
$$+ (X_{N-2} + 1)/N\}. \qquad (18.5.6)$$

(iii) By backward induction, we define for all $1 \leq n \leq N - 2$,

$$\rho^{(n)}(X_{N-n-1}) = \max \{(n + 1)\lambda, [(X_{N-n-1} + 1)/(N - n + 1)]\rho^{(n-1)}(X_{N-n-1} + 1)$$
$$+ [(N - n - X_{N-n-1})/(N - n + 1)]\rho^{(n-1)}(X_{N-n-1})$$
$$+ (X_{N-n-1} + 1)/(N - n + 1)\} \qquad (18.5.7)$$

and,

$$\rho^{(N-2)}(X_1) = \max \{(N - 1)\lambda, [(X_1 + 1)/3]\rho^{(N-2)}(X_1 + 1) + [(2 - X_1)/3]\rho^{(N-2)}(X_1)$$
$$+ (X_1 + 1)/3\}. \qquad (18.5.8)$$

Finally, since the procedure starts at arm B, and $P\{X_1 = 0\} = P\{X_1 = 1\} = 1/2$, we get that the maximal expected reward is $\rho^{(N-1)} = (\rho^{(N-2)}(0) + \rho^{(N-2)}(1))/2$.

Example 18.2. In the following example, we illustrate the values of $\rho^{(n)}(X_{N-n-1})$, for given (N, λ), where n designates the number of available future trials. These values are computed by program "DynOAB".
Since Table 18.12 is too big, we illustrate these for $N = 10$. We get the following values:

Table 18.12   Values of $\rho^{(n)}(X_{N-n-1})$, for $N = 10, n = 0, ..., 9$

n	0	1	2	3	4	5	6	7	8
0	0.5	0.5	0.5	0.5	0.5	0.545	0.636	0.727	0.818
1	1.0	1.0	1.0	1.0	1.023	1.200	1.400	1.600	1.800
2	1.5	1.5	1.5	1.5	1.677	2.000	2.333	2.667	3.000
3	2.0	2.0	2.0	2.088	2.504	3.000	3.500	4.000	
4	2.5	2.5	2.5	2.897	3.573	4.286	5.000		
5	3.0	3.0	3.199	4.014	5.000	6.000			
6	3.5	3.5	4.288	5.603	7.000				
7	4.0	4.394	6.024	8.000					
8	4.5	6.147	9.000						
9	5.382	10.000							

n	9	10
0	0.909	1.000
1	2.000	

Accordingly, the maximal predictive reward in following this strategy is $(5.382 + 10.000)/2 = 7.691$. For each n we stay in arm A as long as the maximal reward is $(n + 1)0.5$. Since Table 18.12 is quite big, the program "DynOAB2" yields just the maximal predictive reward. To compare this with the case of Table 18.11, Program "DynOAB2" yields for $N = 50$ and $\lambda = 0.5$ the maximal reward of 37.507.

18.5.2   Two-armed Bernoulli bandits

The two-armed bandit (TAB) is the case when the probabilities of reward at the two arms are unknown. We consider here the truncated game, in which N trials are allowed at both arms. The sufficient statistic after n trials are $(N^{(1)}(n), X_{N^{(1)}}, N^{(2)}(n), Y_{N^{(2)}})$ where $N^{(i)}(n), i = 1, 2$ are the number of trials on arm A and arm B, correspondingly, among the first n trials. $X_{N^{(1)}}$ and $Y_{N^{(2)}}$ are the number of successes at arms A and B. The optimal strategy is much more complicated than in the OAB case. If the two arms act independently, the situation is less complicated. The rewards from the two arms are cumulative. In principle one could find an optimal strategy by dynamic programming. It is however much more complicated than in the OAB case. Gittins and Jones (1974) proved that one can compute an index for each arm separately, and apply in the next trial the arm having the maximal index value. We provide here a strategy which yields very good results for large total number of trials N. This strategy starts with k trials on any arm, say arm A. If all the k trials yield k successes, i.e., $X_k = k$, we continue all the rest $N - k$ trials on arm A. If the first $m = [k/2]$ trials are all failures, i.e., $X_m = 0$, we immediately switch to arm B. We then compute the posterior estimator of the probability of success in arm A, p_A and use it in the dynamic programming for one arm known probability (program "DynOAB") starting at arm B.

Example 18.3. Consider the case of TAB with $N = 50$ and $k = 10$. If $X_5 = 0$, we switch to arm B. According to the beta(1,1) prior for p_A, the Bayesian estimator is $\hat{p}_A = 1/7 = 0.143$ with predictive expected reward of DynOAB2(45, 0.143) = 33.87. On the other hand, if $X_{10} = 10$, we stay at arm A till the end, with expected reward of $10 + 40 * (11/12) = 46.67$. In Table 18.13, we present all the other possibilities:

Table 18.13 *Expected reward when* $N = 50, k = 10$

X_{10}	$E\{\text{Re } ward\}$
1	31.150
2	32.413
3	33.819
4	35.380
5	37.093
6	38.970
7	41.033
8	43.249
9	45.667

There is much more literature on the armed-bandits problem. The interested reader should see Chapter 8 of Zacks (2009).

18.6 Acceptance sampling plans for variables

It is sometimes possible to determine whether an item is defective or not by performing a measurement on the item which provides a value of a continuous random variable X and comparing it to specification limits. For example, in Chapter 2, we discussed measuring the strength of yarn. In this case, a piece of yarn is deemed defective if its strength X is less than ξ where ξ is the required minimum strength, i.e., its lower specification limit. The proportion of defective yarn pieces in the population (or very large lot) is the probability that $X \leq \xi$.

Suppose now that X has a normal distribution with mean μ and variance σ^2. (If the distribution is not normal, we can often reduce it to a normal one by a proper transformation.) Accordingly, the proportion of defectives in the population is

$$p = \Phi\left(\frac{\xi - \mu}{\sigma}\right). \tag{18.6.1}$$

We have to decide whether $p \leq p_0$ (= AQL), or $p \geq p_t$ (= LQL), in order to accept or reject the lot.

Let x_p represent the pth quantile of a normal distribution with mean μ and standard deviation σ. Then,

$$x_p = \mu + z_p\sigma. \tag{18.6.2}$$

If it were the case that $x_{p_0} \geq \xi$, we should accept the lot since the proportion of defectives is less than p_0. Since we do not know μ and σ, we must make our decision on the basis of estimates from a sample of n measurements. We decide to reject the lot if

$$\overline{X} - kS < \xi$$

and **accept** the lot if

$$\overline{X} - kS \geq \xi.$$

Here, $\overline{X}$ and S are the usual sample mean and standard deviation, respectively. The factor k is chosen so that the producer's risk (the risk of rejecting a good lot) does not exceed α. Values of the factor k are given approximately by the formula

$$k \doteq t_{1-\alpha, p_0, n}$$

where

$$t_{1-a,b,n} = \frac{z_{1-b}}{1 - z_{1-a}^2/2n} + \frac{z_{1-a}\left(1 + \frac{z_b^2}{2} - \frac{z_{1-a}^2}{2n}\right)^{1/2}}{\sqrt{n}(1 - z_{1-a}^2/2n)}. \qquad (18.6.3)$$

The OC function of such a test is given approximately (for large samples), by

$$\mathrm{OC}(p) \approx 1 - \Phi\left(\frac{(z_p + k)/\sqrt{n}}{(1 + k^2/2)^{1/2}}\right), \qquad (18.6.4)$$

where $k = t_{1-\alpha, p, n}$. We can thus determine n and k so that

$$\mathrm{OC}(p_0) = 1 - \alpha$$

and

$$\mathrm{OC}(p_t) = \beta.$$

These two conditions yield the equations

$$(z_{p_t} + k)\sqrt{n} = z_{1-\beta}(1 + k^2/2)^{1/2}$$

and $\qquad\qquad\qquad\qquad\qquad\qquad\qquad\qquad\qquad\qquad\qquad (18.6.5)$

$$(z_{p_0} + k)\sqrt{n} = z_\alpha(1 + k^2/2)^{1/2}.$$

The solution for n and k yields:

$$n = \frac{(z_{1-\alpha} + z_{1-\beta})^2(1 + k^2/2)}{(z_{p_t} - z_{p_0})^2}, \qquad (18.6.6)$$

and

$$k = (z_{p_t}z_\alpha + z_{p_0}z_\beta)/(z_{1-\alpha} + z_{1-\beta}). \qquad (18.6.7)$$

In other words, if the sample size n is given by the above formula, we can replace $t_{1-\alpha, p, n}$ by the simpler term k, and accept the lot if

$$\overline{X} - kS \geq \xi.$$

The statistic $\overline{X} - kS$ is called a **lower tolerance limit**.

Example 18.4. Consider the example of testing the compressive strength of concrete cubes presented in Chapter 2. It is required that the compressive strength be larger than 240 [kg/cm^2]. We found that $Y = \ln X$ had an approximately normal distribution. Suppose that it is required to decide whether to accept or reject this lot with the following specifications: $p_0 = 0.01$, $p_t = 0.05$, and $\alpha = \beta = 0.05$. According to the normal distribution

$$z_{p_0} = -2.326, \quad z_{p_t} = -1.645$$

and

$$z_{1-\alpha} = z_{1-\beta} = 1.645.$$

Thus, according to the above formulas, we find $k = 1.9855$ and $n = 70$. Hence with a sample size of 70, we can accept the lot if $\overline{Y} - 1.9855S \geq \xi$ where $\xi = \ln(240) = 5.48$.   □

The sample size required in this single-stage sampling plan for variables is substantially smaller than the one we determined for the single-stage sampling plan for attributes (which was $n = 176$). However, the sampling plan for attributes is free of any assumption about the distribution of X, while in the above example we had to assume that $Y = \ln X$ is normally distributed. Thus, there is a certain trade-off between the two approaches. In particular, if our assumptions concerning the distribution of X are erroneous, we may not have the desired producer's and consumer's risks.

The above procedure of acceptance sampling for variables can be generalized to **upper and lower tolerance limits**, double sampling, and sequential sampling. The interested reader can find more information on the subject in Duncan (1986, Chapters 12–15). Fuchs and Kenett (1988) applied tolerance limits for the appraisal of ceramic substrates in the multivariate case.

18.7   Rectifying inspection of lots

Rectifying inspection plans are those plans which call for a complete inspection of a **rejected lot** for the purpose of replacing the defectives by nondefective items. (Lots which are accepted are not subjected to rectification.) We shall assume that the tests are nondestructive, that all the defective items in the sample are replaced by good ones, and that the sample is replaced in the lot.

If a lot contains N items and has a proportion p of defectives before the inspection, the proportion of defectives in the lot after inspection is

$$p' = \begin{cases} 0, & \text{if lot is rejected,} \\[2mm] p(N - X)/N, & \text{if lot is accepted,} \end{cases} \tag{18.7.1}$$

where X is the number of defectives in the sample. If the probability of accepting a lot by a given sampling plan, is OC(p), then, the expected proportion of outgoing defectives is, when sampling is single stage by attribute,

$$E\{p'\} = p\,\text{OC}(p)\left(1 - \frac{n}{N}R_s^*\right), \tag{18.7.2}$$

where

$$R_s^* = \frac{H(c - 1; N - 1, [Np] - 1, n - 1)}{H(c; N, [Np], n)}.$$ (18.7.3)

If n/N is small then

$$E\{p'\} \cong p\,\mathrm{OC}(p).$$ (18.7.4)

The expected value of p' is called the **average outgoing quality**, and is denoted by AOQ.

The formula for R_s^* depends on the method of sampling inspection. If the inspection is by double sampling the formula is considerably more complicated.

In Table 18.14, we present the AOQ values corresponding to a rectifying plan, when $N = 1000$, $n = 250$, and $c = 5$.

The average outgoing quality limit (AOQL) of a rectifying plan is defined as the maximal value of AOQ. Thus, the AOQL corresponding to the plan of Table 18.14 is approximately 0.01. The AOQ given is presented graphically in Figure 18.2.

We also characterize a rectifying plan by the **average total inspection** (ATI) associated with a given value of p. If a lot is accepted, only n items (the sample size) have been inspected, while if it is rejected, the number of items inspected is N. Thus,

$$\mathrm{ATI}(p) = n\,\mathrm{OC}(p) + N(1 - \mathrm{OC}(p))$$

$$= n + (N - n)(1 - \mathrm{OC}(p)).$$ (18.7.5)

This function is increasing from n (when $p = 0$) to N (when $p = 1$).

In our example, the lot contains $N = 1000$ items and the sample size is $n = 250$. The graph of the ATI function is presented in Figure 18.3.

Dodge and Romig (1959) published tables for the design of single and double sampling plans for attributes, for which the AOQL is specified and the ATI is minimized at a specified value of p. In Table 18.15, we provide a few values of n and c for such a single sampling plan, for which the AOQL = 0.01.

According to Table 18.15, for a lot of size 2000, to guarantee an AOQL of 1% and minimal ATI at $p = 0.01$ one needs a sample of size $n = 180$, with $c = 3$. For another method of determining n and c, see Duncan (1986, Ch. 16).

Rectifying sampling plans with less than 100% inspection of rejected lots have been developed and are available in the literature.

Table 18.14 *AOQ values for rectifying plan N = 1000, n = 250, c = 5*

p	$\mathrm{OC}(p)$	R_s^*	AOQ
0.005	1.000	1.0000	0.004
0.010	0.981	0.9710	0.007
0.015	0.853	0.8730	0.010
0.020	0.618	0.7568	0.010
0.025	0.376	0.6546	0.008
0.030	0.199	0.5715	0.005
0.035	0.094	0.5053	0.003

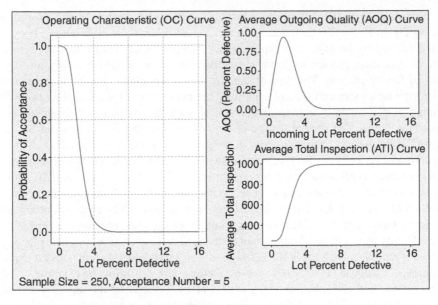

Figure 18.3 *ATI curve for single sampling plan with N = 1000, n = 250 and c = 5 (MINITAB).*

Table 18.15 *Selected values of (n, c) for a single sample plan with AOQL = 0.01 and ATI minimum at p*

p		0.004–0.006		0.006–0.008		0.008–0.01	
N		n	c	n	c	n	c
101–200		32	0	32	0	32	0
201	– 300	33	0	33	0	65	1
501	– 600	75	1	75	1	75	1
1001	– 2000	130	2	130	2	180	3

18.8 National and international standards

During World War II, the US Army developed standards for sampling acceptance schemes by attributes. Army ordinance tables were prepared in 1942 and the Navy issued its own tables in 1945. Joint Army and Navy standards were issued in 1949. These standards were superseded in 1950 by the common standards, named MIL-STD-105A. The MIL-STD-105D was issued by the US Government in 1963 and slightly revised as MIL-STD-105E in 1989. These standards however, are gradually being phased out by the Department of Defense. The American National Standards Institute, ANSI, adopted the military standards with some minor modifications, as ANSI Z1.4 standards. These were adopted in 1974 by the International Organization for Standardization as ISO 2859. In 1981, ANSI Z1.4 were adopted by the American Society for Quality Control with some additions, and the standards issued were named ANSI/ASQC Z1.4.

The military standards were designed to inspect incoming lots from a variety of suppliers. The requirement from all suppliers is to satisfy specified quality levels for the products. These quality levels are indexed by the **AQL**. It is expected that a supplier sends in continuously series of lots (shipments). All these lots are subjected to quality inspection. At the beginning, an AQL value is specified for the product. The type of sampling plan is decided (single, double, sequential, etc.) For a given lot size and type of sampling, the parameters of the sampling procedure are determined. For example, if the sampling is single stage by attribute, the parameter (n, c) are read from the tables. The special feature of the MIL-STD-105E is that lots can be subjected to **normal, tightened**, or **reduced** inspection. Inspection starts at a normal level. If two out of five consecutive lots have been rejected a switch to tightened inspection level takes place. Normal inspection is reinstituted if five consecutive lots have been accepted. If 10 consecutive lots remain under tightened inspection an action may take place to discontinue the contract with the supplier. On the other hand, if the last 10 lots have all been accepted at a normal inspection level, and the total number of defective units found in the samples from these 10 lots is less than a specified value then a switch from normal to reduced inspection level can take place.

We do not reproduce here the MIL-STD-105 tables. The reader can find detailed explanation and examples in Duncan (1986, Ch. 10). We conclude with the following example.

Suppose that for a given product AQL = 0.01 (1%). The size of the lots is $N = 1000$. The military standard specifies that a single stage sampling for attributes, under normal inspection, has the parameters $n = 80$ and $c = 2$. Applying MINITAB yields the following OC values for this plan.

p	0.01	0.02	0.03	0.04	0.05
OC(p)	0.961	0.789	0.564	0.365	0.219

Thus, if the proportion of nonconforming, p, of the supplier is less than AQL, the probability of accepting a lot is larger than 0.961. A supplier which continues to ship lots with $p = 0.01$, has a probability of $(0.961)^{10} = 0.672$ that all the 10 lots will be accepted, and the inspection level will be switched to a reduced one. Under the reduced level, the sample size from the lot is reduced to $n = 32$. The corresponding acceptance number is $c = 1$. Thus, despite the fact that the level of quality of the supplier remains good, there is a probability of 0.33 that there will be no switch to reduced inspection level after the 10th lot. On the other hand, the probability that there will be no switch to tightened level of inspection before the sixth lot is inspected is 0.9859. This is the probability that after each inspection the next lot will continue to be inspected under normal level. If there is no deterioration in the quality level of the supplier, and $p = 0.03$, the probability that the inspection level will be switched to "tightened" after five inspections is 0.722.

18.9   Skip-lot sampling plans for attributes

We have seen in the previous section that according to the MIL-STD-105E, if a supplier keeps shipping high quality lots, then after a while his lots are subjected to inspection under reduced level. All lots are inspected under a reduced level inspection scheme, as long as their quality level remains high. The **Skip Lot Sampling Plans** (SLSP), which was proposed by Liebesman and Saperstein (1983), introduces a new element of savings if lots continue to have very low proportions of nonconforming items. As we will see below, instead of just reduced level of inspection of high quality lots, the SLSP plans do not necessarily inspect such lots. If the lots coming in from a given supplier qualify for skipping, then they are inspected only with probability 0.5. This probability is later reduced to 0.33 and to 0.2, if the inspected lots continue to be almost free of nonconforming items. Thus, suppliers which continue to manufacture their product, with proportion defectives p, considerably smaller than the specified AQL stand a good chance to have only a small fraction of their lots inspected. The SLSP which will be specified below was adopted as the ISO2859/3 standard in 1986.

18.9.1 The ISO 2859 skip-lot sampling procedures

A SLSP has to address three main issues:

(1) What are the conditions for beginning or reinstating the Skip-Lot (SL) state?
(2) What is the fraction of lots to be skipped?
(3) Under what conditions should one stop skipping lots, on a temporary or permanent basis?

The fraction of lots to be skipped is the probability that a given lot will not be inspected. If this probability for example is 0.8, we generate a random number, U, with uniform distribution on $(0, 1)$. If $U < 0.8$ inspection is skipped; otherwise the lot is inspected.

We define three states:

State 1. Every lot is inspected.
State 2. Some lots are skipped and not inspected.
State 3. All lots are inspected, pending a decision of disqualification (back to State 1) or resumption of SL (back to State 2).

Lot by lot inspection is performed during State 3, but the requirements to requalify for skip lot inspection are less stringent than the initial qualification requirements.

Switching rules apply to four transitions between states: Qualification (State 1 to State 2), Interruption (State 2 to State 3), Resumption (State 3 to State 2), Disqualification (State 3 to State 1).

The switching rules for the SLSP procedure are listed below.

Skip lot switching rules

We specify here the rules appropriate for single sampling by attributes. Other rules are available for other sampling schemes.

A. **Qualification** (State 1 → State 2).
   1. Ten consecutive lots are accepted.
   2. Total number of defective items in the samples from the 10 lots is smaller than critical level given in Table 18.16.
   3. Number of defective items in each one of the last two lots is smaller than the values specified in Table 18.17.
   4. Supplier has a stable manufacturing organization, continuous production, and other traits which qualify him to be high quality stable manufacturer.
B. **Interruption** (State 2 → State 3)
   1. An inspected lot has in the sample more defectives than specified in Table 18.17.

C. **Resumption** (State 3 → State 2)
  1. Four consecutive lots are accepted.
  2. The last two lots satisfy the requirements of Table 18.17.
D. **Disqualifications** (State 3 → State 1)
  1. Two lots are rejected within 10 consecutively inspected lots; or
  2. Violation of the supplier qualification criteria (item A4 above).

We have seen in the previous section that, under normal inspection, MIL-STD-105E specifies that, for AQL = 0.01 and lots of size $N = 1000$ random samples of size $n = 80$ should be drawn. The critical level was $c = 2$. If 10 lots have been accepted consecutively, the total number of observed defectives is $S_{10} \le 20$. The total sample size is 800 and according to Table 18.16, S_{10} should not exceed 3 to qualify for a switch to State 2. Moreover, according to Table 18.17, the last two samples should each have less than 1 defective item. Thus, the probability to qualify for State 2, on the basis of the last 10 samples, when $p = $ AQL = 0.01 is

$$QP = b^2(0; 80, 0.01)B(3; 640, 0.01)$$

$$= (0.4475)^2 \times 0.1177 = 0.0236.$$

Thus, if the fraction defectives level is exactly at the AQL value, the probability for qualification is only 0.02. On the other hand, if the supplier maintains the production at fraction defective of

Table 18.16   *Minimum cumulative sample size in 10 lots for skip-lot qualifications*

Cumulative No. of defectives	AQL(%)						
	0.65	1.0	1.5	2.5	4.0	6.5	10.0
0	400	260	174	104	65	40	26
1	654	425	284	170	107	65	43
2	883	574	383	230	144	88	57
3	1098	714	476	286	179	110	71
4	1306	849	566	340	212	131	85
5	1508	980	653	392	245	151	98
6	1706	1109	739	444	277	171	111
7	1902	1236	824	494	309	190	124
8	2094	1361	907	544	340	209	136
9	2285	1485	990	594	371	229	149
10	2474	1608	1072	643	402	247	161
11	2660	1729	1153	692	432	266	173
12	2846	1850	1233	740	463	285	185
13	3031	1970	1313	788	493	303	197
14	3214	2089	1393	836	522	321	209
15	3397	2208	1472	883	552	340	221
16	3578	2326	1550	930	582	358	233
17	3758	2443	1629	977	611	376	244
18	3938	2560	1707	1024	640	394	256
19	4117	2676	1784	1070	669	412	268
20	4297	2793	1862	1117	698	430	279

Table 18.17   *Individual lot acceptance numbers for skip-lot qualification*

Sample size	AQL(%)						
	0.65	1.0	1.5	2.5	4.0	6.5	10.0
2	–	–	–	–	–	0	0
3	–	–	–	–	0	0	0
5	–	–	–	0	0	0	1
8	–	–	0	0	0	1	1
13	–	0	0	0	1	1	2
20	0	0	0	1	1	2	3
32	0	0	1	1	2	3	5
50	0	1	1	2	3	5	7
80	1	1	2	3	5	7	11
125	1	2	3	4	7	11	16
200	2	3	4	7	11	17	25
315	3	4	7	11	16	25	38
500	5	7	10	16	25	39	58
800	7	11	16	25	38	60	91
1250	11	16	23	38	58	92	138
2000	17	25	36	58	91	144	217

$p = 0.001$, then the qualification probability is

$$QP = b^2(0; 80, 0.001)B(3; 640, 0.001)$$

$$= (0.9231)^2 \times 0.9958 = 0.849.$$

Thus, a supplier who maintains a level of $p = 0.001$, when the AQL $= 0.01$, will probably be qualified after the first 10 inspections and will switch to State 2 of skipping lots. Eventually only 20% of his lots will be inspected, under this SLSP standard, with high savings to both producer and consumer. This illustrates the importance of maintaining high quality production processes. In Chapters 10 and 11, we discussed how to statistically control the production processes, to maintain stable processes of high quality. Generally, for the SLSP to be effective, the fraction defectives level of the supplier should be smaller than half of the AQL. For p level close to the AQL, the SLSP and the MIL-STD 105E are very similar in performance characteristics.

18.10   The Deming inspection criterion

Deming (1982) has derived a formula to express the expected cost to the firm caused by sampling of lots of incoming material. Let us define

N = number of items in a lot
k_1 = cost of inspecting one item at the beginning of the process
q = probability of a conforming item
p = probability of a nonconforming item
Q = OC(p) = probability of accepting a lot
k_2 = cost to the firm when one nonconforming item is moved downstream to a customer or to the next stage of the production process

p'' = the probability of nonconforming items being in an accepted lot

n = the sample size inspected from a lot of size N.

Thus, the total expected cost per lot is

$$EC = (Nk_1/q)[1 + Qq\{(k_2/k_1)p'' - 1\}\{1 - n/N\}] \tag{18.10.1}$$

If $(k_2/k_1)p'' > 1$ then any sampling plan increases the cost to the firm and $n = N$ (100% inspection) becomes the least costly alternative.

If $(k_2/k_1)p'' < 1$ then the value $n = 0$ yields the minimum value of EC so that no inspection is the alternative of choice.

Now p'' can be only somewhat smaller than p. For example if $N = 50$, $n = 10$, $c = 0$, and $p = 0.04$ then $p'' = 0.0345$. Substituting p for p'' gives us the following rule

If $(k_2/k_1)p > 1$ inspect every item in the lot

If $(k_2/k_1)p < 1$ accept the lot without inspection.

The Deming assumption is that the process is under control and that p is known. Sampling plans such as MIL-STD-105D do not make such assumptions and, in fact, are designed for catching shifts in process levels.

To keep the process under control Deming suggests the use of control charts and Statistical Process Control (SPC) procedures which are discussed in Chapters 10 and 11.

The assumption that a process is under control means that the firm has absorbed the cost of SPC as internal overhead or as a piece-cost. Deming's assertion then is that assuming up front the cost of SPC implementation is cheaper, in the long run, than doing business in a regime where a process may go out of control undetected until its output undergoes acceptance sampling.

The next chapter will introduce the reader to basic tools and principles of Statistical Process Control.

18.11   Published tables for acceptance sampling

In this section, we list some information on published tables and schemes for sampling inspection by attribute and by variables. The material given here follows Chapters 24 and 25 of Juran (1979). We shall not provide explanation here concerning the usage of these tables. The interested practitioner can use the instructions attached to the tables and/or read more about the tables in Juran (1979), Ch. 24-25, or in Duncan (1986).

I. **Sampling by Attributes**
   1. **MIL-STD-105E**
      Type of sampling: Single, double and multiple.
      Type of application: General.
      Key features: Maintains average quality at a specified level. Aims to minimize rejection of good lots. Provides single sampling plans for specified AQL and producer's risk.
      Reference: MIL-STD-105E, Sampling Procedures and Tables for Inspection by Attributes, Government Printing Office, Washington, DC
   2. **Dodge-Romig**
      Type of sampling: Single and double.
      Type of application: Where 100% rectifying of lots is applicable.

Key features: One type of plan uses a consumer's risk of $\beta = 0.10$. Another type limits the AOQL. Protection is provided with minimum inspection per lot.

Reference: Dodge and Romig (1959)

3. **H107**

Type of sampling: Continuous single stage.

Type of application: When production is continuous and inspection is nondestructive.

Key features: Plans are indexed by AQL, which generally start with 100% inspection until some consecutive number of units free of defects are found. Then inspection continues on a sampling basis until a specified number of defectives are found.

Reference: H-107, Single-Level Continuous Sampling Procedures and Tables For Inspection by Attribute, Government Printing Office, Washington, DC

II. **Sampling by Variables**

1. **MIL-STD-414**

Assumed distribution: Normal.

Criteria specified: AQL.

Features: Lot evaluation by AQL. Includes tightened and reduced inspection.

Reference: Sampling Procedures and Tables for Inspection by Variables for Percent Defectives, MIL-STD-414, Government Printing Office, Washington, DC

2. **H-108**

Assumed distribution: Exponential.

Criteria specified: Mean Life (MTBF).

Features: Life testing for reliability specifications.

Reference: H-108, Sampling Procedures and Tables for Life and Reliability Testing (Based on Exponential Distribution), U.S. Department of Defense, Quality Control and Reliability Handbook, Government Printing Office, Washington, DC

18.12   Chapter highlights

Traditional supervision consists of keeping close control of operations and progress. The focus of attention being the product or process outputs. A direct implication of this approach is to guarantee product quality through inspection screening. The chapter discusses sampling techniques and measures of inspection effectiveness. Performance characteristics of sampling plans are discussed and guidelines for choosing economic sampling plans are presented. The basic theory of single-stage acceptance sampling plans for attributes is first presented including the concepts of Acceptable Quality Level and Limiting Quality Level. Formulas for determining sample size, acceptance levels, and Operating Characteristic functions are provided. Moving on from single-stage sampling the chapter covers double-sampling and sequential sampling using Wald's sequential probability ratio test. One section deals with acceptance sampling for variable data. Other topics covered include computations of Average Sample Numbers and Average Total Inspection for rectifying inspection plans. Modern Skip-Lot sampling procedures are introduced and compared to the standard application of sampling plans where every lot is inspected. The Deming "all or nothing" inspection criterion are presented and the connection between sampling inspection and statistical process control is made. Throughout the chapter we refer MINITAB and R applications which are used to perform various calculations and generate appropriate tables and graphs. The main concepts and definitions introduced in this chapter include:

- Lot
- Acceptable Quality Level
- Limiting Quality Level
- Producer's Risk
- Consumer's Risk
- Single Stage Sampling
- Acceptance Number
- Operating Characteristic
- Double-Sampling Plan
- ASN-Function
- Sequential Sampling
- Sequential Probability Ratio Test (SPRT)
- Rectifying Inspection
- Average Outgoing Quality (AOQ)
- Average Total Inspection (ATI)
- Tightened, Normal or Reduced Inspection Levels
- Skip Lot Sampling Plans

18.13   Exercises

18.1 Determine single sampling plans for attributes, when the lot is $N = 2500$, $\alpha = \beta = 0.01$, and

(i) AQL $= 0.005$, LQL $= 0.01$
(ii) AQL $= 0.01$, LQL $= 0.03$
(iii) AQL $= 0.01$, LQL $= 0.05$

18.2 Investigate how the lot size, N, influences the single sampling plans for attributes, when $\alpha = \beta = 0.05$, AQL $= 0.01$, LQL $= 0.03$, by computing the plans for $N = 100$, $N = 500$, $N = 1000$, $N = 2000$.

18.3 Compute the OC(p) function for the sampling plan computed in Exercise 18.1 (iii). What is the probability of accepting a lot having 2.5% of nonconforming items?

18.4 Compute the large sample approximation to a single sample plan for attributes (n^*, c^*), with $\alpha = \beta = 0.05$ and AQL $= 0.025$, LQL $= 0.06$. Compare these to the exact results. The lot size is $N = 2000$.

18.5 Repeat the previous Exercise with $N = 3000$, $\alpha = \beta = 0.10$, AQL $= 0.01$ and LQL $= 0.06$.

18.6 Obtain the OC and ASN functions of the double sampling plan, with $n_1 = 200$, $n_2 = 2n_1$ and $c_1 = 5$, $c_2 = c_3 = 15$, when $N = 2000$.

(i) What are the attained α and β when AQL $= 0.015$ and LQL $= 0.05$?
(ii) What is the ASN when $p = $ AQL?
(iii) What is a single sampling plan having the same α and β? How many observations we expect to save if $p = $ AQL? Notice that if $p = $ LQL the present double sampling plan is less efficient than the corresponding single sampling plan.

18.7 Compute the OC and ASN values for a double sampling plan with $n_1 = 150$, $n_2 = 200$, $c_1 = 5$, $c_2 = c_3 = 10$, when $N = 2000$. Notice how high β is when LQL $= 0.05$. The present plan is reasonable if LQL $= 0.06$. Compare this plan to a single sampling one for $\alpha = 0.02$ and $\beta = 0.10$, AQL $= 0.02$, LQL $= 0.06$.

18.8 Use the R-program "simOAB" to simulate the expected rewards, for $p = 0.4(0.05)0.8$, when $N = 75$, $\lambda = 0.6$, $k = 15$, $\gamma = 0.95$, $Ns = 1000$.

18.9 Use the R-program "DynOAB2" to predict the expected reward under the optimal strategy, when $N = 75$, $\lambda = 0.6$.

18.10 Consider the TAB with $N = 40$ and $K = 10$. Make a table of all the possible predicted rewards.

18.11 Determine n and k for a continuous variable size sampling plan, when $(p_0) = \text{AQL} = 0.01$ and $(p_t) = \text{LQL} = 0.05$, $\alpha = \beta = 0.05$.

18.12 Consider data file **ALMPIN.csv**. An aluminum pin is considered as defective if its cap diameter is smaller than 14.9 [mm]. For the parameters $p_0 = 0.01$, $\alpha = 0.05$, compute k and decide whether to accept or reject the lot, on the basis of the sample of $n = 70$ pins. What is the probability of accepting a lot with proportion defectives of $p = 0.03$?

18.13 Determine the sample size and k for a single sampling plan by a normal variable, with the parameters $\text{AQL} = 0.02$, $\text{LQL} = 0.04$, $\alpha = \beta = 0.10$.

18.14 A single sampling plan for attributes, from a lot of size $N = 500$, is given by $n = 139$ and $c = 3$. Each lot which is not accepted is rectified. Compute the AOQ, when $p = 0.01$, $p = 0.02$, $p = 0.03$ and $p = 0.05$. What are the corresponding ATI values?

18.15 A single sampling plan, under normal inspection has probability $\alpha = 0.05$ of rejection, when $p = \text{AQL}$. What is the probability, when $p = \text{AQL}$ in five consecutive lots, that there will be a switch to tightened inspection? What is the probability of switching to a tightened inspection if p increases so that $\text{OC}(p) = 0.7$?

18.16 Compute the probability for qualifying for State 2, in a SKLP, when $n = 100$, $c = 1$. What is the upper bound on S_{10}, in order to qualify for State 2, when $\text{AQL} = 0.01$? Compute the probability QP for State 2 qualification.

Appendix

These are three R functions referred to in Section 18.5

(1) The first function "simOAB" is programed to simulate the expected number of trials on Arm B before switching to the known arm A, and the expected reward.

The input variables are:

N = Number of trials; p = the probability of reward on arm B (unknown); al = The known probability of reward on arm A; K = the initial sample size on arm B; gam = Bayesian confidence level, Ns=number of runs in the simulation.

Output variables: Estimates of Expected stopping time M; Standard deviation of M; Expected total reward; Standard deviation of the rewards.

```
   simOAB<-
function(N,p,al,k,gam,Ns){
res<-jay(Ns,2)
for(i in 1:Ns){
n<-k
X<-rbinom(1,k,p)
repeat{
cr<-pbeta(al,X+1,n+1-X)
if((cr>gam)||(n==N)) break
n<-n+1
X<-X+rbinom(1,1,p)
}
res[i,1]<-n
res[i,2]<-X+(N-n)*al
}
AV<-mean(res[,1])
sda<-std(res[,1])
Arew<-mean(res[,2])
sdr<-std(res[,2])
out<-list(AV,sda,Arew,sdr)
out
}
```

(2) The R function "DynOAB" for the dynamic programming of the optimal OAB.

Input variables:

N = Number of trials; al = Known probability of success in arm A.

Output: Table of maximal predicted rewards

```
   DynOAB<-
function(N,al){
```

```
Rew<-0*jay(N,N+1)
 for(i in 1:(N+1)){
 X<-i-1
if(X<(al*(N+1)-1)){ Rew[1,i]<-al}
 else {Rew[1,i]<-(X+1)/(N+1)}
}
for(j in 2:N){
n<-N+1-j
for(i in 1:(n+1)){
X<-i-1
cr1<-(X+1)*Rew[j-1,X+2]/(N-j+2)
cr2<-(N-j+1-X)*Rew[j-1,X+1]/(N-j+2)
cr3<-(X+1)/(N-j+2)
cr<-cr1+cr2+cr3
if(cr<(j*al)){Rew[j,i]<-j*al}
else {Rew[j,i]<-cr}
}
}
Rew
}
```

(3) R function "DynOAB2" for an output of the optimal predicted reward

```
DynOAB2<-
function(N,al){
Rew<-0*jay(N,N+1)
 for(i in 1:(N+1)){
 X<-i-1
if(X<(al*(N+1)-1)){ Rew[1,i]<-al}
 else {Rew[1,i]<-(X+1)/(N+1)}
}
for(j in 2:N){
n<-N+1-j
for(i in 1:(n+1)){
X<-i-1
cr1<-(X+1)*Rew[j-1,X+2]/(N-j+2)
cr2<-(N-j+1-X)*Rew[j-1,X+1]/(N-j+2)
cr3<-(X+1)/(N-j+2)
cr<-cr1+cr2+cr3
if(cr<(j*al)){Rew[j,i]<-j*al}
else {Rew[j,i]<-cr}
}
}
out<-(Rew[N,1]+Rew[N,2])/2
out
}
```

List of R Packages

AcceptanceSampling
Kiermeier A (2008). "Visualizing and Assessing Acceptance Sampling Plans: The R Package AcceptanceSampling." *Journal of Statistical Software*, *26*(6), http://www.jstatsoft.org/v26/i06/.

boot
Canty A and Ripley BD (2020). *boot: Bootstrap R (S-Plus) Functions*. R package version 1.3-7.
Davison AC and Hinkley DV (1997). *Bootstrap Methods and Their Applications*. Cambridge University Press, Cambridge. ISBN 0-521-57391-2, http://statwww.epfl.ch/davison/BMA/.

car
Fox J and Weisberg S (2019). *An R Companion to Applied Regression*, Third edition. Sage, Thousand Oaks CA. http://socserv.socsci.mcmaster.ca/jfox/Books/Companion.

cluster
Maechler M, Rousseeuw P, Struyf A, Hubert M and Hornik K (2019). *cluster: Cluster Analysis Basics and Extensions*. R package version 2.1.0, http://CRAN.R-project.org/package=cluster.

DiceDesign and DiceEval
Dupuy D, Helbert C and Franco J (2015). "DiceDesign and DiceEval: Two R Packages for Design and Analysis of Computer Experiments." *Journal of Statistical Software*, *65*(11), http://www.jstatsoft.org/v65/i11/.

DiceKriging
Roustant O, Ginsbourger D and Deville Y (2012). "DiceKriging, DiceOptim: Two R Packages for the Analysis of Computer Experiments by Kriging-Based Metamodeling and Optimization." *Journal of Statistical Software*, *51*(1), http://www.jstatsoft.org/v51/i01/.

DiceView
Richet Y, Deville Y and Chevalier C (2018). *DiceView: Plot Methods for Computer Experiments Design and Surrogate*. R package version 1.3-2, http://CRAN.R-project.org/package=DiceView.

Modern Industrial Statistics: With Applications in R, MINITAB and JMP, Third Edition.
Ron S. Kenett and Shelemyahu Zacks.
© 2021 John Wiley & Sons, Ltd. Published 2021 by John Wiley & Sons, Ltd.

Dodge
Godfrey AJR and Govindaraju K (2018). *Dodge: Functions for Acceptance Sampling Ideas Originated by H.F. Dodge*. R package version 0.9-2, http://CRAN.R-project.org/package=Dodge.

DoE.base
Groemping U (2018). "R Package DoE.base for Factorial Experiments." *Journal of Statistical Software*, *85*(5), http://www.jstatsoft.org/v85/i05/.

e1071
Meyer D, Dimitriadou E, Hornik K, Weingessel A and Leisch F (2019). *e1071: Misc Functions of the Department of Statistics, Probability Theory Group (Formerly: E1071), TU Wien*. R package version 1.7-3, http://CRAN.R-project.org/package=e1071.

fdapace
Carroll C et al. (2020). *fdapace: Functional Data Analysis and Empirical Dynamics*. R package version 0.5.5, http://CRAN.R-project.org/package=fdapace.

FrF2
Groemping U (2014). "R Package FrF2 for Creating and Analyzing Fractional Factorial 2-Level Designs." *Journal of Statistical Software*, *56*(1), http://www.jstatsoft.org/v56/i01/.

ggplot2
Wickham H (2016). *ggplot2: Elegant Graphics for Data Analysis*. Springer, New York. ISBN 978-3-319-24277-4, https://ggplot2.tidyverse.org.

grid
R Core Team (2020). *R: A Language and Environment for Statistical Computing*. R Foundation for Statistical Computing, Vienna, Austria. http://www.R-project.org/.

lattice
Sarkar D (2008). *Lattice: Multivariate Data Visualization with R*. Springer, New York. ISBN 978-0-387-75968-5, http://lmdvr.r-forge.r-project.org.

LearnBayes
Albert J (2018). *LearnBayes: Functions for Learning Bayesian Inference*. R package version 2.15.1, http://CRAN.R-project.org/package=LearnBayes.

lhs
Carnell R (2020). *lhs: Latin Hypercube Samples*. R package version 1.0.2, http://CRAN.R-project.org/package=lhs.

mistat
Amberti D (2020). *mistat: Data Sets, Functions and Examples from the Book: "Modern Industrial Statistics" by Kenett, Zacks and Amberti*. R package version 2.0, http://CRAN.R-project.org/package=mistat.

mvtnorm
Genz A, Bretz F, Miwa T, Mi X, Leisch F, Scheipl F and Hothorn T (2020). *mvtnorm: Multivariate Normal and t Distributions*. R package version 1.1-0, http://CRAN.R-project.org/package=mvtnorm.

Genz A and Bretz F (2009).
Computation of Multivariate Normal and t Probabilities, Lecture Notes in Statistics. Springer-Verlag, Heidelberg. ISBN 978-3-642-01688-2.

qcc
Scrucca L (2004). "qcc: an R package for quality control charting and statistical process control." *R News*, *4/1*, http://CRAN.R-project.org/doc/Rnews/.

randomForest
Liaw A and Wiener M (2002). "Classification and Regression by randomForest." *R News*, *2/3*, http://CRAN.R-project.org/doc/Rnews/.

rpart
Therneau T and Atkinson B (2019). *rpart: Recursive Partitioning and Regression Trees*. R package version 4.1-15, http://CRAN.R-project.org/package=rpart.

rsm
Lenth RV (2009). "Response-Surface Methods in R, Using rsm." *Journal of Statistical Software*, *32*(7), http://www.jstatsoft.org/v32/i07/.

survival
Therneau T (2020). *A Package for Survival Analysis in R*. R package version 3.1-12, http://CRAN.R-project.org/package=survival.
Therneau T and Grambsch P (2000). *Modeling Survival Data: Extending the Cox Model*. Springer, New York. ISBN 0-387-98784-3.

tseries
Trapletti A and Hornik K (2019). *tseries: Time Series Analysis and Computational Finance*. R package version 0.10-47, http://CRAN.R-project.org/package=tseries.

xgboost
Chen T et al. (2020). *xgboost: Extreme Gradient Boosting*. R package version 1.1.1.1, http://CRAN.R-project.org/package=xgboost.

Solution Manual

Chapter 2: Analyzing Variability: Descriptive Statistics

2.1   The following are the MINITAB commands and output:

```
MTB> random 50 C1;
SUBC> integer 1 6.
MTB> table C1
```

```
Tabulated Statistics

x      Count
1       11
2       10
3        9
4       10
5        4
6        6
ALL     50
```

The expected frequency in each cell, under randomness is $50/6 = 8.3$. You will get different numerical results, due to randomness.

2.2   Following the instructions in the exercise, we obtained (Figure 2.2.1)

Modern Industrial Statistics: With Applications in R, MINITAB and JMP, Third Edition.
Ron S. Kenett and Shelemyahu Zacks.
© 2021 John Wiley & Sons, Ltd. Published 2021 by John Wiley & Sons, Ltd.

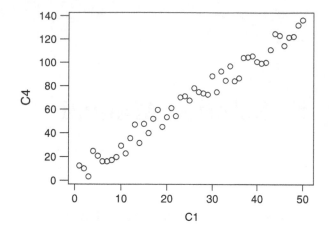

Figure 2.2.1   *Random variability around a straight line*

2.3   Using the MINITAB commands given in the exercise, we obtained

p	0.1	0.3	0.7	0.9
Sum	2	9	36	44

Notice that the expected values of the sums are 5, 15, 35, and 45.

2.4   As shown in Figure 2.4.1, the measurements on Instrument 1, ⊙, seem to be accurate but less precise than those on Instrument 2, *. Instrument 2 seems to have an upward bias (inaccurate). Quantitatively, the mean of the measurements on Instrument 1 is $\bar{X}_1 = 10.034$ and its standard deviation is $S_1 = 0.871$. For Instrument 2, we have $\bar{X}_2 = 10.983$ and $S_2 = 0.569$.

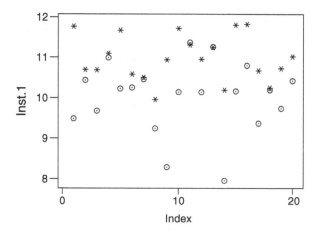

Figure 2.4.1   *Scatterdiagram of measurements on two instruments*

2.5 If the scale is inaccurate, it will show on the average a deterministic component different than the nominal weight. If the scale is imprecise, different weight measurements will show a high degree of variability around the correct nominal weight. Problems with stability arise when the accuracy of the scale changes with time, and the scale should be recalibrated.

2.6 The MINITAB commands to obtain a random sample with replacement (RSWR) from the set $\{1, 2, \ldots, 100\}$ of integers stored in $C1$ are

```
MTB> set C1
DATA> 1(1:100/1)1
DATA> end.
MTB> sample 20 C1 C2;
SUBC> replace.
```

2.7 Use the following MINITAB commands:

```
MTB> set C1
DATA> 1(11:30/1)1
DATA> end.
MTB> sample 10 C1 C2
```

2.8 (i) $26^5 = 11,881,376$; (ii) $7,893,600$; (iii) $26^3 = 17,576$; (iv) $2^{10} = 1024$; (v) $\binom{10}{5} = 252$.

2.9 (i) discrete; (ii) discrete; (iii) continuous; (iv) continuous.

2.10 The histogram in Figure 2.10.1 was created by using the MINITAB graphics menu.

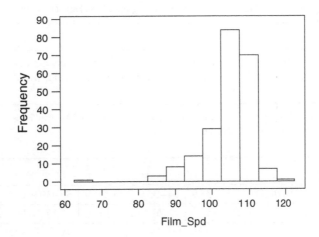

Figure 2.10.1 *Histogram of film speed data*

2.11 The following are MINITAB commands and output. Column C1 contains the yearly number of disasters. Adapt the path of the data file to your computer set up by indicating the folder where the data files are stored.

```
MTB> read 'C:\COAL.csv' C1.
MTB> table C1
```

```
Tabulated Statistics
x           Count
0            33
1            28
2            18
3            17
4             6
5             5
6             3
7             1
ALL         111
```

2.12   (i) Frequency distribution of number of cylinders:

Nu_Cyl	4	6	8	Total
Frequency	66	30	13	109

(ii) Frequency distribution of car's origin:

Origin	United States	Europe	Asia	Total
Frequency	58	14	37	109

The tables for (i) and (ii) are obtained by using the MINITAB command

```
MTB> table C1 (C2, respectively).
```

(iii) Frequency distribution of Turn diameter: We determine the frequency distribution on eight intervals of length 2, from 28 to 44, by using the CODE command. If the variable is in $C3$, we write

```
MTB> code (28:29.9)1 (30:31.9)2 (32:33.9)3 (34:35.9)4 (36:37.9)5
(38:39.9)6 (40:41.9)7 (42:43.9)8 C3 C6
MTB> table C6
```

to obtain

Turn D	28–29.9	30–31.9	32–33.9	34–35.9	36–37.9	38–39.9	40–41.9	42–43.9
Frequency	3	16	15	26	19	17	11	2

(iv) Frequency distribution of horsepower:

HP	50–75	76–100	101–125	126–150	151–175	176–200	201–225	226–250	Total
Frequency	7	34	22	18	17	6	4	1	109

(v) Frequency distribution of MPG:

MPG	14–19	20–24	25–29	30–34	Total
Frequency	43	41	22	3	109

2.13 (i) $X_{(1)} = 66$; (ii) $Q_1 = 102$; (iii) $M_e = X_{(109)} = 105$; (iv) $Q_3 = 109$; (v) $X_{(217)} = 118$; (vi) $x_{0.8} = 110$, (vii) $x_{0.9} = 111$, (viii) $x_{0.99} = 115.64$.

2.14 $\beta_3 = -1.7974$, $\beta_4 = 8.9315$. The distribution of film speed is negatively skewed and much steeper than the normal distribution. Notice that MINITAB defines kurtosis as $\beta_4 - 3$.

2.15

Origin	Mean	Std
United States (1)	20.931	3.598
Europe (2)	19.500	2.624
Asia (3)	23.108	4.280

2.16 Coefficient of variation = 0.084.

2.17

Origin	$\bar{X}$	G
United States	37.203	37.069
Asia	33.046	32.976

We see that $\bar{X}$ is greater than G. The cars from Asia have smaller mean turn diameter.

2.18 $\bar{X} = 104.59$; $S = 6.55$.

Interval	Actual Frequencies	Predicted Frequencies
$\bar{X} \pm S$	173	147.56
$\bar{X} \pm 2S$	205	206.15
$\bar{X} \pm 3S$	213	216.35

The discrepancies between the actual frequencies to the predicted frequencies are due to the fact that the distribution of film speed is neither symmetric nor bell-shaped.

2.19 See Figure 2.19.1

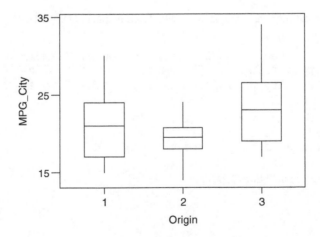

Figure 2.19.1   *Side-by-side boxplots of MPG by car origin*

2.20 The stem-and-leaf diagram obtained from MINITAB is

```
Character Stem-and-Leaf Display
Stem-and-leaf of OTURB N = 100 Leaf Unit = 0.010
     4   2   3444
    18   2   55555666677789
    39   3   00000011111122223334
  (16)   3   5566677788899999
    45   4   00022334444
    34   4   566888999
    25   5   0112333
    18   5   6789
    14   6   01122233444
     3   6   788.
```

$X_{(1)} = 0.23,$

$Q_1 = X_{(25.25)} = X_{(25)} + 0.25(X_{(26)} - X_{(25)}) = 0.31,$

$M_3 = X_{(50.5)} = 0.385,$

$Q_3 = X_{(75.75)} = 0.49 + 0.75(0.50 - 0.49) = 0.4975,$

$X_{(n)} = 0.68.$

2.21 $\bar{T}_\alpha = 0.4056$ and $S_\alpha = 0.0988$, where $\alpha = 0.10$.

2.22 Acceleration statistics in seconds:

Origin	n	$X_{(1)}$	Q_1	Me	Q_3	$X_{(n)}$
German	15	4.8	6.0	7.1	8.7	10.9
Japanese	20	5.7	7.125	8.4	9.475	12.5

2.23 The following statistics are obtained by the MINITAB command DESCRIBE:

```
Variable    N      Mean    Median   TrMean   StDev   SEMean
Res 3      192    1965.2   1967.0   1967.4   163.5    11.8
Res 7      192    1857.8   1880.0   1861.2   151.5    10.9


Variable    Min       Max        Q1        Q3
Res 3      1587.0    2427.0    1860.0    2092.2
Res 7      1420.0    2200.0    1768.7    1960.0
```

The histograms are given in Figures 2.23.1 a,b.

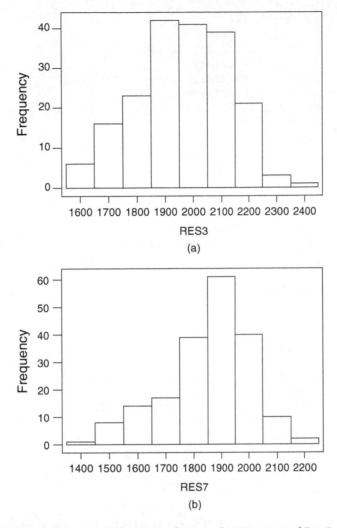

Figure 2.23.1 *(a)* Histogram of Res 3. *(b) Histogram of Res 7*

The stem-leaf diagrams are as follows:

```
Character Stem-and-Leaf Display
Stem-and-leaf of Res 3 N = 192 Leaf Unit = 10
    1   15   8
    6   16   01124
   14   16   56788889
   22   17   00000234
   32   17   5566667899
```

```
  45    18    0011112233444
  60    18    556666677888899
  87    19    00000011111122333334444444444
 (18)   19    566666666667888889
  87    20    00000000122222333334444
  64    20    5555666666667778899999
  44    21    0000011222233344444
  25    21    566667788888
  13    22    000111234
   4    22    668
   1    23
   1    23
   1    24    2
```

```
Character Stem-and-Leaf Display
Stem-and-leaf of Res 7 N = 192 Leaf Unit = 10
    1    14    2
    1    14
    9    15    11222244
   15    15    667789
   23    16    00012334
   30    16    5566799
   40    17    0022233334
   54    17    66666666777999
   79    18    00002222222222233344444444
  (28)   18    5555556666666778888888999999
   85    19    0000000011111122222222233333444444
   52    19    56666666777788888889999
   28    20    0000111222333444
   12    20    678
    9    21    1123344
    2    21    8
    1    22    0
```

2.24 Lower whisker starts at max $(1587, 1511.7) = 1587 = X_{(1)}$; upper whisker ends at min $(2427, 2440.5) = 2427 = X_{(n)}$. There are no outliers.

Chapter 3: Probability Models and Distribution Functions

3.1   (a) $S = \{(w_1, \ldots, w_{20}); w_j = G, D, j = 1, \ldots, 20\}$.

      (b) $2^{20} = 1,048,576$.

      (c) $A_n = \left\{(w_1, \ldots, w_{20}) : \sum_{j=1}^{20} I\{w_j = G\} = n\right\}, n = 0, \ldots, 20,$

where $I\{A\} = 1$ if A is true and $I\{A\} = 0$ otherwise. The number of elementary events in A_n is $\binom{20}{n} = \dfrac{20!}{n!(20-n)!}$.

3.2 $S = \{(\omega_1, \ldots, \omega_{10}) : 10 \le \omega_i \le 20, i = 1, \ldots, 10\}$. Looking at the (ω_1, ω_2) components of A and B, we have the following graphical representation:

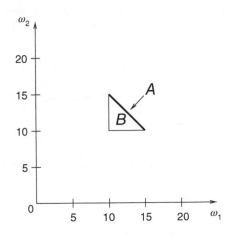

If $(\omega_1, \ldots, \omega_{10}) \in A$, then $(\omega_1, \ldots, \omega_{10}) \in B$. Thus $A \subset B$. $A \cap B = A$.

3.3   (a) $S = \{(i_1, \ldots, i_{30}) : i_j = 0, 1, j = 1, \ldots, 30\}$.

      (b) $A_{10} = \{(1, 1, \ldots, 1, i_{11}, i_{12}, \ldots, i_{30}) : i_j = 0, 1, \quad j = 11, \ldots, 30\}$. $|A_{10}| = 2^{20} = 1,048,576$.

        ($|A_{10}|$ denotes the number of elements in A_{10}.)

      (c) $B_{10} = \{(i_1, \ldots, i_{30}) : i_j = 0, 1 \text{ and } \sum_{j=1}^{30} i_j = 10\}$, $|B_{10}| = \binom{30}{10} = 30,045,015$. $A_{10} \not\subset B_{10}$, in

        fact, A_{10} has only one element belonging to B_{10}.

3.4 $S = (A \cap B) \cup (A \cap B^c) \cup (A^c \cap B) \cup (A^c \cap B^c)$, a union of mutually disjoint sets.

      (i) $A \cup B = (A \cap B) \cup (A \cap B^c) \cup (A^c \cap B)$. Hence, $(A \cup B)^c = A^c \cap B^c$.

      (ii) $(A \cap B)^c = (A \cap B^c) \cup (A^c \cap B) \cup (A^c \cap B^c)$

$$= (A \cap B^c) \cup A^c$$

$$= A^c \cup B^c.$$

3.5   As in Exercise 3.1, $A_n = \left\{(\omega_1, \ldots, \omega_{20}) : \sum_{i=1}^{20} I\{\omega_i = G\} = n\right\}, n = 0, \ldots, 20$. Thus, for any $n \ne n'$, $A_n \cap A_{n'} = \emptyset$, moreover $\bigcup_{n=0}^{20} A_n = S$. Hence, $\{A_0, \ldots, A_{20}\}$ is a partition.

3.6 $\bigcup_{i=1}^{n} A_i = S$ and $A_i \cap A_j = \emptyset$ for all $i \ne j$.

$$B = B \cap S = B \cap \left(\bigcup_{i=1}^{n} A_i\right)$$

$$= \bigcup_{i=1}^{n} A_i B.$$

3.7

$$\Pr\{A \cup B \cup C\} = \Pr\{(A \cup B) \cup C\}$$

$$= \Pr\{(A \cup B)\} + \Pr\{C\} - \Pr\{(A \cup B) \cap C\}$$

$$= \Pr\{A\} + \Pr\{B\} - \Pr\{A \cap B\} + \Pr\{C\}$$

$$- \Pr\{A \cap C\} - \Pr\{B \cap C\} + \Pr\{A \cap B \cap C\}$$

$$= \Pr\{A\} + \Pr\{B\} + \Pr\{C\} - \Pr\{A \cap B\}$$

$$- \Pr\{A \cap C\} - \Pr\{B \cap C\} + \Pr\{A \cap B \cap C\}.$$

3.8   We have shown in Exercise 3.6 that $B = \bigcup_{i=1}^{n} A_i B$. Moreover, since $\{A_1, \ldots, A_n\}$ is a partition, $A_i B \cap A_j B = (A_i \cap A_j) \cap B = \emptyset \cap B = \emptyset$ for all $i \neq j$. Hence, from Axiom 3, $\Pr\{B\} = \Pr\{\bigcup_{i=1}^{n} A_i B\} = \sum_{i=1}^{n} \Pr\{A_i B\}$.

3.9

$$S = \{(i_1, i_2) : i_j = 1, \ldots, 6, \ j = 1, 2\}$$

$$A = \{(i_1, i_2) : i_1 + i_2 = 10\} = \{(4,6), (5,5), (6,4)\}$$

$$\Pr\{A\} = \frac{3}{36} = \frac{1}{12}.$$

3.10 $\Pr\{B\} = \Pr\{A_{150}\} - \Pr\{A_{280}\} = \exp\left(-\frac{150}{200}\right) - \exp\left(-\frac{280}{200}\right) = 0.2258.$

3.11 $\dfrac{\binom{10}{2}\binom{10}{2}\binom{15}{2}\binom{5}{2}}{\binom{40}{8}} = 0.02765.$

3.12   (i) $100^5 = 10^{10}$; (ii) $\binom{100}{5} = 75,287,520.$

3.13 $N = 1000, M = 900, n = 10.$

(i) $\Pr\{X \geq 8\} = \sum_{j=8}^{10} \binom{10}{j} (0.9)^j (0.1)^{10-j} = 0.9298.$

(ii) $\Pr\{X \geq 8\} = \sum_{j=8}^{10} \dfrac{\binom{900}{j}\binom{100}{10-j}}{\binom{1000}{10}} = 0.9308.$

3.14 $1 - (0.9)^{10} = 0.6513.$

3.15 $\Pr\{T > 300 \mid T > 200\} = 0.6065.$

3.16   (i) $\Pr\{D \mid B\} = \frac{1}{4}$; (ii) $\Pr\{C \mid D\} = 1.$

3.17   Since A and B are independent, $\Pr\{A \cap B\} = \Pr\{A\}\Pr\{B\}$. Using this fact and DeMorgan's law,

$$\Pr\{A^c \cap B^c\} = \Pr\{(A \cup B)^c\}$$

$$= 1 - \Pr\{A \cup B\}$$

$$= 1 - (\Pr\{A\} + \Pr\{B\} - \Pr\{A \cap B\})$$

$$= 1 - \Pr\{A\} - \Pr\{B\} + \Pr\{A\}\Pr\{B\}$$

$$= \Pr\{A^c\} - \Pr\{B\}(1 - \Pr\{A\})$$

$$= \Pr\{A^c\}(1 - \Pr\{B\})$$

$$= \Pr\{A^c\}\Pr\{B^c\}.$$

Since $\Pr\{A^c \cap B^c\} = \Pr\{A^c\}\Pr\{B^c\}$, A^c and B^c are independent.

3.18   We assume that $\Pr\{A\} > 0$ and $\Pr\{B\} > 0$. Thus, $\Pr\{A\}\Pr\{B\} > 0$. On the other hand, since $A \cap B = \emptyset$, $\Pr\{A \cap B\} = 0.$

3.19
$$\Pr\{A \cup B\} = \Pr\{A\} + \Pr\{B\} - \Pr\{A \cap B\}$$
$$= \Pr\{A\} + \Pr\{B\} - \Pr\{A\}\Pr\{B\}$$
$$= \Pr\{A\}(1 - \Pr\{B\}) + \Pr\{B\}$$
$$= \Pr\{B\}(1 - \Pr\{A\}) + \Pr\{A\}.$$

3.20 By Bayes' formula,

$$\Pr\{D \mid A\} = \frac{\Pr\{A \mid D\}\Pr\{D\}}{\Pr\{A \mid D\}\Pr\{D\} + \Pr\{A \mid G\}\Pr\{G\}} = \frac{0.10 \times 0.01}{0.10 \times 0.01 + 0.95 \times 0.99} = 0.0011.$$

3.21 Let n be the number of people in the party. The probability that all their birthdays fall on different days is $\Pr\{D_n\} = \prod_{j=1}^{n} \left(\frac{365 - j + 1}{365} \right)$.

    (i) If $n = 10$, $\Pr\{D_{10}\} = 0.8831$.

    (ii) If $n = 23$, $\Pr\{D_{23}\} = 0.4927$. Thus, the probability of at least two persons with the same birthday, when $n = 23$, is $1 - \Pr\{D_{23}\} = 0.5073 > \frac{1}{2}$.

3.22 $\left(\frac{7}{10} \right)^{10} = 0.02825$.

3.23 $\prod_{j=1}^{10} \left(1 - \frac{1}{24 - j + 1} \right) = 0.5833$.

3.24 (i) $\frac{\binom{4}{1}\binom{86}{4}}{\binom{100}{5}} = 0.1128$; (ii) $\frac{\binom{4}{1}\binom{10}{1}\binom{86}{3}}{\binom{100}{5}} = 0.0544$; (iii) $1 - \frac{\binom{86}{5}}{\binom{100}{5}} = 0.5374$.

3.25 $\frac{4}{\binom{10}{2}} = 0.0889$.

3.26 The sample median is $X_{(6)}$, where $X_{(1)} < \cdots < X_{(11)}$ are the ordered sample values. $\Pr\{X_{(6)} = k\} = \frac{\binom{k-1}{5}\binom{20-k}{5}}{\binom{20}{11}}$, $k = 6, \ldots, 15$. This is the probability distribution of the sample median. The probabilities are

k	6	7	8	9	10	11	12	13	14	15
Pr	0.01192	0.04598	0.09902	0.15404	0.18905	0.18905	0.15404	0.09902	0.04598	0.01192

3.27 Without loss of generality, assume that the stick is of length 1. Let x, y, and $(1 - x - y)$, $0 < x, y < 1$, be the length of the three pieces. Obviously, $0 < x + y < 1$. All points in $S = \{(x, y) : x, y > 0, x + y < 1\}$ are uniformly distributed. In order that the three pieces can form a triangle, the following three conditions should be satisfied:

    (i) $x + y > (1 - x - y)$

    (ii) $x + (1 - x - y) > y$

    (iii) $y + (1 - x - y) > x$.

The set of points (x, y) satisfying (i), (ii), and (iii) is bounded by a triangle of area 1/8. S is bounded by a triangle of area 1/2. Hence, the required probability is 1/4.

3.28 Consider Figure 3.28.1. Suppose that the particle is moving along the circumference of the circle in a counterclockwise direction. Then, using the notation in the diagram, $\Pr\{\text{hit}\} = \phi/2\pi$. Since $OD = 1 = OE$, $OB = \sqrt{a^2 + h^2} = OC$ and the lines $\overline{DB}$ and $\overline{EC}$ are tangential to the circle, it follows that the triangles $\triangle ODB$ and $\triangle OEC$ are congruent. Thus, $m(\angle DOB) = m(\angle EOC)$, and it is easily seen that $\phi = 2\alpha$. Now $\alpha = \tan^{-1}\left(\frac{h}{a} \right)$, and hence, $\Pr\{\text{hit}\} = \frac{1}{\pi}\tan^{-1}\left(\frac{h}{a} \right)$.

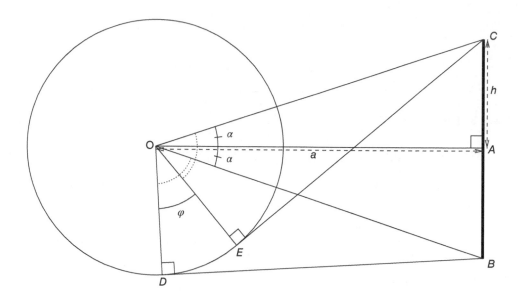

Figure 3.28.1   *Geometry of the solution*

3.29 $1 - (0.999)^{100} - 100 \times (0.001) \times (0.999)^{99} - \binom{100}{2} \times (0.001)^2 (0.999)^{98} = 0.0001504.$

3.30   The probability that n tosses are required is $p(n) = \binom{n-1}{1}\left(\frac{1}{2}\right)^n$, $n \geq 2$. Thus, $p(4) = 3 \cdot \frac{1}{2^4} = \frac{3}{16}.$

3.31 $S = \{(i_1, \ldots, i_{10}) : i_j = 0, 1, \ j = 1, \ldots, 10\}.$ One random variable is the number of 1's in an element, i.e., for $\omega = (i_1, \ldots, i_{10})$ $X_1(\omega) = \sum_{j=1}^{10} i_j.$ Another random variable is the number of zeros to the left of the 1st one, i.e., $X_2(\omega) = \sum_{j=1}^{10} \prod_{k=1}^{j} (1 - i_k).$ Notice that $X_2(\omega) = 0$ if $i_1 = 1$ and $X_2(\omega) = 10$ if $i_1 = i_2 = \ldots = i_{10} = 0.$ The probability distribution of X_1 is $\Pr\{X_1 = k\} = \binom{10}{k}/2^{10}, k = 0, 1, \ldots, 10.$ The probability distribution of X_2 is

$$\Pr\{X_2 = k\} = \begin{cases} \left(\frac{1}{2}\right)^{k+1}, & k = 0, \ldots, 9 \\ \\ \left(\frac{1}{2}\right)^{10}, & k = 10. \end{cases}$$

3.32   (i) Since $\sum_{x=0}^{\infty} \frac{5^x}{x!} = e^5$, we have $\sum_{x=0}^{\infty} p(x) = 1.$; (ii) $\Pr\{X \leq 1\} = e^{-5}(1 + 5) = 0.0404$; (iii) $\Pr\{X \leq 7\} = 0.8666.$

3.33   (i) $\Pr\{X = -1\} = 0.3$; (ii) $\Pr\{-0.5 < X < 0\} = 0.1$; (iii) $\Pr\{0 \leq X < 0.75\} = 0.425$; (iv) $\Pr\{X = 1\} = 0$; (v) $E\{X\} = -0.25, V\{X\} = 0.4042.$

3.34 $E\{X\} = \int_0^{\infty} (1 - F(x))dx = \int_0^{\infty} e^{-x^2/2\sigma^2} dx = \sigma\sqrt{\frac{\pi}{2}}\ .$

3.35 $E\{X\} = \frac{1}{N}\sum_{i=1}^{N} i = \frac{N+1}{2}; \quad E\{X^2\} = \frac{1}{N}\sum_{i=1}^{N} i^2 = \frac{(N+1)(2N+1)}{6}$

$$V\{X\} = E\{X^2\} - (E\{X\})^2 = \frac{2(N+1)(2N+1) - 3(N+1)^2}{12} = \frac{N^2 - 1}{12}.$$

3.36 $\Pr\{8 < X < 12\} = \Pr\{|X - 10| < 2\} \geq 1 - \frac{V\{X\}}{4} = 1 - \frac{0.25}{4} = 0.9375.$

3.37   Notice that $F(x)$ is the standard Cauchy distribution. The pth quantile, x_p, satisfies the equation $\frac{1}{2} + \frac{1}{\pi}\tan^{-1}(x_p) = p$, hence $x_p = \tan\left(\pi\left(p - \frac{1}{2}\right)\right)$. For $p = 0.25, 0.50, 0.75$ we get $x_{0.25} = -1$, $x_{0.50} = 0$, $x_{0.75} = 1$, respectively.

3.38

$$\mu_l^* = E\{(X - \mu_1)^l\} = \sum_{j=0}^{l}(-1)^j\binom{l}{j}\mu_1^j\mu_{l-j}.$$

When $j = l$, the term is $(-1)^l\mu_1^l$. When $j = l - 1$, the term is $(-1)^{l-1}l\mu_1^{l-1}\mu_1 = (-1)^{l-1}l\mu_1^l$. Thus, the sum of the last two terms is $(-1)^{l-1}(l - 1)\mu_1^l$ and we have

$$\mu_l^* = \sum_{j=0}^{l-2}(-1)^j\binom{l}{j}\mu_1^j\mu_{l-j} + (-1)^{l-1}(l - 1)\mu_1^l.$$

3.39   We saw in the solution of Exercise 3.33 that $\mu_1 = -0.25$. Moreover,

$$\mu_2 = V\{X\} + \mu_1^2 = 0.4667.$$

3.40

$$M_X(t) = \frac{1}{t(b - a)}(e^{tb} - e^{ta}), \quad -\infty < t < \infty, \quad a < b.$$

$$E\{X\} = \frac{a + b}{2}; \quad V\{X\} = \frac{(b - a)^2}{12}.$$

3.41   (i) For $t < \lambda$, we have $M(t) = \left(1 - \frac{t}{\lambda}\right)^{-1}$,

$$M'(t) = \frac{1}{\lambda}\left(1 - \frac{t}{\lambda}\right)^{-2}, \qquad \mu_1 = M'(0) = \frac{1}{\lambda}$$

$$M''(t) = \frac{2}{\lambda^2}\left(1 - \frac{t}{\lambda}\right)^{-3}, \qquad \mu_2 = M''(0) = \frac{2}{\lambda^2}$$

$$M^{(3)}(t) = \frac{6}{\lambda^3}\left(1 - \frac{t}{\lambda}\right)^{-4}, \qquad \mu_3 = M^{(3)}(0) = \frac{6}{\lambda^3}$$

$$M^{(4)}(t) = \frac{24}{\lambda^4}\left(1 - \frac{t}{\lambda}\right)^{-5}, \qquad \mu_4 = M^{(4)}(0) = \frac{24}{\lambda^4}.$$

(ii) The central moments are $\mu_1^* = 0$, $\mu_2^* = \frac{1}{\lambda^2}$, $\mu_3^* = \mu_3 - 3\mu_2\mu_1 + 2\mu_1^3 = \frac{6}{\lambda^3} - \frac{6}{\lambda^3} + \frac{2}{\lambda^3} = \frac{2}{\lambda^3}$,

$\mu_4^* = \mu_4 - 4\mu_3\mu_1 + 6\mu_2(\mu_1)^2 - 3\mu_1^4 = \frac{1}{\lambda^4}(24 - 4 \cdot 6 + 6 \cdot 2 - 3) = \frac{9}{\lambda^4}.$

(iii) The index of kurtosis is $\beta_4 = \frac{\mu_4^*}{(\mu_2^*)^2} = 9.$

3.42   Using the commands

```
MTB> set C1
DATA> 1(0:20/1)1
DATA> end.
MTB> pdf C1 C2;
SUBC> binomial 20 0.17.
MTB> cdf C1 C3;
SUBC> binomial 20 0.17.
MTB> print C1-C3
```

we obtained the p.d.f., C2, and c.d.f., C3, of $B(20, 0.17)$:

```
C1      C2          C3
 0    0.024075   0.02407
 1    0.098619   0.12269
 2    0.191892   0.31459
 3    0.235819   0.55041
 4    0.205276   0.75568
 5    0.134543   0.89022
 6    0.068892   0.95912
 7    0.028221   0.98734
 8    0.009393   0.99673
 9    0.002565   0.99930
10    0.000578   0.99987
11    0.000108   0.99998
12    0.000017   1.00000
13    0.000002   1.00000
14    0.000000   1.00000
```

3.43 $Q_1 = 2$, Med $= 3$, $Q_3 = 4$.

3.44 $E\{X\} = 15.75$, $\sigma = 3.1996$.

3.45   Pr {no defective chip on the board} $= p^{50}$. Solving $p^{50} = 0.99$ yields $p = (0.99)^{1/50} = 0.999799$.

3.46   Notice first that $\lim\limits_{\substack{n\to\infty \\ np\to\lambda}} b(0;n,p) = \lim\limits_{n\to\infty}\left(1 - \dfrac{\lambda}{n}\right)^n = e^{-\lambda}$. Moreover, for all $j = 0, 1, \ldots,$

$n - 1$, $\dfrac{b(j+1;n,p)}{b(j;n,p)} = \dfrac{n-j}{j+1} \cdot \dfrac{p}{1-p}$. Thus, by induction on j, for $j > 0$

$$\lim_{\substack{n\to\infty \\ np\to\lambda}} b(j;n,p) = \lim_{\substack{n\to\infty \\ np\to\lambda}} b(j-1;n,p)\frac{n-j+1}{j} \cdot \frac{p}{1-p}$$

$$= e^{-\lambda}\frac{\lambda^{j-1}}{(j-1)!}\lim_{\substack{n\to\infty \\ np\to\lambda}}\frac{(n-j+1)p}{j\left(1-\frac{\lambda}{n}\right)}$$

$$= e^{-\lambda}\frac{\lambda^{j-1}}{(j-1)!} \cdot \frac{\lambda}{j} = e^{-\lambda}\frac{\lambda^j}{j!}.$$

3.47   Using the Poisson approximation, $\lambda = n \cdot p = 1000 \cdot 10^{-3} = 1$.

$$\Pr\{X < 4\} = e^{-1}\sum_{j=0}^{3}\frac{1}{j!} = 0.9810.$$

3.48 $E\{X\} = 20 \cdot \dfrac{350}{500} = 14$; $V\{X\} = 20 \cdot \dfrac{350}{500} \cdot \dfrac{150}{500}\left(1 - \dfrac{19}{499}\right) = 4.0401$.

3.49   Let X be the number of defective items observed.

Pr $\{X > 1\} = 1 - \Pr\{X \leq 1\} = 1 - H(1; 500, 5, 50) = 0.0806$.

3.50 $\Pr\{R\} = 1 - H(3; 100, 10, 20) + \sum_{i=1}^{3} h(i; 100, 10, 20)[1 - H(3 - i; 80, 10 - i, 40)]$

$= 0.87395.$

3.51   The m.g.f. of the Poisson distribution with parameter λ, $P(\lambda)$, is $M(t) = \exp\{-\lambda(1 - e^t)\}$, $-\infty < t < \infty$. Accordingly,

$$M'(t) = \lambda M(t)e^t,$$

$$M''(t) = (\lambda^2 e^{2t} + \lambda e^t)M(t),$$

$$M^{(3)}(t) = (\lambda^3 e^{3t} + 3\lambda^2 e^{2t} + \lambda e^t)M(t),$$

$$M^{(4)}(t) = (\lambda^4 e^{4t} + 6\lambda^3 e^{3t} + 7\lambda^2 e^{2t} + \lambda e^t)M(t).$$

The moments and central moments are

$$\mu_1 = \lambda \qquad\qquad\qquad \mu_1^* = 0$$
$$\mu_2 = \lambda^2 + \lambda \qquad\qquad \mu_2^* = \lambda$$
$$\mu_3 = \lambda^3 + 3\lambda^2 + \lambda \qquad \mu_3^* = \lambda$$
$$\mu_4 = \lambda^4 + 6\lambda^3 + 7\lambda^2 + \lambda. \qquad \mu_4^* = 3\lambda^2 + \lambda.$$

Thus, the indexes of skewness and kurtosis are $\beta_3 = \lambda^{-1/2}$ and $\beta_4 = 3 + \dfrac{1}{\lambda}$.

For $\lambda = 10$, we have $\beta_3 = 0.3162$ and $\beta_4 = 3.1$.

3.52   Let X be the number of blemishes observed. $\Pr\{X > 2\} = 0.1912$.

3.53   Using the Poisson approximation with $N = 8000$ and $p = 380 \times 10^{-6}$, we have $\lambda = 3.04$ and $\Pr\{X > 6\} = 0.0356$, where X is the number of insertion errors in 2 hours of operation.

3.54   The distribution of N is geometric with $p = 0.00038$. $E\{N\} = 2631.6$, $\sigma_N = 2631.08$.

3.55   Using R we obtain that for the $NB(p, k)$ with $p = 0.01$ and $k = 3$, $Q_1 = 170$, Me $= 265$, and $Q_3 = 389$.

Unfortunately, the negative binomial distribution is not given by MINITAB. We can get the results indirectly, however, by using the relationship $NB(j; p, k) = I_p(k, j + 1)$, where $NB(j; p, k)$ is the c.d.f. of $NB(p, k)$ and $I_p(k, j + 1)$ is the c.d.f. of $B(k, j + 1)$ at p. Using the commands

```
MTB> cdf 0.01 k1;
SUBC> beta 3 170.
MTB> print k1
```

we get $k1 = 0.247697$ and for $B(3, 171)$ we get 0.25035. Thus, the first value of j for which $G(j; .01, 3)$ is greater than 0.25 is 170. Thus, $Q_1 = 170$. In a similar manner the other quantiles can be determined. The student should tabulate the c.d.f. of the NB $(0.01, 3)$ and find Q_1, Me, and Q_3 from the table.

3.56   By definition, the m.g.f. of $NB(p, k)$ is $M(t) = \sum_{i=0}^{\infty} \binom{k+i-1}{k-1} p^k((1-p)e^t)^i$, for $t < -\log(1-p)$.

Thus, $M(t) = \dfrac{p^k}{(1-(1-p)e^t)^k} \sum_{i=0}^{\infty} \binom{k+i-1}{k-1} (1-(1-p)e^t)^k((1-p)e^t)^i$.

Since the last infinite series sums to one, $M(t) = \left[\dfrac{p}{1-(1-p)e^t}\right]^k$, $t < -\log(1-p)$.

3.57 $M(t) = \dfrac{pe^t}{1 - (1 - p)e^t}$, for $t < -\log(1 - p)$. The derivatives of $M(t)$ are

$$M'(t) = M(t)(1 - (1 - p)e^t)^{-1}$$

$$M''(t) = M(t)(1 - (1 - p)e^t)^{-2}(1 + (1 - p)e^t)$$

$$M^{(3)}(t) = M(t)(1 - (1 - p)e^t)^{-3} \cdot [(1 + (1 - p)e^t)^2 + 2(1 - p)e^t]$$

$$M^{(4)}(t) = M(t)(1 - (1 - p)e^t)^{-4}[1 + (1 - p)^3 e^{3t} + 11(1 - p)e^t + 11(1 - p)^2 e^{2t}].$$

The moments are

$$\mu_1 = \frac{1}{p}$$

$$\mu_2 = \frac{2 - p}{p^2}$$

$$\mu_3 = \frac{(2 - p)^2 + 2(1 - p)}{p^3} = \frac{6 - 6p + p^2}{p^3}$$

$$\mu_4 = \frac{11(1 - p)(2 - p) + (1 - p)^3 + 1}{p^4} = \frac{24 - 36p + 14p^2 - p^3}{p^4}.$$

The central moments are

$$\mu_1^* = 0$$

$$\mu_2^* = \frac{1 - p}{p^2},$$

$$\mu_3^* = \frac{1}{p^3}(1 - p)(2 - p),$$

$$\mu_4^* = \frac{1}{p^4}(9 - 18p + 10p^2 - p^3).$$

Thus, the indices of skewness and kurtosis are $\beta_3^* = \dfrac{2 - p}{\sqrt{1 - p}}$ and $\beta_4^* = \dfrac{9 - 9p + p^2}{1 - p}$.

3.58   If there are n chips, $n > 50$, the probability of at least 50 good ones is $1 - B(49; n, 0.998)$. Thus, n is the smallest integer > 50 for which $B(49; n, 0.998) < 0.05$. It is sufficient to order 51 chips.

3.59   If X has a geometric distribution then, for every $j, j = 1, 2, \ldots$
$\Pr\{X > j\} = (1 - p)^j$. Thus,

$$\Pr\{X > n + m \mid X > m\} = \frac{\Pr\{X > n + m\}}{\Pr\{X > m\}}$$

$$= \frac{(1 - p)^{n+m}}{(1 - p)^m}$$

$$= (1 - p)^n$$

$$= \Pr\{X > n\}.$$

3.60   For $0 < y < 1$, $\Pr\{F(X) \le y\} = \Pr\{X \le F^{-1}(y)\} = F(F^{-1}(y)) = y$. Hence, the distribution of $F(X)$ is uniform on $(0, 1)$. Conversely, if U has a uniform distribution on $(0, 1)$, then $\Pr\{F^{-1}(U) \le x\} = \Pr\{U \le F(x)\} = F(x)$.

3.61 $E\{U(10, 50)\} = 30$; $V\{U(10, 50)\} = \dfrac{1600}{12} = 133.33$;

$$\sigma\{U(10, 50)\} = \frac{40}{2\sqrt{3}} = 11.547.$$

3.62   Let $X = -\log(U)$ where U has a uniform distribution on $(0,1)$.

$$\Pr\{X \leq x\} = \Pr\{-\log(U) \leq x\}$$
$$= \Pr\{U \geq e^{-x}\}$$
$$= 1 - e^{-x}.$$

Therefore, X has an exponential distribution $E(1)$.

3.63   (i) $\Pr\{92 < X < 108\} = 0.4062$;   (ii) $\Pr\{X > 105\} = 0.3694$;   (iii) $\Pr\{2X + 5 < 200\} = \Pr\{X < 97.5\} = 0.4338$.

3.64   Let z_α denote the α quantile of a $N(0, 1)$ distribution. Then, the two equations $\mu + z_{0.9}\sigma = 15$ and $\mu + z_{0.99}\sigma = 20$ yield the solution $\mu = 8.8670$ and $\sigma = 4.7856$.

3.65   Due to symmetry, $\Pr\{Y > 0\} = \Pr\{Y < 0\} = \Pr\{E < v\}$, where $E \sim N(0, 1)$. If the probability of a bit error is $\alpha = 0.01$, then $\Pr\{E < v\} = \Phi(v) = 1 - \alpha = 0.99$. Thus $v = z_{0.99} = 2.3263$.

3.66   Let X_p denote the diameter of an aluminum pin and X_h denote the size of a hole drilled in an aluminum plate. If $X_p \sim N(10, 0.02)$ and $X_h \sim N(\mu_d, 0.02)$, then the probability that the pin will not enter the hole is $\Pr\{X_h - X_p < 0\}$. Now $X_h - X_p \sim N(\mu_d - 10, \sqrt{0.02^2 + 0.02^2})$ and for $\Pr\{X_h - X_p < 0\} = 0.01$, we obtain $\mu_d = 10.0658$ mm. (The fact that the sum of two independent normal random variables is normally distributed should be given to the student since it has not yet been covered in the text.)

3.67   For $X_1, \ldots, X_n$ i.i.d. $N(\mu, \sigma^2)$, $Y = \sum_{i=1}^n iX_i \sim N(\mu_Y, \sigma_Y^2)$ where

$$\mu_Y = \mu \sum_{i=1}^n i = \mu\frac{n(n+1)}{2} \text{ and } \sigma_Y^2 = \sigma^2 \sum_{i=1}^n i^2 = \sigma^2\frac{n(n+1)(2n+1)}{6}.$$

3.68 $\Pr\{X > 300\} = \Pr\{\log X > 5.7038\} = 1 - \Phi(0.7038) = 0.24078.$

3.69   For $X \sim e^{N(\mu,\sigma)}$, $X \sim e^Y$ where $Y \sim N(\mu, \sigma)$, $M_Y(t) = e^{\mu t + \sigma^2 t^2/2}$.

$$\xi = E\{X\} = E\{e^Y\}$$
$$= M_Y(1)$$
$$= e^{\mu + \sigma^2/2}.$$

Since $E\{X^2\} = E\{e^{2Y}\} = M_Y(2) = e^{2\mu + 2\sigma^2}$ we have

$$V\{X\} = e^{2\mu + 2\sigma^2} - e^{2\mu + \sigma^2}$$
$$= e^{2\mu + \sigma^2}(e^{\sigma^2} - 1)$$
$$= \xi^2(e^{\sigma^2} - 1).$$

3.70   The quantiles of $E(\beta)$ are $x_p = -\beta \log(1 - p)$. Hence,

$$Q_1 = 0.2877\beta, \text{ Me} = 0.6931\beta, \ Q_3 = 1.3863\beta.$$

3.71   If $X \sim E(\beta)$, $\Pr\{X > \beta\} = e^{-\beta/\beta} = e^{-1} = 0.3679.$

3.72   The m.g.f. of $E(\beta)$ is

$$M(t) = \frac{1}{\beta}\int_0^\infty e^{tx - x/\beta}dx$$
$$= \frac{1}{\beta}\int_0^\infty e^{-\frac{(1-t\beta)}{\beta}x}dx$$
$$= (1 - t\beta)^{-1}, \quad \text{for } t < \frac{1}{\beta}.$$

3.73 By independence,

$$M_{(X_1+X_2+X_3)}(t) = E\{e^{t(X_1+X_2+X_3)}\}$$

$$= \prod_{i=1}^{3} E\{e^{tX_i}\}$$

$$= (1-\beta t)^{-3}, \quad t < \frac{1}{\beta}.$$

Thus, $X_1 + X_2 + X_3 \sim G(3, \beta)$.
   Using the formula of the next exercise,

$$\Pr\{X_1 + X_2 + X_3 \geq 3\beta\} = \Pr\{\beta G(3, 1) \geq 3\beta\}$$

$$= \Pr\{G(3, 1) \geq 3\}$$

$$= e^{-3} \sum_{j=0}^{2} \frac{3^j}{j!}$$

$$= 0.4232.$$

3.74 Using integration by parts on the p.d.f. repeatedly,

$$G(t; k, \lambda) = \frac{\lambda^k}{(k+1)!} \int_0^t x^{k-1} e^{-\lambda x} dx$$

$$= \frac{\lambda^k}{k!} t^k e^{-\lambda t} + \frac{\lambda^{k+1}}{k!} \int_0^t x^k e^{-\lambda x} dx$$

$$= \frac{\lambda^k}{k!} t^k e^{-\lambda t} + \frac{\lambda^{k+1}}{(k+1)!} t^{k+1} e^{-\lambda t} + \frac{\lambda^{k+2}}{(k+1)!} \int_0^t x^{k+1} e^{-\lambda x} dx$$

$$= \cdots$$

$$= e^{-\lambda t} \sum_{j=k}^{\infty} \frac{(\lambda t)^j}{j!}$$

$$= 1 - e^{-\lambda t} \sum_{j=0}^{k-1} \frac{(\lambda t)^j}{j!}.$$

3.75 $\Gamma(1.17) = 0.9267$, $\Gamma\left(\frac{1}{2}\right) = 1.77245$, $\Gamma\left(\frac{3}{2}\right) = \frac{1}{2}\Gamma\left(\frac{1}{2}\right) = 0.88623$.

3.76 The moment generating function of the sum of independent random variables is the product of their respective m.g.f.'s. Thus, if $X_1, \ldots, X_k$ are i.i.d. $E(\beta)$, using the result of Exercise 3.72, $M_S(t) = \prod_{i=1}^{k}(1-\beta t)^{-1} = (1-\beta t)^{-k}$, $t < \frac{1}{\beta}$, where $S = \sum_{i=1}^{k} X_i$. On the other hand, $(1-\beta t)^{-k}$ is the m.g.f. of $G(k, \beta)$.

3.77 The expected value and variance of $W(2, 3.5)$ are

$$E\{W(2, 3.5)\} = 3.5 \times \Gamma\left(1 + \frac{1}{2}\right) = 3.1018,$$

$$V\{W(2, 3.5)\} = (3.5)^2 \left[\Gamma\left(1 + \frac{2}{2}\right) - \Gamma^2\left(1 + \frac{1}{2}\right)\right] = 2.6289.$$

3.78 Let T be the number of days until failure. $T \sim W(1.5, 500) \sim 500W(1.5, 1)$.

$$\Pr\{T \geq 600\} = \Pr\left\{W(1.5, 1) \geq \frac{6}{5}\right\} = e^{-(6/5)^{1.5}} = 0.2686.$$

3.79 Let $X \sim B\left(\frac{1}{2}, \frac{3}{2}\right)$.

$$E\{X\} = \frac{1/2}{\frac{1}{2} + \frac{3}{2}} = \frac{1}{4}, \quad V\{X\} = \frac{\frac{1}{2} \cdot \frac{3}{2}}{2^2 \cdot 3} = \frac{1}{16} \text{ and } \sigma\{X\} = \frac{1}{4}.$$

3.80 Let $X \sim B(v, v)$. The first four moments are

$$\mu_1 = v/2v = \frac{1}{2}$$

$$\mu_2 = \frac{B(v+2, v)}{B(v, v)} = \frac{v+1}{2(2v+1)}$$

$$\mu_3 = \frac{B(v+3, v)}{B(v, v)} = \frac{(v+1)(v+2)}{2(2v+1)(2v+2)}$$

$$\mu_4 = \frac{B(v+4, v)}{B(v, v)} = \frac{(v+1)(v+2)(v+3)}{2(2v+1)(2v+2)(2v+3)}.$$

The variance is $\sigma^2 = \dfrac{1}{4(2v+1)}$ and the fourth central moment is

$$\mu_4^* = \mu_4 - 4\mu_3 \cdot \mu_1 + 6\mu_2 \cdot \mu_1^2 - 3\mu_1^4 = \frac{3}{16(3 + 8v + 4v^2)}.$$

Finally, the index of kurtosis is $\beta_2 = \dfrac{\mu_4^*}{\sigma^4} = \dfrac{3(1+2v)}{3+2v}$.

3.81 Let (X, Y) have a joint p.d.f. $f(x, y) = \begin{cases} \dfrac{1}{2}, & (x, y) \in S \\ 0, & \text{otherwise.} \end{cases}$

(i) The marginal distributions of X and Y have p.d.f.'s
$$f_X(x) = \frac{1}{2} \int_{-1+|x|}^{1-|x|} dy = 1 - |x|, -1 < x < 1, \text{ and by symmetry,}$$

$$f_Y(y) = 1 - |y|, -1 < y < 1.$$

(ii) $\qquad E\{X\} = E\{Y\} = 0, \quad V\{X\} = V\{Y\} = 2 \int_0^1 y^2(1-y)dy = 2B(3, 2) = \dfrac{1}{6}.$

3.82 The marginal p.d.f. of Y is $f(y) = e^{-y}$, $y > 0$, that is, $Y \sim E(1)$. The conditional p.d.f. of X, given $Y = y$, is $f(x \mid y) = \dfrac{1}{y}e^{-x/y}$ which is the p.d.f. of an exponential with parameter y. Thus, $E\{X \mid Y = y\} = y$, and $E\{X\} = E\{E\{X \mid Y\}\} = E\{Y\} = 1$. Also,

$$E\{XY\} = E\{YE\{X \mid Y\}\}$$
$$= E\{Y^2\}$$
$$= 2.$$

Hence, $\text{cov}(X, Y) = E\{XY\} - E\{X\}E\{Y\} = 1$. The variance of Y is $\sigma_Y^2 = 1$. The variance of X is

$$\sigma_X^2 = E\{V\{X \mid Y\}\} + V\{E\{X \mid Y\}\}$$
$$= E\{Y^2\} + V\{Y\}$$
$$= 2 + 1 = 3.$$

The correlation between X and Y is $\rho_{XY} = \dfrac{1}{\sqrt{3}}$.

3.83   Let (X, Y) have joint p.d.f. $f(x, y) = \begin{cases} 2, & \text{if } (x, y) \in T \\ 0, & \text{otherwise.} \end{cases}$

The marginal densities of X and Y are

$$f_X(x) = 2(1 - x), \quad 0 \le x \le 1$$
$$f_Y(y) = 2(1 - y), \quad 0 \le y \le 1.$$

Notice that $f(x, y) \ne f_X(x)f_Y(y)$ for $x = \dfrac{1}{2}, y = \dfrac{1}{4}$. Thus, X and Y are dependent.

$$\text{cov}(X, Y) = E\{XY\} - E\{X\}E\{Y\} = E\{XY\} - \frac{1}{9}.$$

$$E\{XY\} = 2 \int_0^1 x \int_0^{1-x} y \, dy \, dx$$
$$= \int_0^1 x(1 - x)^2 \, dx$$
$$= B(2, 3) = \frac{1}{12}.$$

Hence, $\text{cov}(X, Y) = \frac{1}{12} - \frac{1}{9} = -\frac{1}{36}$.

3.84 $J \mid N \sim B(N, p);\; N \sim P(\lambda).\; E\{N\} = \lambda, V\{N\} = \lambda, E\{J\} = \lambda p.$

$$\begin{aligned} V\{J\} &= E\{V\{J \mid N\}\} + V\{E\{J \mid N\}\} & E\{JN\} &= E\{NE\{J \mid N\}\} \\ &= E\{Np(1 - p)\} + V\{Np\} & &= pE\{N^2\} \\ &= \lambda p(1 - p) + p^2\lambda = \lambda p. & &= p(\lambda + \lambda^2) \end{aligned}$$

Hence, $\text{cov}(J, N) = p\lambda(1 + \lambda) - p\lambda^2 = p\lambda$ and $\rho_{JN} = \dfrac{p\lambda}{\lambda\sqrt{p}} = \sqrt{p}$.

3.85   Let $X \sim G(2,100) \sim 100G(2, 1)$ and $Y \sim W(1.5, 500) \sim 500W(1.5, 1)$. Then, $XY \sim 5 \times 10^4 G(2, 1) \cdot W(1.5, 1)$ and $V\{XY\} = 25 \times 10^8 \cdot V\{GW\}$, where $G \sim G(2, 1)$ and $W \sim W\left(\dfrac{3}{2}, 1\right)$.

$$V\{GW\} = E\{G^2\}V\{W\} + E^2\{W\}V\{G\}$$

$$= 6\left(\Gamma\left(1 + \frac{4}{3}\right) - \Gamma^2\left(1 + \frac{2}{3}\right)\right) + 2 \cdot \Gamma^2\left(1 + \frac{2}{3}\right)$$

$$= 3.88404.$$

Thus, $V\{XY\} = 9.7101 \times 10^9$.

3.86   Using the notation of Example 4.33,

(i) $\Pr\{J_2 + J_3 \le 20\} = B(20; 3500, 0.005) = 0.7699.$

(ii) $J_3 \mid J_2 = 15 \sim$ Binomial $B\left(3485, \dfrac{0.004}{0.999}\right)$.

(iii) $\lambda = 3485 \times \dfrac{0.004}{0.999} = 13.954$, $\Pr\{J_2 \le 15 \mid J_3 = 15\} \approx P(15; 13.954) = 0.6739$.

3.87   Using the notation of Example 4.34, the joint p.d.f. of J_1 and J_2 is

$$p(j_1, j_2) = \dfrac{\binom{20}{j_1}\binom{50}{j_2}\binom{30}{20-j_1-j_2}}{\binom{100}{20}}, \quad 0 \le j_1, j_2; \; j_1 + j_2 \le 20.$$

The marginal distribution of J_1 is $H(100, 20, 20)$. The marginal distribution of J_2 is $H(100, 50, 20)$. Accordingly,

$$V\{J_1\} = 20 \times 0.2 \times 0.8 \times \left(1 - \dfrac{19}{99}\right) = 2.585859,$$

$$V\{J_2\} = 20 \times 0.5 \times 0.5 \times \left(1 - \dfrac{19}{99}\right) = 4.040404.$$

The conditional distribution of J_1, given J_2, is $H(50, 20, 20 - J_2)$. Hence,

$$E\{J_1 J_2\} = E\{E\{J_1 J_2 \mid J_2\}\}$$

$$= E\left\{J_2(20 - J_2) \times \dfrac{2}{5}\right\}$$

$$= 8E\{J_2\} - 0.4E\{J_2^2\}$$

$$= 80 - 0.4 \times 104.040404 = 38.38381$$

and $\mathrm{cov}(J_1, J_2) = -1.61616$.

   Finally, the correlation between J_1 and J_2 is $\rho = \dfrac{-1.61616}{\sqrt{2.585859 \times 4.040404}} = -0.50$.

3.88 $V\{Y \mid X\} = 150$, $\quad V\{Y\} = 200$, $\quad V\{Y \mid X\} = V\{Y\}(1 - \rho^2)$. Hence, $|\rho| = 0.5$. The sign of ρ cannot be determined.

3.89   (i) $X_{(1)} \sim E\left(\dfrac{100}{10}\right)$, $E\{X_{(1)}\} = 10$; (ii) $E\{X_{(10)}\} = 100 \sum_{i=1}^{10} \dfrac{1}{i} = 292.8968$.

3.90 $J \sim B(10, 0.95)$. If $\{J = j\}, j > 1$, $X_{(1)}$ is the minimum of a sample of j i.i.d. $E(10)$ random variables. Thus, $X_{(1)} \mid J = j \sim E\left(\dfrac{10}{j}\right)$.

(i) $\Pr\{J = k, X_{(1)} \le x\} = b(k; 10, 0.95)(1 - e^{-\frac{kx}{10}})$, $k = 1, 2, \ldots, 10$.

(ii) First note that $\Pr\{J \ge 1\} = 1 - (0.05)^{10} \approx 1$.

$$\Pr\{X_{(1)} \le x \mid J \ge 1\} = \sum_{k=1}^{10} b(k; 10, 0.95)(1 - e^{-\frac{kx}{10}})$$

$$= 1 - \sum_{k=1}^{10} \binom{10}{k} (0.95 e^{-\frac{x}{10}})^k (0.05)^{(10-k)}$$

$$= 1 - [0.05 + 0.95 e^{-x/10}]^{10} + (0.05)^{10}$$

$$= 1 - (0.05 + 0.95 e^{-x/10})^{10}.$$

3.91   The median is $\mathrm{Me} = X_{(6)}$.

(i) The p.d.f. of Me is $f_{(6)}(x) = \dfrac{11!}{5!5!} \lambda(1 - e^{-\lambda x})^5 e^{-6\lambda x}$, $x \ge 0$.

(ii) The expected value of Me is

$$E\{X_{(6)}\} = 2772\lambda \int_0^\infty x(1 - e^{-\lambda x})^5 e^{-6\lambda x}\, dx$$

$$= 2772\lambda \sum_{j=0}^{5}(-1)^j \binom{5}{j} \int_0^\infty xe^{-\lambda x(6+j)}dx$$

$$= 2772 \sum_{j=0}^{5}(-1)^j \binom{5}{j} \frac{1}{\lambda(6+j)^2}$$

$$= 0.73654/\lambda.$$

3.92 Let X and Y be i.i.d. $E(\beta)$, $T = X + Y$ and $W + X - Y$.

$$V\left\{T + \frac{1}{2}W\right\} = V\{\frac{3}{2}X + \frac{1}{2}Y\}$$

$$= \beta^2\left(\left(\frac{3}{2}\right)^2 + \left(\frac{1}{2}\right)^2\right)$$

$$= 2.5\beta^2.$$

3.93 $cov(X, X + Y) = cov(X, X) + cov(X, Y) = V\{X\} = \sigma^2$.

3.94 $V\{\alpha X + \beta Y\} = \alpha^2\sigma_X^2 + \beta^2\sigma_Y^2 + 2\alpha\beta cov(X, Y) = \alpha^2\sigma_X^2 + \beta^2\sigma_Y^2 + 2\alpha\beta\rho_{XY}\sigma_X\sigma_Y$.

3.95 Let $U \sim N(0, 1)$ and $X \sim N(\mu, \sigma)$. We assume that U and X are independent. Then, $\Phi(X) = \Pr\{U < X \mid X\}$ and therefore

$$E\{\Phi(X)\} = E\{\Pr\{U < X \mid X\}\}$$

$$= \Pr\{U < X\}$$

$$= \Pr\{U - X < 0\}$$

$$= \Phi\left(\frac{\mu}{\sqrt{1 + \sigma^2}}\right).$$

The last equality follows from the fact that $U - X \sim N(-\mu, \sqrt{1 + \sigma^2})$.

3.96 Let U_1, U_2, X be independent random variables; U_1, U_2 i.i.d. $N(0, 1)$. Then, $\Phi^2(X) = \Pr\{U_1 \leq X, U_2 \leq X \mid X\}$. Hence,

$$E\{\Phi^2(X)\} = \Pr\{U_1 \leq X, U_2 \leq X\} = \Pr\{U_1 - X \leq 0, U_2 - X \leq 0\}.$$

Since $(U_1 - X, U_2 - X)$ have a bivariate normal distribution with means $(-\mu, -\mu)$ and variance-covariance

matrix $V = \begin{bmatrix} 1 + \sigma^2 & \sigma^2 \\ \sigma^2 & 1 + \sigma^2 \end{bmatrix}$, it follows that

$$E\{\Phi^2(X)\} = \Phi_2\left(\frac{\mu}{\sqrt{1 + \sigma^2}}, \frac{\mu}{\sqrt{1 + \sigma^2}}; \frac{\sigma^2}{1 + \sigma^2}\right).$$

3.97 Since X and Y are independent, $T = X + Y \sim P(12)$, $\Pr\{T > 15\} = 0.1556$.

3.98 Let $F_2(x) = \int_{-\infty}^x f_2(z)dz$ be the c.d.f. of X_2. Since $X_1 + X_2$ are independent

$$\Pr\{Y \leq y\} = \int_{-\infty}^\infty f_1(x)\Pr\{X_2 \leq y - x\}dx$$

$$= \int_{-\infty}^\infty f_1(x)F_2(y - x)dx.$$

Therefore, the p.d.f. of Y is

$$g(y) = \frac{d}{dy} \Pr\{Y \le y\}$$

$$= \int_{-\infty}^{\infty} f_1(x) f_2(y - x) dx.$$

3.99 Let $Y = X_1 + X_2$ where X_1, X_2 are i.i.d. uniform on $(0,1)$. Then the p.d.f.'s are

$$f_1(x) = f_2(x) = I\{0 < x < 1\}$$

$$g(y) = \begin{cases} \int_0^y dx = y, & \text{if } 0 \le y < 1 \\ \int_{y-1}^1 dx = 2 - y, & \text{if } 1 \le y \le 2. \end{cases}$$

3.100 X_1, X_2 are i.i.d. $E(1)$. $U = X_1 - X_2$. $\Pr\{U \le u\} = \int_0^\infty e^{-x} \Pr\{X_1 \le u + x\} dx$. Notice that $-\infty < u < \infty$ and $\Pr\{X_1 \le u + x\} = 0$ if $x + u < 0$. Let $a^+ = \max(a, 0)$. Then,

$$\Pr\{U \le u\} = \int_0^\infty e^{-x}(1 - e^{-(u+x)^+}) dx$$

$$= 1 - \int_0^\infty e^{-x-(u+x)^+} dx$$

$$= \begin{cases} 1 - \frac{1}{2} e^{-u}, & \text{if } u \ge 0 \\ \frac{1}{2} e^{-|u|}, & \text{if } u < 0 \end{cases}$$

Thus, the p.d.f. of U is $g(u) = \frac{1}{2} e^{-|u|}$, $-\infty < u < \infty$.

3.101 $T = X_1 + \cdots + X_{20} \sim N(20\mu, \sqrt{20\sigma^2})$. $\Pr\{T \le 50\} \approx \Phi\left(\frac{50 - 40}{44.7214}\right) = 0.5885$.

3.102 $X \sim B(200, 0.15)$. $\mu = np = 30$, $\sigma = \sqrt{np(1 - p)} = 5.0497$.

$$\Pr\{25 < X < 35\} \approx \Phi\left(\frac{34.5 - 30}{5.0497}\right) - \Phi\left(\frac{25.5 - 30}{5.0497}\right) = 0.6271.$$

3.103 $X \sim P(200)$, $\Pr\{190 < X < 210\} \approx 2\Phi\left(\frac{9.5}{\sqrt{200}}\right) - 1 = 0.4983$.

3.104 $X \sim B(3, 5)$, $\mu = E\{X\} = \frac{3}{8} = 0.375$, $\sigma = \sqrt{V\{X\}} = \left(\frac{3 \cdot 5}{8^2 \cdot 9}\right)^{1/2} = 0.161374$.

$$\Pr\{|\bar{X}_{200} - 0.375| < 0.2282\} \approx 2\Phi\left(\frac{\sqrt{200} \cdot 0.2282}{0.161374}\right) - 1 = 1.$$

3.105 $t_{0.95}[10] = 1.8125$, $t_{0.95}[15] = 1.7531$, $t_{0.95}[20] = 1.7247$.

3.106 $F_{0.95}[10, 30] = 2.1646$, $F_{0.95}[15, 30] = 2.0148$, $F_{0.95}[20, 30] = 1.9317$.

3.107 The solution to this problem is based on the fact, which is not discussed in the text, that $t[v]$ is distributed like the ratio of two independent random variables, $N(0, 1)$ and $\sqrt{\chi^2[v]/v}$. Accordingly, $t[n] \sim \dfrac{N(0,1)}{\sqrt{\dfrac{\chi^2[n]}{n}}}$, where $N(0, 1)$ and $\chi^2[n]$ are independent. $t^2[n] \sim \dfrac{(N(0,1))^2}{\dfrac{\chi^2[n]}{n}} \sim F[1, n]$. Thus, since $\Pr\{F[1, n] \leq F_{1-\alpha}[1, n]\} = 1 - \alpha$.

$\Pr\{-\sqrt{F_{1-\alpha}[1, n]} \leq t[n] \leq \sqrt{F_{1-\alpha}[1, n]}\} = 1 - \alpha$. It follows that $\sqrt{F_{1-\alpha}[1, n]} = t_{1-\alpha/2}[n]$, or $F_{1-\alpha}[1, n] = t^2_{1-\alpha/2}[n]$. (If you assign this problem, please inform the students of the above fact.)

3.108 A random variable $F[v_1, v_2]$ is distributed like the ratio of two independent random variables $\chi^2[v_1]/v_1$ and $\chi^2[v_2]/v_2$. Accordingly, $F[v_1, v_2] \sim \dfrac{\chi^2[v_1]/v_1}{\chi^2[v_2]/v_2}$ and

$$1 - \alpha = \Pr\{F[v_1, v_2] \leq F_{1-\alpha}[v_1, v_2]\}$$

$$= \Pr\left\{\frac{\chi_1^2[v_1]/v_1}{\chi_2^2[v_2]/v_2} \leq F_{1-\alpha}[v_1, v_2]\right\}$$

$$= \Pr\left\{\frac{\chi_2^2[v_2]/v_2}{\chi_1^2[v_1]/v_1} \geq \frac{1}{F_{1-\alpha}[v_1, v_2]}\right\}$$

$$= \Pr\left\{F[v_2, v_1] \geq \frac{1}{F_{1-\alpha}[v_1, v_2]}\right\}$$

$$= \Pr\{F[v_2, v_1] \geq F_\alpha[v_2, v_1]\}.$$

Hence, $F_{1-\alpha}[v_1, v_2] = \dfrac{1}{F_\alpha[v_2, v_1]}$.

3.109 Using the fact that $t[v] \sim \dfrac{N(0,1)}{\sqrt{\dfrac{\chi^2[v]}{v}}}$, where $N(0, 1)$ and $\chi^2[v]$ are independent,

$$V\{t[v]\} = V\left\{\frac{N(0, 1)}{\sqrt{\chi^2[v]/v}}\right\}$$

$$= E\left\{V\left\{\frac{N(0, 1)}{\sqrt{\dfrac{\chi^2[v]}{v}}}\, \middle|\, \chi^2[v]\right\}\right\} + V\left\{E\left\{\frac{N(0, 1)}{\sqrt{\dfrac{\chi^2[v]}{v}}}\, \middle|\, \chi^2[v]\right\}\right\}.$$

By independence, $V\left\{\dfrac{N(0, 1)}{\sqrt{\dfrac{\chi^2[v]}{v}}}\, \middle|\, \chi^2[v]\right\} = \dfrac{v}{\chi^2[v]}$, and $E\left\{\dfrac{N(0, 1)}{\sqrt{\dfrac{\chi^2[v]}{v}}}\, \middle|\, \chi^2[v]\right\} = 0$.

Thus, $V\{t[v]\} = vE\left\{\dfrac{1}{\chi^2[v]}\right\}$. Since $\chi^2[v] \sim G\left(\dfrac{v}{2}, 2\right) \sim 2G\left(\dfrac{v}{2}, 1\right)$,

$$E\left\{\frac{1}{\chi^2[v]}\right\} = \frac{1}{2} \cdot \frac{1}{\Gamma\left(\frac{v}{2}\right)} \int_0^\infty x^{v-2} e^{-x} dx$$

$$= \frac{1}{2} \cdot \frac{\Gamma\left(\frac{v}{2} - 1\right)}{\Gamma\left(\frac{v}{2}\right)} = \frac{1}{2} \cdot \frac{1}{\frac{v}{2} - 1} = \frac{1}{v - 2}.$$

Finally, $V\{t[v]\} = \dfrac{v}{v - 2}$, $v > 2$.

3.110 $E\{F[3, 10]\} = \dfrac{10}{8} = 1.25$, $V\{F[3, 10]\} = \dfrac{2 \cdot 10^2 \cdot 11}{3 \cdot 8^2 \cdot 6} = 1.9097$.

Chapter 4: Statistical Inference and Bootstrapping

4.1 By the weak law of large numbers (WLLN), for any $\epsilon > 0$, $\lim_{n\to\infty} \Pr\{|M_l - \mu_l| < \epsilon\} = 1$. Hence, M_l is a consistent estimator of the lth moment.

4.2 Using the central limit theorem (CLT), $\Pr\{|\bar{X}_n - \mu| < 1\} \approx 2\Phi\left(\dfrac{\sqrt{n}}{\sigma}\right) - 1$. To determine the sample size n so that this probability is 0.95, we set $2\Phi\left(\dfrac{\sqrt{n}}{\sigma}\right) - 1 = 0.95$ and solve for n. This gives $\dfrac{\sqrt{n}}{\sigma} = z_{0.975} = 1.96$. Thus, $n \geq \sigma^2 (1.96)^2 = 424$ for $\sigma = 10.5$.

4.3 $\hat{\xi}_p = \bar{X}_n + z_p \hat{\sigma}_n$, where $\hat{\sigma}_n^2 = \dfrac{1}{n} \sum (X_i - \bar{X}_n)^2$.

4.4 Let $(X_1, Y_1), \ldots, (X_n, Y_n)$ be a random sample from a bivariate normal distribution with density $f(x, y; \mu, \eta, \sigma_X, \sigma_Y, \rho)$ as given in equation (3.6.6). Let $Z_i = X_i Y_i$ for $i = 1, \ldots, n$. Then, the first moment of Z is given by

$$
\mu_1(F_Z) = E\{Z\}
$$
$$
= \int_{-\infty}^{\infty} \int_{-\infty}^{\infty} xy f(x, y; \mu, \eta, \sigma_X, \sigma_Y, \rho)\, dx\, dy
$$
$$
= \mu\eta + \rho\sigma_X\sigma_Y.
$$

Using this fact, as well as the first two moments of X and Y, we get the following moment equations:

$$
\frac{1}{n} \sum_{i=1}^{n} X_i Y_i = \mu\eta + \rho\sigma_X\sigma_Y
$$

$$
\frac{1}{n} \sum_{i=1}^{n} X_i = \mu
$$

$$
\frac{1}{n} \sum_{i=1}^{n} Y_i = \eta
$$

$$
\frac{1}{n} \sum_{i=1}^{n} X_i^2 = \sigma_X^2 + \mu^2
$$

$$
\frac{1}{n} \sum_{i=1}^{n} Y_i^2 = \sigma_Y^2 + \eta^2.
$$

Solving these equations for the five parameters gives

$$
\hat{\rho}_n = \frac{\frac{1}{n}\sum_{i=1}^{n} X_i Y_i - \bar{X}\bar{Y}}{\left[\left(\frac{1}{n}\sum_{i=1}^{n} X_i^2 - \bar{X}^2\right)\cdot\left(\frac{1}{n}\sum_{i=1}^{n} Y_i^2 - \bar{Y}^2\right)\right]^{1/2}}, \text{ or equivalently,}
$$

$$
\hat{\rho}_n = \frac{\frac{1}{n}\sum_{i=1}^{n}(X_i - \bar{X}_n)(Y_i - \bar{Y}_n)}{\left(\frac{1}{n}\sum_{i=1}^{n}(X_i - \bar{X})^2 \cdot \frac{1}{n}\sum_{i=1}^{n}(Y_i - \bar{Y})^2\right)^{1/2}}.
$$

4.5 Let $M_1 = \dfrac{1}{n}\sum_{i=1}^{n} X_i$, $M_2 = \dfrac{1}{n}\sum_{i=1}^{n} X_i^2$ and $\hat{\sigma}_n^2 = M_2 - M_1^2$. Then, the moment equation estimators are $\hat{\nu}_1 = M_1(M_1 - M_2)/\hat{\sigma}_n^2$ and $\hat{\nu}_2 = (M_1 - M_2)(1 - M_1)/\hat{\sigma}_n^2$.

4.6 $V\{\bar{Y}_w\} = \left(\sum_{i=1}^{k} w_i^2 \frac{\sigma_i^2}{n_i}\right) / \left(\sum_{i=1}^{k} w_i\right)^2$. Let $\lambda_i = \frac{w_i}{\sum_{j=1}^{k} w_j}$, $\sum_{i=1}^{k} \lambda_i = 1$. We find weights λ_i which

minimize $V\{\bar{Y}_w\}$, under the constraint $\sum_{i=1}^{k} \lambda_i = 1$. The Lagrangian is $L(\lambda_1, \ldots, \lambda_k, \rho) = \sum_{i=1}^{k} \lambda_i^2 \frac{\sigma_i^2}{n_i} +$

$\rho\left(\sum_{i=1}^{k} \lambda_i - 1\right)$. Differentiating with respect to λ_i, we get $\frac{\partial}{\partial \lambda_i} L(\lambda_1, \ldots, \lambda_k, \rho) = 2\lambda_i \frac{\sigma_i^2}{n_i} + \rho$, $i = 1, \ldots, k$

and $\frac{\partial}{\partial \rho} L(\lambda_1, \ldots, \lambda_k, \rho) = \sum_{i=1}^{k} \lambda_i - 1$. Equating the partial derivatives to zero, we get $\lambda_i^0 = -\frac{\rho}{2} \frac{n_i}{\sigma_i^2}$ for

$i = 1, \ldots, k$ and $\sum_{i=1}^{k} \lambda_i^0 = -\frac{\rho}{2} \sum_{i=1}^{k} \frac{n_i}{\sigma_i^2} = 1$.

Thus, $-\frac{\rho}{2} = \frac{1}{\sum_{i=1}^{k} n_i/\sigma_i^2}$, $\lambda_i^0 = \frac{n_i/\sigma_i^2}{\sum_{j=1}^{k} n_j/\sigma_j^2}$, and therefore $w_i = n_i/\sigma_i^2$.

4.7 Since the Y_i are uncorrelated,

$$V\{\hat{\beta}_1\} = \sum_{i=1}^{n} w_i^2 V\{Y_i\} = \sigma^2 \sum_{i=1}^{n} \frac{(x_i - \bar{x}_n)^2}{SS_x^2} = \frac{\sigma^2}{SS_x}, \text{ where } SS_x = \sum_{i=1}^{n} (x_i - \bar{x}_n)^2.$$

4.8 Let $w_i = \frac{x_i - \bar{x}_n}{SS_x}$ for $i = 1, \ldots, n$ where $SS_x = \sum_{i=1}^{n} (x_i - \bar{x}_n)^2$. Then we have

$$V\{\hat{\beta}_0\} = V\{\bar{Y}_n - \hat{\beta}_1 \bar{x}_n\}$$

$$= V\left\{\bar{Y}_n - \left(\sum_{i=1}^{n} w_i Y_i\right) \bar{x}_n\right\}$$

$$= V\left\{\sum_{i=1}^{n} \left(\frac{1}{n} - w_i \bar{x}_n\right) Y_i\right\}$$

$$= \sum_{i=1}^{n} \left(\frac{1}{n} - w_i \bar{x}_n\right)^2 \sigma^2$$

$$= \sigma^2 \sum_{i=1}^{n} \left(\frac{1}{n^2} - \frac{2 w_i \bar{x}_n}{n} + w_i^2 \bar{x}_n^2\right)$$

$$= \sigma^2 \left(\frac{1}{n} - \frac{2}{n} \bar{x}_n \sum_{i=1}^{n} w_i + \bar{x}_n^2 \sum_{i=1}^{n} w_i^2\right)$$

$$= \sigma^2 \left(\frac{1}{n} + \frac{\bar{x}_n^2}{SS_x}\right).$$

Also

$$\text{cov}(\hat{\beta}_0, \hat{\beta}_1) = \text{cov}\left(\sum_{i=1}^{n} \left(\frac{1}{n} - w_i \bar{x}_n\right) Y_i, \sum_{i=1}^{n} w_i Y_i\right)$$

$$= \sigma^2 \sum_{i=1}^{n} \left(\frac{1}{n} - w_i \bar{x}_n\right) w_i$$

$$= -\sigma^2 \frac{\bar{x}_n}{SS_x}.$$

4.9 The correlation between $\hat{\beta}_0$ and $\hat{\beta}_1$ is

$$\rho_{\hat{\beta}_0, \hat{\beta}_1} = -\frac{\sigma^2 \bar{x}_n}{\sigma^2 SS_x \left[\left(\frac{1}{n} + \frac{\bar{x}_n^2}{SS_x}\right)\frac{1}{SS_x}\right]^{1/2}}$$

$$= -\frac{\bar{x}_n}{\left(\frac{1}{n}\sum_{i=1}^{n} x_i^2\right)^{1/2}}.$$

4.10 $X_1, X_2, \ldots, X_n$ i.i.d., $X_1 \sim P(\lambda)$. The likelihood function is $L(\lambda; X_1, \ldots, X_n) = e^{-n\lambda}\frac{\lambda^{\sum_{i=1}^{n} X_i}}{\prod_{i=1}^{n} X_i!}$.

Thus, $\frac{\partial}{\partial \lambda} \log L(\lambda; X_1, \ldots, X_n) = -n + \frac{\sum_{i=1}^{n} X_i}{\lambda}$. Equating this to zero and solving for λ, we get $\hat{\lambda}_n = \frac{1}{n}\sum_{i=1}^{n} X_i = \bar{X}_n$.

4.11 Since v is known, the likelihood of β is $L(\beta) = C_n \frac{1}{\beta^{nv}} e^{-\sum_{i=1}^{n} X_i/\beta}$, $0 < \beta < \infty$ where C_n does not

depend on β. The log-likelihood function is $l(\beta) = \log C_n - nv \log \beta - \frac{1}{\beta}\sum_{i=1}^{n} X_i$. The score function is

$l'(\beta) = -\frac{nv}{\beta} + \frac{\sum_{i=1}^{n} X_i}{\beta^2}$. Equating the score to 0 and solving for β, we obtain the maximum likelihood

estimator (MLE) $\hat{\beta} = \frac{1}{nv}\sum_{i=1}^{n} X_i = \frac{1}{v}\bar{X}_n$. The variance of the MLE is $V\{\hat{\beta}\} = \frac{\beta^2}{nv}$.

4.12 We proved in Exercise 3.56 that the m.g.f. of $NB(2, p)$ is

$M_X(t) = \frac{p^2}{(1 - (1-p)e^t)^2}$, $t < -\log(1-p)$. Let $X_1, X_2, \ldots, X_n$ be i.i.d. $NB(2, p)$, then the m.g.f. of $T_n = $

$\sum_{i=1}^{n} X_i$ is $M_{T_n}(t) = \frac{p^{2n}}{(1 - (1-p)e^t)^{2n}}$, $t < -\log(1-p)$. Thus, $T_n \sim NB(2n, p)$. According to Example 6.4,

the MLE of p, based on T_n (which is a sufficient statistic) is $\hat{p}_n = \frac{2n}{T_n + 2n} = \frac{2}{\bar{X}_n + 2}$, where $\bar{X}_n = T_n/n$ is

the sample mean.

(i) According to the WLLN, $\bar{X}_n$ converges in probability to $E\{X_1\} = \frac{2(1-p)}{p}$. Substituting $2\frac{1-p}{p}$

for $\bar{X}_n$ in the formula of $\hat{p}_n$, we obtain $p^* = \frac{2}{2 + 2\frac{1-p}{p}} = p$. This shows that the limit in probability

as $n \to \infty$, of $\hat{p}_n$ is p.

(ii) Substituting $k = 2n$ in the formulas of Example 6.4 we obtain $\text{Bias}(\hat{p}_n) \approx \frac{3p(1-p)}{4n}$ and $V\{\hat{p}_n\} \approx$

$\frac{p^2(1-p)}{2n}$.

4.13 The likelihood function of μ and β is

$$L(\mu, \beta) = I\{X_{(1)} \geq \mu\}\frac{1}{\beta^n} \exp\{-\frac{1}{\beta}\sum_{i=1}^{n}(X_{(i)} - X_{(1)}) - \frac{n}{\beta}(X_{(1)} - \mu)\},$$

for $-\infty < \mu \leq X_{(1)}$, $0 < \beta < \infty$.

(i) $L(\mu, \beta)$ is maximized by $\hat{\mu} = X_{(1)}$, that is,

$$L^*(\beta) = \sup_{\mu \leq X_{(1)}} L(\mu, \beta) = \frac{1}{\beta^n} \exp\left\{-\frac{1}{\beta}\sum_{i=1}^{n}(X_{(i)} - X_{(1)})\right\} \quad \text{where} \quad X_{(1)} < X_{(2)} < \cdots < X_{(n)} \text{ are}$$

the ordered statistics.

(ii) Furthermore, $L^*(\beta)$ is maximized by $\hat{\beta}_n = \frac{1}{n}\sum_{i=2}^{n}(X_{(i)} - X_{(1)})$. The MLEs are $\hat{\mu} = X_{(1)}$, and

$\hat{\beta}_n = \frac{1}{n}\sum_{i=2}^{n}(X_{(i)} - X_{(1)})$.

(iii) $X_{(1)}$ is distributed like $\mu + E\left(\dfrac{\beta}{n}\right)$, with p.d.f. $f_{(1)}(x; \mu, \beta) = I\{x \geq \mu\}\dfrac{n}{\beta}e^{-\frac{n}{\beta}(x-\mu)}$. Thus, the joint

p.d.f. of $(X_1, \ldots, X_n)$ is factored to a product of the p.d.f. of $X_{(1)}$ and a function of $\hat{\beta}_n$, which does not depend on $X_{(1)}$ (nor on μ). This implies that $X_{(1)}$ and $\hat{\beta}_n$ are independent. $V\{\hat{\mu}\} = V\{X_{(1)}\} = \dfrac{\beta^2}{n^2}$. It can be shown that $\hat{\beta}_n \sim \frac{1}{n}G(n-1, \beta)$. Accordingly, $V\{\hat{\beta}_n\} = \dfrac{n-1}{n^2}\beta^2 = \dfrac{1}{n}\left(1 - \dfrac{1}{n}\right)\beta^2$.

4.14 In sampling with replacement, the number of defective items in the sample, X, has the binomial distribution $B(n,p)$. We test the hypotheses $H_0 : p \leq 0.03$ against $H_1 : p > 0.03$. H_0 is rejected if $X > B^{-1}(1 - \alpha, 20, 0.03)$. For $\alpha = 0.05$ the rejection criterion is $k_\alpha = B^{-1}(0.95, 20, 0.03) = 2$. Since the number of defective items in the sample is $X = 2$, H_0 is not rejected at the $\alpha = 0.05$ significance level.

4.15 The OC function is $OC(p) = B(2; 30, p)$, $0 < p < 1$. A plot of this OC function is given in Figure 4.15.1.

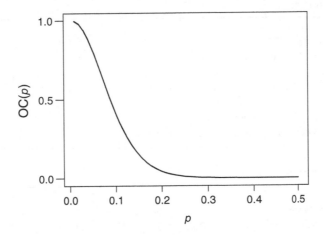

Figure 4.15.1 The OC function $B(2; 30, p)$

4.16 Let $p_0 = 0.01$, $p_1 = 0.03$, $\alpha = 0.05$, $\beta = 0.05$. According to equation (4.3.12), the sample size n should satisfy $1 - \Phi\left(\dfrac{p_1 - p_0}{\sqrt{p_1 q_1}}\sqrt{n} - z_{1-\alpha}\sqrt{\dfrac{p_0 q_0}{p_1 q_1}}\right) = \beta$ or, equivalently, $\dfrac{p_1 - p_0}{\sqrt{p_1 q_1}}\sqrt{n} - z_{1-\alpha}\sqrt{\dfrac{p_0 q_0}{p_1 q_1}} = z_{1-\beta}$. This gives

$$n \approx \frac{(z_{1-\alpha}\sqrt{p_0 q_0} + z_{1-\beta}\sqrt{p_1 q_1})^2}{(p_1 - p_0)^2}$$

$$= \frac{(1.645)^2(\sqrt{0.01 \times 0.99} + \sqrt{0.03 \times 0.97})^2}{(0.02)^2} = 494.$$

For this sample size, the critical value is $k_\alpha = np_0 + z_{1-\alpha}\sqrt{np_0 q_0} = 8.58$. Thus, H_0 is rejected if there are more than 8 "successes" in a sample of size 494.

4.17
$$\bar{X}_n \sim N(\mu, \frac{\sigma}{\sqrt{n}}).$$

(i) If $\mu = \mu_0$, the probability that $\bar{X}_n$ will be outside the control limits is

$$\Pr\left\{\bar{X}_n < \mu_0 - \frac{3\sigma}{\sqrt{n}}\right\} + \Pr\left\{\bar{X}_n > \mu_0 + \frac{3\sigma}{\sqrt{n}}\right\} = \Phi(-3) + 1 - \Phi(3) = 0.0027.$$

(ii) $(1 - 0.0027)^{20} = 0.9474$.

(iii) If $\mu = \mu_0 + 2(\sigma/\sqrt{n})$, the probability that $\bar{X}_n$ will be outside the control limits is

$$\Phi(-5) + 1 - \Phi(1) = 0.1587.$$

(iv) $(1 - 0.1587)^{10} = 0.1777$.

4.18 The MINITAB commands and output are as follows:

```
MTB> Test 4.0 't1';
SUBC> Alternative -1.
```

```
T-Test of the Mean
Test of mu = 4.000 vs mu < 4.000
Variable    N    Mean    StDev    SE Mean    T       P-Value
t1          16   3.956   0.440    0.110      -0.40   0.35
```

The hypothesis $H_0 : \mu \geq 4.0$ amps is not rejected.

4.19 The MINITAB commands and output are as follows:

```
MTB> TTest 4.0 't2';
SUBC> Alternative 1.
```

```
T-Test of the Mean
Test of mu = 4.000 vs mu > 4.000
Variable    N    Mean    StDev    SE Mean    T      P-Value
t2          16   4.209   0.418    0.104      2.01   0.032
```

The hypothesis $H_0 : \mu \leq 4$ is rejected at a 0.05 level of significance.

4.20 Let $n = 30$, $\alpha = 0.01$. The OC(δ) function for a one-sided t-test is

$$\text{OC}(\delta) = 1 - \Phi\left(\frac{\delta\sqrt{30} - 2.462 \times \left(1 - \frac{1}{232}\right)}{\left(1 + \frac{6.0614}{58}\right)^{1/2}}\right)$$

$$= 1 - \Phi(5.2117\delta - 2.3325).$$

Values of OC(δ) for $\delta = 0, 1(0.1)$ are given in the following table.

δ	OC(δ)
0.0	0.990163
0.1	0.964878
0.2	0.901155
0.3	0.778159
0.4	0.596315
0.5	0.390372
0.6	0.211705
0.7	0.092969
0.8	0.032529
0.9	0.008966
1.0	0.001931

This table was computed using the following MINITAB commands:

```
MTB> set C1
DATA> 1(0:1/.1)1
DATA> end.
MTB> let C2 = 5.2217 * C1 - 2.3325
MTB> cdf C2 C3;
MTB> normal 0 1.
MTB> let C4 = 1 - C3
```

The values of δ and OC(δ) are in $C1$ and $C4$, respectively.

4.21 Let $n = 31$, $\alpha = 0.10$. The OC function for testing $H_0 : \sigma^2 \le \sigma_0^2$ against $H_1 : \sigma^2 > \sigma_0^2$ is

$$OC(\sigma^2) = \Pr \left\{ S^2 \le \frac{\sigma_0^2}{n-1} \chi_{0.9}^2[n-1] \right\}$$

$$= \Pr \left\{ \chi^2[30] \le \frac{\sigma_0^2}{\sigma^2} \chi_{0.9}^2[30] \right\}.$$

The values of OC(σ^2) for $\sigma^2 = 1, 2(0.1)$ are given in the following table (here $\sigma_0^2 = 1$):

σ^2	OC(σ^2)
1.0	0.900005
1.1	0.810808
1.2	0.700688
1.3	0.582931
1.4	0.469472
1.5	0.368202
1.6	0.282782
1.7	0.213696
1.8	0.159541
1.9	0.118065
2.0	0.086837

4.22 The OC function, for testing $H_0 : p \leq p_0$ against $H_1 : p > p_0$ is approximated by $OC(p) = 1 - \Phi\left(\frac{p-p_0}{\sqrt{pq}} \sqrt{n} - z_{1-\alpha}\sqrt{\frac{p_0 q_0}{pq}}\right)$, for $p \geq p_0$. In Figure 4.22.1, we present the graph of $OC(p)$, for $p_0 = 0.1$, $n = 100$, $\alpha = 0.05$.

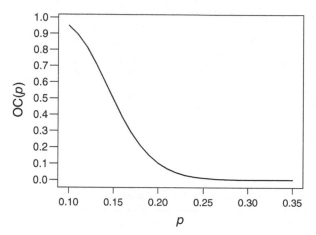

Figure 4.22.1   *OC function*

4.23 The power function is

$$\psi(\sigma^2) = \Pr\left\{S^2 \geq \frac{\sigma_0^2}{n-1}\chi^2_{1-\alpha}[n-1]\right\}$$

$$= \Pr\left\{\chi^2[n-1] \geq \frac{\sigma_0^2}{\sigma^2}\chi^2_{1-\alpha}[n-1]\right\}.$$

4.24 The power function is $\psi(\rho) = \Pr\left\{F[n_1 - 1, n_2 - 1] \geq \frac{1}{\rho}F_{1-\alpha}[n_1 - 1, n_2 - 1]\right\}$, for $\rho \geq 1$, where $\rho = \frac{\sigma_1^2}{\sigma_2^2}$.

4.25   (i) Using the following MINITAB command, we get a 99% confidence interval (CI) for μ:

```
    MTB> TInterval 99.0 'Sample'.
```

```
Confidence Intervals
Variable    N    Mean    StDev   SE Mean        99.0% CI
Sample     20   20.760   0.975    0.218    (20.137, 21.384)
```

(ii) A 99% CI for σ^2 is $(0.468, 2.638)$.
(iii) A 99% CI for σ is $(0.684, 1.624)$.

4.26 Let $(\underline{\mu}_{0.99}, \bar{\mu}_{0.99})$ be a confidence interval for μ, at level 0.99. Let $(\underline{\sigma}_{0.99}, \bar{\sigma}_{0.99})$ be a confidence interval for σ at level 0.99. Let $\xi = \underline{\mu}_{0.99} + 2\underline{\sigma}_{0.99}$ and $\bar{\xi} = \bar{\mu}_{0.99} + 2\bar{\sigma}_{0.99}$. Then, $\Pr\{\xi \leq \mu + 2\sigma \leq \bar{\xi}\} \geq \Pr\{\underline{\mu}_{0.99} \leq \mu \leq \bar{\mu}_{0.99}, \underline{\sigma}_{0.99} \leq \sigma \leq \bar{\sigma}_{0.99}\} \geq 0.98$. Thus, $(\xi, \bar{\xi})$ is a confidence interval for $\mu + 2\sigma$, with confidence level greater or equal to 0.98. Using the data of the previous problem, a 98% CI for $\mu + 2\sigma$ is $(21.505, 24.632)$.

4.27 Let $X \sim B(n, \theta)$. For $X = 17$ and $n = 20$, a confidence interval for θ, at level 0.95, is $(0.6211, 0.9679)$.

4.28 From the data, we have $n = 10$ and $T_{10} = 134$. For $\alpha = 0.05$, $\lambda_L = 11.319$ and $\lambda_U = 15.871$.

4.29 For $n = 20$, $\sigma = 5$, $\bar{Y}_{20} = 13.75$, $\alpha = 0.05$, and $\beta = 0.1$, the tolerance interval is $(3.33, 24.17)$.

4.30 From the data, we have $\bar{X}_{100} = 2.9238$ and $S_{100} = 0.9378$.

$$t(0.025, 0.025, 100) = \frac{1.96}{1 - 1.96^2/200} + \frac{1.96\left(1 + \dfrac{1.96^2}{2} - \dfrac{1.96^2}{200}\right)^{1/2}}{10\left(1 - \dfrac{1.96^2}{200}\right)}$$

$$= 2.3388.$$

The tolerance interval is $(0.7306, 5.1171)$.

4.31 From the data, $Y_{(1)} = 1.151$ and $Y_{(100)} = 5.790$. For $n = 100$ and $\beta = 0.10$, we have $1 - \alpha = 0.988$. For $\beta = 0.05$, $1 - \alpha = 0.847$, the tolerance interval is $(1.151, 5.790)$. The nonparametric tolerance interval is shorter and is shifted to the right with a lower confidence level.

4.32 The following is a normal probability plot of ISC t_1:

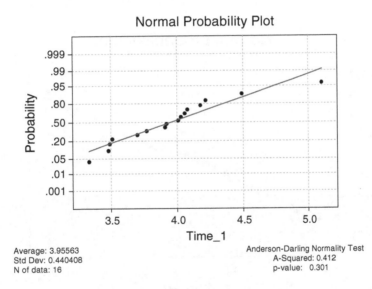

Figure 4.32.1 *Normal probability plot*

According to Figure 4.32.1, the hypothesis of normality is not rejected.

4.33 As is shown in the normal probability plots below, the hypothesis of normality is not rejected in either case (Figure 4.33.1).

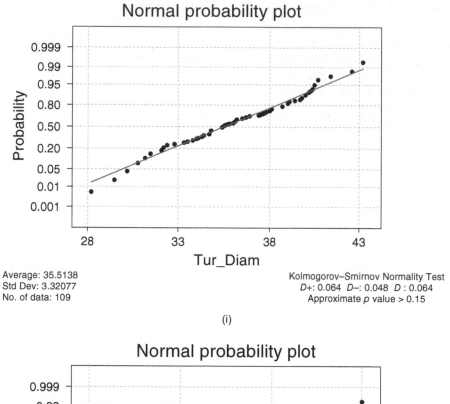

Normal probability plot

Average: 35.5138
Std Dev: 3.32077
No. of data: 109

Kolmogorov–Smirnov Normality Test
$D+$: 0.064 $D-$: 0.048 D : 0.064
Approximate p value > 0.15

(i)

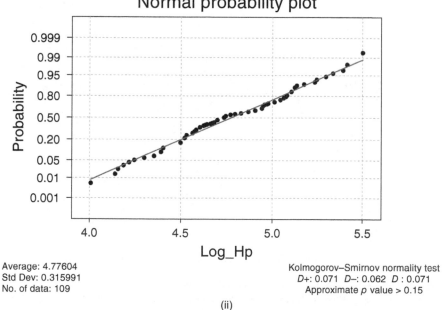

Normal probability plot

Average: 4.77604
Std Dev: 0.315991
No. of data: 109

Kolmogorov–Smirnov normality test
$D+$: 0.071 $D-$: 0.062 D : 0.071
Approximate p value > 0.15

(ii)

Figure 4.33.1 *(i) Normal probability plot for turn diameter. (ii) Normal probability plot of log-horse power*

4.34 Frequency distribution for turn diameter:

Interval	Observed	Expected	$(0-E)^2/E$
-31	11	8.1972	0.9583
$31-32$	8	6.3185	0.4475
$32-33$	9	8.6687	0.0127
$33-34$	6	10.8695	2.1815
$34-35$	18	12.4559	2.4677
$35-36$	8	13.0454	1.9513
$36-37$	13	12.4868	0.0211
$37-38$	6	10.9234	2.2191
$38-39$	9	8.7333	0.0081
$39-40$	8	6.3814	0.4106
$40-$	13	9.0529	1.7213
Total	109	$-$	12.399

The expected frequencies were computed for $N(35.5138, 3.3208)$. Here $\chi^2 = 12.4$, d.f. $= 8$ and the P value is 0.135. The differences from normal are not significant.

4.35 $D_{109}^* = 0.0702$. For $\alpha = 0.05$ $k_\alpha^* = 0.895/\left(\sqrt{109} - 0.01 + \dfrac{0.85}{\sqrt{109}}\right) = 0.0851$. The deviations from

the normal distribution are not significant.

4.36 For $X \sim P(100)$, $n^0 = P^{-1}\left(\frac{0.2}{0.3}, 100\right) = 100 + z_{0.67} \times 10 = 105$.

4.37 Given $X = 6$, the posterior distribution of p is $B(9, 11)$.

4.38 $E\{p \mid X = 6\} = \frac{9}{20} = 0.45$ and

$$V\{p \mid X = 6\} = \frac{99}{20^2 \times 21} = 0.0118 \text{ so } \sigma_{p|X=6} = 0.1086.$$

4.39 Let $X \mid \lambda \sim P(\lambda)$ where $\lambda \sim G(2, 50)$.

   (i) The posterior distribution of $\lambda \mid X = 82$ is $G\left(84, \frac{50}{51}\right)$.

   (ii) $G_{0.025}\left(84, \frac{50}{51}\right) = 65.6879$ and $G_{0.975}\left(84, \frac{50}{51}\right) = 100.873$.

4.40 The posterior probability for $\lambda_0 = 70$ is $\pi(72) = \dfrac{\frac{1}{3}p(72; 70)}{\frac{1}{3}p(72; 70) + \frac{2}{3}p(72; 90)} = 0.771$.

$H_0 : \lambda = \lambda_0$ is accepted if $\pi(X) > \dfrac{r_0}{r_0 + r_1} = 0.4$. Thus, H_0 is accepted.

4.41 The credibility interval for μ is $(43.235, 60.765)$. Since the posterior distribution of μ is symmetric, this credibility interval is also a highest posterior density (HPD) interval.

4.42 The macro for this problem is

```
sample 64 C5 C6;
replace.
let k1 = mean(C6)
stack C7 k1 C7
end
```

Executing the macro 200 times, we obtained $SE\{\bar{X}\} = \dfrac{S}{8} = 0.48965$. The standard deviation of column $C7$ is 0.4951. This is a resampling estimate of $SE\{\bar{X}\}$. The normal probability plotting of the means in $C7$ is shown in Figure 4.42.1.

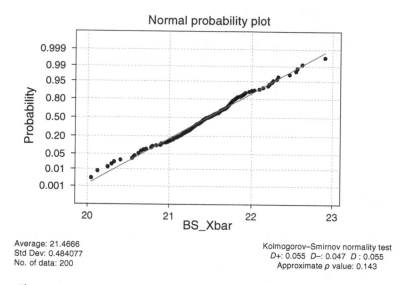

Average: 21.4666
Std Dev: 0.484077
No. of data: 200

Kolmogorov–Smirnov normality test
$D+$: 0.055 $D-$: 0.047 D : 0.055
Approximate p value: 0.143

Figure 4.42.1   *Normal probability plotting of resampling means, $n = 64$*

The resampling distribution is approximately normal.

4.43 In our particular execution with $M = 500$, we have a proportion $\hat{\alpha} = 0.07$ of cases in which the bootstrap confidence intervals do not cover the mean of $C1$, $\mu = 2.9238$. This is not significantly different from the nominal $\alpha = 0.05$. The determination of the proportion $\hat{\alpha}$ can be done by using the following MINITAB commands:

```
MTB> code (1:2.92375)0 (2.92385:5)1 C3 C5
MTB> code (1:2.92375)1 (2.92385:5)0 C4 C6
MTB> rmax C5 C6 C7
MTB> mean C7
```

4.44 We obtained $\tilde{P} = 0.25$. The mean $\bar{X} = 37.203$ is **not** significantly larger than 37.

4.45 Let X_{50} be the number of non-conforming units in a sample of $n = 50$ items. We reject H_0, at level of $\alpha = 0.05$, if $X_{50} > B^{-1}(0.95, 50, 0.03) = 4$. The criterion k_a is obtained by using the MINITAB commands:

```
MTB> invcdf 0.95;
SUBC> binomial 50 0.03.
```

4.46 Generate 1000 bootstrap samples, we obtained the following results:
   (i) A 95% bootstrap CI for the mean is $(0.557, 0.762)$, and a 95% bootstrap CI for the standard deviation is $(0.338, 0.396)$.
   (ii) See Figure 4.46.1a,b.

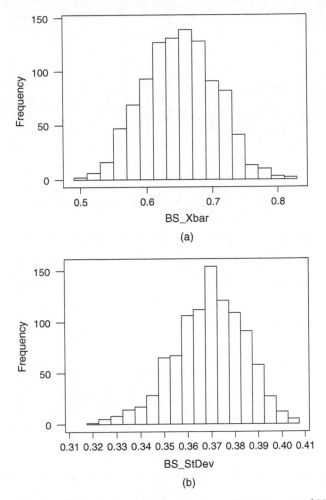

Figure 4.46.1 *(a) Histogram of EBD of 1000 sample means. (b) Histogram of EBD of 1000 sample standard deviations*

4.47 Executing the macro **BOOTPERC.MTB** to generate 1000 bootstrap quartiles, we obtained the following results:

    (i) The bootstrap 0.95 confidence intervals obtained are

	Lower Limit	Upper Limit
Q_1	0.2710	0.3435
Me	0.4340	0.6640
Q_3	1.0385	1.0880

    (ii) The histograms of Q_1 and Q_3 are (Figure 4.47.1)

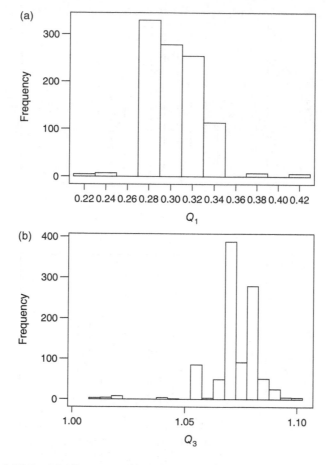

Figure 4.47.1   *(a) Histogram of bootstrap Q_1. (b) Histogram of bootstrap Q_3*

4.48 Running a bootstrapping program with $M = 1000$ yields an EBD, whose histogram is given in Figure 4.48.1. The 0.95 bootstrap CI is $(0.931, 0.994)$.

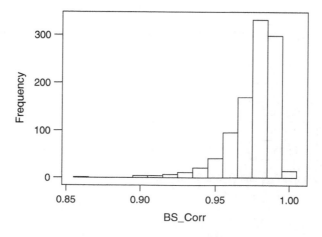

Figure 4.48.1   *Histogram of the bootstrap correlations $M = 1000$*

4.49 Executing the macro **BOOTREGR.MTB** 1000 times, we obtained the following results:
   (i) The 0.95 level bootstrap confidence interval for the intercept a is (29.249,32.254).
   (ii) The 0.95 level bootstrap confidence interval for the slope b is ($-0.0856,-0.0629$).
   (iii) The bootstrap SE of a and b are 0.7930 and 0.00604, respectively. The standard errors of a and b, according to the formulas of Section 4.4.2.1 are 0.8099 and 0.00619, respectively. The bootstrap estimates are quite close to the correct values.

4.50 The mean of the sample is $\bar{X}_{50} = 0.652$. The studentized difference from $\mu_0 = 0.55$ is $t = 1.943$.
   (i) We obtained a bootstrap P-level of $P^* = 0.057$.
   (ii) No, but μ is very close to the lower bootstrap confidence limit (0.557). The null hypothesis $H_0 : \mu = 0.55$ is rejected.
   (iii) No, but since P^* is very close to 0.05, we expect that the bootstrap confidence interval will be very close to μ_0.

4.51 Using an ANOVA bootstrapping program with $M = 500$, we obtained
   (i) First sample variance = 0.0003, second sample variance = 0.0003, variance ratio = 1.2016, and the bootstrap P value is $P^* = 0.896$. The variances are not significantly different.
   (ii) The two boxplots are shown in Figure 4.51.1:

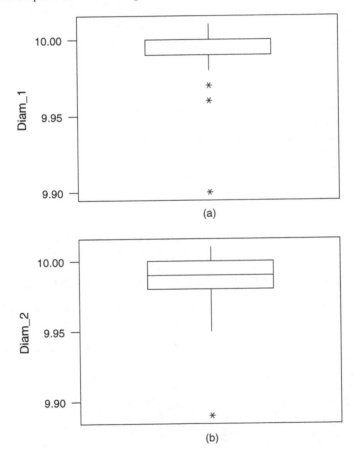

Figure 4.51.1 *(a) Box plot of Sample 1 (b) Box plot of Sample 2*

4.52 The mean of Sample 1 (Diameter1) is $\bar{X}_1 = 9.993$; the mean of Sample 2 (Diameter2) is $\bar{X}_2 = 9.987$. The studentized difference is $t = 1.912$. The P^* level is 0.063 (for $M = 1000$). The variance of Sample 1 is $S_1^2 = 0.000270$. The variance of Sample 2 is $S_2^2 = 0.000324$. The variance ratio is $F = S_2^2/S_1^2 = 1.20$. The bootstrap level for variance ratios is $P^* = 0.756$.

4.53 With $M = 500$, we obtained the following results:

First sample variance $= 12.9425$,

Second sample variance $= 6.8846$,

Third sample variance $= 18.3213$,

$F_{max/min} = 2.6612$ and the bootstrap P value is $P^* = 0.080$.

The bootstrap test does not reject the hypothesis of equal variances at the 0.05 significance level.

4.54 With $M = 500$, we obtained

$$\bar{X}_1 = 20.931 \qquad S_1^2 = 12.9425$$
$$\bar{X}_2 = 19.5 \qquad S_2^2 = 6.8846$$
$$\bar{X}_3 = 23.1081 \qquad S_3^2 = 18.3213$$

$F = 6.0756$, $P^* = 0.02$, and the hypothesis of equal means is rejected.

4.55 The tolerance intervals of the number of defective items in future batches of size $N = 50$, with $\alpha = 0.05$ and $\beta = 0.05$ are as follows.

	Limits	
p	Lower	Upper
0.20	2.5	18
0.10	0	15
0.05	0	10

(Remember to correctly initiate the constants $k1$, $k3$, $k4$, $k7$.)

4.56 A $(0.95, 0.95)$ tolerance interval for OTURB.csv is $(0.238, 0.683)$.

4.57 A $(0.95, 0.95)$ tolerance interval for OTURB.csv is $(0.238, 0.683)$.

4.58 By using the MINITAB commands

```
MTB> read 'C:\CYCLT.csv' C1
MTB> code (0:0.7)0 (0.71:10)1 C1 C2
MTB> sum C2
```

we find that in the sample of $n = 50$ cycle times, there are $X = 20$ values greater than 0.7. If the hypothesis is $H_0 : \xi_{0.5} \leq 0.7$, the probability of observing a value smaller than 0.7 is $p \geq \frac{1}{2}$. Thus, the sign test rejects H_0 if $X < B^{-1}(\alpha; 50, \frac{1}{2})$. For $\alpha = 0.10$ the critical value is $k_\alpha = 20$. H_0 is not rejected.

4.59 We put the sample from file OELECT.csv into column $C1$. The MINITAB commands and the output are

```
MTB> WTest 220.0 C1;
SUBC> Alternative 0.
```

```
Wilcoxon Signed Rank Test
TEST OF MEDIAN = 220.0 VERSUS MEDIAN N.E. 220.0
            N FOR  WILCOXON                ESTIMATED
      N    TEST   STATISTIC  P-VALUE        MEDIAN
C1 99   99      1916.0     0.051          219.2
```

The null hypothesis is rejected with P value equal to 0.051.

4.60 There are 33 US-made cars with four cylinders, and 33 foreign cars with four cylinders. The turn diameter of these cars is put in a data file using the following commands:

```
MTB> read 'C:\CAR.csv' C1 - C5
MTB> sort C1 C2 C3 C6 C7 C8;
SUBC> by C2.
MTB> sort C6 C7 C8 C9 C10 C11;
SUBC> by C6.
MTB> copy C11 C12;
SUBC> use 1:66.
MTB> write 'C:\samplec.csv' C12
```

Using the program RANDTEST , we obtained the following results:

Mean of Sample 1 (US made) = 35.539,

Mean of Sample 2 (foreign) = 33.894,

P level for means = 0.03.

The difference between the means of the turn diameters is significant. Foreign cars have on the average a smaller turn diameter.

Chapter 5: Variability in Several Dimensions and Regression Models

5.1

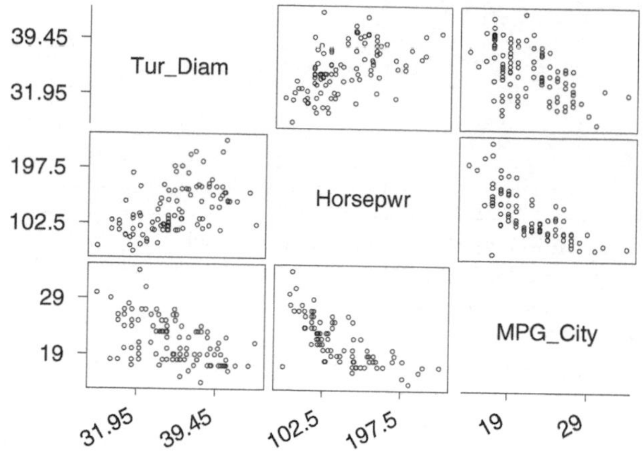

Figure 5.1.1   *Matrix plot of turn diameter versus horsepower versus miles per gallon*

One sees that horsepower and miles per gallon are inversely proportional. Turn diameter seems to increase with horsepower (Figure 5.1.1).

5.2

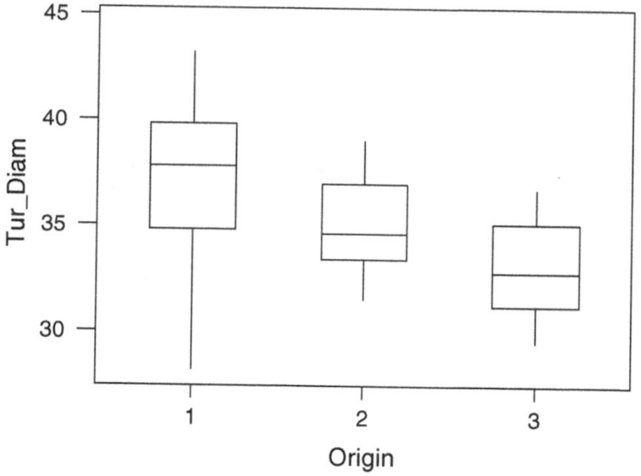

Figure 5.2.1   *Side-by-side boxplots of turn diameter versus car origin*

Figure 5.2.1 shows that cars from Asia generally have the smallest turn diameter. The maximal turn diameter of cars from Asia is smaller than the median turn diameter of US cars. European cars tend to have larger turn diameter than those from Asia, but smaller than those from the United States.

5.3   Figure 5.3.1a,b present the multiple boxplots of Res 3 by Hybrid and a matrix plot of all Res variables. From Figure 5.3.1a we can learn that the conditional distributions of Res at different hybrids are different. Figure 5.3.1b reveals that Res 3 and Res 7 are positively correlated. Res 20 is generally larger than the corresponding Res 14. Res 18 and Res 20 seem to be negatively associated.

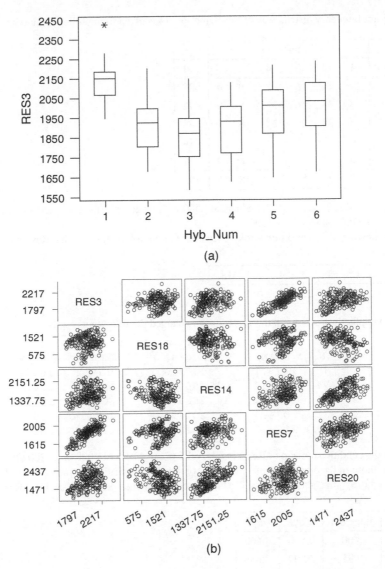

Figure 5.3.1   (a) Multiple box plot of Res 3 by Hybrid. (b) Matrix plot of Res variables

5.4   The joint frequency distribution of horsepower versus miles per gallon is

MPG	HP								All
	1	2	3	4	5	6	7	8	
1	0	0	0	0	0	0	1	0	1
2	1	0	4	10	18	3	3	3	42
3	0	16	17	5	2	1	0	0	41
4	3	13	4	2	0	0	0	0	22
5	3	0	0	0	0	0	0	0	3
All	7	29	25	17	20	4	4	3	109

The intervals for HP are from 50 to 250 at fixed length of 25. The intervals for MPG are from 10 to 35 at length 5. Students may get different results by defining the intervals differently.

5.5   The joint frequency distribution of Res 3 and Res 14 is given in the following table:

CodeRes3	CodeRes14						All
	1	2	3	4	5	6	
1	11	11	5	2	0	0	29
2	3	33	16	11	9	0	72
3	3	28	24	12	8	1	76
4	0	2	6	5	1	0	14
5	0	0	0	1	0	0	1
ALL	17	74	51	31	18	1	192

The intervals for Res 3 start at 1580 and end at 2580 with length of 200. The intervals of Res 14 start at 900 and end at 2700 with length of 300.

5.6   The following is the conditional frequency distribution of Res 3, given that Res 14 is between 1300 and 1500 Ω:

Res 3	1	2	3	4	5	ALL
Frequency	8	21	25	2	0	56

5.7   Following the instructions in the question, we obtained the following results:

```
                      Descriptive Statistics
Variable   C9    N    Mean      Median     TrMean    StDev    SEMean
 Res 3     1    17   1779.1     1707.0     1767.9    162.3     39.4
           2    74   1952.2     1950.5     1955.8    154.7     18.0
           3    51   1997.2     2000.0     2003.2    151.6     21.2
           4    31   2024.8     2014.0     2021.5    156.7     28.2
           5    19   1999.7     2027.0     1998.1    121.5     27.9
```

We see that the conditional mean of Res 3 in interval 1 (Res 14 between 900 and 1200 Ω) is considerably smaller than those of the other four intervals. The conditional standard deviation of interval 5 seems to be smaller than those of the other intervals. This indicates dependence between Res 3 and Res 14.

5.8

Data Set	Intercept	Slope	R^2
1	3.00	0.500	0.667
2	2.97	0.509	0.658
3	2.98	0.501	0.667
4	3.02	0.499	0.667

Notice the influence of the point (19,12.5) on the regression in Data Set 4. Without this point the correlation between x and y is zero.

5.9   The correlation matrix:

	Turn D	HP	MPG
Turn D	1.000	0.508	−0.541
HP	0.508	1.000	−0.755
MPG	−0.541	−0.755	1.000

5.10

$$\text{SSE} = \sum_{i=1}^{n} (Y_i - \beta_0 - \beta_1 X_{i1} - \beta_2 X_{i2})^2$$

(i)

$$\frac{\partial}{\partial \beta_0} \text{SSE} = -2 \sum_{i=1}^{n} (Y_i - \beta_0 - \beta_1 X_{i1} - \beta_2 X_{i2})$$

$$\frac{\partial}{\partial \beta_1} \text{SSE} = -2 \sum_{i=1}^{n} X_{i1}(Y_i - \beta_0 - \beta_1 X_{i1} - \beta_2 X_{i2})$$

$$\frac{\partial}{\partial \beta_2} \text{SSE} = -2 \sum_{i=1}^{n} X_{i2}(Y_i - \beta_0 - \beta_1 X_{i1} - \beta_2 X_{i2}).$$

Equating these partial derivatives to zero and arranging terms, we arrive at the following set of linear equations:

$$\begin{bmatrix} n & \sum_{i=1}^{n} X_{i1} & \sum_{i=1}^{n} X_{i2} \\ \sum_{i=1}^{n} X_{i1} & \sum_{i=1}^{n} X_{i1}^2 & \sum_{i=1}^{n} X_{i1} X_{i2} \\ \sum_{i=1}^{n} X_{i2} & \sum_{i=1}^{n} X_{i1} X_{i2} & \sum_{i=1}^{n} X_{i2}^2 \end{bmatrix} \begin{bmatrix} \beta_0 \\ \beta_1 \\ \beta_2 \end{bmatrix} = \begin{bmatrix} \sum_{i=1}^{n} Y_i \\ \sum_{i=1}^{n} X_{i1} Y_i \\ \sum_{i=1}^{n} X_{i2} Y_i \end{bmatrix}.$$

(ii) Let b_0, b_1, and b_2 be the (unique) solution. From the first equation we get, after dividing by n, $b_0 = \bar{Y} - \bar{X}_1 b_1 - \bar{X}_2 b_2$, where $\bar{Y} = \frac{1}{n} \sum_{i=1}^{n} Y_i$, $\bar{X}_1 = \frac{1}{n} \sum_{i=1}^{n} X_{i1}$, $\bar{X}_2 = \frac{1}{n} \sum_{i=1}^{n} X_{i2}$. Substituting b_0 in the second and third equations and arranging terms, we obtain the reduced system of equations:

$$\begin{bmatrix} (Q_1 - n\bar{X}_1^2) & (P_{12} - n\bar{X}_1\bar{X}_2) \\ (P_{12} - n\bar{X}_1\bar{X}_2) & (Q_2 - n\bar{X}_2^2) \end{bmatrix} \begin{bmatrix} b_1 \\ b_2 \end{bmatrix} = \begin{bmatrix} P_{1y} - n\bar{X}_1\bar{Y} \\ P_{2y} - n\bar{X}_2\bar{Y} \end{bmatrix}$$

where $Q_1 = \sum X_{i1}^2$, $Q_2 = \sum X_{i2}^2$, $P_{12} = \sum X_{i1}X_{i2}$ and $P_{1y} = \sum X_{i1}Y_i$, $P_{2y} = \sum X_{i2}Y_i$. Dividing both sides by $(n-1)$, we obtain equation (5.6.3), and solving we get b_1 and b_2.

5.11 The following is MINITAB output for the command

```
MTB> regr C5 on 2 C3 C4
```

Regression Analysis
The regression equation is MPG = 38.3 - 0.251 TurnD - 0.0631 HP.

Predictor	Coef	Stdev	t-ratio	p
Constant	38.264	2.654	14.42	0.000
TurnD	-0.25101	0.8377	-3.00	0.003
HP	-0.063076	0.006926	-9.11	0.000

s = 2.491 R-sq = 60.3% R-sq(adj) = 59.6%

We see that only 60% of the variability in MPG is explained by the linear relationship with TurnD and HP. Both variables contribute significantly to the regression.

5.12 The partial correlation is -0.27945.

5.13 The partial regression equation is $\hat{e}_1 = -0.251\hat{e}_2$.

5.14 The regression of MPG on HP is MPG $= 30.6633 - 0.07361$HP. The regression of TurnD on HP is TurnD $= 30.2813 + 0.041971$HP. The regression of the residuals $\hat{e}_1$ on $\hat{e}_2$ is $\hat{e}_1 = -0.251 \cdot \hat{e}_2$. Thus,

$$\text{Const.} : \quad b_0 = 30.6633 + 30.2813 \times 0.251 = 38.2639$$

$$\text{HP} : \quad b_1 = -0.07361 + 0.041971 \times 0.251 = -0.063076$$

$$\text{TurnD} : \quad b_2 = -0.251.$$

5.15 The regression of Cap Diameter on Diam2 and Diam3 is

```
MTB> regr C5 2 C3-C4
```

Regression Analysis
```
The regression equation is CapDiam = 4.76 + 0.504 Diam2 + 0.520 Diam3.
Predictor     Coef     Stdev    t-ratio        p
Constant     4.7565    0.5501      8.65     0.000
Diam2        0.5040    0.1607      3.14     0.003
Diam3        0.5203    0.1744      2.98     0.004
s = 0.007568 R-sq = 84.6% R-sq(adj)  = 84.2%.
```

The dependence of CapDiam on Diam2, without Diam1 is significant. This is due to the fact that Diam1 and Diam2 are highly correlated ($\rho = 0.957$). If Diam1 is in the regression, then Diam2 does not furnish additional information on CapDiam. If Diam1 is not included then Diam2 is very informative.

5.16 The regression of yield (y) on the four variables $x_1, x_2, x_3, and x_4$ is

```
MTB> regr C5 4 C1-C4
```

Regression Analysis
```
The regression equation is
y = -6.8 + 0.227 x1 + 0.554 x2 - 0.150 x3 + 0.155 x4.
Predictor        Coef         Stdev       t-ratio          p
Constant        -6.82         10.12         -0.67       0.506
x1             0.22725       0.09994          2.27       0.031
x2              0.5537        0.3698          1.50       0.146
x3            -0.14954       0.02923         -5.12       0.000
x4            0.154650       0.006446        23.99       0.000
s = 2.234 R-sq = 96.2% R-sq(adj)= 95.7%
```

   (i) The regression equation is $\hat{y} = -6.8 + 0.227x_1 + 0.554x_2 - 0.150x_3 + 0.155x_4$.
   (ii) $R^2 = 0.962$.
   (iii) The regression coefficient of x_2 is not significant.
   (iv) Running the multiple regression again, without x_2, we obtain the equation
          $\hat{y} = 4.03 + 0.222x_1 - 0.187x_3 + 0.157x_4$, with $R^2 = 0.959$. Variables x_1, x_3, and x_4 are important.
   (v) Normal probability plotting of the residuals $\hat{e}$ from the equation of (iv) shows that they are normally distributed. See the graph below.

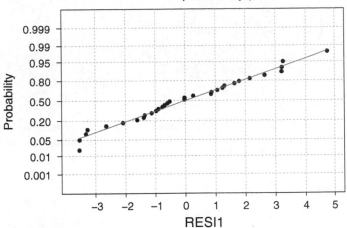

Normal probability plot

Average: 0.0000001
Std Dev: 2.17019
No. of data: 32

W-Test for normality
R: 0.9903
p value (approx): >0.1000

5.17 (i) $(H) = (X)(B) = (X)[(X)'(X)]^{-1}(X)'$

$$H^2 = (X)[(X)'(X)]^{-1}(X)'(X)[(X)'(X)]^{-1}(X)'$$

$$= (X)[(X)'(X)]^{-1}(X)'$$

$$= H.$$

(ii) $(Q) = I - (H)$

$$(Q)^2 = (I - (H))(I - (H))$$

$$= I - (H) - (H) + (H)^2$$

$$= I - (H)$$

$$= Q.$$

$$s_e^2 = \mathbf{y}'(Q)(Q)\mathbf{y}/(n - k - 1)$$

$$= \mathbf{y}'(Q)\mathbf{y}/(n - k - 1).$$

5.18 We have $\hat{\mathbf{y}} = (X)\hat{\beta} = (X)(B)\mathbf{y} = (H)\mathbf{y}$ and $\hat{\mathbf{e}} = Q\mathbf{y} = (I - (H))\mathbf{y}$.

$$\hat{\mathbf{y}}'\hat{\mathbf{e}} = \mathbf{y}'(H)(I - (H))\mathbf{y}$$

$$= \mathbf{y}'(H)\mathbf{y} - \mathbf{y}'(H)^2\mathbf{y}$$

$$= 0.$$

5.19 $1 - R_{y(x)}^2 = \dfrac{\text{SSE}}{\text{SSD}_y}$ where $\text{SSE} = \hat{\mathbf{e}}'\hat{\mathbf{e}} = ||\hat{\mathbf{e}}||^2$.

5.20 From the basic properties of the $\text{cov}(X, Y)$ operator,

$$\text{cov}\left(\sum_{i=1}^{n}\beta_i X_i, \sum_{j=1}^{n}\gamma_j X_j\right) = \sum_{i=1}^{n}\sum_{j=1}^{n}\beta_i\gamma_j\text{cov}(X_i, X_j)$$

$$= \sum_{i=1}^{n}\sum_{j=1}^{n}\beta_i\gamma_j \, \Sigma_{ij}$$

$$= \beta'(\Sigma)\gamma.$$

5.21 $\mathbf{W} = (W_1, \ldots, W_m)'$ where $W_i = \mathbf{b}_i'\mathbf{X}$ $(i = 1, \ldots, m)$. $\mathbf{b}_i'$ is the ith row vector of B. Thus, by the previous exercise, $\text{cov}(W_i, W_j) = \mathbf{b}_i'(\Sigma)\mathbf{b}_j$. This is the (i, j) element of the covariance matrix of $\mathbf{W}$. Hence, the covariance matrix of $\mathbf{W}$ is $C(\mathbf{W}) = (B)(\Sigma)(B)'$.

5.22 From the model, $\Sigma(\mathbf{Y}) = \sigma^2 I$ and $\mathbf{b} = (B)\mathbf{Y}$.

$$\Sigma(\mathbf{b}) = (B)\Sigma(\mathbf{Y})(B)' = \sigma^2(B)(B)'$$

$$= \sigma^2[(\mathbf{X})'(\mathbf{X})]^{-1} \cdot \mathbf{X}'\mathbf{X}[(\mathbf{X})'(\mathbf{X})]^{-1}$$

$$= \sigma^2[(\mathbf{X})'(\mathbf{X})]^{-1}.$$

5.23 The simple linear regression of MPG on horsepower for US- made cars is MPG_US = 30.1 - 0.0697 HP_US.

```
Predictor        Coef        Stdev    t-ratio       p
Constant       30.0637      0.9482      31.71    0.000
HP_US         -0.069660    0.006901    -10.09    0.000
s = 2.161 R-sq = 64.5% R-sq(adj) = 63.9%.
```

For Japanese made cars, the simple linear regression is MPG_Jap = 31.8 - 0.0799 HP_Jap.

```
Predictor        Coef       Stdev    t-ratio       p
Constant       31.833      1.828       17.41    0.000
HP_Jap        -0.07988    0.01598      -5.00    0.000
s = 3.316 R-sq = 41.7% R-sq(adj) = 40.0%.
```

The combined multiple regression of MPG on HP, z and w (where z and w are

```
MPG = 30.1 - 0.0697 HP + 1.77 z - 0.0102 w.
Predictor        Coef       Stdev    t-ratio       p
Constant       30.064      1.169       25.71    0.000
HP            -0.069660   0.008509     -8.19    0.000
z               1.769      1.878        0.94    0.349
w             -0.01022    0.01541      -0.66    0.509
s = 2.665 R-sq = 57.0% R-sq(adj) = 55.6%
```

The P-values corresponding to z and w are 0.349 and 0.509, respectively. Accordingly, we can conclude that the slopes and intercepts of the two simple linear regressions given above are not significantly different. Combining the data we have the following regression line for both US and Japanese cars:

Regression Analysis
```
The regression equation is MPG = 30.9 - 0.0747 HP.
Predictor        Coef        Stdev    t-ratio        p
Constant       30.9330      0.8788      35.20     0.000
HP            -0.074679    0.006816    -10.96     0.000
s = 2.657 R-sq = 56.3% R-sq(adj) = 55.9%.
```

5.24

(a) The regression of Y on X_1 is

```
MTB> regr C5 1 C1
```

Regression Analysis
The regression equation is Y = 81.0 + 1.86 X1.

Predictor	Coef	Stdev	t-ratio	p
Constant	81.042	4.955	16.36	0.000
X1	1.8603	0.5294	3.51	0.005

s = 10.79 R-sq = 52.9% R-sq(adj) = 48.6%.
Analysis of Variance

SOURCE	DF	SS	MS	F	p
Regression	1	1437.0	1437.0	12.35	0.005
Error	11	1279.9	116.4		
Total	12	2716.9			

$F = 12.35$ (In the 1st stage, F is equal to the partial$-F$.) $SSE_1 = 1279.9$, $R^2_{Y|(X_1)} = 0.529$.

(b) The regression of Y on X_1 and X_2 is

```
MTB> regr C5 2 C1 C2
```

Regression Analysis
The regression equation is Y = 51.3 + 1.45 X1 + 0.678 X2.

Predictor	Coef	Stdev	t-ratio	p
Constant	51.278	2.576	19.90	0.000
X1	1.4526	0.1338	10.85	0.000
X2	0.67804	0.05171	13.11	0.000

s = 2.652 R-sq = 97.4% R-sq(adj) = 96.9%

Analysis of Variance

SOURCE	DF	SS	MS	F	p
Regression	2	2646.6	1323.3	188.08	0.000
Error	10	70.4	7.0		
Total	12	2716.9			

SOURCE	DF	SEQ SS
X1	1	1437.0
X2	1	1209.5

$R^2_{Y|(X_1,X_2)} = 0.974$, $SSE_2 = 70.4$, $s^2_{e_2} = 7.04$, Partial $-F = \dfrac{2716.9(0.974 - 0.529)}{7.04} = 171.735$.

Notice that SEQ SS for $X_2 = 2716.9(0.974 - 0.529) = 1209.3$.

(c) The regression of Y on X_1, X_2, X_3 is

```
MTB> regr C5 3 C1 C2 C3
```

Regression Analysis
The regression equation is Y = 45.7 + 1.74 X1 + 0.673 X2 + 0.316 X3.

Predictor	Coef	Stdev	t-ratio	p
Constant	45.658	4.237	10.78	0.000
X1	1.7389	0.2172	8.00	0.000
X2	0.67321	0.04814	13.98	0.000
X3	0.3158	0.9165	1.61	0.142

s = 2.465 R-sq = 98.0% R-sq(adj) = 97.3%

Analysis of Variance

SOURCE	DF	SS	MS	F	p
Regression	3	2662.26	887.42	146.10	0.000
Error	9	54.67	6.07		
Total	12	2716.92			

SOURCE	DF	SEQ SS
X1	1	1437.02
X2	1	1209.54
X3	1	15.69

$R^2_{Y|(X_1,X_2,X_3)} = 0.980$. Partial $-F = \dfrac{2716.92(0.980 - 0.974)}{6.07} = 2.686$. The SEQ SS of X_3 is 15.69.
The 0.95-quantile of $F[1,9]$ is 5.117. Thus, the contribution of X_3 is not significant.

(d) The regression of Y on X_1, X_2, X_3, X_4 is

```
MTB> regr C5 4 C1-C4
```

Regression Analysis
NOTE X2 is highly correlated with other predictor variables
NOTE X4 is highly correlated with other predictor variables
The regression equation is
Y = 84.8 + 1.32 X1 + 0.275 X2 - 0.123 X3 - 0.384 X4.

Predictor	Coef	Stdev	t-ratio	p
Constant	84.78	53.19	1.59	0.150
X1	1.3166	0.6142	2.14	0.064
X2	0.2747	0.5424	0.51	0.626
X3	-0.1230	0.6280	-0.20	0.850
X4	-0.3840	0.5204	-0.74	0.482

s = 2.529 R-sq = 98.1% R-sq(adj) = 97.2%

Analysis of Variance

SOURCE	DF	SS	MS	F	p
Regression	4	2665.74	666.43	104.16	0.000
Error	8	51.18	6.40		
Total	12	2716.92			

SOURCE	DF	SEQ SS
X1	1	1437.02
X2	1	1209.54
X3	1	15.69
X4	1	3.48

Partial $-F = 3.48/6.40 = 0.544$. The effect of X_4 is not significant.

5.25 The following is a MINITAB step-wise regression:

```
MTB> Stepwise 'Y' 'x1' 'x2' 'x3' 'x4';
SUBC> FEnter 4.0;
SUBC> FRemove 4.0.
```

```
Stepwise Regression
F-to-Enter: 4.00 F-to-Remove: 4.00
Response is Y on 4 predictors, with N = 13
Step              1       2       3       4
Constant     117.12  102.76   74.71   51.28

x4           -0.740  -0.617  -0.287
T-Ratio       -4.81  -12.40   -1.82

x1                     1.43    1.43    1.45
T-Ratio               10.10   11.85   10.85

x2                             0.375   0.678
T-Ratio                         2.16   13.11

S              8.92    2.80    2.39    2.65
R-sq          67.77   97.12   98.11   97.41
```

According to this analysis, the suggested model is $Y = 102.76 - 0.617x_4 + 1.43x_1$ with $R^2 = 0.9712$. The following regression of Y on x_1, x_4 yields

```
MTB> regr C5 2 C1 C4
```

Regression Analysis
```
The regression equation is Y = 103 + 1.43 x1 - 0.617 x4.
Predictor        Coef     Stdev   t-ratio        p
Constant      102.758     2.172     47.30    0.000
x1             1.4296    0.1416     10.10    0.000
x4           -0.61672   0.04975    -12.40    0.000
s = 2.797 R-sq = 97.1% R-sq(adj) = 96.5%
```

```
Analysis of Variance
SOURCE        DF        SS        MS        F        p
Regression     2    2638.7    1319.4   168.69    0.000
Error         10      78.2       7.8
Total         12    2716.9

SOURCE    DF    SEQ SS
x1         1    1437.0
x4         1    1201.7
```

We see that both x_1 and x_4 have significant contribution. An alternative model is $Y = 51.28 + 1.45x_1 + 0.678x_2$ with $R^2 = 0.974$. Both models are good for predicting Y.

5.26 The following are MINITAB commands and output:

```
MTB> read 'C:\CAR.csv' C1-C5
MTB> sort C2 C4 C5 C6 C7 C8;
SUBC> by C2.
MTB> copy C7 C8 C9 C10;
SUBC> use 73:109.
MTB> name C9 = 'HP_3' C10 = 'MPG_3'
MTB> regr C10 1 C9;
SUBC> SResiduals C11;
SUBC> Constant;
SUBC> Residuals C12;
SUBC> Hi C13;
SUBC> Cookd C14.
```

Regression Analysis
The regression equation is MPG_3 = 31.8 - 0.0799 HP_3.

Predictor	Coef	Stdev	t-ratio	p
Constant	31.833	1.828	17.41	0.000
HP_3	-0.07988	0.01598	-5.00	0.000

s = 3.316   R-sq = 41.7%   R-sq(adj) = 40.0%

Analysis of Variance

SOURCE	DF	SS	MS	F	p
Regression	1	274.79	274.79	25.00	0.000
Error	35	384.77	10.99		
Total	36	659.57			

Unusual Observations R denotes an obs. with a large st. resid.
X denotes an obs. whose X value gives it large influence.

Obs.	HP_3	MPG_3	Fit	Stdev.Fit	Residual	St.Resid
3	55	17.000	27.439	1.024	-10.439	-3.31R
27	66	34.000	26.560	0.880	7.440	2.33R
33	190	19.000	16.655	1.401	2.345	0.78X
35	200	18.000	15.856	1.550	2.144	0.73X

	C9	C10	C11	C12	C13	C14
	HP_3	MPG_3	SRES1	RESI1	HI1	COOK1
1	118	25	0.79375	2.5936	0.028819	0.009348
2	161	18	-0.30699	-0.9714	0.089302	0.004621
3	55	17	-3.31011	-10.4392	0.095289	0.577018
4	98	23	-0.30748	-1.0041	0.029949	0.001459
5	92	27	0.77220	2.5166	0.033910	0.010465
6	92	29	1.38590	4.5166	0.033910	0.033709
7	104	20	-1.07809	-3.5248	0.027659	0.016531
8	68	27	0.18709	0.5993	0.066478	0.001246
9	70	31	1.48261	4.7591	0.062742	0.073574
10	110	20	-0.93120	-3.0455	0.027041	0.012050
11	121	19	-0.96988	-3.1668	0.030252	0.014672

12	82	24	-0.39558	-1.2823	0.044229	0.003621
13	110	22	-0.31967	-1.0455	0.027041	0.001420
14	158	19	-0.06644	-0.2110	0.082295	0.000198
15	92	26	0.46536	1.5166	0.033910	0.003801
16	102	22	-0.51540	-1.6846	0.028236	0.003859
17	81	27	0.50561	1.6378	0.045516	0.006095
18	142	18	-0.77105	-2.4892	0.051987	0.016301
19	107	18	-1.61608	-5.2852	0.027141	0.036431
20	160	19	-0.01618	-0.0513	0.086920	0.000012
21	90	24	-0.19754	-0.6432	0.035603	0.000720
22	90	26	0.41670	1.3568	0.035603	0.003205
23	97	21	-0.94465	-3.0840	0.030493	0.014033
24	106	18	-1.64062	-5.3650	0.027267	0.037725
25	140	20	-0.20071	-0.6489	0.049034	0.001039
26	165	18	-0.20714	-0.6518	0.099294	0.002365
27	66	34	2.32718	7.4396	0.070400	0.205072
28	97	23	-0.33204	-1.0840	0.030493	0.001734
29	100	25	0.35371	1.1557	0.029000	0.001868
30	115	24	0.41414	1.3539	0.027804	0.002453
31	115	26	1.02591	3.3539	0.027804	0.015050
32	90	27	0.72381	2.3568	0.035603	0.009670
33	190	19	0.78045	2.3453	*0.178583	0.066212
34	115	25	0.72002	2.3539	0.027804	0.007413
35	200	18	0.73148	2.1441	*0.218426	0.074766
36	78	28	0.74195	2.3982	0.049657	0.014382
37	64	28	0.40122	1.2798	0.074507	0.006480

Notice that points 3 and 27 have residuals with large magnitude. Points 3 and 27 have also the largest Cook's distance. Points 33 and 35 have high HI values (leverage).

5.27 The results of the simulations are

	Piston Weight		
30 Kg	40 Kg	50 Kg	60 Kg
1.00838	1.16425	1.30158	1.42575
1.09801	1.26842	1.41851	1.55416
1.12007	1.29327	1.44587	1.58384
0.42310	0.48856	0.54623	0.59837
1.02116	1.17867	1.31748	1.44300

```
One-Way Analysis of Variance
Analysis of Variance on C6
Source    DF     SS      MS      F       p
C5         3   0.416   0.139   1.10   0.378
Error     16   2.014   0.126
Total     19   2.430
```

```
                              Individual 95% CIs for Mean
                              Based on Pooled StDev
Level   N    Mean    StDev    ----------+---------+---------+------
30      5   0.9341   0.2897   (-----------*----------)
40      5   1.0786   0.3345      (----------*----------)
50      5   1.2059   0.3740         (----------*----------)
60      5   1.3210   0.4097            (----------*----------)
                              ----------+---------+---------+------
        Pooled StDev = 0.3421          0.90      1.20      1.50
```

We see that the differences between the sample means are not significant in spite of the apparent upward trend in cycle times.

5.28

(a) The MINITAB command and results are as follows:

```
MTB> aovoneway 'Exp1' 'Exp2' 'Exp3'
```

```
One-Way Analysis of Variance
Analysis of Variance
Source   DF      SS       MS       F       p
Factor    2    3.3363   1.6682   120.92   0.000
Error    27    0.3725   0.0138
Total    29    3.7088
```

```
                              Individual 95% CIs for Mean
                              Based on Pooled StDev
Level   N    Mean    StDev    -------+---------+---------+---------
Exp1   10   2.5170   0.0392                  (--*-)
Exp2   10   2.7740   0.1071                            (-*--)
Exp3   10   1.9740   0.1685    (--*-)
                              -------+---------+---------+---------
        Pooled StDev = 0.1175         2.10      2.40      2.70
```

(b) The bootstrap program for ANOVA table gave, for $M = 1000$, $F^* = 120.917$, with P-level $P^* = 0$. This is identical to the result in (a).

5.2

(i) Ran a randomization test with $M = 500$, gave a P value of 0.230. The difference between the means is not significant.

(ii) One-Way Analysis of Variance

```
Analysis of Variance on C4
Source   DF     SS      MS      F       p
C3        1    28.8    28.8    1.56    0.228
Error    18   333.2    18.5
Total    19   362.0
```

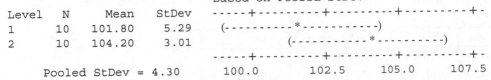

```
                                   Individual 95% CIs For Mean
                                   Based on Pooled StDev
Level   N     Mean    StDev    -----+---------+---------+---------+-
1      10    101.80    5.29    (----------*-----------)
2      10    104.20    3.01            (-----------*----------)
                               -----+---------+---------+---------+-
       Pooled StDev = 4.30     100.0       102.5      105.0     107.5
```

The ANOVA also shows no significant difference in the means. The *P* value is 0.228. Remember that the *F*-test in the ANOVA is based on the assumption of normality and equal variances. The randomization test is nonparametric.

5.30 The results are similar to those obtained in Exercise 5.28.

5.31 The ANOVA of the 26 boards is as follows:

```
One-Way Analysis of Variance
Analysis of Variance on x-dev
Source   DF        SS          MS         F       p
Board    25    0.0011281   0.0000451   203.29   0.000
Error    390   0.0000866   0.0000002
Total    415   0.0012147
```

```
                                     Individual 95% CIs For Mean
                                     Based on Pooled StDev
Level    N      Mean       StDev    -+---------+---------+---------+-----
1       16    -1.4E-03   7.76E-04    (*-)
2       16    -1.3E-03   6.91E-04    (-*)
3       16    -1.2E-03   7.01E-04     (-*)
4       16    -1.0E-03   3.28E-04      (-*)
5       16    -1.0E-03   3.33E-04      (-*)
6       16    -8.1E-04   2.42E-04        (*)
7       16    -9.0E-04   7.30E-04       (*-)
8       16    -1.0E-03   8.11E-04       (*-)
9       16    -8.8E-04   3.01E-04        (*-)
10      16    -9.9E-05   3.18E-04           (*-)
11      16    3.63E-05   3.94E-04            (*-)
12      16    8.25E-05   4.17E-04            (-*)
13      16    2.02E-03   3.35E-04                    (-*)
14      16    2.08E-03   5.09E-04                     (*)
15      16    2.09E-03   6.28E-04                     (*)
16      16    2.48E-03   3.10E-04                      (-*)
17      16    2.59E-03   3.54E-04                       (*-)
18      16    2.56E-03   3.56E-04                       (*)
19      16    2.58E-03   3.54E-04                       (*-)
20      16    3.78E-03   3.98E-04                         (-*)
21      16    2.27E-03   2.81E-04                      (*-)
22      16    2.11E-03   3.42E-04                      (*-)
23      16    2.05E-03   2.55E-04                     (-*)
24      16    2.17E-03   4.16E-04                      (-*)
25      16    2.19E-03   4.19E-04                      (-*)
26      16    2.28E-03   4.54E-04                      (*-)
                                    -+---------+---------+---------+-----
       Pooled StDev = 4.71E-04      -0.0016   0.0000    0.0016    0.0032
```

There seems to be four homogeneous groups: $G_1 = \{1, 2, \ldots, 9\}$, $G_2 = \{10, 11, 12\}$, $G_3 = \{13, \ldots, 19, 21, \ldots, 26\}$, $G_4 = \{20\}$. In multiple comparisons, we use the Scheffé coefficient $S_{0.05} = (25 \times F_{0.95}[25, 390])^{1/2} = (25 \times 1.534)^{1/2} = 6.193$. The group means and standard errors are as follows:

Group	Mean	SE	n	Diff	CR^*
G_4	0.00378	0.000995	16		
G_3	0.002268	0.0000303	208	0.001510	0.000757
G_2	0.0000065	0.0000547	48	0.0022614	0.000467
G_1	−0.00106	0.0000501	144	0.0010683	0.000486

$^*CR = S_{0.05} \times SE\{\text{diff.}\}$.

The differences between the means of the groups are all significant.

5.32 The chi-square test statistic is $X^2 = 2.440$ with d.f. $= 2$ and P value $= 0.296$. The null hypothesis that the number of cylinders a car has is independent of the origin of the car is not rejected.

5.33 The chi-square test statistic is $X^2 = 34.99$ with d.f. $= 6$ and P value $= 0$. The dependence between turn diameter and miles per gallon is significant.

5.34 Since the frequencies in rows 1, 2 and columns 1, 2 of both tables are very small, we collapse the tables to 3×3 ones. The collapsed tables and results are given below.

Table 1. (Question 3 × Question 1)

Q_3	Q_1			
	3	4	5	
3	12	6	1	19
4	13	23	13	49
5	2	15	100	117
	27	44	114	185

For Table 1, we obtain the following values:
    Chi-squared statistic $X^2 = 96.624$,
    Mean squared contingency $\Phi^2 = 0.5223$,
    Tschuprow's index $T = 0.3613$,
    Cramér's index $C = 0.5110$.

Table 2. (Question 3 × Question 2)

Q3	Q_1			
	3	4	5	
3	10	6	1	27
4	12	7	5	24
5	1	30	134	165
	29	47	140	206

For Table 2, we obtain the following values:
    Chi-squared statistic $X^2 = 108.282$,
    Mean squared contingency $\Phi^2 = 0.5256$,
    Tschuprow's index $T = 0.3625$,
    Cramér's index $C = 0.5126$.

Chapter 6: Sampling for Estimation of Finite Population Quantities

6.1 Define the binary random variables

$$
I_{ij} = \begin{cases} 1, & \text{if the } j\text{th element is selected at the } i\text{th sampling} \\ \\ 0, & \text{otherwise.} \end{cases}
$$

The random variables $X_1, \ldots, X_n$ are given by $X_i = \sum_{j=1}^{N} x_j I_{ij}$, $i = 1, \ldots, n$. Since sampling is RSWR, $\Pr\{X_i = x_j\} = \dfrac{1}{N}$ for all $i = 1, \ldots, n$ and $j = 1, \ldots, N$. Hence, $\Pr\{X_i \le x\} = F_N(x)$ for all x, and all $i = 1, \ldots, n$. Moreover, by definition of RSWR, the vectors $\mathbf{I}_i = (I_{i1}, \ldots, I_{iN})$, $i = 1, \ldots, n$ are mutually independent. Therefore, $X_1, \ldots, X_n$ are i.i.d., having a common c.d.f. $F_N(x)$.

6.2 In continuation of the previous exercise, $E\{X_i\} = \dfrac{1}{N} \sum_{i=1}^{n} x_i = \mu_N$. Therefore, by the weak law of large numbers, $\lim_{n \to \infty} P\{|\bar{X}_n - \mu_N| < \epsilon\} = 1$.

6.3 By the CLT $(0 < \sigma_N^2 < \infty)$, $\Pr\{\sqrt{n}|\bar{X}_n - \mu_N| < \delta\} \approx 2\Phi\left(\dfrac{\delta}{\sigma_N}\right) - 1$, as $n \to \infty$.

6.4 We first use the command

```
MTB> read 'C:\PLACE.csv' C1-C4
```

The following macro computes the correlation coefficients of random sample without replacement (RSWOR) of size n from x-dev. and y-dev. The variable x-dev. is in column C2 and y-dev. is in C3. We set $k1 = 20$ which is the sample size.

macro 'CORR.MTB'

```
sample k1 C2 C3 C5 C6
let k2 = sum((C5 - mean(C5))*(C6 - mean(C6)))/
   ((k1-1)*stan(C5)*stan(C6))
stack C7 k2 C7
end
```

In Figure 6.4.1, we present the histogram of these correlations, for samples of size 20.

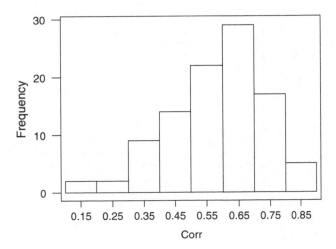

Figure 6.4.1 *Histogram of correlations*

6.5   Use the following macro:

```
sample 50 C3 C9
let k1 = median(C9)
stack C6 k1 C6
sample 50 C4 C9
let k1 = median(C9)
stack C7 k1 C7
sample 50 C5 C9
let k1 = median(C9)
stack C8 k1 C8
end
```

Import the data into columns $C1 - C5$. Then, fill $C6(1)$, $C7(1)$, $C8(1)$ by the medians of $C3$, $C4$, $C5$. Figure 6.5.1a–c present the histograms of the sampling distributions.

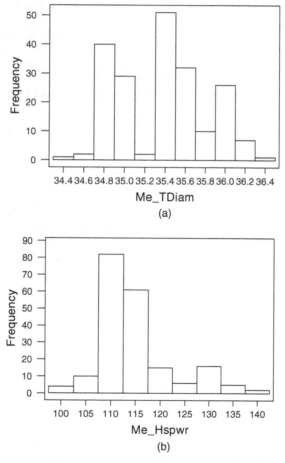

Figure 6.5.1   (a) Histogram of Turn diameter medians. (b) Histogram of horsepower medians. (c) Histogram of MPG medians

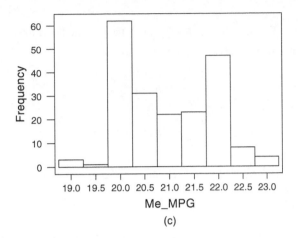

Figure 6.5.1 *(Continued)*

6.7 The required sample size is a solution of the equation

$$0.002 = 2 \cdot 1.96 \cdot \sqrt{\frac{P(1-P)}{n}} \left(1 - \frac{n-1}{N}\right).$$ The solution is $n = 1611$.

6.8 The following are MINITAB commands and a macro to estimate the mean of all $N = 416$ x-dev. values by stratified sampling with proportional allocation. The total sample size is $n = 200$ and the weights are $W_1 = 0.385$, $W_2 = 0.115$, $W_3 = 0.5$. Thus, $n_1 = 77$, $n_2 = 23$, and $n_3 = 100$.

```
MTB> read 'C:\PLACE.csv' C1-C4
MTB> code (1:10)1 (11:13)2 (14:26)3 C1 C5
MTB> copy C2 C6;
SUBC> use 1:160.
MTB> copy C2 C7;
SUBC> use 161:208.
MTB> copy C2 C8;
SUBC> use 209:416.
MTB> let k1 = 77
MTB> let k2 = 23
MTB> let k3 = 100
MTB> stor 'strat.mtb'
STOR> sample k1 C6 C9
STOR> sample k2 C7 C10
STOR> sample k3 C8 C11
STOR> let k4 = 0.385 * mean(C9) + 0.115 * mean(C10) + 0.5 * mean(C11)
STOR> stack C12 k4 C12
STOR> end
MTB> exec 'strat.mtb' 100
```

The standard deviation of column $C12$ is an estimate of $SE(\hat{\mu}_N)$. The true value of this SE is 0.000034442.

6.9 $L(n_1, \ldots, n_k; \lambda) = \sum_{i=1}^{k} W_i^2 \dfrac{\tilde{\sigma}_{N_i}^2}{n_i} - \lambda \left(n - \sum_{i=1}^{k} n_i \right)$. Partial differentiation of L w.r.t. $n_1, \ldots, n_k$ and λ and equating the result to zero yields the following equations:

$$\frac{W_i^2 \tilde{\sigma}_{N_i}^2}{n_i^2} = \lambda, \quad i = 1, \ldots, k$$

$$\sum_{i=1}^{k} n_i = n.$$

Equivalently, $n_i = \dfrac{1}{\sqrt{\lambda}} W_i \tilde{\sigma}_{N_i}$, for $i = 1, \ldots, k$ and $n = \dfrac{1}{\sqrt{\lambda}} \sum_{i=1}^{k} W_i \tilde{\sigma}_{N_i}$.

Thus, $n_i^0 = n \dfrac{W_i \tilde{\sigma}_{N_i}}{\sum_{j=1}^{k} W_j \tilde{\sigma}_{N_j}}, \quad i = 1, \ldots, k.$

6.10 The prediction model is $y_i = \beta + e_i, i = 1, \ldots, N, E\{e_i\} = 0, V\{e_i\} = \sigma^2, \text{cov}(e_i, e_j) = 0$ for all $i \neq j$.

$$E\{\bar{Y}_n - \mu_N\} = E \left\{ \beta + \frac{1}{n} \sum_{i=1}^{N} I_i e_i - \beta - \frac{1}{N} \sum_{i=1}^{N} e_i \right\}$$

$$= \frac{1}{n} \sum_{i=1}^{N} E\{I_i e_i\},$$

where $I_i = \begin{cases} 1, & \text{if } i\text{th population element is sampled} \\ 0, & \text{otherwise.} \end{cases}$

Notice that $I_1, \ldots, I_N$ are independent of $e_1, \ldots, e_N$. Hence, $E\{I_i e_i\} = 0$ for all $i = 1, \ldots, N$. This proves that $\bar{Y}_n$ is prediction unbiased, irrespective of the sample strategy. The prediction MSE of $\bar{Y}_n$ is

$$\text{PMSE}\{\bar{Y}_n\} = E\{(\bar{Y}_n - \mu_N)^2\}$$

$$= V \left\{ \frac{1}{n} \sum_{i=1}^{N} I_i e_i - \frac{1}{N} \sum_{i=1}^{N} e_i \right\}$$

$$= V \left\{ \left(\frac{1}{n} - \frac{1}{N} \right) \sum_{i=1}^{N} I_i e_i - \frac{1}{N} \sum_{i=1}^{N} (1 - I_i) e_i \right\}.$$

Let $\mathbf{s}$ denote the set of units in the sample. Then,

$$\text{PMSE}\{\bar{Y}_n \mid \mathbf{s}\} = \frac{\sigma^2}{n} \left(1 - \frac{n}{N} \right)^2 + \frac{1}{N} \left(1 - \frac{n}{N} \right) \sigma^2$$

$$= \frac{\sigma^2}{n} \left(1 - \frac{n}{N} \right).$$

Notice that $\text{PMSE}\{\bar{Y}_n \mid \mathbf{s}\}$ is independent of $\mathbf{s}$, and is equal for all samples.

6.11 The model is $y_i = \beta_0 + \beta_1 x_i + e_i, i = 1, \ldots, N. E\{e_i\} = 0, V\{e_i\} = \sigma^2 x_i, i = 1, \ldots, N.$ Given a sample $\{(X_1, Y_1), \ldots, (X_n, Y_n)\}$, we estimate β_0 and β_1 by the weighted LSE because the variances of y_i depend on $x_i, i = 1, \ldots, N.$ These weighted

LSE values $\hat{\beta}_0$ and $\hat{\beta}_1$ minimizing $Q = \sum_{i=1}^{n} \frac{1}{X_i}(Y_i - \beta_0 - \beta_1 X_i)^2$, are given by

$$\hat{\beta}_1 = \frac{\bar{Y}_n \cdot \frac{1}{n}\sum_{i=1}^{n}\frac{1}{X_i} - \frac{1}{n}\sum_{i=1}^{n}\frac{Y_i}{X_i}}{\bar{X}_n \cdot \frac{1}{n}\sum_{i=1}^{n}\frac{1}{X_i} - 1} \quad \text{and} \quad \hat{\beta}_0 = \frac{1}{\sum_{i=1}^{n}\frac{1}{X_i}}\left(\sum_{i=1}^{n}\frac{Y_i}{X_i} - n\hat{\beta}_1\right).$$

It is straightforward to show that $E\{\hat{\beta}_1\} = \beta_1$ and $E\{\hat{\beta}_0\} = \beta_0$. Thus, an unbiased predictor of μ_N is $\hat{\mu}_N = \hat{\beta}_0 + \hat{\beta}_1 \bar{x}_N$.

6.12 The predictor $\hat{Y}_{RA}$ can be written as $\hat{Y}_{RA} = \bar{x}_N \cdot \frac{1}{n}\sum_{i=1}^{N} I_i \frac{y_i}{x_i}$, where

$$I_i = \begin{cases} 1, & \text{if } i\text{th population element is in the samples} \\ 0, & \text{otherwise.} \end{cases}$$

Recall that $y_i = \beta x_i + e_i, i = 1, \ldots, N$, and that for any sampling strategy, $e_1, \ldots, e_N$ are independent of $I_1, \ldots, I_N$. Hence, since $\sum_{i=1}^{N} I_i = n$,

$$E\{\hat{Y}_{RA}\} = \bar{x}_N \cdot \frac{1}{n}\sum_{i=1}^{N} E\left\{I_i \frac{\beta x_i + e_i}{x_i}\right\}$$

$$= \bar{x}_N\left(\beta + \frac{1}{n}\sum_{i=1}^{N} E\left\{I_i \frac{e_i}{x_i}\right\}\right)$$

$$= \bar{x}_N \beta,$$

because $E\left\{I_i \frac{e_i}{x_i}\right\} = E\left\{\frac{I_i}{x_i}\right\}E\{e_i\} = 0, i = 1, \ldots, N$. Thus, $E\{\hat{Y}_{RA} - \mu_N\} = 0$ and $\hat{Y}_{RA}$ is an unbiased predictor.

$$\text{PMSE}\{\hat{Y}_{RA}\} = E\left\{\left(\frac{\bar{x}_N}{n}\sum_{i=1}^{N} I_i \frac{e_i}{x_i} - \frac{1}{N}\sum_{i=1}^{N} e_i\right)^2\right\} = V\left\{\sum_{i\in s_n}\left(\frac{\bar{x}_N}{nx_i} - \frac{1}{N}\right)e_i - \sum_{i'\in r_n}\frac{e_{i'}}{N}\right\},$$

where s_n is the set of elements in the sample and r_n is the set of elements in P but not in s_n, $r_n = P - s_n$. Since $e_1, \ldots, e_N$ are uncorrelated,

$$\text{PMSE}\{\hat{Y}_{RA} \mid s_n\} = \sigma^2 \sum_{i\in s_n}\left(\frac{\bar{x}_N}{nx_i} - \frac{1}{N}\right)^2 + \sigma^2 \frac{N-n}{N^2}$$

$$= \frac{\sigma^2}{N^2}\left[(N-n) + \sum_{i\in s_n}\left(\frac{N\bar{x}_N}{nx_i} - 1\right)^2\right].$$

A sample s_n which minimizes $\sum_{i\in s_n}\left(\frac{N\bar{x}_N}{nx_i} - 1\right)^2$ is optimal.

The predictor $\hat{Y}_{\mathrm{RG}}$ can be written as

$$\hat{Y}_{\mathrm{RG}} = \bar{x}_N \frac{\sum_{i=1}^{N} I_i x_i y_i}{\sum_{i=1}^{N} I_i x_i^2} = \bar{x}_N \left(\frac{\sum_{i=1}^{N} I_i x_i (\beta x_i + e_i)}{\sum_{i=1}^{N} I_i x_i^2} \right) = \beta \bar{x}_N + \bar{x}_N \frac{\sum_{i=1}^{N} I_i x_i e_i}{\sum_{i=1}^{N} I_i x_i^2}.$$

Hence, $E\{\hat{Y}_{\mathrm{RG}}\} = \beta \bar{x}_N$ and $\hat{Y}_{\mathrm{RG}}$ is an unbiased predictor of μ_N. The conditional prediction MSE, given $\mathbf{s}_n$, is $\mathrm{PMSE}\{\hat{Y}_{\mathrm{RG}} \mid \mathbf{s}_n\} = \dfrac{\sigma^2}{N^2} \left[N + \dfrac{N^2 \bar{x}_N^2}{\sum_{i \in s_n} x_i^2} - 2N \bar{x}_N \dfrac{n \bar{X}_n}{\sum_{i \in s_n} x_i^2} \right].$

Chapter 7: Time Series Analysis and Prediction

7.2   We are first fitting the seasonal trend to the data in SeasCom data set. We use the linear model
$Y = X\beta + \varepsilon$, where the Y vector has 102 elements, which are in data set. X is a 102×4 matrix of four column vectors.

Let tx be a set of 102 integers $c(1:102)$. The first vector of X is the column of 1's, i.e., $jay(102,1)$. The second column consists of the elements of $(tx-51)/102$. The third column consists of $\cos(pi*tx/6)$, and the fourth column consists of $\sin(tx*pi/6)$. The least squares estimates of β is $b = (101.08252, 106.81809, 10.65397, 10.13015)'$. The fitted trend is $Y_t = Xb$. (see Figure 7.14).

The deviations of the data from the cyclical trend is in the file $Dev = Y - Yt$. A plot of this deviations shows something like randomness. Indeed the correlations between adjacent data points are $corr(X_t, X_{t-1}) = -0.191$, and $corr(X_t, X_{t-2}) = 0.132$.

7.3   According to equation (7.2.2), the auto-correlation in an MA(q) is (for $0 \le h \le q$) is

$$\rho(h) = \sum_{j=0}^{q-h} \beta_j \beta_{j+h} / \sum_{j=0}^{q} \beta_j^2.$$

Notice that $\rho(h) = \rho(-h)$, and $\rho(h) = 0$ for all $h > q$.

7.4   The covariances and correlations are given in the following table:

h	0	1	2	3	4	5
$K(h)$	3.228	2.460	0.542	0.541	1.028	0.550
$\rho(h)$	1		0.762	0.168	0.167	0.380	0.170

7.5   We consider the AQ(∞), given by with coefficients $\beta_j = q^j$, with $0 < q < 1$. In this case, $E\{X_t\} = 0$, and $V\{X_t\} = \sigma^2 \sum_{j=0}^{\infty} q^{2j} = \sigma^2/(1 - q^2)$.

7.6   We consider the AR(1) $X_t = 0.75X_{t-1} + \varepsilon_t$.
  (i)   This time series is equivalent to $X_t = \sum_{j=0}^{\infty} (-0.75)^j \varepsilon_{t-j}$. Hence, it is covariance stationary.
  (ii) $E\{X_t\} = 0_-$
  (iii)   According to the Yule–Walker equations,
$$\begin{bmatrix} 1 & -0.75 \\ -0.75 & 1 \end{bmatrix} \begin{bmatrix} K(0) \\ k(1) \end{bmatrix} = \begin{bmatrix} \sigma^2 \\ 0 \end{bmatrix}$$
  It follows that $K(0) = 2.285714\sigma^2$ and $K(1) = 1.714286\sigma^2$.

7.7   The given AR(2) can be written as $X_t - 0.5X_{t-1} + 0.3X_{t-2} = \varepsilon_t$.
  (1)   The corresponding characteristic polynomial is $P_2(z) = 0.3 - 0.5z + z^2$. The two characteristic roots are $z_{1,2} = 0.5/2 \pm (1.07)^{1/2}i/2$. These two roots belong to the unit circle. Hence, this AR(2) is covariance stationary.
  (2)   We can write $A_2(z)X_t = \varepsilon_t$, where $A_2(z) = 1 - 0.5z^{-1} + 0.3z^{-2}$. Furthermore, $\phi_{1,2}$ are the two roots of $A_2(z) = 0$.
It follows that $X_t = (A_2(z))^{-1}\varepsilon_t = \varepsilon_t + 0.5\varepsilon_{t-1} - 0.08\varepsilon_{t-2} - 0.205\varepsilon_{t-3} - 0.0761\varepsilon_{t-4} + 0.0296\varepsilon_{t-5} + \cdots$

7.8   The Yule–Walker equations are:

$$\begin{bmatrix} 1 & -0.5 & 0.3 & -0.2 \\ -0.5 & 1.3 & -0.2 & 0 \\ 0.3 & -0.7 & 1 & 0 \\ -0.2 & 0.3 & -0.5 & 1 \end{bmatrix} \cdot \begin{bmatrix} K(0) \\ K(1) \\ K(2) \\ K(3) \end{bmatrix} = \begin{bmatrix} 1 \\ 0 \\ 0 \\ 0 \end{bmatrix}$$

The solution is $K(0) = 1.2719, K(1) = 0.4825, K(2) = -0.0439, K(3) = 0.0877$.

7.9   The Toeplitz matrix is

$$
R_4 = \begin{bmatrix}
1.0000 & 0.3794 & -0.0235 & 0.0690 \\
0.3794 & 1.0000 & 0.3794 & -0.0235 \\
-0.0235 & 0.3794 & 1.0000 & 0.3794 \\
0.0690 & -0.0235 & 0.3794 & 1.0000
\end{bmatrix}
$$

7.10 This series is an ARMA(2,2), given by the equation

$$
(1 - z^{-1} + 0.25z^{-2})X_t = (1 + 0.4z^{-1} - 0.45z^{-2})\varepsilon_t.
$$

Accordingly,

$X_t = (1 + 0.4z^{-1} - 0.45z^{-2})(1 - z^{-1} + 0.25z^{-2})^{-1}\varepsilon_t = \varepsilon_t + 1.4\varepsilon_{t-1} + 0.7\varepsilon_{t-2} + 0.35\varepsilon_{t-3} + 0.175\varepsilon_{t-4} + 0.0875\varepsilon_{t-5} + 0.0438\varepsilon_{t-6} + 0.0219\varepsilon_{t-7} + 0.0109\varepsilon_{t-8} + \cdots$

7.11 Let X denote the DOW1941 data set. We create a new set, Y of second order difference, i.e., $Y_t = X_t - 2X_{t-1} + X_{t-2}$.

In the following table, we present the autocorrelations, acf, and the partial autocorrelations, pacf, of Y

h	acf	S/NS	pacf	S/NS
1	-0.342	S	-0.342	S
2	-0.179	S	-0.334	S
3	0.033	NS	-0.207	S
4	0.057	NS	-0.097	NS
5	-0.080	NS	-0.151	S
6	-0.047	NS	-0.187	S
7	-0.010	NS	-0.227	S
8	0.128	NS	-0.066	NS
9	-0.065	NS	-0.177	S
10	-0.076	NS	-0.188	S
11	0.053	NS	-0.171	S

S denotes significantly different from 0. NS denotes not significantly different from 0.

All other correlations are not significant. It seems that the ARIMA(11,2,2) is a good approximation.

7.12 In Figure 7.15, we present the seasonal data SeasCom, and the one-day ahead predictions. We see an excellent prediction.

Chapter 8: Modern Analytic Methods

8.1

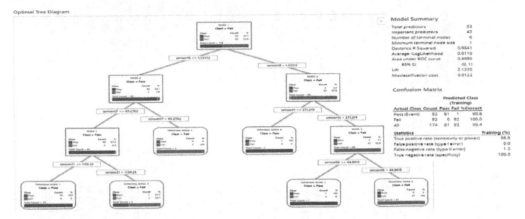

Figure 8.1.1 *CART analysis of Stats and 63 sensors using MINITAB v17*

8.2

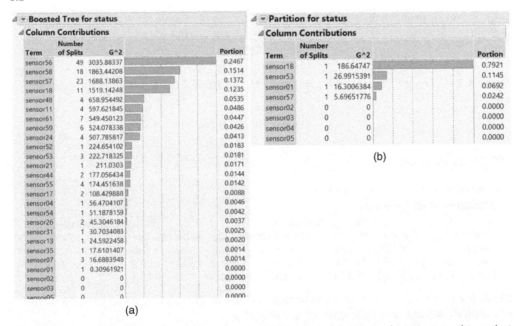

(a)

(b)

Figure 8.2.1 *Column contribution from boosted tree (a) and partition- decision tree (b) analysis of Status and 63 sensors using JMP v. 15.2*

8.3

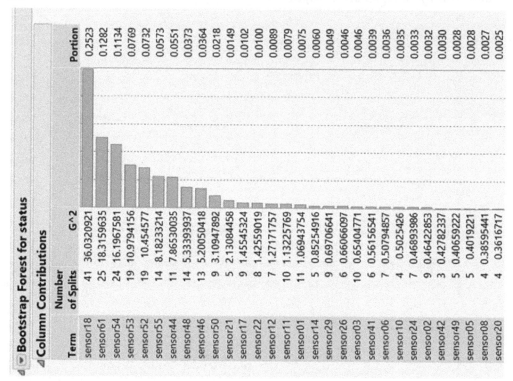

Figure 8.3.1   *Column contribution from boosted forest (random forest) analysis of Status and 63 sensors using JMP v. 15.2*

8.4

Model Comparison

Target status missing a predictor for category Fail

Measures of Fit for status

Creator	.2 .4 .6 .8	Entropy RSquare	Generalized RSquare	Mean -Log p	RMSE	Mean Abs Dev	Misclassification Rate	N
Partition		0.9642	0.9830	0.0248	0.0709	0.0200	0.0057	174
Boosted Tree		0.9958	0.9981	0.0029	0.0057	0.0029	0.0000	174
Bootstrap Forest		0.8974	0.9489	0.071	0.1136	0.0624	0.0057	174

Figure 8.4.1   *Comparison of partition-decision tree, boosted tree, and boosted forest (random forest) analysis of Status and 63 sensors using JMP v. 15.2*

8.5

Model Comparison

Measures of Fit for testResult

Creator		Entropy RSquare	Generalized RSquare	Mean-Log p	RMSE	Mean Abs Dev	Misclassification Rate	N
Partition		0.8571	0.9766	0.234	0.2860	0.1508	0.0977	174
Bootstrap Forest		0.8091	0.9659	0.3127	0.3169	0.2232	0.0517	174

Figure 8.5.1 *Comparison of partition-decision tree and boosted forest (random forest) analysis of testResult and 63 sensors using JMP v. 15.2 (boosted tree in JMP apply to only binary responses)*

8.6

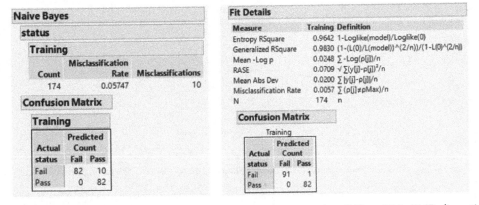

Naive Bayes

status

Training

	Misclassification	
Count	Rate	Misclassifications
174	0.05747	10

Confusion Matrix

Training

Actual status	Predicted Count	
	Fail	Pass
Fail	82	10
Pass	0	82

Fit Details

Measure	Training	Definition
Entropy RSquare	0.9642	1-Loglike(model)/Loglike(0)
Generalized RSquare	0.9830	(1-(L(0)/L(model))^(2/n))/(1-L(0)^(2/n))
Mean -Log p	0.0248	$\sum$ -Log(p[j])/n
RASE	0.0709	$\sqrt{\sum(y[j]-p[j])^2/n}$
Mean Abs Dev	0.0200	$\sum$ |y[j]-p[j]|/n
Misclassification Rate	0.0057	$\sum$ (p[j]≠pMax)/n
N	174	n

Confusion Matrix

Training

Actual status	Predicted Count	
	Fail	Pass
Fail	91	1
Pass	0	82

Figure 8.6.1 *Naïve Bayes analysis of Status and 63 sensors using JMP v. 15.2 (JMP discretizes continuous variables using normal distributions and prior bias*

8.7

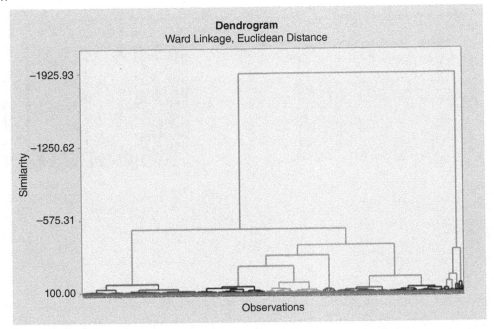

Figure 8.7.1 *Dendrogram of hierarchical clustering of 63 sensors using MINITAB v. 17*

8.8

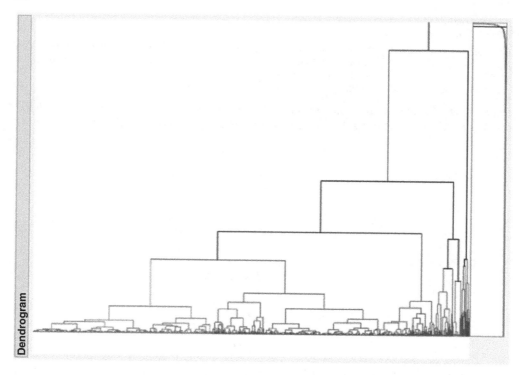

Figure 8.8.1   *Dendrogram of hierarchical clustering of 63 sensors using JMP v.15.2*

8.9

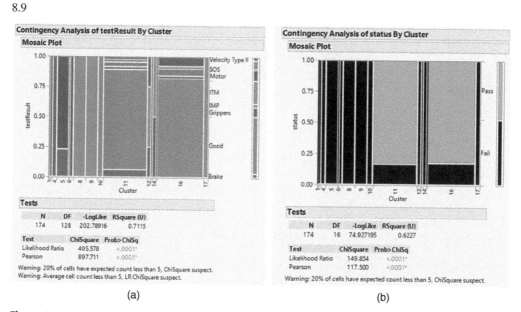

(a)

(b)

Figure 8.9.1   *Comparing of cluster and testResult and status with clusters determined by K-means 17 clustering algorithm, using JMP 15.2*

8.10

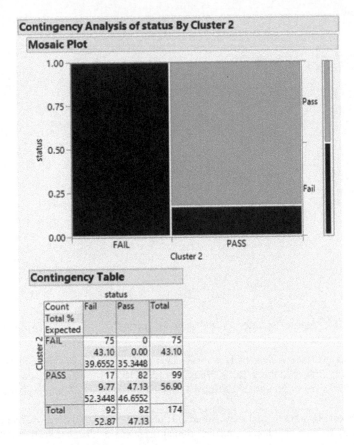

Figure 8.10.1   *Confusion matrix derived from assigning PASS to clusters 11 and 16, using JMP 15.2*

8.11 Supervised application: "modeling the number of cases per hospital over time and the number of days a hospital is at peak capacity."

   Unsupervised application: "Account for the social and geographic connections."

8.12

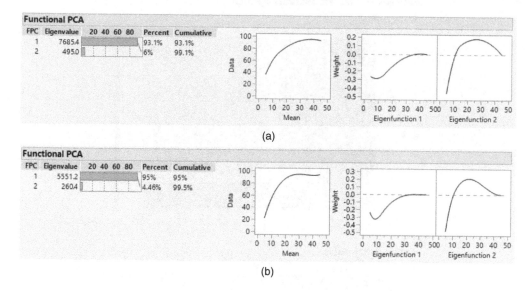

(a)

(b)

Figure 8.12.1   *Mean dissolution of reference (a) and test (b) samples using JMP v.15.2*

8.13 Article 1: An evidence review of face masks against COVID-19, PNAS January 26, 2021, 118 (4),   e2014564118;   https://doi.org/10.1073/pnas.2014564118.   https://www.pnas.org/content/118/4/e2014564118#:~:text=Their%20models%20show%20that%20mask,the%20public%20wear%20face%20masks.

Article 2: Face masks: what the data say, Nature, October 6th, 2020, https://www.nature.com/articles/d41586-020-02801-8. https://media.nature.com/original/magazine-assets/d41586-020-02801-8/d41586-020-02801-8.pdf.

8.14 Article 1: https://www.nytimes.com/2020/05/30/opinion/sunday/coronavirus-globalization.html.

Article 2: https://www.palmbeachpost.com/story/opinion/columns/more-voices/2020/09/24/thomas-friedman-donald-trump-and-covid-19/3504363001.

Chapter 9: The Role of Statistical Methods in Modern Industry and Services

9.1 Examples of 100% inspection:

(a) Airplane pilots rely heavily on checklists before takeoff. These lists provide a systematic check on safety and functional items that can be verified by relatively simple pilot inspection. The lists are used in every flight.

(b) The education system relies on tests to evaluate what students learn in various courses and classes. Tests are administered to all students attending a class. The grades received on these tests reflect the student's ability, the course syllabus and training material, the instructor's ability, the classroom environment, etc.

(c) Production lines of electronic products typically include automatic testers that screen products as PASS or FAIL. Failed products are directed to rework departments where the problems are diagnosed and fixed before the products are retested.

Questions for discussion:

(a) How well does 100% inspection work? Provide examples where it did not work.

(b) What are the alternatives to 100% inspection?

(c) What needs to complement 100% inspection for long-term improvements?

9.2 Search of information on quality initiatives:

Internet searches provide easy access to huge amounts of data on quality initiatives. The journal *Quality Progress*, published by the American Society for Quality, has a special section listing of up-to-date sites on the World Wide Web where such information is available.

Suggestions for discussion: Ask students to present their findings by focusing on business impact of quality initiatives. For example, have them track stock values of companies with quality initiatives relative to the S&P and Dow Jones index.

9.3 Examples of production systems:

(a) *Continuous flow production*: Polymerization is used in the petroleum industry and in the manufacture of synthetic rubber. In polymerization, a reaction occurs in which two or more molecules of the same substance combine to form a compound from which the original substance may or may not be regenerated. Typical parameters affecting the process are feed rate, polymerizer temperature, and sludge levels in the separator kettles. Typical measured responses are yield, concentration, color, and texture.

(b) *Job shop*: Modern print houses are typical job shop operations. Customers provide existing material, computer files, or just an idea. The print house is responsible for producing hard copies in various sizes, colors, and quantities. The process involves several steps combining human labor with machine processing.

(c) *Discrete mass production*: A college or university is, in fact, a discrete mass production system where students are acquiring knowledge and experience through various combinations of classes, laboratories, projects, and homework assignments.

Suggestions for discussion: Arrange for a trip to various organizations such as the local print shop, a bank, a cafeteria, a manufacturing plant, or the central post office processing center and have the students map the processes they observed in the trip. Discuss in class the possible parameters affecting the process and how the process should be controlled.

9.4 Continuous flow production cannot be controlled by 100% product inspection.

Suggestions for discussion: List various continuous flow systems and discuss the parameters that affect them. Show, through examples, that 100% inspection is impractical and that alternative process control approaches do work.

9.5 What management approach characterizes:
  (a) *A school system?*: In most cases, school systems rely on individual learning combined with 100% inspection. Many schools also encourage cooperative learning efforts such as team projects.
  (b) *A military unit?*: Traditionally, this is a classical organization operating with strict regulations and 100% inspection. Improvements are achieved through training and exercises with comprehensive performance reviews.
  (c) *A football team?*: Serious sports teams invest huge efforts in putting together a winning team combination. This includes screening and selection, team building exercises, on-going feedback during training and games, and detailed analysis and review of self and other teams performance.

Suggestions for discussion: Discuss additional organizations you are familiar with focusing on how they are currently managed and possible alternatives to that management approach.

9.6 Examples of how you personally apply the four management approaches:
  (a) *As a student*: With an old car that you maintain with a "don't fix what ain't broke" strategy you are fire fighting. Reviewing your bank account statements is typically done on a 100% basis. Serious learning involves on going process control to assure success in the final exam. Proper preparation, ahead of time, through self study and learning from the experience of others is a form of ensuring quality by design.
  (b) *In your parent's home*: This is a personal exercise for self thinking.
  (c) *With your friends*: This is also a personal exercise for self thinking.

Chapter 10: Basic Tools and Principles of Process Control

10.1

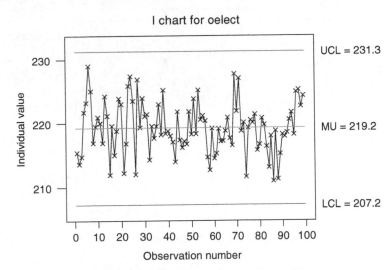

I chart for oelect

UCL = 231.3

MU = 219.2

LCL = 207.2

Figure 10.1.1 *Individual chart for OELECT data*

It seems that the data are randomly distributed around a mean value of 219.25 (Figure 10.1.1).

10.2   See Figure 10.2.1

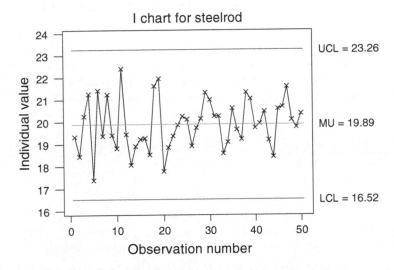

I chart for steelrod

UCL = 23.26

MU = 19.89

LCL = 16.52

Figure 10.2.1 *Individual chart for steel rods data*

It seems that the data are randomly distributed around a mean value of 19.89.

10.3   No patterns of nonrandomness are apparent.

10.4

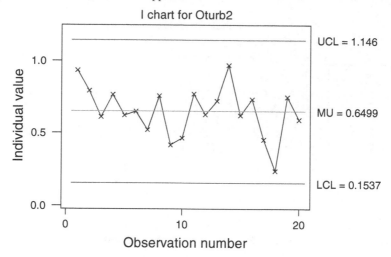

Figure 10.4.1   *Individual chart for means*

No pattern of nonrandomness is apparent (Figure 10.4.1).

10.5   If $\bar{X}_n \sim N\left(\mu - \sigma, \dfrac{\sigma}{\sqrt{n}}\right)$, the proportion of points expected to fall outside the control limits is $\pi_n \equiv$

$\text{Pr}\left\{\bar{X}_n < \mu - \dfrac{3\sigma}{\sqrt{n}}\right\} + \text{Pr}\left\{\bar{X}_n > \mu + \dfrac{3\sigma}{\sqrt{n}}\right\}$. (a) For $n = 4$, $\pi_n = 0.159$; (b) for $n = 6$, $\pi_n = 0.291$; (c) for $n = 9$, $\pi_n = 0.500$.

10.6   Performing a MINITAB capability analysis, we found that $C_p = 0.93$. Although $C_p = 0.93$, there is room for improvement by designing the process with reduced variability. See Chapter 12.

10.7   With $\bar{X} = 219.25$, $S = 4.004$, $n = 99$, $\xi_U = 230$, and $\xi_L = 210$, we obtained $\hat{C}_{pl} = 0.77$, $\hat{C}_{pu} = 0.895$ and $\hat{C}_{pk} = 0.77$. Since $F_{0.975}[1, 98] = 5.1818$, confidence intervals for C_{pl} and C_{pu}, at 0.975 confidence level, are

	Lower Limit	Upper Limit
C_{pl}	0.6434	0.9378
C_{pu}	0.7536	1.0847

Finally, a 0.95-confidence interval for C_{pk} is $(0.6434, 0.9378)$.

10.8   Using the STEELROD.csv data with $\xi_L = 19$ and $\xi_U = 21$, we obtained $\hat{C}_p = 0.297$, $\hat{C}_{pu} = 0.329$, $\hat{C}_{pl} = 0.264$, and $\hat{C}_{pk} = 0.264$. A 0.95-confidence interval for C_{pk} is $(0.1623, 0.3958)$.

10.9   If all the seven control parameters are at their low level, the mean cycle time of 20 observations is $\bar{X}_{20} = 0.7745$. This value is outside the specification limits, 0.1 and 0.5. Thus, the process at the low parameter values is incapable of satisfying the specs. The standard deviation is 0.3675.

    (a) Thus, if the process mean can be moved to 0.3 the C_p value is 0.18, which is low because σ is too high. There is no point in computing.

    (b) Under the high values of the seven control parameters, we get

    (c) $\bar{X}_{20} = 0.452$, $S_{20} = 0.1197$. For these values, $C_p = 0.557$ and $C_{pk} = 0.133$.

    (d) The confidence interval for C_{pk}, at level 0.95, is $(0.0602, 0.2519)$.

(e) As mentioned above, the gas turbine at low control levels is incapable of satisfying the specifications of 0.3 ± 0.2 seconds. At high levels its C_{pk} is not greater than 0.252. In Chapter 11, we study experimental methods for finding the combination of control levels, which maximizes the capability.

10.10 (a) 1.8247　　(b) 1.8063 and 1.8430　　(c) $C_{pk} = 0.44$.

10.11 (a) $\bar{\bar{X}} = \frac{10,950}{30} = 365$, $\bar{S} = 1.153$, $n = 4$, $\hat{\sigma} = \bar{S}/c(n) = 1.252$. The control limits for $\bar{X}_4$ are

$$UCL = 365 + 3 \times \frac{1.252}{\sqrt{4}} = 366.878 \text{ and } LCL = 365 - 3 \times \frac{1.252}{\sqrt{4}} = 363.122.$$

(b) When $X \sim N(\mu = 365, \sigma = 1.252)$, $1 - \Pr\{365 < X < 375\} = 1 - \Phi\left(\frac{375 - 365}{1.252}\right) + \Phi\left(\frac{365 - 365}{1.252}\right) = 0.5$.

(c) When $X \sim N(\mu = 370, \sigma = 1.252)$, $1 - \Pr\{365 < X < 375\} = 2 \times \left(1 - \Phi\left(\frac{5}{1.252}\right)\right) = 0.000065$.

(d) $UCL = 370 + 3 \times \frac{1.252}{\sqrt{4}} = 371.878$ and $LCL = 370 - 3 \times \frac{1.252}{\sqrt{4}} = 368.122$.

(e) When $\bar{X}_4 \sim N(\mu = 365, \sigma = 1.252/\sqrt{4})$, $\Pr\{368.122 < \bar{X}_4 < 371.878\} = \Phi\left(\frac{371.878 - 365}{1.252/\sqrt{4}}\right) - \Phi\left(\frac{368.122 - 365}{1.252/\sqrt{4}}\right) = 0$.

(f) The process capability before and after the adjustment is

	Before	After
C_p	1.331	1.331
C_{pk}	0	1.331

10.12 (a) In Figure 10.12.1a,b, we see the Pareto charts for the month of October and the second week in November, respectively.

(b) Using the test described in Section 10.6,we find that the improvement team has produced significant differences in the type of nonconformities. In particular, the difference in the "Insufficient solder category is significant at the 1% level and the difference in the "Too much solder category is significant at the 10% level (see Table 10.5). The computations involved are summarized in the table below:

Category	October	p	November	Expected	(Obs. - Exp.)	Z
Missing	293	0.253	34	41.239	−7.239	−1.304
Wrong	431	0.372	52	60.636	−8.636	−1.399
Too Much	120	0.103	25	16.789	8.211	2.116
Insufficient	132	0.114	34	18.582	15.418	3.800
Failed	183	0.158	18	25.754	−7.754	−1.665
Total	1159		163			

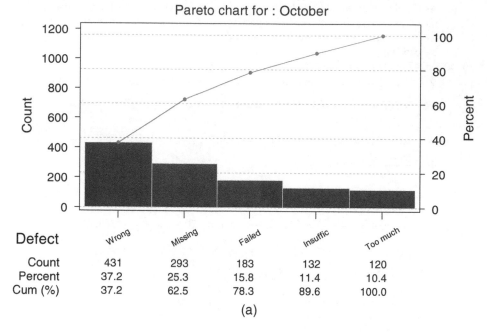

Pareto chart for : October

Defect	Wrong	Missing	Failed	Insuffic	Too much
Count	431	293	183	132	120
Percent	37.2	25.3	15.8	11.4	10.4
Cum (%)	37.2	62.5	78.3	89.6	100.0

(a)

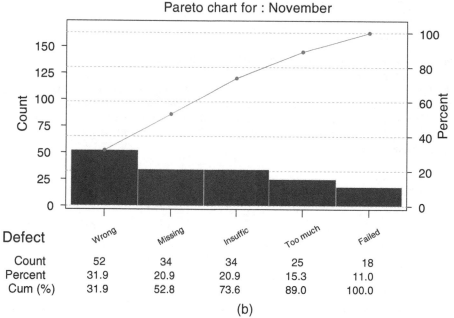

Pareto chart for : November

Defect	Wrong	Missing	Insuffic	Too much	Failed
Count	52	34	34	25	18
Percent	31.9	20.9	20.9	15.3	11.0
Cum (%)	31.9	52.8	73.6	89.0	100.0

(b)

Figure 10.12.1   *(a) Pareto chart for October. (b) Pareto chart for second week of November*

10.13 From the data, $\bar{\bar{X}} = 15.632$ ppm and $\bar{R} = 3.36$. An estimate of the process standard deviation is $\hat{\sigma} = \bar{R}/d(5) = 1.4446$.

(a) $C_{pu} = \dfrac{18 - 15.632}{3 * \hat{\sigma}} = 0.546.$

(b) The proportion expected to be out of spec is 0.051.

(c) The control limits for $\bar{X}$ are LCL = 13.694 and UCL = 17.570. The control limits for R_5 are LCL = 0 and UCL = 7.106.

10.14 Part I: See the control charts in Figure 10.14.1a. The change after the 50th cycle time is so big that the means of the subgroup are immediately out of the control limits. The shift has no influence on the ranges within subgroups.

Part II: See the control charts in Figure 10.14.1b. The effect of adding the random number is to increase the mean of all cycle times by 0.5. The variability is increased too.

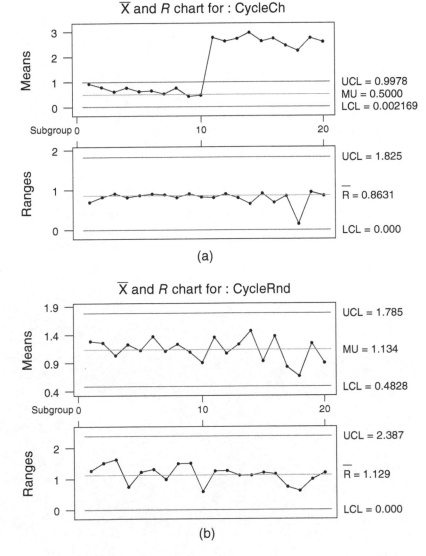

Figure 10.14.1   *(a) $\bar{X}$ and R control charts. (b) $\bar{X}$ and R control charts*

10.15 Part I: Setting all seven control parameters at lower values we obtained $\bar{\bar{X}} = 0.6499$ and $\bar{S} = 0.3587$.
  (a) The control limits for $\bar{X}_5$ are LCL $= 0.138$ and UCL $= 1.162$. The control limits for S_5 are LCL $= 0$ and UCL $= 0.7634$.
      Part II: We obtained $\bar{\bar{X}} = 0.6689$, $\bar{S} = 0.3634$.
  (b) The control limits for $\bar{X}_{10}$ are LCL $= 0.3145$ and UCL $= 1.0232$. The control limits for S_{10} are LCL $= 0.0992$ and UCL $= 0.6275$.
  (c) We see that the control limits of $\bar{X}_{10}$ and of S_{10} are closer to the centerline than those of $\bar{X}_5$ and S_5, respectively.

Chapter 11: Advanced Methods of Statistical Process Control

11.1   (i) Using MINITAB, we obtained $Q_1 = 10$, $M_e = 12$, and $Q_3 = 14$.
   (ii) From equations (11.1.3) and (11.1.4), we get $\mu_R = 13$ and $\sigma_R = 2.3452$.
   (iii) $\Pr\{10 \le R \le 16\} = 0.8657$.
   (iv) Using the normal approximation $\Pr\{10 \le R \le 16\} \approx 0.8644$.

11.2   We obtained the following output with the MINITAB command:

```
MTB> Runs C1.
```

```
Runs Test
Cyclt
K = 0.6525
The observed no. of runs = 26
The expected no. of runs = 25.6400
22 Observations above K 28 below
The test is significant at 0.9169
Cannot reject at alpha = 0.05
```

Additional Run tests are obtained by performing a Run Chart under SPC. This is shown in Figure 11.2.1. All run tests show no significant deviations from randomness.

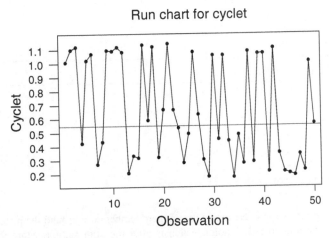

Run chart for cyclet

Number of runs about median:	23.0000	Number of runs up or down:	34.0000
Expected number of runs:	26.0000	Expected number of runs:	33.0000
Longest run about median:	6.0000	Longest run up or down:	4.0000
Approx p-value for Clustering:	0.1956	Approx p-value for Trends:	0.6337
Approx p-value for Mixtures:	0.8044	Approx p-value for Oscillation:	0.3663

Figure 11.2.1   *Run chart for cycle times*

11.3   (i) For $n = 50$, $E\{R^*\} = 33$.
   (ii) Using the cycle time data, $R^* = 34$ (See Figure 11.2.1).
   (iii) We have $\sigma^* = 2.9269$, $\alpha_u^* = 1 - \Phi\left(\dfrac{1}{2.9269}\right) = 0.3663$. The deviation is not significant.

   (iv) The probability that a sample of size 50 will have at least one run of length 5 or longer is 0.102.

11.4

```
MTB> Runs 'Ln_YarnS'.
```

```
Runs Test
Ln_YarnS
K = 2.9238
The observed no. of runs = 49
The expected no. of runs = 50.9200
48 Observations above K 52 below
The test is significant at 0.6992
Cannot reject at alpha = 0.05
```

11.5 The following is the MINITAB output for a runs test: K for XBAR is the grand mean and K for STD is the mean of the 50 standard deviations.

```
MTB> Runs 'XBAR'.
```

```
Runs Test
XBAR
K = 0.4295
The observed no. of runs = 26
The expected no. of runs = 25.9600
26 Observations above K 24 below
The test is significant at 0.9909
Cannot reject at alpha = 0.05
```

```
MTB> Runs 'STD'.
```

```
Runs Test
STD
K = 0.1261
The observed no. of runs = 26
The expected no. of runs = 25.6400
28 Observations above K 22 below
The test is significant at 0.9169
Cannot reject at alpha = 0.05
```

11.6 (i) In simulating the piston cycle times, the ambient temperature around the piston is increased 10% per cycle after the shift point, which is after the 16th sample. Store the data in file XBAR21.MTW. We see in Figure 11.6.1a,b the effect of this change in temperature on the means $\bar{X}$ and standard deviations. The run tests reflect the nonrandomness of the sequence, after the change point.

```
MTB> Runs 'XBAR21'.
```

```
Runs Test
XBAR21
K = 0.4642
The observed no. of runs = 7
The expected no. of runs = 24.0400
32 Observations above K 18 below
The test is significant at 0.0000
```

```
MTB> Runs 'STD21'.
```

```
Runs Test
STD21
K = 0.0604
The observed no. of runs = 16
The expected no. of runs = 26.0000
25 Observations above K 25 below
The test is significant at 0.0044
```

Run chart for X̄

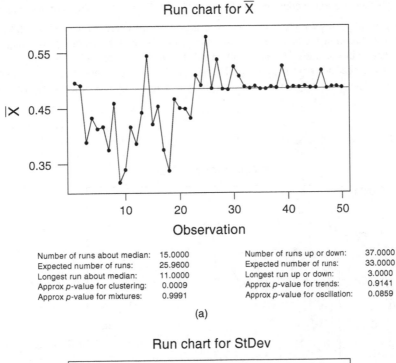

Number of runs about median:	15.0000	Number of runs up or down:	37.0000
Expected number of runs:	25.9600	Expected number of runs:	33.0000
Longest run about median:	11.0000	Longest run up or down:	3.0000
Approx *p*-value for clustering:	0.0009	Approx *p*-value for trends:	0.9141
Approx *p*-value for mixtures:	0.9991	Approx *p*-value for oscillation:	0.0859

(a)

Run chart for StDev

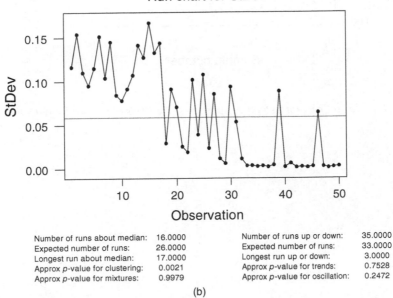

Number of runs about median:	16.0000	Number of runs up or down:	35.0000
Expected number of runs:	26.0000	Expected number of runs:	33.0000
Longest run about median:	17.0000	Longest run up or down:	3.0000
Approx *p*-value for clustering:	0.0021	Approx *p*-value for trends:	0.7528
Approx *p*-value for mixtures:	0.9979	Approx *p*-value for oscillation:	0.2472

(b)

Figure 11.6.1 (a) Run chart for sample means. (b) Run chart for sample standard deviations

(ii) In this part, the shift is in the standard deviation of the spring coefficient, after $k = 25$ samples. We see that after 35 samples the piston does not function properly. This is shown by a run chart of the means or the standard deviations.

(iii) The JMP and R piston simulators can simulate the function of the piston when the ambient temperature drifts from t to $1.1t$ and then the temperature is corrected back to t. Simulate this and look at the run charts both for $\bar{X}$ and for S.

(iv) When changes were introduced into the factors that affect piston performance, it was detectable in the run charts and the hypothesis of randomness was rejected in every case. However, when the factors are kept constant as in Exercise 11.5, the hypothesis of randomness was not rejected.

11.7 In Figure 11.7.1a, we present the p-chart of the data. This chart is obtained by MINITAB, under control charts, by inserting the number of defectives to $C1$ and setting the subgroup size equal to 1000. As we see, the data points of week 16 and 18 fall outside the upper control limit. Also, the last eight weeks show a significant trend down (improvement). In Figure 11.7.1b, we present the p-chart without the 16 and 18 week points. The points are now all in control, but the pattern of the last eight weeks remains.

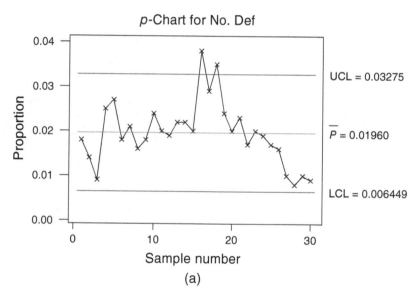

Figure 11.7.1 (a) p-Chart for defectives substrates. (b) Revised p-chart for defective substrates

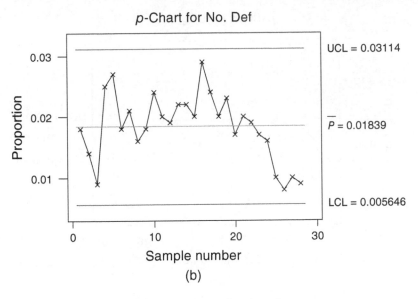

p-Chart for No. Def

(b)

Figure 11.7.1   *(Continued)*

11.8   The average proportion of defectives for line 1 is $\hat{p}_1 = 0.005343$, while that for line 2 is $\hat{p}_2 = 0.056631$. The difference is very significant. Indeed, $Z = \dfrac{\hat{p}_2 - \hat{p}_1}{\sqrt{\dfrac{\hat{p}_1(1 - \hat{p}_1)}{N_1} + \dfrac{\hat{p}_2(1 - \hat{p}_2)}{N_2}}} = 34.31$,

where $N_1 = \sum_{i=1}^{12} n_1(i)$ and $N_2 = \sum_{i=1}^{12} n_2(i)$.

In Figure 11.8.1a,b, we present the *p*-charts for these two production lines. We see that the chart for line 1 reveals that the process was at the beginning (point 2), out of control.

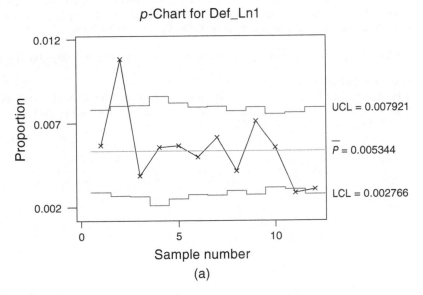

p-Chart for Def_Ln1

(a)

Figure 11.8.1   *(a) p-Chart for Line 1 defects. (b) p-Chart for Line 2 defects*

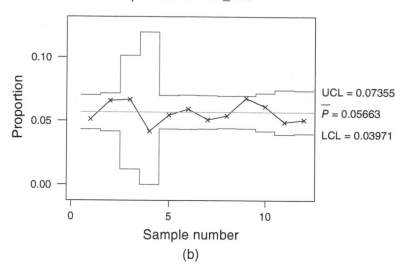

Figure 11.8.1   *(Continued)*

11.9   Let θ be the probability of not detecting the shift in a given day. Solving $\theta^5 = 0.2$ we get $\theta = 0.72478$. We find the smallest n for which

$$\theta \geq \Phi\left(\frac{0.01 + 3\sqrt{\dfrac{0.01 \times 0.99}{n}} - 0.05}{\sqrt{\dfrac{0.05 \times 0.95}{n}}}\right) - \Phi\left(\frac{0.01 - 3\sqrt{\dfrac{0.01 \times 0.99}{n}} - 0.05}{\sqrt{\dfrac{0.05 \times 0.95}{n}}}\right).$$

The solution is $n = 16$. Using equation (11.3.7), we get $n \approx 18$. This solution and the one shown above both use the normal approximation to the binomial. Another approach, which may be preferable for small sample sizes, is to use binomial distribution directly. We find the smallest n for which

$$B(n \times 0.01 + 3\sqrt{n \times 0.01 \times 0.99}; n, 0.05) - B(n \times 0.01 - 3\sqrt{n \times 0.01 \times 0.99}; n, 0.05) \leq \theta.$$

In this case, the solution is $n = 7$.

11.10   (i) In Figure 11.10.1 we see the $\bar{X}$ and S control charts. Points for days 21 and 22 are outside the UCL for $\bar{X}$. Excluding these points, the recalculated control chart limits for $\bar{X}$ are LCL = 29.547 and UCL = 45.966. Also, $\bar{\bar{X}} = 37.757$.

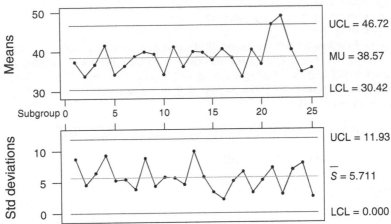

Figure 11.10.1 *X-bar and S chart of dock to stock cycle times*

(ii) The following is the MINITAB output for testing the significance of the difference between 45 and the means of days 21 and 22. We find that only $\bar{X}_{22}$ is significantly larger than 45.

```
MTB> TTest 45.0 'Day 21';
SUBC> Alternative 1.
```

```
T-Test of the Mean
Test of mu = 45.00 vs mu > 45.00
Variable    N    Mean    StDev    SE Mean     T     p-Value
Day 21      5    46.80   7.26     3.25       0.55    0.30
```

```
MTB> TTest 45.0 'Day 22';
SUBC> Alternative 1.
```

```
T-Test of the Mean
Test of mu = 45.00 vs mu > 45.00
Variable    N    Mean    StDev    SE Mean     T     p-Value
Day 22      5    49.00   3.00     1.34       2.98    0.020
```

(iii) Unusual long cycle times can be due to
   (a) Missing or misplaced information in accompanying paperwork
   (b) Missing or misplaced marks on package
   (c) Defective package
   (d) Non standard package
   (e) Wrong information on package destination
   (f) Overloaded stock room so that packages cannot be accepted
   (g) Misplaced packages.
   Additional data that can be collected to explain long cycle times:
   (a) Package destination
   (b) Stock room

    (c) Package size

    (d) Package weight

    (e) Package origin (country, supplier)

    (f) Package courier.

11.11 The average run length (ARL) of the modified Shewhart control chart with $a = 3$, $w = 2$, and $r = 4$ when $n = 10$ and $\delta = 0$ is 370.3. When $\delta = 1$, the ARL is 1.75 and when $\delta = 2$ the ARL is 1.0.

11.12 For $a = 3$, $w = 1$, and $r = 15$ when $n = 5$ the ARL is 33.4.

11.13 Samples should be taken every $h^0 = 34$ [minutes] = 0.57 [hours].

11.14 With $n = 20$ and $p_0 = 0.10$,

$$\text{OC}(p) = B(np_0 + 3\sqrt{np_0(1-p_0)}; n, p) - B(np_0 - 3\sqrt{np_0(1-p_0)}; n, p).$$

The values of OC(p), for $p = 0.05, 0.5 \; (0.05)$ are

p	OC(p)
0.05	0.999966
0.10	0.997614
0.15	0.978065
0.20	0.913307
0.25	0.785782
0.30	0.608010
0.35	0.416625
0.40	0.250011
0.45	0.129934
0.50	0.057659

11.15 The sample size n_0 is the smallest n for which

$$B(n \times 0.01 + 0.2985\sqrt{n}; n, 0.05) - B(n \times 0.01 - 0.2985\sqrt{n}; n, 0.05) \le 0.1.$$

From this we obtain, $n_0 = 184$.

Using equation (11.3.6), which is based on the normal approximation to the binomial, gives $n_0 \approx 209$.

11.16 (i) 4.5 [hours]; (ii) 7; (iii) For a shift of size $\delta = 1$, option 1 detects it on the average after 4.5 [hours]. Option 2 detects it on the average after $2 \times 1.77 = 3.5$ [hours]. Option 2 is preferred from the point of view of detection speed.

11.17 Here $K^+ = 221$ and $h^+ = 55.3724$. Figure 11.17.1 shows the "up" and "down" CUSUM charts. We see that the upper or lower limits are not crossed. There is no signal of change.

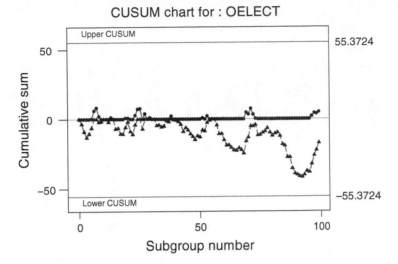

Figure 11.17.1 *CUSUM chart for OELECT*

11.18 Using $K^+ = 19.576$, $h^+ = 13.523$, $K^- = 10.497$, and $h^- = -9.064$, we obtained PFA = 0.014 and CED = 3.01 ± 0.235. Here the value of λ changed from 15 to 25.

11.19 The modified Shewhart chart with $a = 3$, $w = 1.5$, $r = 2.2$, and $n = 7$ yields ARL(0) = 136. This is close to the ARL(0) of the CUSUM. It is interesting that for this control scheme, ARL(0.52) = 6.96 and ARL(1.57) = 1.13. These are significantly smaller than those of the CUSUM.

11.20 According to equation (11.6.6), in the Poisson case

$$
\begin{aligned}
W_m &= \sum_{i=1}^{m-1} \prod_{j=i+1}^{m} R_j \\
&= \sum_{i=1}^{m-1} \exp\left\{ -(m-i)\delta + \sum_{j=i+1}^{m} X_j \log(\rho) \right\} \\
&= \sum_{i=1}^{m-2} \exp\left\{ -(m-1-i)\delta + \sum_{j=i+1}^{m-1} X_j \log(\rho) \right\} \cdot \exp\{-\delta + X_m \log(\rho)\} \\
&\quad + \exp\{-\delta + X_m \log \rho\} \\
&= (1 + W_{m-1})e^{-\delta + X_m \log \rho} \\
&= (1 + W_{m-1})R_m.
\end{aligned}
$$

11.21 The mean of OELECT.csv is $\bar{X} = 219.25$, and its standard deviation is $S = 4.004$. Plotting by MINITAB the EWMA, we obtain Figure 11.21.1, which shows that there is no significant shift in the data.

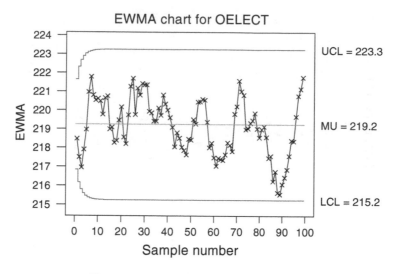

Figure 11.21.1   *EWMA chart for OELECT*

11.22 In Figure 11.22.1, we see the EWMA control chart for Diameter 1. The mean and standard deviation are $\bar{X} = 9.993$ and $S = 0.0164$. The target value for Diameter 1 is 10 mm. Notice in the chart that after a drop below 9.98 mm, the machine corrects itself automatically, and there is a significant run upwards toward the target value. Using the EWMA chart, an automatic control can be based on equation (9.8.11).

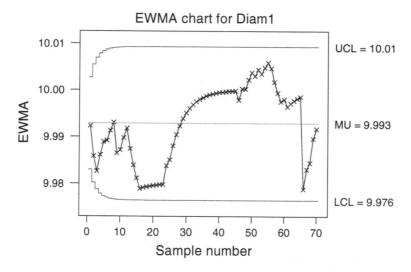

Figure 11.22.1   *EWMA chart for Diameter1*

11.23 Using the first 50 values of DOW1941 and the regression method outlined in the text, the least squares estimates of the initial parameter values are $\hat{\mu}_0 = 132.809$, $\hat{\delta} = -0.2556$ and $\hat{\sigma}_\varepsilon^2 + \hat{\sigma}_2^2 = 0.2995$. For $\hat{\sigma}_\varepsilon^2$, we chose the value of 0.15, and for w_0^2, the value 0.0015. The MINITAB commands used to obtain the Kalman filter estimates of the DOW1941 data are as follows:

```
MTB> read 'c:\dow1941.csv' C1
MTB> let k1 = 132.809
MTB> let k2 = 0.15
MTB> let k3 = 0.1495
MTB> let k4 = 0.0015
MTB> let k5 = -0.2556
MTB> let k6 = 2
MTB> name C2 'B(t)' C3 'Wsq(t)' C4 'Mu(t)'
MTB> let C2(1) = k2/(k2 + k3 + k4)
MTB> let C3(1) = C2(1) * (k3 + k4)
MTB> let C4(1) = C2(1) * (k1 + k5) + (1-C2(1)) * C1(1)
MTB> store 'Kalman'
STOR> let C2(k6) = k2/(k2 + k3 + C3(k6-1))
STOR> let C3(k6) = C2(k6) * (k3+C3(k6-1))
STOR> let C4(k6) = C2(k6) * (C4(k6-1) + k5) + (1 - C2(k6)) * C1(k6)
STOR> let k6 = k6 + 1
STOR> end
MTB> exec 'Kalman' 301
```

The table below shows the first 25 values of the DOW 1941 data as well as the values of B_t, w_t^2, and $\hat{\mu}_t$. Starting with $w_0^2 = 0.0015$, we see that w_t^2 converges fast to 0.09262. The values of $\hat{\mu}_t$ are very close to the data values. In Figure 11.23.1, we show the values of Y_t (circles) and of $\hat{\mu}_t$ (pluses).

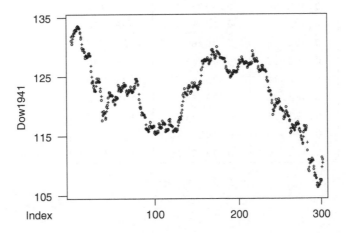

Figure 11.23.1 *Dow-Jones daily, 1941, and the Kalman filter*

C1	C2	C3	C4
Dow1941	B(t)	Wsq(t)	Mu(t)
131.13	0.498339	0.0752492	131.839
130.57	0.400268	0.0899598	130.976
132.01	0.385149	0.0922277	131.513
132.40	0.382919	0.0925621	131.963
132.83	0.382592	0.0926111	132.400
133.02	0.382545	0.0926183	132.685
133.02	0.382538	0.0926194	132.794
133.39	0.382537	0.0926195	133.064
133.59	0.382536	0.0926195	133.291
133.49	0.382536	0.0926195	133.316
133.25	0.382536	0.0926196	133.178
132.44	0.382536	0.0926196	132.624
131.51	0.382536	0.0926196	131.839
129.93	0.382536	0.0926196	130.562
129.54	0.382536	0.0926196	129.833
129.75	0.382536	0.0926196	129.684
129.24	0.382536	0.0926196	129.312
128.20	0.382536	0.0926196	128.528
128.65	0.382536	0.0926196	128.505
128.34	0.382536	0.0926196	128.306
128.52	0.382536	0.0926196	128.340
128.96	0.382536	0.0926196	128.625
129.03	0.382536	0.0926196	128.777
128.60	0.382536	0.0926196	128.570
126.00	0.382536	0.0926196	126.885

Chapter 12: Multivariate Statistical Process Control

12.1 Figure 12.1.1a shows a plot of the T^2 values, and in Figure 12.1.1b, we see the side-by-side boxplots of the TSQ values, by board number. The values of T^2 were computed, with **m** being the mean of the first 48 vectors. The UCL, is UCL = 17.1953. We see that even on the first nine cards there are a few outliers, that is, points whose T^2 is outside the control limits. All points from card 13 on and a majority of points from boards 10, 11, and 12 have T^2 values greater than UCL.

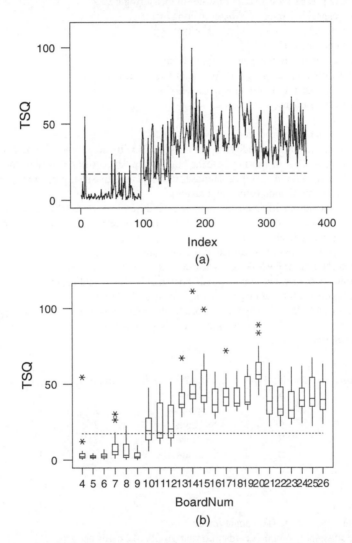

(a)

(b)

Figure 12.1.1 *(a) Plot of T^2 values with control limit. (b) Boxplots of T^2 values by board number*

Chapter 13: Classical Design and Analysis of Experiments

13.1   Preparing pancakes is a production process readily available for everyone to experiment with. The steps are as follows:

(a) Open readymix box

(b) Measure prespecified amount of readymix into measuring cup

(c) Pour prespecified amount of readymix into container

(d) Measure prespecified amount of water in measuring cup

(e) Add prespecified amount of water to container

(f) Mix material in container with one or two eggs

(g) Warm cooking pan

(h) Add oil to pan in liquid, solid or spray form

(i) Pour material into cooking pan for one or more pancakes

(j) Wait for darkening signs on the pancakes rim

(k) Turn over pancakes

(l) Wait again

(m) Remove pancakes from cooking pan.

Examples of response variables include subjective taste testing by household members with classification on a 1–5 scale, drop test for measuring pancake body composition (drop pancake from the table to the floor and count the number of pieces it decomposes into), quantitative tests performed in the laboratory to determine pancake chemical composition and texture.

Noise variables include environmental temperature and humidity, variability in amounts of readymix, water, oil and cooking time, stove heating capacity, and differences in quality of raw materials.

Examples of control variables include:

(a) Amounts of Readymix – less and more than present value

(b) Preheating time of cooking pan – less and more than present value

(c) Pouring speed of water into readymix container – slower and faster

(d) Waiting time before turn over – not just ready and overdone.

13.2   Response variable: Bond strength test. Controllable factors: Adhesive type, Curing pressure. Factor levels: Adhesive type – A, B, C, D, Curing pressure – low, nominal, high. Experimental layout: 4×3 full factorial experiment with four replications.

Experimental Run	Adhesive Type	Curing Pressure	Replication			
1	A	low	1	2	3	4
2	A	nominal	1	2	3	4
3	A	high	1	2	3	4
4	B	low	1	2	3	4
5	B	nominal	1	2	3	4
⋮	⋮	⋮	⋮	⋮	⋮	⋮

Experiment protocol:

(a) Prepare 48 $(3 \times 4 \times 4)$ computer cards.

(b) Randomly split computer cards into four groups, 12 cards per group.

(c) Randomly assign adhesive type to each group, 1 type per group.

(d) Randomly assign curing pressure level to group of 12 cards, 1 level for 4 cards.

(e) Attach to each computer card a sticker indicating experimental run, replication number, adhesive type, and curing pressure.

(f) Randomize the order of the 48 cards.

(g) Run experiment according to randomized order and factor levels indicated on sticker by keeping factors not participating in experiment fixed (e.g., amount of water).

(h) Perform bond strength test and record data.

(i) Analyze data with statistical model including main effects and interaction terms.

13.3 Blocking reduces the variability of the product relative to factors being studied. Examples include:

(a) Track and field athletes on a college team decided to investigate the effect of sleeping hours on athletic performance. The experiment consisted of sleeping a controlled amount of time prior to competitions. Individual athletes are natural experimental blocks.

(b) A natural extension of the example in Section 13.2 on testing shoe sole materials is an experiment designed to test car tires. A natural block consists of the four wheels of a car, with an additional blocking variable determined by the position of the tires (front or rear).

(c) Experiments performed on plants are known to be sensitive to environmental conditions. Blocking variables in such experiments consist of neighboring plots of land where soil, humidity, and temperature conditions are similar.

13.4 Main effects: τ_i^A, τ_j^B, τ_k^C each at three levels. First order interactions: τ_{ij}^{AB}, τ_{ik}^{AC}, τ_{jk}^{BC} each at nine levels. Second order interaction: τ_{ijk}^{ABC} at 27 levels.

13.5 $Y = \beta_0 + \beta_1 x_1 + \beta_2 x_2 + \beta_{12} x_1 x_2 + e$. If $\beta_{12} = 0$, then the model is additive. The regression of Y on x_1 (x_2) are parallel for different values of x_2 (x_1). On the other hand, if $\beta_{12} \neq 0$, then the regression of Y on x_1, for two values of x_2, say $x_2^{(1)}$ and $x_2^{(2)}$, are $Y = \beta_0 + (\beta_1 + \beta_{12} x_2^{(1)}) x_1 + \beta_2 x_2^{(1)} + e$ and $Y = \beta_0 + (\beta_1 + \beta_{12} x_2^{(2)}) x_1 + \beta_2 x_2^{(2)} + e$.

These are regression lines with **different** slopes. This means that the model is nonadditive and the extent of the interaction depends on β_{12}.

13.6 The following output for a paired comparison is obtained by using the MINITAB commands:

```
MTB> let C4 = C2 - C1
MTB> TTest 0.0 C4;
SUBC> Alternative 0.
```

T-Test of the Mean
Test of mu = 0.0000 vs mu not = 0.0000

Variable	N	Mean	StDev	SE Mean	T	p-Value
Diff.	16	0.2537	0.0898	0.0225	11.30	0.0000

The difference between the two means is significant.

13.7 The randomization paired comparison with $M = 100$ yielded a P-value estimate $\hat{P} = 0.11$. The t-test yields a P-value of $P = 0.12$. The two methods give the same result.

13.8 The ANOVA was obtained using the following MINITAB commands:

```
MTB> name C4 = 'ResI1'
MTB> Twoway 'Yield' 'Blends' 'Treatm.' 'RESI1';
SUBC> Means 'Blends' 'Treatm.'.
```

Two-way Analysis of Variance
Analysis of Variance for Yield

Source	DF	SS	MS	F	p
Blends	4	264.0	66.0	3.51	0.041
Treatm.	3	70.0	23.3	1.23	0.342
Error	12	226.0	18.8		
Total	19	560.0			

Individual 95% CI

Blends	Mean	
1	92.0	(---------*---------)
2	83.0	(--------*--------)
3	85.0	(--------*--------)
4	88.0	(--------*--------)
5	82.0	(--------*--------)

```
------+---------+---------+---------+-----
     80.0      85.0      90.0      95.0
```

Individual 95% CI

Treatm.	Mean	
1	84.0	(-----------*-----------)
2	85.0	(-----------*-----------)
3	89.0	(-----------*-----------)
4	86.0	(-----------*-----------)

```
---+---------+---------+---------+--------
  80.5      84.0      87.5      91.0
```

There are no significant differences between the treatments. There are significant differences between blends, the most extreme difference being between blend 1 and blend 5.

13.9 The ANOVA is

Source	d.f.	SS	MS	F	p
Blocks	27	15,030.50	556.685	5.334	0.00011
Treat(˙ adj.)	7	1,901.88	271.697	2.603	0.04220
Error	21	2,191.62	104.363	–	–
Total	55	19,124.00	–	–	–

This table was created using the following calculations:

$$Y_{..} = \sum_{i=1}^{28} \sum_{l\in B_i} Y_{il} = 1931 \qquad \bar{Y} = 34.4821$$

$$Q_{..} = \sum_{i=1}^{28} \sum_{l\in B_i} Y_{li}^2 = 85,709 \qquad SST = Q_{..} - \frac{1}{56} Y_{..}^2 = 19,124$$

$$SSBL = \frac{1}{2} \sum_{i=1}^{28} \left(\sum_{l\in B_i} Y_{il} \right)^2 - \frac{1}{56} Y_{..}^2 = 15,030.5.$$

$W_1 = 213$	$W_5 = 247$	$W_1^* = 447$	$W_5^* = 471$
$W_2 = 213$	$W_6 = 240$	$W_2^* = 504$	$W_6^* = 483$
$W_3 = 217$	$W_7 = 376$	$W_3^* = 436$	$W_7^* = 611$
$W_4 = 204$	$W_8 = 221$	$W_4^* = 467$	$W_8^* = 443$

$Q_1 = -21$	$Q_5 = 23$	$\bar{Y}_1^* = 31.857$	$\bar{Y}_5^* = 37.357$
$Q_2 = -78$	$Q_6 = -3$	$\bar{Y}_2^* = 24.732$	$\bar{Y}_6^* = 34.107$
$Q_3 = -2$	$Q_7 = 141$	$\bar{Y}_3^* = 34.232$	$\bar{Y}_7^* = 52.107$
$Q_4 = -59$	$Q_8 = -1$	$\bar{Y}_4^* = 27.107$	$\bar{Y}_8^* = 34.357$

$$SSTR = \frac{1}{\lambda kt} \sum_{j=1}^{8} Q_j^2 = 1901.88.$$

Both the effects of the treatments and the blocks are significant at the $\alpha = 0.05$ level. The Scheffé coefficient for $\alpha = 0.05$ is $S_{0.05} = (7 \cdot F_{0.95}[7, 21])^{1/2} = 4.173$ and $\hat{\sigma}_p = \sqrt{104.363} = 10.216$. Thus, the treatments can be divided into two homogenous groups, $G_1 = \{A, B, C, D, E, F, H\}$ and $G_2 = \{G\}$. The 0.95 confidence interval for the difference of the group means is 20.413 ± 17.225, which shows that the group means are significantly different.

13.10 The ANOVA was created using the following numbers:

Treatments	Totals
$A = 312, 317, 312, 300$	$T_A = 1241$
$B = 299, 300, 295, 299$	$T_B = 1193$
$C = 315, 295, 315, 314$	$T_C = 1239$
$D = 290, 313, 298, 313$	$T_D = \underline{1214}$
	4887

Totals for Batches and Totals for Days:

B_1 :	1215	D_1 :	1216
B_2 :	1228	D_2 :	1225
B_3 :	1239	D_3 :	1220
B_4 :	1205	D_4 :	1226

ANOVA of Latin Square

Source	d.f.	SS	MS	F	p
Treatments	3	388.75	129.583	1.155	0.40
Batches	3	165.75	55.25	–	–
Days	3	16.25	5.417	–	–
Error	6	673.25	112.208	–	–
Total	15	1244	–	–	–

The differences between treatments, batches, or days are not significant.

13.11 We repeated the experiments of Example 13.7 with the specified factors set at the low levels. The ANOVA was obtained by using the MINITAB commands:

```
MTB> name C4 = 'RESI1'
MTB> Twoway 'Cyclet' 'Pist_Wg' 'Spring_C' 'RESI1';
SUBC> Means 'Pist_Wg' 'Spring_C'.
```

```
Two-way Analysis of Variance
Analysis of Variance for Cyclet
Source       DF      SS       MS       F       p
Pist_Wg       2    0.4749   0.2374   2.929   0.07
Spring_C      2    0.0430   0.0215   0.266   0.78
Interaction   4    0.2858   0.0715   0.895   0.48
Error        36    2.8760   0.0799
Total        44    3.6798
```

The results are similar to those of Example 12.7.

13.12 The regression of $\bar{X}$ on the seven factors is

$$\text{Xbar} = 0.639 + 0.109 \, A - 0.109B + 0.234C - 0.216D - 0.0044E$$

$$- 0.0019F + 0.0023G$$

```
Predictor          Coef        Stdev       t-ratio          p
Constant        0.63923     0.01237         51.66      0.000
A               0.10881     0.01237          8.79      0.000
B              -0.10916     0.01237         -8.82      0.000
C               0.23372     0.01237         18.89      0.000
D              -0.21616     0.01237        -17.47      0.000
E              -0.00437     0.01237         -0.35      0.724
F              -0.00186     0.01237         -0.15      0.881
G               0.00233     0.01237          0.19      0.851
s = 0.1400 R-sq = 87.2% R-sq(adj) = 86.4%.
```

Only factors $A-D$ have significant main effects on the means.
   The regression of the standard deviations on factors $A-G$ is

$$\text{STD} = 0.157 + 0.0283A - 0.0290B - 0.0887C - 0.0751D + 0.00252E$$

$$+ 0.00083F - 0.00025G$$

```
Predictor          Coef        Stdev       t-ratio          p
Constant       0.156805    0.006082         25.78      0.000
A              0.028302    0.006082          4.65      0.000
B             -0.028973    0.006082         -4.76      0.000
C             -0.088706    0.006082        -14.58      0.000
D             -0.075053    0.006082        -12.34      0.000
E              0.002516    0.006082          0.41      0.680
F              0.000825    0.006082          0.14      0.892
G             -0.000252    0.006082         -0.04      0.967
s = 0.06881 R-sq = 77.3% R-sq(adj) = 76.9%.
```

Only factors $A-D$ have significant main effects on the standard deviations.

13.13   (i) The LSE of the main effects are $\hat{A} = -4.0625$, $\hat{B} = 11.8125$, $\hat{C} = -1.5625$, and $\hat{D} = -2.8125$.
   (ii) An estimate of σ^2 with 11 d.f. is $\hat{\sigma}^2 = 9.2898$.
   (iii) A 0.99 level confidence interval for σ^2 is $(3.819, 39.254)$.

13.14 The ANOVA was obtained by using the MINITAB command.

```
MTB> Twoway 'Y' 'B' 'A';
SUBC> Means 'B' 'A'.
```

```
Two-way Analysis of Variance
Analysis of Variance for Y
Source          DF          SS          MS          F          p
B                2       54.170      27.085      88.513        0
A                2       43.081      21.540      70.392        0
Interaction      4        4.346       1.086       3.549      0.027
Error           18        5.500       0.306        ---        ---
Total           26      107.096        ---        ---         ---
```

```
                          Individual 95% CI
B      Mean    ----+---------+---------+---------+--------
-1     18.16   (---*-)
 0     20.82                                (---*---)
 1     21.41                                    (---*---)
               ----+---------+---------+---------+--------
               18.00       19.00      20.00      21.00
```

```
                          Individual 95% CI
A      Mean    ---------+---------+---------+---------+--
-1     20.23                     (---*----)
 0     18.53   (---*---)
 1     21.62                              (---*----)
               ---------+---------+---------+---------+--
                      19.00      20.00      21.00      22.00
```

The factors A and B as well as their interaction are significant.

In order to break the SS due to A and B to 1 d.f. components, we consider the orthogonal matrix $\Psi_{9\times9}$ and the vector of means $\mathbf{m}$, where

$$
\Psi_{9\times9} =
\begin{array}{ccccccccc}
M & A & A^2 & B & AB & A^2B & B^2 & AB^2 & A^2B^2 \\
\end{array}
$$

$$
\Psi_{9\times9} =
\left[
\begin{array}{ccc|ccc|ccc}
1 & -1 & 1 & -1 & 1 & -1 & 1 & -1 & 1 \\
1 & 0 & -2 & -1 & 0 & 2 & 1 & 0 & -2 \\
1 & 1 & 1 & -1 & -1 & -1 & 1 & 1 & 1 \\
\hline
1 & -1 & 1 & & & & -2 & 2 & -2 \\
1 & 0 & -2 & & 0 & & -2 & 0 & 4 \\
1 & 1 & 1 & & & & -2 & -2 & -2 \\
\hline
1 & -1 & 1 & 1 & -1 & 1 & 1 & -1 & 1 \\
1 & 0 & -2 & 1 & 0 & -2 & 1 & 0 & -2 \\
1 & 1 & 1 & 1 & 1 & 1 & 1 & 1 & 1 \\
\end{array}
\right]
$$

and $\mathbf{m} = (18.233, 17.233, 19.000, 20.767, 18.867, 22.833, 21.700, 19.500, 23.033)'$.

The columns corresponding to A, A^2, B, and B^2 are the 2nd, 3rd, 4th, and 7th. To obtain the linear effects of A and B, $\hat{\alpha}_L$, $\hat{\beta}_L$, we multiply the corresponding columns by $\mathbf{m}$ and divide by 6. For A^2 and B^2, we multiply the 3rd and 7th columns, respectively, by $\mathbf{m}$ and divide by 18. The SS linear components are $\text{SSAL} = 18 \times \hat{\alpha}_L^2$ and $\text{SSBL} = 18 \times \hat{\beta}_L^2$. The quadratic components are $\text{SSAQ} = 54\hat{\alpha}_Q^2$ and $\text{SSBQ} = 54\hat{\beta}_Q^2$.

Thus, the partition is

	Source	d.f.	SS	MS	F
A	Linear	1	8.6777	8.6777	28.396
	Quad.	1	34.3970	34.3970	112.556
	Total	2	43.0747	–	–
B	Linear	1	47.6972	47.6972	156.077
	Quad.	1	6.4792	6.4792	21.202
	Total	2	54.1764	–	–
	Error	18	5.5000	0.3056	–

The Scheffé coefficient for $\alpha = 0.05$, is $S_\alpha = (4 \cdot F_{0.95}[4, 18])^{1/2} = 3.4221$. Thus, all four main effects are significant.

13.15 The subgroup of defining parameters is $\{\mu, ABCDG, ABEFH, CDEFGH\}$. The design is of resolution V. The aliases to the main effects and to the first-order interactions with factor A are

Factors	Aliases		
A	$-BCDG$	$-BEFH$	$+ACDEFGH$
B	$-ACDG$	$-AEFH$	$+BCDEFGH$
C	$-ABDG$	$-ABCEFH$	$+DEFGH$
D	$-ABCG$	$-ABDEFH$	$+CEFGH$
E	$-ABCDEG$	$-ABFH$	$+CDFGH$
F	$-ABCDFG$	$-ABEH$	$+CDEGH$
G	$-ABCD$	$-ABEFGH$	$+CDEFH$
H	$-ABCDGH$	$-ABEF$	$+CDEFG$
$A * B$	$-CDG$	$-EFH$	$+ABCDEFGH$
$A * C$	$-BDG$	$-BCEFH$	$+ADEFGH$
$A * D$	$-BCG$	$-BDEFH$	$+ACEFGH$
$A * E$	$-BCDEG$	$-BFH$	$+ACDFGH$
$A * F$	$-BCDFG$	$-BEH$	$+ACDEGH$
$A * G$	$-BCD$	$-BEFGH$	$+ACDEFH$
$A * H$	$-BCDGH$	$-BEF$	$+ACDEFG$

The first block ($v = 0$) of the 2^{8-2} is

```
1  1  1  1  1  1  1  1      2  1  2  1  1  1  1  2
2  2  1  1  1  1  1  1      1  2  2  1  1  1  1  2
1  1  2  2  1  1  1  1      2  1  1  2  1  1  1  2
2  2  2  2  1  1  1  1      1  2  1  2  1  1  1  2
2  1  2  1  2  1  1  1      1  1  1  1  2  1  1  2
1  2  2  1  2  1  1  1      2  2  1  1  2  1  1  2
2  1  1  2  2  1  1  1      1  1  2  2  2  1  1  2
1  2  1  2  2  1  1  1      2  2  2  2  2  1  1  2
2  1  2  1  1  2  1  1      1  1  1  1  1  2  1  2
1  2  2  1  1  2  1  1      2  2  1  1  1  2  1  2
2  1  1  2  1  2  1  1      1  1  2  2  1  2  1  2
1  2  1  2  1  2  1  1      2  2  2  2  1  2  1  2
1  1  1  1  2  2  1  1      2  1  2  1  2  2  1  2
2  2  1  1  2  2  1  1      1  2  2  1  2  2  1  2
1  1  2  2  2  2  1  1      2  1  1  2  2  2  1  2
2  2  2  2  2  2  1  1      1  2  1  2  2  2  1  2
1  1  2  1  1  1  2  1      2  1  1  1  1  1  2  2
2  2  2  1  1  1  2  1      1  2  1  1  1  1  2  2
1  1  1  2  1  1  2  1      2  1  2  2  1  1  2  2
2  2  1  2  1  1  2  1      1  2  2  2  1  1  2  2
2  1  1  1  2  1  2  1      1  1  2  1  2  1  2  2
1  2  1  1  2  1  2  1      2  2  2  1  2  1  2  2
2  1  2  2  2  1  2  1      1  1  1  2  2  1  2  2
1  2  2  2  2  1  2  1      2  2  1  2  2  1  2  2
2  1  1  1  1  2  2  1      1  1  2  1  1  2  2  2
1  2  1  1  1  2  2  1      2  2  2  1  1  2  2  2
2  1  2  2  1  2  2  1      1  1  1  2  1  2  2  2
1  2  2  2  1  2  2  1      2  2  1  2  1  2  2  2
1  1  2  1  2  2  2  1      2  1  1  1  2  2  2  2
2  2  2  1  2  2  2  1      1  2  1  1  2  2  2  2
1  1  1  2  2  2  2  1      2  1  2  2  2  2  2  2
2  2  1  2  2  2  2  1      1  2  2  2  2  2  2  2
```

13.16 The following parameters are not estimable: μ, ACE, $ABEF$, $ABCD$, BCF, BDE, $CDEF$, and AFD.

13.17 (i) The response function is $\hat{Y} = 62.0 + 2.39X_1 + 4.16X_2 - 0.363X_1X_2$, with $R^2 = 0.99$. The contour plot is given in Figure 13.17.1.

(ii) An estimate of the variance is $\hat{\sigma}^2 = 0.13667$, 3 d.f. The variance around the regression is $s^2_{y|(x)} = 0.2357$, 8 d.f. The ratio $F = \dfrac{s^2_{y|(x)}}{\hat{\sigma}^2} = 1.7246$ gives a p-value of $P = 0.36$. This shows there is no significant differences between the variances.

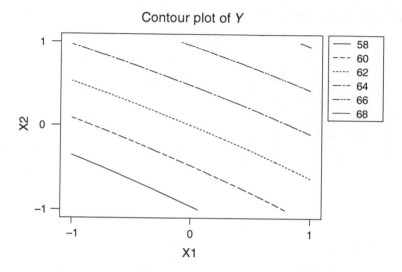

Figure 13.17.1   *Contour plot of equal responses*

13.18   (i) The response function is

$$Y = 95.3 + 16.5X_1 + 3.20X_2 - 20.3X_1^2 - 21.3X_2^2 - 7.79X_1 * X_2.$$

Predictor	Coef	Stdev	t-ratio	p
Constant	95.333	3.349	28.47	0.000
X_1	16.517	3.349	4.93	0.016
X_2	3.204	3.349	0.96	0.409
X_1^2	-20.283	5.294	-3.83	0.031
X_2^2	-21.318	5.295	-4.03	0.028
$X_1 * X_2$	-7.794	6.697	-1.16	0.329

s = 5.800 R-sq = 94.6% R-sq(adj) = 85.5%

(ii) See Figure 13.18.1

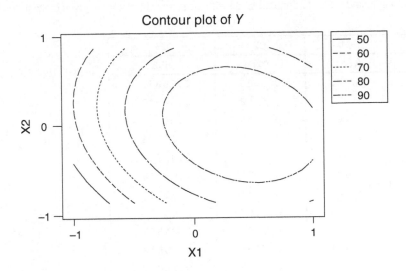

Figure 13.18.1   *Contour plot of equal response*

(iii) Analysis of variance

SOURCE	DF	SS	MS	F	p
Regression	5	1760.85	352.17	10.47	0.041
Error	3	100.91	33.64		
Total	8	1861.76			

SOURCE	DF	SEQ SS	
X_1	1	818.40	
X_2	1	30.80	(n.s.)
X_1^2	1	320.80	
X_2^2	1	545.28	
$X_1 * X_2$	1	45.56	(n.s.)

Chapter 14: Quality by Design

14.1   The output and MSE of a simulation run with the piston simulator are given in the following table:

	C1	C2	C3
	Xbar	STD	MSE
1	0.638	0.3653	0.16879
2	0.917	0.5398	0.50947
3	0.244	0.1639	0.06930
4	0.371	0.3342	0.11793
5	1.045	0.0752	0.35968
6	1.481	0.0923	1.07148
7	0.906	0.1244	0.22341
8	1.260	0.1607	0.68192
9	0.317	0.1604	0.04342
10	0.448	0.2195	0.04818
11	0.143	0.0778	0.10030
12	0.187	0.1036	0.07990
13	0.482	0.0081	0.00109
14	0.681	0.0107	0.05348
15	0.464	0.0196	0.00058*
16	0.657	0.0260	0.04353
17	0.648	0.3924	0.19318
18	0.890	0.5304	0.47492
19	0.224	0.1564	0.07554
20	0.366	0.2971	0.09532
21	1.055	0.0530	0.36883
22	1.474	0.1067	1.05996
23	0.913	0.1220	0.22925
24	1.301	0.1791	0.75628
25	0.326	0.1516	0.03836
26	0.443	0.2248	0.05058
27	0.141	0.0923	0.10400
28	0.233	0.1615	0.07317
29	0.483	0.0086	0.00116
30	0.683	0.0087	0.05436
31	0.464	0.0188	0.00055*
32	0.660	0.0241	0.04468

(i) The treatment combinations yielding minimal MSE are for 15 or 31, namely: $(0, 1, 1, 1, 0, *, *, *)$ or $(0, 1, 1, 1, 1, *, *, *)$, where 0 denotes low level and 1 denotes the high level of the factor.

(ii) For the SN ratio η, we have

	C1 Xbar	C2 STD	C3 SNR
1	0.638	0.3653	4.8292
2	0.917	0.5398	4.5877
3	0.244	0.1639	3.4366
4	0.371	0.3342	0.8720
5	1.045	0.0752	22.8577
6	1.481	0.0923	24.1069
7	0.906	0.1244	17.2453
8	1.260	0.1607	17.8864
9	0.317	0.1604	5.9060
10	0.448	0.2195	6.1864
11	0.143	0.0778	5.2743
12	0.187	0.1036	5.1163
13	0.482	0.0081	35.4912
14	0.681	0.0107	36.0753
15	0.464	0.0196	27.4852
16	0.657	0.0260	28.0518
17	0.648	0.3924	4.3410
18	0.890	0.5304	4.4803
19	0.224	0.1564	3.0990
20	0.366	0.2971	1.7829
21	1.055	0.0530	25.9794
22	1.474	0.1067	22.8064
23	0.913	0.1220	17.4814
24	1.301	0.1791	17.2228
25	0.326	0.1516	6.6410
26	0.443	0.2248	5.8809
27	0.141	0.0923	3.6617
28	0.233	0.1615	3.1628
29	0.483	0.0086	34.9890
30	0.683	0.0087	37.8980*
31	0.464	0.0188	27.8471
32	0.660	0.0241	28.7505

The treatment combination which maximizes the SN ratio is the 30th one, namely $(1, 0, 1, 1, 1, *, *, *)$. This treatment combination has an MSE of 0.05436, which is 94 times bigger than the minimal MSE. This exercise demonstrates the need for a full analysis of the effects and the dangers of relying on a simplistic observation of the experiment's outcomes.

14.2   The regression analysis of SN on the seven factors (only main effects) is

```
SN = 4.06 + 0.004 x1 - 0.407 x2 + 2.318 x3 + 0.675 x4 - 0.176 x5
        + 0.125 x6 + 0.058 x7.
```

```
Predictor         Coef      Stdev     t-ratio        p
Constant        4.0567     0.1172       34.60     0.000
x1              0.0037     0.1172        0.03     0.975
x2             -0.4066     0.1172       -3.47     0.001
x3              2.3176     0.1172       19.77     0.000
x4              0.6754     0.1172        5.76     0.000
x5             -0.1757     0.1172       -1.50     0.137
x6              0.1257     0.1172        1.07     0.286
x7              0.0576     0.1172        0.49     0.624
```

```
s = 1.326 R-sq = 78.6% R-sq(adj) = 77.3%
```

We see that only factors 2, 3, and 4 (piston surface area, initial gas volume and spring coefficient, respectively) have a significant effect. The R^2 is only 78.6%. Adding to the regression equation the first order interactions between X_2, X_3, and X_4, we get

```
SN = 4.06 - 0.407 x2 + 2.318 x3 + 0.675 x4 - 0.677 x2 * x3
        - 0.152 x2 * x4 + 0.630 x3 * x4.
```

```
Predictor         Coef      Stdev     t-ratio        p
Constant      4.05670    0.08236       49.26     0.000
x2           -0.40663    0.08236       -4.94     0.000
x3            2.31758    0.08236       28.14     0.000
x4            0.67535    0.08236        8.20     0.000
x2 * x3      -0.67662    0.08236       -8.22     0.000
x2 * x4      -0.15230    0.08236       -1.85     0.067
x3 * x4       0.63045    0.08236        7.66     0.000
```

```
s = 0.9318 R-sq = 89.3% R-sq(adj) = 88.8%
```

We see that the interactions $X_2 * X_3$ and $X_3 * X_4$ are very significant. The interaction $X_2 * X_4$ is significant at 6.7% level. The regression equation with the interaction terms predicts the SN better, $R^2 = 89.3\%$.

14.3   The approximations to the expected values and variances are as follows:

(i)

$$E\left\{\frac{X_1}{X_2}\right\} \approx \frac{\xi_1}{\xi_2} - \sigma_{12}\frac{1}{\xi_2^2} + \sigma_2^2\frac{\xi_1}{\xi_2^3} \ ; \ V\left\{\frac{X_1}{X_2}\right\} \approx \frac{\sigma_1^2}{\xi_2^2} + \sigma_2^2\frac{\xi_1^2}{\xi_2^4} - 2\sigma_{12}\frac{\xi_1}{\xi_2^3}.$$

(ii)

$$E\left\{\log\frac{X_1^2}{X_2^2}\right\} \approx \log\left(\frac{\xi_1}{\xi_2}\right)^2 - \frac{\sigma_1^2}{\xi_1^2} + \frac{\sigma_2^2}{\xi_2^2} \ ; \ V\left\{\log\left(\frac{X_1}{X_2}\right)^2\right\} \approx \frac{4\sigma_1^2}{\xi_1^2} + \frac{4\sigma_2^2}{\xi_2^2} - 8\frac{\sigma_{12}}{\xi_1\xi_2}.$$

(iii)

$$E\{(X_1^2 + X_2^2)^{1/2}\} \approx (\xi_1^2 + \xi_2^2)^{1/2} + \frac{\sigma_1^2}{2(\xi_1^2 + \xi_2^2)^{1/2}} \left(1 - \frac{\xi_1^2}{\xi_1^2 + \xi_2^2}\right)$$

$$+ \frac{\sigma_2^2}{2(\xi_1^2 + \xi_2^2)^{1/2}} \left(1 - \frac{\xi_2^2}{\xi_1^2 + \xi_2^2}\right) - \frac{\sigma_{12}\xi_1\xi_2}{(\xi_1^2 + \xi_2^2)^{3/2}};$$

$$V\{(X_1^2 + X_2^2)^{1/2}\} \approx \frac{\sigma_1^2\xi_1^2}{(\xi_1^2 + \xi_2^2)} + \frac{\sigma_2^2\xi_2^2}{(\xi_1^2 + \xi_2^2)} + 2\frac{\sigma_{12}\xi_1\xi_2}{(\xi_1^2 + \xi_2^2)}.$$

14.4   Approximation formulas yield: $E\{Y\} \approx 0.5159$ and $V\{Y\} \approx 0.06762$. Simulation with 5000 runs yields the estimates $E\{Y\} \approx 0.6089$, $V\{Y\} \approx 0.05159$. The first approximation of $E\{Y\}$ is significantly lower than the simulation estimate.

14.5   We have that $E\{\bar{X}_n\} = \mu$, $V\{\bar{X}_n\} = \dfrac{\sigma^2}{n}$, $E\{S_n^2\} = \sigma^2$, and $V\{S_n^2\} = \dfrac{2\sigma^4}{n-1}$. The first and second order partial derivatives of $f(\mu, \sigma^2) = 2\log(\mu) - \log(\sigma^2)$ are

$$\frac{\partial}{\partial\mu}f = \frac{2}{\mu}, \; \frac{\partial}{\partial\sigma^2}f = -\frac{1}{\sigma^2}, \; \frac{\partial^2}{\partial\mu\partial\sigma^2}f = 0, \; \frac{\partial^2}{\partial\mu^2}f = -\frac{2}{\mu^2} \text{ and } \frac{\partial^2}{\partial(\sigma^2)^2}f = \frac{1}{\sigma^4}.$$

Thus, the approximations to the expected value and variance of $Y = \log\left(\frac{\bar{X}_n^2}{S_n^2}\right)$ are

$$E\left\{\log\left(\frac{\bar{X}_n^2}{S_n^2}\right)\right\} \approx \log\left(\frac{\mu^2}{\sigma^2}\right) - \frac{\sigma^2}{n\mu^2} + \frac{1}{n-1} \text{ and } V\left\{\log\left(\frac{\bar{X}_n^2}{S_n^2}\right)\right\} \approx \frac{4\sigma^2}{n\mu^2} + \frac{2}{n-1}.$$

14.6   In the regression analysis, we transform in column 1, $1 \rightarrow -1$ and $2 \rightarrow 1$; in columns 2–8 we transform $1 \rightarrow -1$, $2 \rightarrow 0$, and $3 \rightarrow 1$. We will account for only the linear effects of the factors. Regression analysis yields the following:

```
SNR = 3.23 + 0.266x1 + 0.0790x2 + 0.138 x3 - 0.143 x4 - 0.312 x5
        + 0.126 x6 + 0.0263 x7 - 0.171x8
```

Predictor	Coef	Stdev	t-ratio	p	
Constant	3.22601	0.07518	42.91	0.000	
x1	0.26565	0.07518	3.53	0.006	
x2	0.07900	0.09208	0.86	0.413	n.s.
x3	0.13848	0.09208	1.50	0.167	n.s.
x4	-0.14339	0.09208	-1.56	0.154	n.s.
x5	-0.31231	0.09208	-3.39	0.008	
x6	0.12630	0.09208	1.37	0.203	n.s.
x7	0.02632	0.09208	0.29	0.781	n.s.
x8	-0.17109	0.09208	-1.86	0.096	

```
s = 0.3190 R-sq=79.5% R-sq(adj) = 61.2%.
```

The linear effects of x_1, x_5, and x_8 are significant. There might be significant interactions or quadratic effects.

14.7   The coefficients in the regression analysis yield estimates of the main effects. Regression analysis of the **full factorial** yields

```
SN = 2.03 + 0.0019 PW - 0.203 PSA + 1.159 IGV + 0.338 SC - 0.0878 AP
        + 0.0629 AT + 0.0288 FGT
```

Predictor	Coef	Stdev	t-ratio	p
Constant	2.02835	0.05862	34.60	0.000
PW	0.00187	0.05862	0.03	0.975
PSA	-0.20331	0.05862	-3.47	0.001
IGV	1.15879	0.05862	19.77	0.000
SC	0.33768	0.05862	5.76	0.000
AP	-0.08785	0.05862	-1.50	0.137
AT	0.06287	0.05862	1.07	0.286
FGT	0.02880	0.05862	0.49	0.624

```
s = 0.6632 R-sq = 78.6% R-sq(adj) = 77.3%.
```

The three significant main effects are those of piston surface area (PSA), initial gas volume (IGV), and spring coefficients (SC).

Regression analysis of **the 1/2 factorial** yields

```
SN = 1.94 + 0.0127 PW - 0.203 PSA + 1.183 IGV + 0.337 SC
        - 0.102 AP + 0.0529 AT + 0.0329 FGT
```

Predictor	Coef	Stdev	t-ratio	p
Constant	1.93648	0.06951	27.86	0.000
PW	0.01266	0.06951	0.18	0.856
PSA	-0.20334	0.06951	-2.93	0.005
IGV	1.18261	0.06951	17.01	0.000
SC	0.33718	0.06951	4.85	0.000
AP	-0.10199	0.06951	-1.47	0.148
AT	0.05294	0.06951	0.76	0.450
FGT	0.03294	0.06951	0.47	0.637

```
s = 0.5561 R-sq = 85.3% R-sq(adj) = 83.4%.
```

The results are similar to those of the full factorial.

Regression analysis of **the 1/4 factorial** gives

```
SN = 1.955 + 0.026 PW - 0.244 PSA + 1.033 IGV + 0.374 SC
        - 0.210 AP - 0.007 AT + 0.014 FGT
```

Predictor	Coef	Stdev	t-ratio	p
Constant	1.9547	0.1413	13.84	0.000
PW	0.0256	0.1413	0.18	0.858
PSA	-0.2443	0.1413	-1.73	0.097
IGV	1.0329	0.1413	7.31	0.000
SC	0.3741	0.1413	2.65	0.014
AP	-0.2104	0.1413	-1.49	0.149
AT	-0.0070	0.1413	-0.05	0.961
FGT	0.0142	0.1413	0.10	0.921

```
s = 0.7991 R-sq = 73.3% R-sq(adj) = 65.5%.
```

Here the effect of PSA is significant only at a 10% *p*-level.

Finally, the regression analysis of **the 1/8 factorial** yields

```
SN = 2.06 - 0.079 PW - 0.183 PSA + 1.078 IGV + 0.454 SC
      + 0.374 AP - 0.014 AT + 0.021 FGT
```

Predictor	Coef	Stdev	t-ratio	p
Constant	2.0623	0.1659	12.43	0.000
PW	-0.0789	0.1659	-0.48	0.647
PSA	-0.1829	0.1659	-1.10	0.302
IGV	1.0783	0.1659	6.50	0.000
SC	0.4537	0.1659	2.74	0.026
AP	0.374	0.1659	0.23	0.827
AT	-0.0136	0.1659	-0.08	0.937
FGT	0.0213	0.1659	0.13	0.901

s = 0.6635 R-sq = 86.5% R-sq(adj) = 74.7%.

The only significant factors here are IGV and SC. It is interesting to notice that PSA, which showed significant effect in the full, 1/2, and 1/4 factorials, is not significant in the 1/8 factorial.

14.8   In the following table, we present the output of the piston smelter with $n = 20$:

	C1	C2			C1	C2
	Xbar	STD			Xbar	STD
1	0.651	0.3779		17	0.720	0.4111
2	0.872	0.5825		18	0.691	0.4942
3	0.268	0.1641		19	0.304	0.2730
4	0.329	0.3268		20	0.370	0.3622
5	1.036	0.0667		21	1.065	0.0516
6	1.433	0.1396		22	1.419	0.1342
7	0.951	0.1357		23	0.840	0.1331
8	1.190	0.2024		24	1.339	0.1372
9	0.340	0.1601		25	0.362	0.1623
10	0.424	0.2170		26	0.505	0.2124
11	0.164	0.1127		27	0.132	0.0338
12	0.193	0.0644		28	0.207	0.1364
13	0.482	0.0080		29	0.484	0.0040
14	0.681	0.0137		30	0.679	0.0118
15	0.468	0.0155		31	0.468	0.0162
16	0.657	0.0258		32	0.654	0.0352

This program was run with tolerances of 5% and 10%. The column of mean $\bar{X}$ is different from that of Table 14.5, where all tolerances are 0%. The regression of $\bar{X}$ on the five factors $A-E$ is

```
Xbar = 0.637 + 0.0909A - 0.103B + 0.229C - 0.206D + 0.0031E
Predictor           Coef      Stdev     t-ratio                  p
Constant          0.63681    0.02617      24.33              0.000
A                 0.09087    0.02617       3.47              0.002
B                -0.10344    0.02617      -3.95              0.001
C                 0.22856    0.02617       8.73              0.000
D                -0.20556    0.02617      -7.85              0.000
E                 0.00313    0.02617       0.12              0.906
s = 0.1480 R-sq = 86.4% R-sq(adj) = 83.8%.
```

We see that the coefficient of determination now is $R^2 = 0.864$ while if the tolerances are 0%, it is $R^2 = 0.997$. Also, the effect of factor E is not significant.

14.9   In the power circuit simulator change, $5 \to 1$ and $10 \to 2$. Then run the simulator with $n = 100$. The results and the MSE are given in the following table:

	Ybar	STD	MSE
1	219.92	0.4035	0.169213
2	220.08	0.6876	0.479194
3	220.03	0.6193	0.384432
4	219.98	0.6775	0.459406
5	220.02	0.5910	0.349681
6	219.87	0.7102	0.521285
7	220.02	0.5964	0.356093
8	220.03	0.5871	0.345586
9	220.04	0.5376	0.290613
10	220.04	0.6722	0.453452
11	220.09	0.3793	0.151968
12	220.07	0.7225	0.526907
13	220.03	0.5583	0.312599
14	220.07	0.6097	0.376635
15	220.06	0.6107	0.376554
16	220.04	0.6156	0.380563
17	220.10	0.5714	0.336499
18	220.07	0.6425	0.417707
19	220.00	0.4986	0.248602
20	220.00	0.6550	0.429025
21	219.97	0.5368	0.289054
22	219.92	0.6052	0.372667
23	220.04	0.5954	0.356101
24	219.93	0.6554	0.434450
25	220.02	0.3776	0.142982
26	219.99	0.6989	0.488561

	Ybar	STD	MSE
27	219.93	0.5504	0.307841
28	220.04	0.6395	0.410560
29	219.97	0.5474	0.300547
30	220.11	0.5821	0.350940
31	219.95	0.5709	0.328427
32	220.00	0.5337	0.284836

Comparing this table to Table 14.8 in the text, we see that by reducing the tolerances to 1% and 2%, the MSE is reduced by a factor of 100. This, however, increases the cost of the product.

Chapter 16: Reliability Analysis

16.1   MTTF = 5.1 [hours]; MTTR = 6.6 [minutes]; intrinsic availability = 0.979.

16.2   (i) Since 4 years = 35,040 hours, the reliability estimate is $R(4) = 0.30$.

   (ii) Of the solar cells that survive 20,000 hours, the proportion expected to survive 40,000 hours is
   $\frac{0.25}{0.60} = 0.42$.

16.3 $F(t) = \begin{cases} t^4/20{,}736 & 0 \leq t < 12 \\ 1 & 12 \leq t. \end{cases}$

   (i) The hazard function is $h(t) = \begin{cases} \dfrac{4t^3}{20{,}736 \cdot \left(1 - \dfrac{t^4}{20{,}736}\right)}, & 0 \leq t < 12 \\[4mm] \infty, & 12 \leq t. \end{cases}$

   (ii) MTTF $= \int_0^{12} \left(1 - \frac{t^4}{20{,}736}\right) dt = 9.6$ [months].

   (iii) $R(4) = 1 - F(4) = 0.9877$.

16.4 $R(t) = \exp\{-2t - 3t^2\}$ (i) $h(t) = 2 + 6t$, $h(3) = 20$. (ii) $R(5)/R(3) = 0$.

16.5   (i) $1 - B(1; 4, 0.95) = 0.9995$. (ii) $(1 - B(0; 2, 0.95))^2 = (1 - 0.05^2)^2 = 0.9950$.

16.6   (i) The block diagram of the system is

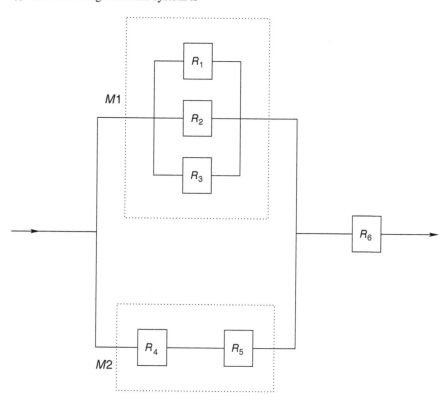

   (ii) $R_{M_1} = 1 - (1 - R_1)(1 - R_2)(1 - R_3)$ and $R_{M_2} = R_4 R_5$. Thus,

$$R_{\text{sys}} = [1 - (1 - R_1)(1 - R_2)(1 - R_3)(1 - R_4 R_5)]R_6$$
$$= R_1 R_6 + R_2 R_6 + R_3 R_6 - R_1 R_2 R_6 + R_4 R_5 R_6$$
$$- R_1 R_3 R_6 - R_2 R_3 R_6 + R_1 R_2 R_3 R_6$$

$$- R_1 R_4 R_5 R_6 - R_2 R_4 R_5 R_6 - R_3 R_4 R_5 R_6$$
$$+ R_2 R_3 R_4 R_5 R_6 + R_1 R_3 R_4 R_5 R_6 + R_1 R_2 R_4 R_5 R_6$$
$$- R_1 R_2 R_3 R_4 R_5 R_6.$$

If all values of $R = 0.8$, then $R_{\text{sys}} = 0.7977$.

16.7 Extending Bonferroni's inequality,

$$\Pr\{C_1 = 1, \ldots, C_n = 1\} = 1 - \Pr\left\{\bigcup_{i=1}^{n}\{C_i = 0\}\right\}$$

$$\geq 1 - \sum_{i=1}^{n} \Pr\{C_i = 0\}$$

$$= 1 - \sum_{i=1}^{n}(1 - R_i);$$

where $\{C_i = 1\}$ is the event that the ith component functions and $\{C_i = 0\}$ is the event that it fails. Since $R_{\text{sys}} = \Pr\{C_1 = 1, \ldots, C_n = 1\}$ for a series structure, the inequality follows.

16.8 The reliability of a component is $\theta(5) = \Pr\left\{W\left(\frac{1}{2}, 100\right) > 5\right\} = 0.79963$.
  Therefore, $R_{\text{sys}} = 1 - B(3; 8, 0.7996) = 0.9895$.

16.9 The life length of the system is the sum of 3 independent $E(100)$, that is $G(3, 100)$. Hence, the MTTF = 300 hours. The reliability function of the system is

$$R_{\text{sys}}(t) = 1 - G(t; 3, 100).$$

16.10 Let C be the length of the renewal cycle. $E\{C\} = \beta\Gamma\left(1 + \frac{1}{\alpha}\right) + e^{\mu + \sigma^2/2}. \ \sigma(C) = \left[\beta^2\left(\Gamma\left(1 + \frac{2}{\alpha}\right) - \Gamma^2\left(1 + \frac{1}{\alpha}\right)\right) + e^{2\mu + \sigma^2}(e^{\sigma^2} - 1)\right]^{1/2}$.

16.11 Let $C_1, C_2, \ldots$ denote the renewal cycle times, $C_i \sim N(100, 10)$. $N_R(200)$ is a random variable representing the number of repairs that occur during the time interval $(0, 200]$. $\Pr\{N_R(200) = 0\} = \Pr\{C_1 > 200\} = 1 - \Phi\left(\frac{200 - 100}{10}\right) = 1 - \Phi(10) = 0$. For $k = 1, 2, 3, \ldots,$ $C_1 + \cdots + C_k \sim N(100k, 10\sqrt{k})$ and so

$$\Pr\{N_R(200) = k\} = \Pr\{N_R(200) \geq k\} - \Pr\{N_R(200) \geq k + 1\}$$

$$= \Pr\{C_1 + \cdots + C_k \leq 200\} - \Pr\{C_1 + \cdots + C_{k+1} \leq 200\}$$

$$= \Phi\left(\frac{200 - 100k}{10\sqrt{k}}\right) - \Phi\left(\frac{200 - 100(k + 1)}{10\sqrt{k + 1}}\right)$$

$$= \Phi\left(\frac{20}{\sqrt{k}} - 10\sqrt{k}\right) - \Phi\left(\frac{20}{\sqrt{k + 1}} - 10\sqrt{k + 1}\right),$$

where $\Phi(\cdot)$ denotes the c.d.f. of a $N(0, 1)$ distribution. Thus, the p.d.f. of $N_R(200)$ is given by the following table.

k	$\Pr\{N_R(200) = k\}$
0	0
1	0.5
2	0.5
3	0
4	0
$\vdots$	$\vdots$

16.12 $V(1000) = \sum_{n=1}^{\infty} \Phi\left(\dfrac{100 - 10n}{\sqrt{n}}\right) \approx 9.501.$

16.13 $v(t) = \dfrac{1}{10\sqrt{n}} \sum_{n=1}^{\infty} \phi\left(\dfrac{t - 100n}{10\sqrt{n}}\right)$, where $\phi(Z)$ is the p.d.f. of $N(0, 1)$.

16.14 Since TTF $\sim \max(E_1(\beta), E_2(\beta))$, $F(t) = (1 - e^{-t/\beta})^2$ and $R(t) = 1 - (1 - e^{-t/\beta})^2$.
Using Laplace transforms, we have

$$R^*(s) = \frac{\beta(3 + s\beta)}{(1 + s\beta)(2 + s\beta)}, \quad f^*(s) = \frac{2}{(1 + \beta s)(2 + \beta s)}, \quad g^*(s) = \frac{1}{(1 + s\gamma)^2}, \quad \text{and}$$

$$A^*(s) = \frac{\beta(3 + s\beta)(1 + \gamma s)^2}{s[3\beta + 4\gamma + (\beta^2 + 6\beta\gamma + 2\gamma^2)s + (2\beta^2\gamma + 3\beta\gamma^2)s^2 + \beta^2\gamma^2 s^3]}.$$

16.15 For our samples, with $M = 500$ runs, we get the following estimates:

$$E\{N_R(2000)\} = V(2000) = 16.574 \quad \text{and} \quad \text{Var}\{N_R(2000)\} = 7.59.$$

16.16 The expected length of the test is 2069.8 [hours].

16.17 Let $T_{n,0} \equiv 0$. Then, $T_{n,r} = \sum_{j=1}^{r}(T_{n,j} - T_{n,j-1}) = \sum_{j=1}^{r} \Delta_{n,j-1}$. Thus,

$$V\{T_{n,r}\} = \beta^2 \sum_{j=1}^{r} \left(\frac{1}{n - j + 1}\right)^2.$$

In Exercise 16.16, $\beta = 2000$ hours, $n = 15$, and $r = 10$. This yields

$$V\{T_{15,10}\} = 4 \times 10^6 \times 0.116829 = 467, 316 \text{ (hours)}^2.$$

16.18 Let $S_{n,r} = \sum_{i=1}^{r} T_{n,i} + (n - r)T_{n,r}$. An unbiased estimator of β is $\hat{\beta} = \dfrac{S_{n,r}}{r}$. $V\{\hat{\beta}\} = \dfrac{\beta^2}{r}$.

16.19 The Weibull QQ plotting is based on the regression of $\ln T_{(i)}$ on $W_i = \ln\left(-\ln\left(1 - \dfrac{i}{n+1}\right)\right)$. The linear relationship is $\ln T_{(i)} = \dfrac{1}{v}W_i + \ln\beta$. The slope of the straight line estimates $\dfrac{1}{v}$ and the intercept estimates $\log\beta$. For our sample of $n = 100$ from $W(2.5, 10)$, the regression of $\ln T_{(i)}$ on W_i is

```
ln(TTF) = 2.32 + 0.464 W
Predictor      Coef        Stdev      t-ratio       p
Constant       2.31968     0.01021    227.15      0.000
W              0.464351    0.007678    60.48      0.000
s = 0.09263 R-sq = 97.4% R-sq(adj) = 97.4%.
```

The graphical estimate of β is $\hat{\beta} = \exp(2.31968) = 10.172$. The graphical estimate of v is $\hat{v} = \frac{1}{0.4643} = 2.154$. Both estimates are close to the nominal values. In Figure 16.9.1, we see the Weibull probability plotting of MINITAB. This graph puts $\ln T_{(i)}$ on the x-axis and $\ln p_i$ on the y-axis where p_i is the calculated probability of occurrence for each observation $T_{(i)}$ assuming a Weibull distribution. The value of p for which $\ln(-\ln(1-p)) = 0$ is $p_0 = 1 - e^{-1} = 0.632$. The value of x, corresponding to p_0 is β.

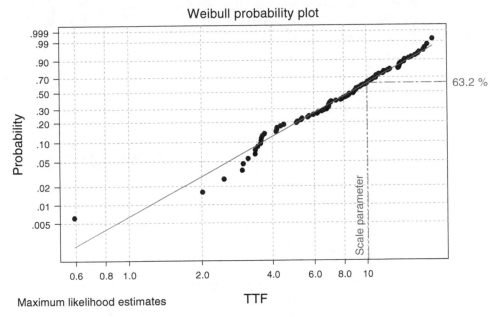

Weibull probability plot

Maximum likelihood estimates

Shape parameter : 2.31146
Scale parameter : 10.1099

Figure 16.19.1 *Weibull probability plot*

16.20 The regression of $Y_i = \ln X_{(i)}, i = 1, \dots, n$ on the normal scores $Z_i = \Phi^{-1}\left(\dfrac{i - 3/8}{n + 1/4}\right)$ is

```
Y = 5.10 + 0.556 Z.
```

Predictor	Coef	Stdev	t-ratio	p
Constant	5.09571	0.02201	231.55	0.000
Z	0.55589	0.02344	23.72	0.000

```
s = 0.09842 R-sq = 96.9% R-sq(adj) = 96.7%.
```

The intercept $\hat{a} = 5.09571$ is an estimate of μ, and the slope $\hat{b} = 0.5559$ is an estimate of σ. The mean of $\text{LN}(\xi, \sigma^2)$ is $\xi = e^{\mu + \sigma^2/2}$ and its variance is $D^2 = e^{2\mu + \sigma^2}(e^{\sigma^2} - 1)$. Thus, the graphical estimates are $\hat{\xi} = 190.608$ and $\hat{D} = 13,154.858$. The mean $\bar{X}$ and variance S^2 of the given sample are 189.04 and 12,948.2. In Figure 16.20.1, we see the normal probability plotting of the data.

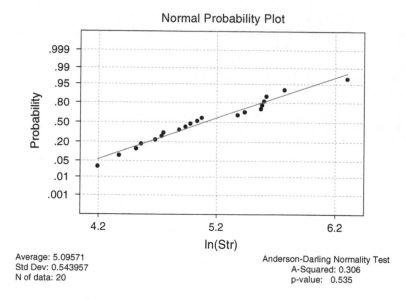

Figure 16.20.1   *Normal probability plot of* ln X

16.21 The *Q-Q* plot is given in Figure 16.21.1.

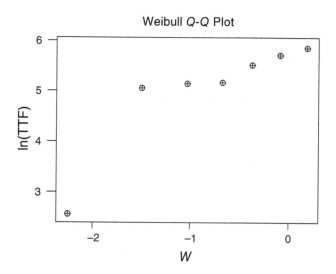

Figure 16.21.1   *Weibull Q–Q plot*

The regression of $\ln T_{(i)}$ on $W_i = \ln\left(-\ln\left(1 - \dfrac{i}{10}\right)\right)$ for $i = 1, \ldots, 7$ is

```
ln(T) = 5.96 + 1.17 W
Predictor        Coef       Stdev      t-ratio          p
Constant       5.9594      0.3068        19.43      0.000
W              1.1689      0.2705         4.32      0.008
s = 0.5627 R-sq = 78.9% R-sq(adj) = 74.7%.
```

According to this regression line, the estimates of β and v are $\hat{\beta} = \exp(5.9594) = 387.378$ and $\hat{v} = 1/1.1689 = 0.855$.

The estimate of the median of the distribution is $\hat{Me} = \hat{\beta}\left(-\ln\left(\frac{1}{2}\right)\right)^{1/\hat{v}} = 252.33$.

16.22 See Figure 16.22.1

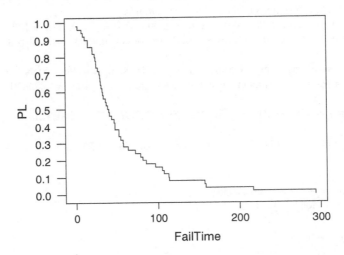

Figure 16.22.1 *Kaplan–Meier estimator of the reliability of an electric device*

16.23 The MLE of β is $\hat{\beta} = \bar{X} = 57.07$. Since $R(t) = \exp(-t/\beta)$, the MLE of R(50) is $\hat{R}(50) = 0.4164$. Note that $\bar{X} \sim \frac{\beta}{2n}\chi^2[2n]$. Hence, confidence limits for β, with $1 - \alpha = 0.95$ are $(44.048, 76.891)$. Substituting these limits into $R(t)$ gives $(0.3214, 0.5219)$ as a 0.95 confidence interval for R(50).

16.24 We have $n = 20$, $K_n = 17$, and $\hat{\beta} = 129.011$. With $M = 500$ runs, the estimated STD of $\hat{\beta}$ is 34.912, with confidence interval $(78.268, 211.312)$.

16.25 For $\beta = 100$, $SE\{\hat{\beta}_n\} = 0.1\beta = 10 = \dfrac{100}{\sqrt{r}}$. Hence, $r = 100$ and

$$n^0 \approx \frac{100}{2}\left(1 + \left(1 + 4\frac{100}{1000}\right)^{1/2}\right) = 109.16 \approx 110. \ E\{T_{n,r}\} = 235.327 \text{ and the expected cost of testing is}$$

$E\{C\} = 10 \times n + 1 \times E\{T_{n,r}\} = \$1335.33$.

16.26 (i) First we put the data into column C1. We then obtained the MLEs of the scale and shape parameters by using the following MINITAB commands:

```
MTB> let k1 = 1
MTB> store 'weibullmles'
STOR> let C2 = C1**k1
STOR> let C3 = loge(C1)
```

```
STOR> let C4 = C2 * C3
STOR> let k2 = 1/((sum(C4)/sum(C2)) - mean(C3))
STOR> let C5 = C1**k2
STOR> let k3 = (mean(C5))**(1/k2)
STOR> let k1 = k2
STOR> end
MTB> exec 'weibullmles' 50
```

The values of $\hat{\nu}$ and $\hat{\beta}$ are stored in $k2$ and $k3$, resp. For the WEIBUL.csv data, we obtained $\hat{\beta} = 27.0789$ and $\hat{\nu} = 1.374$.

(ii) The EBD estimates are SE$\{\hat{\beta}_{50}\} = 3.047$ and SE$\{\hat{\nu}_{50}\} = 0.1568$. The asymptotic estimates are SE$\{\hat{\beta}_{50}\} = 2.934$ and SE$\{\hat{\nu}_{50}\} = 0.1516$. The EBD estimates and the asymptotic estimates of the standard deviations of $\hat{\beta}$ and $\hat{\nu}$ are very similar.

16.27 To discriminate between $R_0 = 0.99$ and $R_1 = 0.90$ with $\alpha = \beta = 0.01$, $n \approx 107$.

16.28 For $R_0 = 0.99$, $R_1 = 0.90$, and $\alpha = \gamma = 0.01$, we get $s = 0.9603$, $h_1 = 1.9163$, $h_2 = 1.9163$, and ASN$(0.9) = 31.17$.

16.29 Without loss of generality, assume that $t = 1$. Thus, $R_0 = e^{-1/\beta_0} = 0.99$, or $\beta_0 = 99.5$. Also, $R_1 = 0.9$, or $\beta_1 = 9.49$. The parameters of the SPRT are $h_1 = 48.205$, $h_2 = 48.205$, and $s = 24.652$. ASN$(0.9) \approx 3$.

16.30 Using $\alpha = \gamma = 0.05$, we have $n = 20$, $H_0 : \beta = 2000$, $H_1 : \beta = 1500$. Thus,

$$\lambda_0 = 0.0005 \quad \text{and} \quad \lambda_1 = 0.00067. \quad \text{The parameters of the SPRT are} \quad h_1 = \frac{\log\left(\frac{95}{5}\right)}{\log\left(\frac{\lambda_1}{\lambda_0}\right)} = 10.061,$$

$$h_2 = 10.061, \text{and } s = \frac{\lambda_1 - \lambda_0}{\log\left(\frac{\lambda_1}{\lambda_0}\right)} = 0.000581.$$

$$E_{\beta_0}\{\tau\} = \frac{2000}{20} E_\beta\{X_{20}(\tau)\}$$

$$\approx 100 \left(\frac{10.061 - 0.95 \times 20.122}{1 - 2000 \times 0.000581}\right)$$

$$= 5589.44 \text{ [hour]}.$$

16.31 With $\beta = 1000$, $p = 0.01$, $\gamma = 500$, and $t^* = 300$, we get

$$F^*(t) = 1 - \frac{0.01e^{-t/500} + 0.99e^{-t/1000}}{0.01e^{-300/500} + 0.99e^{-300/1000}}, \quad t \geq 300.$$

The expected life length of units surviving the burn-in is

$$\beta^* = 300 + \frac{1}{0.738898} \int_{300}^{\infty} (0.01 \cdot e^{-t/500} + 0.99 \cdot e^{-t/1000}) dt$$

$$= 300 + \frac{1}{0.738898}(5e^{-3/5} + 990e^{-3/10})$$

$$= 1296.29.$$

A unit surviving 300 hours of burn-in is expected to operate an additional 996.29 hours. The expected life of these units without the burn-in is $0.01 \times 500 + 0.99 \times 1000 = 995$ hours. The burn-in of 300 hours in the plant only yields an increase of 1.29 hours in the mean life of the product in the field. Whether this very small increase in the MTTF is justified depends on the relative costs involved.

Chapter 18: Sampling Plans for Batch and Sequential Inspection

18.1   The single-sampling plans for attributes with $N = 2500$, $\alpha = \beta = 0.01$, and AQL and LQL as specified are (i) $n = 1878$, $c = 13$; (ii) $n = 702$, $c = 12$; (iii) $n = 305$, $c = 7$.

18.2   For $\alpha = \beta = 0.05$, AQL $= 0.01$, and LQL $= 0.03$, the single-sampling plans for attributes are as follows.

N	n	c
100	87	1
500	254	4
1000	355	6
2000	453	8

We see that as the lot size, N, increases then the required sample size increases, but n/N decreases from 87% to 22.6%. The acceptance number c increases very slowly.

18.3   For the sampling plan in Exercise 18.1(iii), $OC(p) = H(7; 2500, M_p, 305)$. When $p = 0.025$, we get $M_p = 62$ and $OC(0.025) = 0.5091$.

18.4   The large sample approximation yields $n^* = 292$, $c^* = 11$. The "exact" plan is $n = 311$, $c = 12$. Notice that the actual risks of the large sample approximation plan (n^*, c^*) are $\alpha^* = 0.0443$ and $\beta^* = 0.0543$. The actual risks of the "exact" plan are $\alpha = 0.037$ and $\beta = 0.0494$.

18.5   The large sample approximation is $n^* = 73$, $c^* = 1$ with actual risks of $\alpha^* = 0.16$, $\beta^* = 0.06$. The exact plan is $n = 87$, $c = 2$ with actual risks of $\alpha = 0.054$ and $\beta = 0.097$.

18.6   The double-sampling plan with $n_1 = 150$, $n_2 = 200$, $c_1 = 5$, $c_2 = c_3 = 10$, and $N = 2000$ yields

p	$OC(p)$	$ASN(p)$
0	1.000	150
0.01	0.999	150.5
0.02	0.966	164.7
0.03	0.755	206.0
0.04	0.454	248.4
0.05	0.227	262.8

p	$OC(p)$	$ASN(p)$
0.06	0.099	247.7
0.07	0.039	218.9
0.08	0.014	191.5
0.09	0.005	171.9
0.10	0.001	160.4

The single sampling plan for $\alpha = 0.02$, $\beta = 0.10$, AQL $= 0.02$, LQL $= 0.06$ is $n = 210$, $c = 8$ with actual $\alpha = 0.02$, $\beta = 0.10$.

18.7   For the sequential plan, $OC(0.02) = 0.95 = 1 - \alpha$, $OC(0.06) = 0.05 = \beta$.
ASN$(0.02) = 140$, ASN$(0.06) = 99$, and ASN$(0.035) = 191$.

18.8   The single-sampling plan for $N = 10,000$, $\alpha = \beta = 0.01$, AQL $= 0.01$, and LQL $= 0.05$ is $n = 341$, $c = 8$. The actual risks are $\alpha = 0.007$, $\beta = 0.0097$. The corresponding sequential plan, for $p = $ AQL $= 0.01$ has ASN$(0.01) = 182$. On the average, the sequential plan saves, under $p = $ AQL, 159 observations. This is an average savings of $159 per inspection.

18.9   For a continuous variable-size sampling plan when $(p_0) = $ AQL $= 0.01$, $(p_t) = $ LQL $= 0.05$, and $\alpha = \beta = 0.05$, we obtain $n = 70$ and $k = 1.986$.

18.10   From the data, we get $\bar{X} = 14.9846$ and $S = 0.019011$. For $p_0 = 0.01$ and $\alpha = 0.05$, we obtain $k = 2.742375$. Since $\bar{X} - kS = 14.9325 > \xi = 14.9$, the lot is **accepted**. $OC(0.03) \approx 0.4812$.

18.11   For AQL $= 0.02$, LQL $= 0.04$ and $\alpha = \beta = 0.1$, $n = 201$, $k = 1.9022$.

18.12 For a single-sampling plan for attributes where $n = 139$, $c = 3$ and $N = 500$, we obtain $OC(p) = H(3; 500, M_p, 139)$, and $R^* = H(2; 499, M_p - 1,138)/H(3; 500, M_p, 139)$.

p	AOQ	ATI
0.01	0.0072	147.2
0.02	0.0112	244.2
0.03	0.0091	369.3
0.05	0.0022	482.0

18.13 A switch to tightened plan, when all five consecutive lots have $p=$ AQL, with $\alpha = 0.05$, is $\sum_{j=2}^{5} b(j; 5, 0.05) = 0.0226$.

The probability of switching to a tightened plan if $\alpha = 0.3$ is 0.4718.

18.14 The total sample size from 10 consecutive lots is 1000. Thus, from Table 18.16 S_{10} should be less than 5. The last 2 samples should each have less than 2 defective items. Hence, the probability for qualification is

$$QP = b^2(1; 100, 0.01)B(2; 800, 0.01)$$

$$+ 2b(1; 100, 0.01)b(0; 100, 0.01)B(3; 800, 0.01)$$

$$+ b^2(0; 100, 0.01)B(4; 800, 0.01)$$

$$= 0.0263.$$

References

Aoki, M. (1989) *Optimization of Stochastic Systems: Topics in Discrete-Time Dynamics* (Second Edition), Academic Press, New York.

Barnard, G.A. (1959) Control charts and stochastic processes, *Journal of the Royal Statistical Society, Series B*, **21**, 239–271.

Barnett, A.G. and Wren, J.D. (2019) Examination of CIs in health and medical journals from 1976 to 2019: an observational study, *BMJ Open*, **9**, e032506. doi: 10.1136/bmjopen-2019-032506.

Berry, D.A. and Friestedt, B. (1985) *Bandit Problems, Sequential Allocation of Experiments*, Chapman and Hall, New York.

Box, G.E.P. and Jenkins, G.M. (1970) *Time Series Analysis, Forecasting and Control*, Holden-Day, Oakland, CA.

Box, G.E.P. and Kramer, T. (1992) Statistical process monitoring and feedback adjustment: a discussion, *Technometrics*, **34**, 251–285.

Box, G.E.P and Tiao, G. (1973) *Bayesian Inference in Statistical Analysis*, Addison-Wesley Pub. Co., Reading, MA.

Box, G.E.P, Bisgaard, S. and Fung, C. (1988) An explanation and critique of Taguchi's contributions to quality engineering, *Quality and Reliability Engineering International*, **4**(2), 123–131.

Box, G.E.P., Hunter, W.G. and Hunter, S.J. (1978) *Statistics for Experimenters*, John Wiley, New York.

Bratley, P., Fox, B.L. and Schrage, L.E. (1983) *A Guide to Simulation*, Springer-Verlag, New York.

Breiman, L. (2001) Statistical modeling: the two cultures, *Statistical Science*, **16**(3), 199–231.

Breiman, L., Friedman, J., Olshen, R. and Stone, C. (1984) *Classification and Regression Trees*, Chapman Hall/CRC (orig. published by Wadsworth), Boca Raton, FL.

Bundesministerium für Bildung und Forschung (BMBF) (2016) Industrie 4.0. https://www.bmbf.de/de/zukunftsprojekt-industrie-4-0-848.html (Accessed 20/12/2018).

Chen, H. (1994) A multivariate process capability index over a rectangular solid tolerance zone, *Statistics Sinica*, **4**, 749–758.

Cochran, W.G. (1977) *Sampling Techniques*, John Wiley, New York.

Daniel, C. and Wood, F.S. (1971) *Fitting Equations to Data: Computer Analysis of Multifactor Data for Scientist and Engineers*. John Wiley, New York.

Dehand, K. (1989) *Quality Control, Robust Design, and the Taguchi Method*, Wadsworth and Brooks/Cole, Pacific Grove, CA.

Deming, W.E. (1967) A tribute to Walter A. Shewhart, *The American Statistician*, **21**(2), 39–40.

Deming, W.E. (1982) *Quality, Productivity and the Competitive Position*, Center for Advanced Engineering Studies, Cambridge, MA.

Deming, W.E. (1986) *Out of The Crisis*, MIT Press, Boston, MA.

Derringer, G. and Suich, R. (1980) Simultaneous optimization of several response variables, *Journal of Quality Technology*, **12**(4), 214–219.

Dodge, H.F. and Romig, H.G. (1929) A method of sampling inspection, *Bell System Technical Journal*, **8**, 613–631.

Modern Industrial Statistics: With Applications in R, MINITAB and JMP, Third Edition.
Ron S. Kenett and Shelemyahu Zacks.
© 2021 John Wiley & Sons, Ltd. Published 2021 by John Wiley & Sons, Ltd.

Dodge, H.F. and Romig, H.G. (1959) *Sampling Inspection Tables* (Second Edition), John Wiley, New York.

Draper, N.R. and Smith, H. (1981) *Applied Regression Analysis* (Second Edition), John Wiley, New York.

Duncan, A.J. (1956) The economic design of X charts used to maintain current control of a process, *Journal of the American Statistical Association*, **51**, 228–242.

Duncan, A.J. (1971) The economic design of X charts when there is a multiplicity of assignable causes, *Journal of the American Statistical Association*, **66**, 107–121.

Duncan, A.J. (1978) The economic design of p-charts to maintain current control of a process: some numerical results, *Technometrics*, **20**, 235–243.

Duncan, A.J. (1986) *Quality Control and Industrial Statistics* (Fifth Edition), Irwin, Homewood, IL.

Efron, B. and Hastie, T. (2016) *Computer Age Statistical Inference: Algorithms, Evidence and Data Science*, Cambridge University Press, Cambridge, UK.

Faltin, F., Kenett, R.S. and Ruggeri, F. (2012) *Statistical Methods in Healthcare*, John Wiley, UK.

Figini, S., Kenett, R.S. and Salini, S. (2010) Integrating operational and financial risk assessments, *Quality and Reliability Engineering International*, 26(8), 887–897.

Friedman, T.L. (2016) *Thank You for Being Late: An Optimist's Guide to Thriving in the Age of Accelerations*, Farrar, Straus, and Giroux, New York.

Fuchs, C. and Kenett, R.S. (1987) Multivariate tolerance regions and F-tests, *Journal of Quality Technology*, **19**, 122–131.

Fuchs, C. and Kenett, R.S. (1988) Appraisal of ceramic substrates by multivariate tolerance regions, *Journal of the Royal Statistical Society, Series D*, **37**, 401–411.

Fuchs, C. and Kenett, R.S. (1998) *Multivariate Quality Control: Theory and Application, Quality and Reliability Series*, Vol. 54, Marcel Dekker, New York.

Gandin, L.S. (1963) *Objective Analysis of Meteorological Fields*, GIMIZ, Leningrad.

Gelman, A., Carlin, J., Stern, H., Dunson, D., Vehtari, A. and Rubin, D. (2013). *Bayesian Data Analysis*, Chapman and Hall/CRC Texts in Statistical Science.

Gertsbakh, I.B. (1989) *Statistical Reliability Theory*, Marcel Dekker, New York.

Ghosh, S. (1990) *Statistical Design and Analysis of Industrial Experiments*, Marcel Dekker, New York.

Gibra, I.N. (1971) Economically optimal determination of the parameters of X-bar control chart, *Management Science*, **17**, 635–646.

Gittins, J.C. (1989), *Multi-Armed Bandit Allocation Indices*, Wiley, New York.

Gittins, J.C. and Jones, D.M. (1974) *A dynamic allocation index for the sequential design of experiments*. In: Gani, J. (ed.) Progress in Statistics, North-Holland, Amsterdam, 241–266.

Goba, F.A. (1969) Biography on thermal aging of electrical insulation, *IEEE Transactions on Electrical Insulation*, **EI-4**, 31–58.

Godfrey, A.B. (1986) The history and evolution of quality in AT&T, *AT&T Technical Journal*, **65**, 9–20.

Godfrey, A.B. and Kenett, R.S. (2007) Joseph M. Juran, a perspective on past contributions and future impact, *Quality and Reliability Engineering International*, **23**, 653–663

Good, I.J. (1965) *The Estimation of Probabilities: An Essay on Modern Bayesian Methods*, M.I.T. Press.

Haridy, S., Wu, Z. and Castagliola, P. (2011) Univariate and multivariate approaches for evaluating the capability of dynamic-behavior processes (case study), *Statistical Methodology*, **8**, 185–203.

Hastie, T., Tibshirani, R. and Friedman, J. (2009) *The Elements of Statistical Learning: Data Mining, Inference, and Prediction*, Springer, (Second Edition), 12th Printing, 2017, Springer.

Hoadley, B. (1981) The quality measurement plan, *Bell System Technical Journal*, **60**, 215–271.

Insua, D., Ruggeri, F. and Wiper, M. (2012) *Bayesian Analysis of Stochastic Process Models*, John Wiley and Sons.

Ishikawa, K. (1986) *Guide to Quality Control* (Second Edition), Asian Productivity Organization, UNIPAB Kraus International Publications, White Plains, NY.

Jalili, M., Bashiri, M. and Amiri, A. (2012) A new multivariate process capability index under both unilateral and bilateral quality characteristics, Wileylibrary.com. doi: 10.1002/qre.1284.

Jensen, F. and Petersen, N.E. (1982) *Burn-In: An Engineering Approach to the Design and Analysis of Burn-In Procedures*, John Wiley, New York.

John, S. (1963) A tolerance region for multivariate normal distributions, *Sankhya, Series A*, **25**, 363–368.

John, P.W.M. (1990) *Statistical Methods in Engineering and Quality Assurance*, John Wiley, New York.

Juran, J.M. Ed. (1979) *Quality Control Handbook* (Third Edition), McGraw-Hill Book Company, New York.

Juran, J.M. (1988) *Juran on Planning for Quality*, The Free Press, New York.

Juran, J.M. Ed. (1995) *A History of Managing for Quality*, ASQ Quality Press, Milwaukee, WI.

Kacker, R.N. (1985) Off-line quality control, parameter design, and the Taguchi method (with discussion), *Journal of Quality Technology*, **17**, 176–209.

Kelly, T., Kenett, R.S., Newton, E., Roodman, G. and Wowk, A. (1991) Total Quality Management Also Applies to a School of Management, *Proccedings of the 9th IMPRO Conference*, Atlanta, GA.

Kenett, R.S. (1983) On an exploratory analysis of contingency tables, *Journal of the Royal Statistical Society, Series D*, **32**, 395–403.

Kenett, R.S. (1991) Two methods for comparing Pareto charts, *Journal of Quality Technology*, **23**, 27–31.

Kenett, R.S. (2012) Applications of Bayesian Networks, http://ssrn.com/abstract=2172713. accessed 2020.

Kenett, R.S. and Baker, E. (2010) *Process Improvement and CMMI for Systems and Software*, Taylor and Francis, Auerbach CRC Publications, Boca Raton, FL.

Kenett, R.S., de Frenne, A., Tort-Martorell, X., McCollin, C. (2008) *The Statistical Efficiency Conjecture in Applying Statistical Methods in Business and Industry–The State of the Art*, Greenfield, T., Coleman, S., Montgomery, D. (eds.). Wiley, Chichester, UK.

Kenett, R.S. and Kenett, D.A. (2008), Quality by design applications in biosimilar technological products, *ACQUAL, Accreditation and Quality Assurance*, **13**(12), 681–690.

Kenett, R.S. and Pollak, M. (1996) Data analytic aspects of the Shiryayev-Roberts control chart, *Journal of Applied Statistics*, **23**, 125–137.

Kenett, R.S. and Raanan, Y. (2010) *Operational Risk Management: A Practical Approach to Intelligent Data Analysis*, John Wiley, UK, Kindle Edition, 2011.

Kenett, R.S. and Salini, S. (2008) Relative linkage disequilibrium applications to aircraft accidents and operational risks, *Transactions on Machine Learning and Data Mining*, **1**(2), 83–96. The procedure is Implemented in *arules* R Package. Version 0.6-6, Mining Association Rules and Frequent Itemsets.

Kenett, R.S. and Salini, S. (2011) *Modern Analysis of Customer Surveys: With Applications Using R*, John Wiley, UK.

Kenett, R.S and Shmueli, G. (2014) On information quality, *Journal of the Royal Statistical Society, Series A* (with discussion), **177**(1), 3–38.

Kenett, R.S. and Shmueli, G. (2016) *Information Quality: The Potential of Data and Analytics to Generate Knowledge*, John Wiley and Sons.

Kenett, R.S. and Zacks, S. (1998) *Modern Industrial Statistics: Design and Control of Quality and Reliability*, Duxbury.

Kenett, R.S., Coleman, S. and Stewardson, D. (2003) Statistical efficiency: the practical perspective, *Quality and Reliability Engineering International*, **19**, 265–272.

Kenett, R.S., Swarz, R., and Zonnenshain, A. (2020) *Systems Engineering in the Fourth Industrial Revolution: Big Data, Novel Technologies, and Modern Systems Engineering*, John Wiley and Sons.

Kohavi, R. and Thomke, S. (2017) *The Surprising Power of Online Experiments*, Harvard Business Review, September-October, https://hbr.org/2017/09/the-surprising-power-of-online-experiments.

Kotz, S. and Johnson, N.L. (1985). *Encyclopedia of Statistical Sciences*, John Wiley, New York.

Kotz, S. and Johnson, N.L. (1994) *Process Capability Indices*, Chapman and Hall, New York.

Liebesman, B.S. and Saperstein, B. (1983) A proposed attribute skip-lot sampling program, *Journal of Quality Technology*, **15**, 130–140.

Lin, K.M. and Kacker, R.N. (1989) Optimizing the Wave Soldering Process, in *Quality Control Robust Design, and The Taguchi Method*, Khorsrow, Khorsrow, Ed., Wadsworth BrooksCole, Pacific Grove, CA, 143–157.

Lucas, J.M. (1982) Combined Shewhart - CUSUM quality control schemes, *Journal of Quality Technology*, **14**, 51–59.

Lucas, J.M. and Crosier, R.B. (1982) Fast initial response for CUSUM quality control scheme: give your CUSUM a headstart, *Technometrics*, **24**, 199–205.

Mann, N.R., Schafer, R.E. and Singpurwala, N.D. (1974) *Methods for Statistical Analysis of Reliability and Life Data*, John Wiley, New York.

Martz, H. and Waller, R. (1982) *Bayesian Reliability Analysis*, John Wiley, New York.

Matheron, G. (1963) Principles of geostatistics, *Economic Geology*, **58**, 1246–1266.

Meeker, W.Q. and Escobar, L.A. (1989) *Statistical Methods for Reliability Data*, John Wiley and Sons.

Nasr, M. (2007) Quality by design (QbD) – a modern system approach to pharmaceutical development and manufacturing – FDA perspective, FDA Quality Initiatives Workshop, Maryland, USA.

Nelson, W. (1992) *Accelerated Testing: Statistical Models, Test Plans and Data Analysis*, John Wiley, New York.

Oikawa, T. and Oka, T. (1987) New techniques for approximating the stress in pad-type nozzles attached to a spherical shell, *Transactions of the American Society of Mechanical Engineers*, 109, 188–192.

Page, E.S. (1954) Continuous inspection schemes, *Biometrika*, **41**, 100–114.

Page, E.S. (1962) A modified control chart with warning limits, *Biometrika*, **49**, 171–176.

Pao, T.W., Phadke, M.S. and Sherrerd, C.S. (1985) Computer response time optimization using orthogonal array experiments, *IEEE International Communication Conference, Chicago, Conference Record*, Vol. **2**, 890–895.

Peck, D.S. and Trapp, O. (1978) *Accelerated Testing Handbook*, Technology Associations, Portola Valley, CA.

Phadke, M.S. (1989) *Quality Engineering Using Robust Design*, Prentice Hall, Englewood Cliffs, NJ.

Phadke, M.S., Kacker, R.N., Speeney, D.V. and Grieco, M.J. (1983) Quality control in integrated circuits using experimental design, *Bell System Technical Journal*, **62**, 1275–1309.

Press, S. (1989) *Bayesian Statistics: Principles, Models and Applications*, John Wiley, New York.

Quinlan, J. (1985) *Product Improvement by Application of Taguchi Methods, Third Suplier Symposium on Taguchi Methods*, American Supplier Institute, Inc., Dearborn, MI.

Rathore, A.S. and Mhatre, R. (2009) *Quality by Design for Biopharmaceuticals*, John Wiley and Sons, Hoboken, NJ.

Reinman, G., Ayer, T., Davan, T., et al. (2012) Design for variation, *Quality Engineering*, **24**, 317–345.

Robert, C.P. and Casella, G. (2010) *Introducing Monte Carlo Methods in R*, Springer.

Romano, D. and Vicario, G. (2002) Reliable estimation in computer experiments on finite element codes, *Quality Engineering*, **14**(2), 195–204.

Ruggeri, F., Kenett, R., and Faltin, F. (2007) *Encyclopedia of Statistics in Quality and Reliability*, John Wiley and Sons, Chichester, UK.

Ryan, B.F., Joiner, B.L. and Ryan, T.P. (1976) *Minitab Handbook*, Duxbury Press, Belmont, CA.

Sacks, J., Welch, W.J., Mitchell, T.J. and Wynn, H.P. (1989), Design and analysis of computer experiments, *Statistical Science*, **4**(4), 409–435.

Sall, J. (2002). *Monte Carlo Calibration of Distributions of Partition Statistics*, SAS Institute Inc., Cary, NC. https://www.jmp.com/content/dam/jmp/documents/en/white-papers/montecarlocal.pdf (Accessed 20/10/2020).

SAS Institute Inc. (1983). *SAS Technical Report A-108: Cubic Clustering Criterion*, SAS Institute Inc., Cary, NC. https://support.sas.com/kb/22/addl/fusion_22540_1_a108_5903.pdf (Accessed 20/10/2019).

Scheffe, H. (1959) *The Analysis of Variance*, John Wiley, New York.

Senge, P. (1990) *The Fifth Discipline: The Art and Practice of the Learning Organization*, Doubleday, New York.

Shewhart, W.A. (1926) Quality control charts, *Bell System Technical Journal*, **5**, 593–603.

Shumway, R.H. and Stoffer, D.S. (2017) *Time Series Analysis and Applications, Springer Texts in Statistics* (Fourth Edition), Springer, New York.

Singpurwalla, N. (2006) *Reliability and Risk: A Bayesian Perspective*, John Wiley and Sons.

SUPAC (1997) Food and Drug Administration, Center for Drug Evaluation and Research (CDER) Scale-Up and Postapproval Changes: Chemistry, Manufacturing, and Controls, In Vitro Release Testing and In Vivo Bioequivalence Documentation, Rockville, MD, USA.

Taguchi, G. (1987) *Systems of Experimental Design*, Vols. 1–2, Clausing, D. Ed., UNIPUB/Kraus International Publications, New York.

Taguchi, G. and Taguchi, S. (1987) *Taguchi Methods: Orthogonal Arrays and Linear Graphs*, American Supplier Institute, Dearborn, MI.

Tsokos, C. and Shimi, I. (1977) *The Theory and Applications of Reliability with Emphasis on Bayesian and Non-Parametric Methods*, Academic Press, New York.

Tsong, Y., Hammerstrom, T., Sathe, P. and Shah, V. (1996) Statistical assessment of mean differences between two dissolution data sets, *Drug Information Journal*, **30**, 1105–1112.

Tukey, J.W. (1962) The future of data analysis, *The Annals of Mathematical Statistics*, **33**, 1–67.

Weindling, J.I. (1967) Statistical Properties of a General Class of Control Charts Treated As a Markov Process, Ph.D. Dissertation, Columbia University, New York.

Weindling, J.I., Littauer, S.B. and De Olivera, J.T. (1970) Mean action time of the X-bar control chart with warning limits, *Jour. Quality Technology*, **2**, 79–85.

Yashchin, E. (1985) On a unified approach to the analysis of two-sided cumulative sum control schemes with headstarts, *Advances in Applied Probability*, **17**, 562–593.

Yashchin, E. (1991) Some aspects of the theory of statistical control schemes, *IBM Journal of Research Development*, **31**, 199–205.

Zacks, S. (1992) *Introduction to Reliability Analysis: Probability Models and Statistical Methods*, Springer-Verlag, New York.

Zacks, S. (2009), *Stage-Wise Adaptive Designs*, Wiley Series in Probability and Statistics, Wiley, New York.

Author Index

Modern Industrial Statistics: With Applications in R, MINITAB and JMP, Third Edition.
Ron S. Kenett and Shelemyahu Zacks.
© 2021 John Wiley & Sons, Ltd. Published 2021 by John Wiley & Sons, Ltd.

Subject Index

accelerated life testing   358, 603, 604, 638, 640, 641, 643, 644, 646

acceptable quality level (AQL)   354, 672, 673, 679, 685, 690–697, 821, 822

acceptance
  number   674, 675, 690, 693, 696, 821
  region   139, 150, 198
  sampling   20, 354, 358, 671, 672, 674–71, 685–687, 694–695, 701, 702

accuracy   5, 11, 18–19, 42, 43, 325, 326, 330, 546, 550, 582, 634, 707

actuarial estimator   624

additive linear models   476–478

alfa-trimmed mean   39

alfa-trimmed standard deviation   41

aliases   523, 524, 543, 802

alternative hypothesis   4, 139, 147, 179, 181, 196, 401, 405, 406

American National Standard Institute (ANSI)   671, 689

American Society for Quality (ASQ)   358, 689, 775

analysis of variance (ANOVA)   184–186, 197, 198, 225, 235, 241, 242, 247–251, 263, 265, 266, 269, 270, 334, 483–485, 487, 489–491, 495–501, 540–544, 571, 743, 753–759, 797–801, 805

ANOVA Table   248, 251, 483, 484, 489, 501, 536, 540, 758

arcsin transformation   264, 266

Arrhenius law   641

ASN function   635–638, 678, 680, 681, 696

assignable causes   363, 372, 381, 396, 462

attained significance level   142, 407

automatic process control   355, 441–444, 448

autoregressive integrated moving average (ARIMA)   306, 307, 317–319, 768

availability   358, 557, 603–606, 610–617, 643, 662–665, 814

availability function   612, 614, 641, 645

average outgoing quality (AOQ)   679, 688, 696, 697, 822

average outgoing quality limit (AOQL)   688, 689, 695

average run length (ARL)   396, 408–410, 414, 423–427, 430–432, 434, 444, 447, 790, 791

average sample number (ASN)   635–638, 643, 646, 678–681, 695, 696, 820, 821

average total inspection (ATI)   688, 689, 695–697, 822

Modern Industrial Statistics: With Applications in R, MINITAB and JMP, Third Edition.
Ron S. Kenett and Shelemyahu Zacks.
© 2021 John Wiley & Sons, Ltd. Published 2021 by John Wiley & Sons, Ltd.